Protein Dynamics

Foundations and Frontiers in Enzymology

Protein Dynamics

Experimental and Computational Approaches

Edited by

Durba Sengupta
Senior Principal Scientist, Professor,
Physical Chemistry Division,
CSIR-National Chemical Laboratory,
Pune, Maharashtra, India

Ana-Nicoleta Bondar
Professor, Physics, Faculty of Physics and
Faculty for Interdisciplinary Sciences,
University of Bucharest,
Bucharest, Romania

ELSEVIER

ACADEMIC PRESS
An imprint of Elsevier
elsevier.com/books-and-journals

Academic Press is an imprint of Elsevier

125 London Wall, London EC2Y 5AS, United Kingdom
50 Hampshire Street, 5th Floor, Cambridge, MA 02139, United States

ISBN: 978-0-443-30042-4

For Information on all Academic Press publications visit our website at https://www.elsevier.com/books-and-journals

Cover art by *Alina-Maria Halaicu, University of Bucharest, Faculty of Physics*

Based on (1) https://en.m.wikipedia.org/wiki/File:Beta-2-adrenergic-receptor.png (2) RCSB PDB - 2RH1: High resolution crystal structure of human B2-adrenergic G protein-coupled receptor (3) Cherezov, V., Rosenbaum, D.M., Hanson, M.A., Rasmussen, S.G.F., Thian, F.S., Kobilka, T.S., Choi, H., Kuhn, P., Weis, W.I., Kobilka, B.K., & Stevens, R.C. (2007).
(4) High-Resolution Crystal Structure of an Engineered Human β 2 -Adrenergic G Protein–Coupled Receptor. Science, 318(5854), 1258–1265. https://doi.org/10.1126/science.1150577 and (5) Tang, X., & Lagerström, M.C. (2002). G protein-coupled receptor signaling in the brain: New insights and implications for therapy. Nature Reviews Drug Discovery, 1(1), 37–41. https://doi.org/10.1038/nrd913

Publisher: Megan Ball
Acquisitions Editor: Michelle Fisher
Editorial Project Manager: Prachi Mittal
Production Project Manager: Fahmida Sultana
Cover Designer: Raman Kumar

Typeset by Aptara, New Delhi, India

Dedication

Dedicated to all students of Biophysics

Contents

Contributors

Magnus Andersson
Department of Chemistry, Umeå University, Umeå, Sweden

Erika Balog
Department of Biophysics and Radiation Biology, Semmelweis University, Budapest, Hungary

Ana-Nicoleta Bondar
Faculty of Physics, University of Bucharest, Bucharest, Romania; Faculty for Interdisciplinary Sciences, University of Bucharest, Bucharest, Romania

Urban Bren
Laboratory of Physical Chemistry and Chemical Thermodynamics, Faculty of Chemistry and Chemical Engineering, University of Maribor, Maribor, Slovenia; Institute of Environmental Protection and Sensors, Maribor, Slovenia; Faculty of Mathematics, Natural Sciences and Information Technologies, University of Primorska, Koper, Slovenia

Suman Chakrabarty
Department of Chemical and Biological Sciences, S. N. Bose National Centre for Basic Sciences, Kolkata, West Bengal, India

Hirak Chakraborty
School of Chemistry, Sambalpur University, Burla, Sambalpur, Odisha, India

Abhilasha A. Doke
Physical and Materials Chemistry Division, CSIR-National Chemical Laboratory, Pune, Maharashtra, India; Academy of Scientific and Innovative Research (AcSIR), Ghaziabad, Uttar Pradesh, India; Department of Biology, Indian Institute of Science Education and Research Pune, Pune, Maharashtra, India

Lauren R. Dorsett
School of Chemistry and Physics, Queensland University of Technology, Brisbane, QLD, Australia; Centre for Genomics and Personalised Health, Queensland University of Technology, Brisbane, QLD, Australia

Balint Dudas
Department of Biophysics and Radiation Biology, Semmelweis University, Budapest, Hungary; Laboratory of Computational Biology, National Heart, Lung, and Blood Institute, National Institutes of Health, Bethesda, MD, United States

Ioana-Daniela Dumitru
Faculty of Physics, University of Bucharest, Bucharest, Romania

Vito Ferro
School of Chemistry and Molecular Biosciences, The University of Queensland, Brisbane, QLD, Australia

Ran Friedman
Department of Chemistry and Biomedical Sciences, Linnaeus University, Kalmar, Sweden

Neha S. Gandhi
School of Chemistry and Physics, Queensland University of Technology, Brisbane, QLD, Australia; Centre for Genomics and Personalised Health, Queensland University of Technology, Brisbane, QLD, Australia; Department of Biotechnology, Manipal Institute of Technology, Manipal Academy of Higher Education, Manipal, Karnataka, India

Indira Ghosh
School of Computational and Integrative Sciences, Jawaharlal Nehru University, New Delhi, India

Luke Guddat
School of Chemistry and Molecular Biosciences, The University of Queensland, Brisbane, QLD, Australia

Alina-Maria Halaicu
Faculty of Physics, University of Bucharest, Bucharest, Romania

Matthias Heyden
School of Molecular Sciences, Arizona State University, Tempe, AZ, United States

Keiichi Inoue
The Institute for Solid State Physics, The University of Tokyo, Kashiwa, Japan; Graduate School of Frontier Sciences, The University of Tokyo, Kashiwa, Japan

Santosh Kumar Jha
Physical and Materials Chemistry Division, CSIR-National Chemical Laboratory, Pune, Maharashtra, India; Academy of Scientific and Innovative Research (AcSIR), Ghaziabad, Uttar Pradesh, India

Marko Jukić
Laboratory of Physical Chemistry and Chemical Thermodynamics, Faculty of Chemistry and Chemical Engineering, University of Maribor, Maribor, Slovenia; Faculty of Mathematics, Natural Sciences and Information Technologies, University of Primorska, Koper, Slovenia

Kota Katayama
Department of Life Science and Applied Chemistry, Nagoya Institute of Technology, OptoBioTechnology Research Center, Nagoya, Japan

Yoshitaka Kato
The Institute for Solid State Physics, The University of Tokyo, Kashiwa, Japan

Amrutha Keshavamurthy
CSIR-National Chemical Laboratory, Pune, Maharashtra, India; Academy of Scientific and Innovative Research, Ghaziabad, Uttar Pradesh, India

Amit Kumawat
Department of Physics, University of Cagliari, Cagliari, Italy

Sarbajit Layek
Department of Biological Sciences, Indian Institute of Science Education and Research (IISER) Kolkata, Mohanpur, Nadia, West Bengal, India

Samo Lešnik
Laboratory of Physical Chemistry and Chemical Thermodynamics, Faculty of Chemistry and Chemical Engineering, University of Maribor, Maribor, Slovenia; Institute of Environmental Protection and Sensors, Maribor, Slovenia

Priyanka Mahapatra
School of Chemistry, Sambalpur University, Burla, Sambalpur, Odisha, India

Sthitadhi Maiti
Department of Physics, University of Genoa, Genoa, Italy

Aditya Manoj
CSIR-National Chemical Laboratory, Pune, Maharashtra, India; Academy of Scientific and Innovative Research, Ghaziabad, Uttar Pradesh, India

Maria Mernea
Department of Anatomy, Animal Physiology and Biophysics, Faculty of Biology, University of Bucharest, Bucharest, Romania

Dan Florin Mihailescu
Faculty of Biology, Doctoral School of Biology, University of Bucharest, Bucharest, Romania; Department of Anatomy, Animal Physiology and Biophysics, Faculty of Biology, University of Bucharest, Bucharest, Romania

Ayori Mitsutake
Department of Physics, School of Science and Technology, Meiji University, Kawasaki, Kanagawa, Japan

Shailesh Kumar Panday
School of Computational and Integrative Sciences, Jawaharlal Nehru University, New Delhi, India

Swaratmika Pandia
School of Chemistry, Sambalpur University, Burla, Sambalpur, Odisha, India

Meenakshi Pillai
Physical and Materials Chemistry Division, CSIR-National Chemical Laboratory, Pune, Maharashtra, India; Academy of Scientific and Innovative Research (AcSIR), Ghaziabad, Uttar Pradesh, India; Department of Biochemistry and Molecular Biology, Bloomberg School of Public Health, Johns Hopkins University, Baltimore, MD, United States

Durba Sengupta
Physical Chemistry Division, CSIR-National Chemical Laboratory, Pune, Maharashtra, India

Neelanjana Sengupta
Department of Biological Sciences, Indian Institute of Science Education and Research (IISER) Kolkata, Mohanpur, Nadia, West Bengal, India

Shun Yokoi
Department of Physics, School of Science and Technology, Meiji University, Kawasaki, Kanagawa, Japan; International Institute for Integrative Sleep Medicine (WPI-IIIS), Tsukuba Institute for Advanced Research (TIAR), University of Tsukuba, Tsukuba, Ibaraki, Japan; Department of Structural Biology, School of Medicine, Stanford University, Stanford, CA, United States

Acknowledgment

We thank *Alina-Maria Halaicu, University of Bucharest, Faculty of Physics* for the cover art.

CHAPTER

Introduction

1

Ana-Nicoleta Bondar[a,b,*] and Durba Sengupta[c,*]

[a]*Faculty of Physics, University of Bucharest, Bucharest, Romania,* [b]*Faculty of Interdisciplinary Sciences, University of Bucharest, Bucharest, Romania,* [c]*Physical Chemistry Division, CSIR-National Chemical Laboratory, Pune, Maharashtra, India*

**Corresponding authors: a.bondar@fz-juelich.de; d.sengupta@ncl.res.in*

1.1 Overview

Protein dynamics is integral to biological function, encompassing a broad spectrum of motion across multiple time and length scales, from fast bond vibrations to slower time-scale organization of protein complexes. The ability of proteins to interact with their environment, bind to ligands, and subsequently associate with other biomolecules is fundamentally governed by their underlying dynamics. For instance, protein–water interactions not just only dictate protein folding but also biomolecular flexibility and have been shown to be essential to function, such as enzyme catalysis and antifreeze protein activity (Laage et al., 2017). Similarly, protein–lipid interactions regulate membrane protein functional dynamics and can help stabilize specific, often transient, conformational states such as active states of ion channels and receptors (Corradi et al., 2019). Importantly, molecular recognition, such as binding to ligands, substrates, or other proteins, is a dynamic process where the protein samples various conformations until an optimal fit to the partner is obtained (Weikl & Paul, 2014). The binding event itself may involve a subtle "induced fit" or a "conformational selection" mechanism, both dependent on the flexibility and plasticity of the protein (Csermely et al., 2010).

A hierarchical energy landscape governs the structural dynamics of proteins across a wide range of timescales. An overview of the timescales associated with these phenomena is shown in Fig. 1.1. At the fastest end of the dynamic spectrum, subpicosecond bond vibrations constitute the fundamental "jiggling and wiggling" of atoms (Durrant & McCammon, 2011; Orellana, 2019). Up the scale, picosecond side-chain rotamer sampling and nanosecond hinge and loop fluctuations emerge as key drivers of protein dynamics. Atomic fluctuations on the picosecond-to-nanosecond timescale within hinge regions link to the slower, large-scale domain rearrangements such as those required to achieve a catalytically competent state in several enzymes (Henzler-Wildman et al., 2007). Intermediate and slow dynamics, spanning the microsecond-to-millisecond range, involve large-scale domain motions that are often the primary drivers of protein function. For instance, domain motion in molecular motors and enzymes occurs within these timescales. Similarly, within the conformational ensemble sampled by membrane receptors such as G protein-coupled receptors (GPCRs), these timescales govern the transitions that allow ligand binding to shift the equilibrium and selectively stabilize active states. Similarly, concerted

Protein Dynamics: Experimental and Computational Approaches. DOI: https://doi.org/10.1016/B978-0-443-30042-4.00019-5

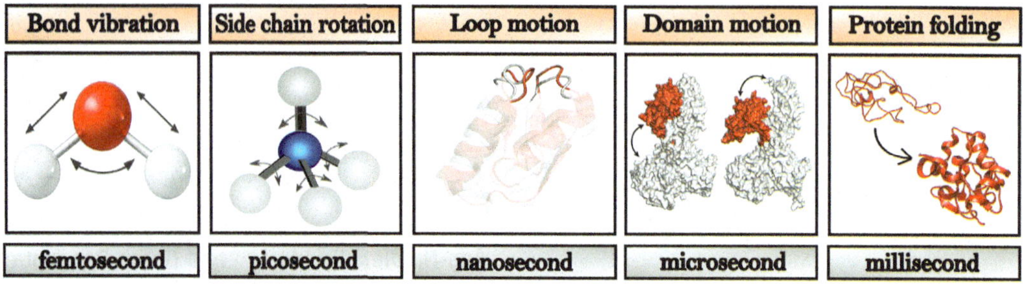

FIGURE 1.1

Timescales associated with protein dynamics: A few representative examples of different timescales sampled, from bond vibration to domain motion and protein folding, are shown.

motion on the millisecond-to-second scale is exhibited by proteins such as membrane transporters, where synchronized, large-scale domain closures drive net, one-directional substrate transport against a gradient. Intrinsically disordered proteins (IDPs) represent a unique set of proteins that, despite their reputation for high flexibility, exhibit significant slow-scale dynamics driven by the transient formation of secondary structures and the binding-induced folding upon partner recognition. Across all scales, protein dynamics is unified by allosteric networks, where a local perturbation shifts the entire conformational ensemble, propagating a signal to a distant site via dynamically correlated motions.

1.2 The free-energy landscape: Governing protein dynamics and function

Understanding protein function necessitates a molecular thermodynamics view, which regards proteins not as rigid structures but as dynamic conformational ensembles diffusing through the basins within the free-energy landscape. The Gibbs free energy (ΔG) is the fundamental thermodynamic determinant of protein dynamics within this landscape, in which the precise interplay between enthalpy (ΔH) and entropy (ΔS) dictates the conformational ensemble. For instance, enthalpy–entropy compensation drives molecular recognition, and high-affinity binding is usually achieved when the entropic penalty arising from the loss of translational/rotational freedom and conformational flexibility is adequately compensated by favorable enthalpic interactions and entropic gain of the solvent.

Chapter 2 (Chakraborty and Kumawat) highlights molecular thermodynamics from the perspective of biomolecular recognition and allostery. By framing proteins as dynamic ensembles on free-energy landscapes the chapter discusses how signals, such as ligands or even posttranslational modifications, shift state populations to drive biological outcomes. In particular, Chapter 2 describes a few of these key concepts with the help of several case studies, including dynamic allostery in PDZ domains, transiently populated metastable states in Rho GTPases and the HipA toxin, and the impact of PTMs and mutations on conformational ensembles of RhoA-RhoGDI and G6PD. Further the chapter discusses how to leverage these allosteric principles for drug discovery targeting protein–protein interactions like PCSK9-LDLR.

Chapter 3 (Ghosh and Panday) discusses the effect of entropy in the context of protein dynamics and interactions. Entropy, a measure of disorder, is a critical (but often antagonistic) determinant of stability and affinity, governing binding affinity through the Gibbs–Helmholtz equation. Experimental approaches such as isothermal titration calorimetry (ITC) can be used to dissect enthalpic and entropic contributions and, in conjunction with computational approaches, can reveal how molecular flexibility drives binding. Accounting for configurational entropy in dynamic targets is essential for rational therapeutic design and accurate affinity predictions. Specific examples of flexible targets, such as PfAMA1 and PfPK5, are used to demonstrate that accurate prediction of binding affinities necessitates incorporating higher order configurational entropy, and neglecting these contributions leads to inaccurate results.

1.3 Role of environment

The dynamics of the surface residues are intimately coupled with the dynamics of the solvent, contributing to the fast, local conformational fluctuations in the protein, which propagate to the entire protein. Further, crowding effects have been reported to restrict the available conformations and favor more compact states, often stabilizing the folded form or promoting the association of protein complexes. In this context, high local viscosity (microviscosity) may slow down the rate of large-scale domain motion, effectively acting as regulator. A critical emerging facet of protein dynamic regulation is the quinary structure, defined by the transient, weak, and nonspecific interactions that occur between macromolecules within the densely crowded cellular interior. Consequently the free-energy landscape of a protein is not just determined by its own intrinsic dynamics but is constantly being reshaped by its neighbors.

1.3.1 Role of water

Water shapes protein dynamics and protein function. Irrespective of where they are in the cell and what their function is, proteins interact with water. Water is not just a "bystander," but it exerts a significant influence on protein dynamics and can play direct roles in protein function. At a hydration level below 3%, for example, protein motions are hindered and enzymes no longer perform their catalytic activity (Kurkal et al., 2005). Membrane proteins that function as transporters or receptors often rely on networks of hydrogen bonds between protein groups and protein-bound water to interconnect functionally important sites. In the case of membrane proteins that transfer protons across membranes, water molecules can be critical to establish transient paths for protons to be transferred.

Chapter 4 (Layek and Sengupta) discusses an overview of the role of protein hydration water from the experimental and computational perspectives. It starts with a brief review of the unique physical–chemical properties of water, including the concept of "hydration shell" of biomolecules, which consists of a shell of a few water layers that surrounds the biomolecule; waters of the hydration shell have properties distinct from bulk water—their motions can be slower, for example, as discussed previously (Lorch et al., 2015). The chapter presents a detailed account of the structure of the water hydration shell and of the experiments and simulations that have illuminated the properties of waters from the vicinity of biomolecules. Key properties of water hydrogen-bond networks in liquid water are introduced, and parameters that can be computed to characterize water structure and dynamics are discussed. The chapter further introduces fundamental aspects of hydration thermodynamics and modern computational approaches to characterize the equilibrium distribution of water near solutes.

It also gives a thorough discussion of water dynamics at biomolecule interfaces, including water orientational dynamics and water behavior in crowded cellular environments.

Chapter 5 (Lešnik, Jukić, and Bren) discusses the importance of water from the perspective of drug design. Water molecules bound to proteins, and especially those conserved in structures of related proteins, are essential for protein structure, dynamics, and function. Of central interest to drug design are the waters found at ligand-binding sites, as these waters may either be replaced or retained when the ligand binds and bring key contributions to the overall free energy of binding.

Structural biology has provided invaluable information about the location and role of water in drug binding. But high resolution structures are required to accurately solve water coordinates; mobile and low-occupancy water molecules could be difficult to solve, and the likelihood that a water molecule occupies a protein cavity can also depend on the temperature (Bondar et al., 2008). Computational approaches are invaluable for the study of the role of water in drug binding. The chapter gives a thorough discussion of conserved waters at protein–protein interfaces and their role. An intriguing example discussed is that of the programmed death receptor (PD1) and its ligand, programmed death ligand 1 (PDL1). Formation of this receptor–ligand complex is implicated in key signaling pathways of the immune system; water participates in the hydrogen-bond network at the interface between the receptor and the ligand. Computational studies of the role of water in drug binding are pursued with specialized tools that are thoroughly reviewed in the chapter. The tools reviewed are grouped according to the methodology used, and specific examples of tools and their usefulness are given. The chapter further explores the current status of the research in the field of conserved water molecules in drug design and the role of water-mediated hydrogen-bond networks in receptor signal propagation across membranes.

1.3.2 Role of lipids

Protein–lipid interactions are fundamentally critical for the function and dynamics of membrane proteins, which constitute a significant fraction of the proteome and mediate essential cellular processes such as signaling and transport. Our understanding of protein–lipid interactions has evolved from a passive, scaffold-like role to a dynamic, bidirectional communication, directly modulating the protein's conformational ensemble. Lipids frequently act as dynamic allosteric regulators by stabilizing or promoting transitions between functional states. The lipids surround a protein can be classified as annular, nonannular, and bulk lipids (Paila, Tiwari, & Chattopadhyay, 2009). The nonannular lipids bind strongly to membrane proteins, exhibit slow exchange with bulk lipids, and contribute to the specific lipid effects. For instance, specific lipids, such as phosphoinositides or cardiolipins, bind to these nonannular sites, inducing global conformational changes required to open or close ion channels. The annular lipids comprise the first lipid shells in the vicinity of the protein and, together with the bulk lipids, modulate protein function through nonspecific effects. In particular the annular lipids surrounding the transmembrane domain can act as a lubricant, enabling rapid structural changes necessary for transporter function, or, conversely, can rigidify specific domains to stabilize inactive conformations. Furthermore, hydrophobic mismatch, the geometrical difference between the protein and bilayer thickness, exerts mechanical stress and lateral pressure on the protein, directly altering its free-energy landscape and biasing the equilibrium toward particular oligomeric (non)-functional states (Prasanna et al., 2014).

The lipid bilayer also defines the essential environment for protein dynamics and localization. Membrane viscosity and fluidity, determined by lipid composition, directly regulate the lateral diffusion of protein, thus controlling the efficiency of signaling complex assembly. Specific lipid components, such as cholesterol and sphingolipidscan organize into microdomains, partitioning proteins to regulate their oligomerization and provide stable signaling platforms. Finally, protein–lipid coupling is necessary for the integrity and assembly of multisubunit complexes. Lipids often stabilize dimer and oligomer interfaces, where their presence is mandatory for achieving the correct quaternary structure and subsequent functional dynamics (Prasanna et al., 2020).

Chapter 6 (Pandia, Mahapatra and Chakraborty) provides a comprehensive overview of membrane effects, both specific and nonspecific effects. A few of the important lipid effects focusing on membrane-active peptides, in particular antimicrobial peptides and fusion peptides, are discussed. An important aspect to highlight in this context is that protein–lipid effects are reciprocal and can modulate each other. Membrane-active proteins involved in processes such as membrane fusion and budding induce local lipid restructuring and high-curvature stress, demonstrating the inextricable link between lipid environment dynamics and the protein function.

1.4 Aggregation dynamics and binding-induced folding of intrinsically disordered proteins and intrinsically disordered regions

The molecular pathology of neurodegenerative disorders, including Alzheimer's and Parkinson's, is fundamentally linked to the aggregation dynamics of IDPs and proteins containing intrinsically disordered regions (IDRs). Unlike their globular counterparts the free-energy landscape of IDPs and IDRs is characteristically flat and highly rugged. This landscape features numerous shallow minima, enabling IDPs and IDRs to exist as a vast, highly flexible conformational ensemble and engage in functional promiscuity. Crucially, this high flexibility imparts only marginal stability, making the ensemble critically sensitive to minor changes in the cellular environment (e.g., pH, mutations). Such perturbations can easily tilt the thermodynamic balance toward the formation of compact, aggregation-prone states, initiating the pathological transition from functional ensemble to ordered, β-sheet-rich amyloid fibrils. The conversion to amyloids is a dynamic process characterized by a significant nucleation barrier. The aggregation pathway is further complicated by the fact that the most cytotoxic species are often transient, kinetically trapped, soluble oligomers, representing shallow, high-energy minima. Moreover, the dynamic interplay of IDPs and IDRs has been shown to drive liquid–liquid phase separation, leading to the formation of dense liquid droplets. Within these concentrated microenvironments the local protein concentration drastically lowers the nucleation barrier, promoting the pathological maturation of the liquid state into irreversible, solid-like fibrillar aggregates, thus linking IDP and IDR dynamics directly to disease progression.

Chapter 7 (Pillai, Doke and Jha) focuses on the example of TAR DNA-binding protein 43 (TDP-43), whose aggregation drives 97% of ALS cases, yet the properties of the full-length protein remain elusive. This chapter explores the shallow energy landscape and structural polymorphism in the context of TDP-43, emphasizing how stabilizing the conformational ensemble of the native states can be a strategy for rational drug design.

Importantly the shallow energy landscape of IDPs and IDRs facilitates a transition toward a more constrained conformational ensemble upon molecular recognition, serving as a primary mechanism for the induction of functional structure through partner interactions. IDPs and IDRs maintain a highly dynamic conformational ensemble in their unbound (apo) state, characterized by rapid interconversion between numerous extended and collapsed structures rather than adopting single rigid structures. This structural plasticity allows it to rapidly encounter and loosely associate with the target partner, accelerating the initial binding step. This transient encounter is then followed by a conformational selection and/or induced fit process, where the interactions with the binding partner stabilize a specific, usually structured, conformation of the IDP from its existing ensemble. The stabilization of the IDB conformation shifts the thermodynamic equilibrium, dramatically lowering the free energy of the bound state, and results in the *apparent* folding of the peptide into a defined structure.

Chapter 8 (Heyden and Maiti) highlights molecular recognition features (MoRFs) that are specific functional regions within IDRs that undergo transitions from disorder to order upon binding, such as the transition from a disordered state to an α-helix structure in α-MoRFs. Using the specific example of α-MoRFs bound to the MCL-1 host protein the chapter addresses the challenge of "undruggability" of IDPs and IDRs by testing and validating improved methodologies to compute the binding free energy of the complex binding-induced folding process.

1.5 Approaches to delineate protein dynamics

A comprehensive understanding of protein dynamics necessitates the use of a complementary toolbox comprising both experimental and computational methodologies. Several well-established approaches have successfully analyzed certain aspects. For instance, for the fast to intermediate timescales (ps to ms), nuclear magnetic resonance (NMR) spectroscopy is important, offering residue-specific information on internal motion. Concurrently, fluorescence spectroscopy (for instance, via anisotropy decay) monitors the rotational mobility and local environment of the probes. To access slower, large-scale dynamics, methods such as hydrogen/deuterium exchange mass spectrometry measure solvent accessibility governed by backbone breathing motions, providing indirect rates of conformational opening. Emerging approaches such as terahertz (THz) spectroscopy probe novel aspects of protein dynamics by directly measuring the collective, low-frequency vibrational modes. To bridge the spatiotemporal gaps inherent in experimental data, computational methods have been extensively used. Molecular dynamics (MD) simulations generate atomic-level trajectories, providing a time-resolved view of dynamics.

An emerging approach in probing protein dynamics is the use of THz spectroscopy (0.1–10 THz) that offers a unique and noninvasive approach to analyze protein dynamics by directly measuring the collective, low-frequency vibrational modes that govern function. The energy scale of the THz range correlates with the weak noncovalent interactions that dictate protein global structure and intrinsic flexibility. By measuring the absorption or transmission of radiation, THz spectroscopy captures the concerted motions of large groups of atoms, domains, or the entire macromolecule and bridges the time gap between ultrafast bond vibrations and slow, large-scale conformational changes. A pivotal application of THz spectroscopy lies in characterizing the protein–solvent coupling. The technique is sensitive to the dynamics of the hydration shell water, which is significantly slower and more structurally restricted compared to bulk water, yet essential for biological activity. Observable changes in the THz spectrum following ligand binding or aggregation provide direct evidence regarding the modulation of this hydration layer, supplying crucial information on the dynamic interplay between the protein and its immediate solvent environment. Furthermore, changes in these low-frequency collective modes upon effector binding are central to understanding allostery, reflecting the ligand-induced stabilization or restriction of the conformational ensemble of the protein. For enzymes, THz modes can be correlated with the collective motions necessary to attain the transition state, thus offering indirect evidence of the dynamic mechanisms employed to lower the activation energy barrier.

Chapter 9 (Mihailescu and Mernea) discusses the use of THz spectroscopic and modeling approaches to elucidate subtle dynamical features, hydration shell behavior, and the effects of posttranslational modifications. A series of case studies, from ligand binding in myoglobin and HIV-1 protease to evolutionary divergence in lysozymes and disease mutations in rhodopsin, is discussed.

Molecular simulations, particularly MD simulations, provide unprecedented insights into molecular-level protein dynamics. Classical simulations can sample up to the submillisecond range with current computing approaches. Enhanced sampling methods (e.g., accelerated MD, metadynamics) extend the sampling to the ms timescale, enabling the accurate mapping of the free-energy landscape and the capture of rare events such as large domain motions.

Chapter 10 (Friedman) details important aspects of MD simulations to study large molecular transitions and elucidate the molecular processes underlying enzyme activation. MD simulations are able to sample the high-dimensional conformational space to characterize the transient, marginally populated precatalytic states required for optimal substrate and residue alignment. Furthermore the role of intrinsic thermal fluctuations and dynamic motions in modulating activation barriers can be elucidated. Enhanced sampling techniques can be used to explicitly compute the free energy profile along a known reaction coordinate, providing quantitative measures of entropic and solvation effects that drive catalytic reactions.

Another major contribution of MD simulations is the capability to model large-scale, cooperative conformational dynamics, which are essential for processes such as transport that occur on micro- to millisecond timescales. **Chapter 11 (Dudas and Balog)** discusses different aspects of MD simulations to sample domain dynamics and how to integrate complementary approaches such as enhanced sampling, alchemical calculations, and coarse-grained modeling to improve the efficiency of sampling at these timescales. The chapter discusses the synergy between computational and experimental frameworks that helps facilitate the discovery of mechanistic pathways, essential for advancing drug discovery and biomolecular research.

1.6 Protein dynamics govern catalysis, misfolding, and pharmacological intervention

The interplay between enzyme mechanism, disease-related mutations, and the development of therapeutics represents an area where protein dynamics serve as the crucial link connecting genetic sequence to functional and pathological outcomes. **Chapter 12 (Gandhi et al.)** takes the example of Sanfilippo syndrome to highlight the development of specific therapeutics that function by altering protein dynamics of mutant proteins. Sanfilippo syndrome, or mucopolysaccharidosis III, is a severe lysosomal storage disorder that typically manifests in young children through cognitive regression and developmental delays. The etiology of the disease lies in mutations of the enzymes heparan-*N*-sulfatase (SGSH) and *N*-acetyl-α-glucosaminidase, which drive the Type A and Type B variants, respectively. The chapter introduces the concept of pharmacological chaperones that are small molecules that act similarly to protein chaperones and selectively stabilize destabilized mutant proteins, shifting conformational ensembles toward functional states. The development of such pharmacological chaperones is described in detail in the chapter.

1.7 Membrane receptors and transporters

Membrane receptors and transporters mediate signaling across cellular membranes. Fundamental common aspects include coupling between protein conformational dynamics and the binding of ligand (receptors) or solute (transporters) and an important role of the fluid lipid membrane environment in protein function (Kharche & Sengupta, 2020).

Chapter 13 (Andersson) presents a comprehensive overview of membrane transporters, with a focus on biophysics studies of intermediate conformational states. To discuss the intermediate states of transporters the chapter first introduces the alternate access model, passive vs. active transport, and primary and secondary transport, and then it discusses the structural dynamics of several key model systems from a structural biology perspective. (1) The apparently simple bacteriorhodopsin protein pump has served as model to understand general physical chemical principles of cellular ion pumps. The chapter summarizes the reaction cycle of bacteriorhodopsin and X-ray crystallography studies that illuminated the structural dynamics of the intermediate states. (2) Transporters of the major facilitator superfamily (MFS) have a typical fold of two bundles of six transmembrane helices each interconnected by a loop. MFS transporters, particularly lactose permease (LacY) from *Escherichia coli* and glucose transporters, are used to discuss the alternating model for solute transport across membranes. (3) P-type ATPases use the energy from the binding and hydrolysis of adenosine triphosphate (ATP) to mediate active transport across membranes. Success stories from trapping intermediate states of P-type ATPases are discussed together with limitations of the approach and advanced computational approaches that can be used for heterogenous refinement. The chapter further includes an overview of modern structural biology techniques—time-resolved cryo-EM, time-resolved serial femtosecond crystallography, and time-resolved solution X-ray scattering, with milestones in solving protein structural dynamics.

Chapter 14 (Inoue and Kato) reviews experimental studies of xenorhodopsins (XeRs) and schizorhodopsins (SzRs), which are inward proton pumps from the large family of microbial rhodopsins. These pumps are all from the same family of proteins as the bacteriorhodopsin outward proton pump . Understanding how microbial proton pumps translocate protons across the membrane (inward or outward) is of general interest to protein sequence–structure–function relationships in mem; some of these pumps are also of interest as tools for optogenetics applications to the pH in specific cells and cell compartments. As discussed in this chapter, inward proton pumps have been discovered relatively recently, and many fundamental open questions remain open. The mechanisms of XeRs and SzRs are discussed based largely on bioinformatics, functional metagenomics, spectroscopy, and structural biology studies. To clarify key aspects pertaining to the inward proton pump rhodopsins the chapter refers to important studies on other microbial pump rhodopsins, including outward sodium-pumping rhodopsins and inward chloride-pumping rhodopsins. Internal hydrogen-bond networks of protein sidechains and protein-bound water molecules are particularly important for the proton-transfer mechanisms discussed. Among the examples that illustrate the key role of hydrogen bonding, the substitution of the helix G counterion (D212 in bacteriorhodopsin and D184 in SzR4) with Pro in XeR disrupts interactions at the extracellular side of the transporter; this difference in sequence at a key functional site of the transporter, and other examples discussed in the chapter, highlight the importance of characterizing the reaction mechanisms of different microbial pump rhodopsins. The chapter discusses the reaction cycles of representative pump rhodopsins, with focus on the similarities and differences between the reaction cycles of XeRs and SzRs. A key structural element is the primary proton donor, the retinal Schiff base, whose orientation toward the extracellular or cytoplasmic side of the membrane depends on the isomeric state and interactions with water molecules and protein groups from the retinal binding site. An intriguing aspect of microbial pump rhodopsins is that, for some of them, functional interconversion could be achieved. The chapter includes a brief discussion of a functional interconversion between an outward and an inward proton pump, as well as applications of inward proton pumps as optogenetic tools.

Chapter 15 (Yokoi and Mitsutake) presents computational studies of GPCRs with a focus on MD and relaxation mode analyses (RMA) of the orexin-2 receptor, OX2R. This is a class A GPCR implicated in sleep regulation. Prolonged, microsecond timescale MD simulations have characterized motions of OX2R in different conformational states—apo wild-type versus mutant, inactive, bound to antagonist or agonist ligands. A wealth of information extracted from this large dataset (36 microseconds in total) is discussed in the chapter. This included detailed structural rearrangements at key functional sites. To discuss findings obtained with RMA the method is first introduced together with specific parameters used for the OX2R simulations. RMA data reviewed in the chapter identified motions of transmembrane helices during receptor activation—which could be characterized by detailed interatomic distance analyses—and changes in the atomic fluctuations of specific protein regions. Activation thus associates with changes in protein flexibility at functionally important sites of the receptor.

Chapter 16 (Katayama) reviews Fourier transform infrared (FTIR) spectroscopy studies of GPCRs. FTIR difference spectroscopy has been invaluable for studies of rhodopsins—these are class A GPCRs that contain retinal, whose Schiff base NH bond provides an invaluable probe of the retinal–counterion interactions. A fundamental question addressed in the chapter is how rhodopsins of different amino acid residue sequences use the same chromophore, the covalently bound retinal, to mediate scotopic vision and to discriminate between colors. The fundamentals of spectral tuning by rhodopsins are discussed by first introducing key structural elements and interactions directly relevant to spectral tuning in rhodopsin and cone pigments (of which there are four groups), the fundamentals of vibrational spectroscopy for a membrane protein, FTIR, and Raman spectroscopy of cone pigments. Compared to rhodopsins, cone pigments have low expression levels, which made it necessary to use large culture plates and reconstitute the proteins into lipid liposomes. The extensive spectroscopy studies that could be performed indicated that the strength of the hydrogen-bonding strength of the Schiff base NH bond is the same in red and green cone pigments—thus, rather counterintuitively, that it does not have a direct influence on the absorption maximum of the retinal chromophore. The retinal hydrogen-bond network is distinct in rhodopsin vs. cone pigments, and dipole moments of protein-bound water influence the spectral tuning—including in red and green cone pigments. The chapter further presents an overview of attenuated total reflection FTIR studies of ion binding to cone pigments; FTIR spectroscopy of the muscarinic M2 receptor, M2R; and modern applications of FTIR in drug discovery.

Chapter 17 (Manoj and Sengupta) discusses recent advances in GPCR biology that challenge the traditional "one receptor–one G protein" dogma, revealing that receptors can actively engage diverse, noncognate G proteins. This functional plasticity is governed by the intricate conformational dynamics of the receptor–transducer complex, which balance structural selectivity with the flexibility required to activate downstream signaling pathways. Further the chapter describes a specific example of how lipid effects may contribute to transmembrane proteins, such as G-protein coupling specificity in GPCRs, by binding at the interface of GPCR–lipid interactions.

1.8 Hands-on student tutorials: Data integration and meta-analyses of protein structure and dynamics

The comprehensive understanding of protein dynamics requires beginners in this area to move beyond single, isolated observations to construct universally applicable models of protein function and dysfunction. One such approach is to use rigorous meta-analysis and integration of existing structural data to understand the emergent protein dynamics. For students, engaging with this integrated data paradigm is a cornerstone of modern scientific training. It fosters the development of critical evaluation skills as they must synthesize information, weigh the limitations of each technique, and use computational tools to manage and analyze large, heterogeneous data sets.

Chapter 18 (Keshavamurthy and Sengupta) provides hands-on exercises that will help students bridge the understanding of protein structures to structural dynamics, highlighting a holistic conceptualization of proteins as dynamic, multiscale entities. The exercises explore structures resolved by different experimental techniques, such as NMR, crystallography, and cryo-EM, to highlight salient features of each approach. The chapter will guide students through their first steps in visualization of protein structures and hands-on coding exercises to compare protein structures.

Chapter 19 (Dumitru, Halaicu, and Bondar) provides a foundational overview of bioinformatics and molecular modeling applied to membrane receptors, using three distinct class A GPCRs, photosensors, chemokine receptors, and proton sensors, as representative models. The chapter is structured for practical computational instruction, guiding students through the essential hierarchy of receptor biology, from primary amino acid sequences to three-dimensional architectures and their resulting structure–function relationships.

In summary, this volume elucidates emerging concepts in protein dynamics, emphasizing novel methodologies and illustrative applications. By balancing cutting-edge research with foundational tutorials the text provides a comprehensive framework for students entering the field. As highlighted in this book, the functional dynamics of biomolecules, in particular proteins, orchestrates cellular function. The classical structure–function paradigm in proteins has been well established, and often a single conformer or topology has been linked to the functional form of that protein. With improved resolution and accuracy of experimental and computational approaches the role of flexibility and dynamics is being highlighted, and a structure-dynamics-function paradigm can be envisaged. The conformational dynamics of proteins occur in a hierarchical fashion at varying time- and length-scales. At one end of the scale, reactions catalyzed by enzymes are characterized by fast dynamics of binding site residues at atomistic level. At the other end of the timescale, molecular recognition between domains from one or more proteins during signal transduction usually occurs over milliseconds and several nanometers. Another aspect of protein conformational dynamics is the plasticity in the folded and nonfolded protein states. In fact, IDPs and domains have been shown to be critical for function and often linked to multiple functions with varying conformational states. Although several facets of protein dynamics at these varying levels have been probed by experimental and computational approaches, the complex interplay at the different levels is only now emerging. Ultimately this volume addresses evolving paradigms in protein conformational dynamics and its importance in function, with focus on modern computational approaches.

Acknowledgments

We thank Amrutha Keshavamurthy and Sowmomita Gharui for help with figures.

References

Bondar, A.-N., Baudry, J., Suhai, S., Fischer, S., & Smith, J. C. (2008). Key role of active-site water molecules in bacteriorhodopsin proton-transfer reactions. *The Journal of Physical Chemistry B, 112*(47), 14729–14741. doi:10.1021/jp801916f.

Corradi, V., Sejdiu, B. I., Mesa-Galloso, H., Abdizadeh, H., Noskov, S. Y., Marrink, S. J., & Tieleman, D. P. (2019). Emerging diversity in lipid–protein interactions. *Chemical Reviews, 119*(9), 5775–5848. doi:10.1021/acs.chemrev.8b00451.

Csermely, P., Palotai, R., & Nussinov, R. (2010). Induced fit, conformational selection and independent dynamic segments: An extended view of binding events. *Trends in Biochemical Sciences, 35*(10), 539–546. doi:10.1016/j.tibs.2010.04.009.

Durrant, J. D., & McCammon, J. A. (2011). Molecular dynamics simulations and drug discovery. *BMC Biology, 9*(1). doi:10.1186/1741-7007-9-71.

Henzler-Wildman, K. A., Lei, M., Thai, V., Kerns, S. J., Karplus, M., & Kern, D. (2007). A hierarchy of timescales in protein dynamics is linked to enzyme catalysis. *Nature, 450*(7171), 913–916. doi:10.1038/nature06407.

Kharche, S. A., & Sengupta, D. (2020). Dynamic protein interfaces and conformational landscapes of membrane protein complexes. *Current Opinion in Structural Biology, 61*, 191–197. doi:10.1016/j.sbi.2020.01.001.

Kurkal, V., Daniel, R. M., Finney, J. L., Tehei, M., Dunn, R. V., & Smith, J. C. (2005). Enzyme activity and flexibility at very low hydration. *Biophysical Journal, 89*(2), 1282–1287. doi:10.1529/biophysj.104.058677.

Laage, D., Elsaesser, T., & Hynes, J. T. (2017). Water dynamics in the hydration shells of biomolecules. *Chemical Reviews, 117*(16), 10694–10725. doi:10.1021/acs.chemrev.6b00765.

Lorch, S., Capponi, S., Pieront, F., & Bondar, A.-N. (2015). Dynamic carboxylate/water networks on the surface of the PsbO subunit of photosystem II. *The Journal of Physical Chemistry B, 119*(37), 12172–12181. doi:10.1021/acs.jpcb.5b06594.

Orellana, L. (2019). Large-scale conformational changes and protein function: Breaking the in silico barrier. *Frontiers in Molecular Biosciences, 6*. doi:10.3389/fmolb.2019.00117.

Paila, Y. D., Tiwari, S., & Chattopadhyay, A. (2009). Are specific nonannular cholesterol binding sites present in G-protein coupled receptors? *Biochimica et Biophysica Acta (BBA)-Biomembranes, 1788*, 295–302. doi:10.1016/j.bbamem.2008.11.020.

Prasanna, X., Chattopadhyay, A., & Sengupta, D. (2014). Role of lipid-mediated effects in β2-adrenergic receptor dimerization. In A. Chakrabarti, & A. Surolia (Eds.), *Biochemical roles of eukaryotic cell surface macromolecules* (pp. 247–261). Springer International Publishing.

Prasanna, X., Mohole, M., Chattopadhyay, A., & Sengupta, D. (2020). Role of cholesterol-mediated effects in GPCR heterodimers. *Chemistry and Physics of Lipids, 227*, 104852. doi:10.1016/j.chemphyslip.2019.104852.

Weikl, T. R., & Paul, F. (2014). Conformational selection in protein binding and function. *Protein Science, 23*(11), 1508–1518. doi:10.1002/pro.2539.

CHAPTER

2 Molecular thermodynamics view of biomolecular recognition, signaling, and allostery

Amit Kumawat[a] and Suman Chakrabarty[b,*]

[a]*Department of Physics, University of Cagliari, Cagliari, Italy,* [b]*Department of Chemical and Biological Sciences, S. N. Bose National Centre for Basic Sciences, Kolkata, West Bengal, India*

**Corresponding author: chakrabarty.suman@gmail.com*

2.1 Introduction

Biomolecular recognition, signaling, and allostery are fundamental processes that underpin all aspects of life, governing cellular functions ranging from metabolism and gene regulation to signal transduction and immune responses (Fenwick et al., 2011; Weis & Kobilka, 2018). These processes involve a complex and organized chain of extra-, intra-, and intercellular communication networks, also referred to as signaling pathways, which comprise interconnected cascades triggered by a wide range of signals. These signals can range from simple physical stimuli (e.g., light, pressure, or heat) to more complex biochemical events such as chemical stimuli (e.g., metal ions, ligands, hormones, neurotransmitters), posttranslational modifications (PTMs), changes in protonation state, or mutations (Artemenko et al., 2016; Du & Lovly, 2018; Duan et al., 2015; Stroka & Konstantopoulos, 2014). A basic biological signaling response involves three distinct steps: activation of a biomolecule in response to a signal, molecular recognition of this active state by a downstream effector, and subsequent propagation of the signal to other cellular components. These intricate steps are orchestrated at the molecular level by a complex interplay of biophysical interactions and conformational dynamics within and between biomolecules (e.g., proteins and nucleic acids, protein–protein, and protein–ligand) that ultimately trigger defined biological responses. The inherent complexity of these signaling processes raises fundamental questions about how these intricate molecular events translate into the observable physiological functions we encounter every day.

A classic overview of this complexity is the example of the G protein-coupled receptor (GPCR), which illustrates the role of signaling pathways in fundamental physiological functions in all domains of life (Weis & Kobilka, 2018). For example, it is remarkable how a seemingly simple signal like light can be sensed and utilized by photoreceptor proteins (a subset of GPCRs) not only for vision but also for nonvisual functions, such as regulating circadian rhythms. The light sensitivity of these proteins arises from bound chromophores (such as retinal, bilin, or flavin) that absorb photons for energy conversion or the initiation of intra- or intercellular signaling. In mammals, rhodopsin is one such membrane protein (photoreceptor) that detects light in the rod photoreceptor cell and is activated by the isomerization of its covalently bound cofactor, retinal, from 11-*cis* to all-*trans* (Hubbard & Kropf, 1958; Palczewski et al., 2000). Like other GPCRs, upon activation, rhodopsin catalyzes the

Protein Dynamics: Experimental and Computational Approaches. DOI: https://doi.org/10.1016/B978-0-443-30042-4.00003-1

replacement of GDP by GTP on the α subunit of a heterotrimeric G protein, transducin, which is bound to rhodopsin. This, in turn, triggers the response of the rod cells in the retina. An essential question is how the isomerization signal propagates efficiently from the middle of rhodopsin, where retinal is located, to the cytoplasmic surface. Moreover, the complexity of biomolecular signaling and recognition is evident in the fact that GPCRs can signal through multiple G-protein subtypes, as well as a single GPCR can bind various agonists and antagonists, leading to diverse cellular responses (Weis & Kobilka, 2018; Zhang et al., 2024b). Consider the μ-opioid receptor, a key target for pain management, which can activate different intracellular pathways depending on the specific ligand bound. For example, morphine binding primarily leads to analgesia, whereas other agonists, such as fentanyl or oxycodone, can have different effects on respiration or reward pathways (Koehl et al., 2018; Manglik et al., 2012). Thus, the signal transduction process involves a multitude of signals that are propagated through numerous protein–protein interactions (PPIs). Understanding how downstream signaling effectors recognize specific activation states triggered by a particular molecule from a large pool of potential agonists, which ultimately leads to precise biological outcomes, is a fascinating and challenging question that requires delving into the intricacies of signaling mechanisms at the molecular level.

Early efforts predominantly aimed to elucidate the molecular nature of the active and inactive states of signaling proteins within signaling pathways. With the advancements of biophysical and biochemical techniques, a structure-based paradigm emerged as a powerful tool to distinguish between these functional states. Long-standing examples, ranging from the classical crystallographic studies of hemoglobin to the detailed structural characterization of the conformational differences between active (GTP-bound) and inactive (GDP-bound) forms of proto-oncogenic Ras proteins, have been instrumental in establishing this paradigm (Ahmed et al., 2020; Milburn et al., 1990; Perrella & Russo, 2003; Simanshu et al., 2017). The availability of these static snapshots of protein structures revealed distinct structural conformations associated with different functional states. Crucially, these structural models provided a foundation for understanding how conformational changes induced by mutations, autophosphorylation, and GTP hydrolysis dictate their functional states (Grant et al., 2010). However, it is evident from several experimental and computational studies that the activation and deactivation mechanisms can no longer be described as a two-state phenomenon (open and closed/active and inactive) (Boehr et al., 2009; Nussinov et al., 2023).

Biomolecular recognition and signaling are increasingly understood as dynamic processes driven by the inherent flexibility and adaptability of biomolecules, where proteins and nucleic acids, in fact, exist as ensembles of interconverting conformations in equilibrium (Boehr et al., 2009; Nussinov, 2025; Nussinov et al., 2023). Biological outcomes often depend on population shifts within the conformational ensemble under different conditions, which can be effectively represented by a conformational free energy landscape (FEL) (Kar et al., 2010; Livesay, 2010). For example, agonists or antagonists binding to GPCRs or phosphorylation events in kinases selectively stabilize otherwise minor conformational states, shifting the ensemble toward functionally active or inactive conformations (Chong et al., 2024; Fleetwood et al., 2020; Latorraca et al., 2017; Lu et al., 2021; Tong & Seeliger, 2015). This thermodynamic perspective provides a crucial foundation for understanding allostery, a fundamental regulatory principle where a perturbation at one site of a biomolecule, such as ligand binding or PTMs (e.g., phosphorylation), induces functional and dynamical changes at a distinct, often distant site within the same molecule (Changeux & Edelstein, 2005; Kumawat et al., 2025; Montserrat-Canals et al., 2025; Perrella & Russo, 2003). Moreover, allosteric effects can be fundamentally understood as a population shift within the FEL where the binding of an effector molecule at one site on the biomolecule alters the relative free energies of the various conformational states, consequently influencing the structure and function of a distant site (Guo & Zhou, 2016; Livesay, 2010; Nussinov et al., 2023; Tsai et al., 2014). Therefore, allostery is not merely a matter of structural changes propagating through a molecule; it is intrinsically linked to changes in the thermodynamic equilibrium of the accessible conformational states (Gunasekaran et al., 2004). Elucidating how these population shifts in signaling pathways can be quantitatively described from a thermodynamic standpoint presents a compelling avenue for future investigation.

This chapter uses the concept of FELs as a unifying framework to interpret these diverse mechanisms of recognition, signaling, and allostery. It illustrates key concepts with case studies from our group, showing how certain conformational states (often more energetic or less populated at rest) are critical for biological function and how perturbations such as ligand binding, protonation, or phosphorylation shift conformational populations toward active or inactive states. Furthermore it will provide a brief introduction to the computational approaches [e.g., molecular dynamics (MD) simulations and data-driven methods] that can capture these dynamics and provide mechanistic insights, setting the stage for understanding the FELs that underlie biomolecular signaling and how they are remodeled by external signals.

2.2 Protein conformational ensembles and free energy landscape

2.2.1 From static structures to dynamic ensembles

Understanding the conformational landscape of proteins is fundamental to unraveling the mechanisms underlying biomolecular signaling and recognition. This concept moves beyond the traditional view of a protein existing as a single, static entity, acknowledging the inherent dynamics of these macromolecules. The conformational landscape of globular protein is often visualized as a folding funnel, where the unfolded states occupy the broad, high-energy top, and the native folded state resides at the narrow, low-energy bottom (Wolynes, 2015). The funnel shape effectively depicts a decrease in both energy and conformational entropy during the folding process (Yang et al., 2013). This means that the unfolded protein exists in a large number of possible configurations, and as the protein folds, it moves down the funnel toward the narrow bottom, representing the native state, which is characterized by low energy and a limited range of closely related conformations. During this process, the protein could become transiently stabilized in metastable conformations, forming localized minima before eventually reaching the global energy minimum (Ghosh & Ranjan, 2020; Rao & Gosavi, 1998). Often, these metastable states are known to play crucial roles in protein function (e.g., processes like folding, aggregation, enzyme activity, and effector recognition) (Biasini & Faccioli, 2025). Thus this concept not only illustrates how proteins can fold through multiple pathways to reach their native conformation but also the dynamics and stability of the intermediate and metastable states, many of which are essential for biological activity.

2.2.2 Understanding the free energy landscape

The FEL provides a thermodynamic framework for comprehending the protein conformational landscape (Maisuradze et al., 2010; Pontiggia et al., 2015; Yang et al., 2013). It assigns a free energy value

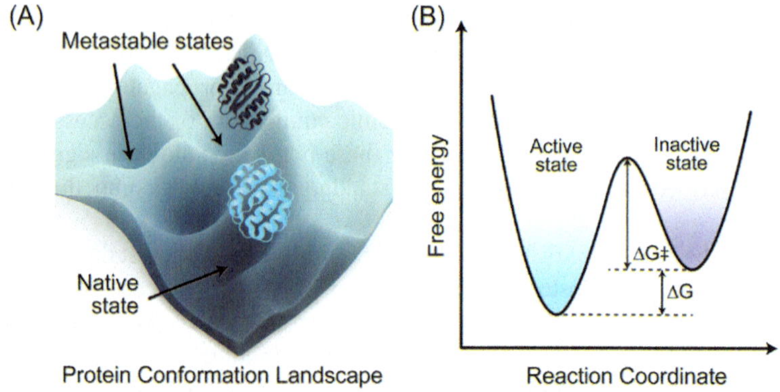

FIGURE 2.1

Conformational free energy landscape and functional states: Multidimensional protein conformational free energy landscape and functional states. (A) Illustration of a multidimensional protein conformational free energy landscape with several local peaks (energy barriers) and valleys (conformational states). The vertical axis represents the Gibbs free energy (*G*), whereas the horizontal axes represent abstract conformational coordinates. The global minimum at the bottom represents the native state ensemble, encompassing the collection of low-energy conformations essential for the biological function. Higher energy local minima correspond to metastable states, which can be transiently populated and play crucial roles in processes such as folding, regulation, or interactions. (B) A simplified one-dimensional projection along a reaction coordinate, showing two functional basins, namely, active and inactive states. The relative stability of these states is reflected by the free energy difference, whereas the interconversion rate is governed by the energy barrier between them.

to each possible conformation within the conformational space. The FEL consists of energy minima (valleys) corresponding to stable or metastable conformations, separated by energy barriers (hills) that regulate transitions between these states (Lois et al., 2008). The global minimum represents the most stable native state, whereas shallower minima correspond to less stable, metastable states (Fig. 2.1A) (Cecchini et al., 2009; Tavernelli et al., 2003). Experimental techniques like X-ray crystallography often capture the most populated state, or so-called native state, which resides at the global minimum of the free energy (Lau & Roux, 2007; Sankar et al., 2015). However, these structures do not capture the inherent dynamics and the ensemble of conformations that a protein can adopt in solution or within the complex environment of a living cell (Wei et al., 2016). In contrast, the FEL reveals the existence of multiple conformational states, taking into account both energetic (enthalpic) and entropic contributions, representing the thermodynamic stability of these conformational states (Peccati & Jiménez-Osés, 2021; Singh & Warshel, 2010). In the context of protein conformation, enthalpy (ΔH) primarily reflects the sum of the internal energy of the protein, arising from intramolecular interactions such as hydrogen bonds, van der Waals forces, hydrophobic interactions, and electrostatic interactions, as well as the energy associated with protein–solvent interactions. Entropy (ΔS), on the other hand, accounts for the number and diversity of microstates accessible to a given conformational state, often linked to the amplitude and nature of atomic fluctuations and backbone flexibility (Frederick et al., 2007). Highly dynamic or disordered states typically have greater conformational entropy than more rigid, ordered states. The Gibbs free energy (ΔG) provides a comprehensive measure of the spontaneity and equilibrium of a biomolecular interaction, encompassing both enthalpic and entropic contributions through the equation $\Delta G = \Delta H - T\Delta S$, where T is the absolute temperature.

Accurately characterizing conformational FEL of biomolecules requires defining suitable reaction coordinates or collective variables. A reaction coordinate is a parameter or set of parameters tracking progression along a biochemical pathway or distinguishing key states in conformational transitions. For relatively simple processes, reaction coordinates, such as interatomic distances, angles, or root-mean-square deviations relative to reference structures, are readily identifiable. For example, the distance between binding sites in protein–ligand interactions commonly serves as an intuitive reaction coordinate, allowing clear visualization of bound, unbound, and intermediate states such as transient encounter complexes.

In the conformational FEL the free energy of a conformation $G(X)$ along the reaction coordinate X is given by the Boltzmann relation as $G(X) = -k_B T \ln P(X)$, where k_B is the Boltzmann constant, T is the absolute temperature, and $P(X)$ is the probability distribution of conformational coordinate X. These states correspond to distinct energy minima or basins on the landscape, each representing an ensemble of conformations. Thus this formulation allows estimation of the FEL from structural ensembles generated through MD simulations or inferred from experimental data. The thermodynamic stability and relative populations of these states are determined by the free energy difference (ΔG) between their minima. At equilibrium the free energy difference (ΔG) between distinct minima on the FEL is directly related to the relative populations of the corresponding conformational states given by the Boltzmann distribution: $\frac{P_{\text{Active}}}{P_{\text{Inactive}}} = e^{-\Delta G/(RT)}$ (Fig. 2.1B). A smaller ΔG between two states indicates that both states will be significantly populated at equilibrium, whereas a larger ΔG signifies that the state with the lower free energy will be much more prevalent in the ensemble. By determining the free energy differences between various conformational states, either through experimental measurements (e.g., ITC and SPR) or computational methods, it becomes possible to predict the likelihood of finding the protein in each of these states under specific conditions (Amaral et al., 2017; Bussi & Laio, 2020; Glöckner & Klebe, 2021; Papadourakis et al., 2024).

Similarly the height of the energy barrier that separates two minima on the FEL dictates the kinetics, or the rate of interconversion, between the corresponding conformational states. According to transition state theory the rate constant (k_{rate}) for conformational exchange is exponentially dependent on this barrier height given by the relationship: $\kappa_{\text{rate}} = \frac{k_B T}{h} e^{-\Delta G^{\ddagger}/(kT)}$. Thus a higher energy barrier implies a slower rate of transition, as the protein requires more energy, typically obtained from thermal fluctuations in the surrounding environment, to overcome the barrier and switch between the two conformations.

In practical terms the FEL captures both thermodynamic and kinetic aspects of biomolecular behavior, providing a powerful tool/paradigm for understanding the dynamic ensemble nature of proteins, beyond the static snapshots provided by traditional structural biology techniques (Frauenfelder et al., 1991; Wang et al., 2012). Ideally, by mapping the conformational FEL, one can gain insights into how different perturbations can modulate the conformational ensembles associated with protein function and stability.

2.3 Perturbations induced population shifts in allosteric mechanisms

Signaling proteins such as enzymes, receptors, or kinases usually exhibit at least two prominent basins on their energy landscape, namely, an inactive state and an active state (Hall, 1990;

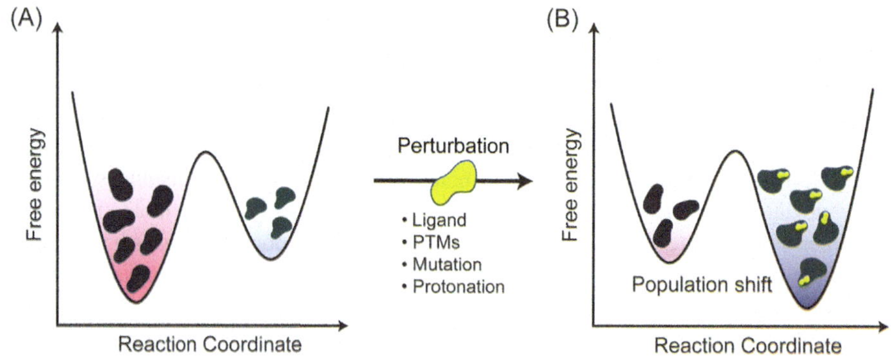

FIGURE 2.2

Conformational population shift on external perturbation: Perturbation-induced population shift on a dynamic energy landscape. The diagram illustrates how the conformational equilibrium of a protein can be altered by external factors, leading to functional modulation. (A) A simplified one-dimensional free energy landscape depicts two stable conformational states. The filled shapes indicate the equilibrium population distribution, showing that the lower energy state (left well) is predominantly populated in the initial condition. (B) Following a perturbation, such as ligand binding, posttranslational modification, mutation, or change in protonation state, the relative free energies of the conformational states are altered. Although the fundamental landscape topology may remain similar, the perturbation selectively stabilizes one state over the other (schematically indicated by the *yellow* markers within the populated shapes). This energetic change results in a redistribution of the conformational ensemble, known as a "population shift."

Johnson et al., 1996; Kobilka & Deupi, 2007; Nussinov et al., 2023; Tsai et al., 2014). The inactive state is often the global minimum under physiological conditions, that is, the most populated state in the absence of input, whereas the active state is a higher energy, less-populated local minimum. These states are generally separated by large energy barriers, which help ensure that the protein remains in the inactive conformation unless an appropriate activation signal is received (Fig. 2.2). External perturbations such as ligand binding, PTMs, genetic mutations, or changes in the physicochemical environment (e.g., pH or ionic strength) can reshape the FEL and affect the relative stability and population of different conformational states (Dutta et al., 2022; Goh et al., 2014; Guarnera & Berezovsky, 2019; Wei et al., 2016).

2.3.1 Evolving models of ligand binding and allostery

Classically, these perturbation-driven conformational changes, particularly ligand binding, were described through two models: induced fit or conformational selection (Boehr et al., 2009; Kumar et al., 2000; Michel, 2016; Whitehead, 1970). The induced fit model, also known as the Koshland–Némethy–Filmer (KNF) model, describe a scenario where the binding of a ligand to a protein induces a conformational change in the protein, leading to altered activity at another site (Koshland et al., 1966). This model emphasizes sequential conformational changes within the protein upon ligand binding. Conversely the conformational selection model, also known as the Monod–Wyman–Changeux (MWC) model, hypothesize that the protein exists in an equilibrium between different preexisting conformational states, such as the tense (T) and relaxed (R) states exemplified by hemoglobin (Monod et al., 1965).

The binding of a ligand then shifts this equilibrium toward the conformation with higher affinity for the ligand, thereby affecting the protein's overall activity. The MWC model provided a foundational framework to understand the concept of allostery, but its two-state view was later recognized as an oversimplification, as it did not account for the full spectrum of conformational dynamics observed in proteins (Perutz, 1970).

2.3.2 Dynamic allostery and the population shift framework

The understanding of allostery has significantly evolved, particularly with the introduction of dynamic allostery by Cooper and Dryden (1984). They proposed that allosteric regulation could arise from alterations in a protein's dynamic properties, such as changes in vibrational modes or thermal fluctuations, rather than solely relying on substantial shifts in the average static structure. This entropy-driven mechanism highlighted the critical role of protein flexibility and thermodynamic factors in allosteric regulation, broadening the scope beyond purely structural interpretations (Lee et al., 2000; Popovych et al., 2006). Building on these insights, Nussinov group introduced a conformational selection and population shift framework in 1999, with an ensemble-centric view that fits naturally with the FEL concept. This model has been widely used to explain allostery, advancing the concept of allostery from two states (MWC model and KNF model) to ensembles of multiple states (Boehr et al., 2009; Nussinov et al., 2023; Tsai et al., 2014; Wei et al., 2016). Furthermore, it emphasizes that the absence of significant differences in static structures between different functional states does not preclude the existence of crucial conformational changes (Gunasekaran et al., 2004; Tsai et al., 2008). Subtle shifts in the populations of energetically less favorable or highly dynamic invisible states within the ensemble can indeed underlie allosteric regulation (Hilser et al., 2012). This mechanism contrasts with the induced fit model, which proposes that the protein undergoes a conformational change only after the initial binding event, adapting its binding site to better accommodate the ligand. Although both models may contribute to biomolecular interactions in different contexts, the conformational selection and population shift model appears to be particularly relevant for explaining phenomena like allostery, where a perturbation at one site can influence activity at a distant site without necessarily involving a large-scale structural change.

2.4 Case studies: Mechanistic insights from biomolecular systems

Having established the theoretical framework of conformational FELs, population shifts, and dynamic allostery, we now turn to selected biomolecular systems to provide concrete mechanistic insights, drawing particularly from the pioneering work of the Chakrabarty group. These case studies, focusing on diverse protein families, showcase how different perturbations drive subtle changes in the balance of conformational states within a protein ensemble and propagate signals across significant distances. Furthermore, through these case studies, we illustrate how understanding the reshaping of the FEL provides a powerful lens for deciphering the intricate molecular choreography of cellular processes.

2.4.1 Dynamic allostery in PDZ domain

Allostery in PDZ domains has been shown to occur dynamically, often in the absence of significant conformational changes detectable by traditional structural methods. Several theoretical and experimental studies propose multiple allosteric pathways for PDZ domains based on evolutionary information, correlated motions, or energy fluctuations (Gerek & Ozkan, 2011; Kong & Karplus, 2009; Lockless & Ranganathan, 1999; Petit et al., 2009). The Chakrabarty group has conducted seminal work in decoding the hidden dynamic allostery in proteins, particularly focusing on PDZ domains (Kumawat & Chakrabarty, 2017, 2020). Their research has elucidated how subtle side-chain motions, often without major global structural changes, can give rise to allosteric effects. A key finding from their work is how electrostatic interactions and their redistribution can induce population shifts within the protein's conformational ensemble, revealing the existence of a potentially universal dynamic response network (Liu & Nussinov, 2017). This has been demonstrated through studies examining two key perturbation effects: ligand binding and the protonation of specific histidine residues. Moreover, the concept of allostery fundamentally involves communication between two distinct sites on a biomolecule, and these studies provide detailed molecular insights into fundamental mechanisms underlying these communication pathways.

2.4.2 Ligand binding–induced population shift in conformational ensemble

The Chakrabarty group's work has investigated how ligand binding at the canonical binding site propagates effects to distant regions through modulation of the protein's dynamic ensemble (Kumawat & Chakrabarty, 2017). Their studies utilizing MD simulations have revealed that while average structures may remain similar upon ligand binding, there are significant population shifts in the conformational substates accessible to the protein (Fig. 2.3). These shifts are linked to the redistribution of electrostatic energies and subtle rearrangements in side-chain interactions throughout the protein, collectively remodeling the FEL. This remodeling leads to increased or decreased populations of certain preexisting or newly stabilized conformational substates characterized by distinct patterns of side-chain orientations and interactions throughout the domain. The allosteric pathway, in this context, can be understood as a network of energetically coupled residues whose dynamic behavior is collectively modulated by the presence of the ligand, enabling signal transmission across the domain.

2.4.3 Protonation–induced population shift in conformational ensemble

Beyond conventional ligand binding the Chakrabarty group's research on PDZ domains has also highlighted the role of other types of perturbations, such as changes in protonation state, in triggering dynamic allostery (Kumawat & Chakrabarty, 2020). They demonstrated that the protonation of specific histidine residues, such as H317 and H372 in the PDZ domain, alters the local electrostatic environment and triggers a cascade of subtle hydrogen bond rearrangements and side-chain repacking events that propagate through the protein's interaction network. Similar to ligand binding, this leads to a redistribution of electrostatic energies and a modulation of the dynamic ensemble, impacting residues distant from the protonation site. The studies revealed that the allosteric response network activated by protonation substantially overlaps with the network involved in ligand-induced allostery, suggesting the existence of a potentially universal response network within the PDZ domain capable of transmitting signals initiated by different types of perturbations. This challenges the traditional view of allostery being solely tied to specific structural changes and emphasizes the critical contribution of the protein's inherent dynamic landscape and the plasticity of electrostatic interactions in mediating allosteric communication.

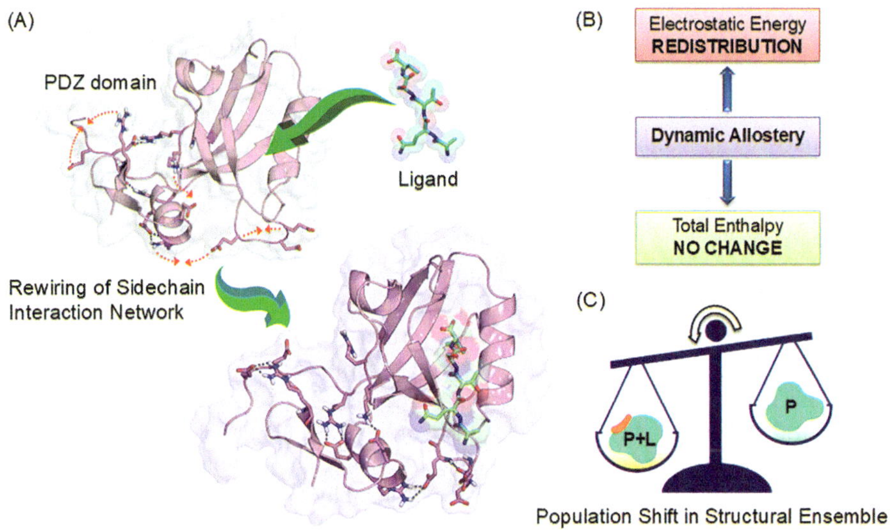

FIGURE 2.3

Dynamic allostery in PDZ domain: Ligand-induced dynamic allostery in PDZ domain. (A) Structural representation of the PDZ domain before and after ligand binding, highlighting the rewiring of the side-chain interaction network. The ligand binds to the canonical binding site, inducing subtle rearrangements in side-chain interactions (*red dashed lines*) that propagate through the protein, despite minimal changes in the overall backbone structure. (B) Schematic of the thermodynamic basis of dynamic allostery, showing how ligand binding leads to a redistribution of electrostatic energy within the PDZ domain. This redistribution drives dynamic allostery without altering the total enthalpy, emphasizing the role of entropy-driven processes in signal propagation. (C) Conceptual depiction of the population shift in the structural ensemble, where ligand binding (P + L) shifts the equilibrium from one conformational substate (P) to another, illustrating how subtle changes in the conformational ensemble underpin allosteric communication in the PDZ domain.

2.4.4 Functional significance of metastable states

Metastable states are intermediate conformations that are transiently populated and energetically less favorable than the global minimum. These states can act as dynamic checkpoints that modulate kinetic barriers and reaction pathways, enabling proteins to explore functionally relevant conformations that might be inaccessible under equilibrium conditions alone. Despite their short-lived nature and typically lower population, these states can significantly influence biological activity, serving as essential intermediates for effector recognition and signal transduction. The following examples illustrate how specific metastable states in different biomolecular systems underpin critical aspects of their functional mechanisms.

2.4.5 Conformational heterogeneity and effector recognition in small GTPases

Rho GTPases, like RhoA, regulate cellular processes via nucleotide-dependent switching, primarily involving conformational changes in the Switch I (SWI) and Switch II (SWII) loops. Although crystal structures often capture dominant "ON" (GTP-bound) or "OFF" (GDP-bound) states, standard MD and metadynamics simulations reveal significant conformational heterogeneity beyond these canonical states (Kumawat et al., 2017). The conformational FEL for these systems suggests the presence of multiple intermediate metastable states. For the GTP-bound form, metastable intermediate states were observed, separated by an interconversion barrier of ~5 kcal/mol. These states, potentially analogous to the "state 1" and "state 2" conformations described for Ras GTPases, exhibit distinct structural signatures in the SWI loop, particularly concerning the orientation of functionally important residues such as Tyr34 and Phe39.

The existence of these metastable states is stabilized by a fine balance between unfavorable solvent exposure of hydrophobic residues and favorable hydrogen-bonded interactions between the γ-phosphate group of GTP with highly conserved Tyr34 and Thr37 residues. These competing molecular interactions lead to a tunable energy landscape of the Switch I conformation. Specifically, in the GDP-bound form, the side chain of Tyr34 points outward, whereas in the GTP-bound state, it flips inward to form a hydrogen bond with the γ-phosphate. Similarly the hydrophobic Phe39 side chain is buried in the GDP form but flips out to become solvent-exposed in the active state. Analysis of SWI residue conformations in the GTP-bound state reveals substantial solvent exposure of hydrophobic residues, where unfavorable solvation energy is counterbalanced by favorable electrostatic interaction with GTP. This results in hydrophobic groups being more exposed in the GTP-bound state and hydrophilic groups more exposed in the GDP-bound state. The free energy surfaces demonstrate that perturbations, such as the oncogenic G14V mutation in RhoA, can shift the identity of the global minimum to an intermediate state. This population shift is also evident in the probability distribution of the Tyr34 dihedral angle upon nucleotide binding and the G14V mutation. The existence of these metastable states, stabilized by a fine balance between unfavorable solvent exposure of hydrophobic residues and favorable interactions with GTP, likely contributes to the dynamic nature of effector recognition and signaling specificity.

2.4.6 HipA toxin—Enabling phosphorylation via transient exposure

The *Escherichia coli* toxin HipA provides a clear example of how a functionally critical metastable state, not observed in static crystal structures, governs its activity (Pandey et al., 2023). The catalytically important residue, Ser150, is buried and inaccessible for autophosphorylation in the dominant "in-state" captured by crystallography. However, for autoinactivation via phosphorylation to occur, Ser150 must become solvent-exposed. Both experimental unfolding studies and MD simulations coupled with FEL analysis revealed the existence of a minorly populated, less stable (approximately 2–4 kcal/mol higher in energy) "out-state" ensemble where Ser150 is transiently exposed. This metastable, molten-globule-like state, characterized by the ejection of the P-loop containing Ser150, is phosphorylation-competent. Its identification underscores that transiently accessible, higher energy conformations on the FEL, potentially missed by crystallography, can be essential for key biological processes like PTM.

MD simulations showed the HipA "in-out" pathway to contain multiple free energy minima with an increasing degree of Ser150 solvent exposure. The free energy difference between the "in-state" (global minimum) and the metastable exposed states is estimated to be approximately 2–2.5 kcal/mol, supported by experimental estimates of approximately 4 kcal/mol. These metastable states are associated with unique sets of hydrogen bonds and salt bridges. In the HipA crystal structure, Ser150 forms two hydrogen bonds: one with carbonyl oxygen atom of Leu64 backbone and another with an internally bound water molecule. The transition to the exposed state involves disrupting specific interactions that stabilize the buried "in-state." Distinct interaction patterns, including hydrogen bonds and salt bridges, stabilize the loop conformations in different clusters on the FEL. For instance the buried state (Conf-WT) is stabilized by a hydrogen bond between Ser150 and Leu64, whereas the exposed state (Conf-SP) involves Ser150 forming a hydrogen bond with Asp309. The metastable "out-state" ensemble is also aggregation-prone, consistent with the transient solvent exposure of Ser150 and its flanking hydrophobic residues (Val/Ile). This highlights how even low-population, higher energy states on the FEL can be functionally critical for key biological processes like PTM.

2.4.7 Impact of mutations and posttranslational modifications on dynamic conformational ensembles

Mutations and PTMs act as perturbations that reshape and modulate the conformational ensemble and dynamics on the protein's FEL. These perturbations, akin to ligand binding, often operate by altering the relative stability and accessibility of preexisting conformational states, redistributing the populations across the landscape. This redistribution can manifest as changes in the depth of energy wells or the height of barriers separating distinct functional states. For example, constitutively activating mutations (as observed in kinases, GTPases, or metabolic enzymes) can stabilize an otherwise transient active state or destabilize an energetically favorable inactive one, effectively reducing energy barriers or shifting basin depths (Arasteh et al., 2019; Pour et al., 2024; Regev et al., 2025; Tsukamoto & Farrens, 2013).

One such example is the regulation of RhoA GTPase dissociation from its inhibitor (RhoGDI) via phosphorylation. MD simulations reveal that RhoA-specific phosphorylation of RhoGDI at Ser34 acts as an allosteric trigger, leading to a significant reduction in the binding affinity between RhoA and RhoGDI (Sinha et al., 2024b). This weakening of the PPI is accompanied by distinct structural changes observed in simulations, including the conformational changes in the N-terminal region of RhoGDI and an outward movement (pulling out) of the geranylgeranyl group of RhoA from its hydrophobic binding cavity within C-terminal β-sandwich of RhoGDI. An important question arising from these observations is how the energetic perturbation created by phosphorylation at Ser34 propagates over such a long distance (>3.5 nm) to affect the polybasic region (PBR) of RhoA and the geranylgeranyl binding site (Fig. 2.4). MD simulations suggest a mechanistic model where the introduction of the negatively charged phosphate group at Ser34 disrupts the local electrostatic environment. This initiates a cascade involving the rearrangement and rewiring of hydrogen bonds and salt bridges (charge–charge interactions) that propagates the signal distally. Specifically, phosphorylation at Ser34 weakens the interaction between the helix–turn–helix region of RhoGDI and the switch regions of RhoA, allowing the N-terminal loop of RhoGDI to become more flexible and move away from the PBR, shifting toward C-terminal region of RhoGDI. This conformational rearrangement facilitates the "pulling out" motion of the buried geranylgeranyl group. These emergent dynamic and structural changes effectively remodel the FEL, shifting the conformational population toward states favoring the dissociation of RhoA from RhoGDI.

Similarly, missense mutations in glucose-6-phosphate dehydrogenase (G6PD), even within the same codon, can induce variant-specific alterations in conformational dynamics, subtly reshaping the FEL (Costa et al., 2024). Computational studies comparing wild-type G6PD with variants affecting Arg454 (Arg454Pro, Arg454Cys, and Arg454His) show that while the overall tetrameric structure remains largely unperturbed, the loss or modification of the positive charge at residue 454 significantly impacts the local electrostatic network and dynamics. MD simulations highlight distinct, variant-specific rearrangements of hydrogen bond networks surrounding the mutation site. These local perturbations lead

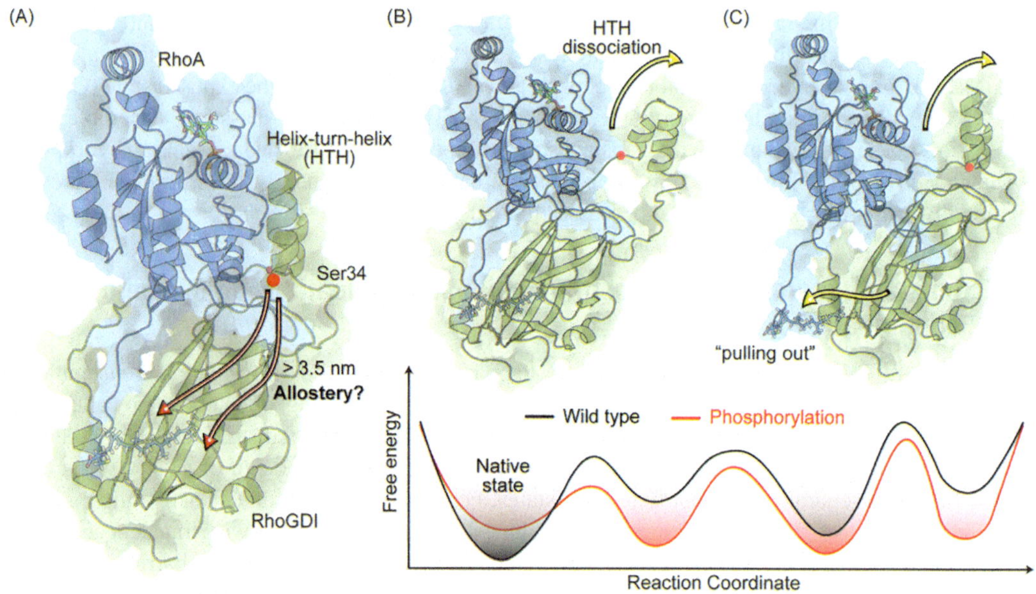

FIGURE 2.4

Phosphorylation-induced conformational population shift: Phosphorylation-induced modulation of the RhoA–RhoGDI complex and its conformational free energy landscape. (A) Representation of the RhoA–RhoGDI complex in its native state (crystal structure), highlighting the phosphorylation site at Ser34 on RhoGDI. *Arrows* suggest possible long-range allosteric communication (>3.5 nm) from the phosphorylation site to the polybasic region (PBR) of RhoA and the hydrophobic cavity of RhoGDI. Parts (B) and (C) illustrate key conformational changes following Ser34 phosphorylation: dissociation of the helix–turn–helix (HTH) motif from RhoA and the outward movement of the geranylgeranyl moiety, reflecting reduced binding affinity and destabilization of the complex. The lower panel illustrates a conceptual free energy landscape (FEL) comparing wild-type (*black curve*) and phosphorylated (*red curve*) systems. Phosphorylation reshapes the FEL, decreasing the depth of the native basin and redistributing populations across alternate states that favor dissociation. This hypothetical landscape does not represent explicit MD simulation data but qualitatively captures the energetic impact of local perturbation on global conformational dynamics and binding stability. *MD*, Molecular dynamics.

to altered collective motions and flexibility, particularly affecting the catalytic region and the binding stability and dynamics of the substrate (G6P) and the catalytic cofactor (NADP+). Such mutation-induced shifts in local interactions, dynamics, and ligand binding reflect subtle modulations of the FEL, potentially leading to broadened or skewed energy wells that correlate with the different pathogenic potentials observed for these variants.

2.4.8 Leveraging allostery for drug discovery

Leveraging allostery for drug discovery presents a promising avenue and offers an alternative strategy to targeting highly conserved or large orthosteric binding sites. The Chakrabarty group's work on the PCSK9–low-density lipoprotein receptor (LDLR) system provides a compelling case study in this regard, illustrating how understanding and exploiting the bidirectional nature of allostery and the dynamics of protein conformational ensembles can lead to novel therapeutic approaches (Sinha et al., 2024a). The interaction between PCSK9 and the LDLR is a critical target for managing hypercholesterolemia. Inhibiting this PPI can lead to increased LDLR levels and consequently lower LDL cholesterol. However, the extensive interface between PCSK9 and LDLR makes designing small molecule inhibitors that directly compete for this site challenging. The Chakrabarty group's research has explored an allosteric strategy, focusing on identifying distal cryptic sites on PCSK9 that can modulate its interaction with LDLR.

Their studies utilized techniques such as MD simulations and Markov state models (MSMs) to show that the binding of LDLR induces significant conformational changes in PCSK9, particularly in regions distant from the direct interaction interface, such as a specific distal loop. These structural changes involve shifts in the population of different conformational substates within the dynamic ensemble of PCSK9. These substates, some of which can be considered metastable, exhibit varying affinities for LDLR. Crucially the research highlights the bidirectional nature of allostery: just as LDLR binding influences the conformation of distant regions, the inherent dynamics and population distributions of these distant regions can in turn modulate the binding affinity at the primary PCSK9–LDLR interface. By identifying and characterizing the metastable states and their associated interaction networks (including hydrogen bonds and salt bridges) in the unbound state of PCSK9 that have lower affinity for LDLR, the Chakrabarty group's work points to the potential of targeting these specific substates with small molecules. An allosteric modulator binding to a cryptic site could stabilize these low-affinity conformations by altering the balance of inter- and intramolecular interactions, thereby shifting the conformational ensemble toward states that are less conducive to LDLR binding. This approach, which leverages the protein's intrinsic dynamics and the thermodynamic principles of population shifts on the FEL, offers a sophisticated strategy for disrupting critical PPIs and presents exciting opportunities for the design of novel allosteric inhibitors.

2.5 Computational approaches to map free energy conformational landscapes

Computational techniques are essential for characterizing conformational FELs, especially for complex biomolecules and processes. Methods like umbrella sampling and metadynamics, which utilize defined reaction coordinates, can enhance the sampling of conformational states and FEL construction. However, many biological processes, including allosteric regulation, PTMs mediated signaling, and complex PPIs, involve multidimensional changes that are poorly described by simple coordinates and often occur over long timescales that require extensive time-taking simulations. To address these complexities, approaches such as MSMs are employed (Bowman et al., 2014; Plattner & Noé, 2015; Sadiq et al., 2012; Shukla et al., 2014; Voelz et al., 2010). MSMs provide a powerful framework by discretizing conformational space into metastable states and describing the transition kinetics between them. Building effective MSMs often relies on advanced dimensionality reduction and data-driven methods to identify the relevant slow modes of motion, such as principal component analysis (PCA) and time-lagged independent component analysis (TICA). PCA is also widely used to simplify MD simulation data by identifying dominant motions based on variance. It captures large-amplitude global motions such as domain movements in proteins like adenylate kinase (Lou & Cukier, 2006). However, it does not explicitly distinguish kinetically important motions from fast fluctuations, potentially missing biologically significant slow transitions. In contrast, TICA identifies the slowest relaxing modes of molecular motion, making it particularly useful for elucidating kinetically relevant transitions essential in protein folding, ligand-induced allostery, and other important events. Beyond PCA and TICA, emerging machine learning methods, including autoencoders, variational approaches for conformational dynamics, and clustering techniques like *k*-means or DBSCAN, have become powerful tools for analyzing complex biomolecular dynamics. These reveal nonlinear and subtle conformational transitions that linear methods might overlook, significantly advancing our understanding of biomolecular function and enabling predictive modeling of signaling pathways, allosteric modulation, and drug-target interactions.

2.6 Conclusion and future perspective

In conclusion the chapter emphasizes the thermodynamic perspective, centered on conformational ensembles and FELs, as a powerful framework for understanding biomolecular recognition, signaling, and allostery. Moving beyond the limitations of static structural snapshots, this dynamic viewpoint illuminates how the inherent flexibility and subtle population shifts within protein conformational ensembles fundamentally govern biological function, as illustrated by case studies such as the dynamic allostery in PDZ domains, metastable states of Rho GTPases, and the therapeutic potential of the PCSK9–LDLR interaction. These examples highlight the versatility and explanatory power of thermodynamics in elucidating complex biomolecular mechanisms. However, the thermodynamic framework is not confined to proteins but extends to other biomolecules, including nucleic acids, lipids, and carbohydrates, where conformational dynamics and energy landscapes similarly govern recognition and function. This broader application is crucial in systems biology, where the crowded cellular environment involves multiple interacting components, such as protein–nucleic acid complexes and lipid–protein interactions. Understanding how thermodynamic principles operate in these contexts is essential for a holistic view of biological processes. In addition, from an evolutionary perspective, the thermodynamic properties of biomolecules have likely been shaped by natural selection to optimize function, offering a rich avenue for exploring how evolution has fine-tuned these systems for adaptability and efficiency.

Although this paradigm offers compelling mechanistic insights, significant challenges remain in fully harnessing its potential. Characterizing the complex, high-dimensional FELs of large biomolecules or multiprotein complexes, especially over biologically relevant timescales, continues to push the boundaries of current computational and experimental capabilities. Defining appropriate reaction coordinates to capture the essence of intricate processes such as allosteric communication or complex binding events remains a nontrivial task. Furthermore bridging the gap between atomistic simulation predictions and quantitative experimental validation of conformational populations, transition kinetics, and thermodynamic parameters requires ongoing innovation in both simulation methodologies and biophysical techniques capable of probing biomolecular dynamics with high resolution. Continued development and application of enhanced sampling techniques, MSMs, and increasingly sophisticated machine learning approaches promise to improve the accuracy and efficiency of mapping FELs and identifying key functional motions. Refining experimental methods, such as time-resolved crystallography, advanced NMR spectroscopy, and single-molecule techniques, will provide crucial data to validate and refine computational models. A major goal will be to extend these detailed dynamic analyses from isolated biomolecules to their behavior within the crowded and complex cellular environment, incorporating factors such as nonspecific interactions and cellular compartmentalization. Ultimately a deeper, quantitative understanding of how perturbations reshape conformational landscapes holds immense promise not only for fundamental biology but also for rational drug design, enabling the development of novel allosteric modulators and providing a clearer picture of the molecular basis of diseases arising from dysregulated protein function. This dynamic ensemble perspective is poised to remain central to unraveling the molecular choreography of life.

References

Ahmed, M. H., Ghatge, M. S., & Safo, M. K. (2020). Hemoglobin: Structure, function and allostery. *Subcellular Biochemistry, 94*, 345–382. http://www.springer.com/series/6515. doi:10.1007/978-3-030-41769-7_14.

Amaral, M., Kokh, D. B., Bomke, J., Wegener, A., Buchstaller, H. P., Eggenweiler, H. M., Matias, P., Sirrenberg, C., Wade, R. C., & Frech, M. (2017). Protein conformational flexibility modulates kinetics and thermodynamics of drug binding. *Nature Communications, 8*(1). doi:10.1038/s41467-017-02258-w.

Arasteh, S., He, P., Haldane, A., & Levy, R. M. (2019). Protein kinase free energy landscapes: The role of the activation loop. *Biophysical Journal, 116*(3), 436a. doi:10.1016/j.bpj.2018.11.2348.

Artemenko, Y., Axiotakis, L., Borleis, J., Iglesias, P. A., & Devreotes, P. N. (2016). Chemical and mechanical stimuli act on common signal transduction and cytoskeletal networks. *Proceedings of the National Academy of Sciences of the United States of America, 113*(47), E7500. http://www.pnas.org/content/113/47/E7500.full.pdf. doi:10.1073/pnas.1608767113.

Biasini, E., & Faccioli, P. (2025). Functional, pathogenic, and pharmacological roles of protein folding intermediates. *Proteins: Structure, Function, and Bioinformatics, 93*, 1299–1307.

Boehr, D. D., Nussinov, R., & Wright, P. E. (2009). The role of dynamic conformational ensembles in biomolecular recognition. *Nature Chemical Biology, 5*(11), 789–796. http://www.nature.com/nchembio. doi:10.1038/nchembio.232.

Bowman, G. R., Pande, V. S., & Noé, F. (2014). *An introduction to Markov state models and their application to long timescale molecular simulation*: 797. Springer.

Bussi, G., & Laio, A. (2020). Using metadynamics to explore complex free-energy landscapes. *Nature Reviews Physics, 2*(4), 200–212. doi:10.1038/s42254-020-0153-0.

Cecchini, M., Krivov, S. V., Spichty, M., & Karplus, M. (2009). Calculation of free-energy differences by confinement simulations. Application to peptide conformers. *The Journal of Physical Chemistry B, 113*(29), 9728–9740. doi:10.1021/jp9020646.

Changeux, J. P., & Edelstein, S. J. (2005). Allosteric mechanisms of signal transduction. *Science, 308*(5727), 1424–1428. doi:10.1126/science.1108595.

Chong, S.-H. o, Oshima, H., & Sugita, Y. (2024). Allosteric changes in the conformational landscape of Src kinase upon substrate binding. *Journal of Molecular Biology*, 168871. doi:10.1016/j.jmb.2024.168871.

Cooper, A., & Dryden, D. T. F. (1984). Allostery without conformational change. *European Biophysics Journal, 11*(2), 103–109. doi:10.1007/bf00276625.

Costa, S., Minucci, A., Kumawat, A., De Bonis, M., Prontera, G., Gelsomino, M., Tana, M., Tiberi, E., Romano, A., Ruggiero, A., Mastrangelo, S., Palumbo, G., Giorgio, V., & Elisabetta, M. (2024). Pathogenic G6PD variants: Different clinical pictures arise from different missense mutations in the same codon. *British Journal of Haematology, 205*(5), 1985–1994. doi:10.1111/bjh.19775.

Du, Z., & Lovly, C. M. (2018). Mechanisms of receptor tyrosine kinase activation in cancer. *Molecular Cancer, 17*(1). http://www.molecular-cancer.com/start.asp. doi:10.1186/s12943-018-0782-4.

Duan, G., Walther, D., & Radivojac, P. (2015). The roles of post-translational modifications in the context of protein interaction networks. *PLOS Computational Biology, 11*(2), e1004049. doi:10.1371/journal.pcbi.1004049.

Dutta, P., Roy, P., & Sengupta, N. (2022). Effects of external perturbations on protein systems: A microscopic view. *ACS Omega, 7*(49), 44556–44572. doi:10.1021/acsomega.2c06199.

Fenwick, R. B., Esteban-Martín, S., & Salvatella, X. (2011). Understanding biomolecular motion, recognition, and allostery by use of conformational ensembles. *European Biophysics Journal, 40*(12), 1339–1355. doi:10.1007/s00249-011-0754-8.

Fleetwood, O., Matricon, P., Carlsson, J., & Delemotte, L. (2020). Energy landscapes reveal agonist control of G protein-coupled receptor activation via microswitches. *Biochemistry, 59*(7), 880–891. doi:10.1021/acs.biochem.9b00842.

Frauenfelder, H., Sligar, S. G., & Wolynes, P. G. (1991). The energy landscapes and motions of proteins. *Science, 254*(5038), 1598–1603. doi:10.1126/science.1749933.

Frederick, K. K., Marlow, M. S., Valentine, K. G., & Wand, A. J. (2007). Conformational entropy in molecular recognition by proteins. *Nature, 448*(7151), 325–329. http://www.nature.com/nature/index.html. doi:10.1038/nature05959.

Gerek, Z. N., & Ozkan, S. B. (2011). Change in allosteric network affects binding affinities of PDZ domains: Analysis through perturbation response scanning. *PLoS Computational Biology, 7*(10). http://www.ploscompbiol.org/article/fetchObjectAttachment.action?uri=info%3Adoi%2F10.1371%2Fjournal.pcbi.1002154&representation=PDF. doi:10.1371/journal.pcbi.1002154.

Ghosh, D. K., & Ranjan, A. (2020). The metastable states of proteins. *Protein Science, 29*(7), 1559–1568. http://onlinelibrary.wiley.com/journal/10.1002/(ISSN)1469-896X. doi:10.1002/pro.3859.

Glöckner, S., & Klebe, G. (2021). Simultaneous determination of thermodynamic and kinetic data by isothermal titration calorimetry. *Biochimica et Biophysica Acta (BBA)—General Subjects, 1865*(2), 129772. doi:10.1016/j.bbagen.2020.129772.

Goh, G. B., Hulbert, B. S., Zhou, H., & Brooks, C. L. (2014). Constant pH molecular dynamics of proteins in explicit solvent with proton tautomerism. *Proteins: Structure, Function and Bioinformatics, 82*(7), 1319–1331. http://www3.interscience.wiley.com/journal/36176/home. doi:10.1002/prot.24499.

Grant, B. J., Gorfe, A. A., & McCammon, J. A. (2010). Large conformational changes in proteins: Signaling and other functions. *Current Opinion in Structural Biology, 20*(2), 142–147. doi:10.1016/j.sbi.2009.12.004.

Guarnera, E., & Berezovsky, I. N. (2019). On the perturbation nature of allostery: Sites, mutations, and signal modulation. *Current Opinion in Structural Biology, 56*, 18–27. http://www.elsevier.com/locate/sbi. doi:10.1016/j.sbi.2018.10.008.

Gunasekaran, K., Ma, B., & Nussinov, R. (2004). Is allostery an intrinsic property of all dynamic proteins? *Proteins: Structure, Function, and Bioinformatics, 57*(3), 433–443. doi:10.1002/prot.20232.

Guo, J., & Zhou, H. X. (2016). Protein allostery and conformational dynamics. *Chemical Reviews, 116*(11), 6503–6515. http://pubs.acs.org/journal/chreay. doi:10.1021/acs.chemrev.5b00590.

Hall, A. (1990). The cellular functions of small GTP-binding proteins. *Science, 249*(4969), 635–640. doi:10.1126/science.2116664.

Hilser, V. J., Wrabl, J. O., & Motlagh, H. N. (2012). Structural and energetic basis of allostery. *Annual Review of Biophysics, 41*(1), 585–609. doi:10.1146/annurev-biophys-050511-102319.

Hubbard, R., & Kropf, A. (1958). The action of light on rhodopsin. *Proceedings of the National Academy of Sciences, 44*(2), 130–139. doi:10.1073/pnas.44.2.130.

Johnson, L. N., Noble, M. E. M., & Owen, D. J. (1996). Active and inactive protein kinases: Structural basis for regulation. *Cell, 85*(2), 149–158. https://www.sciencedirect.com/journal/cell. doi:10.1016/S0092-8674(00)81092-2.

Kar, G., Keskin, O., Gursoy, A., & Nussinov, R. (2010). Allostery and population shift in drug discovery. *Current Opinion in Pharmacology, 10*(6), 715–722. doi:10.1016/j.coph.2010.09.002.

Kobilka, B. K., & Deupi, X. (2007). Conformational complexity of G-protein-coupled receptors. *Trends in Pharmacological Sciences, 28*(8), 397–406. doi:10.1016/j.tips.2007.06.003.

Koehl, A., Hu, H., Maeda, S., Zhang, Y., Qu, Q., Paggi, J. M., Latorraca, N. R., Hilger, D., Dawson, R., Matile, H., Schertler, G. F. X., Granier, S., Weis, W. I., Dror, R. O., Manglik, A., Skiniotis, G., & Kobilka, B. K. (2018). Structure of the μ-opioid receptor–Gi protein complex. *Nature, 558*(7711), 547–552. doi:10.1038/s41586-018-0219-7.

Kong, Y., & Karplus, M. (2009). Signaling pathways of PDZ2 domain: A molecular dynamics interaction correlation analysis. *Proteins: Structure, Function, and Bioinformatics, 74*(1), 145–154. doi:10.1002/prot.22139.

Koshland, D. E., Némethy, G., & Filmer, D. (1966). Comparison of experimental binding data and theoretical models in proteins containing subunits. *Biochemistry, 5*(1), 365–385. doi:10.1021/bi00865a047.

Kumar, S., Ma, B., Tsai, C. J., Sinha, N., & Nussinov, R. (2000). Folding and binding cascades: Dynamic landscapes and population shifts. *Protein Science, 9*(1), 10–19. http://onlinelibrary.wiley.com/journal/10.1002/(ISSN)1469-896X. doi:10.1110/ps.9.1.10.

Kumawat, A., & Chakrabarty, S. (2017). Hidden electrostatic basis of dynamic allostery in a PDZ domain. *Proceedings of the National Academy of Sciences, 114*(29). doi:10.1073/pnas.1705311114.

Kumawat, A., & Chakrabarty, S. (2020). Protonation-induced dynamic allostery in PDZ domain: Evidence of perturbation-independent universal response network. *The Journal of Physical Chemistry Letters, 11*(21), 9026–9031. doi:10.1021/acs.jpclett.0c02885.

Kumawat, A., Chakrabarty, S., & Kulkarni, K. (2017). Nucleotide dependent switching in Rho GTPase: Conformational heterogeneity and competing molecular interactions. *Scientific Reports, 7*(1). doi:10.1038/srep45829.

Kumawat, A., Tavazzani, E., Lentini, G., Trancuccio, A., Kukavica, D., Oldani, A., Denegri, M., Priori, S. G., & Camilloni, C. (2025). Molecular insights into the rescue mechanism of an HERG activator against severe LQT2 mutations. *Journal of Biomedical Science, 32*(1). doi:10.1186/s12929-025-01134-w.

Latorraca, N. R., Venkatakrishnan, A. J., & Dror, R. O. (2017). GPCR dynamics: Structures in motion. *Chemical Reviews, 117*(1), 139–155. http://pubs.acs.org/journal/chreay. doi:10.1021/acs.chemrev.6b00177.

Lau, A. Y., & Roux, B. (2007). The free energy landscapes governing conformational changes in a glutamate receptor ligand-binding domain. *Structure, 15*(10), 1203–1214. doi:10.1016/j.str.2007.07.015.

Lee, A. L., Kinnear, S. A., & Wand, A. J. (2000). Redistribution and loss of side chain entropy upon formation of a calmodulin-peptide complex. *Nature Structural Biology, 7*(1), 72–77. doi:10.1038/71280.

Liu, J., & Nussinov, R. (2017). Energetic redistribution in allostery to execute protein function. *Proceedings of the National Academy of Sciences, 114*, 7480–7482 29. doi:10.1073/pnas.1709071114.

Livesay, D. R. (2010). Protein dynamics: Dancing on an ever-changing free energy stage. *Current Opinion in Pharmacology, 10*(6), 706–708. doi:10.1016/j.coph.2010.09.015.

Lockless, S. W., & Ranganathan, R. (1999). Evolutionarily conserved pathways of energetic connectivity in protein families. *Science, 286*(5438), 295–299. doi:10.1126/science.286.5438.295.

Lois, G., Blawzdziewicz, J., & O'Hern, C. S. (2008). Reliable protein folding on complex energy landscapes: The free energy reaction path. *Biophysical Journal, 95*(6), 2692–2701. http://download.cell.com/biophysj/pdf/PIIS0006349508784145.pdf?intermediate=true. doi:10.1529/biophysj.108.133132.

Lou, H., & Cukier, R. I. (2006). Molecular dynamics of apo-adenylate kinase: A principal component analysis. *Journal of Physical Chemistry B, 110*(25), 12796–12808. http://pubs.acs.org/journal/jpcbfk. doi:10.1021/jp061976m.

Lu, S., He, X., Yang, Z., Chai, Z., Zhou, S., Wang, J., Ur Rehman, A., Ni, D., Pu, J., Sun, J., & Zhang, J. (2021). Activation pathway of a G protein-coupled receptor uncovers conformational intermediates as targets for allosteric drug design. *Nature Communications, 12*(1). doi:10.1038/s41467-021-25020-9.

Maisuradze, G. G., Liwo, A., & Scheraga, H. A. (2010). Relation between free energy landscapes of proteins and dynamics. *Journal of Chemical Theory and Computation, 6*(2), 583–595. doi:10.1021/ct9005745.

Manglik, A., Kruse, A. C., Sun Kobilka, T., Sun Thian, F., Mathiesen, J. M., Sunahara, R. K., Pardo, L., Weis, W. I., Kobilka, B. K., & Granier, S. (2012). Crystal structure of the μ-opioid receptor bound to a morphinan antagonist. *Nature, 485*(7398), 321–326. doi:10.1038/nature10954.

Michel, D. (2016). Conformational selection or induced fit? New insights from old principles. *Biochimie, 128–129*, 48–54. doi:10.1016/j.biochi.2016.06.012.

Milburn, M. V., Tong, L., Devos, A. M., Brünger, A., Yamaizumi, Z., Nishimura, S., & Kim, S. H. (1990). Molecular switch for signal transduction: Structural differences between active and inactive forms of protooncogenic ras proteins. *Science, 247*(4945), 939–945. Www.sciencemag.org.

Monod, J., Wyman, J., & Changeux, J. P. (1965). On the nature of allosteric transitions: A plausible model. *Journal of Molecular Biology, 12*(1), 88–118. doi:10.1016/S0022-2836(65)80285-6.

Montserrat-Canals, M., Cordara, G., & Krengel, U. (2025). Allostery. *Quarterly Reviews of Biophysics, 58*, e5.

Nussinov, R. (2025). Pioneer in molecular biology: Conformational ensembles in molecular recognition, allostery, and cell function. *Journal of Molecular Biology, 437*(11), 169044. doi:10.1016/j.jmb.2025.169044.

Nussinov, R., Liu, Y., Zhang, W., & Jang, H. (2023). Protein conformational ensembles in function: Roles and mechanisms. *RSC Chemical Biology, 4*(11), 850–864. doi:10.1039/d3cb00114h.

Palczewski, K., Kumasaka, T., Hori, T., Behnke, C. A., Motoshima, H., Fox, B. A., Le Trong, I., Teller, D. C., Okada, T., Stenkamp, R. E., Yamamoto, M., & Miyano, M. (2000). Crystal structure of rhodopsin: A G protein-coupled receptor. *Science, 289*(5480), 739–745. doi:10.1126/science.289.5480.739.

Pandey, B., Sinha, K., Dev, A., Ganguly, H. K., Polley, S., Chakrabarty, S., & Basu, G. (2023). Phosphorylation-competent metastable state of *Escherichia coli* toxin HipA. *Biochemistry, 62*(5), 989–999. doi:10.1021/acs.biochem.2c00614.

Papadourakis, M., Cournia, Z., Mey, A. S. J. S., & Michel, J. (2024). Comparison of methodologies for absolute binding free energy calculations of ligands to intrinsically disordered proteins. *Journal of Chemical Theory and Computation, 20*(21), 9699–9707. http://pubs.acs.org/journal/jctcce. doi:10.1021/acs.jctc.4c00942.

Peccati, F., & Jiménez-Osés, G. (2021). Enthalpy–entropy compensation in biomolecular recognition: A computational perspective. *ACS Omega, 6*(17), 11122–11130. doi:10.1021/acsomega.1c00485.

Perrella, M., & Russo, R. (2003). Allosteric proteins: Lessons to be learned from the hemoglobin intermediates. *Physiology, 18*(6), 232–236. doi:10.1152/nips.01451.2003.

Perutz, M. F. (1970). Stereochemistry of cooperative effects in haemoglobin: Haem–haem interaction and the problem of allostery. *Nature, 228*(5273), 726–734. doi:10.1038/228726a0.

Petit, C. M., Zhang, J., Sapienza, P. J., Fuentes, E. J., & Lee, A. L. (2009). Hidden dynamic allostery in a PDZ domain. *Proceedings of the National Academy of Sciences, 106*(43), 18249–18254. doi:10.1073/pnas.0904492106.

Plattner, N., & Noé, F. (2015). Protein conformational plasticity and complex ligand-binding kinetics explored by atomistic simulations and Markov models. *Nature Communications, 6*(1). doi:10.1038/ncomms8653.

Pontiggia, F., Pachov, D. V., Clarkson, M. W., Villali, J., Hagan, M. F., Pande, V. S., & Kern, D. (2015). Free energy landscape of activation in a signalling protein at atomic resolution. *Nature Communications, 6*(1). doi:10.1038/ncomms8284.

Popovych, N., Sun, S., Ebright, R. H., & Kalodimos, C. G. (2006). Dynamically driven protein allostery. *Nature Structural and Molecular Biology, 13*(9), 831–838. doi:10.1038/nsmb1132.

Pour, N. G., Wei, S., Krishnan, S. T., Ray, A. P., Thakur, N., Eddy, M. T., & Lamichhane, R. (2024). Uncovering the structural mechanisms of how constitutively activating mutations alter GPCR function. *Biophysical Journal, 123*(3), 65a. doi:10.1016/j.bpj.2023.11.472.

Rao, G., & Gosavi, V. V. H. (1998). On the folding of a structurally complex protein to its metastable active state. *Proceedings of the National Academy of Sciences, 115*, 1998–2003.

Regev, C., Jang, H., & Nussinov, R. (2025). ERK allosteric activation: The importance of two ordered phosphorylation events. *Journal of Molecular Biology*, 169130. doi:10.1016/j.jmb.2025.169130.

Sadiq, S. K., Noé, F., & De Fabritiis, G. (2012). Kinetic characterization of the critical step in HIV-1 protease maturation. *Proceedings of the National Academy of Sciences, 109*(50), 20449–20454. doi:10.1073/pnas.1210983109.

Sankar, K., Liu, J., Wang, Y., & Jernigan, R. L. (2015). Distributions of experimental protein structures on coarse-grained free energy landscapes. *Journal of Chemical Physics, 143*(24). http://scitation.aip.org/content/aip/journal/jcp. doi:10.1063/1.4937940.

Shukla, D., Meng, Y., Roux, B., & Pande, V. S. (2014). Activation pathway of Src kinase reveals intermediate states as targets for drug design. *Nature Communications, 5*. http://www.nature.com/ncomms/index.html. doi:10.1038/ncomms4397.

Simanshu, D. K., Nissley, D. V., & McCormick, F. (2017). RAS proteins and their regulators in human disease. *Cell, 170*(1), 17–33. https://www.sciencedirect.com/journal/cell. doi:10.1016/j.cell.2017.06.009.

Singh, N., & Warshel, A. (2010). A comprehensive examination of the contributions to the binding entropy of protein–ligand complexes. *Proteins: Structure, Function, and Bioinformatics, 78*(7), 1724–1735. doi:10.1002/prot.22689.

Sinha, K., Basu, I., Shah, Z., Shah, S., & Chakrabarty, S. (2024a). Leveraging bidirectional nature of allostery to inhibit protein–protein interactions (PPIs): A case study of PCSK9–LDLR interaction. *Journal of Chemical Information and Modeling, 64*(9), 3923–3932. doi:10.1021/acs.jcim.4c00294.

Sinha, K., Kumawat, A., Jang, H., Nussinov, R., & Chakrabarty, S. (2024b). Molecular mechanism of regulation of RhoA GTPase by phosphorylation of RhoGDI. *Biophysical Journal, 123*(1), 57–67. doi:10.1016/j.bpj.2023.11.018.

Stroka, K. M., & Konstantopoulos, K. (2014). Physical biology in cancer. 4. Physical cues guide tumor cell adhesion and migration. *American Journal of Physiology—Cell Physiology, 306*(2), C98. http://ajpcell.physiology.org/content/306/2/C98.full-text.pdf+html. doi:10.1152/ajpcell.00289.2013.

Tavernelli, I., Cotesta, S., & Di Iorio, E. E. (2003). Protein dynamics, thermal stability, and free-energy landscapes: A molecular dynamics investigation. *Biophysical Journal, 85*(4), 2641–2649. http://www.sciencedirect.com/science/journal/00063495. doi:10.1016/S0006-3495(03)74687-6.

Tong, M., & Seeliger, M. A. (2015). Targeting conformational plasticity of protein kinases. *ACS Chemical Biology, 10*(1), 190–200. http://pubs.acs.org/journal/acbcct. doi:10.1021/cb500870a.

Tsai, C. J., del Sol, A., & Nussinov, R. (2008). Allostery: Absence of a change in shape does not imply that allostery is not at play. *Journal of Molecular Biology, 378*(1), 1–11. https://www.journals.elsevier.com/journal-of-molecular-biology. doi:10.1016/j.jmb.2008.02.034.

Tsai, C.-J., Nussinov, R., & Livesay, D. R. (2014). A unified view of "how allostery works". *PLoS Computational Biology, 10*(2), e1003394. doi:10.1371/journal.pcbi.1003394.

Tsukamoto, H., & Farrens, D. L. (2013). A constitutively activating mutation alters the dynamics and energetics of a key conformational change in a ligand-free G protein-coupled receptor. *Journal of Biological Chemistry, 288*(39), 28207–28216. http://www.jbc.org/content/288/39/28207.full.pdf+html. doi:10.1074/jbc.M113.472464.

Voelz, V. A., Bowman, G. R., Beauchamp, K., & Pande, V. S. (2010). Molecular simulation of ab initio protein folding for a millisecond folder NTL9(1-39). *Journal of the American Chemical Society, 132*(5), 1526–1528. http://pubs.acs.org/doi/pdfplus/10.1021/ja9090353. doi:10.1021/ja9090353.

Wang, J., Oliveira, R. J., Chu, X., Whitford, P. C., Chahine, J., Han, W., Wang, E., Onuchic, J. N., & Leite, V. B. P. (2012). Topography of funneled landscapes determines the thermodynamics and kinetics of protein folding. *Proceedings of the National Academy of Sciences, 109*(39), 15763–15768. doi:10.1073/pnas.1212842109.

Wei, G., Xi, W., Nussinov, R., & Ma, B. (2016). Protein ensembles: How does nature harness thermodynamic fluctuations for life? The diverse functional roles of conformational ensembles in the cell. *Chemical Reviews, 116*(11), 6516–6551. doi:10.1021/acs.chemrev.5b00562.

Weis, W. I., & Kobilka, B. K. (2018). The molecular basis of G protein-coupled receptor activation. *Annual Review of Biochemistry, 87*, 897–919. http://www.annualreviews.org/journal/biochem. doi:10.1146/annurev-biochem-060614-033910.

Whitehead, E. (1970). The regulation of enzyme activity and allosteric transition. *Progress in Biophysics and Molecular Biology, 21*(C), 321–397. doi:10.1016/0079-6107(70)90028-3.

Wolynes, P. G. (2015). Evolution, energy landscapes and the paradoxes of protein folding. *Biochimie, 119*, 218–230. Www.elsevier.com/locate/biochimie. doi:10.1016/j.biochi.2014.12.007.

Yang, L. Q., Ji, X. L., & Liu, S. Q. (2013). The free energy landscape of protein folding and dynamics: A global view. *Journal of Biomolecular Structure and Dynamics, 31*(9), 982–992. http://www.tandfonline.com/loi/tbsd20. doi:10.1080/07391102.2012.748536.

Zhang, M., Chen, T., Lu, X., Lan, X., Chen, Z., & Lu, S. (2024). G protein-coupled receptors (GPCRs): Advances in structures, mechanisms and drug discovery. *Signal Transduction and Targeted Therapy, 9*(1). https://doi.org/10.1038/s41392-024-01803-6.

CHAPTER

3 Influence of entropy in protein dynamics and interactions

Shailesh Kumar Panday and Indira Ghosh*
School of Computational and Integrative Sciences, Jawaharlal Nehru University, New Delhi, India
**Corresponding author: indira0654@gmail.com*

3.1 Introduction

The concept of entropy, a cornerstone of modern science, has undergone a fascinating evolution from its initial formulation within classical thermodynamics to its profound interpretation in statistical mechanics (Sethna, 2021). Born out of the need to understand the directionality of energy transformations, entropy has grown from an abstract thermodynamic quantity to a fundamental measure of disorder and a key principle governing phenomena across physics, chemistry, biology, and even information theory (Attard, 2002). The first law of thermodynamics, often called the law of energy conservation, states that energy cannot be created or destroyed, only transferred or changed from one form to another. Formally, for a closed system that exchanges heat (δQ) with and performs work (δW) on its surroundings during a process, the change in its internal energy (ΔU) is given by $\Delta U = \delta Q - \delta W$ (Shell, 2015). This law provides a fundamental accounting principle for energy transformations, stating that in any process, the total energy remains constant. However, the first law fails to explain the directionality of spontaneous processes. For instance, it does not forbid heat from spontaneously flowing from cold to hot bodies (Green & Naghdi, 1977). Such processes, while conserving energy, are never observed to occur in reality according to macroscopic experience. This inherent limitation of the first law necessitated another fundamental principle governing the feasibility and direction of physical and chemical transformations, motivating the formulation of the second law of thermodynamics.

3.1.1 Second law of thermodynamics and the concept of entropy

The fundamental idea of the second law of thermodynamics, and consequently the concept of entropy, lies in the study of heat engines. In the early 19th century engineers and scientists were striving to improve the efficiency of these devices, which convert thermal energy into mechanical work. Nicolas Léonard Sadi Carnot, in his seminal work "Reflections on the Motive Power of Fire" (1824), laid the groundwork for understanding the theoretical limits of heat engine efficiency (Carnot et al., 1960).

Carnot conceived the idea of an idealized reversible engine operating in a cycle, now known as the Carnot cycle. This cycle consists of four reversible processes: isothermal expansion, adiabatic expansion, isothermal compression, and adiabatic compression. Carnot demonstrated that the efficiency of such an engine depends only on the temperatures of the hot and cold reservoirs between which it operates and that no engine operating between the same two reservoirs can be more efficient than

Protein Dynamics: Experimental and Computational Approaches. DOI: https://doi.org/10.1016/B978-0-443-30042-4.00016-X

a reversible Carnot engine (Feidt, 2017). Carnot's work, though initially based on the flawed caloric theory of heat, implicitly suggested that not all thermal energy can be converted into useful work. Some energy is inevitably transferred to a lower temperature reservoir, implying a fundamental asymmetry in energy transformations (Callendar, 1910).

Rudolf Clausius in the mid-19th century rigorously formulated the second law of thermodynamics and introduced the concept of entropy as a state function. The significance of Carnot's ideas and the need to reconcile them with the emerging understanding of heat as a form of energy were recognized (Carnot et al., 1960; Magie et al., 1899). Clausius focused on the concept of "uncompensated transformation." He realized that while work can be completely converted into heat, the reverse process—the complete conversion of heat into work—is impossible without some other change occurring in the system or its surroundings (Dunning-Davies, 1969; Green & Naghdi, 1977). This led him to formulate the second law statement that heat cannot spontaneously flow from a colder body to a hotter body. For a reversible process the ratio of the infinitesimal heat transferred (δQ_{rev}) to the absolute temperature (T) at which the transfer occurs is an exact differential. This exact differential defines the change in a new state function called entropy (S), Eq. (3.1):

$$dS = \frac{\delta Q_{rev}}{T} \tag{3.1}$$

The total change in entropy (ΔS) for a reversible process going from state 1 to state 2 is then given by Eq. (3.2):

$$\Delta S = \int_1^2 \frac{\delta Q_{rev}}{T} \tag{3.2}$$

For irreversible processes, Clausius showed that the change in entropy of the system and its surroundings (the total entropy change) is always greater than or equal to zero, Eq. (3.3):

$$\Delta S_{total} = \Delta S_{system} + \Delta S_{surroundings} \geq 0 \tag{3.3}$$

This inequality is a fundamental statement of the second law of thermodynamics regarding entropy. It implies that spontaneous processes in isolated systems always lead to an increase in entropy or remain constant at equilibrium (the equality holds for reversible processes, which are theoretical limits) (Greiner et al., 2012). The introduction of entropy provided a powerful tool for predicting the direction of spontaneous processes. The second law, expressed in terms of entropy, states that the total entropy of an isolated system can never decrease but rather only increase over time or remain constant in ideal cases (reversible processes or at equilibrium).

This principle explains why certain processes occur spontaneously, whereas others do not. For example, when a hot object is placed in contact with a cold object, heat flows from the hot object to the cold object, increasing the entropy of the system as a whole. The reverse process, where heat spontaneously flows from the cold object to the hot object, would decrease the total entropy and is therefore forbidden by the second law (Müller, 2003). Furthermore the second law dictates that an isolated system will spontaneously evolve toward a state of maximum entropy, which corresponds to thermodynamic equilibrium. At equilibrium, no further spontaneous changes occur, and the entropy of the system remains constant (Grandy, 2008).

3.1.2 Statistical mechanics and microscopic overview of entropy

Although Clausius provided a macroscopic thermodynamic definition of entropy, Ludwig Boltzmann offered a profound microscopic interpretation based on statistical mechanics. Boltzmann recognized that the entropy (S) of a macroscopic state is related to the number of accessible microscopic arrangements (microstates, Ω) that can give rise to that macroscopic state, proposing a fundamental relationship involving a proportionality constant (Eq. 3.4) (Lebowitz, 1993).

$$S = k\ln\Omega \tag{3.4}$$

This constant, now known as the Boltzmann constant (k_B), was later quantified and named in his honor by Max Planck (Lebowitz, 1993). Boltzmann's statistical interpretation provided a deeper understanding of the second law, explaining that the tendency of isolated systems to evolve toward higher entropy states reflects the statistical probability of transitioning from a less probable (fewer microstates, more ordered) state to a more probable (more microstates, more disordered) state.

In thermodynamics, macroscopic variables are experimentally measurable quantities that define the state of a system. Examples include the number of particles in the system (N), the pressure (P), the volume (V), the temperature (T), and the chemical potential (μ) (Stølen & Grande, 2004). Typically a small set of independent macroscopic variables is sufficient for a complete description of the macroscopic state of the system, defining what is called a thermodynamic ensemble. Common ensembles, defined by the variables that remain constant, include the microcanonical (constant N, V, E), canonical (constant N, V, T), grand canonical (constant μ, V, T), and isothermal–isobaric (constant N, P, T). The isothermal–isobaric ensemble is often most relevant for biological processes, which typically occur at constant temperature and pressure (Kurzynski, 2006).

A function defining the probability distribution of all possible energy states consistent with the system's constraints in a given ensemble is related to the partition function (Kurzynski, 2006). Starting from the partition function, one can compute key thermodynamic quantities such as internal energy, enthalpy, entropy, and free energy, which are crucial for characterizing biological processes (Shell, 2015). This section briefly describes these quantities and their relation, choosing the canonical ensemble (N, V, T) for simplicity, as the concepts extend to the isothermal–isobaric ensemble (N, P, T) relevant to biological systems.

In the NVT ensemble the system can occupy various discrete microstates. Each microstate i has a particular energy E_i. The probability of finding the system in microstate i is given by the Boltzmann distribution (Eq. 3.5).

$$p_i = \frac{e^{-\beta E_i}}{Z_{NVT}} \tag{3.5}$$

where $\beta = \frac{1}{k_B T}$, and the denominator Z_{NVT} is the canonical partition function, representing the sum over all accessible microstates j (Eq. 3.6):

$$Z_{NVT} = \sum_j e^{-\beta E_j} \tag{3.6}$$

This dimensionless quantity encodes the system's thermodynamic information in the NVT ensemble. The internal energy (U) is the average energy of the system, calculated as the sum of the energies of each microstate weighted by their probability (Eq. 3.7):

$$U = \langle E \rangle = \sum_i p_i E_i = \sum_i \frac{E_i e^{-\beta E_i}}{Z_{NVT}} \tag{3.7}$$

This can be related to the partition function by taking the derivative of $\ln(Z_{NVT})$ with respect to β (Eq. 3.8):

$$U = -\left(\frac{\partial \ln Z_{NVT}}{\partial \beta}\right)_{N,V} = k_B T^2 \left(\frac{\partial \ln Z_{NVT}}{\partial T}\right)_{N,V} \tag{3.8}$$

Entropy (S_{NVT}), a measure of disorder or the number of accessible microstates, is related to the probabilities of the microstates by the Gibbs entropy formula (Eq. 3.9):

$$S_{NVT} = -k_B \sum_i p_i \ln(p_i) \tag{3.9}$$

Substituting the expression for p_i and simplifying leads to the relationship (Eq. 3.10):

$$S_{NVT} = \frac{U}{T} + k_B \ln(Z_{NVT}) \tag{3.10}$$

The Helmholtz free energy (A_{NVT}) is defined as $A_{NVT} = U - TS_{NVT}$. Using the expression for S_{NVT} (Eq. 3.11):

$$A_{NVT} = U - T\left(\frac{U}{T} + k_B \ln(Z_{NVT})\right) = -k_B T \ln(Z_{NVT}) \tag{3.11}$$

For processes occurring at constant pressure the Gibbs free energy ($G = H - TS$) is the relevant potential, where H is the enthalpy. The Gibbs free energy is related to the partition function in the *NPT* ensemble.

3.1.3 Diverse application domains of entropy

The concept of entropy has transcended the boundaries of classical thermodynamics and statistical mechanics, finding applications in diverse fields. For example, in cosmology, the second law of thermodynamics has profound implications for the evolution of the universe. The overall entropy of the universe is believed to be constantly increasing, leading to the concept of the "heat death" of the universe, a hypothetical future state where energy is uniformly distributed, and no further thermodynamic work can be done (Frautschi, 1982; Tolman, 1931).

In material science, entropy plays a crucial role in understanding phase transitions, the formation of alloys, and the stability of different material structures (Guo & Liu, 2011; Otto et al., 2013). The balance between minimizing energy (enthalpy) and maximizing disorder (entropy) determines the favored phase or structure at a given temperature and pressure (Johari, 2021).

In information theory, Shannon entropy quantifies the uncertainty or randomness in a set of data, sharing formal similarities with thermodynamic entropy (Shannon, 1948). Formally the information contained in a possibility is the negative of the logarithm of the possibility at base 2, and the base 2 used in the formulation corresponds to the number of states in the binary number system used for binary digitization. This implies that multiplying I by the probability of the possibility p and summing over all such possibilities quantifies the total information entropy H for this set of possibilities (Eq. 3.12) (Tse, 2022).

$$H = \sum_i p_i I_i = \sum_i p_i \log_2(p_i) \tag{3.12}$$

Detailed reviews regarding the description and application of various entropies in the context of data analysis and machine learning are available (Sepúlveda-Fontaine & Amigó, 2024).

In biology, although living organisms maintain highly ordered structures, they do so by increasing the entropy of their surroundings. Biological processes involve converting ordered energy (e.g., chemical energy in food) into less ordered forms (e.g., heat and waste products), consistent with the second law when considering the organism and its environment as a system. Erwin Schrödinger, in his book "What is Life?: The Physical Aspect of the Living Cell," reflects on the nature of life itself, using thermodynamic principles (Schrödinger, 2025).

He hypothesizes a scenario where an organism functions as a closed system—unable to exchange matter or energy with its surroundings—its internal entropy would progressively increase. This continual rise in disorder would lead to thermodynamic equilibrium, the point of maximum entropy, beyond which biological function would cease and the organism would die. In this conceptual framing, entropy can be viewed as a representation of unavailable energy, and sustaining biological activity requires a constant input of usable energy, typically from food or other structured energy sources.

To resist the inevitable increase in internal entropy and remain far from equilibrium, living organisms must act as open systems, continually importing low-entropy energy or matter. Heterotrophs achieve this by ingesting ordered, energy-rich substances (e.g., food), whereas autotrophs rely on environmental inputs such as sunlight and water, which they convert into biochemical order through photosynthesis.

The internal entropy dynamics of an organism are not uniform across its lifespan. During early developmental stages, such as infancy and childhood, the organism is in a phase of growth and structural organization. In this phase the spontaneous entropy produced by metabolic processes is relatively low, and the rate of entropy reduction—because of incoming low-entropy energy—can exceed the rate of increase. This results in a net decrease in biological entropy and a corresponding increase in complexity and order.

As the organism ages, however, entropy production rises and repair mechanisms decline in efficiency. The balance shifts toward a net increase in internal disorder, contributing to biological aging and eventual systemic breakdown.

Thus, from this conceptual viewpoint, life is sustained by a dynamic equilibrium between entropy generation and the continuous intake of negentropic resources. The degree to which organisms can maintain internal order over time depends on this balance, which varies with physiological and developmental context (Schrödinger, 2025).

3.1.4 Applications of entropy in biology

The application of entropy and a thermodynamic understanding of binding have become increasingly valuable in drug discovery and pharmacology. Although intuitively binding might seem driven purely by favorable interactions (enthalpy), the entropic contributions, often subtle and complex, play a critical role in determining overall binding affinity (ΔG) and can differentiate successful drugs from less effective ones (Cooper et al., 2001). The binding free energy is related to enthalpy and entropy by the Gibbs–Helmholtz (Eq. 3.13):

$$\Delta G = \Delta H - T\Delta S \tag{3.13}$$

where ΔG is the change in Gibbs free energy, ΔH is the change in enthalpy, T is the absolute temperature, and ΔS is the change in entropy. A favorable binding event is characterized by a negative ΔG.

3.1.4.1 *Characterizing binding mechanisms through enthalpy–entropy decomposition*

Numerous studies use experimental techniques, primarily isothermal titration calorimetry (ITC), to dissect the overall binding free energy (ΔG) into its enthalpic (ΔH) and entropic ($-T\Delta S$) components (Krimmer & Klebe, 2015; Velazquez-Campoy et al., 2004). This decomposition provides insights into the primary driving forces behind the binding event. For example, a binding process dominated by favorable ΔH (negative ΔH) might be driven by strong specific interactions such as hydrogen bonds, van der Waals forces, and electrostatic interactions (Bronowska, 2011; Fox et al., 2018). In contrast an entropy-driven process (favorable $-T\Delta S$, which means positive ΔS) often suggests a significant contribution from the hydrophobic effect, where ordered water molecules are released from the surfaces of the interacting molecules upon binding, increasing the overall entropy of the system. Desolvation of polar or charged groups, however, can lead to unfavorable (negative) entropy changes because of the increased ordering of released solvent molecules around newly exposed surfaces (Privalov & Crane-Robinson, 2016).

3.1.4.2 *Identifying enthalpy–entropy compensation*

A frequently observed phenomenon in biological binding is enthalpy–entropy compensation, where a favorable change in enthalpy is partially or fully offset by an unfavorable change in entropy, or vice versa, resulting in a smaller net change in free energy (ΔG) than might be expected from the individual components (Krimmer & Klebe, 2015; Velazquez-Campoy et al., 2004). This means that improvements made to a drug molecule to enhance favorable enthalpic interactions might be counteracted by an entropic penalty (e.g., increased rigidity upon binding and unfavorable desolvation), limiting the gain in binding affinity (Freire, 2008). Understanding the extent and origin of this compensation is crucial for rational drug design. The phenomenon is often observed as a linear correlation between ΔH and $-T\Delta S$ for a series of related ligands binding to the same target or for a single ligand binding to mutated targets (Dragan et al., 2016).

3.1.4.3 *Selected applications of entropy in drug discovery and pharmacology*

In this section, we summarize some selected cases where the individual measurements or estimations of enthalpy and entropy components have provided an insight into the enthalpy–entropy compensation and its physical origins and opened avenues to better handle it and the development of optimized inhibitors.

3.1.4.3.1 HIV protease inhibitors

The development of HIV-1 protease inhibitors is a classic example where thermodynamic studies, including the analysis of entropy, played a significant role (Freire, 2008). Early inhibitors often achieved high affinity primarily through favorable entropic contributions, largely driven by their hydrophobic nature and preorganization to fit the binding site (minimizing conformational entropy loss upon binding). Later generations of inhibitors were designed with improved enthalpic contributions through optimized interactions, leading to even higher affinities and sometimes a better resistance profile (Núñez et al., 2012). Studies have investigated the thermodynamic profiles of various HIV protease inhibitors binding to wild-type and drug-resistant variants, highlighting how shifts in enthalpy and entropy contributions relate to binding potency and resistance (Freire, 2008; Núñez et al., 2012).

For example, studies using ITC have provided detailed thermodynamic parameters for the binding of various inhibitors to HIV-1 protease. For the binding of the inhibitor amprenavir to wild-type HIV-1 protease at 25°C, reported values are approximately $\Delta H = -6.9$ kcal/mol and $-T\Delta S = -6.3$ kcal/mol, resulting in a $\Delta G = -13.2$ kcal/mol (Ohtaka et al., 2002). This shows both favorable enthalpic and entropic contributions. In contrast the binding of indinavir to wild-type HIV-1 protease at 25°C has been reported with values around $\Delta H = 1.7$ kcal/mol and $-T\Delta S = -14.0$ kcal/mol, giving a $\Delta G = -12.3$ kcal/mol (King et al., 2012). This illustrates a case where binding is driven primarily by a large favorable entropy change, compensating for an unfavorable enthalpy change.

Drug resistance mutations can significantly alter these thermodynamic profiles. For instance the Ile50Val (I50V) mutation in HIV-1 protease affects amprenavir binding. Ohtaka et al. (2002) reported that amprenavir binds to HIV-1 protease I50V mutant at 25°C with $\Delta H = -4.2$ kcal/mol and $-T\Delta S = -6.1$ kcal/mol, leading to $\Delta G = -10.3$ kcal/mol. Comparing these to the wild-type values ($\Delta H = -6.9$ kcal/mol, $-T\Delta S = -6.3$ kcal/mol) the resistance is primarily associated with a less favorable enthalpy change ($\Delta\Delta H = +2.7$ kcal/mol), whereas the entropic contribution changes only slightly ($\Delta(T\Delta S) = +0.2$ kcal/mol) (Ohtaka et al., 2002). This demonstrates how changes in thermodynamic components contribute to reduced inhibitor affinity in resistant strains.

3.1.4.3.2 Influenza neuraminidase inhibitors

Similar thermodynamic analyses have been applied to the development of influenza neuraminidase inhibitors, such as oseltamivir. Research has explored the binding thermodynamics of inhibitors to both wild-type and resistant strains of neuraminidase (Albiñana et al., 2016; Li et al., 2012). These studies use a combination of experimental methods (like ITC) and computational approaches to understand how mutations in the enzyme affect the enthalpic and entropic contributions to inhibitor binding, providing insights into the mechanisms of resistance (Singh & Soliman, 2015). The interplay between inhibitor flexibility, solvation effects, and protein conformational changes, all of which influence entropy, is a key area of investigation.

For example, the H275Y mutation in influenza A neuraminidase is a common oseltamivir resistance mutation known to significantly reduce oseltamivir affinity (increase in ΔG) (Masukawa et al., 2003). Experimental ITC $\Delta H = -2.1$ kcal/mol and $-T\Delta S = -4.6$ kcal/mol with $\Delta G = -6.7$ kcal/mol values for oseltamivir binding specifically to H275Y mutant compared to wild-type $\Delta H = -10.8$ kcal/mol and $-T\Delta S = 1.5$ kcal/mol result $\Delta G = -9.4$ kcal/mol, the observed loss of affinity is a result of changes in the balance between enthalpic and entropic contributions (Pokorná et al., 2018). Often, resistance mutations lead to less favorable interactions that can manifest as changes in both ΔH (e.g., reduced favorable contacts) and ΔS (e.g., altered protein or solvent dynamics) compared to the wild-type enzyme, contributing to the increased ΔG and reduced binding affinity and the activity of neuraminidase (Pokorná et al., 2018; Wang et al., 2023).

3.1.4.4 Challenges and application scopes

3.1.4.4.1 Interpreting simulation results

A significant body of work focuses on the challenges of accurately calculating entropy from simulations, particularly the issue of sufficient sampling of conformational space (Hnizdo et al., 2008; Kassem et al., 2015; King et al., 2012; Panday & Ghosh, 2020; Suárez & Díaz, 2014). Literature discusses methods to improve sampling and validate computational entropy estimates against experimental data. The interpretation of computational results helps to link specific molecular interactions and conformational changes to their entropic consequences (Caro et al., 2017; Suárez & Díaz, 2014).

3.1.4.4.2 Entropy in understanding drug resistance and selectivity

Changes in binding thermodynamics, including altered entropy contributions, can underlie the development of drug resistance. Mutations in a target protein can affect its flexibility, interaction profile, and solvation properties, leading to changes in both ΔH and $-T\Delta S$ of drug binding, ultimately reducing affinity. Thermodynamic analysis helps to elucidate the mechanisms by which resistance emerges. Such a case of HIV-1 protease inhibitors' such points has been underscored above (Freire, 2008; Núñez et al., 2012).

Although often discussed in terms of favorable interactions with the target, drug selectivity over off-targets can also have entropic components (Claveria-Gimeno et al., 2017). Differences in the conformational flexibility of the target versus off-targets, or variations in how solvent molecules are displaced from their binding sites, can contribute to differential entropic penalties or gains, influencing selective binding (Frutiger et al., 2021).

3.1.5 Methods for measurement or estimation of entropy

As underscored earlier the importance of measurement or estimation of the entropy lies in gaining insight into the study of biomolecular processes and utilizing it for designing high-affinity drug molecules or circumventing the effects of mutations in the target receptor. Therefore several experimental and computational methods have been developed over the years; some of these key methods are outlined briefly in the following subsections.

3.1.5.1 Experimental measurement of entropy for biological processes

Experimentally determining the entropy change (ΔS) associated with a biological process, such as the binding of a drug molecule to its target, is often achieved indirectly by measuring other thermodynamic quantities (Decherchi & Cavalli, 2020). Although few techniques measure total entropy change directly (Leavitt & Freire, 2001; Paketurytė et al., 2018), several methods provide data from which entropy can be calculated or offer insights into the entropic contributions (Baiz et al., 2020; Juhász et al., 2016; Kumar & Nussinov, 2001).

3.1.5.1.1 Isothermal titration calorimetry

As discussed earlier, ITC is a powerful tool for characterizing binding thermodynamics. It directly measures the heat evolved or absorbed upon binding (ΔH). By analyzing the binding isotherm (the plot of heat per injection versus the molar ratio of ligand to protein) the binding stoichiometry (n), the association constant (K_a), and thus the binding free energy (ΔG) can also be determined

(Leavitt & Freire, 2001). The entropy change (ΔS) is then calculated using the Gibbs–Helmholtz (Eq. 3.14):

$$\Delta S = \frac{\Delta H - \Delta G}{T} \tag{3.14}$$

where $\Delta G = -RT\ln(K_a)$. Thus, although ITC directly measures ΔH, the entropy change ΔS is derived from the measured ΔH and the calculated ΔG (Leavitt & Freire, 2001; Paketurytė et al., 2018).

3.1.5.1.2 Differential scanning calorimetry

Differential scanning calorimetry (DSC) measures the heat capacity change (ΔC_p) associated with temperature-induced transitions, such as protein unfolding or ligand binding coupled to conformational changes (Kumar & Nussinov, 2001). Although not a direct measure of total entropy change upon binding at a single temperature, ΔC_p is critical because it describes how both enthalpy and entropy change with temperature:

$$\Delta H(T) = \Delta H(T_0) + \textstyle\int_{T_0}^{T} \Delta C_p(T')dT' \tag{3.15}$$

$$\Delta S(T) = \Delta S(T_0) + \textstyle\int_{T_0}^{T} \frac{\Delta C_p(T')}{T'}dT' \tag{3.16}$$

A significant ΔC_p upon binding typically indicates changes in solvation (e.g., exposure or burial of hydrophobic surface area) or changes in molecular vibrations and conformational flexibility (Raudino et al., 2013). As changes in solvation and dynamics are major contributors to binding entropy, measuring ΔC_p by DSC provides crucial insights into the nature and temperature dependence of the entropic contribution (Kuril, 2024).

3.1.5.1.3 Binding assays (e.g., surface plasmon resonance and fluorescence spectroscopy)

Techniques such as surface plasmon resonance (SPR), fluorescence spectroscopy, and other equilibrium binding assays measure the binding affinity (association constant Ka or dissociation constant K_d), from which the binding free energy (ΔG) can be calculated directly using $\Delta G = -RT\ln K_a = RT\ln K_d$ (Juhász et al., 2016). If the binding affinity is measured at multiple temperatures, a van't Hoff analysis can be performed by plotting $\ln K_a$ versus $1/T$. The slope of this plot is related to $\Delta H/R$, allowing for the determination of ΔH, from which ΔS can then be calculated using the Gibbs–Helmholtz equation (Varani et al., 2015).

$$\left(\frac{\partial \ln K_a}{\partial(1/T)}\right)_P = -\frac{\Delta H}{R} \tag{3.17}$$

However, van't Hoff analysis assumes that ΔH and ΔS are constant over the temperature range studied (i.e., $\Delta C_p = 0$), which is often not true for biological systems. This can lead to inaccurate determinations of ΔH and ΔS, and can sometimes give a misleading appearance of enthalpy–entropy compensation (Varani et al., 2015). Although these methods are excellent for determining affinity (ΔG), they are less direct and potentially less accurate for obtaining the individual ΔH and ΔS components compared to ITC, especially when ΔC_p is significant (Perozzo et al., 2004).

3.1.5.1.4 Spectroscopic methods (e.g., nuclear magnetic resonance relaxation and vibrational spectroscopy)

Methods such as nuclear magnetic resonance (NMR) relaxation and vibrational spectroscopy (infrared and Raman) can provide information related to molecular dynamics (MD), flexibility, and vibrational states (Reif et al., 2021). Changes in NMR relaxation parameters upon binding can indicate changes in the fast picosecond to nanosecond timescale dynamics, which contribute to conformational entropy. Changes in vibrational spectra can, in principle, reflect changes in vibrational entropy (Baiz et al., 2020). Although these techniques do not typically yield a total ΔS value directly, they provide experimental data that can be used in conjunction with theoretical models or simulations to estimate specific components of the entropy change (e.g., changes in conformational or vibrational entropy) (Azevedo & Bonagamba, 2006).

3.1.5.1.5 Osmotic stress and preferential hydration

Techniques involving osmotic stress explore the role of solvent (water) in binding. By measuring binding affinity under varying concentrations of an inert cosolute (osmolyte), one can determine the number of water molecules released or taken up during the binding process (Timasheff, 1998). Changes in the hydration shell significantly contribute to the overall entropy change, particularly through the hydrophobic effect and changes in the ordering of water molecules around polar/charged groups (Bellissent-Funel et al., 2016). This method provides an estimate of the hydration component of the binding thermodynamics, which is a crucial part of the total entropy change (Athawale et al., 2008).

In summary, although ITC remains the primary method for experimentally obtaining both ΔH and ΔG (and thus deriving ΔS) in a single experiment (Leavitt & Freire, 2001), other techniques like DSC provide essential information about the temperature dependence of thermodynamics and contribute to understanding entropic contributions. Binding assays measure ΔG, which can be coupled with ΔH measurements (preferably from ITC or accurately derived from van't Hoff with caution) to determine ΔS. Spectroscopic methods and osmotic stress offer valuable insights into specific entropic components related to dynamics, vibrations, and hydration.

3.1.5.2 Computational methods and challenges to entropy estimation

Estimating entropy computationally from molecular simulations is a challenging but crucial aspect of understanding biomolecular processes, particularly in drug discovery (Durrant & McCammon, 2011; Panday & Ghosh, 2019). Because of the difficulty in directly computing entropy from the definition involving the partition function, various methods have been developed to estimate it, either directly or as part of free energy calculations (Zhou & Gilson, 2009). These approaches often focus on different components of entropy, including translational, rotational, vibrational, and conformational contributions (Yu et al., 2001).

In this section, we outline the principles, applications, and limitations of key computational methods for entropy estimation using simulations. Methods like normal mode analysis (NMA) employ a harmonic approximation, treating the system as harmonic oscillators based on minimized structures to calculate vibrational frequencies and estimate vibrational entropy, often combined with rigid-rotor and ideal-gas approximations for translational and rotational entropy (Brooks & Karplus, 1983). NMA is sometimes used with end-state free energy methods such as Molecular Mechanics/Generalized-Born Surface Area (MM/GBSA) or Molecular Mechanics/Poisson-Boltzmann Surface Area (MM/PBSA) to assess vibrational entropy contributions to binding free energies. However, its strict harmonic assumption and reliance on energy minima make it sensitive to the starting structure and unable to account for anharmonicity or conformational transitions, limiting its accuracy for flexible biomolecules and making it computationally expensive for large systems because of Hessian matrix diagonalization (Case, 1994; Dykeman & Sankey, 2010; Ghysels et al., 2009). Extending NMA, quasiharmonic analysis (QHA) attempts to incorporate some anharmonicity by using the covariance matrix from an MD simulation to approximate the conformational distribution as a multivariate Gaussian. This allows for a more realistic estimate of vibrational and low-frequency motion entropy within a single energy basin compared to strict NMA (Levy et al., 1984). Nevertheless, QHA's reliance on the Gaussian approximation may not be accurate for highly flexible molecules or systems with multiple distinct conformational states and requires extensive MD sampling for converged covariance matrices, potentially overestimating entropy if transitions between multiple wells are not fully sampled (Suárez & Díaz, 2014). MD simulation of large biomolecular complexes for entropy estimation often becomes computationally very expensive to address such application cases. Recently approximate methods that rely on change in localized atomic packing via Gaussian density change to estimate entropy of binding for protein–protein complex formation have also been reported in conjunction with MM/PBSA-like end-state free energy methods (Panday & Alexov, 2022).

Other approaches, such as information theory-based methods, leverage concepts from information theory to quantify the disorder in the conformational ensemble sampled by simulation, aiming to calculate configurational entropy by analyzing the probability distribution in conformational space, potentially involving partitioning the space (e.g., via clustering) and estimating cluster probabilities (Hnizdo et al., 2008; King et al., 2012; Panday & Ghosh, 2020; Panday et al., 2019). These are valuable for analyzing the complexity of conformational ensembles, but accurately estimating high-dimensional probability distributions requires extensive sampling, careful definition of collective variables, and sensitivity to partitioning choices (Kassem et al., 2015).

Alchemical free energy methods, including free energy perturbation and thermodynamic integration, provide a rigorous thermodynamic framework primarily for calculating the binding free energy (ΔG) by simulating a nonphysical pathway between states. Although they directly yield ΔG, obtaining accurate individual ΔH and ΔS components from these methods for mechanistic insight is less direct and sensitive to sampling along the alchemical pathway, despite these methods often being more robust for ΔG calculation than direct entropy estimation (Hansen & van Gunsteren, 2014).

Finally, methods based on explicitly calculating configurational entropy directly attempt to compute it from sampled configurations, for instance, by clustering conformations and estimating their probabilities or directly estimating probability densities (Singh et al., 2003). These methods are useful for understanding the entropic landscape of flexible molecules, but defining appropriate clusters, estimating probabilities accurately, and overcoming the computational cost and "curse of dimensionality" in high-dimensional density estimation remain significant challenges (Hensen et al., 2010).

Despite the development of these various computational approaches, estimating entropy from simulations presents inherent challenges, particularly regarding adequate sampling of the vast conformational space available to biomolecules (Suárez & Díaz, 2014). Biomolecules often have complex and rugged free energy landscapes with many local minima, making exhaustive sampling computationally prohibitive for accurate entropy determination (Wei et al., 2016). Furthermore the decomposition of the total entropy into physically meaningful components (translational, rotational, vibrational, and conformational) can be complex and depends heavily on the chosen method and underlying approximations

(Kassem et al., 2015). Accounting for anharmonic effects is crucial for accurate entropy calculations, and purely harmonic methods (like basic NMA) have significant limitations in capturing the full dynamic nature of biological systems. Ultimately, many of the more rigorous methods for entropy or free energy calculation remain computationally very expensive, limiting their applicability in high-throughput virtual screening (Panday, 2018; Panday & Ghosh, 2019, 2020). Despite these challenges the continuous development of more sophisticated sampling techniques, improved force fields, and advanced free energy methods is steadily improving the ability to estimate entropy computationally, providing valuable insights for rational drug design (King et al., 2021).

3.2 Case studies with application to malarial targets: Flexibility matters

Previous studies have emphasized the importance of both enthalpic and entropic contributions to molecular recognition. Binding enthalpy represents only one component of the overall binding free energy; entropy, often underappreciated, can act to counterbalance enthalpic changes when diverse ligands bind to a common receptor (Olsson et al., 2011b). The extent of this enthalpy–entropy compensation varies across ligands and systems (Chodera & Mobley, 2013). Therefore accurate characterization of the thermodynamic profile of ligand–receptor interactions requires the estimation of both enthalpic and entropic components. Therefore, in this section, we present two case studies including our previous works (Panday, 2018) involving flexible protein interactions with other molecules: (1) a protein–peptide binding scenario in which both the receptor protein and the peptides exhibit varying degrees of flexibility, and (2) small-molecule binding to a flexible receptor, where the modulation of receptor conformational dynamics is studied in relation to conformational entropy and binding free energy. In the context of drug resistance and the need for drug development aimed at novel targets in *Plasmodium falciparum*, a leading pathogen causing severe malaria.

3.2.1 The persistent challenge of antimalarial drug resistance

Malaria control has been repeatedly challenged by the emergence of drug-resistant parasites. Historically, quinine, extracted from cinchona bark, was the primary antimalarial until the mid-20th century. It was later replaced by chloroquine, synthesized in the 1930s, which became the frontline treatment during the 1960s and 1970s. However, resistance to chloroquine first appeared along the Thailand–Cambodia border (Mita et al., 2009) and subsequently spread to South America and sub-Saharan Africa. This resistance led to a resurgence in malaria morbidity and mortality.

In response, alternative drugs such as pyrimethamine, sulfadoxine, and mefloquine were introduced, but resistance to these also developed quickly (Abdul-Ghani et al., 2013), prompting a transition toward combination therapies. Although these therapies improved efficacy, they also increased treatment costs and the risk of side effects (Packard, 2014).

A significant breakthrough was the discovery of artemisinin by Chinese scientist Tu Youyou in 1972, derived from *Artemisia annua*. Artemisinin showed exceptional efficacy, earning Tu the Nobel Prize in Physiology or Medicine in 2015. However, its global availability remained limited until the 1990s (Packard, 2014). To mitigate the risk of resistance the World Health Organization recommended artemisinin use in artemisinin-based combination therapies (ACTs). Early ACTs, such as Coartem (a combination of artemether and lumefantrine), were initially restricted to delay resistance, but increasing mortality led to broader adoption in 2010. Despite this, artemisinin resistance was reported in Southeast Asia by 2014 (Ashley et al., 2014; Imwong et al., 2017).

The emergence of resistance reflects a complex interplay of biological, environmental, and socioeconomic factors. Unregulated and incomplete antimalarial uses have significantly contributed to resistance (Packard, 2014). For instance, chloroquine resistance emerged in Pailin, Cambodia—a gem mining region with high population mobility and poor vector control. The distribution of chloroquine-medicated salt for mass drug administration inadvertently led to subtherapeutic dosing, exacerbating resistance (Wongsrichanalai et al., 2002).

Ultimately, resistance is shaped by a convergence of drug pharmacokinetics (e.g., long half-lives), reduced host immunity, parasite genetic plasticity, and vector–parasite interactions (Dondorp et al., 2017; Wongsrichanalai et al., 2002). Addressing resistance requires identifying new targets and developing drugs with novel mechanisms of action. Although numerous potential drug targets in *Plasmodium* have been proposed (Muregi, 2012), many remain to be fully validated.

3.2.1.1 Apical membrane antigen 1 (PfAMA1) as a target

Targeting essential steps in the parasite's erythrocytic life cycle is key to malaria intervention (Meis & Verhave, 1988). A critical event is the invasion of red blood cells by merozoites, which depends on an apical complex and the coordinated secretion of micronemal and rhoptry proteins (Counihan et al., 2013). The interaction between apical membrane antigen 1 (PfAMA1) and the rhoptry neck protein PfRON2 is central to the formation of a "tight junction" that enables erythrocyte invasion (Besteiro et al., 2011; Riglar et al., 2011; Srinivasan et al., 2011). This protein–protein interaction is an attractive therapeutic target, and inhibitors of the PfAMA1–PfRON2 interaction have shown promise in blocking merozoite invasion (Harris et al., 2005, 2014; Tan et al., 2019).

3.2.1.2 Protein kinase 5 (PfPK5) as a target

P. falciparum protein kinase 5 (PfPK5) is a Ser/Thr kinase homologous to human cyclin-dependent kinase 1 (CDK1) (Holton et al., 2003). PfPK5 is highly expressed during the schizont stage (approximately 30 hours postinfection), when multiple rounds of nuclear division occur (Graeser et al., 1996). Given that CDKs are essential regulators of eukaryotic cell division, PfPK5 is a compelling target for antimalarial drug development (Holton et al., 2003).

PfPK5 can be activated by human cyclin H and p25 (an activator of human CDK5), and its activity is inhibited by known CDK inhibitors (Le Roch et al., 2000). Unlike its human counterparts, PfPK5 can autophosphorylate in the presence of a cyclin, whereas Pfmrk (a putative *P. falciparum* CDK-activating kinase) does not activate PfPK5 (Le Roch et al., 2000). These differences in regulation suggest that selective inhibition of PfPK5 is possible.

Detailed challenges associated with these two targets and how the current study attempts to enrich scientific understanding via characterization of enthalpy–entropy and binding free energy thermodynamic parameters and aid in further drug development efforts are discussed for each case.

In summary, both PfAMA1 and PfPK5 represent promising yet challenging targets for antimalarial drug development. Targeting flexible regions and understanding ligand-induced conformational dynamics are crucial for developing effective and selective inhibitors.

3.2.2 Methods

The results presented in this section are obtained using similar approaches and methods; therefore we present the methods in a common subsection, whereas materials will be provided in dedicated subsections for each case study.

3.2.2.1 Molecular dynamics simulation setup

MD simulations were performed using the AMBER14 package (Case et al., 2005) with the Amber ff14SB force field (Maier et al., 2015). Each system was placed in a cuboidal box with a 10 Å padding from the solute in all directions. The system was neutralized with Na^+/Cl^- ions to enable the use of periodic boundary conditions. Solvation was performed using the TIP3P water model (Mark & Nilsson, 2001). In case of simulations involving small molecules, the Generalized Amber Force Field was used (Sprenger et al., 2015), whereas ff14SB was used for protein (Maier et al., 2015).

3.2.2.1.1 Minimization and equilibration

Initial energy minimization was conducted in two stages. First, 2700 steps of steepest descent followed by 300 steps of conjugate gradient minimization were performed, with a 100 kcal/mol restraint applied to nonhydrogen atoms of the solute. This was followed by a second minimization stage (3000 steps) with gradually decreasing restraints to allow solvent relaxation.

The system was heated from 0 to 300 K in six 10 ps heating runs under the *NVT* ensemble, each increasing the temperature by 50 K. After reaching 300 K a further minimization phase was carried out using decreasing restraints (2.0, 0.1, and 0.05 kcal/mol) over 8000 steps of steepest descent and 2000 steps of conjugate gradient minimization under the *NVE* ensemble, following the published protocol (Panday & Ghosh, 2020).

A second heating stage followed, in which the system was gradually heated from 0 to 300 K over 100 ps under *NVT*. Bond lengths involving hydrogen atoms were constrained using the SHAKE algorithm (Forester & Smith, 1998). Equilibration was performed in the *NPT* ensemble with decreasing positional restraints (0.5, 0.1, 0.04, and 0.01 kcal/mol) applied over 50, 100, 400, and 400 ps, respectively. Subsequently a 10 ns unrestrained simulation was carried out with a timestep of 2 fs.

3.2.2.1.2 Production simulation settings

Production simulations were performed under the *NPT* ensemble with a 2 fs timestep. Hydrogen-involving bond lengths were constrained using SHAKE (Forester & Smith, 1998). Temperature was regulated using Langevin dynamics (Grest & Kremer, 1986) with a collision frequency (γ) of 2.0 ps^{-1}. Pressure was maintained using isotropic position scaling with a Berendsen barostat (Yoneya et al., 1994) and a relaxation time of 1 ps. Trajectory snapshots (coordinates and energies) were saved every 1 ps. Each simulation ran for 254 ns, with the first 4 ns discarded to eliminate potential equilibration artifacts. For each system, 20 such simulations were performed, totaling 5 μs of production MD data.

3.2.2.2 Binding enthalpy calculation

We have utilized endpoint free-energy methods such as MM–GBSA and MM–PBSA for estimating binding enthalpy (Genheden & Ryde, 2015). It is important to note that these methods do not explicitly account for binding configurational entropy (Genheden & Ryde, 2015). The enthalpy is calculated using MMPBSA.py tool available in Amber simulation package (Miller et al., 2012). The binding configurational entropy (BCE) is calculated separately as detailed in the following subsection and subsequently combined with the enthalpy estimations to determine the total binding free energy.

3.2.2.3 Binding configurational entropy calculation

We estimated binding configurational entropy using the *Neighbor-Approximated Maximum Information Spanning Tree* (A-MIST) method (Panday & Ghosh, 2020). A-MIST is an information-theoretic approach that estimates entropy by transforming the Cartesian coordinate trajectory of a molecular system into an internal coordinate representation, comprising bonds, angles, and torsions. This transformation eliminates overall translational and rotational motions, which are accounted for analytically.

For a molecular system consisting of N atoms the method constructs a unique set of $3N-6$ internal degrees of freedom (DOFs), representing the system as a connected graph composed of $N-1$ bond lengths, $N-2$ bond angles, and $N-3$ torsional angle DOFs. Under the first-order approximation, all DOFs are assumed to be statistically independent. The entropy contribution of each degree of freedom (DOF) is then computed via histogramming its observed values along the trajectory, using the following expression:

$$S_1(r_i) = \sum\nolimits_{k=1}^{p} v_k(r_i) \ln\left(\frac{v_k(r_i)}{J(r_j)[k] \cdot hw(r_i)}\right) \tag{3.18}$$

where $v_k(r_i)$ denotes the normalized frequency in the kth histogram bin for the ith DOF, $hw(r_i)$ is the bin width, and $J(r_i)[k]$ is the Jacobian term accounting for the coordinate transformation. The Jacobian takes the form $b_i[k]^2$, $\sin(\theta_i[k])$, or 1, for bond length, bond angle, and torsional DOFs, respectively (Panday & Ghosh, 2020). Summing over all DOFs yields the first-order entropy estimate $S_{(1)}$, defined as $S_{(1)} = \sum_{i=1}^{3N-6} S_1(r_i)$.

However, in biomolecular systems, internal DOFs often exhibit correlated fluctuations. To account for such correlations, second-order (and, in principle, higher order) mutual information terms can be included. Because of the high dimensionality and sampling limitations inherent to biomolecules, higher order correlations are computationally challenging to converge reliably. Therefore we restrict our analysis to second-order correlations, as also recommended in previous studies (Panday & Ghosh, 2020).

For the second-order terms, joint histograms are constructed for each pair of DOFs (r_i, r_j), and the joint entropy $S_2(r_i, r_j)$ is computed. The mutual information for each DOF pair is given by

$$I_2(r_i, r_j) = S_1(r_i) + S_1(r_j) - S_2(r_i, r_j) \tag{3.19}$$

To reduce computational cost and focus on physically relevant correlations, a distance-based cutoff C (in Å) is applied, restricting the computation of mutual information to pairs of DOFs that are within this spatial proximity. The second-order entropy using the A-MIST framework is then computed as

$$S_{\mathrm{MIST}_2}^{N_b(C)}(\overline{r_f}) = \sum_{r_j \in \overline{r_f}} \left[S_1(r_j) - \max_{(r_i, r_j) \in N_b(C);\, j>i} I_2(r_i, r_j) \right] \tag{3.20}$$

where $N_b(C)$ denotes the set of DOF pairs separated by less than the cutoff distance C, and $\overline{r_f}$ represents the full set of DOFs considered in the system, excluding DOFs involving hydrogen atoms that were restrained using SHAKE during MD simulation. Previous studies have demonstrated that this cutoff-based approximation yields entropy estimates comparable to the full, noncutoff methods across various biomolecular systems (Panday & Ghosh, 2020; Panday et al., 2019).

3.2.3 Role of entropy in flexible protein–flexible peptide binding

In this subsection, we focus on the binding of apical membrane antigen 1 (AMA1), a crucial vaccine target in malaria-causing apicomplexan parasites such as *P. falciparum. Plasmodium* species, including *P. falciparum*, are the etiological agents of malaria, a global health burden transmitted by *Anopheles* mosquitoes, with substantial socioeconomic consequences (Doerig et al., 2010; WHO, 2023). Among them, *P. falciparum* is the most virulent, responsible for the majority of severe malaria cases and deaths, particularly affecting young children and pregnant women in sub-Saharan Africa (Doerig et al., 2010; WHO, 2023).

The parasite's complex life cycle spans both mosquito and human hosts (Cowman et al., 2016; Miller et al., 2013). In the human host, sporozoites invade hepatocytes, subsequently releasing merozoites that infect erythrocytes and cause the symptomatic phase of the disease (Miller et al., 2013). Interrupting essential stages of the parasite's intraerythrocytic life cycle is therefore key to therapeutic intervention (Meis & Verhave, 1988).

A critical event in the blood stage is the rapid invasion of red blood cells by merozoites, which relies on a specialized apical complex and coordinated secretion of proteins from micronemes and rhoptries (Counihan et al., 2013). Central to this process is the interaction between the micronemal protein apical membrane antigen 1 (PfAMA1) and rhoptry neck proteins, notably PfRON2, to form a "tight junction" that enables invasion (Besteiro et al., 2011; Srinivasan et al., 2011). Disruption of the PfAMA1–PfRON2 interaction has been shown to inhibit invasion, underscoring its therapeutic potential (Harris et al., 2005, 2009; Tan et al., 2019).

Peptides and small molecules that interfere with this protein–protein interaction have shown efficacy in blocking red blood cell invasion, making it an attractive target for antimalarial drug design (Harris et al., 2005; Tan et al., 2019).

3.2.3.1 *PfAMA1–peptide binding and strain-specific inhibition*

Phage display screening by Li et al. (2002) identified three peptides that bind PfAMA1, potentially at the PfRON2 binding site. One peptide, F1 (GWRLLGFGPASSFSM), effectively blocked invasion in 3D7 and 3D10 strains, but not in HB3. Sequence analysis revealed key residue differences—particularly charge alterations—likely responsible for strain-specific binding affinity (Li et al., 2002).

In a subsequent study, Harris et al. (2005) identified a 20-residue peptide, R1 (VFAEFLPLFSK-FGSRMHILK), as a PfAMA1 inhibitor. Like F1, R1 exhibited strain-specific inhibition, acting against 3D7 and 3D10 but not against HB3 or W2mef. NMR spectroscopy revealed two structured regions in R1: a turn-like motif (Phe5–Ser10) and a distorted type II β-turn (Gly13–Met16) (Lee et al., 2011). Alanine scanning mutagenesis and ELISA-based binding assays further identified Phe5, Pro7, Leu8, and Phe9 as critical residues for PfAMA1 binding (Lee et al., 2011).

In this study, we explore the PfAMA1–PfRON2 and PfAMA1–R1 interactions using MD simulations. Our aim is to dissect the structural and energetic basis of these interactions to guide rational ligand design. In particular, we assess the contributions of individual binding-site residues to the thermodynamic components of binding—enthalpy, entropy, and consequently free energy—to understand how flexibility and complementarity shape peptide binding to PfAMA1.

3.2.3.2 Materials

In the present study the receptor is *P. falciparum* apical membrane antigen-1 (PfAMA1), a 622-residue type I integral membrane protein (PF3D7_1133400) (Waters et al., 1990). Interacting partners of PfAMA1 include the transmembrane segments of RON—specifically PfRON2sp1, a truncated variant (lacking segments from both N-terminal and C-terminal ends) denoted as PfRON2sp2, and the R1 peptide identified via phage display screening. All these molecules exhibit considerable flexibility in their unbound states (Bai et al., 2005). The binding of any of these peptides (PfRON2sp1, PfRON2sp2, and R1) to PfAMA1 induces conformational changes in PfAMA1, including the loss of helical structure in a region within the D-II loop upon complex formation (Vulliez-Le Normand et al., 2012). These subtle modifications in structural flexibility upon binding are anticipated to contribute significantly to the overall entropy (Panday, 2018). Therefore we will quantify the entropy of binding, particularly the binding configurational entropy (BCE), in addition to enthalpy, to comprehensively study and understand the underlying thermodynamics of the interaction mechanism (Panday, 2018).

Crystal structures of PfAMA1 from the 3D7 strain [apo form, Protein Data Bank (PDB): 1Z40] and in complex with PfRON2sp1 (3D7 strain, PDB: 3ZWZ), PfRON2sp2 (CAMP strain, PDB: 3SRI), and R1-peptide (3D7 strain, PDB: 3SRJ) were utilized. These structures primarily cover domains I and II (sequence residues Asn108–Glu438) of PfAMA1 (Bai et al., 2005; Vulliez-Le Normand et al., 2012). Domain I is crucial for inhibitory antibody binding and contains the important C1L region (residues 196–207) (Harris et al., 2014).

The available crystal structures of *P. falciparum* apical membrane antigen 1 (PfAMA1) reveal unresolved loop regions that were necessary to model for comparative analysis. The apo-PfAMA1 structure (PDB ID: 1Z40) is missing residues 172–176, 264–273, and 383–387. These regions were modeled using MODELLER v9.15 (Šali et al., 1995), and the model quality was assessed using PROCHECK (Laskowski et al., 1993). In the peptide-bound complex structures (PDB IDs: 3ZWZ, 3SRI, 3SRJ) the PfAMA1 segment from residues 351 to 387 is disordered and not resolved (Vulliez-Le Normand et al., 2012). The structures 3SRI and 3SRJ contain mutations N162K, T288V, S373D, N422D, and S423K, which were reverted to the wild-type sequence for subsequent analysis (Vulliez-Le Normand et al., 2012).

Peptide-binding interfaces were identified for PfRON2sp1, PfRON2sp2, and the R1 peptide. A common region encompassing residues Asn108 to Glu438 was used across all structural and MD analyses for consistency. Interface mapping revealed a set of residues on PfAMA1 consistently involved in binding across all three peptides. These include Leu131, Tyr142, Arg143, Phe183, Pro184, Pro185, Thr186, Glu187, Pro188, Gly222, Asn223, Met224, Ile225, Pro226, Ser232, Tyr234, Lys235, Tyr236, Tyr251, Ala254, Glu256, Gln349, and Pro350. These conserved contacts likely define a core recognition surface on PfAMA1.

A summary of the structural data and listing of only the peptide-specific binding residues exclusive to each complex, after exclusion of the shared interface residues mentioned above, is provided in Table 3.1.

Table 3.1 Summary of PfAMA1 crystal structures and exclusive peptide-binding residues. Residues common to all three peptide interfaces are described in the main text.

PDB	Molecule	Strain	Res. (Å)	pH	Mutations	Missing regions	Exclusive binding residues	References
1Z40	apo-AMA1	3D7	1.9	6.0	–	172–176, 264–273, 383–387	–	Bai et al. (2005)
3ZWZ	AMA1-RON2sp1	3D7	2.1	8.5	–	351–387	Asn271, Met273, Tyr390	Vulliez-Le Normand et al. (2012)
3SRI	AMA1-RON2sp2	CAMP	1.6	8.6	N162K, T288V, S373D, N422D, S423K	261–272, 351–387	Val137, Gln141, Phe201, Tyr202, Asn225	Vulliez-Le Normand et al. (2012)
3SRJ	AMA1-R1	3D7	2.15	8.5	N162K, T288V, S373D, N422D, S423K	261–270, 353–387	Asp134, Val137, Thr140, Gln141, Ala170, Tyr175, Phe201, Asn228, Lys230	Vulliez-Le Normand et al. (2012)

Table 3.2 Structure comparison of AMA1 (Asn108–Glu438) and AMA1 complexed with RON2sp1, RON2sp2, and R1 peptides. RMSD is measured over the aligned CA atoms.

Structure comparison	RMSD (Å)	Number of atoms
1Z40 vs. 3ZWZ	0.488	268
1Z40 vs. 3SRI	0.407	266
1Z40 vs. 3SRJ	0.470	274
3ZWZ vs. 3SRI	0.465	264
3ZWZ vs. 3SRJ	0.524	261
3SRI vs. 3SRJ	0.464	267

3.2.3.2.1 Structural analysis of PfAMA1 and its complexes with peptides

The available crystal structures of *P. falciparum* apical membrane antigen 1 (PfAMA1) encompass only domains I and II, corresponding to sequence residues Asn108 to Glu438. To ensure consistency across structural and MD analyses, this specific region (Asn108–Glu438) was selected for all comparative studies.

Structural comparisons were performed by superimposing the PfAMA1 crystal structures—both unbound and in complex with various peptides—and calculating the root mean square deviation (RMSD) values. These comparisons (Table 3.2) reveal that while the overall structures are similar, notable differences arise in the Domain II (D-II) loop region (residues Lys351 to Ala387). This loop is absent in peptide-bound PfAMA1 structures, leading to its exclusion from RMSD computations.

The unbound PfAMA1 structure (PDB ID: 1Z40) lacks three loop regions: L1 (residues 172–176), L2 (264–273), and L3 (383–387). These loops were modeled using MODELLER v9.15 with default parameters (Šali et al., 1995). The quality of the modeled structures was assessed using PROCHECK (Laskowski et al., 1993), and the best model, based on zDOPE scores, was selected for further analysis. The modeled loops are depicted in *red* in Fig. 3.1A.

Upon binding of PfRON2-derived peptides and the R1 peptide the D-II loop region (residues 351–387) of PfAMA1 becomes disordered and is not resolved in the crystal structures of the complexes (PDB IDs: 3ZWZ, 3SRI, and 3SRJ). Interestingly the helical structure present in the unbound PfAMA1's D-II loop is mimicked by the N-terminal helix of the PfRON2sp1 peptide, as illustrated in Fig. 3.1B.

3.2.3.2.2 Comparative analysis of PfRON2sp1 and PfRON2sp2 complexes

The PfRON2sp2 peptide, a truncated variant of PfRON2sp1, lacks both the N-terminal helical region and the C-terminal segment that interacts with the hydrophobic groove of PfAMA1. Despite these differences, PfRON2sp2 occupies the same binding region on PfAMA1 as PfRON2sp1. To facilitate comparative analyses the PfRON2sp2-PfAMA1 complex structure (originally from the CAMP strain) was superimposed onto the PfRON2sp1-PfAMA1 complex (3D7 strain). These structures are presented in Fig. 3.1B and C.

The R1 peptide binds to the hydrophobic groove of PfAMA1, similar to PfRON2-derived peptides. In the crystal structure, two R1 peptide molecules are observed: R1-major, which binds deeply within the groove, and R1-minor, which overlays R1-major and covers the remaining groove surface (Fig. 3.1D). The R1-major peptide shares five identical and two similar residues with PfRON2sp1, indicating structural mimicry (Fig. 3.1F).

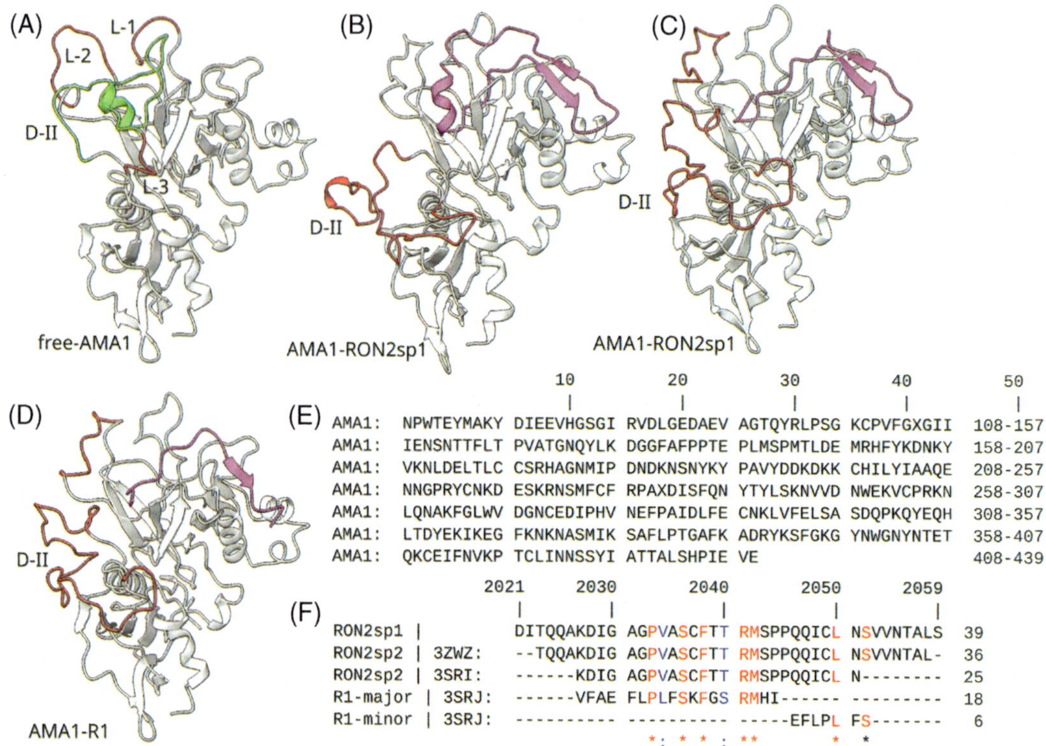

FIGURE 3.1

Structures of unbound-PfAMA1 and bound with inhibitory peptides. Structures of (A) unbound PfAMA1 (PDB: 1Z40), (B) PfAMA1 bound to PfRON2sp1 (PDB: 3ZWZ), (C) PFAMA1 bound to PfRON2sp2 (PDB: 3SRI), (D) PfAMA1 bound to R1-peptides are shown, (E) the sequence of the simulated structure of the AMA1, and (F) aligned sequences of the peptides binding to PfAMA1 are shown. The three loops, L1 (residues 172–176), L2 (264–273), and L3 (383–387), are shown with *red color*. The D-II loop, which is structured in an unbound structure, is shown with *green*. In peptide bound structures this loop becomes unordered; the modeled loop for these is shown in *red color*. All binding peptides are shown in *purple*. Two molecules of R1 mimic the binding of PfRON2 peptides; one crystallized, covering a longer sequence, is named R1-major, and the other, R1-minor; however, assay stoichiometry suggests R1-minor binding is very weak.

To correlate residue numbering between structural data and the full-length protein sequences: (1) For PfAMA1 (UniProtKB: Q7KQK5), add 107 to the residue numbers to map residues mentioned in the enthalpy and entropy results to protein sequence numbering. (2) For PfRON2sp1, add 2022 to the structure's residue numbers, as the crystallized segment starts from residue 2023. (3) For PfRON2sp2, add 2026 to the structure's residue numbers, accounting for the truncation at both termini.

The binding affinities (K_d values) of PfRON2sp1, PfRON2sp2, and R1 peptides to PfAMA1 (3D7 strain) were determined using SPR, as summarized in Table 3.3. These values serve as benchmarks for computational predictions of binding enthalpy, configurational entropy, and free energy.

Table 3.3 Binding affinities of peptides to PfAMA1 (3D7 strain).

Peptide	K_d (nM)	ΔG (kcal/mol)	References
PfRON2sp1	20.3 ± 6.3	−10.57	Vulliez-Le Normand et al. (2012)
R1	80.3 ± 5.5	−9.74	Harris et al. (2009)
PfRON2sp2	520 ± 74	−8.63	Vulliez-Le Normand et al. (2012)

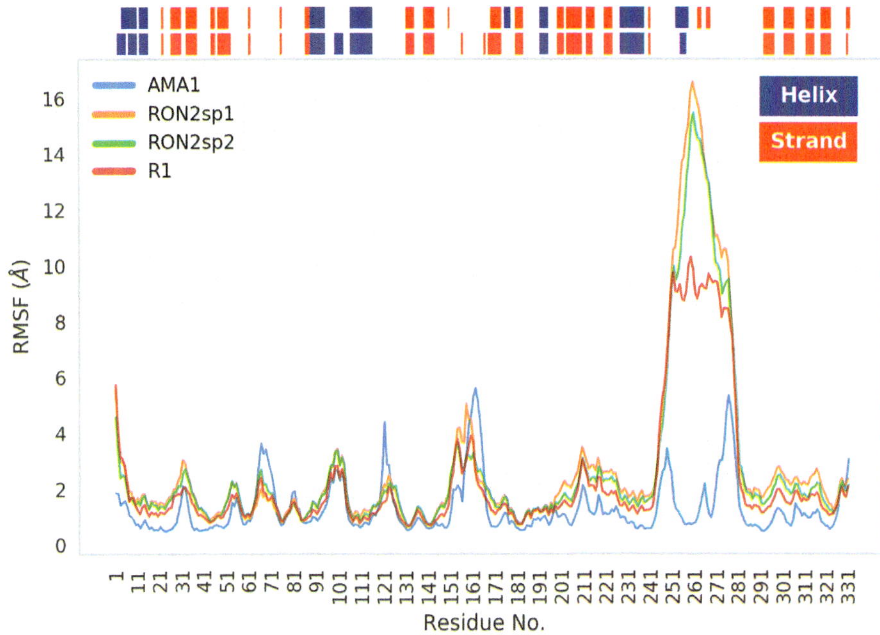

FIGURE 3.2

Root mean square fluctuations of Cα atoms of unbound PfAMA1 and PfAMA1 in complexes: root mean square fluctuations of Cα atoms of unbound PfAMA1 and PfAMA1 in complex with PfRON2sp1, PfRON2sp2, and R1 peptides. Secondary structure elements (SSEs) in free and bound from PfAMA1 are annotated at top of the plot; helix is represented with *blue rectangular box* and strand with *red rectangular box*. Missing loops, L1, L2, and L3, are modeled in free-PfAMA1, D-II is missing in PfRON2sp1-bound PfAMA1, and it is modeled; all modeled regions are annotated with magenta rectangular bars over associated residues.

3.2.3.3 Results and discussion

3.2.3.3.1 Analysis of dynamics

The root mean square fluctuations (RMSF) analysis reveals that, as expected, residues within the modeled regions exhibit higher fluctuations than those in structured regions defined by the crystal structure (Fig. 3.2). Across all systems, domain II of PfAMA1 demonstrates higher fluctuations compared to domain I. Particularly the large D-II loop becomes disordered upon complex formation with RON2 and R1 peptides and shows notable flexibility. Interestingly, in the R1-PfAMA1 complex, this region

Table 3.4 Details of calculated binding energies using MM/GBSA and MM/PBSA (indi = 1) for PfRON2sp1, PfRON2sp2, and R1-peptide binding to PfAMA1. All energy values are in kcal/mol.

		E_{MM}		MM/GBSA (indi = 1)			MM/PBSA (indi = 1)		
Peptide	ΔG_{expr}	vdW	EEL	GB	Surf.	Total	PB	$E_{nonpolar}$	Total
PfRON2sp1	−10.57	−178.0	−293.7	347.1	−23.0	−147.6 ± 13.3	353.7	−17.7	−135.7 ± 14.2
R1	−9.74	−112.7	−308.0	343.2	−15.9	−93.4 ± 10.2	346.0	−12.7	−87.4 ± 12.2
PfRON2sp2	−8.63	−130.3	−211.8	254.3	−17.0	−104.7 ± 11.9	261.9	−13.5	−93.6 ± 13.2

vdW, *van der Waals.*

Table 3.5 Details of calculated ΔH using MM/PBSA with two different internal dielectric constants (indi = 2 and indi = 4) for PfRON2sp1, PfRON2sp2, and R1-peptide binding to PfAMA1. All energy values are in kcal/mol.

		E_{MM}			MM/PBSA (indi = 2)		MM/PBSA (indi = 4)	
Peptide	ΔG_{expr}	vdW	EEL	$E_{nonpolar}$	PB	Total	PB	Total
PfRON2sp1	−10.57	−178.0	−146.8	−17.7	174.2	−168.3 ± 11.8	84.5	−184.6 ± 11.8
R1	−9.74	−112.7	−154.0	−12.7	170.5	108.9 ± 9.1	82.7	−119.7 ± 8.6
PfRON2sp2	−8.63	−130.3	−105.9	−13.5	129.6	−120.1 ± 10.9	63.2	−133.5 ± 11.2

vdW, van der Waals.

exhibits reduced fluctuation compared to the corresponding region in PfRON2sp1- and PfRON2sp2-bound complexes (Fig. 3.2). Overall, all peptide-bound PfAMA1 structures display higher RMSF values than the unbound protein. These calculations were based on 5 μs MD simulation trajectories for each system.

3.2.3.3.2 Binding enthalpy for PfAMA1 surface interactions with peptides

Binding enthalpy for the three peptides—PfRON2sp1, PfRON2sp2, and R1—was calculated using both MM/GBSA and MM/PBSA methods. For each system, 50,000 frames were extracted at intervals of 100 ps from the full 5 μs MD trajectory. The calculated binding enthalpies are summarized in Table 3.4. These data indicate that the predicted binding enthalpies do not reproduce the experimental binding affinity trends. Specifically, PfRON2sp2, which is experimentally the weakest binder, exhibits a more favorable predicted binding affinity than R1. This discrepancy underscores that enthalpy alone may not be sufficient to determine the correct affinity ranking.

Given that binding enthalpies from MM/PBSA calculations are influenced by the choice of the internal dielectric constant (indi), additional calculations were performed using indi values of 1, 2, and 4. The results for indi = 2 and indi = 4 are shown in Table 3.5. Even with adjusted dielectric constants, the predicted affinities do not align with experimental trends. PfRON2sp1 consistently appears as the strongest binder across methods and parameter sets, but the ranking between R1 and PfRON2sp2 remains inconsistent with experimental data.

For the purposes of this analysis, MM/PBSA results with indi = 2 are adopted as the reference values for binding enthalpy. Further insight into peptide binding is obtained by decomposing the

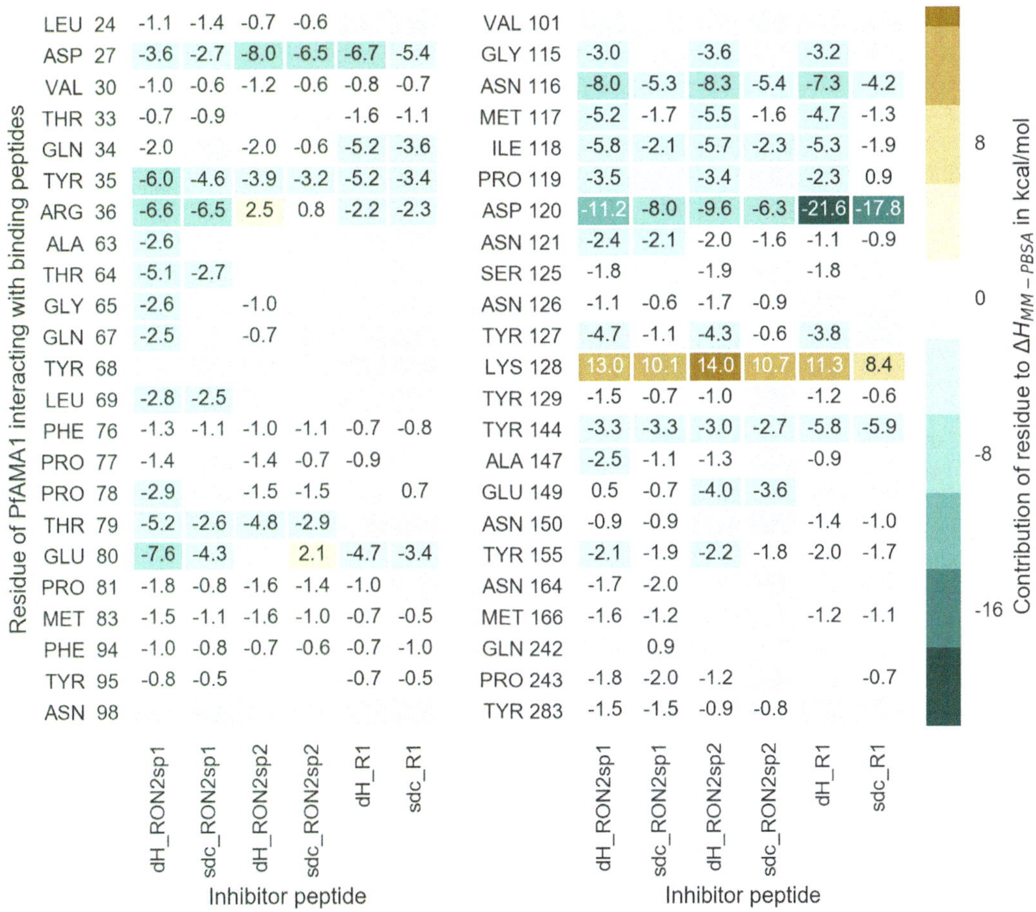

FIGURE 3.3

Residue-wise enthalpy contributions from PfAMA1 to peptide binding were calculated using the MM/PBSA. Residue-wise enthalpy contributions from PfAMA1 to peptide binding, calculated using the MM/PBSA method with internal dielectric constant indi = 2. ΔH_AAA and sdc_AAA represent the total and sidechain contributions to the binding enthalpy for peptide AAA from the receptor residue indicated on the *y*-axis. Heatmap cells are annotated with enthalpy values in kcal/mol, omitting values with magnitudes less than 0.5 for clarity.

binding enthalpy into residue-wise contributions from the receptor (PfAMA1). Note that residue numbering starts from 1 in the simulation setup; thus 107 must be added to align with sequence-based numbering.

Fig. 3.3 shows the residue-wise binding enthalpy contributions of PfAMA1, computed using MM/PBSA with an internal dielectric constant of 2. The quantities dH_*AAA* and sdc_*AAA* denote the total and sidechain-specific enthalpic contributions, respectively, from each receptor residue to

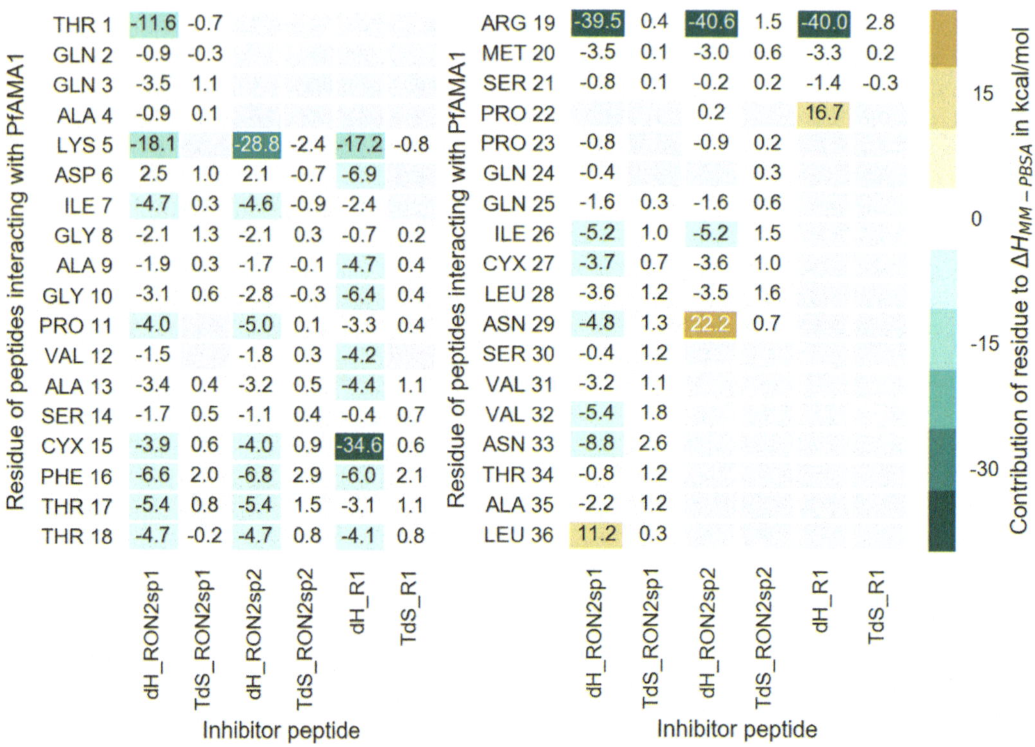

FIGURE 3.4

Residue-wise enthalpy contributions from the peptide to binding with PfAMA1, computed using MM/PBSA. Residue-wise enthalpy contributions from the peptide to binding with PfAMA1, computed using MM/PBSA with indi = 2. The *y*-axis labels correspond to the crystal structure sequence of PfRON2sp1. In PfRON2sp2 the first four N-terminal and last seven C-terminal residues are truncated. For R1, residues are aligned to the PfRON2sp1 sequence for comparative purposes.

the binding of peptide *AAA*. Only values with magnitude greater than 0.5 kcal/mol are annotated for readability.

Several key residues exhibit differential contributions across peptides. For instance, Asp27 contributes more favorably to the binding of RON2sp2 and R1 than to RON2sp1. Tyr35 shows a strong favorable contribution specifically in the RON2sp1 complex. Arg36 displays contrasting behavior: it contributes favorably in RON2sp1 (−6.6 kcal/mol) but opposes binding in RON2sp2 (+2.5 kcal/mol). The segment spanning Pro78 to Glu80 consistently favors interaction with RON2sp1. Asp120 supports binding across all peptides, with the most favorable contribution observed in the R1 complex. Lys128 is the only residue that consistently disfavors binding across all three peptides.

Fig. 3.4 illustrates the enthalpy contributions from peptide residues to PfAMA1 binding. In panel A the peptide sequences of PfRON2sp1, PfRON2sp2, and R1 are aligned according to the PfRON2sp1 crystal structure. Panel B displays the corresponding enthalpy contributions along this alignment. Notably the first four N-terminal residues of PfRON2sp1, which are absent in PfRON2sp2, contribute

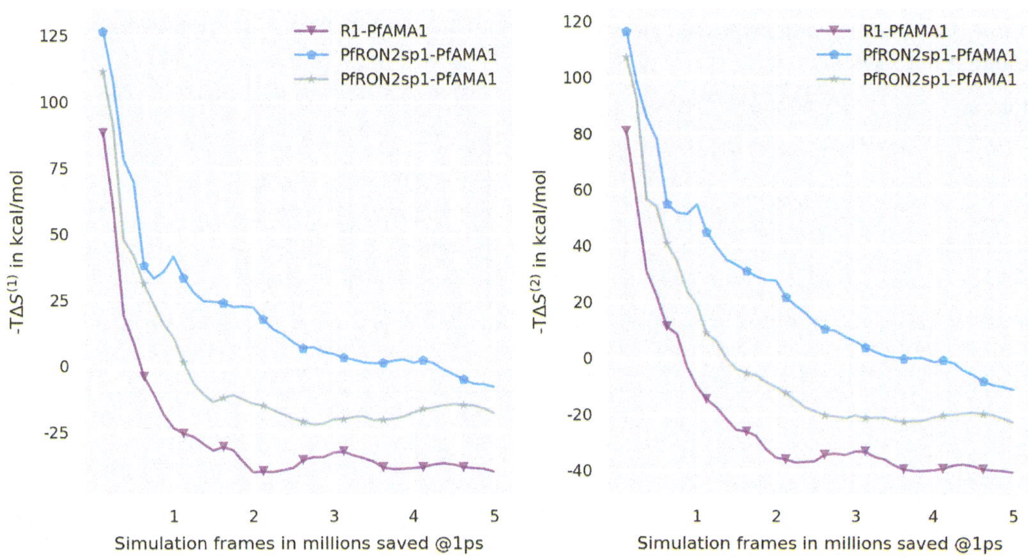

FIGURE 3.5

Convergence of binding configurational entropy (BCE) estimated using A-MIST method. Convergence of first order ($-T\Delta S_{(1)}$) and second order ($-T\Delta S_{(2)}$) binding configurational entropy (BCE) is estimated using A-MIST method, where histogram is used for discretization with parameter values as described in the main text.

approximately −17 kcal/mol to the overall binding enthalpy. This substantial loss of favorable interactions upon truncation may partly explain the reduced binding affinity of PfRON2sp2 compared to PfRON2sp1.

3.2.3.3.3 Binding configurational entropy for surface interactions

This study investigates the binding configurational entropy (BCE) of peptides (PfRON2sp1, PfRON2sp2, and R1) binding to PfAMA1. The BCE is estimated using Neighbor A-MIST. A consistent set of parameters was applied across all systems examined in this case: a distance cutoff of 20 Å, using the Miller–Madow entropy estimator, and histogram-based discretization. For one-dimensional (1D) probability density function (PDF) estimation, 30 bins were used, whereas for two-dimensional (2D) PDFs, a 30 × 30 bin histogram was employed to capture fluctuations in degrees of freedom (DOFs) during the simulations, as used in previous works (Panday & Ghosh, 2020). To avoid repetition, these parameters will be omitted in subsequent discussions unless a specific variation is highlighted.

Fig. 3.5 illustrates the convergence of the first-order $-T\Delta S_{(1)}$ and second-order $-T\Delta S_{(2)}$ BCE estimates obtained using the A-MIST method. The convergence is shown for three peptides (PfRON2sp1, PfRON2sp2, and R1) binding to PfAMA1 as a function of simulation time over 5 million frames (5 μs at 1 ps per frame). The total BCE (derived from the final step of entropy estimation) for all cases is summarized in Table 3.6.

The first-order BCE estimate $S^{(1)}$ for R1-PfAMA1 and PfRON2sp1-PfAMA1 complexes decreases rapidly during the initial 1 million frames. Subsequently the rate of decrease diminishes, with

Table 3.6 Binding configurational entropy (BCE) for peptides PfRON2sp1, PfRON2sp2, and R1 binding to PfAMA1.

Entropy contribution	PfAMA1-PfRON2sp1		PfAMA1-PfRON2sp1		PfAMA1-R1	PfAMA1-R1
	$S^{(1)}$	$S^{(2)}$	$S^{(1)}$	$S^{(2)}$	$S^{(1)}$	$S^{(2)}$
Receptor	−36.63	−32.48	−33.19	−34.09	−57.45	−51.71
Ligand	30.14	22.38	15.67	11.50	17.23	9.98
Rot./Trans.	−1.35	−1.07	−0.33	−0.23	0.29	1.03
Total	−7.83	−11.17	−17.85	−22.82	−39.93	−40.70

BCE is calculated using the A-MIST method, with parameter values as described in the main text. All values are rounded to two decimal places and given in kcal/mol, where temperature is 300 K. $S^{(1)}$ *represents the first-order estimate (assuming independence of DOFs), and* $S^{(2)}$ *represents the second-order estimate (accounting for correlations between pairs of DOFs).*

convergence observed after 2 million frames, at which point the values fluctuate around −27 and −23 kcal/mol, respectively (see Fig. 3.5). In contrast, for PfRON2sp2 binding, $S^{(2)}$ continues to decrease even beyond 3 million frames. Similarly the second-order BCE estimate $S^{(2)}$ for R1-PfAMA1 and PfRON2sp1-PfAMA1 binding also appears to converge fairly after 3 million frames, while for PfRON2sp2-PfAMA1, it continues to decrease even after 4 million frames. Considering the standard deviations of BCE estimates for the last 5 steps (each step includes additional frames, covering the whole data in 40 steps), values are 0.83 kcal/mol for R1-PfAMA1, 1.37 kcal/mol for PfRON2sp1-PfAMA1, and 2.00 kcal/mol for PfRON2sp2-PfAMA1. These standard deviation values suggest that the BCE estimates have converged within approximately 2 kcal/mol for all cases, indicating an estimation error within this range. Further analysis can be performed to elucidate the contributions of the receptor, ligand, and rotational/translational entropy to the total BCE; these contributions are presented in Table 3.6 for all three peptides binding to PfAMA1.

The binding configurational entropy (BCE) contributions from the ligands PfRON2sp1 (36 residues), PfRON2sp2 (25 residues), and R1 (18 residues) are 22.38, 11.50, and 9.98 kcal/mol, respectively. These positive values indicate that the ligands oppose binding by losing flexibility in their bound state compared to their free state. Furthermore the magnitude of this opposing BCE contribution from the peptides decreases with increasing ligand size (Table 3.6). The rotational/translational component of BCE is approximately 1 kcal/mol in magnitude across all cases. However, it is positive for R1-PfAMA1 binding, while negative for PfRON2sp1-PfAMA1 and PfRON2sp2-PfAMA1 binding, suggesting higher mobility of the peptide relative to the PfAMA1 protein in the latter two cases.

The binding of PfRON2sp1 and PfRON2sp2 to PfAMA1 results in a similar extent of modulation in the receptor's flexibility. In contrast the binding of R1 leads to a more negative contribution to BCE from the receptor (−51.71 kcal/mol) compared to PfRON2sp1 (−32.48 kcal/mol) and PfRON2sp2 (−34.09 kcal/mol) binding to PfAMA1 (Table 3.6). This result suggests that the interaction hotspots of PfRON2sp1 and PfRON2sp2 on the PfAMA1 receptor are very similar, and they stabilize/destabilize (positive/negative BCE contributing residues) similar regions of the receptor. However, PfRON2sp1 exhibits a significantly higher affinity than PfRON2sp2 despite a similar entropy profile, indicating that the additional 11 residues of PfRON2sp1 likely form enthalpically favorable interactions with the receptor.

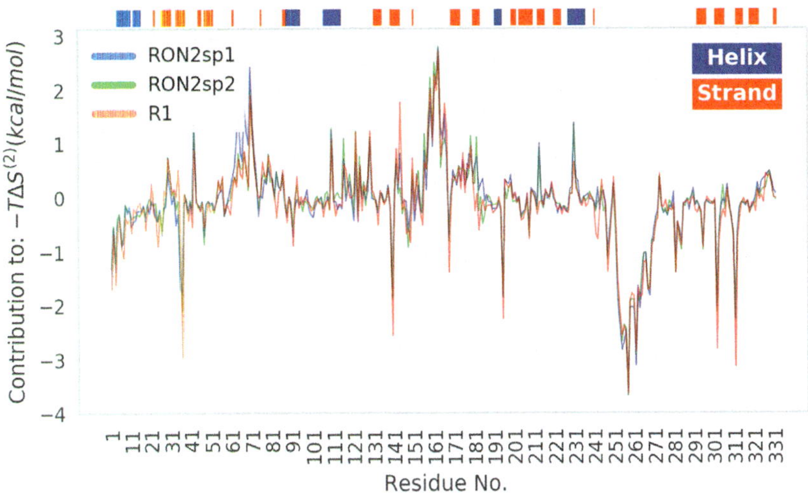

FIGURE 3.6

Residue-wise contribution of receptor PfAMA1 residues to binding configurational entropy for three peptides binding to the common receptor PfAMA1. Residue-wise contribution of receptor PfAMA1 residues to binding configurational entropy $-T\Delta S^{(2)}$ for three peptides: PfRON2sp1 (*blue*), PfRON2sp2 (*green*), and R1 (*red*) binding to the common receptor PfAMA1. All values are given in kcal/mol. The binding configurational entropy was estimated using the Neighbor Approximated Maximum Information Spanning Tree (A-MIST) method.

3.2.3.3.4 Comparative analysis of binding configurational entropy of PfAMA1 interaction with PfRON2sp1, PfRON2sp2, and R1 peptides

A comparative analysis of the binding of all three peptides (PfRON2sp1, PfRON2sp2, and R1) to PfAMA1 has been performed. Identifying individual residues that favor or oppose binding is crucial for the pharmacological optimization of inhibitor molecules targeting PfAMA1. We compared the entropy contribution profiles of all PfAMA1 residues across the three peptide-binding cases to pinpoint structural regions that contribute differentially. Residue-wise binding configurational entropy $S^{(2)}$ has been estimated for peptide binding and is shown in Fig. 3.6.

Among the three peptides, R1-peptide exhibits the most favorable binding configurational entropy (BCE) for binding to PfAMA1, with a value of −40.70 kcal/mol. In contrast, for PfRON2sp1 and PfRON2sp2, the BCE values are −11.17 and −22.82 kcal/mol, respectively (Table 3.6). The binding of R1 is entropically more favored by specific PfAMA1 residues: 36, 140, 195, 253 to 257, 302, and 311. The sum of BCE contributions from these residues is −23.70, −23.64, and −27.0 kcal/mol for PfRON2sp1, PfRON2sp2, and R1 peptides, respectively (based on Table 3.7 summing only for these specific residues). Notably the sum for R1 is approximately 3.3 kcal/mol more negative (more favorable) than for PfRON2sp1 (−27.0 − −23.70 = −3.3 kcal/mol). Conversely, residues contributing the highest magnitude of unfavorable entropy are 42, 61–65, 70, 144, 161–167, and 230. The sum of BCE contributions from these opposing residues are 23.33, 18.62, and 19.59 kcal/mol for PfRON2sp1, PfRON2sp2, and R1-peptide, respectively (Fig. 3.7). Here, R1 shows a 3.74 kcal/mol smaller opposing

Table 3.7 Binding configurational entropy (BCE) decomposed by residue: top-15 favoring and opposing residues of the receptor (PfAMA1) for three peptides binding to PfAMA1.

No.	PfRON2sp1		PfRON2sp2		R1	
	Favoring ResID/ΔS_{res}	Opposing ResID/ΔS_{res}	Favoring ResID/ΔS_{res}	Opposing ResID/ΔS_{res}	Favoring ResID/ΔS_{res}	Opposing ResID/ΔS_{res}
1	257/−3.549	163/2.768	257/−3.673	163/2.805	257/−3.632	163/2.715
2	261/−3.118	70/2.419	261/−2.934	161/2.511	311/−3.141	161/2.278
3	254/−2.835	162/2.194	254/−2.668	159/2.238	36/−2.975	159/2.015
4	255/−2.688	161/2.011	256/−2.419	162/2.098	261/−2.849	162/1.983
5	256/−2.461	159/1.772	255/−2.355	70/1.975	302/−2.815	70/1.899
6	253/−2.374	64/1.769	253/−2.283	158/1.410	140/−2.560	144/1.770
7	311/−2.247	67/1.727	311/−2.281	42/1.399	254/−2.559	166/1.751
8	263/−2.059	166/1.553	302/−2.134	230/1.367	256/−2.534	158/1.433
9	260/−2.057	71/1.414	36 /−2.122	110/1.294	255/−2.390	167/1.269
10	258/−2.047	230/1.385	263/−1.938	122/1.236	195/−2.254	129/1.232

BCE was estimated using the A-MIST method with histogram-based discretization [30 bins for 1D and 30 × 30 bins for 2D probability density functions (PDFs)] and the MM entropy estimator. Second-order entropy estimates, S(2), are in kcal/mol. The temperature was 300 K. ΔSres is the per-residue BCE contribution.

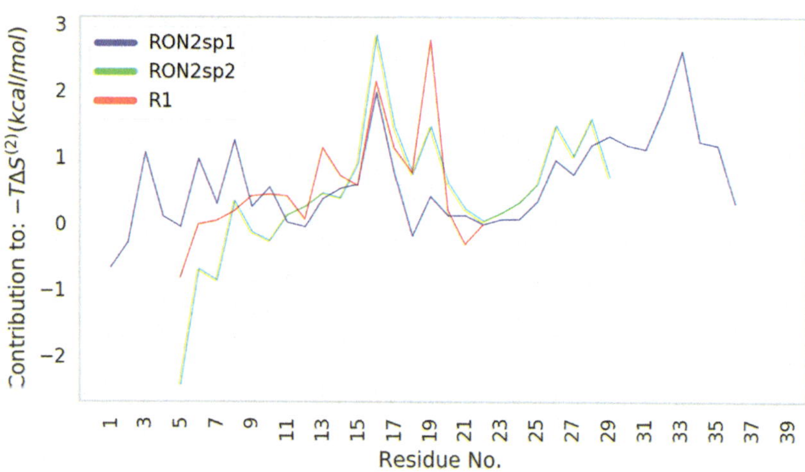

FIGURE 3.7

Residue-wise contribution of peptide residues to binding configurational entropy for binding to the common receptor PfAMA1. Residue-wise contribution of peptide residues to binding configurational entropy $-T\Delta S^{(2)}$, for binding to the common receptor PfAMA1. All values are given in kcal/mol. The binding configurational entropy was estimated using the Neighbor Approximated Maximum Information Spanning Tree (A-MIST) method. The residue numbers for the three peptides start from 1, 5, and 5, as entropy contributions are shown with reference to PfRON2sp1 residues.

Table 3.8 Binding configurational entropy (BCE) decomposed by residue: top-5 binding-favoring and top-5 binding-opposing residues of the peptides binding to PfAMA1.

	PfRON2sp1		PfRON2sp2		R1	
No.	Favoring ResID/ΔS_{res}	Opposing ResID/ΔS_{res}	Favoring ResID/ΔS_{res}	Opposing ResID/ΔS_{res}	Favoring ResID/ΔS_{res}	Opposing ResID/ΔS_{res}
1	1/−0.659	33/2.607	1/−2.420	12/2.851	1/−0.813	15/2.776
2	2/−0.281	16/1.979	3/−0.852	24/1.586	17/−0.300	12/2.150
3	18/−0.183	32/1.780	2/−0.682	22/1.489	2/−0.009	9/1.150
4	5/−0.050	29/1.326	6/−0.258	15/1.472	18/−0.006	13/1.147
5	12/−0.046	8/1.262	5/−0.123	13/1.463	–	14/0.774

BCE is estimated using the A-MIST method with histogram-based discretization. Second-order entropy estimates (S(2)) are given in kcal/mol. Temperature is 300 K. The column ΔSres represents the contribution to BCE from a residue of the binding peptide.

contribution than PfRON2sp1 (23.33 − 19.59 = 3.74 kcal/mol). This implies that a small subset of receptor residues (the aforementioned favoring/opposing residues) collectively contributes approximately 7 kcal/mol (3.3+3.74) more favorable (negative) binding configurational entropy for the R1 peptide than for PfRON2sp1. Therefore enhancing the enthalpy of binding for residues within this list could potentially increase the affinity of the R1-peptide toward PfAMA1.

The entropy contribution of PfRON2sp1 residues is positive for all residues except N-terminal residues 1–2 and residue 19. In the case of PfRON2sp2, five residues exhibit favorable (negative) binding configurational entropy; however, a number of residues show higher opposing entropy compared to PfRON2sp1 (Fig. 3.7). Entropy contributions from R1-peptide residues are generally similar but mostly show higher unfavorable entropy than aligned PfRON2sp2 residues and slightly smaller unfavorable entropy than aligned PfRON2sp1 residues.

As noted in the previous section, only a small subset of receptor and peptide residues makes a significant contribution to the configurational entropic difference between the three peptides. Therefore, entropically, the top-10 binding-favoring and top-10 binding-opposing residues of the PfAMA1 receptor (selected based on magnitude of BCE contribution) are presented in Table 3.7.

Similarly, entropically, the top-5 binding-favoring and top-5 binding-opposing residues of the peptides binding to PfAMA1 are provided in Table 3.8. In the case of R1 (18 residues), only four residues entropically favor binding; therefore the space for the fifth entry is filled with "–."

Among the top-5 binding-favoring residues, PfRON2sp1 and PfRON2sp2 share only one common residue (residue 1 of PfRON2sp2 aligns with residue 5 of PfRON2sp1, based on the alignment shown in Fig. 3.1F). Similarly, they share only one common residue among the top-5 opposing residues (residue 12 of PfRON2sp2 aligns with residue 16 of PfRON2sp1). In contrast, R1 and PfRON2sp1 share two common residues (by position, see Fig. 3.1F) among the top-4 favoring residues and three common residues among the top-5 opposing residues. This analysis suggests that the entropic profile of R1 is more similar to PfRON2sp2 than to PfRON2sp1.

3.2.3.3.5 Binding free energy analysis

Binding free energy (ΔG) arises from both enthalpic (ΔH) and entropic ($-T\Delta S$) components. In this study, we calculate ΔG computationally by combining enthalpy obtained using the MM/PBSA method

Table 3.9 Experimental and computed binding affinities of PfRON2sp1, PfRON2sp2, and R1 peptides to PfAMA1.

Peptide–PfAMA1	K_d nM	ΔG_{exp} kcal/mol	ΔH_{calc} kcal/mol	$-T\Delta S_{calc}$ kcal/mol	ΔG_{calc} kcal/mol
PfRON2sp1	20.3	−10.57	-168.3 ± 11.8	−11.17	-180.10 ± 11.8
R1	80.3	−9.74	-108.9 ± 9.1	−40.70	-149.60 ± 9.1
PfRON2sp2	520.0	−8.63	-120.1 ± 10.9	−22.82	-142.92 ± 10.9

Enthalpy (ΔHcalc) was calculated using the MM-PBSA method with an internal dielectric constant of 2, based on 50,000 frames sampled every 100 ps from 5 μs MD simulations of each peptide–PfAMA1 complex.

and entropy estimated using the A-MIST method (Panday & Ghosh, 2020). These predictions are compared with experimentally determined binding free energies derived from equilibrium dissociation constants (K_d), using the thermodynamic relation: $\Delta G_{\text{bind}} = RT\ln(K_d)$ where $R = 1.987 \times 10^{-3}$ kcal/mol/K and $T = 300$ K. Table 3.9 lists the experimentally determined K_d and ΔG_{exp} values for three peptides (PfRON2sp1, PfRON2sp2, and R1) binding to PfAMA1, alongside the corresponding computational estimates of ΔH_{calc}, $-T\Delta S_{\text{calc}}$, and ΔG_{calc}, all in kcal/mol units.

Enthalpy-based ranking alone fails to reproduce the experimentally observed order of peptide affinities. For instance, PfRON2sp2 shows more favorable enthalpy than R1 despite having a significantly weaker experimental binding affinity. When binding entropy is incorporated, the calculated free energy rankings align with experimental trends: PfRON2sp1 shows the strongest affinity, followed by R1 and then PfRON2sp2. These findings highlight the indispensable role of configurational entropy in capturing the thermodynamics of peptide–receptor binding, especially for systems involving conformationally flexible partners.

3.2.3.4 *Summary and conclusions for the case of flexible protein–flexible peptide binding*

This study investigated binding thermodynamics in peptide–receptor systems involving large, flexible molecules at a surface interface. We focused on the interaction between PfAMA1, a key receptor in *P. falciparum*, and three peptides: PfRON2sp1 and PfRON2sp2 (derived from the parasite RON2 protein) and R1 (a nonplasmodium inhibitory peptide). The biological relevance of these peptides lies in their ability to inhibit erythrocyte invasion by parasite merozoites, although with varying efficacy across parasite strains because of sequence polymorphisms in PfAMA1.

We used crystal structures of PfAMA1 (domains I and II) in its free and peptide-bound forms to initialize all-atom MD simulations, generating extensive structural ensembles (5 μs per system). MM–PBSA and A-MIST methods were applied to extract per-residue enthalpy and configurational entropy profiles, respectively. These analyses offer insights into individual residue contributions to binding thermodynamics and help identify hotspots that could be targeted for rational inhibitor design.

A key outcome is the demonstration that enthalpy-based binding affinity predictions can be misleading when receptor or ligand flexibility is substantial. In contrast the combination of enthalpic and entropic contributions yields predictions that correctly reproduce experimental affinity rankings. The incorporation of entropy is thus essential for understanding binding in flexible biomolecular complexes, where configurational freedom plays a significant thermodynamic role.

The residue-level thermodynamic maps generated here can guide future experimental validation via mutagenesis studies. Once validated, these results can inform the design of optimized peptide inhibitors capable of broad efficacy against multiple PfAMA1 variants. Ultimately, our findings underscore the importance of comprehensive thermodynamic analysis—including both enthalpic and entropic components—in the development of effective therapeutics targeting dynamic protein–protein interfaces.

3.2.4 Role of entropy in flexible protein–small molecule binding

In this case, we study *P. falciparum* protein kinase 5 (PfPK5) as an antimalarial target. As discussed earlier, PfPK5, a flexible Ser/Thr protein kinase, is a potential target for a drug discovery campaign against malaria caused by *P. falciparum*. However, there are challenges that need to be considered and better understood for effective strategic development.

3.2.4.1 Selectivity challenges for PfPK5

Achieving selectivity for PfPK5 over human CDKs remains a major challenge because of high sequence similarity. PfPK5 shares approximately 60% identity with human CDK1 and CDK5, and about 35% with CDK7 (Le Roch et al., 2000). Notably, structural studies have revealed greater similarity between PfPK5 and CDK5 than previously assumed based on sequence alone.

The regulatory mechanisms of CDKs offer potential avenues for selective inhibition. For instance the CDK5 activator p25 resembles cyclin A in structure but not in sequence (Tang et al., 1997). p25 selectively activates CDK5 but not other CDKs, whereas cyclin H activates CDK7 but not CDK5 (Sharma et al., 1999). These unique regulatory patterns suggest that exploiting PfPK5's specific activation requirements could enable the design of selective inhibitors.

3.2.4.2 Materials

Available crystal structures of PfPK5 provide valuable insights into its conformation and ligand-binding characteristics. As of July 6, 2018, multiple PfPK5 structures were available in the PDB (Table 3.10). For this study, we primarily focused on chain A of these structures because of fewer missing regions compared to chain B.

Binding site residues were initially identified using SITE records in the PDB files and cross-validated with predictions from LigPlot and information reported in the literature. Unbound structures 1V0B and 1OB3 are shown in Fig. 3.8A and B, respectively. The PfPK5 sequence used for simulations is depicted in Fig. 3.8E.

Structures 1V0P and 1V0O represent PfPK5 bound to the known inhibitors Purvalanol B and indirubin-5-sulfonate, respectively (Fig. 3.8C and D). These complexes reveal key features of inhibitor binding modes.

For structure 1V0P, which contains a V161I mutation, the isoleucine was computationally mutated back to valine using the rotamer wizard in UCSF Chimera and the Dunbrack rotamer library (Shapovalov & Dunbrack, 2011), selecting the most probable χ_1 angle. A missing loop segment (residues 153–160) was modeled using MODELLER v9.15 (Šali et al., 1995), using 1V0P as the template. A short C-terminal stretch (residues 287–288) was not modeled, as it lies far from the binding site. The quality of the final model was assessed using PROCHECK v3.5.4 (Laskowski et al., 1993).

Structure 1V0O lacked residues at the N-terminus (1–2) and within a nonterminal loop (residues 37–45). These regions were also modeled using MODELLER v9.15 with 1V0O as the template. A

Table 3.10 Summary of relevant features of available crystal structures of PfPK5 in the PDB.

PDB ID	1OB3	1V0B	1V0P	1V0O
Molecule	PfPK5	PfPK5 (T198A mutant)	PfPK5-Purvalanol B	PfPK5-indirubin-5-sulfonate
Res (Å)	1.9	2.2	2.0	1.9
pH	7.5	7.4	7.4	7.4
Mutation	V161I	T198A	V161I	None
Missing regions	152–160, 288	287–288	152–160, 287–288	1–2, 37–44, 288
Binding residues (SITE)	N/A	N/A	Ile10, Val18, Ala30, Lys32, Phe79, Glu80, His81, Leu82, Asp83, Asp85, Leu132, Asp143	Ile10, Gly11, Glu12, Gly13, Ala30, Glu80, Leu82, Asp83, Asp85, Gln129, Leu132, Ala142
Binding site predicted by LigPlot (underlined)			Ile10, Gly11, Glu12, Gly13, Val18, Ala30, Val63, Phe79, Glu80, His81, Asp83, Asp85, Lys88, Gln129, Leu132, Ala142	Ile10, Val18, Ala30, Lys32, Val63, Phe79, Glu80, His81, Leu82, Asp83, Asp85, Leu132, Asp143

All missing region information corresponds to chain A; chain B has more missing regions and was not considered for further study. Binding site residues are collected from the SITE record of corresponding PDB files. The last row contains residues from literature and LigPlot predictions, overlapping residues are underlined. Data from PDB accessed on July 6, 2018.

single missing C-terminal residue was left unmodeled. The model quality was similarly evaluated using PROCHECK v3.5.4 (Laskowski et al., 1993).

The energy-minimized and equilibrated models of the ligand-bound structures are shown in Fig. 3.8C and D, with modeled loops highlighted in yellow. Fig. 3.8C illustrates key hydrogen bonds between indirubin-5-sulfonate and PfPK5 residues—two with Lys32 and one each with Glu80, His81, Leu82, and Asp143 (*black lines*). In Fig. 3.8D, Purvalanol B is seen forming two hydrogen bonds with the backbone atoms of Leu82.

3.2.4.2.1 Selection of known PfPK5 binders

To identify suitable inhibitors for this study, we compiled a dataset of PfPK5 assay records from the ChEMBL database, accessed on October 11, 2015. A total of 99 assay entries were retrieved, including measurements of % inhibition (2 records), general activity (1 record), IC_{50} (24 records), and K_d (72 records) (see Table 3.11). Based on these data, we selected seven small molecules with experimentally determined binding affinities for PfPK5. Four inhibitors were chosen based on their reported K_d values being less than 6000 nM. In addition, three more molecules were included, two that had been cocrystallized with PfPK5 (indirubin-5-sulfonate and Purvalanol B), for which IC_{50} values were available, and one additional compound from the literature with an IC_{50} less than 15000 nM. Table 3.12 summarizes the key chemical properties of these selected inhibitors, including the number of rotatable bonds, heavy atoms, total atoms, hydrogen bond donors and acceptors, and formal molecular charge.

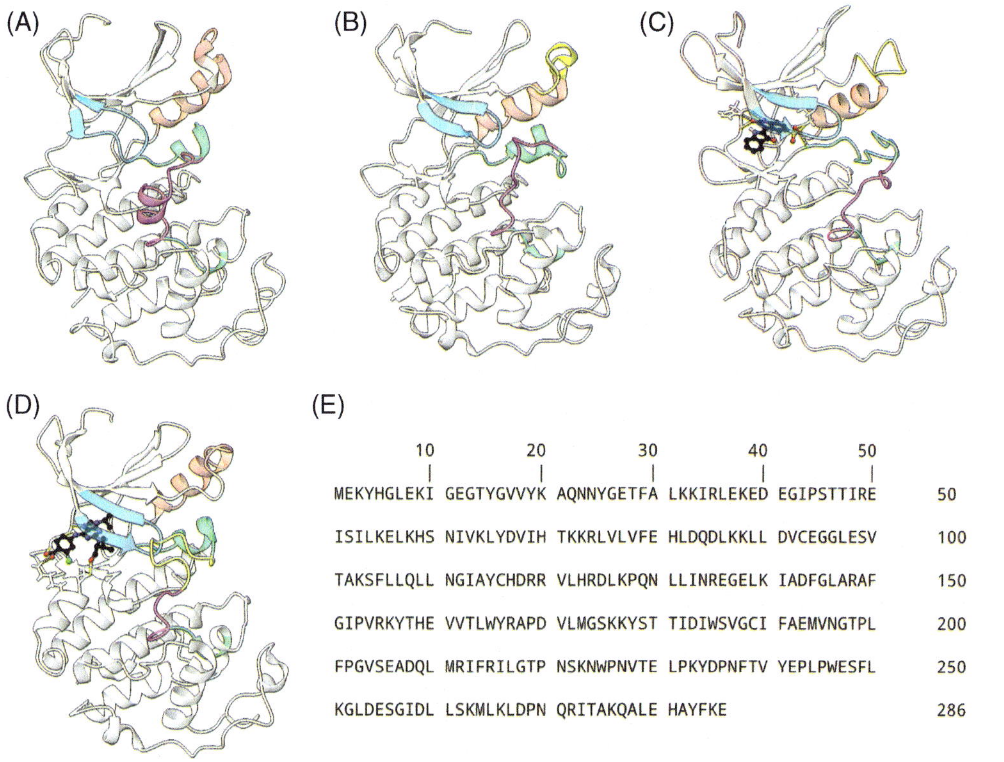

FIGURE 3.8

Structure of apo and two ligand-bound PfPK5. Crystal structures of PfPK5. Panels (A) and (B) show unbound PfPK5 from PDB IDs 1OB3 and 1V0B, respectively. Panels (C) and (D) show PfPK5 in complex with indirubin-5-sulfonate (PDB: 1V0O) and Purvalanol B (PDB: 1V0P), respectively. Panel (E) displays the PfPK5 sequence used in the simulated models. Structural regions are colored as follows: N-lobe (residues 1–82) in *ghost white*, C-lobe (83–286) in *white*, C-helix (39–56) in *coral*, glycine-rich loop (10–19) in *sky blue*, activation segment (14–170) in *green*, and the helix/loop region (157–166) in *magenta*. Segments corresponding to missing residues in the crystal structures are modeled as described in the main text and shown in *yellow*.

Table 3.11 Summary of PfPK5 assay information records available in ChEMBL.

Assay type	Number of assays with exact values	Number of assays with threshold-defined values
Activity	1	0
IC_{50}	7	16
% Inhibition	2	0
K_d	4	68

All data were obtained from the ChEMBL database (accessed October 11, 2015).

Table 3.12 Summary of seven small molecules with known experimental binding affinity for PfPK5 and their chemical properties.

Inhibitor	ChEMBL ID[a]	PbCh CID[b]	ZINC ID[c]	RtB[d]	Hv Atm[e]	Tot Atm[f]	Hb Dnr[g]	Hb Acc[h]	Chrg[i]
Olomoucine (OLM)	280074	4592	01641925	6	22	40	3	6	0
Lestaurtinib/CEP701 (LTU)	603469	126565	03781738	1	33	54	3	4	0
Indirubin-5-sulfonate (INR)	1207227	5288641	01661760	1	24	33	3	6	−1
Flavopiridol (FLV)	428690	5287969	3873287	2	28	48	3	6	0
Staurosporine (STA)	388978	6711321	3814434	2	35	61	2	4	0
NU6102 (NU6)	319467	4566	3873287	6	28	50	3	8	0
Purvalanol B	23254	448991	582590	8	30	55	4	8	0

Identifiers sourced from their respective databases.
[a]*Prefix CHEMBL omitted for brevity in ChEMBL IDs.*
[b]*PubChem CID.*
[c]*ZINC_ID; Chemical property data sourced from PubChem.*
[d]*No. rotatable bonds.*
[e]*No. Heavy Atoms (nonhydrogen atoms).*
[f]*No. of total atoms.*
[g]*hydrogen bond donors.*
[h]*hydrogen bond acceptors.*
[i]*molecular charge.*

Experimental binding affinities were reported either as IC_{50} or K_d values, which are not directly interchangeable. The Cheng–Prusoff equation (Brandt et al., 1987) provides a theoretical framework for converting IC_{50} values to inhibition constants (K_i):

$$K_i = \frac{IC_{50}}{1 + \frac{[S]}{K_m}}$$

where K_i represents the true binding affinity, IC_{50} is the inhibitor concentration required to reduce enzyme activity by 50%, $[S]$ is the substrate concentration used in the assay, and K_m is the Michaelis constant of the substrate. This equation is derived from competitive inhibition models in which an unlabeled ligand displaces a radiolabeled tracer from its binding site. However, its applicability is limited by the requirement for specific assay conditions and inhibition mechanisms. Notably, several of the compounds examined in this study exhibit sigmoid dose–response curves, suggesting noncompetitive or allosteric inhibition behavior (Le Roch et al., 2000), thereby limiting the reliability of this conversion for all selected inhibitors.

3.2.4.2.2 System preparation

Two C-terminal asparagine residues, which were absent in most of the selected crystal structures and located far from the binding site, were excluded from the models. Consequently, all subsequent analyses were performed using residues 1–286. The crystal structures of PfPK5 in complex with indirubin-5-sulfonate (PDB ID: 1V0O, chain A) and Purvalanol B (PDB ID: 1V0P, chain A) were resolved at pH 7.4, consistent with the pH conditions of the experimental assays (Holton et al., 2003). Accordingly, these structures were prepared using protonation states of titratable residues predicted at pH 7.4 with PROPKA3 (Olsson et al., 2011a).

Protonation states of histidine residues were determined for both the apo (PDB IDs: 1OB3, 1V0B) and inhibitor-bound (PDB IDs: 1V0O, 1V0P) forms of PfPK5 at pH 7.4. For His59, His70, His81, His117, His123, and His281, the same tautomeric form either HIE (ε-protonated histidine) or HID (δ-protonated histidine) was consistently predicted across all structures. In contrast, His5 and His159 exhibited different tautomeric assignments between the apo and bound forms. Structural inspection revealed that the side chains of His5 and His159 did not participate in hydrogen bonding or van der Waals interactions in any of the structures. Given that His5 is solvent-exposed and displays similar predicted pK_a values across all structures, its tautomer was uniformly assigned as HIE. Likewise, His159, which is also solvent-exposed and located outside the binding site, was fixed in the HIE tautomeric form.

3.2.4.3 Results and discussion

3.2.4.3.1 Creating complex structures and docking analysis

Molecular docking is a computational method used to predict the binding poses and affinities of small molecules to their target receptors (Panday & Ghosh, 2019). For this study the seven selected inhibitors Olomoucine (OLM), Lestaurtinib (LTU), Indirubin-5-sulfonate (INR), Flavopiridol (FLV), Staurosporine (STA), NU6102 (NU6) and Purvalanol B (PVB) were geometry-optimized and docked into the binding sites of PfPK5 derived from crystal structures complexed with indirubin-5-sulfonate and Purvalanol B. The ability of the docking methods to reproduce the experimentally determined binding poses for the cocrystallized ligands (INR and PVB) was assessed to validate the docking protocols (Panday & Ghosh, 2019). Initial structures for the complexes between the remaining five inhibitors and

Table 3.13 Experimental IC_{50}, K_d values of selected inhibitors for PfPK5 and docking scores from GOLD, GLIDE, and DOCK 6 using binding sites from 1V0O and 1V0P.

	IC_{50}	K_d	$RT\ln(K_i)$	Gold		Glide		Dock 6	
Inhibitor	nM	nM	kcal/mol	1V0O	1V0P	1V0O	1V0P	1V0O	1V0P
OLM	15,000	7500	−7.034	72.98	55.56	−6.77	−5.75	−47.67	−56.46
LTU	NA	5600	−7.208	85.98	60.53	−7.24	0.67	−61.73	−62.64
FLV	NA	3300	−7.524	70.55	65.74	−8.55	−8.17	−56.34	−56.59
INR	5500	2230	−7.757	76.38	64.56	−13.57	−8.65	−52.52	−55.04
STA	1000	800	−8.368	72.31	60.98	−11.48	−4.99	−64.67	−64.55
NU6	215	107	−9.567	73.93	75.56	−6.96	−8.23	−60.12	−61.67
PVB	130	65	−9.865	87.55	78.40	−8.98	−7.98	−66.02	−71.41

PfPK5 were generated through these docking simulations. The most consistent poses predicted by the different docking programs were selected as starting points for further analysis. The docking results obtained from DOCK 6 (Allen et al., 2015), GLIDE (Friesner et al., 2004), and GOLD (Verdonk et al., 2003) programs, including scores for docking into both the 1V0O (INR-bound) and 1V0P (PVB-bound) binding sites, are detailed in previous work (Panday & Ghosh, 2019) and summarized in Table 3.13, alongside the experimental IC_{50} and K_d values.

Docking scores were normalized for ranking purposes to a consistent scale from 0 to 1 using the formula score − min(score))/(max(score) − min(score)), where 0 represents the weakest predicted binder and 1 the strongest. Experimental binding free energies, calculated as $RT\ln(K_d)$ or converted $RT\ln(K_i)$ values, were similarly normalized. The standard error between the normalized predicted scores and normalized experimental binding free energies was calculated using the following equation:

$$\text{Standard error} = \sqrt{\frac{1}{n}\sum_{i=1}^{n}(\text{norm\_score}_i - \text{norm\_experimental}_{\Delta G_i})^2} \tag{3.21}$$

where norm_score$_i$ is the normalized ith docking score and norm_experimentalΔ_{Gi} is the ith normalized experimental binding free energy.

We analyzed the correlation (R^2) between the predicted normalized docking scores and the normalized experimental binding free energies, along with the number of predicted points falling outside the one-standard deviation (1σ) range. When docking into the 1V0O binding site (Fig. 3.9A), Glide predicted three out of seven points beyond the 1σ range and showed the poorest correlation ($R^2 = 0.011$). GOLD predicted two points outside the 1σ range with a slightly better correlation ($R^2 = 0.078$). DOCK 6 predicted only one point outside the 1σ range and exhibited the best correlation ($R^2 = 0.358$) among the three programs for the 1V0O site.

When docking into the 1V0P binding site (Fig. 3.9B), Glide again predicted three points outside the 1σ range, with an $R^2 = 0.221$, which was the lowest among the three programs in this case. DOCK 6's R^2 improved to 0.512, with only one predicted point falling outside the 1σ range. GOLD showed

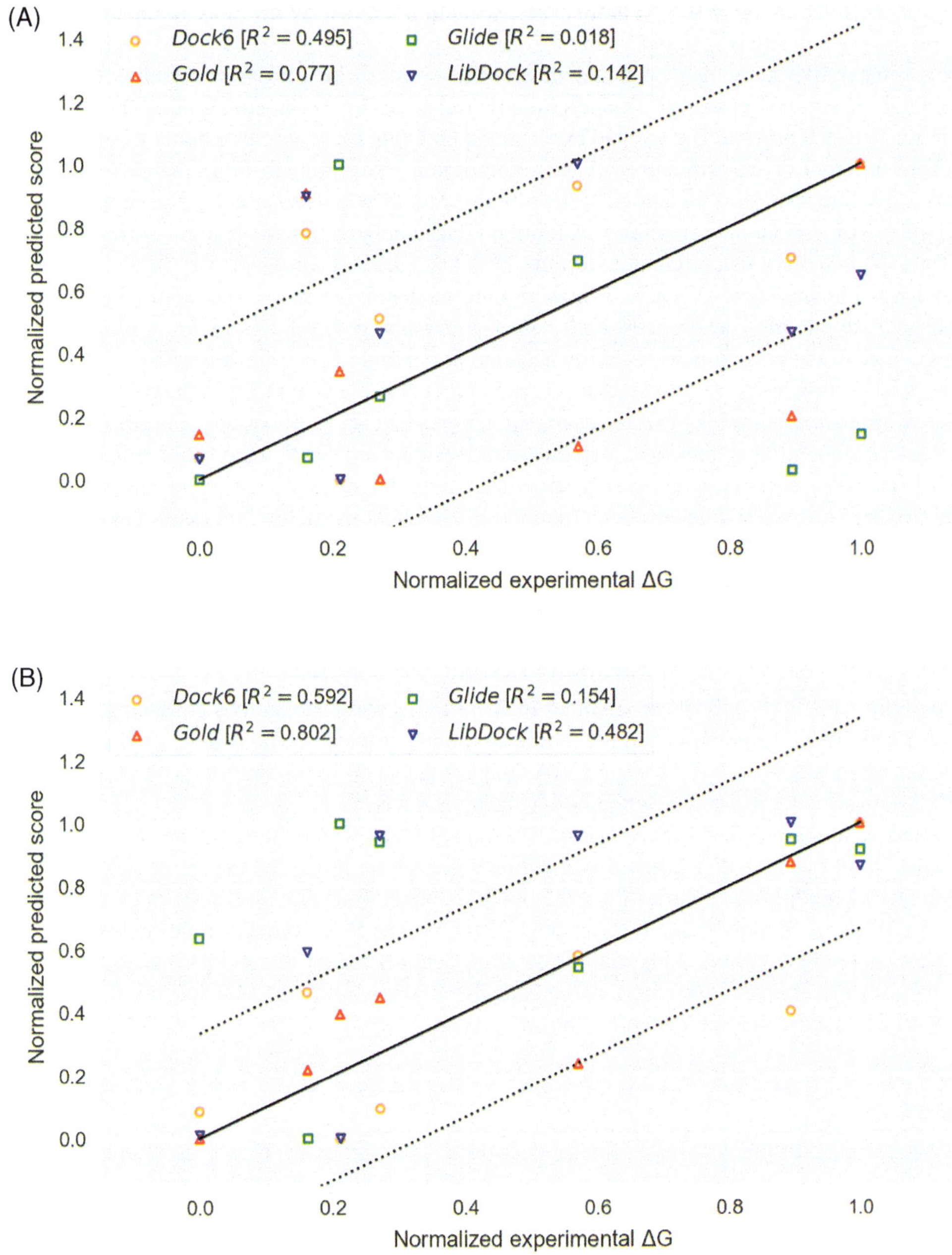
(A)
Dock6 [$R^2 = 0.495$]
Glide [$R^2 = 0.018$]
Gold [$R^2 = 0.077$]
LibDock [$R^2 = 0.142$]
Normalized predicted score
1.4
1.2
1.0
0.8
0.6
0.4
0.2
0.0
0.0
0.2
0.4
0.6
0.8
1.0
Normalized experimental ΔG
(B)
Dock6 [$R^2 = 0.592$]
Glide [$R^2 = 0.154$]
Gold [$R^2 = 0.802$]
LibDock [$R^2 = 0.482$]
Normalized predicted score
1.4
1.2
1.0
0.8
0.6
0.4
0.2
0.0
0.0
0.2
0.4
0.6
0.8
1.0
Normalized experimental ΔG

the highest correlation ($R^2 = 0.834$) in this case, with all predicted points close to the ideal correlation line.

This comparative study highlights that none of the tested docking programs [DOCK 6 (Allen et al., 2015), GOLD (Verdonk et al., 2003), and Glide (Friesner et al., 2004)] consistently provided a strong correlation between predicted scores and experimental binding free energies when docking into binding sites from different crystal structures of the same protein. This suggests that the docking scores are sensitive to subtle structural variations in the binding site, as was seen in crystal structures (PDB IDs: 1VOO and 1VOP) (Holton et al., 2003). Although GOLD demonstrated excellent correlation with the 1VOP site ($R^2 = 0.834$), its correlation with the 1VOO site was very poor ($R^2 = 0.078$), indicating high sensitivity and inconsistency. DOCK 6 showed more moderate but more consistent correlations ($R^2 \approx 0.5$) across both binding sites, suggesting it was qualitatively more reliable in this specific application, although none of the programs were highly accurate predictors of absolute affinity.

The relative flexibility of corresponding structural regions in the 1V0B and 1OB3 structures was estimated by calculating the RMSF for each residue over the MD simulations (Fig. 3.10). The RMSF plot indicates higher fluctuations in the region around residues 140–165 in the 1OB3 structure compared to 1V0B. This observation aligns with the structural difference in the 157–166 regions, which adopts a helical structure in 1V0B but appears disordered in 1OB3. Other structural regions show similar RMSF values between the two structures, with minor differences observed around residues 12–13 and 48–55. Secondary structure elements (SSEs) for both structures are annotated at the top of the plot, with helices and strands represented by *red* and *blue rectangular boxes*, respectively.

3.2.4.3.2 Analysis of protein flexibility in inhibitor complexes

The flexibility of PfPK5 in the presence of various inhibitors was evaluated by computing the RMSF of each residue from MD simulations of the inhibitor-bound complexes. The results are illustrated in Fig. 3.11A for complexes with INR, LTU, and STA, and in Fig. 3.11B for complexes with OLM, FLV, NU6, and PVB. Annotated SSEs are shown in both panels for reference.

Among the first group of complexes (Fig. 3.11A) the PfPK5-STA complex (*green line*) demonstrates the highest level of fluctuations, broadly distributed across the protein structure, indicating increased conformational mobility. In contrast the LTU-bound complex (*orange line*) exhibits pronounced flexibility specifically in the region spanning residues 221–231. The INR-bound complex (*blue line*) displays the lowest overall fluctuations, with minor exceptions near residues 130 and 160. The reduced mobility in the INR-PfPK5 complex may reflect a greater degree of cooperative or collective motions, potentially contributing to a more favorable binding entropy profile.

FIGURE 3.9

Correlation plot of normalized docking scores for docking into the binding site of two bound structures of PK5 versus normalized experimental binding free energy. Correlation plot of normalized docking scores for docking into the (A) 1VOO binding site and (B) 1VOP binding site versus normalized experimental binding free energy [ΔG, derived from $RT\ln(K_i)$ or $RT\ln(K_d)$]. The *black solid line* represents perfect correlation. *Dotted lines* indicate the ±1 standard error range. Points for LTU, STA, and FLV (where K_i is experimentally known) are shown, with marker shapes indicating the docking program. At least one prediction from GOLD and DOCK 6 falls outside the 1σ range.

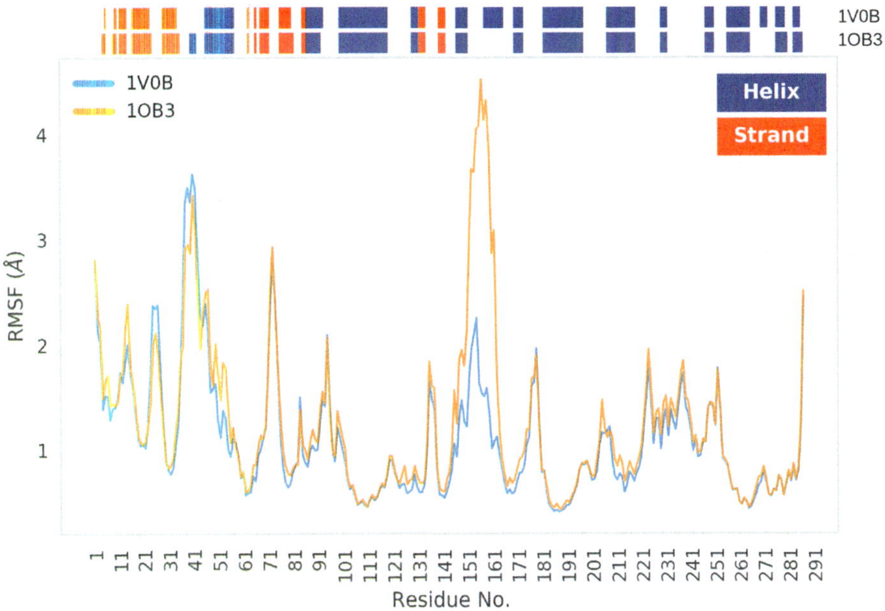

FIGURE 3.10

Residue-wise root mean square fluctuation (RMSF) from MD simulations of 1V0B (*blue line*) and 1OB3 (*orange line*). Residue-wise RMSF from MD simulations of 1V0B (*blue line*) and 1OB3 (*orange line*). Secondary structure elements (SSEs) are annotated at the top, with the top annotation corresponding to 1V0B and the bottom to 1OB3. Helices are shown as *red boxes* and strands as *blue boxes*.

Table 3.14 Decomposition of binding configurational entropy estimated using the A-MIST method for seven inhibitors bound to PfPK5.

Contribution to ΔS_{conf}	PfPK5 in complex with inhibitors						
	OLM	LTU	INR	FLV	STA	NU6	PVB
Receptor	9.36	10.49	16.77	2.93	13.33	−14.20	−2.11
Ligand	2.26	−0.39	0.51	−1.57	−1.23	1.40	4.92
Rot./Trans.	1.49	3.94	4.48	1.90	1.18	2.39	0.10
Total	**13.11**	**14.03**	**21.76**	**3.27**	**13.28**	**−10.40**	**2.92**

A 14 Å distance cutoff, 30 bins for 1D histograms, 30 × 30 bins for 2D histograms, and the Miller–Madow entropy estimator were used. Values are in kcal/mol at 300 K and rounded to two decimal places.

In the second group (Fig. 3.11B) the NU6-bound complex (*green line*) exhibits greater structural fluctuations relative to those bound to OLM, FLV, and PVB. This elevated flexibility suggests that NU6 imposes fewer conformational constraints on the receptor, which is consistent with the more favorable (i.e., more negative) configurational entropy of binding calculated for NU6 (−10.40 kcal/mol), compared to OLM (13.11 kcal/mol), FLV (3.27 kcal/mol), and PVB (2.92 kcal/mol) (see Table 3.14).

(A)

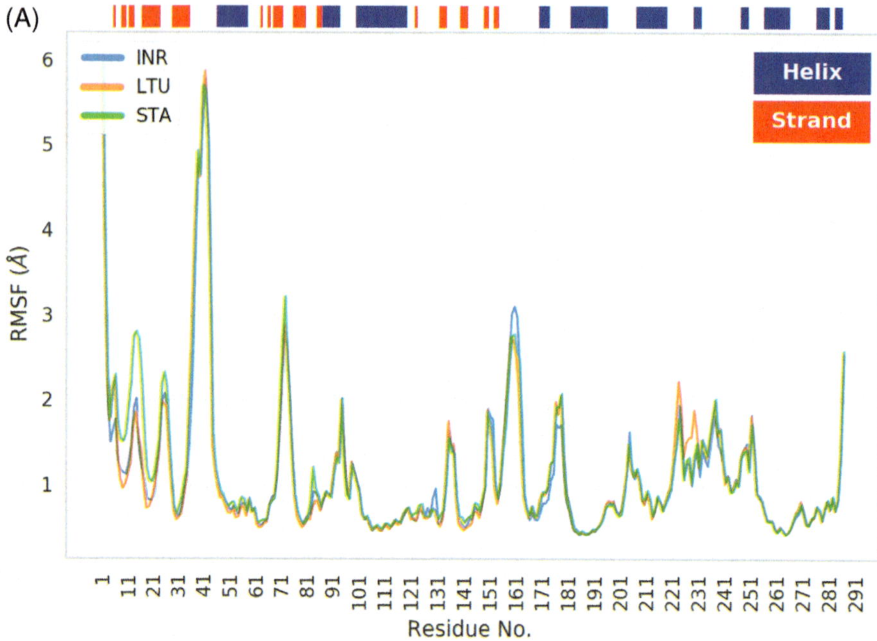
INR
LTU
STA
Helix
Strand
RMSF (Å)
6
5
4
3
2
1
1 11 21 31 41 51 61 71 81 91 101 111 121 131 141 151 161 171 181 191 201 211 221 231 241 251 261 271 281 291
Residue No.

(B)

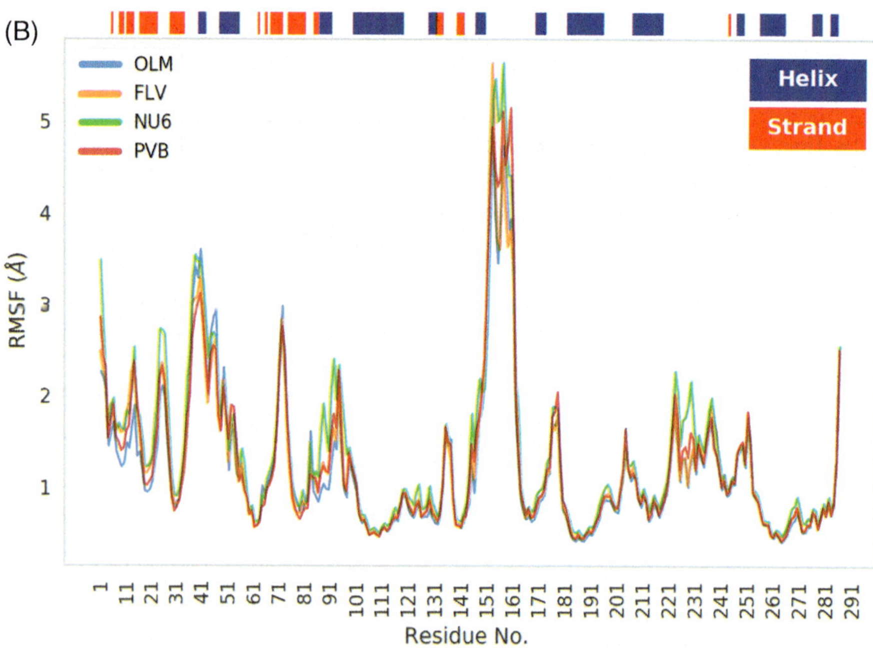
OLM
FLV
NU6
PVB
Helix
Strand
RMSF (Å)
5
4
3
2
1
1 11 21 31 41 51 61 71 81 91 101 111 121 131 141 151 161 171 181 191 201 211 221 231 241 251 261 271 281 291
Residue No.

Notably, residues within the T-loop region (150–160) display particularly high fluctuations in the NU6-bound complex. These observations support a mechanistic link between changes in receptor flexibility upon inhibitor binding and the associated entropic contributions to binding affinity.

3.2.4.3.3 Conformational analysis of Apo PfPK5 simulation trajectories

MD simulations provide an ensemble of structures representing the system's dynamic behavior, where different conformations can exist in equilibrium. To characterize these conformational states, we clustered simulation trajectories based on structural similarity. We resampled the 5 μs trajectory of 1OB3, extracting 1000 frames every 5 ns.

These ensembles of 1000 structures were clustered using the *k*-means algorithm with two different similarity metrics. The first metric was the widely used RMSD, which quantifies global structural changes but does not differentiate between rigid and flexible regions like loops. Recognizing that localized changes involving topologically important SSEs might be missed by global RMSD, we also employed an in-house developed topological descriptor called ContactString (Khan & Ghosh, 2015).

A ContactString represents the SSEs of a protein and their relative orientations as a string. First, all SSEs in a structure are identified and indexed from N- to C-terminus (Khan & Ghosh, 2015). A ContactMatrix is constructed where each cell (i, j) (with $i < j$) records the relative orientation (antiparallel, parallel, orthogonal, or none) between SSE_i and SSE_j. The ContactString is generated by traversing this matrix diagonally and concatenating the entries, separated by symbols (Khan et al., 2018).

In the second step an SSE-fingerprint is derived from the ContactString. Each entry in the upper diagonal of the ContactMatrix is encoded into a three-bit string ("100" for antiparallel, "010" for parallel, "001" for orthogonal, and "000" for none). These bit triplets are concatenated to form the SSE fingerprint (Khan et al., 2018). To apply this to trajectories, a reference dictionary of SSEs (ref_{SSEs}) is built from the first frame, updated with new SSEs from subsequent frames. SSE boundaries are extended to cover their full residue range across the trajectory. SSEs in each frame are mapped to ref_{SSEs} before generating the fingerprint.

Similarity between any pair of structures is then measured using the Tanimoto distance (Lipkus, 1999) between their SSE fingerprints:

$$\text{Tanimoto distance}(A, B) = 1 - \frac{\sum_{i=1}^{L} A[i] \cdot B[i]}{\sum_{i=1}^{L} (A[i] + B[i] - A[i] \cdot B[i])} \tag{3.22}$$

where A and B are SSE fingerprints of length L, and $A[i]$ and $B[i]$ are the ith bits of the fingerprints.

FIGURE 3.11

Root mean square fluctuation (RMSF) profiles from MD simulations of PfPK5 complexes. RMSF profiles from MD simulations of PfPK5 complexes. (A) Complexes with INR (*blue*), LTU (*orange*), and STA (*green*). (B) Complexes with OLM (*blue*), FLV (*orange*), NU6 (*green*), and PVB (*red*). Annotated secondary structure elements (SSEs) are shown at the top: helices are depicted as *red boxes* and *β*-strands as *blue boxes*.

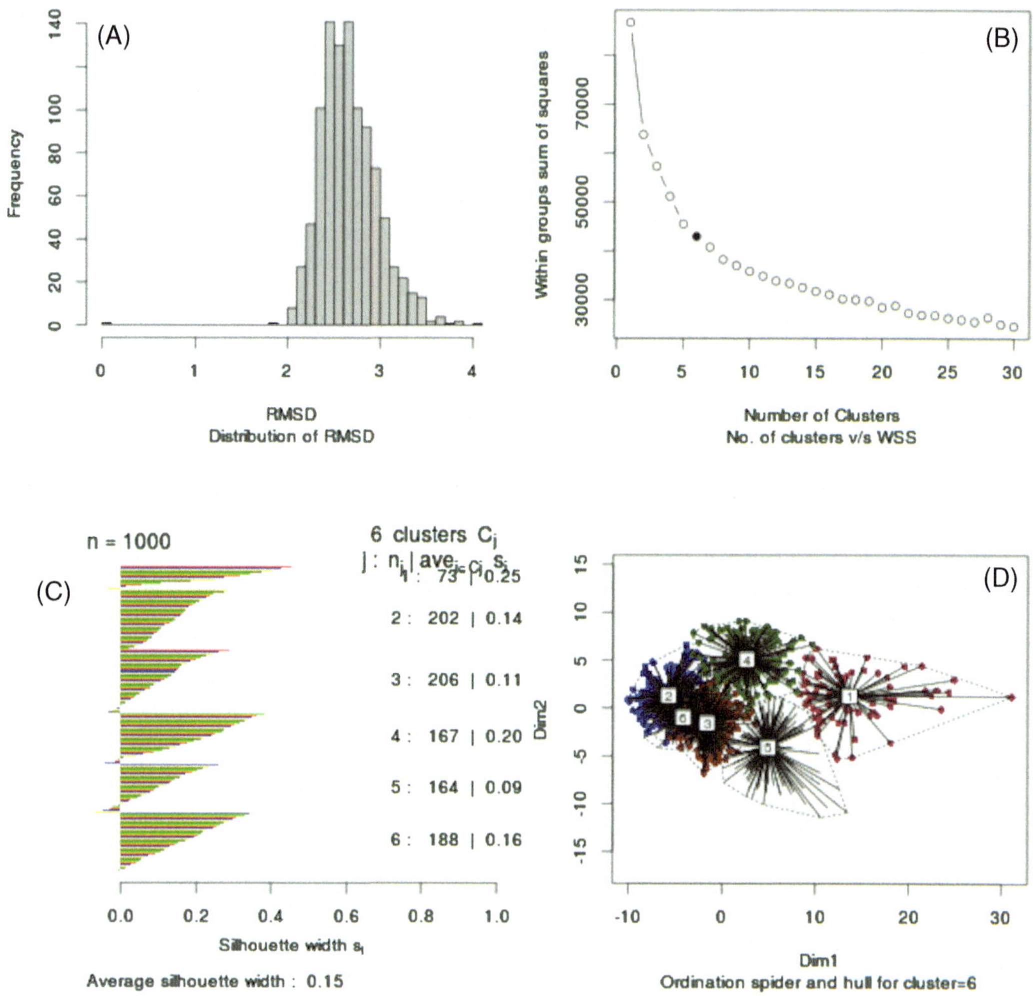

FIGURE 3.12

Clustering of an ensemble of 1000 frames of 1OB3 trajectory using *k*-means and RMSD as a similarity metric for structure and Euclidean distance for clustering. Clustering of an ensemble of 1000 frames of 1OB3 trajectory using *k*-means and RMSD as a similarity metric for structure and Euclidean distance for clustering. (A) Histogram of RMSDs with reference to first structure in ensemble, (B) elbow plot of within group sum of squares for clustering using 2–30 clusters, (C) silhouette plot for clustering using six clusters, and (D) spider plot for clustering. *RMSD*, Root mean square deviation.

3.2.4.3.4 Cluster analysis of simulation of apo-PfPK5 (Protein Data Bank: 1OB3)

The same clustering protocol using RMSD was applied to the 1000 frames of the 1OB3 trajectory (Fig. 3.12). This yielded six clusters, with a between-group sum of squares to total sum of squares ratio of 50.4%. Five clusters contained over 160 frames, with cluster 1 being smaller. Clusters 1, 4, and 5

Table 3.15 Calculated binding enthalpy using MM/GBSA and MM/PBSA with indi = 1 for small-molecule binders of PfPK5. All energy values are in kcal/mol.

		EMM		MM/GBSA (indi = 1)			MM/PBSA (indi = 1)		
Inhibitor	ΔG_{expr}	vdW	EEL	GB	Surf.	Total	PB	$E_{nonpolar}$	Total
OLM	−6.75	−39.1	−37.9	59.6	−5.4	−22.8 ± 5.6	58.2	−3.9	−22.6 ± 7.2
LTU	−7.21	−57.7	−26.2	44.6	−6.4	−45.7 ± 4.2	54.1	−4.6	−34.3 ± 4.8
INR	−7.35	−42.2	−118.8	134.0	−5.3	−32.4 ± 4.4	132.5	−3.6	−32.1 ± 5.1
FLV	−7.52	−37.8	−29.1	45.0	−4.9	−26.8 ± 6.0	49.5	−3.8	−21.2 ± 6.2
STA	−8.37	−55.9	−28.2	43.9	−6.2	−46.4 ± 5.4	52.0	−4.8	−36.9 ± 6.0
NU6	−9.29	−43.8	−18.1	46.0	−5.1	−21.0 ± 4.9	42.2	−4.2	−23.9 ± 6.6
PVB	−9.59	−49.3	−33.7	54.2	−6.1	−34.9 ± 5.8	54.8	−4.6	−32.8 ± 6.5

vdW, *van der Waals.*

were well-separated in the 2D principal component space (Fig. 3.12D), whereas clusters 2, 3, and 6 were denser and closer together. Structural inspection of cluster representatives showed major changes primarily located in the activation segment loop, which is disordered in the 1OB3 crystal structure.

When the same ensemble of 1OB3 structures was clustered using the Tanimoto distance of SSE fingerprints (Fig. 3.12), comparable between-group to total sum of squares ratios, and analysis using the elbow and Silhouette methods suggested 12 clusters (Fig. 3.12B and C). The 2D principal component space showed distinct separation for most clusters (Fig. 3.12D). This higher number of clusters further reflects the increased sensitivity of the SSE-fingerprint metric to subtle, localized conformational changes within the protein, particularly in flexible regions like the activation segment.

3.2.4.3.5 Enthalpy of inhibitor binding to PfPK5

The binding enthalpy (ΔH) for the seven selected inhibitors binding to PfPK5 was calculated using endpoint free energy methods, specifically MM/GBSA and MM/PBSA. A single trajectory protocol was employed, utilizing 50,000 snapshots collected every 100 ps from the 5 μs MD simulations of the complexes. For MM/PBSA calculations the effect of varying the internal dielectric constant (indi) for the protein was investigated by performing analyses with indi values of 1, 2, and 4 to assess its impact on the correlation with experimental binding affinity.

The predicted enthalpy values from MM/GBSA and MM/PBSA methods spanned a range significantly different from the range of experimental binding affinities. For ease of comparison, both predicted enthalpy and experimental affinity values were normalized to a range of 0–1, as detailed previously.

Analysis of the calculated MM/GBSA enthalpy values revealed that STA showed the highest magnitude of favorable binding enthalpy (−46.4 ± 5.4 kcal/mol), whereas NU6 showed the smallest magnitude (−21.0 ± 4.9 kcal/mol; Table 3.15). This predicted trend for binding strength (STA > NU6) is opposite to the experimentally known affinity order (NU6 is a stronger binder than STA). This discrepancy was examined by looking at the components of the binding enthalpy. For STA, favorable electrostatic (EEL) and van der Waals (vdW) contributions were high (−55.9 and −28.2 kcal/mol, respectively), with a relatively smaller unfavorable polar component of solvation free energy (GB) (43.9 kcal/mol). In contrast, NU6 had less favorable (higher) EEL and vdW contributions (−43.8 and −18.1 kcal/mol) and smaller (less unfavorable) solvation penalties, resulting in a less favorable total

Table 3.16 Calculated binding enthalpy using MM/PBSA with indi = 2, 4 for small-molecule binders of PfPK5.

				MM/PBSA (indi = 2)		MM/PBSA (indi = 4)	
Inhibitor	ΔG_{expr}	E_{MM}	$E_{nonpolar}$	PB	Total	PB	Total
OLM	−6.75	−39.1	−3.9	28.8	−33.1 ± 4.1	14.0	−38.4 ± 3.4
LTU	−7.21	−57.7	−4.6	26.8	−48.6 ± 3.5	13.0	−55.7 ± 3.4
INR	−7.35	−42.2	−3.6	65.6	−39.6 ± 4.3	32.0	−43.5 ± 4.2
FLV	−7.52	−37.8	−3.8	24.5	−31.6 ± 4.9	12.0	−36.9 ± 5.0
STA	−8.37	−55.9	−4.8	25.7	−49.1 ± 4.6	12.6	−55.2 ± 4.5
NU6	−9.29	−43.8	−4.2	21.1	−35.9 ± 4.7	10.5	−42.0 ± 4.7
PVB	−9.59	−49.3	−4.6	27.2	−43.6 ± 4.8	13.3	−49.0 ± 4.5

E_{MM} is the molecular mechanical component of enthalpy. PB is the polar component of solvation free energy and $E_{nonpolar}$ is the nonpolar component. All energy values are in kcal/mol.

enthalpy compared to STA according to MM/GBSA. Similar trends were observed with MM/PBSA (Table 3.16), although PBSA generally predicted higher solvation penalties, leading to a smaller overall range of binding enthalpies.

Increasing the internal dielectric constant in MM/PBSA (indi = 2 and indi = 4, Table 3.16) reduced the magnitude of the positive (unfavorable) polar component of solvation free energy, resulting in larger favorable binding enthalpies.

Correlation analysis between normalized predicted enthalpy (from MM/GBSA or MM/PBSA with different indi values) and normalized experimental affinity (Fig. 3.13) showed that MM/GBSA predictions had poor correlation ($R^2 = 0.011$). MM/PBSA predictions, while still exhibiting small R^2 values, showed slightly better correlations (indi = 1: $R^2 = 0.024$; indi = 2: $R^2 = 0.023$; indi = 4: $R^2 = 0.022$). However, in all cases, neither MM/GBSA nor MM/PBSA showed substantial correlation with experimental binding affinity, indicating that enthalpy alone is not sufficient to accurately represent the binding affinity for these inhibitors to PfPK5.

To gain further insight into the contributions to binding enthalpy a residue-wise decomposition was performed using MM/PBSA with indi = 2 (Fig. 3.14). Analysis of individual residue contributions highlighted variations in how different residues interact with the inhibitors. Lys32, for example, showed opposing enthalpy contributions depending on the ligand, being unfavorable for LTU, NU6, and PVB but favorable for OLM, INR, and FLV, with a particularly high favorable contribution for INR (−29.7 kcal/mol). Glu80 selectively opposed binding for FLV and NU6, albeit weakly. The binding of INR was distinct, with several residues (Glu12, Glu80, Asp83, Asp85, Asp118, and Asp143) strongly opposing binding (positive enthalpy), whereas two Lys residues (Lys32 and Lys88) strongly favored it.

3.2.4.3.6 Configurational entropy of inhibitor binding to PfPK5

Accurate estimation of binding configurational entropy requires sufficient convergence of the calculated values. The convergence of the first-order and second-order binding configurational entropy, estimated using the A-MIST method with a 14 Å distance cutoff, Miller–Madow entropy estimator, and histogram-based discretization (30 bins for 1D, 30 × 30 for 2D), was monitored over the 5 μs MD simulations for all seven inhibitors (Fig. 3.15). Both first-order and second-order entropy estimates were found to converge

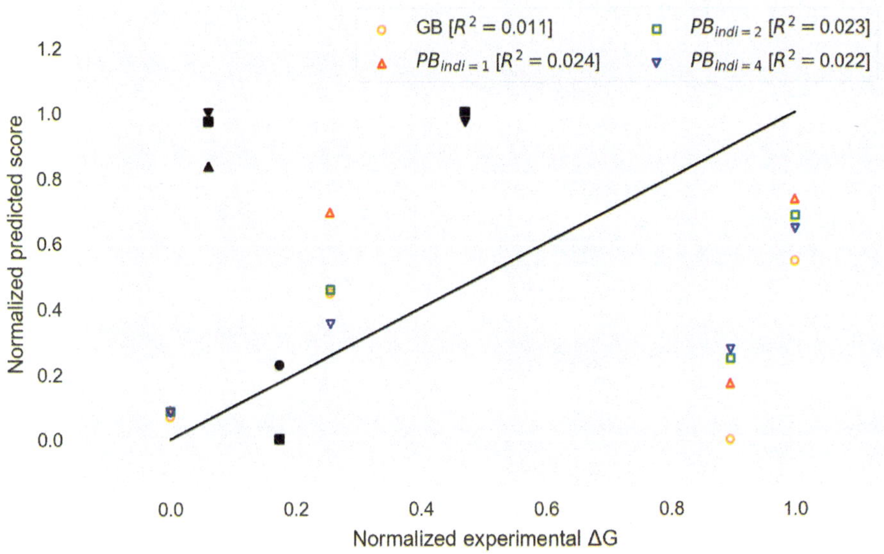

FIGURE 3.13

Normalized enthalpy (using MM/GBSA and MM/PBSA) versus normalized experimental is shown. MM/PBSA has been shown to correspond to internal dielectric constants 1, 2, and 4. The solid *black line* represents a perfect correlation, whereas deviation of points from the line indicates a large error in predicted values. The points and R^2 of predicted simple linear regression for each case are annotated in the legend label. LTU, STA, and FLV, for which K_i is experimentally known, are shown with *black color,* and corresponding marker shape represents the associated normalized score docking program. At least two points of the three predictions are very different from the perfect correlation line. LTU, STA, and FLV for which K_i is experimentally known are shown in *black*, and the corresponding marker shape represents the associated docking program; at least two points of the three predictions are very different from the perfect correlation line.

after approximately 4 μs of simulation.

Analysis of the total binding configurational entropy values (Table 3.14) indicated that binding was entropically favorable (negative contribution) for NU6, whereas it was unfavorable (positive contribution) for the other six inhibitors. Considering the first-order entropy estimates alone (Fig. 3.15A), NU6 was the most entropically favorable binder, and OLM was the most unfavorable. Including first-order correlations (second-order estimate, Fig. 3.15B) maintained NU6 as the most favorable but shifted INR to being the most entropically unfavorable binder.

The total binding configurational entropy change can be decomposed into contributions from the receptor, the ligand, and changes in rotational/translational constraints upon complex formation (Table 3.14). This partitioning provides insight into how the flexibility of the receptor and ligand is modulated upon complex formation. The decomposition highlights that the contribution from the ligand (small molecule) to the total configurational entropy is typically small in magnitude compared to the receptor's contribution. Even small changes in ligand flexibility can significantly influence the receptor's dynamics by affecting the interaction network within the complex. For most ligands studied

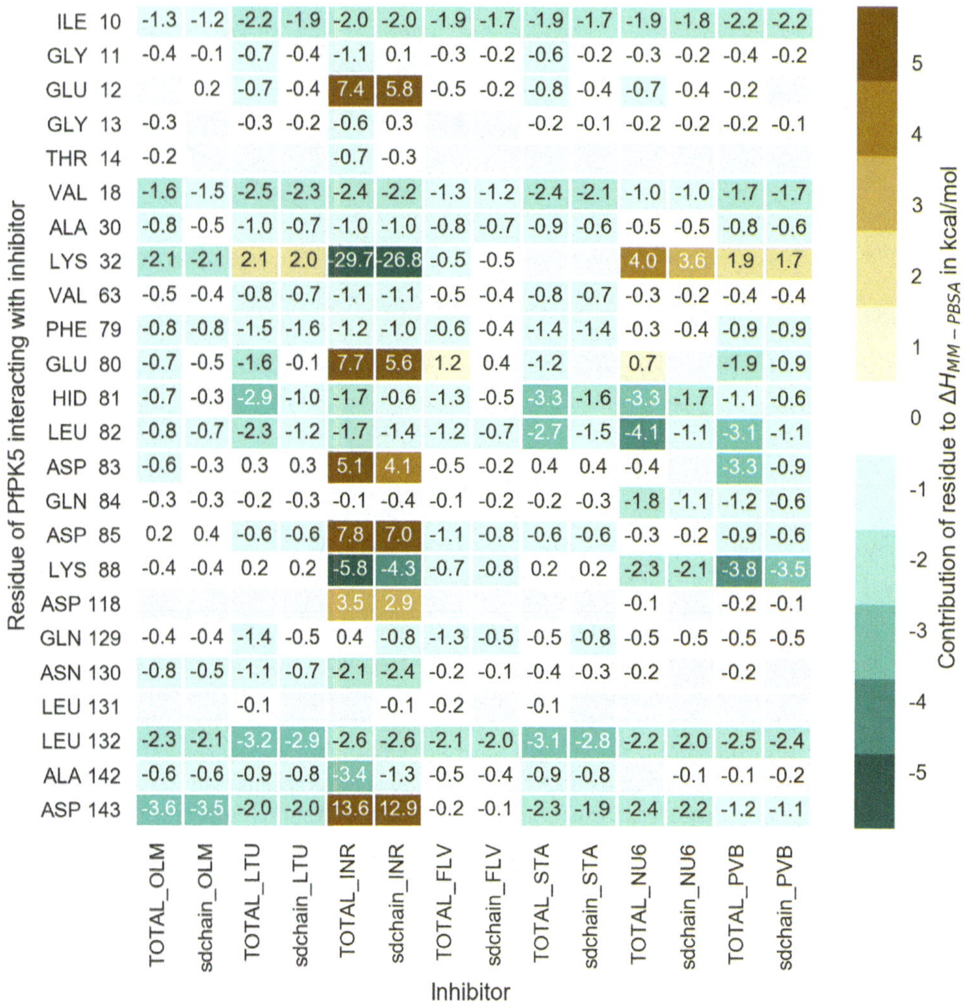

FIGURE 3.14

Residues of PfPK5 interacting with chosen small molecules and their contribution to binding enthalpy. Residues of PfPK5 interacting with chosen small molecules and their contribution to binding enthalpy. Enthalpy in this case is calculated using MM/PBSA method with an internal dielectric constant value (indi = 2). TOTAL_AAA, and sdchain_AAA represent total and sidechain contributions to the binding enthalpy for the small molecule AAA.

(OLM, LTU, INR, STA, and NU6), the ligand's contribution was less than 2.3 kcal/mol in magnitude, whereas the receptor's contribution was around 10 kcal/mol. The rotational/translational entropy of binding generally opposes complex formation and is more significant for rigid molecules. For instance, LTU and INR, with only one rotatable bond each, had higher rotational/translational contributions

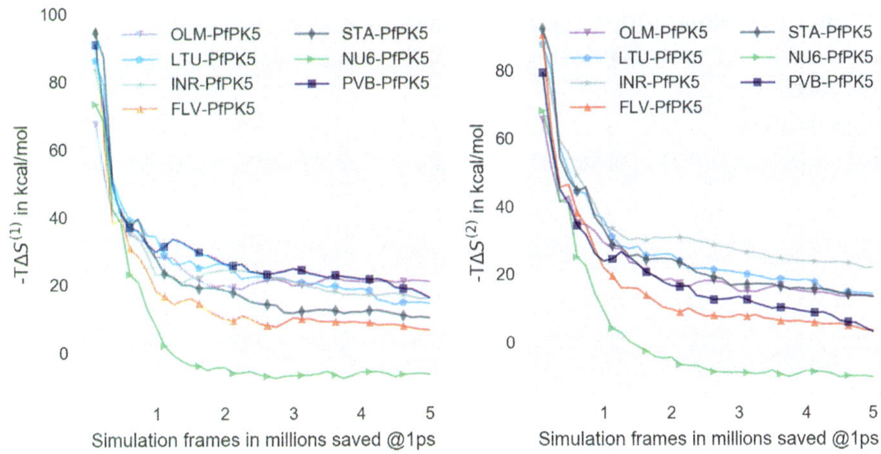

FIGURE 3.15

Convergence of binding configurational entropy estimated using the A-MIST method. Convergence of first-order $S^{(1)}$ and second-order $S^{(2)}$ binding configurational entropy estimated using the A-MIST method, where a histogram is used for discretization with 30 bins for 1D and 30×30 for 2D, using the Miller–Madow entropy estimator and a distance cutoff of 14 Å for seven small-molecule binders of PfPK5 with simulation time (number of frames included 5 million @ 1 ps = 5 μs).

(3.94 and 4.48 kcal/mol) compared to molecules with more rotatable bonds such as OLM (six bonds, 1.49 kcal/mol), NU6 (six bonds, 2.39 kcal/mol), and PVB (eight bonds, 0.10 kcal/mol).

A residue-wise decomposition of binding configurational entropy provides a means to assess the entropic changes of individual residues upon binding and identify residues with potentially significant contributions for lead optimization. The residue-wise binding configurational entropy decomposition for the seven inhibitors. Residues contributing favorably (causing an increase in affinity) to the binding configurational entropy were identified, and the top-10 such residues for each inhibitor are listed in Table 3.17.

Table 3.18 lists top 10 residues contributing negatively to the binding affinity (i.e., opposing residues) by increasing the configurational entropy penalty upon ligand binding to PfPK5. All energetic contributions are in kcal/mol. For each small-molecule inhibitor, two columns are shown: one listing the residue number (Res.) and the other the corresponding first-order configurational entropy contribution (ΔS_{conf}). Residues with higher positive values contribute more unfavorably to binding, thus reflecting an entropic penalty.

3.2.4.3.7 Enthalpic and entropic contributions to binding free energy

Previously, we examined the enthalpic component of binding free energy using molecular docking scores and MM/GBSA and MM/PBSA methods. However, correlations between predicted and experimentally determined binding affinities were generally poor using these approaches. Independently, we also estimated the configurational entropy of binding using the information-theoretic method known as the A-MIST.

Table 3.17 Top 10 residues with favorable (i.e., negative) configurational entropy contributions (ΔS_{conf}) to binding for each PfPK5–ligand complex. Values are in kcal/mol. "Res." indicates residue number.

Rank	ResidueID/ΔS_{conf}						
	OLM	**LTU**	**INR**	**FLV**	**STA**	**NU6**	**PVB**
1	81/−0.468	73/−1.485	73/−1.516	160/−0.975	73/−1.519	155/−0.983	160/−0.816
2	163/−0.432	3/−1.434	3/−1.298	85/−0.518	3/−1.311	160/−0.828	85/−0.736
3	160/−0.429	72/−0.974	167/−1.137	163/−0.446	71/−1.026	85/−0.742	167/−0.593
4	167/−0.369	2/−0.929	72/−0.907	95/−0.435	72/−1.007	14/−0.716	155/−0.569
5	12/−0.349	178/−0.893	2/−0.88	86/−0.433	2/−0.892	167/−0.559	159/−0.536
6	153/−0.329	166/−0.764	4/−0.579	34/−0.393	167/−0.808	34/−0.557	158/−0.534
7	155/−0.271	34/−0.736	130/−0.578	90/−0.392	122/−0.772	153/−0.554	163/−0.511
8	37/−0.255	4/−0.711	71/−0.547	80/−0.344	178/−0.724	86/−0.54	153/−0.457
9	130/−0.235	130/−0.684	164/−0.535	159/−0.315	34/−0.7	158/−0.46	156/−0.449
10	38/−0.223	158/−0.656	158/−0.511	167/−0.307	130/−0.667	129/−0.44	150/−0.436

Table 3.18 Top 10 residues opposing binding via first-order configurational entropy (kcal/mol).

Rank	ResidueID/ΔS_{conf}						
	OLM	**LTU**	**INR**	**FLV**	**STA**	**NU6**	**PVB**
1	156/0.983	154/1.945	140/2.017	141/1.461	154/1.958	141/1.123	140/1.364
2	141/0.718	141/1.805	154/1.823	156/1.116	155/1.767	212/0.453	141/0.792
3	140/0.664	150/1.673	155/1.779	157/0.755	141/1.714	156/0.452	156/0.775
4	148/0.557	49/1.479	32/1.763	150/0.651	49/1.536	129/0.425	80/0.623
5	212/0.494	51/1.418	144/1.343	148/0.504	150/1.473	84/0.399	79/0.526
6	157/0.46	50/1.36	50/1.329	10/0.432	51/1.409	138/0.394	212/0.371
7	79/0.422	155/1.326	51/1.291	138/0.412	50/1.268	134/0.346	157/0.341
8	2/0.391	54/1.279	49/1.245	129/0.39	144/1.262	67/0.325	81/0.333
9	129/0.369	144/1.265	54/1.186	152/0.387	156/1.149	54/0.246	124/0.318
10	124/0.334	119/1.205	10/1.097	134/0.307	54/1.138	136/0.242	32/0.286

The absence of entropy contributions in predictive models may explain the poor correlation with experimental affinities. Therefore, to improve accuracy, it is essential to incorporate configurational entropy into the total binding free energy and reassess its correlation with experimental data.

In this section, we analyze correlations that include first-order configurational entropy, which captures modulations in individual degrees of freedom (DOFs) but not their pairwise correlations, and second-order configurational entropy, which accounts for both individual fluctuations and correlated fluctuations between pairs of DOFs.

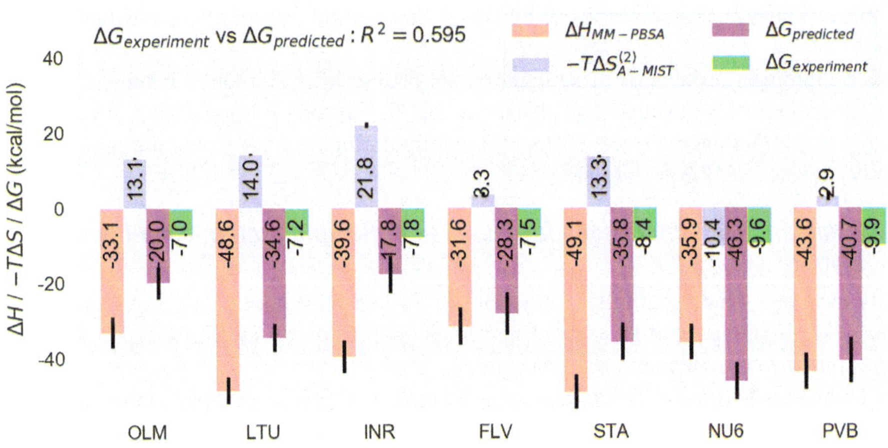

FIGURE 3.16

The binding enthalpy (MM/PBSA), configurational entropy, combined binding free energy, and experimental free energy. The binding enthalpy (MM/PBSA), second-order configurational entropy (A-MIST), combined binding free energy, and experimental free energy are shown as bar plots. Incorporating second-order entropy improved the predictive correlation significantly ($R^2 = 0.595$), compared to $R^2 = 0.045$ when only enthalpy was considered.

3.2.4.3.8 Binding thermodynamics of inhibitors to PfPK5

Among the inhibitors for which experimental K_i values are available (e.g., LTU, STA, and FLV), the predicted binding affinities, when incorporating first-order entropy, follow the experimental affinity trend for at least two compounds (FLV and STA).

The inclusion of first-order entropy improved correlation with experimental values significantly, increasing R^2 from 0.045 (MM/PBSA only) to 0.282. However, this improvement is not yet sufficient for robust prediction of binding affinity. The exclusion of DOF pairwise correlations may be contributing to the limited accuracy. Therefore we extend the analysis to include second-order entropy contributions.

3.2.4.3.9 Enhanced prediction with second-order entropy (PfPK5)

As shown in Fig. 3.16, although LTU and NU6 remain outliers and reduce overall correlation, the inclusion of pairwise correlated DOF fluctuations substantially enhances predictive accuracy.

This $R^2 = 0.595$ correlation outperforms all docking scoring functions and MM/GBSA and MM/PBSA methods, with the exception of GOLD when docked to the 1V0P receptor. However, the superior correlation from GOLD/1V0P is questionable, as performance degrades significantly when docking is performed using the INR-bound PfPK5 structure (PDB: 1V0O), indicating high sensitivity to receptor conformation and possible overfitting.

3.2.4.3.10 Entropy's role in differential inhibitor binding

Our results clearly show that configurational entropy contributions vary significantly in both magnitude and direction among different inhibitors binding to the same target. The assumption that binding entropy is negligible or uniform across a series of ligands is thus invalid, especially in flexible systems such as PfPK5.

Notably, inhibition profiles of these compounds demonstrate sigmoidal behavior, consistent with flexible kinase binding, and suggest a substantial role for entropy in modulating affinity. This insight underscores the importance of accounting for configurational entropy in rational inhibitor design, particularly for selective kinase targeting.

3.2.4.4 *Summary and conclusions for the case of flexible protein–small molecule binding*

This study evaluates binding affinity prediction for PfPK5 inhibitors using three molecular docking programs—GOLD, Glide, and DOCK 6—and assesses their performance with known experimental IC_{50} values ranging from nanomolar to micromolar.

Two inhibitors (INR and PVB) have experimentally resolved complex structures with PfPK5 (PDB: 1V0O and 1V0P). These structures were used for docking, with docking accuracy evaluated by correlating predicted scores with experimental binding free energies.

Docking-based virtual screening was employed to estimate the relative binding affinities of small-molecule inhibitors targeting PfPK5, using two experimentally determined crystal structures as receptor models. Normalized docking scores were compared against experimental binding affinities to assess predictive performance. Key observations indicate that docking can serve as a useful initial filter for prioritizing compounds, provided that receptor conformation and ligand diversity are carefully considered. Among the tools tested, DOCK 6 consistently yielded the highest correlation between normalized docking scores and experimental affinities across both receptor structures. In contrast, GOLD exhibited high sensitivity to receptor structure, achieving good correlation for the 1V0P conformation but poor results for 1V0O. Glide did not produce meaningful correlations for either receptor, highlighting variability in performance across docking algorithms.

These findings underscore both the potential and the limitations of structure-based docking for affinity prediction in flexible kinase targets. Although DOCK 6 outperformed the other programs for this particular receptor-ligand system, caution is warranted in generalizing these results. The ligand set is relatively small, and receptor flexibility—a critical factor in kinase-ligand recognition—was not explicitly modeled in the docking workflow. Therefore docking should be viewed as a rapid, low-cost screening strategy whose results require further refinement through dynamics and thermodynamic modeling.

To improve the predictive power beyond static scoring functions, MM/GBSA and MM/PBSA methods were employed using trajectories from MD simulations. However, the resulting binding free energy estimates showed poor agreement with experimental values, indicating limitations in enthalpy-only models.

Configurational entropy was subsequently evaluated using the A-MIST method, which allows quantification of both first-order (individual DOFs) and second-order (correlated DOFs) entropy contributions from MD simulations. The inclusion of first-order entropy alone improved the correlation between predicted and experimental binding free energies ($R^2 = 0.282$), but only marginally. When second-order entropy was included, predictive performance improved significantly ($R^2 = 0.595$), demonstrating that entropy plays a substantial and nonuniform role in modulating inhibitor binding to PfPK5.

These results emphasize that configurational entropy cannot be neglected in systems such as protein kinases, where flexible binding pockets and conformational adaptation play critical roles in ligand recognition. The assumption that entropy is constant across ligand series is not supported by our data. Instead, entropy emerges as a key determinant of differential binding affinity and specificity.

In conclusion the integration of entropic contributions—especially higher order configurational entropy—into free energy calculations is essential for accurate binding affinity prediction in flexible targets such as PfPK5. This work provides a framework for combining enthalpic and entropic data in kinase-inhibitor studies and supports the broader application of information-theoretic entropy estimators in computational drug design.

3.3 Summary

Entropy, a concept extending beyond its thermodynamic origins, is increasingly recognized as fundamental in understanding biological systems and is a critical factor in rational drug discovery. Although living organisms maintain internal order, they do so by continuously increasing the entropy of their surroundings, consuming low-entropy energy to sustain life and counteract internal disorder. This dynamic balance between negentropic input and entropy generation governs biological processes, from development to aging.

In pharmacology, entropy intrinsically influences molecular binding affinity (ΔG), alongside enthalpy (ΔH), as dictated by the Gibbs–Helmholtz equation ($\Delta G = \Delta H - T\Delta S$). Experimental techniques like ITC are crucial for dissecting these enthalpic and entropic contributions, revealing the true driving forces behind binding events. A significant challenge in drug design is enthalpy–entropy compensation, where favorable changes in one component are often offset by unfavorable changes in the other. This phenomenon, often linked to the flexibility and dynamic alterations of biomolecules upon binding, can limit the net gain in binding affinity despite efforts to optimize interactions. Studies on HIV protease and influenza neuraminidase inhibitors exemplify how understanding these thermodynamic shifts is vital for overcoming drug resistance and improving potency.

Our investigation into the binding thermodynamics of peptide-receptor systems, specifically involving *P. falciparum* apical membrane antigen 1 (PfAMA1) and inhibitory peptides (PfRON2sp1, PfRON2sp2, and R1), critically underscores the indispensable role of entropy in antimalarial drug discovery, particularly for flexible targets. These peptides are vital in inhibiting merozoite invasion, a key step in malaria infection. We found that when dealing with large, flexible biomolecules such as PfAMA1 and its peptide binders, predictions based solely on enthalpy proved misleading and failed to accurately rank experimental binding affinities. However, by integrating both enthalpic and configurational entropic contributions, our predictions accurately reproduced the experimentally observed affinity rankings. This forcefully demonstrates that configurational entropy is not merely a minor factor but an essential component for understanding binding in highly dynamic and flexible biomolecular complexes, where the conformational freedom of both the receptor and ligand significantly influences the overall thermodynamics. The residue-level thermodynamic maps generated from our analyses offer targeted insights, directly informing the design of optimized peptide inhibitors capable of broad efficacy against diverse PfAMA1 variants. This work establishes a robust framework for enhancing rational inhibitor design against crucial antimalarial targets by emphasizing a comprehensive thermodynamic approach.

Our second in-house study delved into the binding affinity prediction for inhibitors of PfPK5, another vital target in *P. falciparum*. We evaluated various molecular docking programs and free energy calculation methods. Although docking offered an initial screening capability, traditional MM/GBSA and

MM/PBSA methods, which primarily account for enthalpic contributions, showed poor correlation with experimental binding affinities. This highlights their limitations in accurately modeling highly flexible protein kinase targets. The notable improvement came with the explicit inclusion of configurational entropy, particularly its higher order contributions, using the A-MIST method from MD simulations. Incorporating even first-order entropy marginally improved predictions, but the inclusion of second-order entropy dramatically enhanced the correlation with experimental values ($R^2 = 0.595$). This unequivocally demonstrated that entropy plays a substantial, nonuniform role in modulating inhibitor binding to PfPK5, challenging the common assumption that entropy is constant across ligand series in such flexible systems. These findings affirm that configurational entropy is a key determinant of differential binding affinity and specificity for flexible targets like protein kinases. This study provides a vital framework for integrating comprehensive enthalpic and entropic data into computational drug design, paving the way for more accurate predictions and the development of potent antimalarial therapeutics.

In conclusion, our in-house research on PfAMA1 and PfPK5 distinctly illustrates that explicitly accounting for biomolecular flexibility, dynamic alterations, and the often-overlooked entropic contributions is paramount for accurate binding affinity prediction and, ultimately, the successful rational design of effective antimalarial drugs targeting dynamic protein–protein interfaces and flexible enzymes.

References

Abdul-Ghani, R., Farag, H. F., & Allam, A. F. (2013). Sulfadoxine-pyrimethamine resistance in *Plasmodium falciparum*: A zoomed image at the molecular level within a geographic context. *Acta Tropica, 125*(2), 163–190. https://doi.org/10.1016/j.actatropica.2012.10.013.

Albiñana, C. B., Machara, A., Řezáčová, P., Pachl, P., Konvalinka, J., & Kožíšek, M. (2016). Kinetic, thermodynamic and structural analysis of tamiphosphor binding to neuraminidase of H1N1 (2009) pandemic influenza. *European Journal of Medicinal Chemistry, 121*, 100–109. https://doi.org/10.1016/j.ejmech.2016.05.016.

Allen, W. J., Balius, T. E., Mukherjee, S., Brozell, S. R., Moustakas, D. T., Therese Lang, P., Case, D. A., Kuntz, I. D., & Rizzo, R. C. (2015). DOCK 6: Impact of new features and current docking performance. *Journal of Computational Chemistry, 36*(15), 1132–1156. https://doi.org/10.1002/jcc.23905.

Ashley, E. A., Dhorda, M., Fairhurst, R. M., Amaratunga, C., Lim, P., Suon, S., Sreng, S., Anderson, J. M., Mao, S., Sam, B., Sopha, C., Chuor, C. M., Nguon, C., Sovannaroth, S., Pukrittayakamee, S., Jittamala, P., Chotivanich, K., Chutasmit, K., Suchatsoonthorn, C., … White, N. J. (2014). Spread of artemisinin resistance in *Plasmodium falciparum* malaria. *New England Journal of Medicine, 371*(5), 411–423. https://doi.org/10.1056/NEJMoa1314981.

Athawale, M. V., Sarupria, S., & Garde, S. (2008). Enthalpy−entropy contributions to salt and osmolyte effects on molecular-scale hydrophobic hydration and interactions. *The Journal of Physical Chemistry B, 112*(18), 5661–5670. https://doi.org/10.1021/jp073485n.

Attard, P. (2002). *Thermodynamics and statistical mechanics: Equilibrium by entropy maximisation*. Elsevier.

Bai, T., Becker, M., Gupta, A., Strike, P., Murphy, V. J., Anders, R. F., & Batchelor, A. H. (2005). Structure of AMA1 from *Plasmodium falciparum* reveals a clustering of polymorphisms that surround a conserved hydrophobic pocket. *Proceedings of the National Academy of Sciences, 102*(36), 12736–12741. https://doi.org/10.1073/pnas.0501808102.

Baiz, C. R., Błasiak, B., Bredenbeck, J., Cho, M., Choi, J.-H., Corcelli, S. A., Dijkstra, A. G., Feng, C.-J., Garrett-Roe, S., Ge, N.-H., Hanson-Heine, M. W. D., Hirst, J. D., Jansen, T. L. C., Kwac, K., Kubarych, K. J., Londergan, C. H., Maekawa, H., Reppert, M., Saito, S., … Zanni, M. T. (2020). Vibrational spectroscopic map, vibrational spectroscopy, and intermolecular interaction. *Chemical Reviews, 120*(15), 7152–7218. https://doi.org/10.1021/acs.chemrev.9b00813.

Bellissent-Funel, M.-C., Hassanali, A., Havenith, M., Henchman, R., Pohl, P., Sterpone, F., Spoel, D., Xu, Y., & Garcia, A. E. (2016). Water determines the structure and dynamics of proteins. *Chemical Reviews, 116*(13), 7673–7697. https://doi.org/10.1021/acs.chemrev.5b00664.

Besteiro, S., Dubremetz, J.-F., & Lebrun, M. (2011). The moving junction of apicomplexan parasites: A key structure for invasion. *Cellular Microbiology, 13*(6), 797–805. https://doi.org/10.1111/j.1462-5822.2011.01597.x.

Brandt, R. B., Laux, J. E., & Yates, S. W. (1987). Calculation of inhibitor Ki and inhibitor type from the concentration of inhibitor for 50% inhibition for Michaelis-Menten enzymes. *Biochemical Medicine and Metabolic Biology, 37*(3), 344–349. https://doi.org/10.1016/0885-4505(87)90046-6.

Bronowska, A. K. (2011). *Thermodynamics of ligand-protein interactions: Implications for molecular design thermodynamics—Interaction studies—Solids, liquids and gases*. InTech https://doi.org/10.5772/19447.

Brooks, B., & Karplus, M. (1983). Harmonic dynamics of proteins: Normal modes and fluctuations in bovine pancreatic trypsin inhibitor. *Proceedings of the National Academy of Sciences, 80*(21), 6571–6575. https://doi.org/10.1073/pnas.80.21.6571.

Callendar, H. L. (1910). The caloric theory of heat and Carnot's principle. *Proceedings of the Physical Society of London, 23*(1), 153–189. https://doi.org/10.1088/1478-7814/23/1/315.

Carnot, S., Clapeyron, E., & Clausius, R. (1960). *Reflections on the motive power of fire*. Springer.

Caro, J. A., Harpole, K. W., Kasinath, V., Lim, J., Granja, J., Valentine, K. G., Sharp, K. A., & Joshua Wand, A. (2017). Entropy in molecular recognition by proteins. *Proceedings of the National Academy of Sciences, 114*(25), 6563–6568. https://doi.org/10.1073/pnas.1621154114.

Case, D. A. (1994). Normal mode analysis of protein dynamics. *Current Opinion in Structural Biology, 4*(2), 285–290. https://doi.org/10.1016/S0959-440X(94)90321-2.

Case, D. A., Cheatham, T. E., Darden, T., Gohlke, H., Luo, R., Merz, K. M., Onufriev, A., Simmerling, C., Wang, B., & Woods, R. J. (2005). The Amber biomolecular simulation programs. *Journal of Computational Chemistry, 26*(16), 1668–1688. https://doi.org/10.1002/jcc.20290.

Chodera, J. D., & Mobley, D. L. (2013). Entropy-enthalpy compensation: Role and ramifications in biomolecular ligand recognition and design. *Annual Review of Biophysics, 42*(1), 121–142. https://doi.org/10.1146/annurev-biophys-083012-130318.

Claveria-Gimeno, R., Vega, S., Abian, O., & Velazquez-Campoy, A. (2017). A look at ligand binding thermodynamics in drug discovery. *Expert Opinion on Drug Discovery, 12*(4), 363–377. https://doi.org/10.1080/17460441.2017.1297418.

Cooper, A., Johnson, C. M., Lakey, J. H., & Nöllmann, M. (2001). Heat does not come in different colours: Entropy–enthalpy compensation, free energy windows, quantum confinement, pressure perturbation calorimetry, solvation and the multiple causes of heat capacity effects in biomolecular interactions. *Biophysical Chemistry, 93*(2–3), 215–230. https://doi.org/10.1016/S0301-4622(01)00222-8.

Counihan, N. A., Kalanon, M., Coppel, R. L., & Koning-Ward, T. F. (2013). *Plasmodium* rhoptry proteins: Why order is important. *Trends in Parasitology, 29*(5), 228–236. https://doi.org/10.1016/j.pt.2013.03.003.

Cowman, A. F., Healer, J., Marapana, D., & Marsh, K. (2016). Malaria: Biology and disease. *Cell, 167*(3), 610–624. https://doi.org/10.1016/j.cell.2016.07.055.

de Azevedo, E. R., & Bonagamba, T. J. (2006). Molecular dynamics and local molecular conformation in solid materials studied by nuclear magnetic resonance. *Brazilian Journal of Physics, 36*(1a). https://doi.org/10.1590/S0103-97332006000100012.

Decherchi, S., & Cavalli, A. (2020). Thermodynamics and kinetics of drug-target binding by molecular simulation. *Chemical Reviews, 120*(23), 12788–12833. https://doi.org/10.1021/acs.chemrev.0c00534.

Doerig, C., Abdi, A., Bland, N., Eschenlauer, S., Dorin-Semblat, D., Fennell, C., Halbert, J., Holland, Z., Nivez, M.-P., Semblat, J.-P., Sicard, A., & Reininger, L. (2010). Malaria: Targeting parasite and host cell kinomes.

Biochimica et Biophysica Acta (BBA)—Proteins and Proteomics, 1804(3), 604–612. https://doi.org/10.1016/j.bbapap.2009.10.009.

Dondorp, A. M., Smithuis, F. M., Woodrow, C., & von Seidlein, L. (2017). How to contain artemisinin- and multidrug-resistant falciparum malaria. *Trends in Parasitology, 33*(5), 353–363. https://doi.org/10.1016/j.pt.2017.01.004.

Dragan, A. I., Read, C. M., & Crane-Robinson, C. (2016). Enthalpy–entropy compensation: The role of solvation. *European Biophysics Journal, 46*(4), 301–308. https://doi.org/10.1007/s00249-016-1182-6.

Dunning-Davies, J. (1969). Connections between the various forms of the second law of thermodynamics. *Il Nuovo Cimento B Series 10, 64*(1), 82–87. https://doi.org/10.1007/BF02710283.

Durrant, J. D., & McCammon, J. A. (2011). Molecular dynamics simulations and drug discovery. *BMC Biology, 9*(1). https://doi.org/10.1186/1741-7007-9-71.

Dykeman, E. C., & Sankey, O. F. (2010). Normal mode analysis and applications in biological physics. *Journal of Physics: Condensed Matter, 22*(42), 423202. https://doi.org/10.1088/0953-8984/22/42/423202.

Feidt, M. (2017). The history and perspectives of efficiency at maximum power of the Carnot engine. *Entropy, 19*(7), 369. https://doi.org/10.3390/e19070369.

Forester, T. R., & Smith, W. (1998). SHAKE, rattle, and roll: Efficient constraint algorithms for linked rigid bodies. *Journal of Computational Chemistry, 19*(1), 102–111. https://doi.org/10.1002/(SICI)1096-987X(19980115)19:1⟨102::AID-JCC9⟩3.0.CO;2-T.

Fox, J. M., Zhao, M., Fink, M. J., Kang, K., & Whitesides, G. M. (2018). The molecular origin of enthalpy/entropy compensation in biomolecular recognition. *Annual Review of Biophysics, 47*(1), 223–250. https://doi.org/10.1146/annurev-biophys-070816-033743.

Frautschi, S. (1982). Entropy in an expanding universe. *Science, 217*(4560), 593–599. https://doi.org/10.1126/science.217.4560.593.

Freire, E. (2008). Do enthalpy and entropy distinguish first in class from best in class? *Drug Discovery Today, 13*(19–20), 869–874. https://doi.org/10.1016/j.drudis.2008.07.005.

Friesner, R. A., Banks, J. L., Murphy, R. B., Halgren, T. A., Klicic, J. J., Mainz, D. T., Repasky, M. P., Knoll, E. H., Shelley, M., Perry, J. K., Shaw, D. E., Francis, P., & Shenkin, P. S. (2004). Glide: A new approach for rapid, accurate docking and scoring. 1. Method and assessment of docking accuracy. *Journal of Medicinal Chemistry, 47*(7), 1739–1749. https://doi.org/10.1021/jm0306430.

Frutiger, A., Tanno, A., Hwu, S., Tiefenauer, R. F., Vörös, J., & Nakatsuka, N. (2021). Nonspecific binding—Fundamental concepts and consequences for biosensing applications. *Chemical Reviews, 121*(13), 8095–8160. https://doi.org/10.1021/acs.chemrev.1c00044.

Genheden, S., & Ryde, U. (2015). The MM/PBSA and MM/GBSA methods to estimate ligand-binding affinities. *Expert Opinion on Drug Discovery, 10*(5), 449–461. https://doi.org/10.1517/17460441.2015.1032936.

Ghysels, A., Speybroeck, V. V., Pauwels, E., Catak, S., Brooks, B. R., Neck, D. V., & Waroquier, M. (2009). Comparative study of various normal mode analysis techniques based on partial Hessians. *Journal of Computational Chemistry, 31*(5), 994–1007. https://doi.org/10.1002/jcc.21386.

Graeser, R., Wernli, B., Franklin, R. M., & Kappes, B. (1996). *Plasmodium falciparum* protein kinase 5 and the malarial nuclear division cycles. *Molecular and Biochemical Parasitology, 82*(1), 37–49. https://doi.org/10.1016/0166-6851(96)02716-8.

Grandy, W. T. (2008). *Entropy and the time evolution of macroscopic systems*: 10. Oxford University Press.

Green, A. E., & Naghdi, P. M. (1977). On thermodynamics and the nature of the second law. *Proceedings of the Royal Society of London. A. Mathematical and Physical Sciences, 357*(1690), 253–270. https://doi.org/10.1098/rspa.1977.0166.

Greiner, W., Rischke, D., Neise, L., & Stöcker, H. (2012). *Thermodynamics and statistical mechanics*. Springer https://books.google.com/books?id=8jrTBwAAQBAJ.

Grest, G. S., & Kremer, K. (1986). Molecular dynamics simulation for polymers in the presence of a heat bath. *Physical Review A, 33*(5), 3628–3631. https://doi.org/10.1103/PhysRevA.33.3628.

Guo, S., & Liu, C. T. (2011). Phase stability in high entropy alloys: Formation of solid-solution phase or amorphous phase. *Progress in Natural Science: Materials International, 21*(6), 433–446. https://doi.org/10.1016/S1002-0071(12)60080-X.

Hansen, N., & van Gunsteren, W. F. (2014). Practical aspects of free-energy calculations: A review. *Journal of Chemical Theory and Computation, 10*(7), 2632–2647. https://doi.org/10.1021/ct500161f.

Harris, K. S., Adda, C. G., Khore, M., Drew, D. R., Valentini-Gatt, A., Fowkes, F. J. I., Beeson, J. G., Dutta, S., Anders, R. F., & Foley, M. (2014). Use of immunodampening to overcome diversity in the malarial vaccine candidate apical membrane antigen 1. *Infection and Immunity, 82*(11), 4707–4717. https://doi.org/10.1128/IAI.02061-14.

Harris, K. S., Casey, J. L., Coley, A. M., Karas, J. A., Sabo, J. K., Yee Tan, Y., Dolezal, O., Norton, R. S., Hughes, A. B., Scanlon, D., & Foley, M. (2009). Rapid optimization of a peptide inhibitor of malaria parasite invasion by comprehensive N-methyl scanning. *Journal of Biological Chemistry, 284*(14), 9361–9371. https://doi.org/10.1074/jbc.M808762200.

Harris, K. S., Casey, J. L., Coley, A. M., Masciantonio, R., Sabo, J. K., Keizer, D. W., Lee, E. F., McMahon, A., Norton, R. S., Anders, R. F., & Foley, M. (2005). Binding hot spot for invasion inhibitory molecules on *Plasmodium falciparum* apical membrane antigen 1. *Infection and Immunity, 73*(10), 6981–6989. https://doi.org/10.1128/IAI.73.10.6981-6989.2005.

Hensen, U., Lange, O. F., & Grubmüller, H. (2010). Estimating absolute configurational entropies of macromolecules: The minimally coupled subspace approach. *PLoS ONE, 5*(2), e9179. https://doi.org/10.1371/journal.pone.0009179.

Hnizdo, V., Tan, J., Killian, B. J., & Gilson, M. K. (2008). Efficient calculation of configurational entropy from molecular simulations by combining the mutual-information expansion and nearest-neighbor methods. *Journal of Computational Chemistry, 29*(10), 1605–1614. https://doi.org/10.1002/jcc.20919.

Holton, S., Merckx, A., Burgess, D., Doerig, C., Noble, M., & Endicott, J. (2003). Structures of *P. falciparum* PfPK5 test the CDK regulation paradigm and suggest mechanisms of small molecule inhibition. *Structure, 11*(11), 1329–1337. https://doi.org/10.1016/j.str.2003.09.020.

Imwong, M., Suwannasin, K., Kunasol, C., Sutawong, K., Mayxay, M., Rekol, H., Smithuis, F. M., Hlaing, T. M., Tun, K. M., Pluijm, R. W., Tripura, R., Miotto, O., Menard, D., Dhorda, M., Day, N. P. J., White, N. J., & Dondorp, A. M. (2017). The spread of artemisinin-resistant *Plasmodium falciparum* in the Greater Mekong subregion: A molecular epidemiology observational study. *The Lancet Infectious Diseases, 17*(5), 491–497. https://doi.org/10.1016/S1473-3099(17)30048-8.

Johari, G. P. (2021). Entropy, enthalpy and volume of perfect crystals at limiting high pressure and the third law of thermodynamics. *Thermochimica Acta, 698*, 178891. https://doi.org/10.1016/j.tca.2021.178891.

Juhász, Á., Csapó, E., Ungor, D., Tóth, G. K., Vécsei, L., & Dékány, I. (2016). Kinetic and thermodynamic evaluation of Kynurenic acid binding to GluR1270–300 polypeptide by surface plasmon resonance experiments. *The Journal of Physical Chemistry B, 120*(32), 7844–7850. https://doi.org/10.1021/acs.jpcb.6b05682.

Kassem, S., Ahmed, M., El-Sheikh, S., & Barakat, K. H. (2015). Entropy in bimolecular simulations: A comprehensive review of atomic fluctuations-based methods. *Journal of Molecular Graphics and Modelling, 62*, 105–117. https://doi.org/10.1016/j.jmgm.2015.09.010.

Khan, T., & Ghosh, I. (2015). Modularity in protein structures: Study on all-alpha proteins. *Journal of Biomolecular Structure and Dynamics, 33*(12), 2667–2681. https://doi.org/10.1080/07391102.2014.1003969.

Khan, T., Kumar Panday, S., & Ghosh, I. (2018). ProLego: Tool for extracting and visualizing topological modules in protein structures. *BMC Bioinformatics, 19*(1). https://doi.org/10.1186/s12859-018-2171-9.

King, B. M., Silver, N. W., & Tidor, B. (2012). Efficient calculation of molecular configurational entropies using an information theoretic approximation. *The Journal of Physical Chemistry B, 116*(9), 2891–2904. https://doi.org/10.1021/jp2068123.

King, E., Aitchison, E., Li, H., & Luo, R. (2021). Recent developments in free energy calculations for drug discovery. *Frontiers in Molecular Biosciences, 8*. https://doi.org/10.3389/fmolb.2021.712085.

King, N. M., Prabu-Jeyabalan, M., Bandaranayake, R. M., Nalam, M. N. L., Nalivaika, E. A., Özen, A., Haliloğlu, T., Yılmaz, N. K., & Schiffer, C. A. (2012). Extreme entropy–enthalpy compensation in a drug-resistant variant of HIV-1 protease. *ACS Chemical Biology, 7*(9), 1536–1546. https://doi.org/10.1021/cb300191k.

Krimmer, S. G., & Klebe, G. (2015). Thermodynamics of protein–ligand interactions as a reference for computational analysis: How to assess accuracy, reliability and relevance of experimental data. *Journal of Computer-Aided Molecular Design, 29*(9), 867–883. https://doi.org/10.1007/s10822-015-9867-y.

Kumar, S., & Nussinov, R. (2001). How do thermophilic proteins deal with heat? *Cellular and Molecular Life Sciences, 58*(9), 1216–1233. https://doi.org/10.1007/PL00000935.

Kuril, A. K. (2024). Differential scanning calorimetry: A powerful and versatile tool for analyzing proteins and peptides. *Journal of Pharmaceutical Research International, 36*(7), 179–187. https://doi.org/10.9734/jpri/2024/v36i77549.

Kurzynski, M. (2006). *Thermodynamic state*. Springer.

Laskowski, R. A., MacArthur, M. W., Moss, D. S., & Thornton, J. M. (1993). PROCHECK: A program to check the stereochemical quality of protein structures. *Journal of Applied Crystallography, 26*(2), 283–291. https://doi.org/10.1107/S0021889892009944.

Leavitt, S., & Freire, E. (2001). Direct measurement of protein binding energetics by isothermal titration calorimetry. *Current Opinion in Structural Biology, 11*(5), 560–566. https://doi.org/10.1016/S0959-440X(00)00248-7.

Lebowitz, J. L. (1993). Boltzmann's entropy and time's arrow. *Physics Today, 46*(9), 32–38. https://doi.org/10.1063/1.881363.

Lee, E. F., Yao, S., Sabo, J. K., Douglas Fairlie, W., Stevenson, R. A., Harris, K. S., Anders, R. F., Foley, M., & Norton, R. S. (2011). Peptide inhibitors of the malaria surface protein, apical membrane antigen 1: Identification of key binding residues. *Biopolymers, 95*(5), 354–364. https://doi.org/10.1002/bip.21582.

Le Roch, K., Sestier, C., Dorin, D., Waters, N., Kappes, B., Chakrabarti, D., Meijer, L., & Doerig, C. (2000). Activation of a *Plasmodium falciparum* cdc2-related kinase by heterologous p25 and cyclin H. *Journal of Biological Chemistry, 275*(12), 8952–8958. https://doi.org/10.1074/jbc.275.12.8952.

Levy, R. M., Karplus, M., Kushick, J., & Perahia, D. (1984). Evaluation of the configurational entropy for proteins: Application to molecular dynamics simulations of an α-helix. *Macromolecules, 17*(7), 1370–1374.

Li, F., Dluzewski, A., Coley, A. M., Thomas, A., Tilley, L., Anders, R. F., & Foley, M. (2002). Phage-displayed peptides bind to the malarial protein apical membrane antigen-1 and inhibit the merozoite invasion of host erythrocytes. *Journal of Biological Chemistry, 277*(52), 50303–50310. https://doi.org/10.1074/jbc.M207985200.

Li, L., Li, Y., Zhang, L., & Hou, T. (2012). Theoretical studies on the susceptibility of oseltamivir against variants of 2009 A/H1N1 influenza neuraminidase. *Journal of Chemical Information and Modeling, 52*(10), 2715–2729. https://doi.org/10.1021/ci300375k.

Lipkus, A. H. (1999). A proof of the triangle inequality for the Tanimoto distance. *Journal of Mathematical Chemistry, 26*(1/3), 263–265. https://doi.org/10.1023/A:1019154432472.

Magie, W. F., Carnot, S., Clausius, R., & Kelvin, W. T. B. (1899). *The second law of thermodynamics: Memoirs by Carnot, Clausius, and Thomson*. Harper & Brothers https://books.google.com/books?id=8jMNAAAAYAAJ.

Maier, J. A., Martinez, C., Kasavajhala, K., Wickstrom, L., Hauser, K. E., & Simmerling, C. (2015). ff14SB: Improving the accuracy of protein side chain and backbone parameters from ff99SB. *Journal of Chemical Theory and Computation, 11*(8), 3696–3713. https://doi.org/10.1021/acs.jctc.5b00255.

Mark, P., & Nilsson, L. (2001). Structure and dynamics of the TIP3P, SPC, and SPC/E water models at 298 K. *The Journal of Physical Chemistry A, 105*(43), 9954–9960. https://doi.org/10.1021/jp003020w.

Masukawa, K. M., Kollman, P. A., & Kuntz, I. D. (2003). Investigation of neuraminidase-substrate recognition using molecular dynamics and free energy calculations. *Journal of Medicinal Chemistry, 46*(26), 5628–5637. https://doi.org/10.1021/jm030060q.

Meis, J. F. G. M., & Verhave, J. P. (1988). Exoerythrocytic development of malarial parasites. *Advances in Parasitology, 27*, 1–61. https://doi.org/10.1016/S0065-308X(08)60352-8.

Miller, B. R., McGee, T. D., Jason, M. S., Nadine, H., Holger, G., & Roitberg, A. E. (2012). MMPBSA.py: An efficient program for end-state free energy calculations. *Journal of Chemical Theory and Computation, 8*(9), 3314–3321. https://doi.org/10.1021/ct300418h.

Miller, L. H., Ackerman, H. C., Su, X.-Z., & Wellems, T. E. (2013). Malaria biology and disease pathogenesis: Insights for new treatments. *Nature Medicine, 19*(2), 156–167. https://doi.org/10.1038/nm.3073.

Mita, T., Tanabe, K., & Kita, K. (2009). Spread and evolution of *Plasmodium falciparum* drug resistance. *Parasitology International, 58*(3), 201–209. https://doi.org/10.1016/j.parint.2009.04.004.

Müller, I. (2003). Entropy: A subtle concept in thermodynamics. *Entropy, 47*, 19.

Muregi, F. (2012). Novel drug targets in malaria parasite with potential to yield antimalarial drugs with long useful therapeutic lives. *Current Pharmaceutical Design*. https://doi.org/10.2174/138161212801327239.

Núñez, S., Venhorst, J., & Kruse, C. G. (2012). Target–drug interactions: First principles and their application to drug discovery. *Drug Discovery Today, 17*(1–2), 10–22. https://doi.org/10.1016/j.drudis.2011.06.013.

Ohtaka, H., Velázquez-Campoy, A., Xie, D., & Freire, E. (2002). Overcoming drug resistance in HIV-1 chemotherapy: The binding thermodynamics of amprenavir and TMC-126 to wild-type and drug-resistant mutants of the HIV-1 protease. *Protein Science, 11*(8), 1908–1916. https://doi.org/10.1110/ps.0206402.

Olsson, M. H. M., Søndergaard, C. R., Rostkowski, M., & Jensen, J. H. (2011a). PROPKA3: Consistent treatment of internal and surface residues in empirical p*K*a predictions. *Journal of Chemical Theory and Computation, 7*(2), 525–537. https://doi.org/10.1021/ct100578z.

Olsson, T. S. G., Ladbury, J. E., Pitt, W. R., & Williams, M. A. (2011b). Extent of enthalpy–entropy compensation in protein–ligand interactions. *Protein Science, 20*(9), 1607–1618. https://doi.org/10.1002/pro.692.

Otto, F., Yang, Y., Bei, H., & George, E. P. (2013). Relative effects of enthalpy and entropy on the phase stability of equiatomic high-entropy alloys. *Acta Materialia, 61*(7), 2628–2638. https://doi.org/10.1016/j.actamat.2013.01.042.

Packard, R. M. (2014). The origins of antimalarial-drug resistance. *New England Journal of Medicine, 371*(5), 397–399. https://doi.org/10.1056/NEJMp1403340.

Paketurytė, V., Linkuvienė, V., Krainer, G., Chen, W.-Y., & Matulis, D. (2018). Repeatability, precision, and accuracy of the enthalpies and Gibbs energies of a protein–ligand binding reaction measured by isothermal titration calorimetry. *European Biophysics Journal, 48*(2), 139–152. https://doi.org/10.1007/s00249-018-1341-z.

Panday, S. K. (2018). Study of Ligands-Receptors Interactions to Estimate Components of Binding Free Energy. PhD dissertation, Jawaharlal Nehru University.

Panday, S. K., & Alexov, E. (2022). protein–protein binding free energy predictions with the MM/PBSA approach complemented with the Gaussian-based method for entropy estimation. *ACS Omega, 7*(13), 11057–11067. https://doi.org/10.1021/acsomega.1c07037.

Panday, S. K., & Ghosh, I. (2019). In silico structure-based prediction of receptor–ligand binding affinity: Current progress and challenges. *Structural bioinformatics: Applications in preclinical drug discovery process* (pp. 109–175). Springer International Publishing. https://doi.org/10.1007/978-3-030-05282-9_5.

Panday, S. K., & Ghosh, I. (2020). Application and comprehensive analysis of neighbor approximated information theoretic configurational entropy methods to protein–ligand binding cases. *Journal of Chemical Theory and Computation, 16*(12), 7581–7600. https://doi.org/10.1021/acs.jctc.0c00764.

Panday, S. K., Sturlese, M., Salmaso, V., Ghosh, I., & Moro, S. (2019). Coupling supervised molecular dynamics (SuMD) with entropy estimations to shine light on the stability of multiple binding sites. *ACS Medicinal Chemistry Letters, 10*(4), 444–449. https://doi.org/10.1021/acsmedchemlett.8b00490.

Perozzo, R., Folkers, G., & Scapozza, L. (2004). Thermodynamics of protein–ligand interactions: History, presence, and future aspects. *Journal of Receptors and Signal Transduction, 24*(1–2), 1–52. https://doi.org/10.1081/RRS-120037896.

Pokorná, J., Pachl, P., Karlukova, E., Hejdánek, J., Řezáčová, P., Machara, A., Hudlický, J., Konvalinka, J., & Kožíšek, M. (2018). Kinetic, thermodynamic, and structural analysis of drug resistance mutations in neuraminidase from the 2009 pandemic influenza virus. *Viruses, 10*(7), 339. https://doi.org/10.3390/v10070339.

Privalov, P. L., & Crane-Robinson, C. (2016). Role of water in the formation of macromolecular structures. *European Biophysics Journal, 46*(3), 203–224. https://doi.org/10.1007/s00249-016-1161-y.

Raudino, A., Sarpietro, M. G., & Pannuzzo, M. (2013). Differential scanning calorimetry (DSC): Theoretical fundamentals. *Drug-biomembrane interaction studies* (pp. 127–168). Elsevier. https://doi.org/10.1533/9781908818348.127.

Reif, B., Ashbrook, S. E., Emsley, L., & Hong, M. (2021). Solid-state NMR spectroscopy. *Nature Reviews Methods Primers, 1*(1). https://doi.org/10.1038/s43586-020-00002-1.

Riglar, D. T., Richard, D., Wilson, D. W., Boyle, M. J., Dekiwadia, C., Turnbull, L., Angrisano, F., Marapana, D. S., Rogers, K. L., Whitchurch, C. B., Beeson, J. G., Cowman, A. F., Ralph, S. A., & Baum, J. (2011). Super-resolution dissection of coordinated events during malaria parasite invasion of the human erythrocyte. *Cell Host & Microbe, 9*(1), 9–20. https://doi.org/10.1016/j.chom.2010.12.003.

Šali, A., Potterton, L., Yuan, F., van Vlijmen, H., & Karplus, M. (1995). Evaluation of comparative protein modeling by MODELLER. *Proteins: Structure, Function, and Bioinformatics, 23*(3), 318–326. https://doi.org/10.1002/prot.340230306.

Schrödinger, E. (2025). *What is life? The physical aspect of the living cell.* Rare Treasure Editions https://books.google.com/books?id=ZIRQEQAAQBAJ.

Sepúlveda-Fontaine, S. A., & Amigó, J. M. (2024). Applications of entropy in data analysis and machine learning: A review. *Entropy, 26*(12), 1126. https://doi.org/10.3390/e26121126.

Sethna, J. P. (2021). *Statistical mechanics: Entropy, order parameters, and complexity*: 14. Oxford University Press.

Shannon, C. E. (1948). A mathematical theory of communication. *Bell System Technical Journal, 27*(3), 379–423. https://doi.org/10.1002/j.1538-7305.1948.tb01338.x.

Shapovalov, M. V., & Dunbrack, R. L. (2011). A smoothed backbone-dependent rotamer library for proteins derived from adaptive kernel density estimates and regressions. *Structure, 19*(6), 844–858. https://doi.org/10.1016/j.str.2011.03.019.

Sharma, P., Sharma, M., Amin, N. D., Wayne Albers, R., & Pant, H. C. (1999). Regulation of cyclin-dependent kinase 5 catalytic activity by phosphorylation. *Proceedings of the National Academy of Sciences, 96*(20), 11156–11160. https://doi.org/10.1073/pnas.96.20.11156.

Shell, M. S. (2015). *Thermodynamics and statistical mechanics: An integrated approach.* Cambridge University Press.

Singh, A., & Soliman, M. E. (2015). Understanding the cross-resistance of oseltamivir to H1N1 and H5N1 influenza A neuraminidase mutations using multidimensional computational analyses. *Drug Design, Development and Therapy*, 4137. https://doi.org/10.2147/DDDT.S81934.

Singh, H., Misra, N., Hnizdo, V., Fedorowicz, A., & Demchuk, E. (2003). Nearest neighbor estimates of entropy. *American Journal of Mathematical and Management Sciences, 23*(3–4), 301–321. https://doi.org/10.1080/01966324.2003.10737616

Sprenger, K. G., Jaeger, V. W., & Pfaendtner, J. (2015). The General AMBER Force Field (GAFF) can accurately predict thermodynamic and transport properties of many ionic liquids. *The Journal of Physical Chemistry B, 119*(18), 5882–5895. https://doi.org/10.1021/acs.jpcb.5b00689.

Srinivasan, P., Beatty, W. L., Diouf, A., Herrera, R., Ambroggio, X., Kathleen Moch, J., Tyler, J. S., Narum, D. L., Pierce, S. K., Boothroyd, J. C., David Haynes, J., & Miller, L. H. (2011). Binding of *Plasmodium* merozoite proteins RON2 and AMA1 triggers commitment to invasion. *Proceedings of the National Academy of Sciences, 108*(32), 13275–13280. https://doi.org/10.1073/pnas.1110303108.

Stølen, S., & Grande, T. (2004). *Chemical thermodynamics of materials: Macroscopic and microscopic aspects.* John Wiley & Sons.

Suárez, D., & Díaz, N. (2014). Direct methods for computing single-molecule entropies from molecular simulations. *WIREs Computational Molecular Science, 5*(1), 1–26. https://doi.org/10.1002/wcms.1195.

Tan, J., Piccoli, L., & Lanzavecchia, A. (2019). The antibody response to *Plasmodium falciparum*: Cues for vaccine design and the discovery of receptor-based antibodies. *Annual Review of Immunology, 37*(1), 225–246. https://doi.org/10.1146/annurev-immunol-042617-053301.

Tang, D., Chun, A. C. S., Zhang, M., & Wang, J. H. (1997). Cyclin-dependent kinase 5 (Cdk5) activation domain of neuronal Cdk5 activator. Evidence of the existence of cyclin fold in neuronal Cdk5a activator. *Journal of Biological Chemistry, 272*(19), 12318–12327. https://doi.org/10.1074/jbc.272.19.12318.

Timasheff, S. N. (1998). In disperse solution, "osmotic stress" is a restricted case of preferential interactions. *Proceedings of the National Academy of Sciences, 95*(13), 7363–7367. https://doi.org/10.1073/pnas.95.13.7363.

Tolman, R. C. (1931). On the problem of the entropy of the universe as a whole. *Physical Review, 37*(12), 1639–1660. https://doi.org/10.1103/PhysRev.37.1639.

Tse, H. T. (2022). The definition and application of entropy. *SHS Web of Conferences, 144*, 1016. https://doi.org/10.1051/shsconf/202214401016.

Varani, K., Gessi, S., Merighi, S., & Borea, P. A. (2015). van't Hoff based thermodynamics. *Thermodynamics and kinetics of drug binding* (pp. 15–35). Wiley. https://doi.org/10.1002/9783527673025.ch2.

Velazquez-Campoy, A., Leavitt, S. A., & Freire, E. (2004). Characterization of protein–protein interactions by isothermal titration calorimetry. In H. Fu (Ed.), *Protein-protein interactions. Methods in molecular biology* (261, pp. 35–54). Humana Press. https://doi.org/10.1385/1-59259-762-9:035.

Verdonk, M. L., Cole, J. C., Hartshorn, M. J., Murray, C. W., & Taylor, R. D. (2003). Improved protein–ligand docking using GOLD. *Proteins: Structure, Function, and Bioinformatics, 52*(4), 609–623. https://doi.org/10.1002/prot.10465.

Vulliez-Le Normand, B., Tonkin, M. L., Lamarque, M. H., Langer, S., Hoos, S., Roques, M., Saul, F. A., Faber, B. W., Bentley, G. A., Boulanger, M. J., & Lebrun, M. (2012). Structural and functional insights into the malaria parasite moving junction complex. *PLoS Pathogens, 8*(6), e1002755. https://doi.org/10.1371/journal.ppat.1002755.

Wang, S., Zhang, T.-H., Hu, M., Tang, K., Sheng, L., Hong, M., Chen, D., Chen, L., Shi, Y., Feng, J., Qian, J., Sun, L., Ding, K., Sun, R., & Du, Y. (2023). Deep mutational scanning of influenza A virus neuraminidase facilitates the identification of drug resistance mutations in vivo. *mSystems, 8*(5). https://doi.org/10.1128/msystems.00670-23.

Waters, A. P., Thomas, A. W., Deans, J. A., Mitchell, G. H., Hudson, D. E., Miller, L. H., F McCutchan, T., & Cohen, S. (1990). A merozoite receptor protein from *Plasmodium knowlesi* is highly conserved and distributed throughout *Plasmodium*. *Journal of Biological Chemistry, 265*(29), 17974–17979. https://doi.org/10.1016/S0021-9258(18)38259-0.

Wei, G., Xi, W., Nussinov, R., & Ma, B. (2016). Protein ensembles: How does nature harness thermodynamic fluctuations for life? The diverse functional roles of conformational ensembles in the cell. *Chemical Reviews, 116*(11), 6516–6551. https://doi.org/10.1021/acs.chemrev.5b00562.

WHO. (2023). *World malaria report*. World Health Organization https://books.google.com/books?id=u6UOEQAAQBAJ.

Wongsrichanalai, C., Pickard, A. L., Wernsdorfer, W. H., & Meshnick, S. R. (2002). Epidemiology of drug-resistant malaria. *The Lancet Infectious Diseases, 2*(4), 209–218. https://doi.org/10.1016/S1473-3099(02)00239-6.

Yoneya, M., Berendsen, H. J. C., & Hirasawa, K. (1994). A non-iterative matrix method for constraint molecular dynamics simulations. *Molecular Simulation, 13*(6), 395–405. https://doi.org/10.1080/08927029408022001.

Yu, Y. B., Privalov, P. L., & Hodges, R. S. (2001). Contribution of translational and rotational motions to molecular association in aqueous solution. *Biophysical Journal, 81*(3), 1632–1642. https://doi.org/10.1016/S0006-3495(01)75817-1.

Zhou, H.-X., & Gilson, M. K. (2009). Theory of free energy and entropy in noncovalent binding. *Chemical Reviews, 109*(9), 4092–4107. https://doi.org/10.1021/cr800551w.

CHAPTER

Water dynamics and thermodynamics in the biological and interfacial contexts

4

Sarbajit Layek and Neelanjana Sengupta*

Department of Biological Sciences, Indian Institute of Science Education and Research (IISER) Kolkata, Mohanpur, Nadia, West Bengal, India

**Corresponding author: n.sengupta@iiserkol.ac.in*

4.1 Introduction

Water is integral to structural and functional viability of biological systems. It serves as a medium for underlying interactions and influences the stabilities of functional assemblies. At biological interfaces, its dynamic and thermodynamic properties become crucial in modulating processes such as protein folding, membrane stability, and molecular recognition, where water structuring and hydrogen bonding play decisive roles (Dargaville & Hutmacher, 2022; Sobhia et al., 2022). The altered density, viscosity, and orientation of interfacial water molecules significantly affect biochemical reactions and signal transduction (Chaplin, 2006; Mondal et al., 2023). A comprehensive understanding of the behavior of water in these environments is essential for elucidating the mechanisms that govern biomolecular function and stability (Fogarty & Laage, 2014; Ghanta et al., 2022; Schotte et al., 2012). This behavior is largely dictated by the polar nature of water, which underpins its unique chemical and physical properties (Brini et al., 2017; Giovambattista et al., 2007). The polarity arises from the significant difference in electronegativity between the hydrogen and oxygen atoms, resulting in strong dipole interactions. These interactions drive the formation of hydrogen bonds (H-bonds), facilitating their extensive networks that underpin the distinctive physicochemical properties (Dargaville & Hutmacher, 2022). Extensive studies, utilizing both computational and experimental approaches, have been conducted to explore the behavior of water. Nonetheless, our understanding is still not fully comprehensive, and the field remains highly active area for further studies. Most studies focus on elucidating the distinctive structural and dynamic properties of water that differentiate it from other molecules (Morawietz et al., 2016; Pettersson et al., 2016). These include its exceptionally high boiling point, anomalous maximum density at 4°C, lower density of ice compared to liquid water, and its nonmonotonic viscosity under pressure (Gallo et al., 2016; Leoni et al., 2021; Pettersson et al., 2016). In addition, unusual high specific heat capacity, strong surface tension, and high dielectric constant of water are crucial in regulating biological functions and interfacial phenomena (Errington, 2001; Altabet et al., 2017; Krott & Barbosa, 2024; Mallamace et al., 2013). These anomalies collectively enable water to stabilize macromolecules, mediate biochemical reactions, and influence interfacial properties in biological systems.

The molecular arrangement of water is highly dynamic, fluctuating continuously because of the thermal excitation of low-frequency modes. A key interaction in this context is the H-bond (Arunan et al., 2011; Chandler, 2005), which forms among water molecules and between water and the polar or ionic

Protein Dynamics: Experimental and Computational Approaches. DOI: https://doi.org/10.1016/B978-0-443-30042-4.00012-2

group resides in biomolecules (Lindow et al., 2013). The structural dynamics of water span multiple time scales (Laage et al., 2017a). At the shortest time scales, intramolecular OH bond stretching and bending vibrations occur within 10 and 20 femtoseconds (Heyden, 2014). On a scale of tens to hundreds of femtoseconds, librational (hindered rotational) motions are observed, initially localized to a few molecules and then delocalized over larger groups. H-bond stretching and bending vibration occur over 200–800 femtoseconds, driving rapid structural reorganization (Asbury et al., 2004; Cowan et al., 2005; Fecko et al., 2003; Kraemer et al., 2008; Møller et al., 2004). A particularly significant process in water is the exchange of H-bond partners, which involves the breaking and reformation of bonds. In bulk water, this process occurs within a time frame of 1–5 picoseconds. At the molecular level, this behavior is explained by the jump mechanism, which involves the reorientation of water molecules primarily through sudden, large-amplitude angular jumps (Das et al., 2024, 2015; Laage & Hynes, 2006; Laage et al., 2012). This occurs when the hydroxyl (OH) group of water exchanges H-bond acceptors, a process facilitated by the formation of transient five-coordinated water defects (Laage & Hynes, 2006; Laage et al., 2011).

These unique properties of water are essential for life and influence the behavior of biomolecules such as proteins, nucleic acids, and membranes. Exploring biomolecular hydration is an active research area that incorporates various experimental techniques, such as X-ray and neutron scattering, solution and solid-state nuclear magnetic resonance (NMR), and ultrafast infrared (IR) spectroscopy, along with electronic and vibrational spectroscopy (Blaffert et al., 2018; Schotte et al., 2012). Complementing these experimental approaches, molecular simulations provide crucial insights, especially when experiments alone are insufficient to fully capture biomolecular hydration (Nibali & Havenith, 2014). The concept of a hydration shell, which has long been used for simple solutes, is now extended to biomolecular ensembles such as membranes (Debnath et al., 2010; Kuntz & Kauzmann, 1974; Wyttenbach & Bowers, 2009). This shell, consisting of the first few layers of water around a biomolecule, exhibits distinct structural and dynamic properties compared to bulk water. However, quantitatively defining this hydration layer remains a challenge, as its characteristics depend on the specific property being examined. The distinction between hydration shell water and bulk water, both in structure and dynamics, remains a complex issue requiring the integration of structure-sensitive and time-resolved spectroscopic and computational techniques for detailed analysis. Ongoing debates persist regarding the extent to which water structure and dynamics are altered near a biomolecule compared to bulk water. At the biomolecule–water interface, interfacial water exhibits pronounced anisotropy, with restricted mobility and directionally dependent dynamics governed by surface interactions, confinement, and chemical heterogeneity (Nandi & Bagchi, 1997; Shafiei et al., 2019; Xiao et al., 2016). A comprehensive understanding of these interfacial properties is essential for elucidating molecular recognition, protein stability, and enzymatic activity, as well as for optimizing artificial interfaces in biomimetic systems.

Furthermore the concept of molecular crowding, prevalent in biological systems, imposes additional complexity on water dynamics. Crowded environments, which are often associated with soft matter systems, significantly affect the movement and structuring of interfacial water (Fuentes-Lemus & Davies, 2023; Harada et al., 2012; Nakano & Sugimoto, 2017). In confined spaces, such as those within biological macromolecules, biomolecular condensates are formed via liquid–liquid phase separation (LLPS) and have been implicated in neurodegenerative disorders such as Alzheimer's and Parkinson's diseases (Pakravan et al., 2021; Wegmann et al., 2018). Molecular dynamics (MDs) simulations reveal an entropy tug-of-war between two distinct classes of hydration water during condensate formation (Ahlers et al., 2021; Mukherjee & Schäfer, 2023; Mukherjee et al., 2024). The release of cavity-wrap

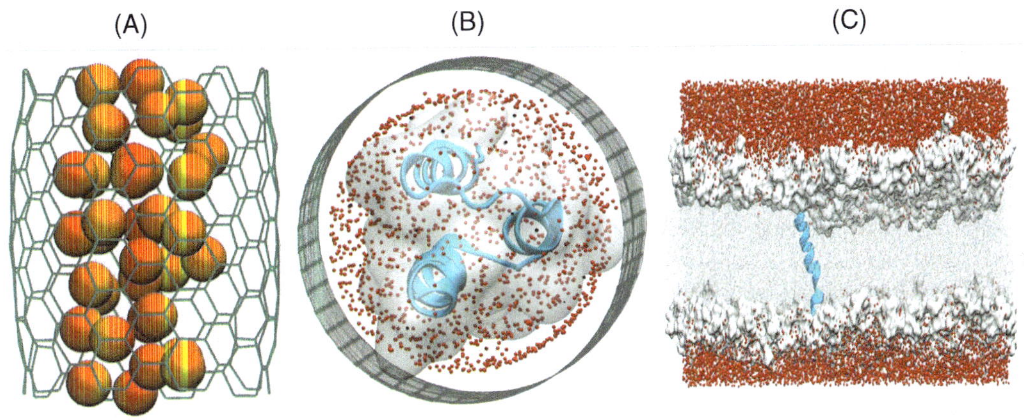

FIGURE 4.1

Schematic representation of interfacial water environments: (A) Single-channel water arrangement within a SWCNT. (B) Hydration layer at the protein–water interface confined inside a SWCNT. (C) Structured interfacial region at the lipid membrane-water boundary.

water from hydrophobic regions increases system entropy, whereas the retention of bound water (BW) in hydrophilic regions is enthalpically favored because of enhanced restriction in the dense protein environment. This interplay of water release and retention governs the thermodynamics of biomolecular phase separation, demonstrating how spatial constraints modulate hydration thermodynamics. In nanoscale spatial confinement, comparable alterations in water structure and dynamics arise, further highlighting the influence of restricted environments on hydration behavior. The phenomenon of water confinement has been extensively studied through both computational and experimental methods. MDs' studies, such as those conducted by Hummer et al. (Hummer, 2001) along with experimental observations, have reported the spontaneous entry of water molecules into hydrophobic confinements, most notably within carbon nanotubes (CNTs). These studies revealed that confined water molecules exhibit greater rotational entropy compared to bulk water, with the effect being most pronounced in single-walled CNTs (SWCNTs) of small diameters, particularly chirality (5,5). Both energy and entropy favor water entry into these hydrophobic confinements (Kumar et al., 2014; Roy & Sengupta, 2021). Upon confinement the intermolecular structure of water deviates significantly from that observed in bulk, indicating substantial modifications to the H-bond network (Joseph & Aluru, 2008; Layek & Sengupta, 2024). This structural rearrangements are closely linked to the rapid transport of water molecules through SWCNTs, where confinement induces the formation of a single chain-like arrangement (see Fig. 4.1A) of water molecules (Chakraborty et al., 2017; Ghosh et al., 2023a; Roy et al., 2019; Su & Guo, 2011; Takaiwa et al., 2008). Understanding these properties is crucial for elucidating their role in biological processes, while also offering insights into material science applications, including nanofluidics, drug delivery, and the design of bioinspired materials.

Building on these insights, this review aims to provide a comprehensive overview of the behavior of water at biomolecular interfaces, integrating findings from both experimental and computational studies. By examining the structural, thermodynamic, and dynamic properties of interfacial water, we seek to elucidate its role in mediating key biological processes such as protein folding, molecular recognition, and membrane stability. Our goal is to highlight the current understanding, address existing challenges, and propose future directions in this rapidly evolving field. Through this review, we aspire to contribute to a deeper understanding of interfacial water, which holds profound implications for both fundamental biophysics and the development of novel therapeutic strategies. This article is organized as follows: Section 4.2 emphasizes the structural features of water based on its hydration shell around biomolecules, order parameters, and spatial anisotropy. Section 4.3 enlightens the thermodynamics of water, whereas Section 4.4 covers the heterogenous water dynamics at the interface, including the importance of the extended jump model (EJM). Section 4.5 explores the behavior of water in molecular crowding environments and LLPS. Each section integrates theoretical frameworks with experimental findings to provide a holistic understanding. Lastly, conclusions are presented in Section 4.6.

4.2 Structural features of water

4.2.1 Importance of hydration shell

To explain the distinct properties of water adjacent to biomolecules the concept of a hydration shell was explored (Bubon et al., 2023; Konstantinovsky et al., 2022; Morón, 2021). However, this is preceded by the concept of biological water by Bagchi, Zewail and co-workers (Pal et al., 2002). The hydration shell, whose physical characteristics differ markedly from bulk water, consists of a few layers of water surrounding the biomolecule, either interacting directly with it or being significantly influenced by its presence (see Fig. 4.1B and C). In proteins the first hydration shell typically extends up to 4 Å, while around DNA, it ranges from 3 to 5 Å but can extend up to 10 Å, with tightly BW interacting with the phosphate backbone and bases through hydrogen bonding (Laage et al., 2017a). Similarly, in phospholipid bilayers, the first hydration shell spans 3–6 Å, where water molecules closely interact with the polar head groups of the lipids (Fogarty & Laage, 2014; Linse & Hub, 2023; Nagle & Tristram-Nagle, 2000).

As hydration water governs biomolecular stability and function, distinguishing it from bulk water remains a considerable challenge. This requires employing structure-sensitive, time-resolved spectroscopic, and computational approaches to obtain quantitative insights. The hydration shell is typically defined from a perturbative perspective, where water molecules display altered dynamics compared to bulk water. MD simulations often use geometric criteria to define this shell. A typical method involves selecting water molecules whose oxygen atoms are within a defined maximum distance, usually around 4 Å, from the closest atom of the biomolecule (Ghosh et al., 2023b; Rudling et al., 2018). This distance usually corresponds to the first hydration layer. The shell thickness can also be determined by locating the first minimum using radial distribution function, which represents the likelihood of finding a particle at a specific distance from a reference particle (Duboué-Dijon & Laage, 2014; Duboué-Dijon et al., 2016; Fogarty & Laage, 2014; Stirnemann et al., 2013). MD simulations enable the calculation of various dynamical properties, such as rotational, translational, and hydrogen-bond dynamics, as well as residence times, which quantify how long a water molecule remains in the shell before diffusing into the bulk. These methods have been widely applied to study hydration dynamics around proteins, DNA, and phospholipids (Fogarty & Laage, 2014; Ghosh et al., 2023b; Gruenbaum & Skinner, 2011; Halle & Nilsson, 2009; Pasenkiewicz-Gierula et al., 1997; Patra et al., 2003; Rossky & Karplus, 1979; Schneider et al., 1998; Sinha & Bandyopadhyay, 2011).

Various experimental techniques provide complementary insights into biomolecular hydration shell dynamics. Dielectric relaxation and NMR spectroscopy have been employed to probe hydration dynamics around biomolecules, whereas ultrafast time-dependent Stokes-shift measurements have enabled time-resolved studies of water interactions. Furthermore, diffraction analysis of crystallized samples has demonstrated hydration architectures around phospholipid membranes, reverse micelles, and protein surfaces as well as major and minor grooves of DNA (Abel et al., 2004; Kermani, 2021; Martin & Matyushov, 2015; Zaki et al., 2022). These studies reveal how biomolecular topology dictates the arrangement of water molecules within this shell. For example, phosphate and choline groups in phospholipids, the distinct hydration structures in the major and minor grooves of DNA (with a rigid "spine of water" in the minor groove), and the heterogeneous hydration patterns of proteins illustrate these specific arrangements (Edwards et al., 1992; Finer-Moore et al., 1992; Foglia et al., 2010; Pearson & Pascher, 1979; Vlieghe et al., 1999). Small-angle X-ray scattering (SAXS) and small-angle neutron scattering provide insights into structural features over nanometer scales (Svergun et al., 1998). SAXS has also been applied to study counterion distributions around DNA (Das et al., 2003; Meisburger et al., 2015; Pabit et al., 2010). More recently, terahertz (THz) spectroscopy has emerged as a powerful tool to probe the collective hydrogen bond network vibrations of water without requiring artificial probes or chromophores (Dexheimer, 2008; Markelz et al., 2000; Novelli et al., 2023; Pyne et al., 2021; Walther et al., 2002). Studies using THz spectroscopy have observed broad absorption features in hydrated biomolecules, such as DNA, bovine serum albumin, and collagen, revealing low-frequency collective modes that have been validated using normal mode analysis (Bhattacharya et al., 2024; Konstantinovsky et al., 2022; Saha & Mitra, 2023; Sobhia et al., 2022; Walther et al., 2002).

A comprehensive understanding of hydration dynamics must encompass the heterogeneity at the biomolecule–water interface. Water actively participates in biomolecular dynamics through both direct short-range interactions in the first hydration shell and long-range interactions mediated by extended H-bond networks. These coupled interactions can facilitate critical biological processes such as the water-mediated induced-fit mechanism in enzyme catalysis. As advanced techniques continue to evolve, they promise to broaden our understanding of biomolecular hydration in complex environments, such as under extreme conditions, at interfaces, and in confined or crowded molecular assemblies.

4.2.2 Water hydrogen-bond network and order parameters

The sp^3-hybridization of oxygen governs the H-bond network of water in both solid and liquid phases. Each water molecule can form up to four H bonds, leading to a tetrahedral geometry (Fransson & Pettersson, 2024; Roberts et al., 2009; Zupančič & Grdadolnik, 2021). In ice, this results in a well-ordered structure (You & Baiz, 2022) (see Fig. 4.2), with each water molecule bonded to four neighbors at an angle of 109.5 degrees (Myneni et al., 2002). In contrast, liquid water retains only partial tetrahedral order, with its H-bond network undergoing continuous dynamic rearrangements. These fluctuations lead to transient "ice-like" configurations, contributing to water's unique properties. Because of the complexity of this geometry, various order parameters have been developed to quantify the local structure, especially in the context of molecular simulations. One widely used metric is the orientational tetrahedral order parameter (q), introduced by Errington and Debenedetti to describe supercooled water (Errington, 2001). It calculates the angles between a central water molecule and its four nearest neighbors, yielding a value of one for perfect tetrahedral ice and zero for a random configuration

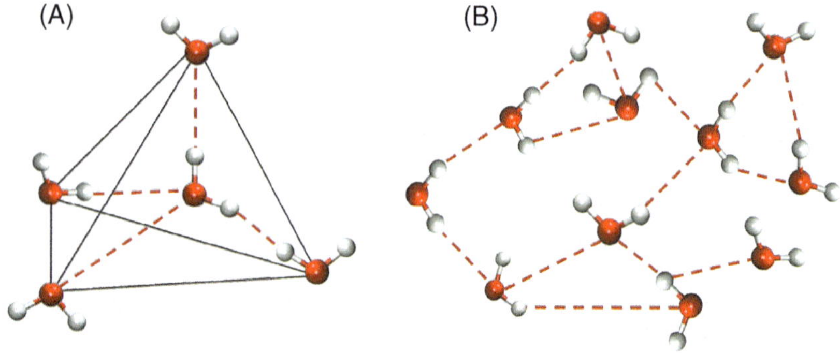

FIGURE 4.2

Diagrams illustrating H-bond arrangements: (A) The tetrahedral arrangement of water H-bond network and (B) The extended H-bond network of liquid water.

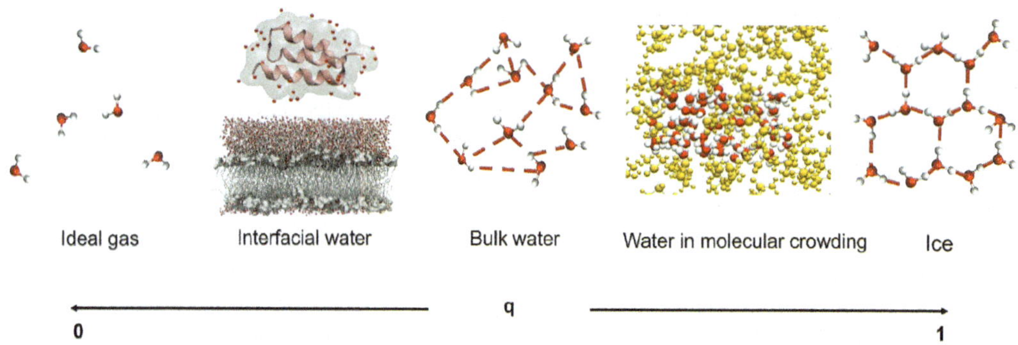

FIGURE 4.3

Schematic view of H-bond network organization in different systems: The tetrahedral order parameter (q) quantifies angular distortions in water, ranging from zero to one in different scenarios.

(e.g., an ideal gas; see Fig. 4.3). The tetrahedral order parameter is defined as

$$q = 1 - \frac{3}{8}\sum_{j=1}^{3}\sum_{k=j+1}^{4}\left(\cos\psi_{jk} + \frac{1}{3}\right)^2 \tag{4.1}$$

where ψ_{jk} is the angle formed by the lines joining the central water oxygen atom and two of its four nearest neighbors j and k. The formulation allows for a quantitative measure of the angular tetrahedrality of local water structure. The q parameter effectively characterizes the H-bond network structure and the impact of cosolvents on water's tetrahedral arrangement (Adhikari & Mondal, 2023; Neophytou et al., 2022; Zimmermann & Jain, 2020). This network spans from a highly ordered state, as observed in ice, to complete randomness akin to an ideal gas. Interestingly, in crowded environments, like in the presence of polyethylene glycol (PEG), water can display greater structural ordering than bulk water. However, near the biomolecular interface, it often shows reduced tetrahedral order because of the imposition of constraints by the local environment.

The presence of biomolecules disrupts the extended H-bond networks of water because of electrostatic interactions and an imbalance of donors and acceptors (Harada et al., 2012; Persson & Halle, 2008; Saenger, 1987). MD simulations reveal that H-bond configurations in heterogeneous aqueous solutions vary locally. Typically, high-concentration solutions exhibit more disordered or unstructured water arrangements compared to pure water (Oh & Baiz, 2018; Oh et al., 2020; You et al., 2021). Interestingly, certain crowders promote long-range ordering, resulting in more structured water beyond the first two to three solvation shells (~1 nm) compared to bulk water (Myneni et al., 2002; Pal & Mitra, 2022). The broader implications of this unexpected ordering effect are still not fully understood.

4.2.3 Spatial anisotropy

Water at interfacial regions exhibits pronounced spatial anisotropy, where its structural and dynamical properties vary along different directions relative to the surface (Nandi et al., 2000; Shafiei et al., 2019; Xiao et al., 2016). Unlike bulk water, where molecular motion is isotropic, interfacial water is restricted by surface roughness, interactions, and confinement, leading to directionally dependent diffusion and relaxation dynamics. In rigid interfaces and sufficiently low temperatures, water primarily undergoes large-angle jumps with mild anisotropy, where the imposed energy landscape limits cooperative rearrangements. In weakly hydrated proteins or narrow confinements, where all water remains interfacial, their dynamics follow the Arrhenius law with an activation energy of ~0.5 eV, but deviations emerge with increasing matrix flexibility, particularly during the glass transition (Cerveny et al., 2004, 2008; Gainaru et al., 2009).

The temperature dependence of these interfacial dynamics can be described by the Arrhenius-type relation (Gainaru et al., 2009):

$$\tau = (2\pi\nu_{max})^{-1} = \nu_0^{-1}\exp\left(\frac{E_a}{T}\right) \tag{4.2}$$

where τ is the relaxation time, ν_{max} is the characteristic peak frequency obtained from spectroscopic measurements, ν_0 is the preexponential factor, E_a is the activation energy, and T is the corresponding temperature. This relationship has been leveraged to highlight how the cooperative rearrangement of hydration water becomes increasingly suppressed under spatial confinement and at lower temperatures, leading to slower reorientation dynamics. In moderately hydrated proteins or completely confined nanoporous environments, water shows a complex dynamical signature. As temperature increases, water transitions from an interface-dominated to a bulk-like dynamical state. Conversely, upon cooling, a fraction of water solidifies at the confinement center, driving the remaining liquid into a fully anisotropic regime, thereby ceasing bulk-like relaxation, respectively. This transition holds biological relevance, as it directly modulates protein flexibility, leading to restricted backbone motions. The attenuation of power-law correlations disappears near 170–190 K, coinciding with the glass transition of interfacial water (Capaccioli et al., 2011; Ngai et al., 2010).

At the protein–water interface, hydration layers exhibit distinct structural and dynamical characteristics compared to bulk water (Heyden, 2019; Ni et al., 2025; Sengupta et al., 2008). MD simulations reveal that hydration water responds anisotropically to applied electric fields (Gekle & Netz, 2012; Nibali et al., 2023). When the electric field is applied parallel to the surface, energy absorption resembles that of bulk water, with a slight shift to lower frequencies (redshift) (Bopp et al., 1998; Chandra & Bagchi, 1989). Conversely a perpendicular field induces weaker absorption and shifts to higher frequencies (blueshift). This anisotropic dielectric response influences molecular processes such as protein folding and function. In addition, near the protein surface, water molecules tend to adopt specific orientations, forming a structured layer that ranges from 10 to 15 Å, influencing the protein energy landscape (Ghosh et al., 2014; Heyden et al., 2012; Martin & Matyushov, 2015). These orientations impose energetic constraints, making molecular reorientation slower compared to bulk water. Within ~10 Å of the protein, hydration water exhibits altered hydrogen bonding and reduced mobility because of pronounced intermolecular forces at the interface. Beyond this region, hydrogen bonding and mobility resemble bulk-like characteristics (Heyden, 2019). The spatial heterogeneity of hydration water modulates protein surface properties, driven by solvation thermodynamics (Chong & Ham, 2015; Timasheff, 2002). When a ligand binds to the protein surface, it induces some thermodynamic changes in the hydration environment at the binding site (Päslack et al., 2021). The protein surface retains tightly BW molecules, known as BW, which are released into the surrounding environment upon ligand binding. This release increases the overall system entropy, compensating for the entropy loss associated with protein–ligand complex formation. But there are few water molecules present on the protein surface, which are comparatively stabilized by the interface-dominated interactions, and they exhibit slower dynamics. Their removal incurs an enthalpic cost, but once liberated into bulk water, their enhanced mobility contributes to entropy gain, and they are classified as highly diffusive water (Mattea et al., 2008; Qin et al., 2017). Overhauser dynamic nuclear polarization provides site-specific insights into hydration water dynamics on nanosecond to picosecond timescales, revealing that BW persists for tens of nanoseconds, whereas DW fluctuates within 1 picoseconds to hundreds of picoseconds (Abhyankar & Szalai, 2021; Brown et al., 2024; Mahanta et al., 2024). Literature suggests that hydration water retardation near protein surfaces is primarily governed by excluded volume effects as well as hydrogen bond formation. Folpini et al. (2017) found that the rate of hydrogen bond breaking and reformation between water and the protein surface depends on the strength of the hydrogen bond. Stronger hydrogen bonds lead to slower exchange dynamics, prolonging water residence time at the interface (Barnes et al., 2017; Mondal & Bandyopadhyay, 2019).

Lipid membranes present natural biological interfaces, not only serving as a barrier between cellular compartments but also mediating hydrophilic and hydrophobic interactions, facilitating protein folding, self-assembly, and functional regulation. Ultrafast vibrational spectroscopy, including two-dimensional IR (2D IR) and vibrational sum-frequency generation, has been used to characterize these interfacial dynamics at picosecond timescales. Existing reports suggest that water dynamics at the lipid interface are significantly slower than in bulk because of disrupted hydrogen-bond networks near the headgroups. The charge of lipid headgroups modulates interfacial water dynamics, with anionic lipids imposing stronger restrictions on the water mobility than cationic ones (Inoue et al., 2017; Singh et al., 2016). In addition, the presence of hydrogen-bond donors, such as primary amines, further impedes water dynamics compared to tertiary amines (Singh et al., 2016). Lipids also undergo multiple phase transitions before adopting a lamellar fluid phase, necessitating distinct characterization of minimally hydrated and fully hydrated membranes. Studies on lipid reverse micelles demonstrate prolonged vibrational lifetime of interfacial water and its restricted spectral diffusion relative to bulk water, highlighting confinement-induced modifications in interfacial dynamics (Costard et al., 2012a, 2012b). The role of cosolvents, particularly dimethyl sulfoxide (DMSO), is notable in modulating hydrogen-bond networks in the membrane interface. Interestingly, DMSO enhances interfacial water dynamics up to 10 mol%

but suppresses them at higher concentrations, driven by competing effects of hydrogen-bond disruption and lipid–lipid repulsion (Oh et al., 2020; Sahu et al., 2024; Venkatraman & Baiz, 2020).

Beyond biological interfaces, chemical heterogeneity plays a crucial role in tuning the thermodynamic and dynamic properties of artificial interfaces (Jin et al., 2023; Skorb & Andreeva, 2013). Graphene, a 2D nanomaterial, has gained significant attention because of its tunability, cost-effectiveness, and ease of production, with applications ranging from water desalination, gas separation, fuel cells, photocatalysis, and (bio)sensors to drug delivery (Ashok Kumar et al., 2022; Majumdar et al., 2024; Raza et al., 2024). Although pristine graphene is largely hydrophobic, functionalization with polar groups renders it amphiphilic, modulating interfacial water structure and dynamics. At the graphene–water interface, water molecules exhibit anisotropic mobility, with restricted perpendicular motion because of strong surface interactions, whereas lateral diffusion remains relatively high (Calero & Franzese, 2020; Majumdar et al., 2023; Ruiz-Barragan et al., 2020; Schaaf & Gekle, 2016). This directional dependence alters translational and rotational dynamics, leading to modified diffusion coefficients and prolonged residence times in the interfacial layer. Furthermore, interfacial water undergoes orientational polarization, which suppresses fluctuations and reduces the dielectric constant compared to bulk water (Ahmadabadi et al., 2021; Moid et al., 2022). Despite these constraints, lateral diffusivity can remain enhanced because of collective motion of water molecules along the interface, often surpassing bulk water diffusivity in specific confinement conditions. Beyond anisotropic mobility and dielectric properties the interfacial hydrogen bond network undergoes structural rearrangements, as reflected by vibrational spectral shifts (Zhang et al., 2020). These insights are crucial for optimizing graphene-based materials in nanofluidics, biomolecular sensing, and other applications where interfacial water plays a crucial role in governing performance.

4.3 Thermodynamic features of water

Hydration thermodynamics is fundamental to a wide range of biological processes, including protein folding, biomolecular recognition, ligand binding, enzyme activity, and self-assembly (Huang & Lau, 2016; Mukherjee & Schäfer, 2022). Despite its significance the molecular-level interactions between water and biomolecules are not fully understood, making it challenging to manipulate these processes in targeted ways. From a thermodynamic perspective the free energy change (ΔG) governs the spontaneity of these processes, with ΔG dependent on enthalpic (ΔH) and entropic (ΔS) contributions, as per the equation $\Delta G = \Delta H - T\Delta S$ (Maiti & Heyden, 2023). Enthalpy, representing interaction energies, is readily computed from potential energy differences. In contrast, entropy, tied to atomic fluctuations, is difficult to quantify because of the complexity of biological macromolecules with many degrees of freedom and nonlocalized water molecules (Lazaridis & Karplus, 2002; Scott & Winzor, 2024). This makes entropy estimation particularly challenging in such systems. In cases where the free energy difference is known, entropy (ΔS) can be determined via methods like thermodynamic integration or free energy perturbation using the relation; $\Delta S = \left(\frac{\Delta H - \Delta G}{T}\right)$. As solvation free energy (ΔG_{solv}) arises because of molecular interactions, it can be decomposed into solute–water (SW) and water–water (WW) interactions. The enthalpic and entropic components of solvation are expressed as, $\Delta H_{solv} = \Delta U_{SW} + \Delta U_{WW}$ and $\Delta S_{solv} = \Delta S_{SW} + \Delta S_{WW}$. However, the WW contributions exactly cancel out each other ($\Delta U_{WW} - T\Delta S_{WW} = 0$) (Chalikian, 2021; Mukherjee & Schäfer, 2022). Thus solvation free energy is driven solely by solute–solvent interactions ($\Delta U_{SW} - T\Delta S_{SW}$), with the enthalpy–entropy compensation being particularly significant in water (Fenley et al., 2014; Gopal et al., 2017; Heinz & Grubmüller, 2023; Kopel & Giovambattista, 2019). To gain spatially resolved insights into the role of water, computational methods, such as grid inhomogeneous solvation theory (GIST), two-phase thermodynamics (2PT), and various cell theories, have been developed (Lin et al., 2003, 2010; Nguyen et al., 2012, 2014, 2016). These methods provide detailed maps of local water dynamics and thermodynamics, yielding deeper mechanistic insights into the hydration effects relevant to biological function.

4.3.1 Grid-inhomogeneous solvation theory

The GIST (grid inhomogeneous solvation theory) method employs MDs or Monte Carlo simulations to characterize the equilibrium distribution of water molecules around a solute (Nguyen et al., 2014, 2016). By discretizing water density and solvation energy onto a 3D grid with cubic voxels, typically spaced at 0.5 Å, this method provides detailed spatial resolution while ensuring statistical convergence. This method focuses on single-body entropies, approximating solvation entropy by considering SW interactions alone ($\Delta S_{\text{solv}} \approx \Delta S_{\text{SW}}$) (Huggins, 2015; Persson et al., 2017). In principle, solvation entropy comprises both SW and WW correlation terms. However, the contribution of the former being negligible (Nguyen et al., 2016), the solvation entropy can be expressed as (Mukherjee & Schäfer, 2022):

$$\Delta S_{\text{solv}} = \Delta S_{\text{SW}} + \Delta S_{\text{WW}} \approx \Delta S_{\text{SW}} \tag{4.3}$$

This enables a tractable but approximate estimation of local thermodynamic quantities. In a study by Ryde et al., this method was applied to investigate solvation entropy contributions to ligand binding in a protein, galectin-3, that plays a multifaceted role in various biological processes, including cell growth, apoptosis, inflammation, and fibrosis (Wallerstein et al., 2021). The binding affinities of three ligands, varying only in fluorine atom position, were analyzed, revealing differences in solvation patterns at the binding site. Although GIST calculations provide valuable microscopic insights that align with experimental findings, the estimated free energy contribution from solvation entropy differences (25–30 kJ mol^{-1}) is likely overestimated. However, this level of overestimation is not typically observed in other cases. Nonetheless, in a related study, smaller discrepancies ($- T\Delta\Delta S_{\text{solv}} \approx 3$ kJ mol^{-1}) were observed for diastereomeric ligand binding to the same protein (Verteramo et al., 2019). Despite its utility a key limitation of this theory is its reliance on restraining solute atoms, which may artificially constrain conformational flexibility and overestimate entropic contributions in dynamic systems.

4.3.2 Two-phase thermodynamics model

The 2PT model, originally developed by Lin et al. (2003, 2010) Lin et al., extracts thermodynamic properties of bulk liquids from MD simulations. A key advantage of the 2PT method is its spectral resolution, which enables the extraction of different entropy contributions by integrating over various frequency domains of the vibrational density of states (VDOS) (Debnath et al., 2013; Fisette et al., 2016; Jana et al., 2006). Fundamentally the 2PT method involves an ad hoc partitioning of the VDOS into two contributions: "gas-like" and "solid-like." The diffusive, gas-like component is modeled by attributing portions of the translational and rotational motions to a hard-sphere fluid and a rigid-rotor framework, respectively. The corresponding entropy terms, $S_{\text{HS}}^{\text{trans}}$ and $S_{\text{RR}}^{\text{rot}}$ are analytically derived using the Carnahan–Starling equation of state (Mukherjee & Schäfer, 2022; Wang & Ma, 2010). The residual contributions of translational and rotational motions are then interpreted as vibrational (solid-like) and treated as harmonic oscillators (HO). Using the frequency-dependent quantum HO partition function the corresponding entropies S_{HO}^{trans} and S_{HO}^{rot} are calculated. The total entropy of the solvent system is thus given by (Mukherjee and Schäfer, 2022)

$$S = S_{HS}^{trans} + S_{HO}^{trans} + S_{RR}^{rot} + S_{HO}^{rot} \tag{4.4}$$

This formulation captures both harmonic and anharmonic contributions as well as higher order couplings between water molecules, which are encoded in translational and rotational velocity correlations. In a recent study by Päslack et al. (2021) the entropy of water molecules in the active site of human carbonic anhydrase II (hCAII) was investigated, focusing on changes upon binding the inhibitor dorzolamide. The apo form of hCAII contains a structurally conserved network of approximately 10 water molecules in its active site, stabilized by hydrogen bonds among themselves with the catalytic zinc ion and with critical amino acids for catalysis (Singh et al., 2019, 2020). Upon binding the inhibitor dorzolamide, three of these active-site water molecules were displaced into the bulk, whereas the remaining ones exhibited reduced conformational entropy. To quantify this effect, VDOS was calculated for the water molecules dynamically selected based on their distance from local maxima in the 3D water density, with a cutoff of 1.4 Å. This method approximates the anisotropic water density distribution effectively. The 2PT results indicate that water molecules in the hCAII active site exhibit lower entropy compared to bulk water, with location-dependent variations. For instance the entropy of the water molecule at site 1 in apo hCAII is 38.8 J mol K^{-1}, which is 17 J mol K^{-1} lower than the bulk value of 55.8 J mol K^{-1}. Binding dorzolamide results in the release of three water molecules into the bulk, increasing their entropy. Conversely the entropy of remaining water molecules decreases, most notably at site W4, where it drops by 5.4 J mol K^{-1} (from 50.7 to 45.3 J mol K^{-1}). Overall the entropy gain from the three released water molecules exceeds the loss of the remaining molecules, yielding a positive total entropy contribution upon inhibitor binding, $\Delta S_{solv,bind}$= 15.6 J mol K^{-1}, which corresponds to a favorable free energy contribution of approximately −5 kJ mol^{-1} at 300 K (Päslack et al., 2021).

This discussion underscores recent methodological advancements in obtaining spatially resolved maps of hydration thermodynamics in biomolecular systems through computer simulations. Notably the focus is on the challenges of computing entropy, which is more complex than enthalpy. The highlighted examples demonstrate that thermodynamic forces associated with hydration significantly influence processes such as protein folding and ligand binding. The microscopic insights derived from these simulations are vital for understanding the underlying mechanisms and can guide targeted modifications in rational ligand or protein design.

4.4 Dynamics of water at interfaces

4.4.1 Site-specific water dynamics at the biomolecular interfaces

The presence of a biomolecular interface significantly alters the dynamics of nearby water molecules compared to the bulk phase, but the precise quantitative characterization of this effect remains an ongoing challenge. Biomolecular surfaces exhibit chemical and topological heterogeneity, incorporating charged, polar, and apolar groups, along with hydrogen-bond donors and acceptors. In addition,

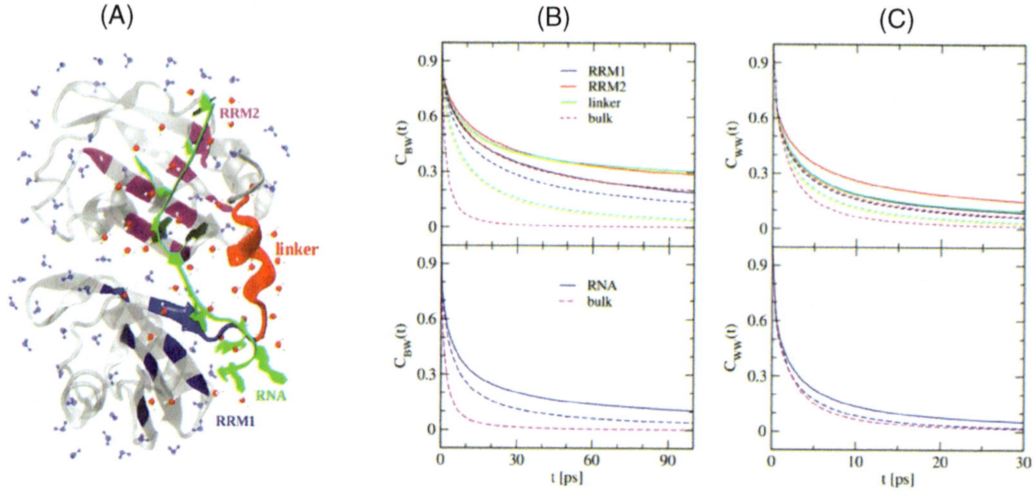

FIGURE 4.4

Structural representation of PABP(poly(A)-binding protein)–poly(A) RNA complex and hydrogen bond dynamics of interfacial water: (A) Schematic representation of the poly(A)-binding protein (PABP) bound to poly(A) RNA. The protein consists of three regions: RRM1 (residues 11–89), a linker helix (residues 90–98), and RRM2 (residues 99–175). Both RRM1 and RRM2 adopt a conserved fold comprising two α-helices and four antiparallel β-strands, whereas the linker region forms a short α-helix that connects the two motifs. Parts (B) and (C) are the intermittent hydrogen bond time correlation functions for backbone water (BW) and water–water (WW) hydrogen bonds, respectively. *Solid* and *dashed lines* denote the complex and free states, whereas the time correlation function of bulk water is included as a reference.

From Chakrabortty, A., & Bandyopadhyay, S. (2025). Probing the degree of restriction in solvent dynamics at the interface of a protein–RNA complex. The Journal of Physical Chemistry B. *https://doi.org/10.1021/acs.jpcb.4c08804.*

the surface is often rough at the molecular level, with features such as pockets, protrusions, and grooves. MD simulations have demonstrated that interfacial water dynamics are highly sensitive to such heterogeneity (Dewan et al., 2014; García & Hummer, 2000; Wang et al., 2022). Specifically, rotational dynamics and residence times of water molecules in the hydration shell reflect these variations. Residence time quantifies how long a water molecule spends in the first hydration layer before transitioning to the bulk phase. The distribution of reorientation times further reveals site-specific variability, providing a measure of how quickly a water molecule loses memory of its initial orientation. Notably, time correlation functions of interfacial water (see Fig. 4.4) feature pronounced nonexponential decay, indicating a broad distribution of relaxation times and confirming the heterogeneous nature of hydration dynamics (Chakrabortty & Bandyopadhyay, 2025; Duboué-Dijon et al., 2016; García & Hummer, 2000). However, elucidating site-specific hydration dynamics poses a significant challenge for experimental techniques. Recent nuclear Overhauser effect–NMR studies have provided spatially resolved maps of water dynamics around protein such as ubiquitin, although it required encapsulating the protein in reverse micelles to slow down water motion enough for nanosecond-scale resolution (Nucci et al., 2011a, 2011b). Although such approaches provide valuable insights, they remain limited in scope. But MD simulations of the subtilisin protein have shown that confinement introduces a strong, yet nonuniform, effect on hydration dynamics, differing significantly from those in free aqueous solutions (Becker-Baldus et al., 2023; Layek & Sengupta, 2024; Roy & Sengupta, 2021). These simulations are particularly well-suited for site-specific analysis of water dynamics. For each protein or DNA site the reorientation times of neighboring water molecules can be precisely measured. The relaxation time distributions have been mapped for a variety of globular proteins, including lysozyme, ubiquitin, subtilisin Carlsberg, acetylcholinesterase, and a hyperactive antifreeze protein, as well as for a DNA dodecamer duplex (Duboué-Dijon & Laage, 2014; Fogarty & Laage, 2014). Despite the chemical and topological heterogeneity of these surfaces, hydration layer dynamics are fairly consistent across most exposed surfaces, only moderately slower than in bulk water. Notably, despite the differing sizes, structures, and functions of the globular proteins studied, including enzymes, ubiquitin, and antifreeze proteins, the distributions of water retardation factors remain strikingly similar.

4.4.2 Extended jump model

Recent theoretical and experimental findings propose a mechanism where water molecules reorient through sudden, large-amplitude jumps as their hydroxyl (OH) groups switch H-bond partners. The EJM provides a theoretical framework to describe these dynamics, particularly within the hydration shells of biomolecules such as proteins and DNA, linking these jumps to the reorientation time of water molecules (Ji et al., 2010; Laage & Hynes, 2008; Laage et al., 2012; Moilanen et al., 2009). Two key factors influence the water jump rate: solute interface topography and initial H-bond strength (Fogarty et al., 2014b; Sterpone et al., 2010). Steric hindrance from solute geometry slows reorientation, with convex sites causing minimal slowdown and concave pockets leading to more significant effects. The strength of the initial H-bond further modulates the jump rate, where stronger initial H-bonds decelerate the rate and weaker bonds accelerate it. This is governed by both enthalpic and entropic factors. Fig. 4.5 highlights that hydrophobic groups slow H-bond jumps by limiting access to new partners via excluded volume effects. H-bond donors, while forming bonds of varying strength, mainly affect water oxygen with negligible impact on OH reorientation (Laage et al., 2017b). In contrast, H-bond acceptors influence water dynamics through both excluded volume and bond strength.

EJM has been widely applied, from small solutes to interfaces and nanopores, but its relevance to biomolecules such as proteins, DNA, and phospholipid bilayers is particularly important. In these systems, large angular jumps remain the dominant reorientation mechanism for water near biomolecular surfaces, with model-predicted water reorientation times aligning well with direct calculations (Duboué-Dijon et al., 2016; Fogarty & Laage, 2014; Fogarty et al., 2014a; Jana et al., 2008; Laage & Thompson, 2012; Sterpone et al., 2012; Stirnemann et al., 2010). Water molecules within protein or DNA hydration shells experience a moderate slowdown compared to the bulk, with a slowdown factor limited to 2–3. This steric slowdown is largely entropic, as supported by the weak temperature dependence of hydration shell dynamics observed in MRD experiments and MD simulations (Duboué-Dijon & Laage, 2014; Fogarty & Laage, 2014; Mattea et al., 2008). This slowdown is caused by the biomolecular interface obstructing the hydrogen bond rearrangements (Capponi et al., 2019; Chakrabortty & Bandyopadhyay, 2025; Mondal & Bandyopadhyay, 2020). For DNA a sharper slowdown (factor 4–10) arises from water near the phosphate backbone, where strong H-bonds decelerate dynamics (Duboué-Dijon & Laage, 2014). The most retarded water molecules are found in minor grooves, where they form

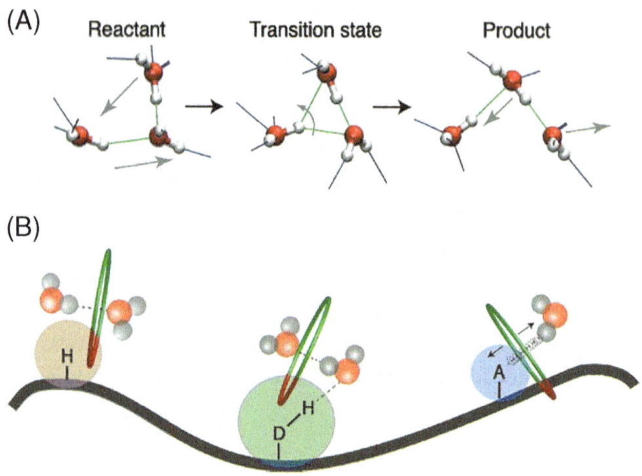

FIGURE 4.5

Water reorientation dynamics and perturbative features on protein interfaces: (A) Molecular jump mechanism via water reorientation. (B) Schematic of protein interfaces highlighting hydrophobic, H-bond donor and H-bond acceptor sites, along with their respective perturbative effects on water dynamics.

From Laage, D., Elsaesser, T., & Hynes, J. T. (2017). Perspective: Structure and ultrafast dynamics of biomolecular hydration shells. Structural Dynamics, *4(4). https://doi.org/10.1063/1.4981019.*

strong H bonds with buried biomolecular sites, particularly in regions like the DNA "spine of hydration" (Khah et al., 2020; Shweta & Sen, 2018). Both excluded volume and H-bond strength contribute to these strongly slowed dynamics.

The jump model emphasizes that the local topography of a biomolecule significantly affects its hydration shell dynamics (Fogarty & Laage, 2014; Sheu et al., 2019). However, because biomolecules are flexible and their conformations fluctuate on varying time scales, these changes influence local water dynamics, leading to variable jump rates at specific sites (King et al., 2014; Pathak & Bandyopadhyay, 2019). If water dynamics are significantly faster than biomolecular fluctuations, the biomolecule appears static, resulting in an inhomogeneous distribution of water dynamics. Conversely, if the biomolecule adopts different local conformations before the hydration shell rearranges, the water dynamics equilibrate and average over these conformations, resulting in no temporal heterogeneity. This interplay introduces a limitation of the model, as it may not fully capture the dynamic relationship between biomolecular fluctuations and hydration shell behavior, necessitating further exploration to better understand these interactions.

4.5 Water in molecular crowding environments

4.5.1 Water confinement and mobility in biomolecular systems

The behavior of water under spatial confinement (such as condensed matter confinement) deviated notably from its bulk properties, affecting essential processes such as ion transport, molecular diffusion, and chemical reactions in aqueous environments (Bampoulis et al., 2018; Tao et al., 2018; Zhu et al., 2019). In this section, we focus on confinement in the context of biomolecules. IR spectroscopy has been widely employed to study water pools in reverse micelles, which serve as model systems for molecular nanoreactors and confinements (Chowdhary & Ladanyi, 2011; Moilanen et al., 2007). By varying the surfactant composition and controlling the micelle size, it has been observed that water relaxation dynamics are primarily influenced by spatial confinement (Geethu et al., 2017; Hanot et al., 2015, 2016; Sharma et al., 2019). Specific interactions between the solvent and the polar surfactant heads are present but have a lesser impact (Borse & Devi, 2006; Shinoda et al., 1996). Computer simulations have provided detailed molecular-level insights into the dynamics and structure of these internal water pools (Faramarzi et al., 2017; Lu et al., 2020). Notably the commonly assumed simplistic spherical model used to derive the water-to-surfactant ratio in experiments does not hold (Martinez et al., 2013). Moreover, simulations underscore the key role of hydrogen bonding between water and the polar or ionic surfactant head groups at the interface in locally perturbing both water reorientation and diffusion (Pal et al., 2018; Parra et al., 2024; Skinner et al., 2024). In the context of protein encapsulation within reverse micelles, hydration dynamics are systematically slower compared to those in dilute solutions (Dodevski et al., 2014; Roche et al., 2013). This kinetic shift can be accessed using NMR techniques (Persson & Halle, 2008). However, it remains unclear whether the extracted information can be directly applied to protein hydration in dilute conditions. One hypothesis is that confinement only globally scales solvent kinetics. Alternatively, specific enclaves of water at the protein–micelle interface may reveal strong residence sites at the protein surface that are otherwise rapidly displaced by interfacial water.

A more intricate form of biological confinement occurs in the cytoplasm, where proteins are situated. NMR studies have been used to track the perturbation of water reorientation within this environment, revealing that water mobility slows by a factor of 30 compared to bulk (Comez et al., 2016; Fleury & Canet, 2014; Galamba, 2014). However, this degree of slowdown is not enough to explain the near cessation of molecular motion required to account for the dormant protein functionality observed within cells. As systems transition from dilute solutions to crowded environments the increasing retardation of water reorientation (from a factor of 5 to as much as 30) appears as a natural consequence of reduced hydration levels. Results from NMR studies partially conflict with findings from neutron scattering experiments (Khodadadi et al., 2008; Rezaei-Ghaleh, 2022). Computer simulations of water in crowded protein solutions have also been reported, shedding light on how solvent-mediated interactions influence protein mobility under crowded conditions (Kleinjung & Fraternali, 2014; Mondal & Bagchi, 2022; Srinivasan et al., 1998). Such interactions are crucial for understanding many in vivo functional processes. In silico models of cellular environments suggest that the observed reduction in protein mobility is not solely because of excluded volume effects but is primarily driven by solvent-mediated interactions (Blinov et al., 2019; Das & Sen, 2020). Skolnick et al. demonstrated that incorporating hydrodynamic interactions in coarse-grained, water-free simulations of proteins (modeled as rigid spheres) in a cell-like environment reproduced the experimental slowdown observed in dilute solutions. Recent efforts have aimed to realistically simulate cellular environments (Frembgen-Kesner & Elcock, 2013). For instance, McGuffee and Elcock (2010) conducted a large-scale simulation of a portion of the *Escherichia coli* cytoplasm, although essential hydrodynamic interactions were not included because of the system's size. A novel multiscale algorithm based on the Lattice Boltzmann methodology has since been developed to address this challenge. As a proof of concept, this method was used to simulate a system of approximately 18,000 proteins at quasiatomistic resolution, including hydrodynamic interactions, allowing for effective investigation of the role of solvent-mediated interactions in protein mobility (Elcock, 2013; Sterpone et al., 2014).

4.5.2 Liquid–liquid phase separation

Recent advancements have deepened our understanding of cellular compartmentalization, particularly through the formation of biomolecular condensates (Brangwynne et al., 2009; Mittag & Pappu, 2022; Pappu et al., 2023). These condensates are dense liquid droplets composed of proteins, RNA, and other biomolecules, exhibiting dynamic properties that facilitate efficient molecular interactions crucial for specific biochemical processes (Banjade & Rosen, 2014; Shin & Brangwynne, 2017; Strom et al., 2017; Woodruff, 2018). A key mechanism driving their formation is LLPS, which has been implicated in regulating various cellular activities and in the development of pathological aggregates associated with certain diseases (Alberti & Dormann, 2019; Alberti et al., 2019). LLPS-derived condensates can act as nucleation sites for protein aggregates linked to neurodegenerative diseases, such as amyotrophic lateral sclerosis, Alzheimer's, and Parkinson's diseases (Pakravan et al., 2021; Wegmann et al., 2018). Water plays a pivotal role in LLPS, actively influencing the reversible compartmentalization of protein solutions by modulating hydration entropy and enthalpy. MD simulations reveal an "entropy tug-of-war" between two classes of hydration water, that is, cavity-wrap and BW (Mukherjee & Schäfer, 2023; Mukherjee et al., 2024). Cavity-wrap water hydrates hydrophobic regions, but it is released during condensate formation, increasing system entropy. In contrast, BW, which hydrates hydrophilic regions, is retained in a significant amount within the condensates (see Fig. 4.6). It experiences enhanced restriction because of the dense protein environment, making its retention enthalpically favorable. This interplay between the release and retention of hydration water molecules highlights the intricate thermodynamic balance that drives phase separation in biomolecular condensates (König et al., 2024; Pezzotti et al., 2023). As proteins coalesce into condensates the population of hydration water surrounding ~0.3 nm of the protein surface decreases with increasing protein concentration, primarily due to the release of cavity-wrapped water into the dilute phase. Meanwhile the retained BW forms hydrogen bonds with proteins, inducing spatial confinement that disrupts extended hydrogen bond networks among water molecules. Conversely the released water molecules, no longer constrained by protein interactions, establish stronger WW hydrogen bonds, leading to an entropy gain. Studies indicate that this entropy-driven release governs the thermodynamic landscape, strengthening the favorable solvation forces that facilitate condensate formation (Mukherjee & Schäfer, 2023; Ribeiro et al., 2019). Furthermore, THz spectroscopy has identified two distinct spectral signatures corresponding to different hydration water categories, providing direct experimental validation of theoretical predictions (Ahlers et al., 2021; König et al., 2024; Mukherjee et al., 2024; Saha & Mitra, 2022). These findings underscore the fundamental role of hydration thermodynamics in protein condensation, refining our understanding of the molecular forces that dictate cellular compartmentalization (Acharya, 2022). Spatial confinement in protein condensates perturbs the tetrahedral arrangement of water, reducing the fraction of molecules in a tetrahedral coordination (Chau & Hardwick, 1998). This effect is evident in phase-separating proteins like the RNA-binding protein. Similar effects have been observed in the fused in sarcoma (FUS protein) that forms intracellular condensates and plays a key role in RNA shearing and transport, DNA repair, microRNA processing, and gene transcription and regulation. The intrinsically disordered N-terminal low-complexity domain (LCD) is responsible for LLPS. A structural comparison of the tetrahedral

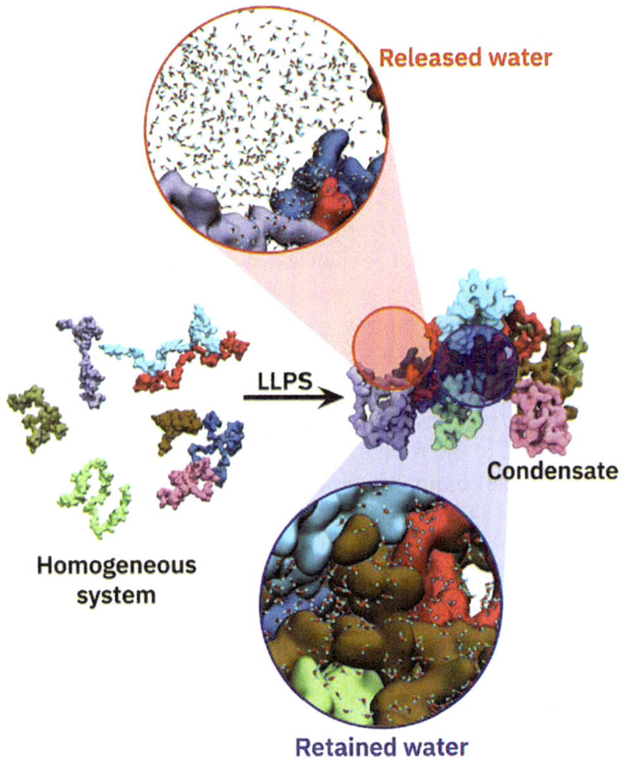

FIGURE 4.6

Illustration of the process of protein condensate formation via liquid–liquid phase separation: Enlarged snapshots depict the release of cavity wrap water into the bulk environment and the sustained retention of interfacial bound water molecules.

From Mukherjee, S., & Schäfer, L. V. (2023). Thermodynamic forces from protein and water govern condensate formation of an intrinsically disordered protein domain. Nature Communications, 14*(1). https://doi.org/10.1038/s41467-023-41586-y.*

order parameter (q) shows that for bulk water, the peak is centered around $q = 0.8$. In the FUS–LCD condensate (350 mg mL^{-1}), its intensity decreases with a concomitant rise near $q = 0.6$, indicating a disruption in tetrahedrality. As a result, water molecules within protein condensates adopt a less tetrahedral local structure, favoring 2D trigonal configurations to optimize intermolecular hydrogen bonding (Mukherjee & Schäfer, 2023).

4.6 Summary and conclusion

Water plays an indispensable role in biological systems, with its unique structural, thermodynamic, and dynamic properties significantly influencing biomolecular function. A central feature in this regard is the hydration shell, the layer of water molecules closely surrounding biomolecules, which differ markedly from bulk water in terms of structure and behavior. Around proteins, this shell typically extends from 3 to 6 Å, whereas, for DNA, it can extend up to 10 Å, with water molecules forming hydrogen bonds with polar or charged functional groups. This structured layer is essential for stabilizing biomolecular conformation and interactions, mediating key processes such as ligand binding, protein folding, and enzymatic catalysis. At the protein–water interface, hydration water exhibits anisotropic dielectric responses under an applied electric field, with parallel fields causing a redshift and perpendicular fields inducing a blueshift in absorption spectra. Ligand binding further alters hydration thermodynamics by releasing tightly BW, increasing system entropy, whereas the removal of slower, interface-stabilized water incurs an enthalpic cost but ultimately enhances entropy upon release into bulk water. Similarly, at graphene–water interfaces, water displays anisotropic mobility, with lateral diffusion remaining relatively high, whereas perpendicular motion is restricted because of strong surface interactions. Experimental techniques such as NMR, SAXS, and THz spectroscopy, alongside MD simulations, have deepened our understanding of hydration shell dynamics, although fully characterizing the extent of biomolecular influence on water remains a challenge.

Hydrogen bonding is central to the role of water in biomolecular systems, where deviations from bulk tetrahedrality arise because of biomolecular interfaces and spatial confinement. The orientational tetrahedral order parameter (q) is a useful metric to quantify these deviations, particularly in molecular simulations. These deviations are most pronounced in confined environments, such as biomolecular condensates formed via LLPS, where the entropic tug-of-war between cavity-wrap and BW governs phase separation. The release of cavity-wrap water increases entropy, whereas retained BW stabilizes the dense protein environment. Such hydration-driven mechanisms contribute to cellular compartmentalization and the regulation of biomolecular interactions. In addition, spatial confinement within condensates perturbs water's tetrahedral structure, further highlighting the structural rearrangement of hydration water in biological phase separation.

Beyond structural aspects, hydration thermodynamics provides crucial insights into biomolecular stability and interactions. Solvation free energy, which influences protein folding, ligand binding, and molecular recognition, arises from a delicate balance of enthalpy and entropy. This free energy can be decomposed into SW and WW interaction contributions, with enthalpy–entropy compensation playing a major role. Computational methods, such as grid-based inhomogeneous solvation theory (GIST) and 2PT models, provide fine spatial resolution on hydration thermodynamics, offering insights into how water–biomolecule interactions modulate key processes. These methods are valuable in drug discovery, enabling predictions about how ligands interact with biomolecules by influencing local water dynamics and entropy.

Dynamically, water at biomolecular interfaces exhibits complex and heterogeneous behavior. MD simulations reveal that hydration water around proteins and DNA experiences slowed rotational and translational dynamics compared to bulk water, primarily because of steric hindrance and hydrogen-bonding interactions imposed by the biomolecular surface. The orientational relaxation of these water molecules is highly dependent on local environmental factors such as steric constraints and hydrogen bond strength. The EJM provides a theoretical framework to describe these dynamics, where water molecules reorient via sudden, large-angle jumps as their OH groups switch hydrogen-bond partners. This model is particularly effective in describing water behavior near biomolecular surfaces, where reorientation times can slow by factors of 2–10 compared to bulk water, especially near the phosphate backbone of DNA or buried sites in proteins.

In conclusion, water's structural, thermodynamic, and dynamic properties at biomolecular interfaces are fundamental to the function and stability of biological macromolecules. The hydration shell, with its distinct structure and behavior, mediates a wide array of biomolecular processes, from protein folding to ligand binding. Advances in spectroscopy and molecular simulations have significantly enhanced our understanding of these interactions, though challenges remain in fully capturing the complexity of hydration dynamics. Future advancements, in experimental and computational methodologies, promise to provide even deeper insights into the role of water in modulating biological function at the molecular level.

Acknowledgment

S.L. acknowledges funding assistance from the Prime Minister's Research Fellowship (PMRF) for the Ph.D. studies. The authors wish to express their appreciation to Hindol Chatterjee for important discussions. N.S. acknowledges support from IISER Kolkata and computational resources from PARAM Rudra and PARAM Shakti, established by MeiTY, Govt. of India. The support of ANRF, Govt. of India, via the ARG grant (ANRF/ARF/2025/000861/CS) is gratefully acknowledged.

References

Abel, S., Sterpone, F., Bandyopadhyay, S., & Marchi, M. (2004). Modeling, and simulations of AOT–water reverse micelles in isooctane: Structural and dynamic properties. *The Journal of Physical Chemistry B, 108*(50), 19458–19466. doi:10.1021/jp047138e.

Abhyankar, N., & Szalai, V. (2021). Challenges and advances in the application of dynamic nuclear polarization to liquid-state NMR spectroscopy. *The Journal of Physical Chemistry B, 125*(20), 5171–5190. doi:10.1021/acs.jpcb.0c10937.

Acharya, S. (2022). Tug-of-war between internal and external frictions and viscosity dependence of rate in biological reactions. *Physical Review Letters, 128*, 108101.

Adhikari, S., & Mondal, J. (2023). Machine learning subtle conformational change due to phosphorylation in intrinsically disordered proteins. *The Journal of Physical Chemistry B, 127*(44), 9433–9449. doi:10.1021/acs.jpcb.3c05136.

Ahlers, J., Adams, E. M., Bader, V., Pezzotti, S., Winklhofer, K. F., Tatzelt, J., & Havenith, M. (2021). The key role of solvent in condensation: Mapping water in liquid-liquid phase-separated FUS. *Biophysical Journal, 120*(7), 1266–1275. http://www.sciencedirect.com/science/journal/00063495. doi:10.1016/j.bpj.2021.01.019.

Ahmadabadi, I., Esfandiar, A., Hassanali, A., & Ejtehadi, M. R. (2021). Structural and dynamical fingerprints of the anomalous dielectric properties of water under confinement. *Physical Review Materials, 5*(2). https://journals.aps.org/prmaterials/abstract/10.1103/PhysRevMaterials.5.024008. doi:10.1103/PhysRevMaterials.5.024008.

Alberti, S., & Dormann, D. (2019). Liquid–liquid phase separation in disease. *Annual Review of Genetics, 53*(1), 171–194. doi:10.1146/annurev-genet-112618-043527.

Alberti, S., Gladfelter, A., & Mittag, T. (2019). Considerations and challenges in studying liquid–liquid phase separation and biomolecular condensates. *Cell, 176*(3), 419–434. doi:10.1016/j.cell.2018.12.035.

Altabet, Y. E., Singh, R. S., Stillinger, F. H., & Debenedetti, P. G. (2017). Thermodynamic anomalies in stretched water. *Langmuir, 33*(42), 11771–11778. doi:10.1021/acs.langmuir.7b02339.

Arunan, E., Desiraju, G. R., Klein, R. A., Sadlej, J., Scheiner, S., Alkorta, I., Clary, D. C., Crabtree, R. H., Dannenberg, J. J., Hobza, P., Kjaergaard, H. G., Legon, A. C., Mennucci, B., & Nesbitt, D. J. (2011). Defining the hydrogen bond: An account (IUPAC Technical Report). *Pure and Applied Chemistry, 83*(8), 1619–1636. http://www.iupac.org/publications/pac/pdf/2011/pdf/8308x1619.pdf. doi:10.1351/PAC-REP-10-01-01India.

Asbury, J. B., Steinel, T., Stromberg, C., Corcelli, S. A., Lawrence, C. P., Skinner, J. L., & Fayer, M. D. (2004). Water dynamics: Vibrational echo correlation spectroscopy and comparison to molecular dynamics simulations. *Journal of Physical Chemistry A, 108*(7), 1107–1119. doi:10.1021/jp036266k.

Ashok Kumar, S. S., Bashir, S., Ramesh, K., & Ramesh, S. (2022). A review on graphene and its derivatives as the forerunner of the two-dimensional material family for the future. *Journal of Materials Science, 57*(26), 12236–12278. Www.springer.com/journal/10853. doi:10.1007/s10853-022-07346-x.

Bampoulis, P., Sotthewes, K., Dollekamp, E., & Poelsema, B. (2018). Water confined in two-dimensions: Fundamentals and applications. *Surface Science Reports, 73*(6), 233–264. doi:10.1016/j.urfrep.2018.09.001.

Banjade, S., & Rosen, M. K. (2014). Phase transitions of multivalent proteins can promote clustering of membrane receptors. *eLife, 3*. doi:10.7554/eLife.04123.

Barnes, R., Sun, S., Fichou, Y., Dahlquist, F. W., Heyden, M., & Han, S. (2017). Spatially heterogeneous surface water diffusivity around structured protein surfaces at equilibrium. *Journal of the American Chemical Society, 139*(49), 17890–17901. http://pubs.acs.org/journal/jacsat. doi:10.1021/jacs.7b08606.

Becker-Baldus, J., Yeliseev, A., Joseph, T. T., Sigurdsson, S. Th., Zoubak, L., Hines, K., Iyer, M. R., Berg, A., Stepnowski, S., Zmuda, J., Gawrisch, K., & Glaubitz, C. (2023). Probing the conformational space of the cannabinoid receptor 2 and a systematic investigation of DNP-enhanced MAS NMR spectroscopy of proteins in detergent micelles. *ACS Omega, 8*(36), 32963–32976. doi:10.1021/acsomega.3c04681.

Bhattacharya, I., Saha, R., Pyne, S., Bera, A., & Mitra, R. K. (2024). Excipient induced unusual phase separation in bovine serum albumin solution: An explicit role played by ion-hydration. *Langmuir, 40*(49), 25822–25833. http://pubs.acs.org/journal/langd5. doi:10.1021/acs.langmuir.4c02802.

Blaffert, J., Haeri, H. H., Blech, M., Hinderberger, D., & Garidel, P. (2018). Spectroscopic methods for assessing the molecular origins of macroscopic solution properties of highly concentrated liquid protein solutions. *Analytical Biochemistry, 561–562*, 70–88. http://www.elsevier.com/inca/publications/store/6/2/2/7/8/1/index.htt. doi:10.1016/j.ab.2018.09.013.

Blinov, N., Wishart, D. S., & Kovalenko, A. (2019). Solvent composition effects on the structural properties of the Aβ42 monomer from the 3D-RISM-KH molecular theory of solvation. *The Journal of Physical Chemistry B, 123*(11), 2491–2506. doi:10.1021/acs.jpcb.9b00480.

Bopp, P. A., Kornyshev, A. A., & Sutmann, G. (1998). Frequency and wave-vector dependent dielectric function of water: Collective modes and relaxation spectra. *Journal of Chemical Physics, 109*(5), 1939–1958. http://scitation.aip.org/content/aip/journal/jcp. doi:10.1063/1.476884.

Borse, M. S., & Devi, S. (2006). Importance of head group polarity in controlling aggregation properties of cationic gemini surfactants. *Advances in Colloid and Interface Science, 123–126*, 387–399.

Brangwynne, C. P., Eckmann, C. R., Courson, D. S., Rybarska, A., Hoege, C., Gharakhani, J., Jülicher, F., & Hyman, A. A. (2009). Germline P granules are liquid droplets that localize by controlled dissolution/condensation. *Science, 324*(5935), 1729–1732. doi:10.1126/science.1172046.

Brini, E., Fennell, C. J., Fernandez-Serra, M., Hribar-Lee, B., Lukšič, M., & Dill, K. A. (2017). How water's properties are encoded in its molecular structure and energies. *Chemical Reviews, 117*(19), 12385–12414. http://pubs.acs.org/journal/chreay. doi:10.1021/acs.chemrev.7b00259.

Brown, D. C. R., Webber, T. R., Casey, T. M., Franck, J., Shell, M. S., & Han, S. (2024). Computation of Overhauser dynamic nuclear polarization processes reveals fundamental correlation between water dynamics, structure, and solvent restructuring entropy. *Physical Chemistry Chemical Physics*, (20), 14637–14650. doi:10.1039/D4CP00030G.

Bubon, T., Zdorevskyi, O., & Perepelytsya, S. (2023). Molecular dynamics study of collective water vibrations in a DNA hydration shell. *European Biophysics Journal, 52*(1–2), 69–79. doi:10.1007/s00249-023-01638-z.

Calero, C., & Franzese, G. (2020). Water under extreme confinement in graphene: Oscillatory dynamics, structure, and hydration pressure explained as a function of the confinement width. *Journal of Molecular Liquids, 317*, 114027. doi:10.1016/j.molliq.2020.114027.

Capaccioli, S., Ngai, K. L., Ancherbak, S., Rolla, P. A., & Shinyashiki, N. (2011). The role of primitive relaxation in the dynamics of aqueous mixtures, nano-confined water and hydrated proteins. *Journal of Non-Crystalline Solids, 357*(2), 641–654. doi:10.1016/j.jnoncrysol.2010.07.054.

Capponi, S., White, S. H., Tobias, D. J., & Heyden, M. (2019). Structural relaxation processes and collective dynamics of water in biomolecular environments. *Journal of Physical Chemistry B, 123*(2), 480–486. http://pubs.acs.org/journal/jpcbfk. doi:10.1021/acs.jpcb.8b12052.

Cerveny, S., Alegría, Á., & Colmenero, J. (2008). Universal features of water dynamics in solutions of hydrophilic polymers, biopolymers, and small glass-forming materials. *Physical Review E, 77*(3). doi:10.1103/physreve.77.031803.

Cerveny, S., Schwartz, G. A., Bergman, R., & Swenson, J. (2004). Glass transition and relaxation processes in supercooled water. *Physical Review Letters, 93*(24). doi:10.1103/PhysRevLett.93.245702.

Chakrabortty, A., & Bandyopadhyay, S. (2025). Probing the degree of restriction in solvent dynamics at the interface of a protein–RNA complex. *The Journal of Physical Chemistry B*. doi:10.1021/acs.jpcb.4c08804.

Chakraborty, S., Kumar, H., Dasgupta, C., & Maiti, P. K. (2017). Confined water: Structure, dynamics, and thermodynamics. *Accounts of Chemical Research, 50*(9), 2139–2146. doi:10.1021/acs.accounts.6b00617.

Chalikian, T. V. (2021). Does the release of hydration water come with a Gibbs energy contribution? *The Journal of Chemical Thermodynamics, 158*, 106409. doi:10.1016/j.jct.2021.106409.

Chandler, D. (2005). Interfaces and the driving force of hydrophobic assembly. *Nature, 437*(7059), 640–647. doi:10.1038/nature04162.

Chandra, A., & Bagchi, B. (1989). Microscopic expression for frequency and wave vector dependent dielectric constant of a dipolar liquid. *The Journal of Chemical Physics, 90*(3), 1832–1840. doi:10.1063/1.456025.

Chaplin, M. (2006). Do we underestimate the importance of water in cell biology? *Nature Reviews Molecular Cell Biology, 7*(11), 861–866. doi:10.1038/nrm2021.

Chau, P., & Hardwick, A. J. (1998). A new order parameter for tetrahedral configurations. *Molecular Physics, 93*(3), 511–518. doi:10.1080/002689798169195.

Chong, S. H., & Ham, S. (2015). Distinct role of hydration water in protein misfolding and aggregation revealed by fluctuating thermodynamics analysis. *Accounts of Chemical Research, 48*(4), 956–965. http://pubs.acs.org/journal/achre4. doi:10.1021/acs.accounts.5b00032.

Chowdhary, J., & Ladanyi, B. M. (2011). Molecular simulation study of water mobility in aerosol-OT reverse micelles. *Journal of Physical Chemistry A, 115*(23), 6306–6316. doi:10.1021/jp201866t.

Comez, L., Paolantoni, M., Sassi, P., Corezzi, S., Morresi, A., & Fioretto, D. (2016). Molecular properties of aqueous solutions: A focus on the collective dynamics of hydration water. *Soft Matter, 12*(25), 5501–5514. doi:10.1039/C5SM03119B.

Costard, R., Greve, C., Heisler, I. A., & Elsaesser, T. (2012a). Ultrafast energy redistribution in local hydration shells of phospholipids: A two-dimensional infrared study. *Journal of Physical Chemistry Letters, 3*(23), 3646–3651. doi:10.1021/jz3018978.

Costard, R., Levinger, N. E., Nibbering, E. T. J., & Elsaesser, T. (2012b). Ultrafast vibrational dynamics of water confined in phospholipid reverse micelles. *Journal of Physical Chemistry B, 116*(19), 5752–5759. http://pubs.acs.org/journal/jpcbfk. doi:10.1021/jp3039016.

Cowan, M. L., Bruner, B. D., Huse, N., Dwyer, J. R., Chugh, B., Nibbering, E. T. J., Elsaesser, T., & Miller, R. J. D. (2005). Ultrafast memory loss and energy redistribution in the hydrogen bond network of liquid H_2O. *Nature, 434*(7030), 199–202. doi:10.1038/nature03383.

Dargaville, B. L., & Hutmacher, D. W. (2022). Water as the often neglected medium at the interface between materials and biology. *Nature Communications, 13*(1). doi:10.1038/s41467-022-31889-x.

Das, B., Ruiz-Barragan, S., Bagchi, B., & Marx, D. (2024). Topological frustration triggers ultrafast dynamics of monolayer water confined in graphene slit pores. *Nano Letters, 24*(49), 15623–15628. doi:10.1021/acs.nanolett.4c04077.

Das, N., & Sen, P. (2020). Shape-dependent macromolecular crowding on the thermodynamics and microsecond conformational dynamics of protein unfolding revealed at the single-molecule level. *The Journal of Physical Chemistry B, 124*(28), 5858–5871. doi:10.1021/acs.jpcb.0c03897.

Das, R., Mills, T. T., Kwok, L. W., Maskel, G. S., Millett, I. S., Doniach, S., Finkelstein, K. D., Herschlag, D., & Pollack, L. (2003). Counterion distribution around DNA probed by solution X-ray scattering. *Physical Review Letters, 90*(18), 4. doi:10.1103/PhysRevLett.90.188103.

Das, S., Biswas, R., & Mukherjee, B. (2015). Reorientational jump dynamics and its connections to hydrogen bond relaxation in molten acetamide: An all-atom molecular dynamics simulation study. *The Journal of Physical Chemistry B, 119*(1), 274–283. doi:10.1021/jp509869d.

Debnath, A., Ayappa, K. G., & Maiti, P. K. (2013). Simulation of influence of bilayer melting on dynamics and thermodynamics of interfacial water. *Physical Review Letters, 110*(1). http://oai.aps.org/filefetch?identifier=10.1103/PhysRevLett.110.018303&component=fulltext&description=markup&format=xml. doi:10.1103/PhysRevLett.110.018303India.

Debnath, A., Mukherjee, B., Ayappa, K. G., Maiti, P. K., & Lin, S. T. (2010). Entropy and dynamics of water in hydration layers of a bilayer. *Journal of Chemical Physics, 133*(17). doi:10.1063/1.3494115.

Dewan, S., Carnevale, V., Bankura, A., Eftekhari-Bafrooei, A., Fiorin, G., Klein, M. L., & Borguet, E. (2014). Structure of water at charged interfaces: A molecular dynamics study. *Langmuir, 30*(27), 8056–8065. doi:10.1021/la5011055.

Dexheimer, S. L. (2008). *Terahertz spectroscopy: Principles and applications* (1st ed.). CRC Press.

Dodevski, I., Nucci, N. V., Valentine, K. G., Sidhu, G. K., O'Brien, E. S., Pardi, A., & Wand, A. J. (2014). Optimized reverse micelle surfactant system for high-resolution NMR spectroscopy of encapsulated proteins and nucleic acids dissolved in low viscosity fluids. *Journal of the American Chemical Society, 136*(9), 3465–3474. http://pubs.acs.org/journal/jacsat. doi:10.1021/ja410716w.

Duboué-Dijon, E., Fogarty, A. C., Hynes, J. T., & Laage, D. (2016). Dynamical disorder in the DNA hydration shell. *Journal of the American Chemical Society, 138*(24), 7610–7620. http://pubs.acs.org/journal/jacsat. doi:10.1021/jacs.6b02715.

Duboué-Dijon, E., & Laage, D. (2014). Comparative study of hydration shell dynamics around a hyperactive antifreeze protein and around ubiquitin. *The Journal of Chemical Physics, 141*(22). doi:10.1063/1.4902822.

Edwards, K. J., Brown, D. G., Spink, N., Skelly, J. V., & Neidle, S. (1992). Molecular structure of the B-DNA dodecamer d(CGCAAATTTGCG)2. An examination of propeller twist and minor-groove water structure at 2.2 Å resolution. *Journal of Molecular Biology, 226*(4), 1161–1173. doi:10.1016/0022-2836(92)91059-X.

Elcock, A. H. (2013). Molecule-centered method for accelerating the calculation of hydrodynamic interactions in Brownian dynamics simulations containing many flexible biomolecules. *Journal of Chemical Theory and Computation, 9*(7), 3224–3239. doi:10.1021/ct400240w.

Errington (2001). *Relationship between structural order and the anomalies of liquid water, 409*, 318–321. doi:10.1038/35053024.

Faramarzi, S., Bonnett, B., Scaggs, C. A., Hoffmaster, A., Grodi, D., Harvey, E., & Mertz, B. (2017). Molecular dynamics simulations as a tool for accurate determination of surfactant micelle properties. *Langmuir, 33*(38), 9934–9943. doi:10.1021/acs.langmuir.7b02666.

Fecko, C. J., Eaves, J. D., Loparo, J. J., Tokmakoff, A., & Geissler, P. L. (2003). Ultrafast hydrogen-bond dynamics in the infrared spectroscopy of water. *Science, 301*(5640), 1698–1702. doi:10.1126/science.1087251.

Fenley, A. T., Henriksen, N. M., Muddana, H. S., & Gilson, M. K. (2014). Bridging calorimetry and simulation through precise calculations of cucurbituril–guest binding enthalpies. *Journal of Chemical Theory and Computation, 10*(9), 4069–4078. doi:10.1021/ct5004109.

Finer-Moore, J. S., Kossiakoff, A. A., Hurley, J. H., Earnest, T., & Stroud, R. M. (1992). Solvent structure in crystals of trypsin determined by X-ray and neutron diffraction. *Proteins: Structure, Function, and Bioinformatics, 12*(3), 203–222. doi:10.1002/prot.340120302.

Fisette, O., Päslack, C., Barnes, R., Isas, J. M., Langen, R., Heyden, M., Han, S., & Schäfer, L. V. (2016). Hydration dynamics of a peripheral membrane protein. *Journal of the American Chemical Society, 138*(36), 11526–11535. http://pubs.acs.org/journal/jacsat. doi:10.1021/jacs.6b07005.

Fleury, M., & Canet, D. (2014). Water orientation in smectites using NMR nutation experiments. *The Journal of Physical Chemistry C, 118*(9), 4733–4740. doi:10.1021/jp4118503.

Fogarty, A. C., Coudert, F. X., Boutin, A., & Laage, D. (2014b). Reorientational dynamics of water confined in zeolites. *ChemPhysChem, 15*(3), 521–529. http://onlinelibrary.wiley.com/journal/10.1002/(ISSN)1439-7641. doi:10.1002/cphc.201300928.

Fogarty, A. C., Duboué-Dijon, E., Laage, D., & Thompson, W. H. (2014a). Origins of the non-exponential reorientation dynamics of nanoconfined water. *Journal of Chemical Physics, 141*(18). http://scitation.aip.org/content/aip/journal/jcp. doi:10.1063/1.4896983.

Fogarty, A. C., & Laage, D. (2014). Water dynamics in protein hydration shells: The molecular origins of the dynamical perturbation. *Journal of Physical Chemistry B, 118*(28), 7715–7729. http://pubs.acs.org/journal/jpcbfk. doi:10.1021/jp409805p.

Foglia, F., Lawrence, M. J., Lorenz, C. D., & McLain, S. E. (2010). On the hydration of the phosphocholine headgroup in aqueous solution. *Journal of Chemical Physics, 133*(14). doi:10.1063/1.3488998.

Folpini, G., Siebert, T., Woerner, M., Abel, S., Laage, D., & Elsaesser, T. (2017). Water librations in the hydration shell of phospholipids. *The Journal of Physical Chemistry Letters, 8*(18), 4492–4497. doi:10.1021/acs.jpclett.7b01942.

Fransson, T., & Pettersson, L. G. M. (2024). TDDFT and the x-ray absorption spectrum of liquid water: Finding the "best" functional. *The Journal of Chemical Physics, 160*, 234105.

Frembgen-Kesner, T., & Elcock, A. H. (2013). Computer simulations of the bacterial cytoplasm. *Biophysical Reviews, 5*(2), 109–119. http://www.springer.com/life+sciences/biochemistry+%26+biophysics/journal/12551. doi:10.1007/s12551-013-0110-6.

Fuentes-Lemus, E., & Davies, M. J. (2023). Effect of crowding, compartmentalization and nanodomains on protein modification and redox signaling – Current state and future challenges. *Free Radical Biology and Medicine, 196*, 81–92. Www.elsevier.com/locate/freeradbiomed. doi:10.1016/j.freeradbiomed.2023.01.011.

Gainaru, C., Fillmer, A., & Böhmer, R. (2009). Dielectric response of deeply supercooled hydration water in the connective tissue proteins collagen and elastin. *The Journal of Physical Chemistry B, 113*(38), 12628–12631. doi:10.1021/jp9065899.

Galamba, N., & Tetrahedrons, W. (2014). Hydrogen-bond dynamics, and the orientational mobility of water around hydrophobic solutes. *The Journal of Physical Chemistry B, 118*(15), 4169–4176. doi:10.1021/jp500067a.

Gallo, P., Amann-Winkel, K., Angell, C. A., Anisimov, M. A., Caupin, F. rédé, Chakravarty, C., Lascaris, E., Loerting, T., Panagiotopoulos, A. Z., Russo, J., Sellberg, J. A., Stanley, H. E., Tanaka, H., Vega, C., Xu, L., & Pettersson, L. G. M. (2016). Water: A tale of two liquids. *Chemical Reviews, 116*(13), 7463–7500. doi:10.1021/acs.chemrev.5b00750.

García, A. E., & Hummer, G. (2000). Water penetration and escape in proteins. *Proteins: Structure, Function, and Genetics, 38*(3), 261–272. doi:10.1002/(SICI)1097-0134(20000215)38:3⟨261::AID-PROT3⟩3.3.CO;2-H.

Geethu, P. M., Yadav, I., Deshpande, S. K., Aswal, V. K., & Satapathy, D. K. (2017). Soft confinement effects on dynamics of hydrated gelatin. *Macromolecules, 50*(17), 6518–6528. doi:10.1021/acs.macromol.7b01521.

Gekle, S., & Netz, R. R. (2012). Anisotropy in the dielectric spectrum of hydration water and its relation to water dynamics. *Journal of Chemical Physics, 137*(10). doi:10.1063/1.4749380.

Ghanta, K. P., Mondal, S., & Bandyopadhyay, S. (2022). Exploring the dynamic heterogeneity at the interface of a protein in aqueous ionic liquid solutions. *Journal of Physical Chemistry B, 126*(38), 7271–7285. http://pubs.acs.org/journal/jpcbfk. doi:10.1021/acs.jpcb.2c03940.

Ghosh, B., Layek, S., Bhattacharyya, D., & Sengupta, N. (2023a). Base pair compositional variability influences DNA structural stability and tunes hydration thermodynamics and dynamics. *The Journal of Chemical Physics, 159*(9). doi:10.1063/5.0154977.

Ghosh, B., Layek, S., & Sengupta, N. (2023b). Hydration behavior under confinement and in the presence of a cosolvent: An exploration based on molecular dynamics simulation. *Journal of Chemical Sciences, 135*(3). doi:10.1007/s12039-023-02201-1.

Ghosh, R., Banerjee, S., Hazra, M., Roy, S., & Bagchi, B. (2014). Sensitivity of polarization fluctuations to the nature of protein-water interactions: Study of biological water in four different protein-water systems. *The Journal of Chemical Physics, 141*(22). doi:10.1063/1.4902821.

Giovambattista, N., Debenedetti, P. G., & Rossky, P. J. (2007). Effect of surface polarity on water contact angle and interfacial hydration structure. *The Journal of Physical Chemistry B, 111*(32), 9581–9587. doi:10.1021/jp071957s.

Gopal, S. M., Klumpers, F., Herrmann, C., & Schafer, L. V. (2017). Solvent effects on ligand binding to a serine protease. *Physical Chemistry Chemical Physics, 19*, 10753–10766.

Gruenbaum, S. M., & Skinner, J. L. (2011). Vibrational spectroscopy of water in hydrated lipid multi-bilayers. I. Infrared spectra and ultrafast pump-probe observables. *The Journal of Chemical Physics, 135*(7), 075101. doi:10.1063/1.3615717.

Halle, B., & Nilsson, L. (2009). Does the dynamic stokes shift report on slow protein hydration dynamics? *The Journal of Physical Chemistry B, 113*(24), 8210–8213. doi:10.1021/jp9027589.

Hanot, S., Lyonnard, S., & Mossa, S. (2015). Water confined in self-assembled ionic surfactant nano-structures. *Soft Matter, 11*(12), 2469–2478. doi:10.1039/C5SM00179J.

Hanot, S., Lyonnard, S., & Mossa, S. (2016). Sub-diffusion and population dynamics of water confined in soft environments. *Nanoscale, 8*(6), 3314–3325. doi:10.1039/C5NR05853H.

Harada, R., Sugita, Y., & Feig, M. (2012). Protein crowding affects hydration structure and dynamics. *Journal of the American Chemical Society, 134*(10), 4842–4849. doi:10.1021/ja211115q.

Heinz, L. P., & Grubmüller, H. (2023). Why solvent response contributions to solvation free energies are compatible with Ben-Naim's theorem. *Journal of Chemical Theory and Computation, 19*(22), 8013–8019. doi:10.1021/acs.jctc.3c00655.

Heyden, M. (2014). Resolving anisotropic distributions of correlated vibrational motion in protein hydration water. *The Journal of Chemical Physics, 141*(22). doi:10.1063/1.4896073.

Heyden, M. (2019). Heterogeneity of water structure and dynamics at the protein-water interface. *The Journal of Chemical Physics, 150*(9). doi:10.1063/1.5081739.

Heyden, M., Tobias, D. J., & Matyushov, D. V. (2012). Terahertz absorption of dilute aqueous solutions. *Journal of Chemical Physics, 137*(23). doi:10.1063/1.4772000.

Huang, R., & Lau, B. L. T. (2016). Biomolecule-nanoparticle interactions: Elucidation of the thermodynamics by isothermal titration calorimetry. *Biochimica et Biophysica Acta—General Subjects, 1860*(5), 945–956. Www.elsevier.com/locate/bbagen. doi:10.1016/j.bbagen.2016.01.027.

Huggins, D. J. (2015). Quantifying the entropy of binding for water molecules in protein cavities by computing correlations. *Biophysical Journal, 108*(4), 928–936. http://www.sciencedirect.com/science/journal/00063495. doi:10.1016/j.bpj.2014.12.035.

Hummer (2001). *Water conduction through the hydrophobic channel of a carbon nanotube, 418*, 188–190. doi:10.1038/35102535.

Inoue, K.-I., Singh, P. C., Nihonyanagi, S., Yamaguchi, S., & Tahara, T. (2017). Cooperative hydrogen-bond dynamics at a zwitterionic lipid/water interface revealed by 2D HD-VSFG spectroscopy. *The Journal of Physical Chemistry Letters, 8*(20), 5160–5165. doi:10.1021/acs.jpclett.7b02057.

Jana, B., Pal, S., & Bagchi, B. (2008). Hydrogen bond breaking mechanism and water reorientational dynamics in the hydration layer of lysozyme. *The Journal of Physical Chemistry B, 112*(30), 9112–9117. doi:10.1021/jp800998w.

Jana, B., Pal, S., Maiti, P. K., Lin, S. T., Hynes, J. T., & Bagchi, B. (2006). Entropy of water in the hydration layer of major and minor grooves of DNA. *Journal of Physical Chemistry B, 110*(39), 19611–19618. http://pubs.acs.org/journal/jpcbfk. doi:10.1021/jp061588k.

Ji, M., Odelius, M., & Gaffney, K. J. (2010). Large angular jump mechanism observed for hydrogen bond exchange in aqueous perchlorate solution. *Science, 328*(5981), 1003–1005. doi:10.1126/science.1187707.

Jin, T., Zhang, Q., & Zhuang, W. (2023). Toward understanding the mechanism of water dynamical onset on the material interfaces. *The Journal of Chemical Physics, 159*(22). doi:10.1063/5.0176362.

Joseph, S., & Aluru, N. R. (2008). Why are carbon nanotubes fast transporters of water? *Nano Letters, 8*(2), 452–458. doi:10.1021/nl072385q.

Kermani, A. A. (2021). A guide to membrane protein X-ray crystallography. *The FEBS Journal, 288*(20), 5788–5804. doi:10.1111/febs.15676.

Khah, A. M., Reinholdt, P., Nuernberger, P., Kongsted, J., & Hättig, C. (2020). Relaxation dynamics of the triazene compound Berenil in DNA-minor-groove confinement after photoexcitation. *Journal of Chemical Theory and Computation, 16*(8), 5203–5211. doi:10.1021/acs.jctc.0c00489.

Khodadadi, S., Pawlus, S., & Sokolov, A. P. (2008). Influence of hydration on protein dynamics: Combining dielectric and neutron scattering spectroscopy data. *The Journal of Physical Chemistry B, 112*(45), 14273–14280. doi:10.1021/jp8059807.

King, J. T., Arthur, E. J., Brooks, C. L., & Kubarych, K. J. (2014). Crowding induced collective hydration of biological macromolecules over extended distances. *Journal of the American Chemical Society, 136*(1), 188–194. doi:10.1021/ja407858c.

Kleinjung, J., & Fraternali, F. (2014). Design and application of implicit solvent models in biomolecular simulations. *Current Opinion in Structural Biology, 25*, 126–134. doi:10.1016/j.sbi.2014.04.003.

König, B., Pezzotti, S., Ramos, S., Schwaab, G., & Havenith, M. (2024). Real-time measure of solvation free energy changes upon liquid-liquid phase separation of α-elastin. *Biophysical Journal, 123*(11), 1367–1375. doi:10.1016/j.bpj.2023.07.023.

Konstantinovsky, D., Perets, E. A., Santiago, T. y, Velarde, L., Hammes-Schiffer, S., & Yan, E. C. Y. (2022). Detecting the first hydration shell structure around biomolecules at interfaces. *ACS Central Science, 8*(10), 1404–1414. doi:10.1021/acscentsci.2c00702.

Kopel, Y., & Giovambattista, N. (2019). Comparative study of water-mediated interactions between hydrophilic and hydrophobic nanoscale surfaces. *The Journal of Physical Chemistry B, 123*(50), 10814–10824. doi:10.1021/acs.jpcb.9b08725.

Kraemer, D., Cowan, M. L., Paarmann, A., Huse, N., Nibbering, E. T. J., Elsaesser, T., & Miller, R. J. D. (2008). Temperature dependence of the two-dimensional infrared spectrum of liquid H_2O. *Proceedings of the National Academy of Sciences, 105*(2), 437–442. doi:10.1073/pnas.0705792105.

Krott, L. B., & Barbosa, M. C. (2024). A simple model for density anomaly and excess thermodynamic properties of alcohol-like in water-like mixture. *Molecular Physics, 122*(21–22). doi:10.1080/00268976.2024.2352084.

Kumar, H., Dasgupta, C., & Maiti, P. K. (2014). Driving force of water entry into hydrophobic channels of carbon nanotubes: Entropy or energy? *Molecular Simulation, 41*(5-6), 504–511. doi:10.1080/08927022.2014.998211.

Kumar, H., Mukherjee, B., Lin, S. T., Dasgupta, C., Sood, A. K., & Maiti, P. K. (2011). Thermodynamics of water entry in hydrophobic channels of carbon nanotubes. *Journal of Chemical Physics, 134*(12). doi:10.1063/1.3571007.

Kuntz, I. D., & Kauzmann, W. (1974). Hydration of proteins and polypeptides. *Advances in Protein Chemistry, 28*(C), 239–345. doi:10.1016/S0065-3233(08)60232-6.

Laage, D., Elsaesser, T., & Hynes, J. T. (2017a). Perspective: Structure and ultrafast dynamics of biomolecular hydration shells. *Structural Dynamics, 4*(4). http://scitation.aip.org/content/aca/journal/sdy. doi:10.1063/1.4981019.

Laage, D., Elsaesser, T., & Hynes, J. T. (2017b). Water dynamics in the hydration shells of biomolecules. *Chemical Reviews, 117*(16), 10694–10725. doi:10.1021/acs.chemrev.6b00765.

Laage, D., & Hynes, J. T. (2006). A molecular jump mechanism of water reorientation. *Science, 311*(5762), 832–835. Www.sciencemag.org. doi:10.1126/science.1122154.

Laage, D., & Hynes, J. T. (2008). On the molecular mechanism of water reorientation. *The Journal of Physical Chemistry B, 112*(45), 14230–14242. doi:10.1021/jp805217u.

Laage, D., Stirnemann, G., Sterpone, F., & Hynes, J. T. (2012). Water jump reorientation: From theoretical prediction to experimental observation. *Accounts of Chemical Research, 45*(1), 53–62. doi:10.1021/ar200075u.

Laage, D., Stirnemann, G., Sterpone, F., Rey, R., & Hynes, J. T. (2011). Reorientation and allied dynamics in water and aqueous solutions. *Annual Reviews Inc., France Annual Review of Physical Chemistry, 62*, 395–416. http://www.annualreviews.org/journal/physchem. doi:10.1146/annurev.physchem.012809.103503.

Laage, D., & Thompson, W. H. (2012). Reorientation dynamics of nanoconfined water: Power-law decay, hydrogen-bond jumps, and test of a two-state model. *Journal of Chemical Physics, 136*(4). doi:10.1063/1.3679404.

Layek, S., & Sengupta, N. (2024). Response of foldable protein conformations to non-physiological perturbations: Interplay of thermal factors and confinement. *ChemPhysChem, 25*(23). doi:10.1002/cphc.202400618.

Lazaridis, T., & Karplus, M. (2002). Thermodynamics of protein folding: A microscopic view. *Biophysical Chemistry, 100*(1-3), 367–395. doi:10.1016/s0301-4622(02)00293-4.

Leoni, F., Calero, C., & Franzese, G. (2021). Nanoconfined fluids: Uniqueness of water compared to other liquids. *ACS Nano, 15*(12), 19864–19876. doi:10.1021/acsnano.1c07381.

Lin, S.-T., Blanco, M., & Goddard, W. A., III (2003). The two-phase model for calculating thermodynamic properties of liquids from molecular dynamics: Validation for the phase diagram of Lennard-Jones fluids. *The Journal of Chemical Physics, 119*, 11792–11805.

Lin, S. T., Maiti, P. K., & Goddard, W. A. (2010). Two-phase thermodynamic model for efficient and accurate absolute entropy of water from molecular dynamics simulations. *Journal of Physical Chemistry B, 114*(24), 8191–8198. http://pubs.acs.org/journal/jpcbfk. doi:10.1021/jp103120q.

Lindow, N., Baum, D., Bondar, A. N., & Hege, H. C. (2013). Exploring cavity dynamics in biomolecular systems. *BMC Bioinformatics, 14*(19). http://www.biomedcentral.com/bmcbioinformatics/. doi:10.1186/1471-2105-14-S19-S5.

Linse, J. B., & Hub, J. S. (2023). Scrutinizing the protein hydration shell from molecular dynamics simulations against consensus small-angle scattering data. *Communications Chemistry, 6*(1). https://www.nature.com/commschem/. doi:10.1038/s42004-023-01067-1.

Lu, Y., Lin, J., Wang, L., Zhang, L., & Cai, C. (2020). Self-assembly of copolymer micelles: Higher-level assembly for constructing hierarchical structure. *Chemical Reviews, 120*(9), 4111–4140. doi:10.1021/acs.chemrev.9b00774.

Mahanta, D. D., Brown, D. R., Webber, T., Pezzotti, S., Schwaab, G., Han, S., Shell, M. S., & Havenith, M. (2024). Bridging the gap in cryopreservation mechanism: Unraveling the interplay between structure, dynamics, and thermodynamics in cryoprotectant aqueous solutions. *Journal of Physical Chemistry B, 128*(15), 3720–3731. http://pubs.acs.org/journal/jpcbfk. doi:10.1021/acs.jpcb.4c00264.

Maiti, S., & Heyden, M. (2023). Model-dependent solvation of the K-18 domain of the intrinsically disordered protein tau. *The Journal of Physical Chemistry B, 127*(33), 7220–7230. doi:10.1021/acs.jpcb.3c01726.

Majumdar, J., Dasgupta, S., Mandal, S., Moid, M., Jain, M., & Maiti, P. K. (2023). Does twist angle affect the properties of water confined inside twisted bilayer graphene? *Journal of Chemical Physics, 158*(3). http://scitation.aip.org/content/aip/journal/jcp. doi:10.1063/5.0139256.

Majumdar, J., Mandal, S., Rajan, A. G., & Maiti, P. K. (2024). Similar structure but different thermodynamic, dielectric, and frictional properties of confined water in twisted 2D materials: MoS_2 vs. graphene. *Nanoscale, 17*(4), 2354–2364. http://pubs.rsc.org/en/journals/journal/nr. doi:10.1039/d4nr03821e.

Mallamace, F., Corsaro, C., Mallamace, D., Vasi, C., & Stanley, H. E. (2013). The thermodynamical response functions and the origin of the anomalous behavior of liquid water. *Faraday Discussions, 167*, 95–108. http://www.rsc.org/faraday_d. doi:10.1039/c3fd00073g.

Markelz, A. G., Roitberg, A., & Heilweil, E. J. (2000). Pulsed terahertz spectroscopy of DNA, bovine serum albumin and collagen between 0.1 and 2.0 THz. *Chemical Physics Letters, 320*(1-2), 42–48. doi:10.1016/s0009-2614(00)00227-x.

Martin, D. R., & Matyushov, D. V. (2015). Dipolar nanodomains in protein hydration shells. *The Journal of Physical Chemistry Letters, 6*(3), 407–412. doi:10.1021/jz5025433.

Martinez, A. V., Dominguez, L., Małolepsza, E., Moser, A., Ziegler, Z., & Straub, J. E. (2013). Probing the structure and dynamics of confined water in AOT reverse micelles. *Journal of Physical Chemistry B, 117*(24), 7345–7351. http://pubs.acs.org/journal/jpcbfk. doi:10.1021/jp402270e.

Mattea, C., Qvist, J., & Halle, B. (2008). Dynamics at the protein-water interface from 17O spin relaxation in deeply supercooled solutions. *Biophysical Journal, 95*(6), 2951–2963. doi:10.1529/biophysj.108.135194.

McGuffee, S. R., & Elcock, A. H. (2010). Diffusion, crowding & protein stability in a dynamic molecular model of the bacterial cytoplasm. *PLoS Computational Biology, 6*(3). http://www.ploscompbiol.org/article/info%3Adoi%2F10.1371%2Fjournal.pcbi.1000694. doi:10.1371/journal.pcbi.1000694.

Meisburger, S. P., Pabit, S. A., & Pollack, L. (2015). Determining the locations of ions and water around DNA from X-ray scattering measurements. *Biophysical Journal, 108*(12), 2886–2895. http://www.sciencedirect.com/science/journal/00063495. doi:10.1016/j.bpj.2015.05.006.

Mittag, T., & Pappu, R. V. (2022). A conceptual framework for understanding phase separation and addressing open questions and challenges. *Molecular Cell, 82*(12), 2201–2214. Www.molecule.org. doi:10.1016/j.molcel.2022.05.018.

Moid, M., Finkelstein, Y., Moreh, R., & Maiti, P. K. (2022). Anisotropy of the proton kinetic energy as a tool for capturing structural transition in water confined in a graphene nanoslit pore. *The Journal of Physical Chemistry Letters, 13*(2), 455–461. doi:10.1021/acs.jpclett.1c03086.

Moilanen, D. E., Levinger, N. E., Spry, D. B., & Fayer, M. D. (2007). Confinement or the nature of the interface? Dynamics of nanoscopic water. *Journal of the American Chemical Society, 129*(46), 14311–14318. doi:10.1021/ja073977d.

Moilanen, D. E., Wong, D., Rosenfeld, D. E., Fenn, E. E., & Fayer, M. D. (2009). Ion-water hydrogen-bond switching observed with 2D IR vibrational echo chemical exchange spectroscopy. *Proceedings of the National Academy of Sciences of the United States of America, 106*(2), 375–380. http://www.pnas.org/content/106/2/375.full.pdf. doi:10.1073/pnas.0811489106.

Møller, K. B., Rey, R., & Hynes, J. T. (2004). Hydrogen bond dynamics in water and ultrafast infrared spectroscopy: A theoretical study. *The Journal of Physical Chemistry A, 108*(7), 1275–1289. doi:10.1021/jp035935r.

Mondal, S., & Bagchi, B. (2022). From structure and dynamics to biomolecular functions: The ubiquitous role of solvent in biology. *Current Opinion in Structural Biology, 77*, 102462. doi:10.1016/j.sbi.2022.102462.

Mondal, S., & Bandyopadhyay, S. (2019). Flexibility of the binding regions of a protein–DNA complex and the structure and ordering of interfacial water. *Journal of Chemical Information and Modeling, 59*(10), 4427–4437. doi:10.1021/acs.jcim.9b00685.

Mondal, S., & Bandyopadhyay, S. (2020). Heterogeneous dynamical environment at the interface of a protein–DNA complex. *Langmuir, 36*(17), 4567–4581. doi:10.1021/acs.langmuir.9b03175.

Mondal, S., Prasad Ghanta, K., Mondal, S., & Bandyopadhyay, S. (2023). Vibrational spectrum and randomness of water at the interface of a protein–DNA complex. *Theoretical Chemistry Accounts, 142*(8). doi:10.1007/s00214-023-03017-5.

Morawietz, T., Singraber, A., Dellago, C., & Behler, J. ö (2016). How van der Waals interactions determine the unique properties of water. *Proceedings of the National Academy of Sciences, 113*(30), 8368–8373. doi:10.1073/pnas.1602375113.

Morón, M. C. (2021). Protein hydration shell formation: Dynamics of water in biological systems exhibiting nanoscopic cavities. *Journal of Molecular Liquids, 337*, 116584. doi:10.1016/j.molliq.2021.116584.

Mukherjee, S., Ramos, S., Pezzotti, S., Kalarikkal, A., Prass, T. M., Galazzo, L., Gendreizig, D., Barbosa, N., Bordignon, E., Havenith, M., & Schäfer, L. V. (2024). Entropy tug-of-war determines solvent effects in the liquid-liquid phase separation of a globular protein. *Journal of Physical Chemistry Letters, 15*(15), 4047–4055. http://pubs.acs.org/journal/jpclcd. doi:10.1021/acs.jpclett.3c03421.

Mukherjee, S., & Schäfer, L. V. (2022). Spatially resolved hydration thermodynamics in biomolecular systems. *The Journal of Physical Chemistry B, 126*(20), 3619–3631. doi:10.1021/acs.jpcb.2c01088.

Mukherjee, S., & Schäfer, L. V. (2023). Thermodynamic forces from protein and water govern condensate formation of an intrinsically disordered protein domain. *Nature Communications, 14*(1). https://www.nature.com/ncomms/. doi:10.1038/s41467-023-41586-y.

Myneni, S., Luo, Y., Naslund, L. A., Cavalleri, M., Ojamäe, L., Ogasawara, H., Pelmenschikov, A., Wernet, P. h, Väterlein, P., Heske, C., Hussain, Z., Pettersson, L. G. M., & Nilsson, A. (2002). Spectroscopic probing of local hydrogen-bonding structures in liquid water. *Journal of Physics: Condensed Matter, 14*(8), L213. doi:10.1088/0953-8984/14/8/106.

Nagle, J. F., & Tristram-Nagle, S. (2000). Structure of lipid bilayers. *Biochimica et Biophysica Acta—Reviews on Biomembranes, 1469*(3), 159–195. doi:10.1016/S0304-4157(00)00016-2.

Nakano, S.-, & Sugimoto, N. (2017). Model studies of the effects of intracellular crowding on nucleic acid interactions. *Molecular BioSystems, 13*(1), 32–41. doi:10.1039/c6mb00654j.

Nandi, N., & Bagchi, B. (1997). Dielectric relaxation of biological water. *The Journal of Physical Chemistry B, 101*(50), 10954–10961. doi:10.1021/jp971879g.

Nandi, N., Bhattacharyya, K., & Bagchi, B. (2000). Dielectric relaxation and solvation dynamics of water in complex chemical and biological systems. *Chemical Reviews, 100*(6), 2013–2046. doi:10.1021/cr980127v.

Neophytou, A., Chakrabarti, D., & Sciortino, F. (2022). Topological nature of the liquid–liquid phase transition in tetrahedral liquids. *Nature Physics, 18*, 1248–1253.

Ngai, K. L., Capaccioli, S., Shahin Thayyil, M., & Shinyashiki, N. (2010). Resolution of problems in soft matter dynamics by combining calorimetry and other spectroscopies. *Journal of Thermal Analysis and Calorimetry, 99*(1), 123–138. doi:10.1007/s10973-009-0500-y.

Nguyen, C. N., Cruz, A., Gilson, M. K., & Kurtzman, T. (2014). Thermodynamics of water in an enzyme active site: Grid-based hydration analysis of coagulation factor Xa. *Journal of Chemical Theory and Computation, 10*(7), 2769–2780. doi:10.1021/ct401110x.

Nguyen, C. N., Kurtzman, T., & Gilson, M. K. (2016). Spatial decomposition of translational water–water correlation entropy in binding pockets. *Journal of Chemical Theory and Computation, 12*(1), 414–429. doi:10.1021/acs.jctc.5b00939.

Nguyen, C. N., Kurtzman Young, T., & Gilson, M. K. (2012). Grid inhomogeneous solvation theory: Hydration structure and thermodynamics of the miniature receptor cucurbit[7]uril. *Journal of Chemical Physics, 137*(4), 973–980. doi:10.1063/1.4733951.

Ni, Z., Tan, J., Luo, Y. i, & Ye, S. (2025). Dynamic protein hydration water mediates the aggregation kinetics of amyloid β peptides at interfaces. *Journal of Colloid and Interface Science, 679*(Pt B), 539–546. doi:10.1016/j.jcis.2024.10.105.

Nibali, V. C., & Havenith, M. (2014). New insights into the role of water in biological function: Studying solvated biomolecules using terahertz absorption spectroscopy in conjunction with molecular dynamics simulations. *Journal of the American Chemical Society, 136*(37), 12800–12807. doi:10.1021/ja504441h.

Nibali, V. C., Maiti, S., Saija, F., Heyden, M., & Cassone, G. (2023). Electric-field induced entropic effects in liquid water. *The Journal of Chemical Physics, 158*(18). doi:10.1063/5.0139460.

Novelli, F., Chen, K., Buchmann, A., Ockelmann, T., Hoberg, C., Head-Gordon, T., & Havenith, M. (2023). The birth and evolution of solvated electrons in the water. *Proceedings of the National Academy of Sciences of the United States of America, 120*(8). https://www.pnas.org/doi/10.1073/pnas.2216480120. doi:10.1073/pnas.2216480120.

Nucci, N. V., Pometun, M. S., & Wand, A. J. (2011a). Mapping the hydration dynamics of ubiquitin. *Journal of the American Chemical Society, 133*(32), 12326–12329. doi:10.1021/ja202033k.

Nucci, N. V., Pometun, M. S., & Wand, A. J. (2011b). Site-resolved measurement of water-protein interactions by solution NMR. *Nature Structural and Molecular Biology, 18*(2), 245–250. doi:10.1038/nsmb.1955.

Oh, K. I., & Baiz, C. R. (2018). Crowding stabilizes DMSO-water hydrogen-bonding interactions. *Journal of Physical Chemistry B, 122*(22), 5984–5990. http://pubs.acs.org/journal/jpcbfk. doi:10.1021/acs.jpcb.8b02739.

Oh, K. I., You, X., Flanagan, J. C., & Baiz, C. R. (2020). Liquid–liquid phase separation produces fast H-bond dynamics in DMSO-water mixtures. *Journal of Physical Chemistry Letters, 11*(5), 1903–1908. http://pubs.acs.org/journal/jpclcd. doi:10.1021/acs.jpclett.0c00378.

Pabit, S. A., Meisburger, S. P., Li, L., Blose, J. M., Jones, C. D., & Pollack, L. (2010). Counting ions around DNA with anomalous small-angle X-ray scattering. *Journal of the American Chemical Society, 132*(46), 16334–16336. doi:10.1021/ja107259y.

Pakravan, D., Michiels, E., Bratek-Skicki, A., De Decker, M., Van Lindt, J., Alsteens, D., Derclaye, S., Van Damme, P., Schymkowitz, J., Rousseau, F., Tompa, P., & Van Den Bosch, L. (2021). Liquid–liquid phase separation enhances TDP-43 LCD aggregation but delays seeded aggregation. *Biomolecules, 11*(4), 548. doi:10.3390/biom11040548.

Pal, S., & Mitra, R. K. (2022). Investigation on the effect of nonpolar amino acids as macromolecular crowders on the stability of globular proteins. *Chemical Thermodynamics and Thermal Analysis, 6*, 100044. doi:10.1016/j.ctta.2022.100044.

Pal, S., Samanta, N., Das Mahanta, D., Mitra, R. K., & Chattopadhyay, A. (2018). Effect of phospholipid headgroup charge on the structure and dynamics of water at the membrane interface: A terahertz spectroscopic study. *Journal of Physical Chemistry B, 122*(19), 5066–5074. http://pubs.acs.org/journal/jpcbfk. doi:10.1021/acs.jpcb.8b01633.

Pal, S. K., Peon, J., Bagchi, B., & Zewail, A. H. (2002). Biological water: Femtosecond dynamics of macromolecular hydration. *Journal of Physical Chemistry B, 106*(48), 12376–12395. doi:10.1021/jp0213506.

Pappu, R. V., Cohen, S. R., Dar, F., Farag, M., & Kar, M. (2023). Phase transitions of associative biomacromolecules. *Chemical Reviews, 123*(14), 8945–8987. http://pubs.acs.org/journal/chreay. doi:10.1021/acs.chemrev.2c00814.

Parra, J.G., Iza, P., Dominguez, H., Schott, E. & Zarat, X. (2024). Unveiling the hydrophilic nature of SDS surfactant through molecular simulations: Exploring the influence of charge distribution on interfacial properties in the vacuum/SDS/water system. Journal of Molecul Liquids. *401*, 124692.

Pasenkiewicz-Gierula, M., Takaoka, Y., Miyagawa, H., Kitamura, K., & Kusumi, A. (1997). Hydrogen bonding of water to phosphatidylcholine in the membrane as studied by a molecular dynamics simulation: Location, geometry, and lipid–lipid bridging via hydrogen-bonded water. *The Journal of Physical Chemistry A, 101*(20), 3677–3691. doi:10.1021/jp962099v.

Päslack, C., Das, C. K., Schlitter, J. ü, & Schäfer, L. V. (2021). Spectrally resolved estimation of water entropy in the active site of human carbonic anhydrase II. *Journal of Chemical Theory and Computation, 17*(8), 5409–5418. doi:10.1021/acs.jctc.1c00554.

Päslack, C., Schäfer, L. V., & Heyden, M. (2021). Protein flexibility reduces solvent-mediated friction barriers of ligand binding to a hydrophobic surface patch. *Physical Chemistry Chemical Physics, 23*(9), 5665–5672. http://pubs.rsc.org/en/journals/journal/cp. doi:10.1039/d1cp00181g.

Pathak, A. K., & Bandyopadhyay, T. (2019). Temperature induced dynamical transition of biomolecules in polarizable and nonpolarizable TIP3P water. *Journal of Chemical Theory and Computation, 15*(4), 2706–2718. http://pubs.acs.org/journal/jctcce. doi:10.1021/acs.jctc.9b00005.

Patra, M., Karttunen, M., Hyvönen, M. T., Falck, E., Lindqvist, P., & Vattulainen, I. (2003). Molecular dynamics simulations of lipid Bilayers: Major artifacts due to truncating electrostatic interactions. *Biophysical Journal, 84*(6), 3636–3645. doi:10.1016/s0006-3495(03)75094-2.

Pearson, R. H., & Pascher, I. (1979). The molecular structure of lecithin dihydrate. *Nature, 281*(5731), 499–501. doi:10.1038/281499a0.

Persson, E., & Halle, B. (2008). Cell water dynamics on multiple time scales. *Proceedings of the National Academy of Sciences, 105*(17), 6266–6271. doi:10.1073/pnas.0709585105.

Persson, R. A. X., Pattni, V., Singh, A., Kast, S. M., & Heyden, M. (2017). Signatures of solvation thermodynamics in spectra of intermolecular vibrations. *Journal of Chemical Theory and Computation, 13*(9), 4467–4481. http://pubs.acs.org/journal/jctcce. doi:10.1021/acs.jctc.7b00184.

Pettersson, L. G. M., Henchman, R. H., & Nilsson, A. (2016). Water—The most anomalous liquid. *Chemical Reviews, 116*(13), 7459–7462. http://pubs.acs.org/journal/chreay. doi:10.1021/acs.chemrev.6b00363.

Pezzotti, S., König, B., Ramos, S., Schwaab, G., & Havenith, M. (2023). Liquid–liquid phase separation? Ask the water!. *The Journal of Physical Chemistry Letters, 14*(6), 1556–1563. doi:10.1021/acs.jpclett.2c02697.

Pyne, P., Samanta, N., Gohil, H., Prabhu, S. S., & Mitra, R. K. (2021). Alteration of water absorption in the THz region traces the onset of fibrillation in proteins. *Chemical Communications, 57*(8), 998–1001. http://pubs.rsc.org/en/journals/journal/cc. doi:10.1039/d0cc06500e.

Qin, Y., Zhang, L., Wang, L., & Zhong, D. (2017). Observation of the global dynamic collectivity of a hydration shell around apomyoglobin. *The Journal of Physical Chemistry Letters, 8*(6), 1124–1131. doi:10.1021/acs.jpclett.7b00205.

Raza, S., Hayat, A., Bashir, T., Chen, C., Shen, L., Orooji, Y., & Lin, H. (2024). Electrochemistry of 2D-materials for the remediation of environmental pollutants and alternative energy storage/conversion materials and devices, a comprehensive review. *Sustainable Materials and Technologies, 40*, e00963. doi:10.1016/j.susmat.2024.e00963.

Rezaei-Ghaleh, N. (2022). Water dynamics in highly concentrated salt solutions: A multi-nuclear NMR approach. *ChemistryOpen, 11*(6). doi:10.1002/open.202200080.

Ribeiro, S. S., Samanta, N., Ebbinghaus, S., & Marcos, J. C. (2019). The synergic effect of water and biomolecules in intracellular phase separation. *Nature Reviews Chemistry, 3*(9), 552–561. https://www.nature.com/natrevchem/. doi:10.1038/s41570-019-0120-4.

Roberts, S. T., Ramasesha, K., & Tokmakoff, A. (2009). Structural rearrangements in water viewed through two-dimensional infrared spectroscopy. *Accounts of Chemical Research, 42*(9), 1239–1249. doi:10.1021/ar900088g.

Roche, C. J., Dantsker, D., Heller, E. R., Sabat, J. E., & Friedman, J. M. (2013). Reverse micelles as a tool for probing solvent modulation of protein dynamics: Reverse micelle encapsulated hemoglobin. *Chemical Physics, 422*, 88–97. http://www.journals.elsevier.com/chemical-physics/. doi:10.1016/j.chemphys.2013.04.006.

Rossky, P. J., & Karplus, M. (1979). Solvation. A molecular dynamics study of a dipeptide in water. *Journal of the American Chemical Society, 101*(8), 1913–1937. doi:10.1021/ja00502a001.

Roy, P., Ghosh, B., Chatterjee, P., & Sengupta, N. (2019). Cosolvent impurities in SWCNT nanochannel confinement: Length dependence of water dynamics investigated with atomistic simulations. *Journal of Chemical Information and Modeling, 59*(5), 2026–2034. doi:10.1021/acs.jcim.8b00889.

Roy, P., & Sengupta, N. (2021). Hydration of a small protein under carbon nanotube confinement: Adsorbed substates induce selective separation of the dynamical response. *The Journal of Chemical Physics, 154*, 204702.

Rudling, A., Orro, A., & Carlsson, J. (2018). Prediction of ordered water molecules in protein binding sites from molecular dynamics simulations: The impact of ligand binding on hydration networks. *Journal of Chemical Information and Modeling, 58*(2), 350–361. doi:10.1021/acs.jcim.7b00520.

Ruiz-Barragan, S., Muñoz-Santiburcio, D., Körning, S., & Marx, D. (2020). Quantifying anisotropic dielectric response properties of nanoconfined water within graphene slit pores. *Physical Chemistry Chemical Physics, 22*(19), 10833–10837. doi:10.1039/d0cp00916d.

Saenger, W. (1987). Structure and dynamics of water surrounding biomolecules. *Annual Review of Biophysics and Biophysical Chemistry, 16*(1), 93–114. doi:10.1146/annurev.bb.16.060187.000521.

Saha, R., & Mitra, R. K. (2022). Trivalent cation-induced phase separation in proteins: Ion specific contribution in hydration also counts. *Physical Chemistry Chemical Physics, 24*(38), 23661–23668. doi:10.1039/d2cp01061e.

Saha, R., & Mitra, R. K. (2023). Thermo-resistive phase behavior of trivalent ion-induced microscopic protein-rich phases: Correlating with ion-specific protein hydration. *Langmuir, 39*(13), 4601–4610. http://pubs.acs.org/journal/langd5. doi:10.1021/acs.langmuir.2c03302.

Sahu, S., Garg, A., Saini, R., & Debnath, A. (2024). Interface water assists in dimethyl sulfoxide crossing and poration in model Bilayer. *Langmuir, 40*(11), 5764–5775. doi:10.1021/acs.langmuir.3c03498.

Schaaf, C., & Gekle, S. (2016). Spatially resolved dielectric constant of confined water and its connection to the non-local nature of bulk water. *The Journal of Chemical Physics, 145*(8). doi:10.1063/1.4960775.

Schneider, B., Patel, K., & Berman, H. M. (1998). Hydration of the phosphate group in double-helical DNA. *Biophysical Journal, 75*(5), 2422–2434. http://www.sciencedirect.com/science/journal/00063495. doi:10.1016/S0006-3495(98)77686-6.

Schotte, F., Cho, H. S., Kaila, V. R. I., Kamikubo, H., Dashdorj, N., Henry, E. R., Graber, T. J., Henning, R., Wulff, M., Hummer, G., Kataoka, M., & Anfinrud, P. A. (2012). Watching a signaling protein function in real time via 100-ps time-resolved Laue crystallography. *Proceedings of the National Academy of Sciences of the United States of America, 109*(47), 19256–19261. http://www.pnas.org/content/109/47/19256.full.pdf+html. doi:10.1073/pnas.1210938109.

Scott, D. J., & Winzor, D. J. (2024). Extreme enthalpy–entropy compensation in the dimerization of small solutes in aqueous solution. *European Biophysics Journal, 53*(7-8), 373–384. doi:10.1007/s00249-024-01722-y.

Sengupta, N., Jaud, S., & Tobias, D. J. (2008). Hydration dynamics in a partially denatured ensemble of the globular protein human α-lactalbumin investigated with molecular dynamics simulations. *Biophysical Journal, 95*(11), 5257–5267. http://download.cell.com/biophysj/pdf/PIIS0006349508789513.pdf?intermediate=true. doi:10.1529/biophysj.108.136531.

Shafiei, M., Domaros, M., Bratko, D., & Luzar, A. (2019). Anisotropic structure and dynamics of water under static electric fields. *The Journal of Chemical Physics, 150*(7). doi:10.1063/1.5079393.

Sharma, V. K., Mitra, S., & Mukhopadhyay, R. (2019). Dynamic landscape in self-assembled surfactant aggregates. *Langmuir, 35*(44), 14151–14172. doi:10.1021/acs.langmuir.8b03596.

Sheu, S.-Y., Liu, Y. u-C., Zhou, J.-K., Schlag, E. W., & Yang, D.-Y. (2019). Surface topography effects of globular biomolecules on hydration water. *The Journal of Physical Chemistry B, 123*(32), 6917–6932. doi:10.1021/acs.jpcb.9b03734.

Shin, Y., & Brangwynne, C. P. (2017). Liquid phase condensation in cell physiology and disease. *Science, 357*(6357). http://science.sciencemag.org/content/sci/357/6357/eaaf4382.full.pdf. doi:10.1126/science.aaf4382.

Shinoda, K., Carlsson, A., & Lindman, B. jö (1996). On the importance of hydroxyl groups in the polar head-group of nonionic surfactants and membrane lipids. *Advances in Colloid and Interface Science, 64*, 253–271. doi:10.1016/0001-8686(95)00287-1.

Shweta, H., & Sen, S. (2018). Dynamics of water and ions around DNA: What is so special about them? *Journal of Biosciences, 43*(3), 499–518. doi:10.1007/s12038-018-9771-4.

Singh, H., Das, C. K., Vasa, S. K., Grohe, K., Schäfer, L. V., & Linser, R. (2020). The active site of a prototypical "rigid" drug target is marked by extensive conformational dynamics. *Angewandte Chemie International Edition, 59*(51), 22916–22921. doi:10.1002/anie.202009348.

Singh, H., Vasa, S. K., Jangra, H., Rovó, P., Päslack, C., Das, C. K., Zipse, H., Schäfer, L. V., & Linser, R. (2019). Fast microsecond dynamics of the protein–water network in the active site of human carbonic anhydrase ii studied by solid-state NMR spectroscopy. *Journal of the American Chemical Society, 141*(49), 19276–19288. doi:10.1021/jacs.9b05311.

Singh, P. C., Inoue, K. I., Nihonyanagi, S., Yamaguchi, S., & Tahara, T. (2016). Femtosecond hydrogen bond dynamics of bulk-like and bound water at positively and negatively charged lipid interfaces revealed by 2D

HD-VSFG spectroscopy. *Angewandte Chemie—International Edition, 55*(36), 10621–10625. http://onlinelibrary.wiley.com/journal/10.1002/(ISSN)1521-3773. doi:10.1002/anie.201603676.

Sinha, S. K., & Bandyopadhyay, S. (2011). Dynamic properties of water around a protein-DNA complex from molecular dynamics simulations. *Journal of Chemical Physics, 135*(13). doi:10.1063/1.3634004.

Skinner, J. L., Pieniazek, P. A., & Gruenbaum, S. M. (2012). Vibrational spectroscopy of water at interfaces. *Accounts of Chemical Research, 45*(1), 93–100.

Skorb, E. V., & Andreeva, D. V. (2013). Surface nanoarchitecture for bio-applications: Self-regulating intelligent interfaces. *Advanced Functional Materials, 23*(36), 4483–4506. doi:10.1002/adfm.201203884.

Sobhia, M. E., Ghosh, K., Kumar, G. S., Sivangula, S., Laddha, K., Kumari, S., & Kumar, H. (2022). The role of water network chemistry in proteins: A structural bioinformatics perspective in drug discovery and development. *Current Topics in Medicinal Chemistry, 22*(20), 1636–1653. http://www.eurekaselect.com/615/journal/current-topics-medicinal-chemistry. doi:10.2174/1568026622666220726114407.

Srinivasan, J., Cheatham, T. E., Cieplak, P., Kollman, P. A., & Case, D. A. (1998). Continuum solvent studies of the stability of DNA, RNA, and phosphoramidate–DNA helices. *Journal of the American Chemical Society, 120*, 9401–9409. doi:10.1021/ja981844%2B.

Sterpone, F., Melchionna, S., Tuffery, P., Pasquali, S., Mousseau, N., Cragnolini, T., Chebaro, Y., St-Pierre, J.-F., Kalimeri, M., Barducci, A., Laurin, Y., Tek, A., Baaden, M., Hoang Nguyen, P., & Derreumaux, P. (2014). The OPEP protein model: From single molecules, amyloid formation, crowding and hydrodynamics to DNA/RNA systems. *Chemical Society Reviews, 43*(13), 4871–4893. doi:10.1039/c4cs00048j.

Sterpone, F., Stirnemann, G., Hynes, J. T., & Laage, D. (2010). Water hydrogen-bond dynamics around amino acids: The key role of hydrophilic hydrogen-bond acceptor groups. *Journal of Physical Chemistry B, 114*(5), 2083–2089. http://pubs.acs.org/journal/jpcbfk. doi:10.1021/jp9119793.

Sterpone, F., Stirnemann, G., & Laage, D. (2012). Magnitude and molecular origin of water slowdown next to a protein. *Journal of the American Chemical Society, 134*(9), 4116–4119. doi:10.1021/ja3007897.

Stirnemann, G., Rossky, P. J., Hynes, J. T., & Laage, D. (2010). Water reorientation, hydrogen-bond dynamics and 2D-IR spectroscopy next to an extended hydrophobic surface. *Faraday Discussions, 146*, 263–281. doi:10.1039/b925673c.

Stirnemann, G., Wernersson, E., Jungwirth, P., & Laage, D. (2013). Mechanisms of acceleration and retardation of water dynamics by ions. *Journal of the American Chemical Society, 135*(32), 11824–11831. doi:10.1021/ja405201s.

Strom, A. R., Emelyanov, A. V., Mir, M., Fyodorov, D. V., Darzacq, X., & Karpen, G. H. (2017). Phase separation drives heterochromatin domain formation. *Nature, 547*(7662), 241–245. http://www.nature.com/nature/index.html. doi:10.1038/nature22989.

Su, J., & Guo, H. (2011). Effect of nanotube-length on the transport properties of single-file water molecules: Transition from bidirectional to unidirectional. *The Journal of Chemical Physics, 134*, 244513.

Svergun, D. I., Richard, S., Koch, M. H. J., Sayers, Z., Kuprin, S., & Zaccai, G. (1998). Protein hydration in solution: Experimental observation by x-ray and neutron scattering. *Proceedings of the National Academy of Sciences of the United States of America, 95*(5), 2267–2272. Www.pnas.org. doi:10.1073/pnas.95.5.2267.

Takaiwa, D., Hatano, I., Koga, K., & Tanaka, H. (2008). Phase diagram of water in carbon nanotubes. *Proceedings of the National Academy of Sciences, 105*(1), 39–43. doi:10.1073/pnas.0707917105.

Tao, J., Song, X., Zhao, T., Zhao, S., & Liu, H. (2018). Confinement effect on water transport in CNT membranes. *Chemical Engineering Science, 192*, 1252–1259. doi:10.1016/j.ces.2018.05.018.

Timasheff, S. N. (2002). Protein hydration, thermodynamic binding, and preferential hydration. *Biochemistry, 41*(46), 13473–13482. doi:10.1021/bi020316e.

Venkatraman, R. K., & Baiz, C. R. (2020). Ultrafast dynamics at the lipid-water interface: DMSO modulates H-bond lifetimes. *Langmuir, 36*(23), 6502–6511. http://pubs.acs.org/journal/langd5. doi:10.1021/acs.langmuir.0c00870.

Verteramo, M. L., Stenström, O., Misini Ignjatović, M., Caldararu, O., Olsson, M. A., Manzoni, F., Leffler, H., Oksanen, E., Logan, D. T., Nilsson, U. J., Ryde, U., & Akke, M. (2019). Interplay between conformational entropy and solvation entropy in protein–ligand binding. *Journal of the American Chemical Society, 141*(5), 2012–2026. doi:10.1021/jacs.8b11099.

Vlieghe, D., Turkenburg, J. P., & Van Meervelt, L. (1999). B-DNA at atomic resolution reveals extended hydration patterns. *Acta Crystallographica Section D: Biological Crystallography, 55*(9), 1495–1502. doi:10.1107/S0907444999007933.

Wallerstein, J., Ekberg, V., Misini Ignjatović, M., Kumar, R., Caldararu, O., Peterson, K., Wernersson, S., Brath, U., Leffler, H., Oksanen, E., Logan, D. T., Nilsson, U. J., Ryde, U., & Akke, M. (2021). Entropy–entropy compensation between the protein, ligand, and solvent degrees of freedom fine-tunes affinity in ligand binding to galectin-3C. *JACS Au, 1*(4), 484–500. doi:10.1021/jacsau.0c00094.

Walther, M., Plochocka, P., Fischer, B., Helm, H., & Jepsen, P. U. (2002). Collective vibrational modes in biological molecules investigated by terahertz time-domain spectroscopy. *Biopolymers, 67*(4–5), 310–313. doi:10.1002/bip.10106.

Wang, J., Xing, Y., Gui, X., Li, G., & Cao, Y. (2022). Experimental and molecular dynamics simulation study on wetting interaction between water droplets and kaolinite surface. *Chemical Physics Letters, 800*, 139659. doi:10.1016/j.cplett.2022.139659.

Wang, X.-, & Ma, H.- (2010). Improvement on the Carnahan-starling equation of state for hard-sphere fluids. *Chinese Journal of Chemical Physics, 23*(6), 675–679. doi:10.1088/1674-0068/23/06/675-679.

Wegmann, S., Eftekharzadeh, B., Tepper, K., Zoltowska, K. M., Bennett, R. E., Dujardin, S., Laskowski, P. R., MacKenzie, D., Kamath, T., Commins, C., Vanderburg, C., Roe, A. D., Fan, Z., Molliex, A. M., Hernandez-Vega, A., Muller, D., Hyman, A. A., Mandelkow, E., Taylor, J. P., & Hyman, B. T. (2018). Tau protein liquid–liquid phase separation can initiate tau aggregation. *EMBO Journal, 37*(7). http://emboj.embopress.org/. doi:10.15252/embj.201798049.

Woodruff, J. B. (2018). Assembly of mitotic structures through phase separation. *Journal of Molecular Biology, 430*(23), 4762–4772. https://www.journals.elsevier.com/journal-of-molecular-biology. doi:10.1016/j.jmb.2018.04.041.

Wyttenbach, T., & Bowers, M. T. (2009). Hydration of biomolecules. *Chemical Physics Letters, 480*(1-3), 1–16. doi:10.1016/j.cplett.2009.08.042.

Xiao, S., Figge, F., Stirnemann, G., Laage, D., & McGuire, J. A. (2016). Orientational dynamics of water at an extended hydrophobic interface. *Journal of the American Chemical Society, 138*(17), 5551–5560. http://pubs.acs.org/journal/jacsat. doi:10.1021/jacs.6b01820.

You, X., & Baiz, C. R. (2022). Importance of hydrogen bonding in crowded environments: A physical chemistry perspective. *Journal of Physical Chemistry A, 126*(35), 5881–5889. http://pubs.acs.org/jpca. doi:10.1021/acs.jpca.2c03803.

You, X., Shirley, J. C., Lee, E., & Baiz, C. R. (2021). Short- and long-range crowding effects on water's hydrogen bond networks. *Cell Reports Physical Science, 2*(5). https://www.cell.com/cell-reports-physical-science/aims. doi:10.1016/j.xcrp.2021.100419.

Zaki, M., Hairat, S., Kamaal, S., Aljarba, N. H., AL-Johani, N. S., & Alkahtani, S. (2022). Synthesis, crystal structure elucidation and DNA/HSA binding profile of Ni(II) complex of Schiff base derived from 3-ethoxy salicylaldehyde and *o*-phenylenediamine. *Journal of Molecular Structure, 1265*. https://www.journals.elsevier.com/journal-of-molecular-structure. doi:10.1016/j.molstruc.2022.133351.

Zhang, Y., de Aguiar, H. B., Hynes, J. T., & Laage, D. (2020). Water structure, dynamics, and sum-frequency generation spectra at electrified graphene interfaces. *The Journal of Physical Chemistry Letters, 11*(3), 624–631. doi:10.1021/acs.jpclett.9b02924.

Zhu, H., Wang, Y., Fan, Y., Xu, J., & Yang, C. (2019). Structure and transport properties of water and hydrated ions in nano-confined channels. *Advanced Theory and Simulations, 2*(6). doi:10.1002/adts.201900016.

Zimmermann, N. E. R., & Jain, A. (2020). Local structure order parameters and site fingerprints for quantification of coordination environment and crystal structure similarity. *RSC Advances, 10*(10), 6063–6081. http://pubs.rsc.org/en/journals/journal/ra. doi:10.1039/c9ra07755c.

Zupančič, B., & Grdadolnik, J.ože (2021). Solute-induced changes in the water H-bond network of different alcohol-aqueous systems. *Journal of Molecular Liquids, 341*, 117349. doi:10.1016/j.molliq.2021.117349.

CHAPTER

5 Conserved waters in protein structure and function with applications to drug design

Samo Lešnik[a,b], Marko Jukić[a,c,*] and Urban Bren[a,b,c,*]

[a]*Laboratory of Physical Chemistry and Chemical Thermodynamics, Faculty of Chemistry and Chemical Engineering, University of Maribor, Maribor, Slovenia,* [b]*Institute of Environmental Protection and Sensors, Maribor, Slovenia,* [c]*Faculty of Mathematics, Natural Sciences and Information Technologies, University of Primorska, Koper, Slovenia*

**Corresponding authors: urban.bren@um.si; marko.jukic@um.si*

5.1 Introduction

Water represents an omnipresent and indispensable component of all biological systems, playing critical roles in maintaining the structural integrity and functionality of proteins (de Beer et al., 2010; Morningstar-Kywi et al., 2022). Its unique properties, including polarity, hydrogen-bonding capabilities, and the ability to form extensive networks, allow water to participate in a wide range of biochemical processes. A critical distinction exists between bulk water, residing in the solvent, and protein-bound water, which interacts closely with protein surfaces and often plays a structural role. Protein-bound waters are strategically positioned, with some acting transiently and others being highly fixed (Maurer & Oostenbrink, 2019; Spyrakis et al., 2017). The water molecules slow down considerably near protein surfaces, extending across multiple hydration layers (Bellissent-Funel et al., 2016), which enhances their ability to mediate protein stability, folding, and function. The residence times of conserved waters can range from a few picoseconds to nanoseconds, depending on their specific role and environment (Chong & Ham, 2017).

Water molecules contribute to both the stability and dynamic flexibility of proteins. Protein-bound water molecules form hydrogen bonds with polar and charged residues, stabilizing structural elements and maintaining protein integrity. These waters, often conserved across related protein structures, (Jukič et al. (2020); Konc and Janežič (2010)), play key roles in stabilizing protein folds, mediating protein–protein interactions (PPI) and protein–ligand interaction, as well as influencing ligand binding specificity and affinity (Spyrakis et al., 2017). In parallel, water molecules facilitate the dynamic flexibility necessary for protein activity by stabilizing transient states and supporting conformational changes. In enzymes, water mediates structural transitions that drive catalysis, whereas, in signaling proteins, it facilitates signal propagation across membranes (Bondar et al., 2008; Maurer & Oostenbrink, 2019). In addition, water molecules act as proton mediators in hydrogen-bonded networks, essential for signaling, transport, and enzymatic reactions (Bondar et al., 2008; Demchenko, 2023; Ishikita & Saito, 2014).

In drug design, water molecules can either be retained during ligand binding—forming hydrogen-bond networks and enhancing binding interactions through enthalpic contributions—or displaced into

Protein Dynamics: Experimental and Computational Approaches. DOI: https://doi.org/10.1016/B978-0-443-30042-4.00015-8

the bulk solvent, contributing entropically to binding (de Beer et al., 2010; Morningstar-Kywi et al., 2022). This enthalpy/entropy balance is crucial in optimizing the overall binding free energy. The phenomenon of enthalpy/entropy compensation, where changes in one component compensate for the other, often involves conserved water molecules and presents both challenges and opportunities in drug design (Breiten et al., 2013).

Accurately characterizing water molecules within protein binding sites remains challenging, particularly because of their dynamic nature. Techniques, such as X-ray crystallography, nuclear magnetic resonance (NMR) spectroscopy, and cryogenic electron microscopy (cryo-EM), offer valuable insights into water location and behavior, but each with specific limitations (Arkhipova et al., 2017; Doyle et al., 2023; Nittinger et al., 2015; Wlodawer et al., 2013). X-ray crystallography can struggle with resolving individual water molecules because of crystallization conditions and weak electron densities, whereas NMR provides dynamic insights but faces challenges such as signal overlap and rapid exchange with bulk water (Ernst et al., 1995; Nucci et al., 2011; Wang et al., 2017). Cryo-EM is emerging as a powerful tool, offering the possibility to solve structures of proteins difficult to crystallize, although it faces its own limitations, especially in resolving smaller proteins and visualizing water molecules, as their small size and diffuse networks are often difficult to capture even at cryogenic temperatures (Chua et al., 2022; Oliveira et al., 2021; Saur et al., 2020; Scapin et al., 2018; Zhao & Tajkhorshid, 2024). Crystallographic resolution is crucial for accurately detecting water molecules in protein structures. Recent research indicates that resolutions around 1.5–1.6 Å are necessary to observe continuous hydration layers and precisely localize water molecules (Carugo, 2016). Lower resolutions result in fewer detectable water molecules and less reliable hydration models, emphasizing the need for high-resolution data in protein–water interaction studies (Gnesi & Carugo, 2017).

Given these experimental limitations, molecular dynamics (MD) simulations serve as a complementary tool for studying the distribution and dynamics of water molecules in protein systems (Zia et al., 2016). Advances in force fields and water models have improved the accuracy of these simulations, though challenges remain, particularly in capturing long-timescale behaviors and specific interactions in complex protein environments (Fischer et al., 2024; Kadaoluwa Pathirannahalage et al., 2021). Despite these hurdles the integration of conserved waters into computational drug discovery workflows, such as virtual screening and molecular docking, continues to yield valuable outcomes (Pavlovicz et al., 2020). Overall, by incorporating conserved waters, researchers can improve predictions of ligand binding modes and affinities, ultimately optimizing drug efficacy (Nittinger et al., 2018, 2019; Ross et al., 2012).

5.2 Conserved waters in protein structure and function

Water plays a role in both facilitating proper protein folding and maintaining protein structure postfolding (Mallamace et al., 2011). During folding, water hydrates polypeptide chains, aiding in the protein adopting its correct functional conformation (Spyrakis et al., 2017). Meanwhile, it simultaneously drives the hydrophobic effect, pushing nonpolar residues into the protein's core, away from the aqueous environment, which promotes the compact structure of proteins.

Although much of the water is excluded from the protein's interior during folding, certain water molecules are selectively retained in specific positions (Bottoms et al., 2006). These conserved waters form hydrogen bonds with polar residues, stabilizing the proteins' folded state. This highlights water's essential dual role: excluding itself to drive folding while being retained within key regions to ensure structural stability.

This stabilization is crucial in enzymes and functional proteins, where conserved water molecules are often located within or near active sites. These waters not only stabilize the protein structure but also participate directly in enzymatic reactions, acting as nucleophiles, proton donors, or acceptors, and ensuring the precise geometry required for catalysis and molecular recognition sites (Nguyen et al., 2014; Pocker, 2000; Prasad & Suguna, 2002).

Beyond its role in maintaining the correct overall protein structure, water is a dynamic component that influences protein flexibility and enables transitions between different conformational states. This flexibility is vital for biological functions, as water molecules can "slave" the motion of proteins, facilitating the conformational changes necessary for activity (Zhao et al., 2013). These coupled motions exhibit the same temperature dependence as solvent fluctuations, linking protein dynamics to the surrounding water (Fenimore et al., 2002). Specifically, α- and β-fluctuations are two distinct types of solvent dynamics that play crucial roles in modulating protein motions through slaving dynamics (Fenimore et al., 2004). α-Fluctuations involve collective, large-scale movements in the bulk water surrounding the protein. These fluctuations are coupled to the collective motions of the protein backbone, such as folding, unfolding, and significant conformational changes essential for functions such as enzyme catalysis and allosteric regulation. In contrast, β-fluctuations are localized movements coupled to motions of side chains and loops, facilitating quick conformational adjustments necessary for processes such as ligand binding and substrate recognition.

MD simulations provide valuable insights into these dynamic processes. However, Rhee et al. (2004) showed in the studies of BBA5 proteins that explicit solvent models, which account for the detailed interactions between water and the protein, capture the folding transition states more accurately than implicit models, particularly during critical transition states. In addition, the choice of water model is critical in MD simulations, as shown by Fischer et al. (2024), who compared different combinations of AMBER force fields and water models. Their findings demonstrated that the choice of water model significantly affects the accuracy of folding simulations, with the OPC model favoring unfolded states and the TIP3P model stabilizing the folded state, regardless of the force field used.

A key concept in understanding water's role in protein stability is *cooperative hydration* (Huggins, 2016). Water molecules often form an interconnected hydrogen-bonding network within proteins, where the presence of one water molecule enhances the stability of neighboring waters. This cooperative hydration network contributes significantly to the stabilization of the protein's native conformation. Enhancing these networks through targeted mutations, such as modifying surface residues to favor extended hydration, could be a potential strategy to increase protein stability. Although cooperative hydration stabilizes protein folds, the transient nature of these hydrogen-bond networks also facilitates protein dynamics. For example, in the study by Bertalan et al. (2020), MD simulations and subsequent network analyses revealed the critical role of dynamic water-mediated hydrogen bonds in regulating the activity of κ-opioid receptors. Specifically the authors demonstrated that conserved water molecules form hydrogen bond networks that are essential for facilitating the functional transitions of this receptor. Upon receptor activation a continuous hydrogen bond network connects key functional motifs via water-mediated interactions, enabling signal propagation from the extracellular to the intracellular regions (Fig. 5.1). This connected dynamic network is largely absent in the inactive state, illustrating how water molecules help mediate the receptor's conformational shifts during activation.

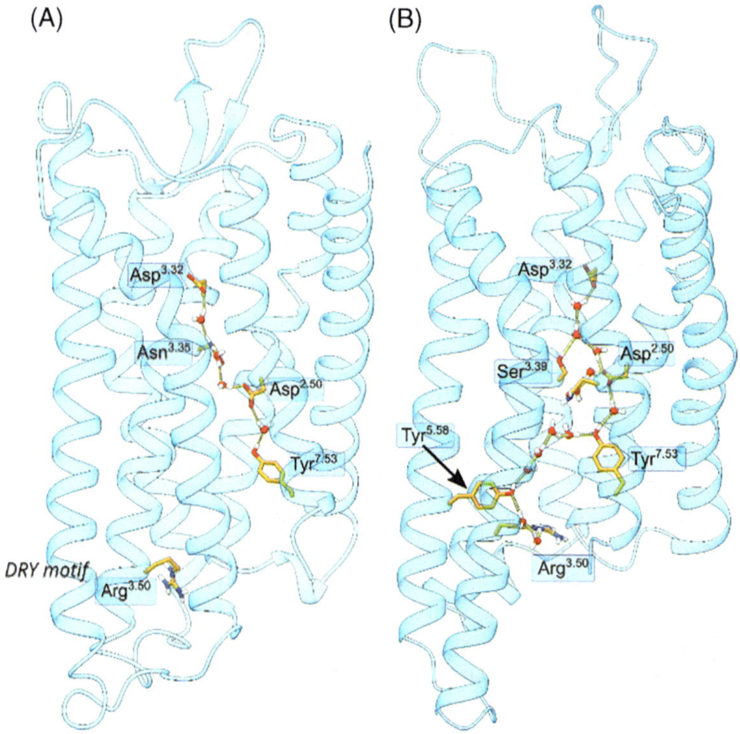

FIGURE 5.1

Structural dynamics of water-mediated hydrogen bonding in κ-opioid receptor: Water-mediated hydrogen bond pathways in the inactive (A) and active (B) states of the κ-opioid receptor after molecular dynamics simulations. The protein is represented with blue cartoons, and key residues are shown as orange sticks. Water molecules are depicted as red spheres for oxygen and white for hydrogen, and hydrogen bonds are indicated by *yellow dashed lines*. In the inactive state (A) the hydrogen bond network from $Asp^{3.32}$, represented at an occupancy of 50%, does not extend fully to the DRY motif residue $Arg^{3.50}$. In contrast the active state (B) shows a complete connection, with the network extending through $Tyr^{5.58}$ (*arrow*), facilitating the full-length interaction from $Asp^{3.32}$ to $Arg^{3.50}$, stabilizing the active conformation.

Based on the PDB structures 4DJH (inactive form) (Wu et al., 2012), 6B73 (active form) (Che et al., 2018), and the work of Bertalan, É, Lešnik, S., Bren, U., & Bondar, A.-N. (2020). Protein-water hydrogen-bond networks of G protein-coupled receptors: Graph-based analyses of static structures and molecular dynamics. Journal of Structural Biology, 212*(3). https://doi.org/10.1016/j.jsb.2020.107634.*

In addition to structural transitions, water facilitates key biochemical processes such as proton transfer, which is vital in photosynthesis, cellular respiration, and enzymatic reactions (Bondar & Smith, 2017; Bondar et al., 2008; Demchenko, 2023; Doyle et al., 2023). Proton transfer relies on transient hydrogen-bond chains, where protons hop between water molecules or titratable groups (Miyake & Rolandi, 2016). These water-mediated networks regulate the speed and efficiency of proton transfer. In photosystem II, for example, dynamic hydrogen-bonded water molecules shape proton-transfer pathways and energetics, thereby supporting efficient photosynthesis (Kemmler et al., 2019).

Water's influence also extends beyond normal folding and function to processes such as misfolding and aggregation, which are implicated in neurodegenerative diseases such as Alzheimer's and Parkinson's (Chong & Ham, 2015; Samantray et al., 2020; Stephens & Kaminski Schierle, 2019; Thirumalai et al., 2012). Water molecules can either stabilize misfolded intermediates, promoting aggregation into amyloid fibrils, or, under different conditions, contribute to disaggregation. Understanding how water mediates these transitions is crucial for developing therapeutic strategies to counter protein misfolding diseases.

5.3 Conserved water in protein–protein interfaces

Conserved water molecules at protein–protein interfaces play a critical role in PPIs that mediate recognition patterns between neighboring surfaces. Clackson et al. postulated that 30% of the total protein–protein interface is hydrated (Clackson & Wells, 1995). Water is ubiquitous and pervasive in biological and crystal nonbiological protein–protein interfaces (Ahmed et al., 2013). In general, waters are conserved at protein binding interfaces even in homologous proteins, suggesting their important interaction and functional and structural roles. Li et al. (2012) demonstrated in a structural alignment-agnostic manner and using a solvated sequence method that hydrophobic residues tend to desolvate the local area, that small residues retain their contact water, and that deeply buried water molecules are mostly conserved. All observations thus point to a flexible organization of the water surface of protein–protein binding interfaces.

When analyzing 4741 water molecules from 179 high-resolution X-ray crystal structures, Ahmed et al. found that water forms bridging interactions between both surfaces (mainly via polar residues) in 21% of cases, favorable hydrogen bonds to only one surface in 53% of cases, and no interactions in 26% of cases (Ahmed et al., 2011). The latter mostly correlated with the presence of nonpolar residues. Interestingly, water interacting with both proteins was found to stabilize interface formation on average, whereas water without interactions was mostly found with very low solvent-accessible surface area (SASA) values and was conserved and buried, forming so-called hydrophobic bubbles, similar to what was observed by Li et al. Akin to the "water-centric" view of protein folding, waters are integral and stabilizing elements of protein–protein interfaces (Mittal & Jayaram, 2011). They fill the voids, provide bridging interactions, and shield the electrostatic repulsions. Comprehensive MD simulations of the interaction of two hydrophilic proteins have shown that the interfacial water between the proteins acts as an adhesive, directional network, resulting in dielectric shielding that causes highly directional electrostatic interactions between the hydrophilic groups of the proteins (Mittal & Jayaram, 2011). Mukherjee et al. (2019) found that there are three times more water-mediated hydrogen bonds at protein–protein interfaces than direct hydrogen bonds. Furthermore the group demonstrated that the density of water desolvation sites is on average the highest, followed by water hydration (water forming hydrogen bonds on contact) and conserved water sites. Similarly, conserved water sites are followed by desolvation and water-hydration sites in the formation of interfaces, and residues near conserved water sites are often themselves conserved.

Beyond general observations on the hydrophobic character of protein–protein interfaces and the occurrence of conserved residues at interaction "hot spots" (Clackson & Wells, 1995), the molecular mechanisms that determine affinity, and particularly specificity, at protein–protein surfaces remain incompletely understood. However, the thermodynamic properties of water at protein interfaces can be investigated in more detail. Ikura et al. (2004) calculated that water-mediated indirect interactions contribute enthalpically to protein–protein complex formation. Specific contributions of individual water molecules were sometimes found to be unfavorable, whereas interface formation was enriched by all waters present. Water entropy at protein–protein interfaces was investigated by Huggins et al. (2011), who used a combination of MD and inhomogeneous fluid solvation theory (IFST) to localize a series of hydration and hydrophobic sites in the crystal structure of human RadA recombinase. The authors identified a correlation between the predicted entropy of water molecules at a studied site and the solvent-accessible surface area and confirmed that the release of water molecules from the hydrophobic region is one of the most important factors for PPIs. Thus water is displaced during the formation of protein–protein interfaces, but conserved water molecules are often found to contribute favorably to specific interactions and total energy in an enthalpy–entropy compensating manner (Huggins et al., 2011). In drug development, targeting conserved water molecules at protein–protein interfaces could represent a strategy for the development of probes, inhibitors, or modulators of PPIs. As the water molecules are actively involved in binding, stability, and recognition, studying and understanding the role of individual water molecules may be crucial also for deciphering protein function.

For example, the human programmed death receptor 1 (PD-1) and its ligand (PD-L1) interact specifically with an interface where conserved water molecules play a crucial role. This is particularly true for immune regulation and checkpoint signaling pathways, where PD-1 and PD-L1 suppress immune responses and allow cancer cells to evade the immune system. The crystal structure (PDB ID: 4ZQK) shows that the receptor–ligand interaction is built around a central hydrophobic core contributed by both partners (Zak et al., 2015). PD-1 contributes Val64, Ile126, Leu128, Ala132, Ile134, and PD-L1 contributes Ile54, Tyr56, Met115, Ala121, and Tyr123. The hydrophobic region is surrounded by a buried region with mixed polar/nonpolar interactions and by a network of polar residues. Water-bridged interactions are present between the backbone amide and carbonyl of Ile134 and ligand Tyr56 hydroxyl and ligand Glu58 carboxyl. Additional water bridges are found on the side chain of Asn66 and the main-chain carbonyl oxygen of ligand Ala121 that stabilize the complex (Fig. 5.2). Additional water molecules could also be positioned at the interface upon analysis of similar structures (Jukič et al., 2020).

A further general exemplary case would be the antibody–antigen complexes. In such systems, water molecules often mediate interactions between the paratope and the epitope and thus contribute to the high specificity and affinity. Braden et al. demonstrated in 1995 that the interface between antibody and lysozyme contains 25 ordered water molecules that are bound directly or via other water molecules to both the antibody and the antigen. In addition to the hydrogen bonds that contribute to the antibody–antigen interaction, the solvent molecules fill many interfacial voids (Braden et al., 1995). Moreover, Nguyen et al. examined a database of 403 antibody–antigen complexes and found 2580 interfacial waters in 242 complexes forming 3863 hydrogen bonds out of all 3700 hydrogen bonds formed by all 8603 contact residues of the antibodies with the antigens. These numbers are informative, and the authors suggest that the antibody may be able to interact specifically with a range of antigens. Furthermore the water molecules appear to form more highly ordered structures at the interface that supplement the lack of shape complementarity and stabilize the paratope–epitope interface (Nguyen et al., 2017).

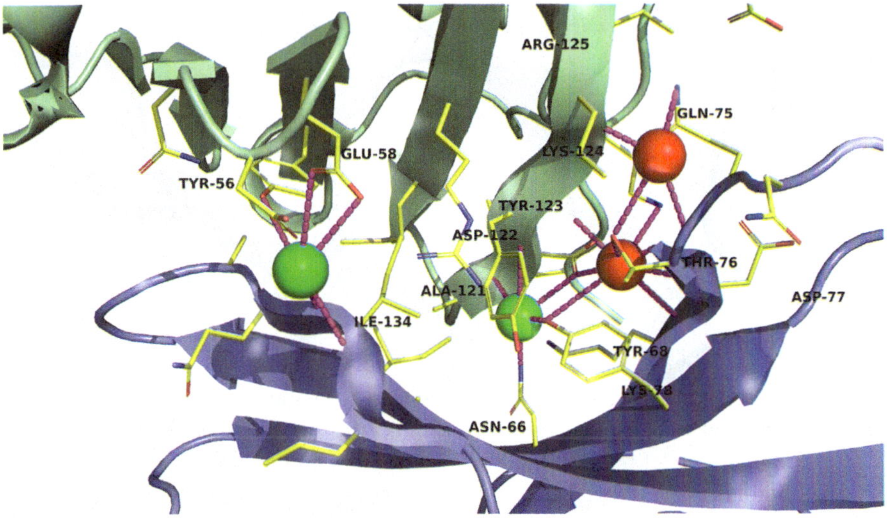

FIGURE 5.2

Complex interface between human programmed death-1 protein (PD-1; *blue cartoon model*) and its ligand (PD-L1; *light green cartoon model*). *Green* CPK models depict two conserved water oxygens highlighted by the crystallographers [PDB ID: 4ZQK (Zak et al., 2015); in contact with Ile134, Tyr56, Glu58, Asn66, Tyr68, and Ala121, respectively (*magenta dotted line*)]. *Red* CPK models depict two conserved water oxygens identified by superposition of similar structures and analysis by ProBiS H_2O and could be of interest as they successfully occupy the interface cavity and bridge the interacting residues (in contact with Thr76, Lys78, Tyr123, Lys124, and Gln75, Thr76, Lys124, and Arg125, respectively, *magenta dotted line*). Interacting residues are emphasized in the yellow stick model with colored heteroatoms. The complex surface is *shaded gray*.

5.4 Tools for the analysis of conserved waters

Tools to analyze single water molecules, water clusters, and networks at the atomic level are crucial for an efficient interactome study. The positioning of water in the context of the intra- or intermacromolecular atomic structure affects the stability and the existing or even possible interaction patterns (specificity, selectivity). In this way the approaches can be roughly divided into six groups, which are explained in more detail in Table 5.1.

The first approach is MD or Monte Carlo (MC)–based sampling using explicit water models with subsequent trajectory analysis (Nguyen et al., 2014). An example is WaterMap (WScore; description of ligand and protein desolvation) by Schrödinger (Yang et al., 2013). WaterMap uses MD with explicit solvent and restrained protein to model solvation effects. WaterMap then clusters and characterizes each identified hydration site energetically using IFST. Using this approach, it has been described that the specific high-energy hydration sites can be used to identify potential binding sites of interest and druggability of targets (Beuming et al., 2012). WaterMap can therefore assess the effects of water displacements and has been used in drug design problems. Another approach based on statistical mechanics to identify and energetically evaluate discrete hydration sites was described by Cui et al. (2013).

Table 5.1 Diverse approaches toward water analysis and exemplary software.

	Methodology	*Examples*
1.	MD, MC-based sampling	WaterMap, SPAM, 3D-RISM, aquaPELE, GIST
2.	Cavity detection, probe placing	DOWSER++, SZMAP, WaterFLAP
3.	Experimental water, clustering, network analysis (+MD)	DRoP, PyWATER, MADE, Bridge (2)
4.	Descriptor-based, scoring	WaterScore, Consolv, PyRod
5.	QM, QM/MM	AIM, Gaussian, Gromacs, Mopac, MobyWat
6.	ML-boosted	GalaxyWater-CNN, 3D-RISM-AI

The SPAM (maps spelled in reverse) software uses explicit solvent MD simulations to capture discrete hydration sites and provides an energetic evaluation (thermodynamic profile) based on the distribution of interaction energies between water and the macromolecular environment. SPAM has been validated on known systems such as HIV1 protease and can identify important water or "hot spot" regions (Michel & Essex, 2010). Sindhikara et al. are authors of the prominent analysis of biomolecular solvation by three-dimensional reference interaction site model (3D-RISM) theory (Sindhikara & Hirata, 2013). The 3D-RISM calculation of a static solute is followed by postprocessing to identify individual solvation sites (placevent algorithm) with final geometric and thermodynamic evaluation (Sindhikara et al., 2012). The authors have also modified 3D-RISM with the Ng-bridge function (3D-RISM-NgB) to quickly and rapidly calculate the free energies of hydration of small molecules for drug design applications (Truchon et al., 2014). 3D-RISM, together with grid inhomogeneous solvation theory (GIST) (Ramsey et al., 2016), is also available in the Flare software package from Cresset LLC. MC methods for identifying conserved water sites have also been used to solve sampling problems, for example, Grand Canonical MC (Ross et al., 2020, 2015) and Just Add Water Molecules (Bodnarchuk, 2016; Bodnarchuk et al., 2020). In addition, protein energy landscape exploration (PELE) tools have been extended to aquaPELE, which uses an additional MC routine to perturb a set of explicit water molecules and assess key hydration sites or water shifts (Municoy et al., 2020). Harrigan et al. (2015) reported on the conserve water protocol, which uses solute coordinate Markov state model analysis to determine the solvent's degrees of freedom (solvent-shells featurization). Ge et al. (2022) reported the calculation of absolute binding free energy for buried waters using the nonequilibrium switching method. The authors identified important sampling improvements for future development of the method (implemented in GROMACS), such as using position restraints on protein atoms to prevent binding site collapse and defining the integration region to use only relevant binding pocket data in the analysis. GIST was also employed by Nguyen et al. (2012) to provide a smoothly varying representation of water as a function of position, rather than focusing on individual hydration sites.

The second approach can be described as cavity detection (Marchand et al., 2021) with hydration (Denisov & Halle, 1996; Wade et al., 1991). An example is the software Dowser, which detects cavities in protein structures, hydrates all pockets and selects the preferred water sites based on energy cut-offs (Zhang & Hermans, 1996). A more recent method is SZMAP/GAMEPLAN from OpenEye Scientific Software Inc. (Bayden et al., 2015). Herein analysis of sites near the protein or ligand surface is performed using semicontinuum solvation theory, where a single explicit water probe is translated through space and interactions are calculated to find 3D grid maps that attempt to probe how water displacement in specific regions affects binding affinity. The software is ultimately used to understand the water structure in the immediate vicinity of the ligand under investigation. There is also WaterFLAP from Molecular Discovery Inc., which is based on the Fingerprints for Ligands and Proteins (FLAP) algorithm (Baroni et al., 2007). The FLAP algorithm analyses macromolecular space using GRID molecular interaction fields obtained by spatially translating a series of chemical probes. The 3D field data are then condensed into pharmacophoric points, which are further enumerated and processed. Waters are then assessed for their structural, displaceable, or bulk character. Similarly, there is the WaterDock protocol for docking water molecules to a site of interest, followed by filtering and clustering steps (Sridhar et al., 2017). Later WaterDocking data were used to develop the improved new method for hydration of protein structures called DOWSER++ software (Morozenko & Stuchebrukhov, 2016). In this context, we would also like to emphasize the importance of including explicit water molecules in docking experiments, a practice that can lead to increased accuracy when quality input data are used (Hu et al., 2020; Kumar & Zhang, 2013; Uehara & Tanaka, 2016). It was reported that modern machine learning-boosted scoring functions can be improved by the inclusion of water molecules (Qu et al., 2022). Moreover, excellent comparative reviews of the described approaches are available to critically select the best option for the particular system under study (Bucher et al., 2018; Nittinger et al., 2019; Samways et al., 2021).

The third approach uses experimental structures to compile water location data. Thus water location data can be collected from accessible and curated databases such as the Research Collaboratory for Structural Bioinformatics (RCSB) PDB, wwPDB, or PDBj (Berman et al., 2003; Kinjo et al., 2017; Rose et al., 2017). In this manner, deposited RCSB PDB data were used in the development of DRoP (Kearney et al., 2014). DRoP collects a set of similar protein structures based on user selection, water-protein contacts are enumerated in a scoring/filtering step, and then all structures are globally aligned using the Ce-align algorithm. The DRoP approach has also been successfully used to detect conserved solvent binding sites on proteins (Kearney et al., 2020). Water data can also be found at the Cambridge Crystallographic Data Centre with its IsoStar (Cole et al., 2017) and SuperStar databases for crystal structures of small molecules and intermolecular interactions (identification of protein "hot" spots) (Radoux et al., 2016), respectively. The IsoStar database with crystal structures of small molecules was used to develop the AcquaAlta approach (Rossato et al., 2011). Data on the geometry of water interactions with generic functional groups were collected to perform ab initio interaction energy calculations and to establish an empirical scale for hydration propensity. PyWATER is a PyMOL plugin (DeLano, 2002) to identify PDB structures with similar sequences (Patel et al., 2014). The plugin follows the standard protocol of performing water filtering to eliminate waters *with* high normalized B-factors prior to a sequence-independent structure-based superposition. The final hierarchical clustering with the calculated degree of conservation between structures provides discrete conserved water location data. Experimental PDB data are also utilized by the ProBiS H_2O approach, where local superposition by ProBiS (Konc & Janežič, 2010) is followed by density clustering (Khan, Rehman, Aziz, Fong, & Sarasvady, 2014) and a water analysis protocol to effectively identify conserved water molecules (Jukič et al., 2020). ProBiS H_2O also bridges the gap between the approaches classified herein, as it can also use MD sampling data (Jukič et al., 2020) and has since been generalized for the analysis of ions and water molecules together with the analysis and comparison of superposition algorithms named the MADE approach (Ravnik et al., 2023). Tools that can utilize experimental and MD sampling data can also effectively analyze water networks in the presence of ligand binding (Tošović et al., 2022). In the latter case, oxygen clustering is followed by hydrogen bond clustering (OPTICS) (Ankerst et al., 1999) with a final network analysis that can detect both water orientations and conservation in closely related networks. In addition, large protein networks can be analyzed using the Bridge approach (Siemers et al., 2019). Herein, graph networks of protein–hydrogen bonds can be effectively analyzed from short to long distances (transient bridging of hydrogen bond partners), and the propagation of hydrogen bond network changes that correlate with structural changes can be visualized and investigated. The algorithm is also available as a PyMOL plugin and has been successfully used to identify hydrogen bond hubs on the severe acute respiratory syndrome coronavirus 2 (SARS-CoV-2) spike protein (Spro) (Siemers & Bondar, 2021). Bridge will be discussed in more detail in the following section.

Descriptor-based approaches are presented as the fourth approach. Rather than identifying conservation trends, descriptor-based approaches evaluate specific waters obtained from experimental data. WaterScore was developed to establish a statistical correlation between the structural properties of water molecules (B-factor, solvent contact surface area, total hydrogen bond energy, and number of atomic protein contacts) in the apo- and holo-protein complexes. It was found that bound and displaceable water could be distinguished based on the descriptors used (García-Sosa et al., 2003). Similarly the Consolv program used a hybrid *k*-nearest neighbor (KNN)/genetic algorithm classifier to predict conserved water molecules by examining the environment (B-factor, number of hydrogen bonds between the water molecule and the protein, density, and hydrophilicity of neighboring protein atoms) of water molecules in experimentally obtained crystal structures (Raymer et al., 1997). Recently a modern consensus approach called PyRod (Schaller et al., 2019) was presented. It uses MD data to identify different waters based on their environment and calculates molecular interaction fields with the aim of improving current pharmacophore-based approaches.

The fifth complementary approach toward water molecule energetic assessment relies on the quantum mechanical (QM, QM/MM) description of systems (Ryde & Söderhjelm, 2016). QM methods are characterized by their ability to accurately account for electronic effects in intermolecular interactions. Even though they are elaborate and computationally intensive, they could be an ideal partner for the water placement and conservation tools described above. The key role of bridging waters in inhibitor design has been demonstrated by ONIOM-based geometry optimization and elegant atom in molecule analysis at the GSK3β-inhibitor interfaces (Lu et al., 2011). Another application was reported by Woods et al. (2015). In it, the authors calculated absolute free energies of binding of water molecules (waters B, C; difference in complex with or without key waters) in prominent neuraminidase–ligand complexes (Barillari et al., 2007). In this manner the waters could be accurately classified as to whether or not they should be displaced during the ligand design campaign. The contributions of the water network to ligand binding were also evaluated using the fragment molecular orbital method (Lukac et al., 2021). In this chapter the authors point out the overall stability of the water network compared to individual water studies. In drug development the interactions formed should naturally outweigh the energy loss caused by the disruption of water network formation. A notable example is the recent study by Szél et al., which presents a method for enthalpic classification of water molecules (Huggins et al., 2011; Szél et al., 2024). This method calculates the changes in bond enthalpy (ΔH_b) for individual water molecules using semiempirical, QM-based calculations and allows the distinction between conserved and displaced water at atomic resolution. The main finding of the study was that water molecules with more negative ΔH_b values are more likely to be conserved, especially in polar protein environments. Conversely, water molecules at nonpolar binding sites tend to have higher ΔH_b values, leading to their displacement. This distinction between polar and nonpolar environments is critical to understanding how water molecules contribute to binding stability and can be strategically utilized in drug development. Here we would also like to demystify the CPU time of QM. Modern scaling and parallelization approaches make QM calculations feasible even on a screening scale, as the example of Raghavan et al. (2023) shows. The authors used the MiMiC-QM/MM-MD method with a parallel efficiency of 70% on up to 80,000 CPU cores and are an exemplary case for modern HPC utilization.

Finally, there is the sixth modern data-driven segment, where machine learning approaches are used to boost existing methods or develop new models for water classification. Bortolato et al. (2013) evaluated water networks using probabilistic classification trees based on GRID-OH2, WaterMap, SZMAP, and GCMC sampling data. A2A receptor antagonists were correctly classified according to the water network contextualized prediction of ligand-free energy of binding in a modest 67% of the test set. Xiao et al. (2017) investigated water positions in binding sites and developed a prediction model for tetrahedron–water clusters based on a series of residue group triplets forming feature triangles of different shapes. The same group later published a machine learning framework to identify conserved water molecules in the binding sites of proteins (Xiao et al., 2022). The group evaluated models such as support vector machine, KNN, decision tree, logistic regression, discriminant analysis, naive Bayes, and ensemble learning (EL) based on water features (atomic density, hydrophilicity, hydrophobicity, solvent-accessible surface area, temperature B-factors, and mobility) to identify the conservation of individual water molecules. EL model was singled out with a prediction accuracy of 82% and validated against the DOWSER++ results. Moreover, a convolutional neural network (CNN) was trained on the PISCES dataset by Park and Seok (2022) to predict the position of water molecules in protein structures. The GalaxyWater-CNN tool is available via a web app (https://galaxy.seoklab.org/cgi-bin/submit.cgi?type=GWCNN), and the authors report that the tool predicts water molecules with an average RMSD of 2.3 Å. A similar approach, but for placing water at the protein–protein interfaces, was reported by Bui et al. (2007). Based on Amber force field energy prediction, authors claim WATGEN algorithm predicts 72% and 88% of water molecule sites within 1.5 and 2.0 Å, respectively. Osaki et al. (2022) attempted to predict free energies of binding for several sets of protein–ligand complexes using a machine learning approach that incorporates the free energies of hydration obtained with 3D-RISM, called 3D-RISM-AI. The input data were ~4000 complexes in the PDBbind dataset (2018) with structural metrics, intra-/intermolecular energies, and 3D-RISM results as features, which were used to develop an extreme gradient boosting regression model. The developed model correlated well with the experimental data (Pearson correlation coefficient of 0.80 and mean square error of 1.91 kcal/mol). As expected the main features identified were the difference in solvent-accessible surface area and the hydration properties of the ligand. At this point, we would like to critically point out the quality of the input data for machine learning methods (Jain et al., 2020). In particular, datasets are available for water analysis in the context of drug discovery, and the reader would do well to utilize such data collections (Huang et al., 2021; Samways et al., 2023).

5.5 Conserved waters in drug design

Conserved water molecules in protein–ligand complexes are critical contributors to the thermodynamics and kinetics of binding (Cui et al., 2018). These water molecules play two crucial roles: replacement and incorporation into hydrogen bond networks. Replacement occurs when a ligand binds to a protein

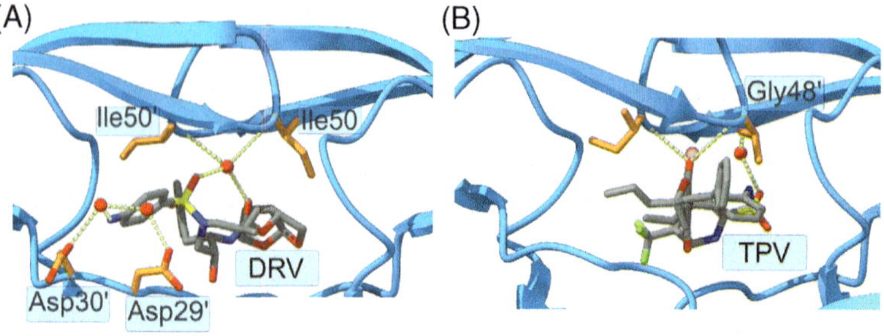

FIGURE 5.3

Comparison of interaction patterns in HIV-1 protease between darunavir (DRV) and tipranavir (TPV) inhibitors: (A) Binding of DRV to HIV-1 protease (protein shown with blue cartoon, sticks with orange carbons, and ligands shown with *grey* carbons), showing extensive water-mediated hydrogen bonds (*yellow dashed lines*) involving key residues Ile50 and Asp29/Asp30. (B) Binding of TPV to the same protease, where direct hydrogen bonds between TPV and the protease replace the water-mediated interactions observed with DRV. Notably, TPV displaces the conserved water molecule bound to Ile50′ (*transparent red sphere*) and forms a water-mediated interaction with Gly48. This displacement results in an entropic gain, which, together with direct hydrogen bonding to the Ile50 residues, contributes to TPV's high binding affinity. The comparison highlights the distinct binding strategies of DRV and TPV, with DRV relying heavily on water-mediated stabilization, whereas TPV achieves binding predominantly through direct interactions. The figure is based on the PDB structure 1T3R (Surleraux et al., 2005) and 2O4P (Muzammil et al., 2007).

and replaces a conserved water molecule from the binding site. This typically involves a functional group of the ligand that is isosteric to the water molecule, mimicking the hydrogen-bonding interactions originally formed by the water molecule (Michel et al., 2009). Displacing water often provides a stronger enthalpic contribution, resulting in a more stable ligand–protein interaction, and leads to a significant entropic gain as the displaced water molecule returns to the bulk solvent, increasing system disorder (Du et al., 2016). In contrast, incorporation involves retaining conserved water molecules within the ligand–protein complex. Instead of being displaced, these water molecules become integral to the interaction by forming hydrogen bonds that stabilize the complex (Lu et al., 2007). Although incorporation yields an enthalpic gain because of the formation of strong hydrogen bonds, it often incurs an entropic penalty because the retained waters are immobilized, reducing their freedom of movement.

Understanding the dual roles of water—whether to replace or incorporate—is pivotal for optimizing ligand design and improving binding affinity. These processes directly influence the overall binding free energy, guiding the rational optimization of ligands for increased potency and selectivity in drug development.

A compelling example of the role of conserved water molecules is the HIV-1 protease enzyme, a key target in AIDS treatment. Inhibiting this protease prevents the virus from processing polyproteins into functional viral proteins, thereby halting replication (Weber et al., 2021). High-resolution crystal structures have provided insight into the roles of conserved water molecules within the protease's active site, particularly the *flap water*, which interacts with the backbone nitrogen atoms of the Ile50 residues (Fig. 5.3) and plays a crucial role in stabilizing the enzyme's functional dimer (Muzammil et al., 2007). This conserved water has been a key focal point in the development of protease inhibitors. Two such inhibitors, tipranavir (TPV) and darunavir (DRV), exhibit distinct binding strategies that are critically shaped by their interactions with these conserved water molecules (Fig. 5.3).

DRV exemplifies water incorporation, relying on multiple water-mediated hydrogen bonds for its binding. It forms water-mediated interactions with the main chains of the Ile50 residues and with Asp29 and Asp30, creating a stabilizing network of hydrogen bonds. This network formation is primarily enthalpy-driven, with strong hydrogen bonds stabilizing the complex, though offset by an entropic penalty because of the immobilization of water molecules (Muzammil et al., 2007).

Conversely, TPV uses a replacement strategy. It forms direct hydrogen bonds with the Ile50 residues, displacing the conserved water molecule. This displacement leads to a significant entropic gain as the released water molecules return to the bulk solvent, complementing the enthalpic contributions from the direct ligand–protein interactions. This scenario exemplifies the enthalpy–entropy compensation effect, with TPV's binding driven more by entropy, whereas DRV's binding is largely enthalpy-driven, despite similar Gibbs free energy of binding for both inhibitors (Muzammil et al., 2007).

A similar phenomenon is observed in A2A adenosine receptor inhibitors, where water displacement can contribute both enthalpically and entropically, depending on the reorganization of the binding site upon ligand binding (Matricon et al., 2021). In cases, such as the binding of a 3-hydroxymethyl adenosine derivative, the replacement of a conserved water molecule results in an entropic gain as the water is released into the bulk solvent. Despite the overall structural similarity in the hydration network the absence of the displaced water molecule leads to significant changes in the interaction dynamics and energies among the remaining water molecules. The term "reorganization" here refers to these changes in interaction dynamics and energies, rather than a wholesome rearrangement of the network, which remains largely similar except for the displaced water molecule. This reorganization leads to more optimal hydrogen bonding, providing an additional enthalpic contribution (Matricon et al., 2021). In these specific cases the enthalpic contributions from reorganized water interactions can surpass the expected entropy gain from water displacement. This challenges the traditional view that water displacement is primarily entropy-driven and provides valuable insights for rational drug design, demonstrating that both enthalpic and entropic benefits can be achieved through careful modulation of water dynamics in the binding site.

In drug design, ligands often incorporate isosteric functional groups specifically designed to mimic the hydrogen-bonding properties of water molecules, replacing them in active sites while maintaining critical interactions (Maurer & Oostenbrink, 2019; Meanwell, 2014). A classic example is the mentioned replacement of a conserved water molecule in HIV protease inhibitors, where the lactone oxygen atom of TPV's dihydropyrone ring takes the place of the flap water molecule (Fig. 5.3) (Lv et al., 2015). Similarly, cyclic ureas have been designed to displace water molecules in the same protease target (De Lucca et al., 1997; Lam et al., 1994, 1996). Other examples include oxime moieties in poly(ADP-ribose) polymerase inhibitors (García-Sosa et al., 2005; Tikhe et al., 2004), nitrile groups in inhibitors of p38α MAP kinase (Chen et al., 1998; Liu et al., 2005), and the cyano group in heat shock protein 90 inhibitors (Davies et al., 2012).

In some cases, water displacement can occur without direct compensatory interactions, particularly when hydrophobic functional groups are involved. This is exemplified by the adamantyl-amine class of inhibitors targeting the M2 proton channel of influenza A. These inhibitors bind to the hydrophobic channel, displacing water molecules that normally stabilize the channel lining. By disrupting this critical water network the inhibitors effectively "dehydrate" the channel, inhibiting proton conduction and thereby blocking viral replication (Fig. 5.4).

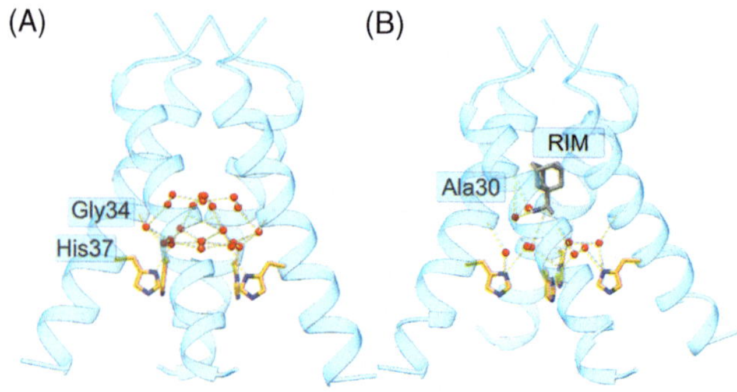

FIGURE 5.4

Water network in the M2 proton channel in the apo- and inhibitor-bound states: (A) The network of hydrogen-bonded waters (*red spheres*) stabilized by Gly34 and His37 residues in the drug-free (apo) state of the M2 channel. (B) Binding of the rimantadine (RIM) inhibitor (*gray*) at the hydrophobic Ala30 site, demonstrating the disruption of the hydrogen-bonded water network because of water displacement, which ultimately inhibits proton conduction through the channel. The M2 proton channel is depicted as *blue cartoons*, with proton-gating histidine residues shown as *orange sticks*. The figure is based on PDBs 5JOO (Thomaston et al., 2017) and 6BOC (Thomaston et al., 2018).

5.6 Dynamic water networks and signal propagation

One of the most compelling examples of the role of dynamic water networks in signal propagation is found in G-protein-coupled receptors (GPCRs) (Fried et al., 2022; Louet et al., 2021). These transmembrane proteins represent a large and diverse family that is essential for transmitting signals from outside the cell to the interior, playing a critical role in various physiological processes such as sensory perception, immune response, and neurotransmission (Cheng et al., 2023). GPCRs undergo significant conformational changes in response to external stimuli, including ligand binding, light, ions, and protons, which trigger the activation of intracellular signaling pathways. These conformational shifts are intricately linked with the reorganization of water molecules within the receptor's structure. Conserved water molecules, particularly those located around key GPCR motifs, often form hydrogen bonded networks that are crucial for maintaining the receptor's structural integrity and facilitating signal transduction (Venkatakrishnan et al., 2019). An analysis of hydrogen bond networks within rhodopsins shows that when accounting for water molecules in the formation of these networks, the size of the hydrogen-bond network can increase by approximately three to four times compared to networks that only consider direct residue-to-residue hydrogen bonds (Bondar & Alfonso-Prieto, 2022).

MD simulations have illuminated the critical role of highly conserved water-mediated interactions near the G protein–coupling interface across various GPCRs, such as the $\beta 2$ adrenergic receptor, M2 muscarinic receptor, μ-opioid receptor (MOR), and adenosine A2A receptor (Venkatakrishnan et al., 2019). These water networks help stabilize crucial structural features, like the helical kink in transmembrane helix 6, which connects different regions of the receptor. Importantly, these networks can be classified into state-independent networks, which are present in both active and inactive receptor states, and state-dependent networks, which undergo reorganization during the receptor's transition between states. For example, the conserved residue $Tyr^{7.53}$ acts as a molecular switch, forming different hydrogen-bonding patterns with water molecules depending on the receptor's activation state. In its inactive state, $Tyr^{7.53}$ interacts with water molecules near the TM1–TM2 interface, whereas, in the active state, it shifts to interact with water molecules connecting TM3, TM5, and TM6. This dynamic rearrangement illustrates the critical role that water plays in maintaining GPCR function and guiding receptor signaling transitions.

The importance of dynamic water networks in receptor function is further illustrated in the binding of ligands to GPCRs. A detailed examination of the water-mediated hydrogen-bond networks formed by fentanyl (FNT) and its fluorinated derivative NFEPP upon binding to the MOR provides a clear example (Lešnik et al., 2023). Despite the near-identical structures of FNT and NFEPP, where NFEPP differs by only a single fluorine substitution, their engagement with water-mediated hydrogen-bonding networks within the receptor is markedly different. FNT tends to lead to a formation of a receptor spanning water-mediated hydrogen-bond networks that propagate conformational changes throughout the receptor. NFEPP, which preferentially binds in acidic environments, establishes a more redundant hydrogen-bond network, with multiple alternative pathways connecting the binding site with the G-protein binding region. This distinction, visualized through graph-based network representations (Fig. 5.5), demonstrates how minor structural modifications of the ligand can drastically alter receptor dynamics and potentially signaling outcomes.

Dynamic water networks not only influence receptor function by manipulating the activation/deactivation states but are also key in modulating how receptors respond to different ligands, playing an important role in biased agonism. Biased agonism refers to the ability of certain ligands to selectively activate specific signaling pathways over others (Rankovic et al., 2016). Leveraging the unique interactions produced by different ligands—such as distinct water-mediated hydrogen-bond networks—enables the development of biased agonists that preferentially activate beneficial signaling pathways while minimizing adverse effects. Although the precise role of water networks in biased agonism is still an emerging field of study, the ability of these networks to influence receptor conformation suggests a potential avenue for designing drugs that achieve targeted therapeutic outcomes (Hu et al., 2023).

For example, in the MOR, the presence of water-mediated networks at allosteric sites has been shown to promote G-protein signaling while minimizing β-arrestin-mediated pathways, which are often associated with adverse effects such as opioid tolerance and addiction (Marmolejo-Valencia et al., 2021). Notably the opioid agonist herkinorin mimics sodium's effect on water-mediated networks by stabilizing the receptor and favoring G-protein signaling. This ability to fine-tune water networks within the receptor therefore suggests potential pathways for designing biased agonists, potentially leading, in this case, to safer, more effective opioid therapies that minimize the risk of adverse effects.

Water-mediated hydrogen-bond networks are also essential for other protein systems, such as microbial rhodopsin pumps, which rely heavily on conserved Asp/Glu side chains that bind protons via water-mediated hydrogen bonds with nearby Asn, Ser, Thr, or Tyr residues (Bertalan et al., 2024; Muroda et al., 2012). These water-mediated interactions are crucial for the functionality of the pumps, enabling them to transport ions efficiently. However, attempts to convert microbial proton pumps into sodium pumps via targeted mutations face significant challenges because of the complexity of the water-mediated interaction networks involved. For example, in gloeobacter rhodopsin, mutations aimed at

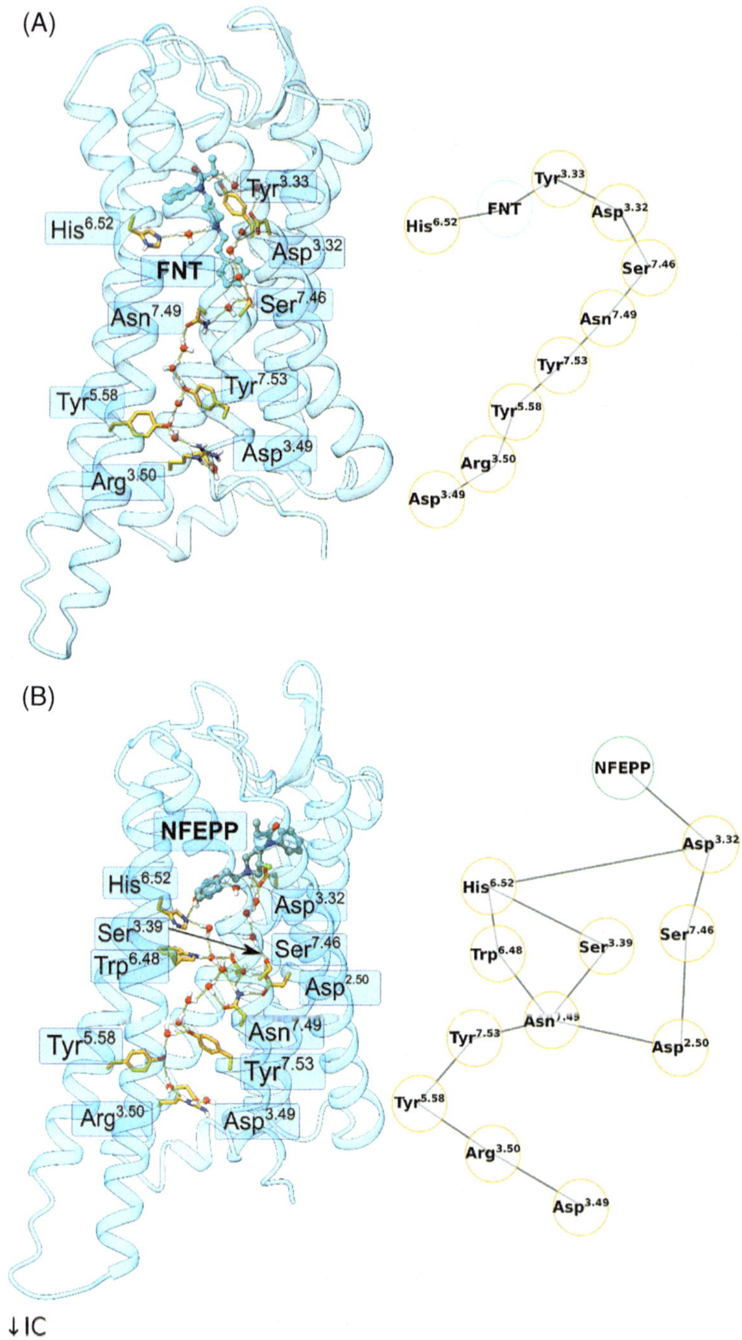
(A)
Tyr3.33
His6.52
Asp3.32
FNT
Ser7.46
Asn7.49
Tyr7.53
Tyr5.58
Asp3.49
Arg3.50
(B)
NFEPP
His6.52
Asp3.32
Ser3.39
Trp6.48
Ser7.46
Asp2.50
Asn7.49
Tyr5.58
Tyr7.53
Arg3.50
Asp3.49
↓IC

transforming the proton pump into a sodium pump were unsuccessful, likely because of the inability to recreate the critical water-mediated interactions near the Schiff base (Bertalan et al., 2024). Pairwise structural analyses suggest that while the key H-bonding aminoacid residues are largely conserved, the hydrogen bonds themselves—especially those mediated by water—undergo local rearrangements, making it difficult to reconstruct functional networks. In these mutants the persistence of H-bonding between the Schiff base and D253 likely alters the energy landscape, disrupting retinal bond torsion and preventing functional ion transport. This highlights the importance of the entire water-mediated hydrogen bond network in shaping the energy landscape and governing ion transport function in microbial rhodopsin pumps.

Dynamic water networks are not only critical in GPCRs and their closely related receptors but also play an essential role in other protein systems, such as carbonic anhydrase II, a zinc-containing enzyme crucial for catalyzing the conversion of carbon dioxide to bicarbonate (Avvaru et al., 2010). This enzyme is a key drug target for conditions like glaucoma, osteoporosis, and certain cancers. The catalytic site of carbonic anhydrase II contains a stable, water-mediated hydrogen-bond network that includes highly conserved water molecules coordinated to the enzyme's zinc ion (Avvaru et al., 2010). Dorzolamide, a well-known inhibitor of carbonic anhydrase, disrupts this water network by displacing water molecules normally coordinated to the zinc ion. Advanced solid-state NMR and MD simulations have shown that this disruption not only alters the water network but also reduces the enzyme's catalytic efficiency by impairing the flexibility and adaptability of the protein structure. This change of the dynamic water network results in diminished substrate turnover and enzyme activity (Fu et al., 2021). The dynamics of this network are closely linked to the enzyme's catalytic process, occurring on a microsecond time scale and driving substrate turnover by facilitating proton transport via His64, a key residue that connects to the active site through the water network (Fig. 5.6). The displacement of water molecules by dorzolamide effectively impairs these crucial dynamics, underscoring the importance of maintaining specific dynamic water networks for enzymatic function.

FIGURE 5.5

Comparative water-mediated hydrogen bond networks formed by fentanyl (FNT) and its fluorinated derivative NFEPP upon binding to the MOR: Panel (A) shows FNT (*cyan*) interacting with key residues within the receptor at 50% occupancy, highlighting the extensive shortest path network of water-mediated hydrogen bonds (*red spheres* and *dashed lines*) residues from the binding site $His^{6.52}$, to the DRY $Asp^{3.49}$. Panel (B) illustrates NFEPP (*green*) binding to the MOR, where the fluorinated derivative forms distinct hydrogen-bonding patterns, forming a more redundant shortest path compared to FNT. Both panels underscore the differences in binding modes between the two ligands, with water molecules playing a crucial role in stabilizing ligand–receptor interactions. The right-hand side of each panel presents a graph-based representation of these networks, created using the Bridge software (Siemers & Bondar, 2021; Siemers et al., 2019), where nodes represent residues and edges indicate the presence of a direct or water-mediated hydrogen bond. The figure is based on the PDB structure 5C1M (Huang et al., 2015) and the work of

Lešnik, S., Bren, U., Domratcheva, T., & Bondar, A. N. (2023). Fentanyl and the fluorinated fentanyl derivative NFEPP elicit distinct hydrogen-bond dynamics of the opioid receptor. Journal of Chemical Information and Modeling, *63(15), 4732–4748. https://doi.org/10.1021/acs.jcim.3c00197.*

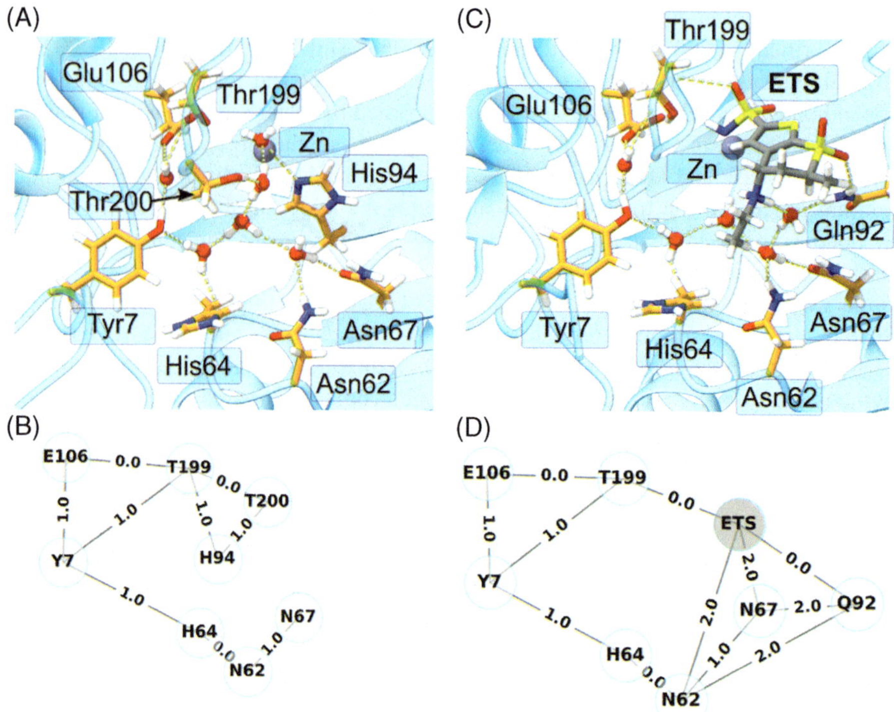

FIGURE 5.6

Structural representation and network analysis of water-mediated interactions in the active site of carbonic anhydrase II: The top panels depict the spatial arrangement of key residues and water molecules in the protein's active site, with the left panel (A) showing the apo state and the right panel (C) illustrating the complex with ETS (dorzolamide). Key residues involved in the water network are shown with orange sticks, and zinc (Zn) is a *purple* sphere. The bottom panels (B) and (D) present the corresponding network graphs, where nodes represent residues and edges indicate the presence of a direct or water-mediated hydrogen bond. The weights on the edges reflect the number of waters connecting the aminoacid residues. The apo structure is based on the PDB ID 4Y0J (Michalczyk et al., 2015), and the holo on PDB ID 6BC9 (Kovalevsky, 2018).

5.6.1 Graph-based water network analysis

One of the key challenges in studying water-mediated hydrogen-bond networks from MD simulations is the sheer volume of data generated. Thousands of simulation frames contain information on residue and water molecule positions, resulting in an overwhelming array of potential H-bond interactions (Bondar, 2022). Compounding this complexity is the dynamic nature of H-bonds, which form and break on the picosecond timescale. This rapid fluctuation makes it difficult to distinguish between critical, persistent interactions and fleeting ones, complicating the identification of water molecules essential for protein function and stability (Adhikari et al., 2020; Qin et al., 2016; Ryan et al., 2024).

Graph-based methods offer a robust solution by simplifying these complex interactions into a graph, where residues and water molecules are treated as nodes, and H-bonds are represented as edges (Bondar, 2022). These methods not only reduce the complexity of large datasets but also allow for the use of metrics like centrality and betweenness to pinpoint key aminoacid residues and critical water molecules that are crucial for protein stability and function (Amitai et al., 2004; Karathanou & Bondar, 2019).

Tracking changes in the graph over time offers insights into the dynamic behavior of H-bonds, helping to differentiate between transient and stable interactions. As mentioned above, this approach is particularly useful in studying proteins like GPCRs, where dynamic water molecules play a critical role in signal transduction (Bertalan et al., 2020). In addition, graph methods enable the mapping of communication pathways within proteins, providing a clearer understanding of how water-mediated interactions facilitate signal propagation across receptors like GPCRs (Lešnik et al., 2023).

Graph-based representations can be seamlessly integrated into machine learning methods, enabling the prediction of hydrogen bond behavior and the classification of such interactions based on their significance (Omwansu et al., 2024; Qu et al., 2022). This integration could enhance the analysis of H-bond networks by providing more accurate models of water-mediated interactions. Such models could be especially useful in drug design, where a better understanding of the role of water molecules in ligand binding can inform the development of drugs that optimize these interactions.

One of the newly developed tools that simplify the analysis of dynamic water networks from MD simulations is Bridge2 (Siemers & Bondar, 2021; Siemers et al., 2019). It uses a graph-based approach to represent atoms as nodes and direct or water-mediated H-bonds as edges. Bridge2 provides an interactive interface, allowing researchers to manipulate these graphs in real-time. This is particularly useful for focusing on specific regions, such as active sites. Its user friendly design supports the exporting of results for further analysis and visualization.

Recently, Bridge2 was further enhanced with a tool called C-Graphs, which significantly improves the analysis of hydrogen-bond networks in proteins (Bertalan et al., 2021). C-Graphs introduces a user-friendly interface for comparing independent trajectories and identifying conserved or unique hydrogen-bond networks. By enabling the simultaneous analysis of multiple structures or MD datasets, the tool facilitates the study of how water networks change under different conditions, including mutations or the binding of distinct ligands. For example, in GPCRs, C-Graphs has revealed an internal protein–water hydrogen-bond network conserved across both visual rhodopsins and adenosine A2A receptors, which is critical for receptor function (Bertalan et al., 2021).

5.7 Conclusion and future topics

Water molecules play a crucial role in protein structure, function, dynamics, stability, folding, and ligand binding, thereby shaping biological processes and informing drug design strategies. This importance is particularly evident in the context of conserved water molecules, which are found across various protein structures. Proteins require a delicate balance between flexibility and structural stability to function properly, and water molecules are key in maintaining this balance. Beyond their role in stabilizing protein structures and assisting in correct protein folding, water molecules regulate dynamic functions by mediating conformational changes necessary for protein activities such as catalysis and signal transduction. In enzymes, conserved water molecules are often located in active sites, where they support catalytic processes, serve as proton carriers, and facilitate proton transfer via transient hydrogen bonds, all of which are crucial for enzymatic reactions and signaling. As emphasized in this chapter, conserved water molecules represent not only passive but also active participants in the functional transitions of proteins.

At protein–protein interfaces, water molecules mediate interactions, often forming bridging hydrogen bonds between adjacent surfaces. These molecules fill voids at the interface, mitigate electrostatic repulsion, and facilitate protein recognition, thereby ensuring a specific and stable binding. The conservation of water molecules and their surrounding residues underscores their essential role in maintaining both the structural and functional integrity of protein complexes, while simultaneously providing the flexibility needed to accommodate variations in binding partners.

The role of water extends beyond structural stability to include dynamic processes such as signal transduction. The reorganization of water-mediated hydrogen bond networks during receptor activation illustrates how water molecules can influence protein function. For example, in GPCRs, state-dependent water networks either stabilize the inactive receptor state or facilitate the transition to its active form. The flexibility and adaptability of water molecules are critical for the precise regulation of signal propagation in these receptors, highlighting the need to account for water dynamics in structural and functional analyses.

To better understand the role of conserved waters, a combination of experimental and computational tools is often needed. Although X-ray crystallography and NMR spectroscopy provide valuable insights into the positioning and dynamics of water molecules, they face limitations, such as resolution constraints, the transient nature of water, and the inability to assess the thermodynamics of a single molecule. MD simulations and computational tools, such as WaterMap, SZMAP, 3D-RISM, Bridge, and MADE, have become indispensable, offering detailed insights into the behavior of water molecules over time. Future advancements in computational methods will be crucial for understanding the energetic contributions of conserved waters and their roles in protein stabilization and ligand binding.

Conserved water molecules present both challenges and opportunities in drug design. Their presence in binding sites can either enhance or hinder ligand binding, depending on whether they are retained or displaced during the interaction. When conserved water molecules are retained, they form stable hydrogen bonds that increase binding affinity via enthalpic contributions. Conversely the displacement of high-energy water molecules into bulk solvent can result in significant entropic gains. Drug design strategies increasingly aim to exploit these thermodynamic principles by either mimicking the role of conserved water molecules or replacing them with isosteric functional groups that replicate the necessary interactions. The balance between enthalpy and entropy, influenced by conserved waters, plays a pivotal role in optimizing drug candidates.

Further advances in experimental techniques and computational modeling will be essential for deepening our understanding of water's role in protein function. Accurate tools for predicting the behavior of water molecules and their thermodynamic contributions will become increasingly important, particularly when linked to comprehensive databases of high-quality protein structural data. In addition, the exploration of dynamic water networks in processes such as signal propagation and ligand binding will provide new insights into drug optimization. Leveraging the unique properties of conserved water molecules offers a powerful approach to improving ligand binding and enhancing drug efficacy. The ability to strategically displace or incorporate these water molecules to modulate both enthalpic and entropic contributions represents an advanced strategy in both drug design and the mechanistic study of biological systems.

References

Adhikari, A., Park, W.-W., & Kwon, O.-H. (2020). Hydrogen-bond dynamics and energetics of biological water. *ChemPlusChem, 85*(12), 2657–2665. doi:10.1002/cplu.202000744.

Ahmed, M. H., Habtemariam, M., Safo, M. K., Scarsdale, J. N., Spyrakis, F., Cozzini, P., Mozzarelli, A., & Kellogg, G. E. (2013). Unintended consequences? Water molecules at biological and crystallographic protein-protein interfaces. *Computational Biology and Chemistry, 47*, 126–141. doi:10.1016/j.compbiolchem.2013.08.009.

Ahmed, M. H., Spyrakis, F., Cozzini, P., Tripathi, P. K., Mozzarelli, A., Scarsdale, J. N., Safo, M. A., Kellogg, G. E., & Deane, C. M. (2011). Bound water at protein-protein interfaces: Partners, roles and hydrophobic bubbles as a conserved motif. *PLoS One, 6*(9), e24712. doi:10.1371/journal.pone.0024712.

Amitai, G., Shemesh, A., Sitbon, E., Shklar, M., Netanely, D., Venger, I., & Pietrokovski, S. (2004). Network analysis of protein structures identifies functional residues. *Journal of Molecular Biology, 344*(4), 1135–1146. doi:10.1016/j.jmb.2004.10.055.

Ankerst, M., Breunig, M. M., Kriegel, H. P., & Sander, J. (1999). OPTICS: Ordering points to identify the clustering structure. *SIGMOD Record (ACM Special Interest Group on Management of Data), 28*(2), 49–60. http://dl.acm.org/citation.cfm?id=J689. doi:10.1145/304181.304187.

Arkhipova, V., Guskov, A., & Slotboom, D. J. (2017). Analysis of the quality of crystallographic data and the limitations of structural models. *Journal of General Physiology, 149*(12), 1091–1103. http://jgp.rupress.org/content/149/12/1091/tab-pdf. doi:10.1085/jgp.201711852.

Avvaru, B. S., Kim, C. U., Sippel, K. H., Gruner, S. M., Agbandje-McKenna, M., Silverman, D. N., & McKenna, R. (2010). A short, strong hydrogen bond in the active site of human carbonic anhydrase II. *Biochemistry, 49*(2), 249–251. http://pubs.acs.org/doi/pdfplus/10.1021/bi902007b. doi:10.1021/bi902007b.

Barillari, C., Taylor, J., Viner, R., & Essex, J. W. (2007). Classification of water molecules in protein binding sites. *Journal of the American Chemical Society, 129*(9), 2577–2587. doi:10.1021/ja066980q.

Baroni, M., Cruciani, G., Sciabola, S., Perruccio, F., & Mason, J. S. (2007). A common reference framework for analyzing/comparing proteins and ligands. Fingerprints for Ligands and Proteins (FLAP): Theory and application. *Journal of Chemical Information and Modeling, 47*(2), 279–294. http://pubs.acs.org/journal/jcisd8. doi:10.1021/ci600253e.

Bayden, A. S., Moustakas, D. T., Joseph-McCarthy, D., & Lamb, M. L. (2015). Evaluating free energies of binding and conservation of crystallographic waters using SZMAP. *Journal of Chemical Information and Modeling, 55*(8), 1552–1565. http://pubs.acs.org/journal/jcisd8. doi:10.1021/ci500746d.

Bellissent-Funel, M. C., Hassanali, A., Havenith, M., Henchman, R., Pohl, P., Sterpone, F., Van Der Spoel, D., Xu, Y., & Garcia, A. E. (2016). Water determines the structure and dynamics of proteins. *Chemical Reviews, 116*(13), 7673–7697. http://pubs.acs.org/journal/chreay. doi:10.1021/acs.chemrev.5b00664.

Berman, H., Henrick, K., & Nakamura, H. (2003). Announcing the worldwide Protein Data Bank. *Nature Structural & Molecular Biology, 10*(12), 980. doi:10.1038/nsb1203-980.

Bertalan, É, Konno, M., del Carmen Marín, M., Bagherzadeh, R., Nagata, T., Brown, L., Inoue, K., & Bondar, A.-N. (2024). Hydrogen-bonding and hydrophobic interaction networks as structural determinants of microbial rhodopsin function. *The Journal of Physical Chemistry B, 128*(30), 7407–7426. doi:10.1021/acs.jpcb.4c02946.

Bertalan, É., Lesca, E., Schertler, G. F. X., & Bondar, A. N. (2021). C-Graphs tool with graphical user interface to dissect conserved hydrogen-bond networks: Applications to visual rhodopsins. *Journal of Chemical Information and Modeling, 61*(11), 5692–5707. http://pubs.acs.org/journal/jcisd8. doi:10.1021/acs.jcim.1c00827.

Bertalan, É., Lešnik, S., Bren, U., & Bondar, A.-N. (2020). Protein-water hydrogen-bond networks of G protein-coupled receptors: Graph-based analyses of static structures and molecular dynamics. *Journal of Structural Biology, 212*(3), 107634. doi:10.1016/j.jsb.2020.107634.

Beuming, T., Che, Y., Abel, R., Kim, B., Shanmugasundaram, V., & Sherman, W. (2012). Thermodynamic analysis of water molecules at the surface of proteins and applications to binding site prediction and characterization. *Proteins: Structure, Function, and Bioinformatics, 80*(3), 871–883. doi:10.1002/prot.23244.

Bodnarchuk, M. S. (2016). Water, water, everywhere. It's time to stop and think. *Drug Discovery Today, 21*(7), 1139–1146. Www.elsevier.com/locate/drugdiscov. doi:10.1016/j.drudis.2016.05.009.

Bodnarchuk, M. S., Packer, M. J., & Haywood, A. (2020). Utilizing grand canonical Monte Carlo methods in drug discovery. *ACS Medicinal Chemistry Letters, 11*(1), 77–82. http://pubs.acs.org/journal/amclct. doi:10.1021/acsmedchemlett.9b00499.

Bondar, A. N. (2022). Graphs of hydrogen-bond networks to dissect protein conformational dynamics. *Journal of Physical Chemistry B, 126*(22), 3973–3984. http://pubs.acs.org/journal/jpcbfk. doi:10.1021/acs.jpcb.2c00200.

Bondar, A. N., & Alfonso-Prieto, M. (2022). Hydrogen-bond networks for proton couplings in G-protein coupled receptors. *Frontiers in Physics, 10*. Journal.frontiersin.org/journal/physics. doi:10.3389/fphy.2022.963716.

Bondar, A. N., Baudry, J., Suhai, S., Fischer, S., & Smith, J. C. (2008). Key role of active-site water molecules in bacteriorhodopsin proton-transfer reactions. *Journal of Physical Chemistry B, 112*(47), 14729–14741. http://pubs.acs.org/journal/jpcbfk. doi:10.1021/jp801916f.

Bondar, A. N., & Smith, J. C. (2017). Protonation–state-coupled conformational dynamics in reaction mechanisms of channel and pump rhodopsins. *Photochemistry and Photobiology, 93*(6), 1336–1344. http://www.blackwellpublishing.com. doi:10.1111/php.12790.

Bortolato, A., Tehan, B. G., Bodnarchuk, M. S., Essex, J. W., & Mason, J. S. (2013). Water network perturbation in ligand binding: Adenosine A2A antagonists as a case study. *Journal of Chemical Information and Modeling, 53*(7), 1700–1713. http://pubs.acs.org/journal/jcisd8. doi:10.1021/ci4001458.

Bottoms, C. A., White, T. A., & Tanner, J. J. (2006). Exploring structurally conserved solvent sites in protein families. *Proteins: Structure, Function and Genetics, 64*(2), 404–421. doi:10.1002/prot.21014.

Braden, B. C., Fields, B. A., & Poljak, R. J. (1995). Conservation of water molecules in an antibody–antigen interaction. *Journal of Molecular Recognition, 8*(5), 317–325. doi:10.1002/jmr.300080505.

Breiten, B., Lockett, M. R., Sherman, W., Fujita, S., Al-Sayah, M., Lange, H., Bowers, C. M., Heroux, A., Krilov, G., & Whitesides, G. M. (2013). Water networks contribute to enthalpy/entropy compensation in protein–ligand binding. *Journal of the American Chemical Society, 135*(41), 15579–15584. doi:10.1021/ja4075776.

Bucher, D., Stouten, P., & Triballeau, N. (2018). Shedding light on important waters for drug design: Simulations versus grid-based methods. *Journal of Chemical Information and Modeling, 58*(3), 692–699. doi:10.1021/acs.jcim.7b00642.

Bui, H. H., Schiewe, A. J., & Haworth, I. S. (2007). WATGEN: An algorithm for modeling water networks at protein-protein interfaces. *Journal of Computational Chemistry, 28*(14), 2241–2251. doi:10.1002/jcc.20751.

Carugo, O. (2016). When proteins are completely hydrated in crystals. *International Journal of Biological Macromolecules, 89*, 137–143. doi:10.1016/j.ijbiomac.2016.04.061.

Che, T., Majumdar, S., Zaidi, S. A., Ondachi, P., McCorvy, J. D., Wang, S., Mosier, P. D., Uprety, R., Vardy, E., Krumm, B. E., Han, G. W., Lee, M.-Y., Pardon, E., Steyaert, J., Huang, X. i-P., Strachan, R. T., Tribo, A. R., Pasternak, G. W., Carroll, F. I., … Roth, B. L. (2018). Structure of the nanobody-stabilized active state of the kappa opioid receptor. *Cell, 172*(1–2), 55. doi:10.1016/j.cell.2017.12.011.

Chen, J. M., Xu, S. L., Wawrzak, Z., Basarab, G. S., & Jordan, D. B. (1998). Structure-based design of potent inhibitors of scytalone dehydratase: Displacement of a water molecule from the active site. *Biochemistry, 37*(51), 17735–17744. doi:10.1021/bi981848r.

Cheng, L., Xia, F., Li, Z., Shen, C., Yang, Z., Hou, H., Sun, S., Feng, Y., Yong, X., Tian, X., Qin, H., Yan, W., & Shao, Z. (2023). Structure, function and drug discovery of GPCR signaling. *Molecular Biomedicine, 4*(1). doi:10.1186/s43556-023-00156-w.

Chong, S. H., & Ham, S. (2015). Distinct role of hydration water in protein misfolding and aggregation revealed by fluctuating thermodynamics analysis. *Accounts of Chemical Research, 48*(4), 956–965. http://pubs.acs.org/journal/achre4. doi:10.1021/acs.accounts.5b00032.

Chong, S. H., & Ham, S. (2017). Dynamics of hydration water plays a key role in determining the binding thermodynamics of protein complexes. *Scientific Reports, 7*(1). Www.nature.com/srep/index.html. doi:10.1038/s41598-017-09466-w.

Chua, E. Y. D., Mendez, J. H., Rapp, M., Ilca, S. L., Tan, Y. Z., Maruthi, K., Kuang, H., Zimanyi, C. M., Cheng, A., Eng, E. T., Noble, A. J., Potter, C. S., Carragher, B., & Better, F. (2022). Cheaper: Recent advances in cryo-electron microscopy. *Annual Review of Biochemistry, 91*, 1–32. http://www.annualreviews.org/journal/biochem. doi:10.1146/annurev-biochem-032620-110705.

Clackson, T., & Wells, J. A. (1995). A hot spot of binding energy in a hormone-receptor interface. *Science, 267*(5196), 383–386. doi:10.1126/science.7529940.

Cole, J. C., Giangreco, I., & Groom, C. R. (2017). Using more than 801 296 small-molecule crystal structures to aid in protein structure refinement and analysis. Addendum. *Acta Crystallographica Section D Structural Biology, 73*(12), 1029. doi:10.1107/s2059798317016904.

Cui, D., Zhang, B. W., Matubayasi, N., & Levy, R. M. (2018). The role of interfacial water in protein-ligand binding: Insights from the indirect solvent mediated potential of mean force. *Journal of Chemical Theory and Computation, 14*(2), 512–526. http://pubs.acs.org/journal/jctcce. doi:10.1021/acs.jctc.7b01076.

Cui, G., Swails, J. M., & Manas, E. S. (2013). SPAM: A simple approach for profiling bound water molecules. *Journal of Chemical Theory and Computation, 9*(12), 5539–5549. doi:10.1021/ct400711g.

Davies, N. G. M., Browne, H., Davis, B., Drysdale, M. J., Foloppe, N., Geoffrey, S., Gibbons, B., Hart, T., Hubbard, R., Jensen, M. R., Mansell, H., Massey, A., Matassova, N., Moore, J. D., Murray, J., Pratt, R., Ray, S., Robertson, A., Roughley, S. D., … Brough, P. (2012). Targeting conserved water molecules: Design of 4-aryl-5-cyanopyrrolo[2,3-d] pyrimidine Hsp90 inhibitors using fragment-based screening and structure-based optimization. *Bioorganic and Medicinal Chemistry, 20*(22), 6770–6789. doi:10.1016/j.bmc.2012.08.050.

de Beer, S., Vermeulen, N., & Oostenbrink, C. (2010). The role of water molecules in computational drug design. *Current Topics in Medicinal Chemistry, 10*(1), 55–66. doi:10.2174/156802610790232288.

DeLano, W. L. (2002). Unraveling hot spots in binding interfaces: Progress and challenges. *Current Opinion in Structural Biology, 12*(1), 14–20. http://www.elsevier.com/locate/sbi. doi:10.1016/S0959-440X(02)00283-X.

De Lucca, G. V., Erickson-Viitanen, S., & Lam, P. Y. S. (1997). Cyclic HIV protease inhibitors capable of displacing the active site structural water molecule. *Drug Discovery Today, 2*(1), 6–18. Www.elsevier.com/locate/drugdiscov. doi:10.1016/S1359-6446(96)10048-9.

Demchenko, A. P. (2023). Proton transfer reactions: From photochemistry to biochemistry and bioenergetics. *BBA Advances, 3*, 100085. doi:10.1016/j.bbadva.2023.100085.

Denisov, V. P., & Halle, B. (1996). Protein hydration dynamics in aqueous solution. *Faraday Discussions, 103*, 227–244. http://pubs.rsc.org/en/journals/journal/fd. doi:10.1039/fd9960300227.

Doyle, M. D., Bhowmick, A., Wych, D. C., Lassalle, L., Simon, P. S., Holton, J., Sauter, N. K., Yachandra, V. K., Kern, J. F., Yano, J., & Wall, M. E. (2023). Water networks in photosystem II using crystalline molecular dynamics simulations and room-temperature XFEL serial crystallography. *Journal of the American Chemical Society, 145*(27), 14621–14635. http://pubs.acs.org/journal/jacsat. doi:10.1021/jacs.3c01412.

Du, X., Li, Y., Xia, Y.-L., Ai, S.-M., Liang, J., Sang, P., Ji, X.-L., & Liu, S.-Q. (2016). Insights into protein–ligand interactions: Mechanisms, models, and methods. *International Journal of Molecular Sciences, 17*, 144.

Ernst, J. A., Clubb, R. T., Zhou, H. X., Gronenborn, A. M., & Clore, G. M. (1995). Demonstration of positionally disordered water within a protein hydrophobic cavity by NMR. *Science, 67*(5205), 1813–1817. doi:10.1126/science.7892604.

Fenimore, P. W., Frauenfelder, H., McMahon, B. H., & Parak, F. G. (2002). Slaving: Solvent fluctuations dominate protein dynamics and functions. *Proceedings of the National Academy of Sciences, 99*(25), 16047–16051. doi:10.1073/pnas.212637899.

Fenimore, P. W., Frauenfelder, H., McMahon, B. H., & Young, R. D. (2004). Bulk-solvent and hydration-shell fluctuations, similar to α- and β-fluctuations in glasses, control protein motions and functions. *Proceedings of the National Academy of Sciences, 101*(40), 14408–14413. doi:10.1073/pnas.0405573101.

Fischer, A. L. M., Tichy, A., Kokot, J., Hoerschinger, V. J., Wild, R. F., Riccabona, J. R., Loeffler, J. R., Waibl, F., Quoika, P. K., Gschwandtner, P., Forli, S., Ward, A. B., Liedl, K. R., Zacharias, M., & Fernández-Quintero, M. L. (2024). The role of force fields and water models in protein folding and unfolding dynamics. *Journal of Chemical Theory and Computation, 20*(5), 2321–2333. http://pubs.acs.org/journal/jctcce. doi:10.1021/acs.jctc.3c01106.

Fried, S. D. E., Hewage, K. S. K., Eitel, A. R., Struts, A. V., Weerasinghe, N., Perera, S. M. D. C., & Brown, M. F. (2022). Hydration-mediated G-protein-coupled receptor activation. *Proceedings of the National Academy of Sciences of the United States of America, 119*(21). https://www.pnas.org/doi/10.1073/pnas.2117349119. doi:10.1073/pnas.2117349119.

Fu, Y., Fan, F., Zhang, Y., Wang, B., & Cao, Z. (2021). Conformational change of H64 and substrate transportation: Insight into a full picture of enzymatic hydration of CO_2 by carbonic anhydrase. *Frontiers in Chemistry, 9*. doi:10.3389/fchem.2021.706959.

García-Sosa, A. T., Firth-Clark, S., & Mancera, R. L. (2005). Including tightly-bound water molecules in de novo drug design exemplification through the in silico generation of poly(ADP-ribose)polymerase ligands. *Journal of Chemical Information and Modeling, 45*(3), 624–633. http://pubs.acs.org/journal/jcisd8. doi:10.1021/ci049694b.

García-Sosa, A. T., Mancera, R. L., & Dean, P. M. (2003). WaterScore: A novel method for distinguishing between bound and displaceable water molecules in the crystal structure of the binding site of protein-ligand complexes. *Journal of Molecular Modeling, 9*(3), 172–182. http://www.springerlink.com/app/home/content.asp?wasp=5hyy105mht6qywb178rn&referrer=contribution&format=2&page=1&pagecount=11. doi:10.1007/s00894-003-0129-x.

Ge, Y., Baumann, H. M., & Mobley, D. L. (2022). Absolute binding free energy calculations for Buried Water Molecules. *Journal of Chemical Theory and Computation, 18*(11), 6482–6499. http://pubs.acs.org/journal/jctcce. doi:10.1021/acs.jctc.2c00658.

Gnesi, M., & Carugo, O. (2017). How many water molecules are detected in X-ray protein crystal structures? *Journal of Applied Crystallography, 50*(1), 96–101. doi:10.1107/s1600576716018719.

Harrigan, M. P., Shukla, D., & Pande, V. S. (2015). Conserve water: A method for the analysis of solvent in molecular dynamics. *Journal of Chemical Theory and Computation, 11*(3), 1094–1101. http://pubs.acs.org/journal/jctcce. doi:10.1021/ct5010017.

Hu, J., Sun, X., Kang, Z., & Cheng, J. (2023). Computational investigation of functional water molecules in GPCRs bound to G protein or arrestin. *Journal of Computer-Aided Molecular Design, 37*(2), 91–105. doi:10.1007/s10822-022-00492-z.

Hu, X., Maffucci, I., & Contini, A. (2020). Advances in the treatment of explicit water molecules in docking and binding free energy calculations. *Current Medicinal Chemistry, 26*(42), 7598–7622. doi:10.2174/0929867325666180514110824.

Huang, K., Fu, T., Gao, W., Zhao, Y., Roohani, Y., Leskovec, J., Coley, C. W., Xiao, C., Sun, J., & Zitnik, M. (2021). Therapeutics data commons: Machine learning datasets and tasks for drug discovery and development. *arXiv*. https://arxiv.org. doi:10.48550/arxiv.2102.09548.

Huang, W., Manglik, A., Venkatakrishnan, A. J., Laeremans, T., Feinberg, E. N., Sanborn, A. L., Kato, H. E., Livingston, K. E., Thorsen, T. S., Kling, R. C., Granier, S., Gmeiner, P., Husbands, S. M., Traynor, J. R., Weis, W. I., Steyaert, J., Dror, R. O., & Kobilka, B. K. (2015). Structural insights into μ-opioid receptor activation. *Nature, 524*(7565), 315–321. doi:10.1038/nature14886.

Huggins, D. J. (2016). Studying the role of cooperative hydration in stabilizing folded protein states. *Journal of Structural Biology, 196*(3), 394–406. http://www.elsevier.com/inca/publications/store/6/2/2/9/0/0/index.htt. doi:10.1016/j.jsb.2016.09.003.

Huggins, D. J., Marsh, M., & Payne, M. C. (2011). Thermodynamic properties of water molecules at a protein-protein interaction surface. *Journal of Chemical Theory and Computation, 7*(11), 3514–3522. doi:10.1021/ct200465z.

Ikura, T., Urakubo, Y., & Ito, N. (2004). Water-mediated interaction at a protein–protein interface. *Chemical Physics, 307*(2-3), 111–119. doi:10.1016/j.chemphys.2004.05.010.

Ishikita, H., & Saito, K. (2014). Proton transfer reactions and hydrogen-bond networks in protein environments. *Journal of the Royal Society Interface, 11*(91), 20130518. doi:10.1098/rsif.2013.0518.

Jain, A., Patel, H., Nagalapatti, L., Gupta, N., Mehta, S., Guttula, S., Mujumdar, S., Afzal, S., Sharma Mittal, R., & Munigala, V. (2020). Overview and importance of data quality for machine learning tasks. In *Proceedings of the ACM SIGKDD International Conference on Knowledge Discovery and Data Mining* (pp. 3561–3562). India: Association for Computing Machinery. 9781450379984. doi:10.1145/3394486.3406477.

Jukič, M., Konc, J., Janežič, D., & Bren, U. (2020). ProBiS H_2O MD approach for identification of conserved water sites in protein structures for drug design. *ACS Medicinal Chemistry Letters, 11*(5), 877–882. http://pubs.acs.org/journal/amclct. doi:10.1021/acsmedchemlett.9b00651.

Kadaoluwa Pathirannahalage, S. P., Meftahi, N., Elbourne, A., Weiss, A. C. G., McConville, C. F., Padua, A., Winkler, D. A., Gomes, M. C., Greaves, T. L., Le, T. C., Besford, Q. A., & Christofferson, A. J. (2021). Systematic comparison of the structural and dynamic properties of commonly used water models for molecular dynamics simulations. *Journal of Chemical Information and Modeling, 61*(9), 4521–4536. http://pubs.acs.org/journal/jcisd8. doi:10.1021/acs.jcim.1c00794.

Karathanou, K., & Bondar, A. N. (2019). Using graphs of dynamic hydrogen-bond networks to dissect conformational coupling in a protein motor. *Journal of Chemical Information and Modeling, 59*(5), 1882–1896. http://pubs.acs.org/journal/jcisd8. doi:10.1021/acs.jcim.8b00979.

Kearney, B. M., Johnson, C. W., Roberts, D. M., Swartz, P., & Mattos, C. (2014). DRoP: A water analysis program identifies Ras-GTP-specific pathway of communication between membrane-interacting regions and the active site. *Journal of Molecular Biology, 426*(3), 611–629. https://www.journals.elsevier.com/journal-of-molecular-biology. doi:10.1016/j.jmb.2013.10.036.

Kearney, B. M., Schwabe, M., Marcus, K. C., Roberts, D. M., Dechene, M., Swartz, P., & Mattos, C. (2020). DRoP: Automated detection of conserved solvent-binding sites on proteins. *Proteins: Structure, Function and Bioinformatics, 88*(1), 152–165. http://onlinelibrary.wiley.com/journal/10.1002/. doi:10.1002/prot.25781. (ISSN)1097-0134.

Kemmler, L., Ibrahim, M., Dobbek, H., Zouni, A., & Bondar, A.-N. (2019). Dynamic water bridging and proton transfer at a surface carboxylate cluster of photosystem II. *Physical Chemistry Chemical Physics, 21*(45), 25449–25466. doi:10.1039/C9CP03926K.

Khan, K., Rehman, S-U., Aziz, K., Fong, S., Sarasvady, S., et al. (2014). DBSCAN: Past, present and future. *The Fifth International Conference on the Applications of Digital Information and Web Technologies (ICADIWT 2014)*, 232–238. doi:10.1109/ICADIWT.2014.6814687.

Kinjo, A. R., Bekker, G. J., Suzuki, H., Tsuchiya, Y., Kawabata, T., Ikegawa, Y., & Nakamura, H. (2017). Protein Data Bank Japan (PDBj): Updated user interfaces, resource description framework, analysis tools for large structures. *Nucleic Acids Research, 45*(1), D282. http://nar.oxfordjournals.org/. doi:10.1093/nar/gkw962.

Konc, J., & Janežič, D. uš (2010). ProBiS algorithm for detection of structurally similar protein binding sites by local structural alignment. *Bioinformatics, 26*(9), 1160–1168. doi:10.1093/bioinformatics/btq100.

Kovalevsky, A. (2018). To be or not to be' protonated: Atomic details of human carbonic anhydrase-clinical drug complexes by neutron crystallography and simulation. *Structure, 26*, 383–390.

Kumar, A., & Zhang, K. Y. J. (2013). Investigation on the effect of key water molecules on docking performance in CSARdock exercise. *Journal of Chemical Information and Modeling, 53*(8), 1880–1892. http://pubs.acs.org/journal/jcisd8. doi:10.1021/ci400052w.

Lam, P. Y. S., Jadhav, P. K., Eyermann, C. J., Hodge, C. N., Ru, Y., Bacheler, L. T., Meek, J. L., Otto, M. J., Rayner, M. M., Wong, Y. N., Chang, C. H., Weber, P. C., Jackson, D. A., Sharpe, T. R., & Erickson-Viitanen, S. (1994). Rational design of potent, bioavailable, nonpeptide cyclic ureas as HIV protease inhibitors. *Science, 263*(5145), 380–384. doi:10.1126/science.8278812.

Lam, P. Y. S., Ru, Y., Jadhav, P. K., Aldrich, P. E., DeLucca, G. V., Eyermann, C. J., Chang, C. H., Emmett, G., Holler, E. R., Daneker, W. F., Li, L., Confalone, P. N., McHugh, R. J., Han, Q., Li, R., Markwalder, J. A., Seitz, S. P., Sharpe, T. R., Bacheler, L. T., … Hodge, C. N. (1996). Cyclic HIV protease inhibitors: Synthesis, conformational analysis, P2/P2′ structure-activity relationship, and molecular recognition of cyclic ureas. *Journal of Medicinal Chemistry, 39*(18), 3514–3525. doi:10.1021/jm9602571.

Lešnik, S., Bren, U., Domratcheva, T., & Bondar, A. N. (2023). Fentanyl and the fluorinated fentanyl derivative NFEPP elicit distinct hydrogen-bond dynamics of the opioid receptor. *Journal of Chemical Information and Modeling, 63*(15), 4732–4748. http://pubs.acs.org/journal/jcisd8. doi:10.1021/acs.jcim.3c00197.

Li, Z., He, Y., Cao, L., Wong, L., & Li, J. (2012). Conservation of water molecules in protein binding interfaces. *International Journal of Bioinformatics Research and Applications, 8*(3/4), 228. doi:10.1504/IJBRA.2012.048968.

Liu, C., Wrobleski, S. T., Lin, J., Ahmed, G., Metzger, A., Wityak, J., Gillooly, K. M., Shuster, D. J., McIntyre, K. W., Pitt, S., Shen, D. R., Zhang, R. F., Zhang, H., Doweyko, A. M., Diller, D., Henderson, I., Barrish, J. C., Dodd, J. H., Schieven, G. L., & Leftheris, K. (2005). 5-Cyanopyrimidine derivatives as a novel class of potent, selective, and orally active inhibitors of p38α MAP kinase. *Journal of Medicinal Chemistry, 48*(20), 6261–6270. doi:10.1021/jm0503594.

Louet, M., Casiraghi, M., Damian, M., Costa, M. G. S., Renault, P., Gomes, A. A. S., Batista, P. R., M'Kadmi, C., Mary, S., Cantel, S., Denoyelle, S., Salah, K. B. H., Perahia, D., Bisch, P. M., Fehrentz, J.-A., Catoire, L. J., Floquet, N., & Banères, J.-L. (2021). Concerted conformational dynamics and water movements in the ghrelin G protein-coupled receptor. *eLife, 10*. doi:10.7554/elife.63201.

Lu, S. Y., Jiang, Y. J., Lv, J., Zou, J. W., & Wu, T. X. (2011). Role of bridging water molecules in GSK3β-inhibitor complexes: Insights from QM/MM, MD, and molecular docking studies. *Journal of Computational Chemistry, 32*(9), 1907–1918. doi:10.1002/jcc.21775.

Lu, Y., Wang, R., Yang, C. Y., & Wang, S. (2007). Analysis of ligand-bound water molecules in high-resolution crystal structures of protein-ligand complexes. *Journal of Chemical Information and Modeling, 47*(2), 668–675. http://pubs.acs.org/journal/jcisd8. doi:10.1021/ci6003527.

Lukac, I., Wyatt, P. G., Gilbert, I. H., & Zuccotto, F. (2021). Ligand binding: Evaluating the contribution of the water molecules network using the fragment molecular orbital method. *Journal of Computer-Aided Molecular Design, 35*(10), 1025–1036. https://www.springer.com/journal/10822. doi:10.1007/s10822-021-00416-3.

Lv, Z., Chu, Y., & Wang, Y. (2015). HIV protease inhibitors: A review of molecular selectivity and toxicity. *HIV/AIDS—Research and Palliative Care, 7*, 95–104. http://www.dovepress.com/getfile.php?fileID=24509. doi:10.2147/HIV.S79956.

Mallamace, F., Corsaro, C., Mallamace, D., Baglioni, P., Stanley, H. E., & Chen, S.-H. (2011). A possible role of water in the protein folding process. *The Journal of Physical Chemistry B, 115*(48), 14280–14294. doi:10.1021/jp205285t.

Marchand, J. R., Pirard, B., Ertl, P., & Sirockin, F. (2021). CAVIAR: A method for automatic cavity detection, description and decomposition into subcavities. *Journal of Computer-Aided Molecular Design, 35*(6), 737–750. https://www.springer.com/journal/10822. doi:10.1007/s10822-021-00390-w.

Marmolejo-Valencia, A. F., Madariaga-Mazón, A., & Martinez-Mayorga, K. (2021). Bias-inducing allosteric binding site in mu-opioid receptor signaling. *SN Applied Sciences, 3*(5). Springer.com/snas. doi:10.1007/s42452-021-04505-8.

Matricon, P., Suresh, R. R., Gao, Z.-G., Panel, N., Jacobson, K. A., & Carlsson, J. (2021). Ligand design by targeting a binding site water. *Chemical Science, 12*(3), 960–968. doi:10.1039/d0sc04938g.

Maurer, M., & Oostenbrink, C. (2019). Water in protein hydration and ligand recognition. *Journal of Molecular Recognition, 32*(12). doi:10.1002/jmr.2810.

Meanwell, N. A. (2014). The influence of bioisosteres in drug design: Tactical applications to address developability problems. *Springer Science and Business Media LLC, 9*, 283–381. doi:10.1007/7355_2013_29.

Michalczyk, R., Unkefer, C. J., Bacik, J.-P., Schrader, T. E., Ostermann, A., Kovalevsky, A. Y., McKenna, R., & Fisher, S. Z. (2015). Joint neutron crystallographic and NMR solution studies of Tyr residue ionization and hydrogen bonding: Implications for enzyme-mediated proton transfer. *Proceedings of the National Academy of Sciences, 112*(18), 5673–5678. doi:10.1073/pnas.1502255112.

Michel, J., & Essex, J. W. (2010). Prediction of protein-ligand binding affinity by free energy simulations: Assumptions, pitfalls and expectations. *Journal of Computer-Aided Molecular Design, 24*(8), 639–658. https://www.springer.com/journal/10822. doi:10.1007/s10822-010-9363-3.

Michel, J., Tirado-Rives, J., & Jorgensen, W. L. (2009). Energetics of displacing water molecules from protein binding sites: Consequences for ligand optimization. *Journal of the American Chemical Society, 131*(42), 15403–15411. http://pubs.acs.org/doi/pdfplus/10.1021/ja906058w. doi:10.1021/ja906058w.

Mittal, A., & Jayaram, B. (2011). Backbones of folded proteins reveal novel invariant amino acid neighborhoods. *Journal of Biomolecular Structure and Dynamics, 28*(4), 443–454. doi:10.1080/073911011010524954.

Miyake, T., & Rolandi, M. (2016). Grotthuss mechanisms: From proton transport in proton wires to bioprotonic devices. *Journal of Physics: Condensed Matter, 28*(2), 023001. doi:10.1088/0953-8984/28/2/023001.

Morningstar-Kywi, N., Wang, K., Asbell, T. R., Wang, Z., Giles, J. B., Lai, J., Brill, D., Sutch, B. T., & Haworth, I. S. (2022). Prediction of water distributions and displacement at protein–ligand interfaces. *Journal of Chemical Information and Modeling, 62*(6), 1489–1497. doi:10.1021/acs.jcim.1c01266.

Morozenko, A., & Stuchebrukhov, A. A. (2016). Dowser++, a new method of hydrating protein structures. *Proteins: Structure, Function, and Bioinformatics, 84*(10), 1347–1357. doi:10.1002/prot.25081.

Mukherjee, S., Nithin, C., Divakaruni, Y., & Bahadur, R. P. (2019). Dissecting water binding sites at protein–protein interfaces: A lesson from the atomic structures in the Protein Data Bank. *Journal of Biomolecular Structure and Dynamics, 37*(5), 1204–1219. http://www.tandfonline.com/loi/tbsd20. doi:10.1080/07391102.2018.1453379.

Municoy, M., Sergi, R., Soler, D., Soutullo, A., & Guallar, V. (2020). aquaPELE: A monte carlo-based algorithm to sample the effects of buried water molecules in proteins. *Journal of Chemical Theory and Computation, 16*(12), 7655–7670. doi:10.1021/acs.jctc.0c00925.

Muroda, K., Nakashima, K., Shibata, M., Demura, M., & Kandori, H. (2012). Protein-bound water as the determinant of asymmetric functional conversion between light-driven proton and chloride pumps. *Biochemistry, 51*(23), 4677–4684. doi:10.1021/bi300485r.

Muzammil, S., Armstrong, A. A., Kang, L. W., Jakalian, A., Bonneau, P. R., Schmelmer, V., Amzel, L. M., & Freire, E. (2007). Unique thermodynamic response of tipranavir to human immunodeficiency virus type 1 protease drug resistance mutations. *Journal of Virology, 81*(10), 5144–5154. doi:10.1128/jvi.02706-06.

Nguyen, C. N., Cruz, A., Gilson, M. K., & Kurtzman, T. (2014). Thermodynamics of water in an enzyme active site: Grid-based hydration analysis of coagulation factor xa. *Journal of Chemical Theory and Computation, 10*(7), 2769–2780. http://pubs.acs.org/journal/jctcce. doi:10.1021/ct401110x.

Nguyen, C. N., Kurtzman Young, T., & Gilson, M. K. (2012). Grid inhomogeneous solvation theory: Hydration structure and thermodynamics of the miniature receptor cucurbit[7]uril. *Journal of Chemical Physics, 137*(4), 973–980. doi:10.1063/1.4733951.

Nguyen, M. N., Pradhan, M. R., Verma, C., & Zhong, P. (2017). The interfacial character of antibody paratopes: Analysis of antibody-antigen structures. *Bioinformatics, 33*(19), 2971–2976. http://bioinformatics.oxfordjournals.org/. doi:10.1093/bioinformatics/btx389.

Nguyen, T. T., Viet, M. H., & Li, M. S. (2014). Effects of water models on binding affinity: Evidence from all-atom simulation of binding of tamiflu to A/H5N1 neuraminidase. *The Scientific World Journal, 2014*. http://www.hindawi.com/journals/tswj/. doi:10.1155/2014/536084.

Nittinger, E., Flachsenberg, F., Bietz, S., Lange, G., Klein, R., & Rarey, M. (2018). Placement of water molecules in protein structures: From large-scale evaluations to single-case examples. *Journal of Chemical Information and Modeling, 58*(8), 1625–1637. doi:10.1021/acs.jcim.8b00271.

Nittinger, E., Gibbons, P., Eigenbrot, C., Davies, D. R., Maurer, B., Yu, C. L., Kiefer, J. R., Kuglstatter, A., Murray, J., Ortwine, D. F., Tang, Y., & Tsui, V. (2019). Water molecules in protein–ligand interfaces. Evaluation of software tools and SAR comparison. *Journal of Computer-Aided Molecular Design, 33*(3), 307–330. doi:10.1007/s10822-019-00187-y.

Nittinger, E., Schneider, N., Lange, G., & Rarey, M. (2015). Evidence of water molecules—A statistical evaluation of water molecules based on electron density. *Journal of Chemical Information and Modeling, 55*(4), 771–783. doi:10.1021/ci500662d.

Nucci, N. V., Pometun, M. S., & Wand, A. J. (2011). Site-resolved measurement of water-protein interactions by solution NMR. *Nature Structural and Molecular Biology, 18*(2), 245–250. doi:10.1038/nsmb.1955.

Oliveira, B., van Beek, L., Shilliday, F., Debreczeni, J. É., & Phillips, C. (2021). Cryo-EM, the resolution revolution and drug discovery. *SLAS Discovery, 26*, 17–31.

Omwansu, W., Musembi, R., & Derese, S. (2024). Graph-based analyses of H-bond networks and unsupervised learning reveal conformational coupling in prion peptide segments. *Physical Chemistry Chemical Physics*. doi:10.1039/d4cp02123a.

Osaki, K., Ekimoto, T., Yamane, T., & Ikeguchi, M. (2022). 3D-RISM-AI: A machine learning approach to predict protein–ligand binding affinity using 3D-RISM. *The Journal of Physical Chemistry B, 126*(33), 6148–6158. doi:10.1021/acs.jpcb.2c03384.

Park, S., & Seok, C. (2022). GalaxyWater-CNN: Prediction of water positions on the protein structure by a 3D-convolutional neural network. *Journal of Chemical Information and Modeling, 62*(13), 3157–3168. doi:10.1021/acs.jcim.2c00306.

Patel, H., Grüning, B. jöA., Günther, S., & Merfort, I. (2014). PyWATER: A PyMOL plug-in to find conserved water molecules in proteins by clustering. *Bioinformatics, 30*(20), 2978–2980. doi:10.1093/bioinformatics/btu424.

Pavlovicz, R. E., Park, H., DiMaio, F., & Wallner, B. jö (2020). Efficient consideration of coordinated water molecules improves computational protein-protein and protein-ligand docking discrimination. *PLOS Computational Biology, 16*(9), e1008103. doi:10.1371/journal.pcbi.1008103.

Pocker, Y. (2000). Water in enzyme reactions: Biophysical aspects of hydration-dehydration processes. *Cellular and Molecular Life Sciences, 57*(7), 1008–1017. doi:10.1007/pl00000741.

Prasad, B. V. L. S., & Suguna, K. (2002). Role of water molecules in the structure and function of aspartic proteinases. *Acta Crystallographica Section D: Biological Crystallography, 58*(2), 250–259. doi:10.1107/s0907444901018327.

Qin, Y., Jia, M., Yang, J., Wang, D., Wang, L., Xu, J., & Zhong, D. (2016). Molecular origin of ultrafast water–protein coupled interactions. *The Journal of Physical Chemistry Letters, 7*(20), 4171–4177. doi:10.1021/acs.jpclett.6b01954.

Qu, X., Dong, L., Zhang, J., Si, Y., & Wang, B. (2022). Systematic improvement of the performance of machine learning scoring functions by incorporating features of protein-bound water molecules. *Journal of Chemical Information and Modeling, 62*(18), 4369–4379. doi:10.1021/acs.jcim.2c00916.

Radoux, C. J., Olsson, T. S. G., Pitt, W. R., Groom, C. R., & Blundell, T. L. (2016). Identifying interactions that determine fragment binding at protein hotspots. *Journal of Medicinal Chemistry, 59*(9), 4314–4325. http://pubs.acs.org/jmc. doi:10.1021/acs.jmedchem.5b01980.

Raghavan, B., Paulikat, M., Ahmad, K., Callea, L., Rizzi, A., Ippoliti, E., Mandelli, D., Bonati, L., Vivo, M., & Carloni, P. (2023). Drug design in the exascale era: A perspective from massively parallel QM/MM simulations. *Journal of Chemical Information and Modeling, 63*(12), 3647–3658. doi:10.1021/acs.jcim.3c00557.

Ramsey, S., Nguyen, C., Salomon-Ferrer, R., Walker, R. C., Gilson, M. K., & Kurtzman, T. (2016). Solvation thermodynamic mapping of molecular surfaces in AmberTools: GIST. *Journal of Computational Chemistry, 37*(21), 2029–2037. doi:10.1002/jcc.24417.

Rankovic, Z., Brust, T. F., & Bohn, L. M. (2016). Biased agonism: An emerging paradigm in GPCR drug discovery. *Bioorganic and Medicinal Chemistry Letters, 26*(2), 241–250. http://www.journals.elsevier.com/bioorganic-and-medicinal-chemistry-letters/. doi:10.1016/j.bmcl.2015.12.024.

Ravnik, V., Jukič, M., & Bren, U. (2023). Identifying metal binding sites in proteins using homologous structures, the MADE approach. *Journal of Chemical Information and Modeling, 63*(16), 5204–5219. doi:10.1021/acs.jcim.3c00558.

Raymer, M. L., Sanschagrin, P. C., Punch, W. F., Venkataraman, S., Goodman, E. D., & Kuhn, L. A. (1997). Predicting conserved water-mediated and polar ligand interactions in proteins using a K-nearest-neighbors genetic algorithm. *Journal of Molecular Biology, 265*(4), 445–464. doi:10.1006/jmbi.1996.0746.

Rhee, Y. M., Sorin, E. J., Jayachandran, G., Lindahl, E., & Pande, V. S. (2004). Simulations of the role of water in the protein-folding mechanism. *Proceedings of the National Academy of Sciences of the United States of America, 101*(17), 6456–6461. doi:10.1073/pnas.0307898101.

Rose, P. W., Prlić, A., Altunkaya, A., Bi, C., Bradley, A. R., Christie, C. H., Di Costanzo, L., Duarte, J. M., Dutta, S., Feng, Z., Green, R. K., Goodsell, D. S., Hudson, B., Kalro, T., Lowe, R., Peisach, E., Randle, C., Rose, A. S., Shao, C., … Burley, S. K. (2017). The RCSB protein data bank: Integrative view of protein, gene and 3D structural information. *Nucleic Acids Research, 45*(1), D271. https://academic.oup.com/nar/issue.

Ross, G. A., Bodnarchuk, M. S., & Essex, J. W. (2015). Water sites, networks, and free energies with grand canonical Monte Carlo. *Journal of the American Chemical Society, 137*(47), 14930–14943. http://pubs.acs.org/journal/jacsat. doi:10.1021/jacs.5b07940.

Ross, G. A., Morris, G. M., & Biggin, P. C. (2012). Rapid and accurate prediction and scoring of water molecules in protein binding sites. *PLoS One, 7*(3). http://www.plosone.org/article/fetchObjectAttachment.action?uri=info%3Adoi%2F10.1371%2Fjournal.pone.0032036&representation=PDF. doi:10.1371/journal.pone.0032036.

Ross, G. A., Russell, E., Deng, Y., Lu, C., Harder, E. D., Abel, R., & Wang, L. (2020). Enhancing water sampling in free energy calculations with grand canonical Monte Carlo. *Journal of Chemical Theory and Computation, 16*(10), 6061–6076. http://pubs.acs.org/journal/jctcce. doi:10.1021/acs.jctc.0c00660.

Rossato, G., Ernst, B., Vedani, A., & Smieško, M. (2011). AcquaAlta: A directional approach to the solvation of ligand–protein complexes. *Journal of Chemical Information and Modeling, 51*(8), 1867–1881. doi:10.1021/ci200150p.

Ryan, M. J., Gao, L., Valiyaveetil, F. I., Kananenka, A. A., & Zanni, M. T. (2024). Water inside the selectivity filter of a K+ ion channel: Structural heterogeneity, picosecond dynamics, and hydrogen bonding. *Journal of the American Chemical Society, 146*(2), 1543–1553. http://pubs.acs.org/journal/jacsat. doi:10.1021/jacs.3c11513.

Ryde, U., & Söderhjelm, P. är (2016). Ligand-binding affinity estimates supported by quantum-mechanical methods. *Chemical Reviews, 116*(9), 5520–5566. doi:10.1021/acs.chemrev.5b00630.

Samantray, S., Yin, F., Kav, B., & Strodel, B. (2020). Different force fields give rise to different amyloid aggregation pathways in molecular dynamics simulations. *Journal of Chemical Information and Modeling, 60*(12), 6462–6475. doi:10.1021/acs.jcim.0c01063.

Samways, M. L., Bruce Macdonald, H. E., Taylor, R. D., & Essex, J. W. (2023). Water networks in complexes between proteins and FDA-approved drugs. *Journal of Chemical Information and Modeling, 63*(1), 387–396. http://pubs.acs.org/journal/jcisd8. doi:10.1021/acs.jcim.2c01225.

Samways, M. L., Taylor, R. D., Bruce Macdonald, H. E., & Essex, J. W. (2021). Water molecules at protein-drug interfaces: Computational prediction and analysis methods. *Chemical Society Reviews, 50*(16), 9104–9120. http://pubs.rsc.org/en/journals/journal/cs. doi:10.1039/d0cs00151a.

Saur, M., Hartshorn, M. J., Dong, J., Reeks, J., Bunkoczi, G., Jhoti, H., & Williams, P. A. (2020). Fragment-based drug discovery using cryo-EM. *Drug Discovery Today, 25*(3), 485–490. doi:10.1016/j.drudis.2019.12.006.

Scapin, G., Potter, C. S., & Carragher, B. (2018). Cryo-EM for small molecules discovery, design, understanding, and application. *Cell Chemical Biology, 25*(11), 1318–1325. doi:10.1016/j.chembiol.2018.07.006.

Schaller, D., Pach, S., & Wolber, G. (2019). PyRod: Tracing water molecules in molecular dynamics simulations. *Journal of Chemical Information and Modeling, 59*(6), 2818–2829. doi:10.1021/acs.jcim.9b00281.

Siemers, M., & Bondar, A. N. (2021). Interactive interface for graph-based analyses of dynamic H-bond networks: Application to spike protein S. *Journal of Chemical Information and Modeling*. http://pubs.acs.org/journal/jcisd8. doi:10.1021/acs.jcim.1c00306.

Siemers, M., Lazaratos, M., Karathanou, K., Guerra, F., Brown, L. S., & Bondar, A.-N. (2019). Bridge: A graph-based algorithm to analyze dynamic H-bond networks in membrane proteins. *Journal of Chemical Theory and Computation, 15*(12), 6781–6798. doi:10.1021/acs.jctc.9b00697.

Sindhikara, D. J., & Hirata, F. (2013). Analysis of biomolecular solvation sites by 3D-RISM theory. *Journal of Physical Chemistry B, 117*(22), 6718–6723. http://pubs.acs.org/journal/jpcbfk. doi:10.1021/jp4046116.

Sindhikara, D. J., Yoshida, N., & Hirata, F. (2012). Placevent: An algorithm for prediction of explicit solvent atom distribution-application to HIV-1 protease and F-ATP synthase. *Journal of Computational Chemistry, 33*(18), 1536–1543. doi:10.1002/jcc.22984.

Spyrakis, F., Ahmed, M. H., Bayden, A. S., Cozzini, P., Mozzarelli, A., & Kellogg, G. E. (2017). The roles of water in the protein matrix: A largely untapped resource for drug discovery. *Journal of Medicinal Chemistry, 60*(16), 6781–6827. doi:10.1021/acs.jmedchem.7b00057.

Sridhar, A., Ross, G. A., & Biggin, P. C. (2017). Waterdock 2.0: Water placement prediction for Holo-structures with a pymol plugin. *PLoS One, 12*(2). http://journals.plos.org/plosone/article/file?id=10.1371/journal.pone.0172743&type=printable. doi:10.1371/journal.pone.0172743.

Stephens, A. D., & Kaminski Schierle, G. S. (2019). The role of water in amyloid aggregation kinetics. *Current Opinion in Structural Biology, 58*, 115–123. http://www.elsevier.com/locate/sbi. doi:10.1016/j.sbi.2019.06.001.

Surleraux, D. L. N. G., Tahri, A., Verschueren, W. G., Pille, G. M. E., De Kock, H. A., Jonckers, T. H. M., Peeters, A., De Meyer, S., Azijn, H., Pauwels, R., De Bethune, M. P., King, N. M., Prabu-Jeyabalan, M., Schiffer, C. A., & Wigerinck, P. B. T. P. (2005). Discovery and selection of TMC114, a next generation HIV-1 protease inhibitor. *Journal of Medicinal Chemistry, 48*(6), 1813–1822. doi:10.1021/jm049560p.

Szél, V., Zsidó, B. Z., & Hetényi, C. (2024). Enthalpic classification of water molecules in target-ligand binding. *Journal of Chemical Information and Modeling, 64*(16), 6583–6595. http://pubs.acs.org/journal/jcisd8. doi:10.1021/acs.jcim.4c00794.

Thirumalai, D., Reddy, G., & Straub, J. E. (2012). Role of water in protein aggregation and amyloid polymorphism. *Accounts of Chemical Research, 45*(1), 83–92. doi:10.1021/ar2000869.

Thomaston, J. L., Polizzi, N. F., Konstantinidi, A., Wang, J., Kolocouris, A., & Degrado, W. F. (2018). Inhibitors of the M2 proton channel engage and disrupt transmembrane networks of hydrogen-bonded waters. *Journal of the American Chemical Society, 140*(45), 15219–15226. http://pubs.acs.org/journal/jacsat. doi:10.1021/jacs.8b06741.

Thomaston, J. L., Woldeyes, R. A., Nakane, T., Yamashita, A., Tanaka, T., Koiwai, K., Brewster, A. S., Barad, B. A., Chen, Y., Lemmin, T., Uervirojnangkoorn, M., Arima, T., Kobayashi, J., Masuda, T., Suzuki, M., Sugahara, M., Sauter, N. K., Tanaka, R., Nureki, O., … DeGrado, W. F. (2017). XFEL structures of the influenza M2 proton channel: Room temperature water networks and insights into proton conduction. *Proceedings of the National Academy of Sciences of the United States of America, 114*(51), 13357–13362. http://www.pnas.org/content/114/51/13357.full.pdf. doi:10.1073/pnas.1705624114.

Tikhe, J. G., Webber, S. E., Hostomsky, Z., Maegley, K. A., Ekkers, A., Li, J., Yu, X. H., Almassy, R. J., Kumpf, R. A., Boritzki, T. J., Zhang, C., Calabrese, C. R., Curtin, N. J., Kyle, S., Thomas, H. D., Wang, L. Z., Calvert, A. H., Golding, B. T., Griffin, R. J., & Newell, D. R. (2004). Design, synthesis, and evaluation of 3,4-dihydro-2*H*-[1,4]diazepino[6,7,1-hi]indol-1-ones as inhibitors of poly(ADP-ribose) polymerase. *Journal of Medicinal Chemistry, 47*(22), 5467–5481. doi:10.1021/jm030513r.

Tošović, J., Fijan, D., Jukič, M., & Bren, U. (2022). Conserved water networks identification for drug design using density clustering approaches on positional and orientational data. *Journal of Chemical Information and Modeling, 62*(23), 6105–6117. doi:10.1021/acs.jcim.2c00801.

Truchon, J. F., Pettitt, B. M., & Labute, P. (2014). A cavity corrected 3D-RISM functional for accurate solvation free energies. *Journal of Chemical Theory and Computation, 10*(3), 934–941. http://pubs.acs.org/journal/jctcce. doi:10.1021/ct4009359.

Uehara, S., & Tanaka, S. (2016). AutoDock-GIST: Incorporating thermodynamics of active-site water into scoring function for accurate protein-ligand docking. *Molecules, 21*(11), 1604. doi:10.3390/molecules21111604.

Venkatakrishnan, A. J., Ma, A. K., Fonseca, R., Latorraca, N. R., Kelly, B., Betz, R. M., Asawa, C., Kobilka, B. K., & Dror, R. O. (2019). Diverse GPCRs exhibit conserved water networks for stabilization and activation. *Proceedings of the National Academy of Sciences, 116*(8), 3288–3293. doi:10.1073/pnas.1809251116.

Wade, R. C., Mazor, M. H., McCammon, J. A., & Quiocho, F. A. (1991). A molecular dynamics study of thermodynamic and structural aspects of the hydration of cavities in proteins. *Biopolymers, 31*(8), 919–931. doi:10.1002/bip.360310802.

Wang, T., Jo, H., DeGrado, W. F., & Hong, M. (2017). Water distribution, dynamics, and interactions with Alzheimer's β-amyloid fibrils investigated by solid-state NMR. *Journal of the American Chemical Society, 139*(17), 6242–6252. http://pubs.acs.org/journal/jacsat. doi:10.1021/jacs.7b02089.

Weber, I. T., Wang, Y.-F., & Harrison, R. W. (2021). HIV protease: Historical perspective and current research. *Viruses, 13*, 839.

Wlodawer, A., Minor, W., Dauter, Z., & Jaskolski, M. (2013). Protein crystallography for aspiring crystallographers or how to avoid pitfalls and traps in macromolecular structure determination. *The FEBS Journal, 280*(22), 5705–5736. doi:10.1111/febs.12495.

Woods, C. J., Shaw, K. E., & Mulholland, A. J. (2015). Combined quantum mechanics/molecular mechanics (QM/MM) simulations for protein-ligand complexes: Free energies of binding of water molecules in influenza neuraminidase. *Journal of Physical Chemistry B, 119*(3), 997–1001. http://pubs.acs.org/journal/jpcbfk. doi:10.1021/jp506413j.

Wu, H., Wacker, D., Mileni, M., Katritch, V., Han, G. W., Vardy, E., Liu, W., Thompson, A. A., Huang, X. i-P., Carroll, F. I., Mascarella, S. W., Westkaemper, R. B., Mosier, P. D., Roth, B. L., Cherezov, V., & Stevens, R. C. (2012). Structure of the human κ-opioid receptor in complex with JDTic. *Nature, 485*(7398), 327–332. doi:10.1038/nature10939.

Xiao, W., He, Z., Sun, M., Li, S., & Li, H. (2017). Statistical analysis, investigation, and prediction of the water positions in the binding sites of proteins. *Journal of Chemical Information and Modeling, 57*(7), 1517–1528. http://pubs.acs.org/journal/jcisd8. doi:10.1021/acs.jcim.6b00620.

Xiao, W., Ren, J., Hao, J., Wang, H., Li, Y., Lin, L., & Chen, L. (2022). Predicting conserved water molecules in binding sites of proteins using machine learning methods and combining features. *Computational and Mathematical Methods in Medicine, 2022*, 1–11. doi:10.1155/2022/5104464.

Yang, Y., Lightstone, F. C., & Wong, S. E. (2013). Approaches to efficiently estimate solvation and explicit water energetics in ligand binding: The use of WaterMap. *Expert Opinion on Drug Discovery, 8*(3), 277–287. doi:10.1517/17460441.2013.749853.

Zak, K. M., Kitel, R., Przetocka, S., Golik, P., Guzik, K., Musielak, B., Dömling, A., Dubin, G., & Holak, T. A. (2015). Structure of the complex of human programmed death 1, PD-1, and its ligand PD-L1. *Structure, 23*(12), 2341–2348. http://www.journals.elsevier.com/structure/. doi:10.1016/j.str.2015.09.010.

Zhang, L., & Hermans, J. (1996). Hydrophilicity of cavities in proteins. *Proteins: Structure, Function, and Genetics, 24*(4), 433–438. doi:10.1002/(sici)1097-0134(199604)24:4⟨433::Aid-prot3⟩3.0.co;2-f.

Zhao, L., Li, W., Tian, P., & Agarwal, P. K. (2013). Reconciling mediating and slaving roles of water in protein conformational dynamics. *PLoS One, 8*(4), e60553. doi:10.1371/journal.pone.0060553.

Zhao, Z., & Tajkhorshid, E. (2024). GOLEM: Automated and robust Cryo-EM-guided ligand docking with explicit water molecules. *Journal of Chemical Information and Modeling, 64*(14), 5680–5690. doi:10.1021/acs.jcim.4c00917.

Zia, S. R., Gaspari, R., Decherchi, S., & Rocchia, W. (2016). Probing hydration patterns in class-A GPCRs via biased MD: The A2A receptor. *Journal of Chemical Theory and Computation, 12*(12), 6049–6061. http://pubs.acs.org/journal/jctcce. doi:10.1021/acs.jctc.6b00475.

CHAPTER

Protein-lipid interactions

6

Swaratmika Pandia, Priyanka Mahapatra and Hirak Chakraborty*
School of Chemistry, Sambalpur University, Burla, Sambalpur, Odisha, India
**Corresponding author: hirak@suniv.ac.in*

6.1 Introduction to membranes

Biological membranes offer morphological compartmentalization to the cell and intracellular organelles to provide selectively permeable barriers that control the free transport or exchange of ions and nutrients across cells and subcellular organelles. The morphological compartmentalization has been recognized as a physical prerequisite for Darwinian evolution (Budin & Szostak, 2010; Deamer et al., 2002). Moreover, membranes provide the platform where proteins can interact with lipids for their necessary function (Escribá et al., 2008). Grossly an organized assembly of the lipid bilayer of around 50–60 Å thickness is considered the membrane; however, the biological membrane is more complex because of the presence of several cytoskeleton, transmembrane, and associated proteins. The complexity further increases with the posttranslational modifications (addition of carbohydrates and other organic moieties) of proteins and lipids. All these partners maintain the dynamic equilibrium as if proteins were floating in a sea of lipids. The semipermeability arises from the amphipathic lipid molecules, which generally contain a hydrophilic headgroup and two fatty acyl hydrophobic tails. The cytoskeleton proteins provide mechanical stability to the membrane while associated and transmembrane proteins communicate between two sides of the membrane in response to external stimuli such as chemicals and ions (Singer & Nicolson, 1972).

The cell membrane was first discovered by botanists Carl Näegeli and C. Cramer in 1855, provoking the existence of an osmotic barrier between the protoplasm of plant cells and their environment (Nägeli, 1858). Overton in 1895 generalized that the living protoplasm of all basic organisms is readily permeable for various organic solutes. Complexities in its structure were added subsequently from various measurements of the physical and chemical properties of the membrane (Overton, 1895). Agnes Pockels spread oil on water using the Langmuir trough and observed that the polar heads are oriented toward the water at the air–water interface, whereas the hydrophobic tails are positioned in the air (Langmuir, 1916; Pockels, 1891). Langmuir (1917) performed a similar experiment with an advanced instrument, where he measured the surface pressure and surface tension of the monolayer at the air–water interface and demonstrated that the hydrophobic tails were flexible. Gorter and Grendel (1925) discovered that the surface area covered by the lipids of the whole erythrocyte in a Langmuir trough is double its actual area. This result provides the concept of the lipid bilayer (Bar et al., 1966). Similar experiments were extended to chromocytes of six mammalian species, and the calculated result showed a 2:1 film to erythrocyte area, which envisages the layer of fatty substances as two

Protein Dynamics: Experimental and Computational Approaches. DOI: https://doi.org/10.1016/B978-0-443-30042-4.00020-1

molecules thick (Bar et al., 1966). Fricke (1925) showed that the plasma membranes are 4 nm thick, and Cole (1932) suggested that the membrane was covered with proteins along with the lipids. Danielli supported the argument of Fricke that the erythrocyte cell contents are separated electrically from the surrounding medium by a film generally in between unimolecular and trimolecular thickness (Danielli & Davson, 1935). Based on surface tension measurement, he proposed that proteins are adsorbed at the lipid-water interface. However, subsequent experiments showed that the oil used in model systems for the lipid-water interface did not exactly represent the properties of the phospholipid-water interface, as phospholipid alone can produce low tension. The model suggests that a protein film is adsorbed on the lipid film, and the combination of these two films distinguishes molecules of different sizes and solubility and ions of different charges (Danielli & Davson, 1935).

Cellular membranes are not exactly dominated by lipids; rather, these are packed with transmembrane proteins. Therefore the plasma membrane is considered more proteinaceous than oil-like protoplasmic surfaces (Clements, 1962; Haydon, 1968; Haydon & Taylor, 1963). The fluidity of membrane lipids was detected by Chapman, showing biological systems operate at different temperatures. Membranes become more unsaturated at lower environmental temperatures and saturated at higher environmental temperatures, retaining a constant fluidity to support diffusion and permeability. The bilayer lipids were in motion and diffused in the plane of the membrane, therefore the function of cells depends on the phase transition temperature (Chapman, 1975). Bretscher (1973) observed the asymmetrical distribution of lipid bilayers in erythrocyte membranes, which implies a differential distribution of lipids in inner and outer leaflets, leading to dissimilar fluidity of the leaflets. Protein diffusion is slower in lipid bilayers compared to buffer medium, as the average viscosity of the bilayer is about 100 times greater than the viscosity of water. However, the lipid bilayer is highly dynamic, and proteins can rotate and diffuse laterally in a two-dimensional space. The diffusion of protein and lipid was random unless it was constrained by the cytoskeleton or by membrane proteins (Edidin, 2003).

Danielli and Davson showed that the plasma membrane consists of two layers of phospholipid molecules, which are arranged in such a way that hydrophilic heads of the phospholipid molecules face outside, and hydrophobic nonpolar lipid chains are located in between two headgroup regions, hiding from water. The model further suggested that the polar ends of lipid molecules are associated with a monomolecular layer of globular proteins. The plasma membrane would thus consist of a double layer of phospholipid molecules sandwiched between two essentially continuous layers (Danielli & Davson, 1935). Green et al. proposed another model for cell membranes in 1970. They showed that the proteins polymerize to form loosely packed globular proteins of around 30–40 Å diameter These proteins have nonpolar as well as polar regions on their surfaces. The nonpolar regions exhibit hydrophobic interaction with nonpolar groups of phospholipid molecules that fill the cavities between the globular protein units, whereas the polar heads of phospholipid molecules face the membrane surface (Vanderkooi & Green, 1970).

Jonathan Singer and Garth Nicolson proposed the complex model of the plasma membrane in 1972, replacing the protein-lipid–protein sandwich concept. Interestingly the Singer–Nicolson model is still considered the most descriptive model of the membrane. Considering all the thermodynamic restrictions, they have proposed a model of the membrane where the proteins that are associated with the membrane are amphipathic and organized the ionic and highly polar groups protruding from the membrane into the aqueous phase and the nonpolar groups are largely buried in the hydrophobic interior of the membrane (Singer & Nicolson, 1972). Overall the globular proteins are partially embedded in the phospholipid matrix, and the bulk lipids are organized as a discontinuous, fluid bilayer, and

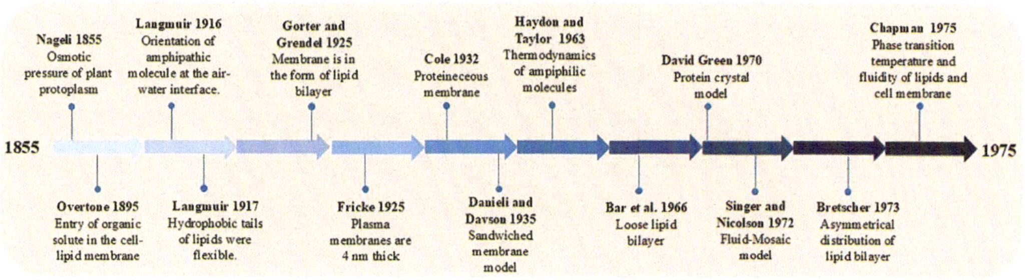

FIGURE 6.1

Timeline of our understanding of the structure of membranes.

some of the lipid molecules may be interacting with the proteins. Therefore the membrane is a two-dimensional oriented solution of proteins and lipids, where proteins are analogous to the mosaic stones in the phospholipid floor. The exposed surface of functional membranes such as plasmalemma, vacuolar, nuclear, chloroplast, mitochondria, and bacterial membranes was also examined. These showed a mosaic-like structure consisting of a smooth matrix that is interrupted by many particles. The model further predicted that some integral proteins having appropriate size and structure may span the entire thickness of the membrane and can be exposed at both membrane surfaces. The membrane proteins as well as lipid components are distributed homogeneously in the long range, whereas specific aggregates can form in the short range, termed lipid rafts or membrane microdomains (Singer & Nicolson, 1972). The timeline of our understanding of membrane structure is shown in Fig. 6.1.

The interaction of cells with their environment is important in several biological processes, such as in the endocytic pathway, which includes receptor signaling, cell adhesion and migration, pathogen entry, membrane protein turnover, and nutrient uptake. These various trafficking processes require motor proteins that transport or tether membranes along the actin and microtubule cytoskeletons. The function of several membrane proteins depends on their interaction with membrane lipids. G protein-coupled receptors (GPCRs), the largest family of drug targets, are known to be stabilized in the presence of membrane cholesterol (CH) (Sengupta & Chattopadhyay, 2015). The molecular modeling study demonstrated that membrane CH makes the serotonin$_{1A}$ receptor more compact (Paila et al., 2011) and provides more stability to the β_2-adrenergic receptor (Zocher et al., 2012). Overall, membrane proteins provide structural support to the membrane and facilitate a multitude of functions for the survival of cells. It is important to mention here that the diversity of lipids creates a dynamic platform for lipid–protein interaction, leading to modulation of protein function. Therefore understanding lipid–protein interaction is important to decipher the effect of lipid diversity on protein function as it houses the membrane proteins.

6.2 Diversity in lipids

Biological membranes are the supramolecular assembly of different lipids and proteins. Lipids are diversified by their backbone chemistry, unsaturation at the acyl chain region, and headgroup variation. Moreover, these lipids are present in different molar ratios in plasma membranes and organelle membranes. The lipidomics data demonstrated that the organelle membranes can be speciated by their

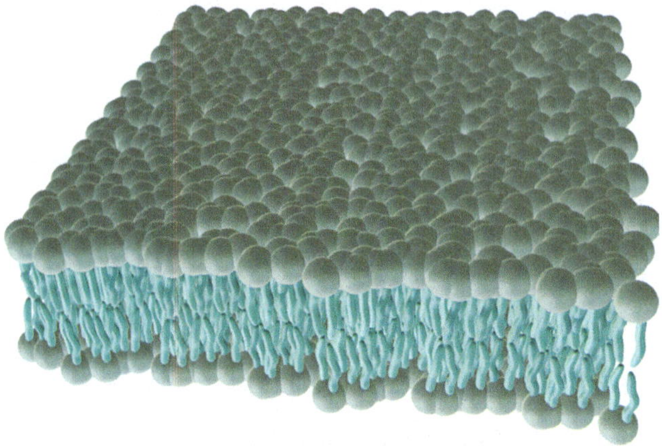

FIGURE 6.2

Structure of lipid bilayer.

lipid composition (Van Meer et al., 2008). Lipids are generally classified into three categories based on their backbone chemistry, namely, glycerophospholipids, sphingolipids, and CH.

6.2.1 Glycerophospholipids

Glycerophospholipids have two hydrophobic fatty acyl chains linked to a glycerol backbone through ester bonds, and the polar headgroup (such as choline, ethanolamine, serine, and glycerol) is attached to the phosphate group through phosphoester linkage. In general the cross-sectional area of the headgroup and fatty acyl tails are similar, and the cylindrical shape of lipid molecules favors the formation of bilayer structure. In a monolayer the polar headgroups are positioned toward water and hydrophobic tails away from water, and two monolayers are mirrored to each other to form the bilayer structure (Fig. 6.2).

However, depending on the size of the headgroup, the lipid can assume a cone or inverted cone shape, and these lipids generate intrinsic local curvature in the lipid bilayer. Glycerophospholipids can be further classified into the following types based on their headgroup chemistry (Fig. 6.3).

6.2.1.1 Phosphatidic acid

Phosphatidic acid (PA) headgroup contains the phosphate group, having a -1 net charge at the physiological pH, whereas it carries 0 charge at less than pH 4 and carries a -2 charge at more than pH 12. It is present in small amounts in biological membranes, and it has a significant role in the binding of positively charged proteins (Kooijman & Burger, 2009), membrane dynamics, membrane curvature, and cellular signaling (Zhukovsky et al., 2019).

6.2.1.2 Phosphatidylcholine

Phosphatidylcholine (PC) contains a choline group, which is bound to the phosphate headgroup of PA via the hydroxyl group of choline, making it an overall neutral charged lipid at the physiological condition. It is the most abundant phospholipid and comprises 60%–80% of total phospholipids

Backbone Types	Backbones	Head Group Variation	Classification
Glycerophospholipids	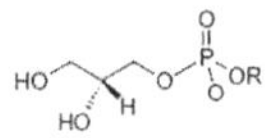	Phosphatidic Acid (PA)	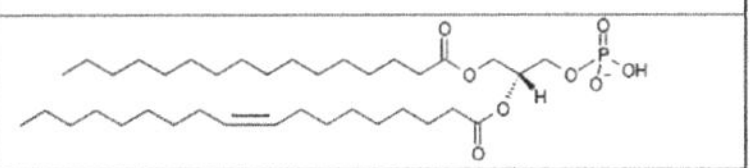
		Phophatidylchoine (PC)	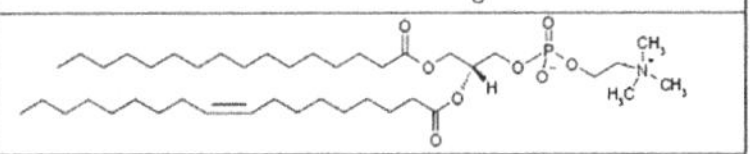
		Phosphatidylethanol emine (PE)	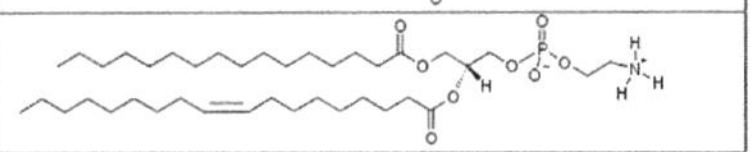
		Phosphatidylglycerol (PG)	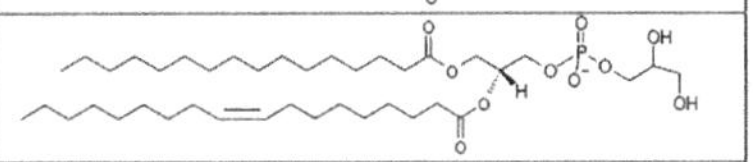
		Phosphatidylserine (PS)	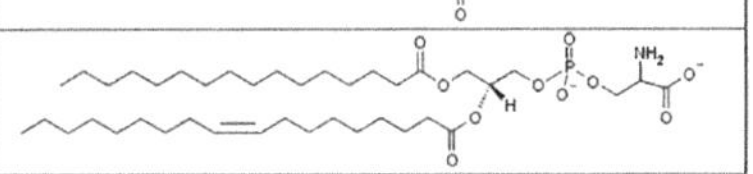
		Phosphatidylinositol (PI)	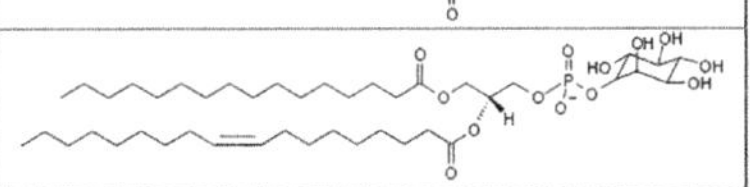

FIGURE 6.3

Classification of glycerophospholipids.

(Cole et al., 2012). It is essential for maintaining membrane fluidity, signaling, and neurotransmission (Roy et al., 2022). Generally, among the two acyl chains, the sn-1 acyl chain contains either saturated or monounsaturated fatty acid, and sn-2 contains either mono or polyunsaturated fatty acid.

6.2.1.3 *Phosphatidylethanolamine*

Phosphatidylethanolamine (PE) contains an additional ethanolamine group, which is bound to the phosphate headgroup of PA via the hydroxyl group of ethanolamine. As the amino group present in PE has a p*Ka* value of 9.5, it is present in zwitterionic form in the physiological pH condition. It prefers to locate at the inner leaflet of the membrane, facing toward the cytosol. Because of lesser cross-sectional area of the PE headgroup as compared to the hydrophobic tail, it gets a conical shape that modulates the membrane curvature and stabilizes the membrane proteins (Vance, 2008).

6.2.1.4 *Phosphatidylglycerol*

Phosphatidylglycerol (PG) contains an additional glycerol group, which is bound to the phosphate headgroup of PA via one hydroxyl group of glycerol. It has a net-1 charge at the physiological pH. It plays an important role in cardiolipin (CL) synthesis and modulates the bacterial membrane function (Short & White, 1972).

6.2.1.5 *Phosphatidylserine*

Phosphatidylserine (PS) is composed of PA, having an amino acid serine group, which is bound to the negatively charged phosphate group of PA via one hydroxyl group of amino acid serine. It plays an important role in blood coagulation (Lentz, 2003) and cell signaling. PS is mainly present in the inner leaflet of the lipid bilayer; when it is present on the outside of the cell membrane, it is a signal that the cell is undergoing apoptosis (Slotte, 2013).

Backbone Types	Backbones	Head Group Variation	Classification
Sphingolipid		Ceramide	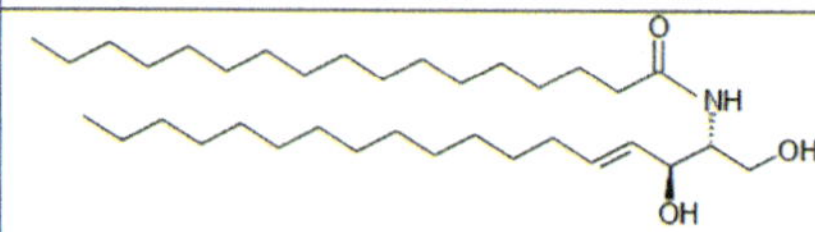
		Sphingomylin	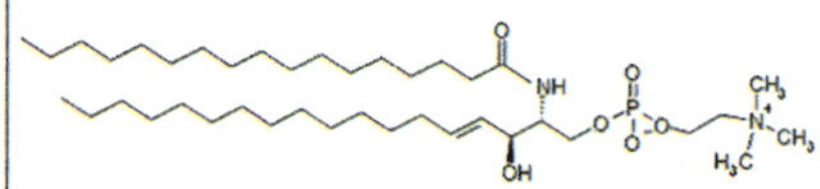

FIGURE 6.4

Classification of sphingolipids.

6.2.1.6 *Phosphatidylinositol*

Phosphatidylinositol (PI) contains a cyclohexane carbohydrate called inositol that is attached to the phosphate moiety at the 1′ carbon of the carbohydrate. The inositol ring can be further phosphorylated at the 3′, 4′, and 5′ carbons of carbohydrate to form phosphate derivatives of PIs, that is, PI phosphate (PIP), PIP2, and PIP3. The phosphoinositol is critical for maintaining cellular signaling, protein binding, and membrane dynamics (Di Paolo & De Camilli, 2006).

6.2.2 Sphingolipids

Sphingolipids differ from glycerophospholipids because of their sphingosine backbone, which forms ceramide when bonded to a fatty acid via an amide bond. Ceramide can further be modified by the addition of a phosphocholine group, resulting in the formation of sphingomyelin (SM) (Slotte, 2013). The chemical structures of ceramide and SM are shown in Fig. 6.4. These molecules play a crucial role in membrane stability and signaling, often concentrating in specialized regions of the membrane known as lipid domains. Sphingolipids also chemically diversified in accordance with their fatty acyl chain length, position, and amount of unsaturation. SM shows a strong affinity toward CH that leads to the segregation of lipids on the membrane and the formation of domains. SM also plays an important role in protein function, especially for G-protein coupled receptors.

6.2.3 Glycolipids

Glycolipid typically represents the glycerophospholipids or SM where the headgroup phosphate is associated with mono- or oligosaccharide moiety (Fig. 6.5). The mono- or oligosaccharide moieties are generally extending to the extracellular space or the lumen of organelles. The mono- or oligosaccharide headgroups are composed of D-glucose, D-galactose, *N*-acetyl-D-galactosamine, and fucose. Glycolipids are primarily found on the outer leaflet of the plasma membrane, and they play important roles in cell recognition, signaling, and adhesion. They contribute to the formation of lipid rafts (specialized membrane microdomains) and are involved in interactions with peptides and proteins, particularly in immune responses and cell communication (Hanafusa et al., 2020).

Backbone Types	Classification
Glycerophospholipid	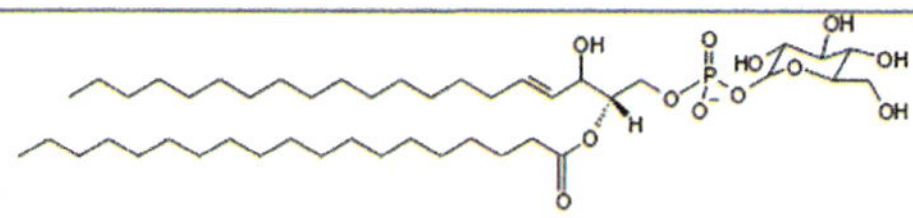
Sphingomyelin	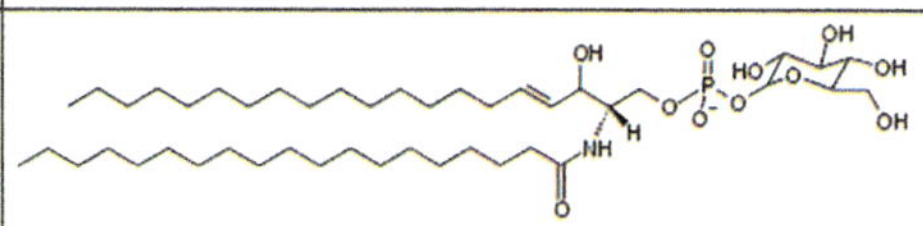

FIGURE 6.5

Classification of glycolipids.

Backbone Types	Backbones	Example
Sterol	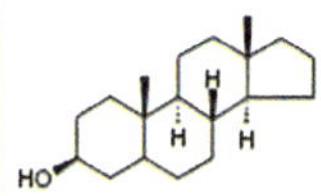	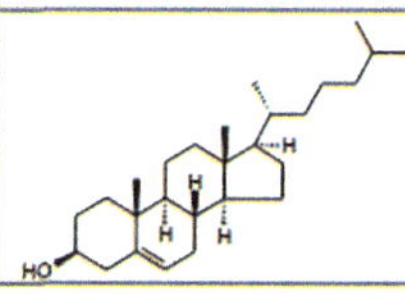

FIGURE 6.6

Structure and example of sterol.

6.2.4 Sterol

Sterols have steroid nucleus that contains four fused rings, out of which three are six-membered and one is a five-membered ring (Fig. 6.6). CH, the most well-known sterol, plays a crucial role in maintaining cell membrane integrity, fluidity, and permeability (Tsai & O'Malley, 1994).

6.2.5 Cardiolipins

Cardiolipins (CLs) have a distinct "double phospholipid" structure, which is composed of four fatty acid chains instead of the typical two, and two phosphatidyl groups connected by a glycerol bridge (Fig. 6.7).

6.3 Effect of fatty acyl tail and degree of unsaturation on membrane properties

The glycerophospholipids contain two esterified fatty acyl tails at positions 1 and 2 of the glycerol backbone, whereas one esterified fatty acyl tail is found with the sphingosine backbone for sphingolipids. The length of the fatty acyl tail and degree of unsaturation affect the membrane packing vis-à-vis fluidity and membrane–protein interaction. The temperature required for the gel phase lipids to transition to the fluid phase (phase transition temperature) reduces with an increase in the amount of

Backbone Types	Structure
Cardiolipin	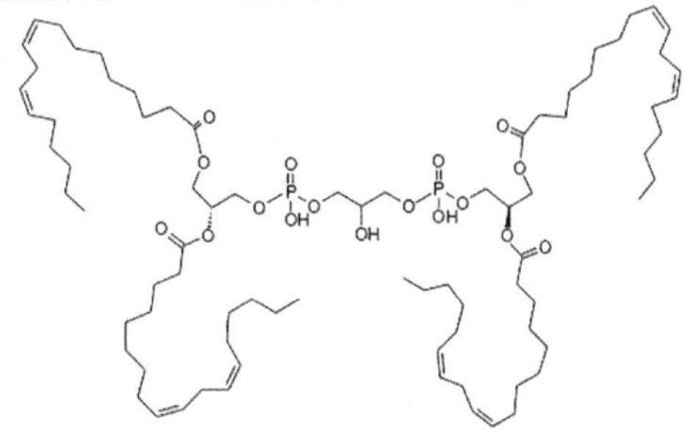

FIGURE 6.7

Structure of cardiolipin.

unsaturation at the fatty acyl tails. For a saturated fatty acyl tail, all carbons are sp (Escribá et al., 2008) hybridized and facilitate the compact packing of the hydrophobic tail region. The addition of sp (Budin & Szostak, 2010) carbons in the tail (because of unsaturation) leads to kink formation that reduces the packing ability of the fatty acyl chains, thereby reducing the phase transition temperature. An increase in lipidic chain length alters the thickness of the membrane and the phase transition temperature of lipids. However, the effect of unsaturation is more pronounced in the phase transition temperature of lipids. Interestingly the distribution of lipids in plasma and organelle membranes is highly heterogeneous (Van Meer et al., 2008). In addition to the lateral heterogeneity, there is lipid diversity between two leaflets of the biological membranes. For example, PC and SM are abundant in the outer leaflet, whereas lipids with smaller headgroups such as PE and negatively charged PS and PI prefer the inner leaflet of the biological membranes (Watson, 2015). The asymmetric distribution of phospholipids between different organelles could be a result of the synthesis and sorting of the lipids in ER and Golgi (Van Meer et al., 2008). The ATP-dependent lipid flip-flop by flippases and floppases is responsible for the change in lipid distribution between two leaflets (Watson, 2015). The composition specificity of membrane bilayer of various organelles has unique importance toward their physiological function (Casares et al., 2019; Shahane et al., 2019). Lack of CH regulates protein trafficking process (Mirdha et al., 2022). Moreover, some enveloped viruses are enriched by specific lipids such as PE (Ivanova et al., 2015), CH, and SM (Aloia et al., 1993).

6.4 Forces responsible for lipid–protein interaction

Membrane proteins are classified based on their degree of insertion in the lipid bilayer. Thus membrane proteins that go transversal to the lipid bilayer are called integral membrane proteins, whereas proteins that are associated with the membrane surface are designated as peripheral proteins. Peripheral membrane proteins are majorly containing polar amino acids, and there exist specific nonbonding interactions between the lipid headgroup and certain amino acids of the protein. The integral proteins are also called transmembrane proteins as they traverse across the membrane and contain hydrophobic amino acids

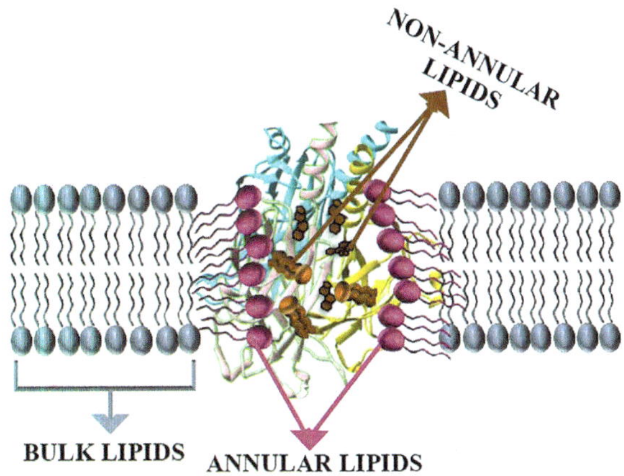

FIGURE 6.8

Schematic representation of bulk, annular, and nonannular lipids.

that are solubilized in the nonpolar environment of membranes. Major GPCRs pass the membrane seven times and are commonly called seven-pass receptors. It has been demonstrated that the transmembrane helices of GPCRs also have an affinity for a special class of lipids such as CH (Sengupta et al., 2018). Based on the involvement of lipids in solubilizing membrane proteins, they are classified into three categories: bulk lipids, annular lipids, and nonannular lipids. Bulk lipids do not interact with proteins directly, but their environment affects the geometry of membrane proteins, thereby helping in signaling, transport, and mortality processes (Blouin et al., 2016; Brown & Rose, 1992; Lin et al., 2017; Sych et al., 2022). Individually they do not have much effect, but collectively they create a heterogeneous environment that regulates protein sorting (Javanainen et al., 2019). However, annular lipids are the immediate environment of the proteins. Their interaction with proteins is often nonspecific, forming a ring around the transmembrane domain of the protein. On the other hand, nonannular lipids have specific interactions with proteins, thus affecting the function of the protein either via allosteric (lipid interaction on one side of the protein causing conformational changes on another side) or oligomerization state of the protein (Fig. 6.8) (Cannarozzo et al., 2021; Chavent et al., 2018; Juhola et al., 2018).

The location of proteins in the membrane is guided through different nonbonding interactions, which we have discussed in the subsequent section.

6.4.1 Electrostatic force

Electrostatic force arises from the attractive or repulsive interaction between two dissimilar or similar charges (or dipoles) on the lipid headgroup and the acidic or basic amino acids present in the protein. The Coulomb equation can be used to assess the electrostatic force's strength. The electrostatic force is the central enthalpic interaction for the binding of proteins to the membrane surface, where charged amino acids interact with the charged headgroups of lipids. The headgroup charge of phospholipids can be varied depending on the modification of the phosphate group, such as PG, PS, and PI, containing a negative charge in their headgroup. The charge–dipole interaction is important for the zwitterionic headgroup of PC, PE, CH, and sphingolipids with the charged amino acids of proteins. Even the lipidic dipoles of zwitterionic lipids can interact with the dipolar amino acids through dipole–dipole interactions. The dipole–dipole and charge–dipole interactions are relatively weaker than the charge–charge interaction; nonetheless, the number of these interactions is much larger than the charge–charge interactions. Thereby the charge–dipole and dipole–dipole interactions play a significant role in solubilizing the membrane proteins in different regions for associated and integral membrane proteins. Biological membranes with compositional heterogeneity of lipid headgroups provide a versatile platform in terms of charge and dipole for the interaction of proteins (Mulgrew-Nesbitt et al., 2006).

The association of peripheral proteins with the surface of intracellular membranes driven by the electrostatic forces is crucial for several cellular functions, including signaling, membrane trafficking, and membrane anchoring of the cytoskeleton proteins (Cho & Stahelin, 2005; Groves & Kuriyan, 2010; Lemmon et al., 2002). The electrostatic interaction makes the lipid–protein interaction flexible and dynamic. Several proteins contain membrane-binding domains that bind to a specific lipid in the membrane, such as the C1 domain of protein kinase C (PKC), which binds to the PS, and the pleckstrin homology domain, which binds to PI (Garcia et al., 1995; Guerrero-Valero et al., 2007; Lemmon, 2008). Conversely, some proteins bind to the membranes through metal ions, such as annexins and synaptotagmin, which interact with the membrane through Ca^{2+} (McLaughlin & Murray, 2005; McMahon et al., 2010). This has been demonstrated that the PS binding site in blood coagulation factor Va regulates the assembly and activity of the prothrombinase complex, an important precursor of thrombin generation (Majumder et al., 2008). The PS binding is so essential for blood clotting that the soluble PS like C_6-PS can activate Va and form the prothrombinase complex (Majumder et al., 2005). The electrostatic interaction has been successfully exploited in designing antimicrobial peptides (AMPs). The overall charge of the bacterial membranes is negative, as they contain majorly negatively charged PG and zwitterionic PE. As a result the bacteria can be killed when a positively charged peptide readily attaches to the bacterial membrane and damages its integrity (Gong et al., 2023; Teixeira et al., 2012).

6.4.2 Hydrophobic effect

The hydrophobic effect is defined as the propensity of nonpolar substances to aggregate in an aqueous medium and to be excluded by water. The effect arises from the high positive enthalpy of solubilization of a nonpolar molecule in an aqueous medium. Micelle formation is a unique example of the hydrophobic effect, where the tails of the surfactant are insoluble in water and form a supramolecular assembly to avoid water. Biologically, membranes offer a platform for the meeting of hydrophobic residues or domains of a protein to lipids to minimize the high positive enthalpy of their solubilization in an aqueous medium. The solubilization of the hydrophobic (nonpolar) domains of the protein to the membrane depends on the membrane thickness and length of the hydrophobic domain. A mismatch in length (hydrophobic mismatch) plays an important role in protein folding, conformation, and activity (Dumas et al., 2000; Kleinschmidt & Tamm, 2002; Williamson et al., 2002). The hydrophobic mismatch can be generated when the hydrophobic domain of the protein is either smaller or larger than the thickness of the membrane. The positive mismatch occurs when the hydrophobic thickness of the membrane protein (d_{TM}) is greater than the hydrophobic thickness of the membrane (d_{MT}), that is, $d_{TM} > d_{MT}$. In this case the surrounding fatty acid chains of the lipids stretch, and the protein tilts or aggregates to match the hydrophobic thickness of the protein and the membrane. Conversely the negative mismatch occurs when the d_{TM} is lower than the d_{MT}; to accommodate the smaller membrane-spanning domain of the protein, the fatty acyl tails of the lipid get compressed, and the proteins either aggregate or change their surface orientation. This minimizes the exposure of their hydrophobic portion of protein to the aqueous environment, which makes it energetically favorable (Rao et al., 2017). Lipid deformation-free energy change (ΔF) is associated with the annular lipidic chain order, which is altered by the hydrophobic mismatch between the lipid and protein (Fattal & Ben-Shaul, 1993). Membrane deformation is influenced by the material properties of the membrane, which in turn depend on the membrane composition. Specifically, CH plays a significant role in determining membrane deformation as it rigidifies the membrane (Needham et al., 1995).

Gramicidin is a peptide-based antibiotic derived from the bacterium *Brevibacillus brevis* that forms ion channels in lipid membranes for monovalent ions (Kelkar and Chattopadhyay, 2007a). However, it assumes a nonchannel conformation in a membrane with hydrophobic mismatch and affects the ion transport (Kelkar and Chattopadhyay, 2007b). Perfringolysin O is a toxin produced by the bacterium *Clostridium perfringens*, which is a soluble, monomeric protein that can oligomerize and form large pores in CH-containing membranes. This CH-dependent peptide has increased affinity for ordered lipid domains because of the hydrophobic matching between the hydrophobic thickness of the protein and the bilayer hydrophobic thickness (Lin & London, 2013). GPCRs are an important class of proteins that transmit external signals to the intracellular compartment and are used as favorable drug targets for many diseases (Chattopadhyay, 2014; Rosenbaum et al., 2009). NMR studies have demonstrated that the increasing bilayer thickness supports the formation of functionally active metarhodopsin II (MII). At the same time, oligomerization leads to functionally inactive metarhodopsin I (MI) conformation (Soubias et al., 2015). β2-adrenergic receptor dimerized in a CH-dependent manner. A higher concentration of CH enhances the thickness of the bilayer by its dimerization, resulting in enhancement of hydrophobic mismatch. In this situation the protein oligomerizes to reduce the extent of mismatch to hide the hydrophobic residues from the aqueous environment (Prasanna et al., 2015; Sengupta & Chattopadhyay, 2015). It is also believed that the hydrophobic mismatch facilitates the protein sorting in different organelles (Hanulová & Weiss, 2012). It is hypothesized that the differential membrane thickness of the organelles supports the insertion of different membrane proteins to minimize the hydrophobic mismatch. Therefore the matching of d_{MP} and d_{TM} has been suggested to act as a physicochemical mechanism of protein sorting.

6.4.3 Membrane curvature

Membrane curvature arises from the geometrical shape of lipids; cylindrical lipids form the flat bilayer with zero curvature. However, the presence of cone-shaped and inverted-cone-shaped lipids in the membrane generates either positive or negative curvature to the membrane. Curvature formation is important for forming complex geometrical shapes of organelles. The membrane also forms curvature during endo- and exocytosis (Cail & Drubin, 2023).

The membrane forms spontaneous positive curvature in the presence of lipids that have a larger headgroup area compared to that of the acyl tail region (Fig. 6.9), such as lysolipids. Conversely, membranes assume intrinsic negative curvature in the presence of lipids having a smaller headgroup area compared to that of the acyl chain region (Fig. 6.9), for example, PE, PA, diacylglycerol (DAG), CH, and CL. Moreover, the acyl chain unsaturation affects the membrane curvature (McMahon & Boucrot, 2015). Lipids with unsaturated acyl chains generate kinks and occupy more space than their saturated counterparts. Therefore, the ratio between the size of the headgroup and acyl chain saturation outlines the geometry of the lipid as well as the curvature of the membrane containing these lipids

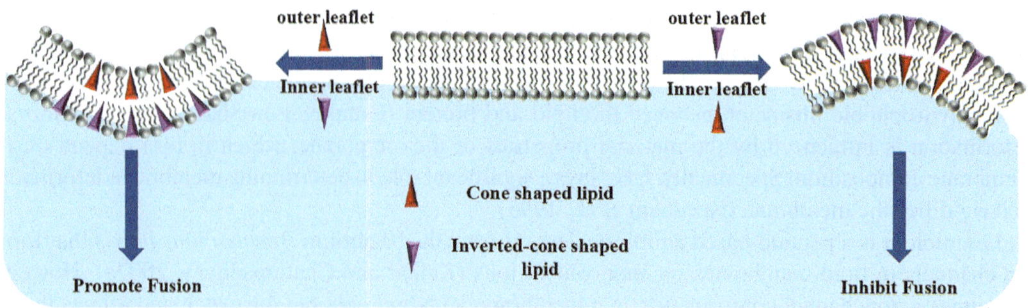

FIGURE 6.9

Representation of different intrinsic curvatures of the membrane in the presence of lipids with various shapes.

(Vanni et al., 2014). In addition to the intrinsic curvature of the lipids, the insertion of amphipathic helices of transmembrane proteins generates curvature in the membrane, whereas the cytoskeleton proteins stabilize the curvature (McMahon & Boucrot, 2015). Coat proteins, such as clathrin, COPI, and COPII, influence the membrane-bending for endocytosis (Jensen & Schekman, 2011; Kirchhausen, 2000; McMahon & Boucrot, 2011). The Bin/amphiphysin/Rvs (BAR) domain is the most conserved region in amphiphysins from lower organisms to humans. The N-terminal region of the BAR domain in amphiphysin drives membrane curvature in vivo and in vitro (Peter et al., 2004). Caveolae are small bulb-shaped invaginations in the plasma membrane (Galbiati et al., 1998) that participate in several homeostatic cellular processes, including mechanosensing, CH homeostasis, and regulation of cell signaling (Ariotti et al., 2014; Couet et al., 1997; Galbiati et al., 1998). Caveolin, an essential structural protein in caveolae, inserts into the membrane and facilitates the formation of curvature in the membrane (Ariotti et al., 2014). The membrane also generates curvature to accommodate the hydrophobic portion of the protein by reducing the hydrophobic mismatch. Nonetheless, it is not sure whether these proteins bind to the curved membrane or whether their binding to the membrane induces curvature (Johnson et al., 2024).

6.5 Lipid domains

In vitro studies with lipid bilayers revealed that the saturated long-chain lipids organize with sterols to form compact domains of nano- to micrometer dimensions in the membrane, leading to organizational heterogeneity. Fluorescence and freeze-fracture electron microscopic measurements demonstrated that the lipids in the domain exist in a liquid-ordered phase (L_o), where fatty acyl chains are packed tightly in the middle of a liquid-disordered phase (L_d). The L_o phase converts into the L_d phase at a relatively higher temperature. The temperature at which the L_o phase transitions to the L_d phase is known as phase transition temperature, and it depends on the length of the unsaturated chain. The CH-SM pair majorly dominates the liquid–liquid immiscibility, leading to subcompartmentalization of the membrane, as they have specific interactions (Simons & Sampaio, 2011). It is challenging to study the domains in the membrane as the size of the domains is beyond the scope of optical imaging. However, the recent development of spectroscopic and microscopic techniques provides detailed information on the membrane domains (Sezgin & Schwille, 2011). These studies have demonstrated the presence of

CH-dependent nanoscale domains of sphingolipids and Glycosylphosphatidylinositol (GPI)-anchored proteins in the plasma membranes of living cells (Sharma et al., 2004). Even the domains are not homogeneous; a domain consists of several nanodomains (Chaudhury et al., 2023).

Domain formation in the membrane is not completely lipid-centric; the presence of proteins also induces phase separation in the membrane (Zhao & Zhang, 2020). Membrane domains play a crucial role in multiple cellular functions of proteins by increasing their local concentration. Usually, protein uses two approaches to form domains in the plasma membrane. The first approach is via the formation of small protein-based membrane domains where the proteins directly interact with the lipids, forming multicomponent complexes as observed in the respiratory chain of prokaryotes. Second is the formation of large membrane domains, which requires a protein scaffold to imprint an ordered structure to the plasma membrane and recruit selected functional membrane domain components. Galectins are soluble and oligomeric extracellular lectins that can assemble membrane domains by cross-bridging subsets of glycoproteins, thus influencing their signaling and sorting. The cytosolic membrane scaffold formation happens during the tubulation induced by BAR domain proteins. Tetraspanins are known to interact with a variety of membrane proteins and can form homo-oligomers and arrange them in tetraspanin-rich membrane domains or tetraspanin webs (Björkholm et al., 2014). It is seen that CH depletion affects the homo-oligomerization and membrane domain formation ability of tetraspanin. Caveolae are another membrane domain formed by intramembrane scaffolds by the arrangement of caveolins and the cavin family of PKC adaptors. Caveolins (CAV1, CAV2, and CAV3) interact with CH, PS, and PIP_2, and it is well established that CH depletion flattens the flask-shaped caveolae and disassembles the caveolins (D'Alessio, 2023). Oreopoulos et al. (2010) have shown that AMPs like PFWRIRIRR-amide can induce the formation of lipid domains in supported phospholipid bilayers containing lipids such as 1-palmitoyl-2-oleoyl-sn-glycero-3-phosphoethanolamine and CL. The introduction of these peptides leads to the clustering of anionic lipids and a reorganization of membrane structure, which is vital for their antimicrobial activity. The myristoylated alanine-rich C-kinase substrate (MARCKS) peptide interacts with negatively charged lipids such as PIP_2 and PS (Al-Mualem et al., 2024). This interaction results in domain formation that alters the local lipid environment, enhancing cell signaling processes. Studies using two-dimensional infrared spectroscopy have demonstrated that MARCKS binding can lead to the colocalization of PIP_2 and PS into distinct domains, indicating a complex interplay between peptide binding and lipid composition. Multiple signaling and trafficking proteins contain several modular domains that display specific lipid binding properties (Stahelin, 2009). Lipidic domains are also utilized by the viral fusion proteins to invade the host cell. The N-terminal fusion peptide from the human immunodeficiency virus binds to the interface of the L_o–L_d phase and reduces the line tension for easy entry (Yang et al., 2016).

6.6 Lipid–peptide interactions and their relevance in different biological processes

6.6.1 Protein trafficking and signal transduction

Lipids recruit proteins to membranes, which helps in the formation of trafficking vesicles. Specific lipids like PIs are considered guiding lights in facilitating protein sorting and membrane–cytosol interactions (Di Paolo & De Camilli, 2006; Stace & Ktistakis, 2006). Proteins that are involved in trafficking undergo modifications, such as palmitoylation (Rocks et al., 2005; Valdez-Taubas & Pelham, 2005) or myristoylation (Magee & Seabra, 2005) for its smooth attachment and sorting to the membrane. These modifications can also regulate protein recycling between the Golgi and the plasma membrane (Table 6.1) (Nakano, 2022). Lipids such as DAG and phosphoinositides regulate the transport of proteins from the trans-Golgi network to the plasma membrane, as well as receptor recycling from endosomes to the surface (von Blume & Hausser, 2019; Xie et al., 2015). Lipids influence membrane dynamics and contribute to the formation of microdomains that concentrate proteins, which impacts sorting and trafficking. Glycerophospholipids influence SNARE-mediated membrane fusion, whereas hydrophobic mismatch during vesicle formation drives protein clustering and membrane curvature. Lipids like PA influence membrane dynamics by altering curvature, which is essential for vesicle formation, budding, and fusion. The PA generated by phospholipase D is especially significant for exocytosis and endocytosis (Haucke & Di Paolo, 2007).

Membrane lipids actively participate in signal transmission, influencing protein–protein interactions and receptor functions. Lipids such as phosphoinositides can serve as secondary messengers, whereas gangliosides and CH modulate integrin and GPCR signaling (Lopez & Schnaar, 2009; Prasanna et al., 2014; Sunshine & Iruela-Arispe, 2017).

The lipid environment determines how proteins, like K-Ras, anchor to the membrane and adopt the specific orientation essential for interacting with effectors and downstream signaling molecules (Lee, 2024). Lipid–protein binding specificity influences receptor localization and cellular responses (Corradi et al., 2019). Lipid microdomains act as platforms that organize signaling proteins that enable endocytosis, exocytosis, and clustering of receptors. CH and sphingolipids assist these domain formations, aiding dynamic protein sorting and signal propagation. The degree of lipid unsaturation impacts membrane fluidity and protein behavior. Polyunsaturated fatty acids were shown to disrupt membrane microdomains and reduce inflammatory signaling through Toll-like receptor 4 (Hwang et al., 2016). CH plays a crucial role in regulating GPCR activity, such as Smoothened (Smo) receptors in Hedgehog (Hh) signaling (Byrne et al., 2016). CH binding acts as an allosteric modulator, thereby promoting receptor activation and impacting downstream pathways (Huang et al., 2016; Luchetti et al., 2016).

6.6.2 Antimicrobial activity

AMPs represent an essential component of innate immunity across living organisms, providing a first line of defense against pathogens such as bacteria, fungi, and viruses (Ahmed et al., 2019). Unlike conventional antibiotics, AMPs disrupt microbial membranes within minutes and minimize the risk of resistance development. Their activity is closely linked to the interaction between the peptides and membrane lipids, which determines both the efficacy and selectivity. Here, we will discuss the various modes by which lipid–peptide interactions enhance antimicrobial activity, focusing on their selective action on bacterial membranes and the molecular mechanisms underlying membrane disruption.

The selective targeting of bacterial membranes by AMPs largely depends on electrostatic interactions. Most AMPs are cationic, which allows them to preferentially bind to negatively charged bacterial membranes containing PG and CL. These lipids are prevalent in bacterial cell walls and are known to provide a strong negative surface charge, which attracts positively charged peptides (Sani & Separovic, 2016). In contrast, mammalian membranes are primarily composed of neutral or zwitterionic phospholipids such as PC and SM that reduce AMP affinity and minimize host cell toxicity. The

Table 6.1 Lipid modifications involved in intracellular protein sorting.

Modifications	Characteristics	Enzymatic regulators	Example substrates	Exemplary functions in sorting	References
N-Myristoylation	Attachment of myristate to N-terminal Gly residue following removal of initiator Met; irreversible	*N*-Myristoyltransferase	Src-family kinases (Fyn, Lck); Gα; Arf1-6; yeast vacuolar protein Vac8p	Weak and reversible membrane attachment, regulating Arf targeting to membranes	Magee and Seabra (2005)
N-Acetylation	Attachment of acetate to N-terminal Met, penultimate Thr, or other residues; cotranslational or posttranslational; presumably irreversible	N-terminal acetyltransferase (Nat A, B, and C in yeast)	β-Endorphin and other hormones; Arl3p; Arl8a/b	Golgi or lysosome targeting of Arls	Behnia et al. (2004), Setty et al. (2003)
N-Palmitoylation	Attachment of palmitate to Cys at protein N-terminus, followed by rearrangement to stable amide linkage; irreversible	PAT	Hedgehog; possibly Spitz (spi)	Spread of Hedgehog signaling within tissues; regulation of Hedgehog targeting and localization	Pepinsky et al. (1998)
S-Palmitoylation	Thioacylation of Cys residue via thioester linkage	PATs, characterized by DHHC motif (aspartate-histidine-histidine-cysteine), are acylprotein thioesterases that carry out depalmitoylation	Ras, human δ-opioid receptor, AMPARs, postsynaptic scaffold postsynaptic density protein (PSD)-95, vacuolar protein Vac8p, SNAP-25, SNAP-23	Anterograde or retrograde trafficking (Ras and AMPARs); targeting to postsynaptic clusters (PSD-95); protection from ubiquitination (Tlg1)	Goodwin et al. (2005), Greaves and Chamberlain (2006, 2007), Han et al. (2002), Hayashi et al. (2005), Prior et al. (2003), Rocks et al. (2005), Roth et al. (2006), Valdez-Taubas and Pelham (2005)

(continued on next page)

Table 6.1 Lipid modifications involved in intracellular protein sorting—cont'd

Modifications	Characteristics	Enzymatic regulators	Example substrates	Exemplary functions in sorting	References
Prenylation	Attachment of farnesyl or geranylgeranyl moiety to CaaX sequence at C-terminal end usually followed by aaX cleavage and carboxymethylation; irreversible	Farnesyltransferase (FTase); geranylgeranyl transferase (GGTase)	Ras (farnesylation); Rabs (geranylgeranylation)	Anterograde and retrograde trafficking of Ras between the Golgi and the PM; membrane recruitment of Rabs to organelles	Magee and Seabra (2005)
Cholesteroylation	Attachment of cholesterol to proteins, that is, hedgehog following autocatalytic cleavage	Cholesterol transferase	Hedgehog	Spread of Hedgehog signaling within tissues; exact relationship to sorting is unknown	Magee and Seabra (2005)
Lipidation	Attachment of amino group of PE (perhaps also PS) to C-terminal Gly of ubiquitin-like proteins	E1-like and E2-like enzymes, that is, Atg7p/hAtg7 and Atg3p/nAtg3	Atg8p in yeast and its mammalian homologs LC3, GABARAP, GATE-16	Regulates autophagy in yeast and presumably mammals	Magee and Seabra (2005)

AMPARs, *AMPA-type glutamate receptors;* PATs, *palmitoyltransferases;* PE, *phosphatidylethanolamine;* PS, *phosphatidylserine.*

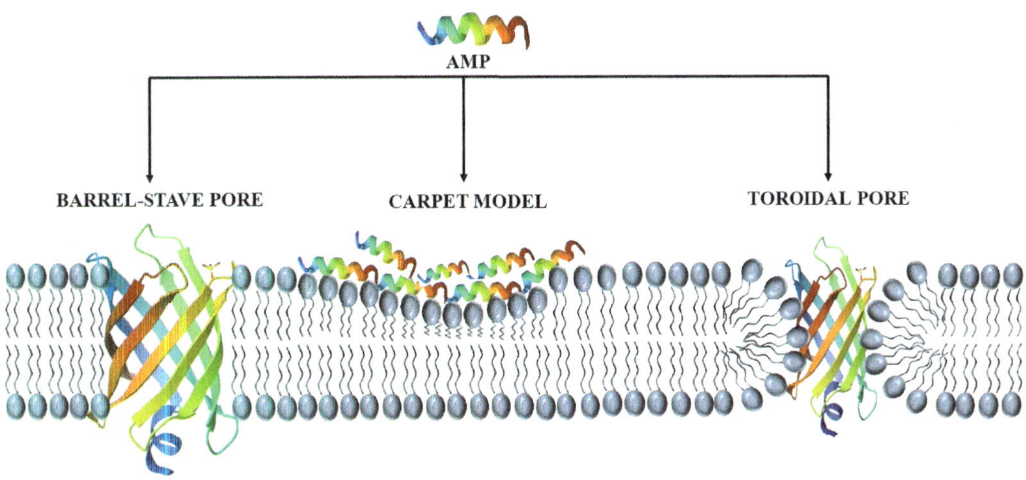

FIGURE 6.10

Schematic representation of modes of activities of antimicrobial peptides. *AMP*, Antimicrobial peptide.

electrostatic attraction initiates AMP binding to bacterial membranes, facilitating membrane insertion or the formation of surface-aligned peptide aggregation. This specific binding disrupts the integrity of bacterial membranes, leading to leakage, pore formation, or cell lysis while sparing mammalian cells, as they have different lipid compositions. AMPs act through multiple mechanisms to disrupt membranes depending on the concentration and lipid composition of the target membrane. Here, we have discussed the key models of membrane interaction.

6.6.2.1 Membrane thinning and pore formation

Membrane thinning is caused when AMPs are inserted into the membrane after binding to the lipid bilayer. This disruption occurs as the peptides position themselves parallel or perpendicular to the bilayer depending on the lipid-to-peptide ratio and types of lipids. As the bilayer thins, pore formation becomes energetically favorable (Grage et al., 2016). Two types of pores are generally formed because of bilayer thinning; toroidal pores are formed when peptides interact with lipid headgroups and create continuous pores lined by both peptides and lipid molecules (Fig. 6.10). This pore type is dynamic and transient; it enhances leakage without requiring complete peptide aggregation. On the contrary, barrel-stave pores are formed when peptides aggregate into cylindrical bundles with their hydrophobic regions aligned along the membrane core (Fig. 6.10). This structure creates stable transmembrane pores that allow ions and molecules to pass freely through the membrane (Lyu & Narsimhan, 2023). Researchers have found that magainin 2, gramicidin S, and BP100 induced membrane thinning for amphiphilic peptides residing in the polar/apolar interface of the bilayer (Grage et al., 2016). Melittin is also a well-known example of AMPs that employ pore-forming mechanisms (Volovik & Batishchev, 2024). Their ability to induce rapid lipid flip-flop between membrane leaflets accelerates membrane destabilization and enhances antimicrobial potency.

6.6.2.2 *Carpet mechanism and detergent-like disruption*

In the carpet model, AMPs accumulate on the membrane surface in a parallel orientation, covering it like a "carpet." This alignment disrupts the lipid bilayer without penetrating deeply into the hydrophobic core (Fig. 6.10). At sufficiently high concentrations, peptides destabilize the membrane, leading to the formation of micelle-like aggregates, which result in complete membrane disintegration (Huan et al., 2020). Melittin also exemplifies this mechanism by acting similarly to detergents, solubilizing membrane lipids into micellar structures. This detergent-like behavior is observed at higher peptide-to-lipid ratios, further contributing to antimicrobial action through nonspecific membrane dissolution (Wimley, 2010).

6.6.3 Membrane fusion

Membrane fusion is a fundamental process for various cellular functions such as intracellular trafficking, neurotransmission, fertilization, and viral infection (Jahn et al., 2003). The fusion process involves the merging of two lipid bilayers, leading to the formation of a continuous bilayer and mixing of the internal contents. According to the stalk hypothesis, membrane fusion occurs through multiple intermediate steps where lipidic rearrangements lead to the fusion of two distinct vesicles (Chernomordik et al., 1987; Kozlov & Markin, 1983). A crucial aspect of this process is the interaction between the N-terminal fusion peptide of the fusion proteins and lipids, which orchestrates membrane destabilization and subsequent fusion. Peptides that interact with membranes are diverse, encompassing fusion peptides from viral proteins, AMPs, and others with unique biophysical properties. Understanding the lipid–peptide interaction mechanisms offers insights into biological processes and provides pathways for developing therapeutic strategies, such as antiviral agents and drug delivery systems.

6.6.3.1 *Mechanism of membrane fusion*

Membrane fusion proceeds through a series of highly regulated steps. The stalk hypothesis gives us information about lipid rearrangement based on mechanical properties and phase behavior of water-lipid mesomorphic systems. According to the hypothesis, two apposed bilayers come to a closer proximity where the outer leaflet of each bilayer fuses to form a semistable stalk-like intermediate. In this stage the lipids undergo rearrangement from a lamellar to a nonlamellar phase. This stalk intermediate then expands longitudinally to form another semistable intermediate called extended transmembrane contact via a transmembrane contact. Finally the process culminates in the formation of a fusion pore. The schematic representation of the stepwise fusion process is shown in Fig. 6.11.

The fusion intermediates involve transient and unfavorable fluctuations in lipid/water rearrangement. The exposure of fatty acyl tails to the aqueous environment provides the activation barrier for membrane fusion. The fusion process requires overcoming the substantial activation energy barriers associated with lipid rearrangement, highlighting the importance of fusion peptides and lipid composition in modulating these transitions (Chernomordik et al., 1987).

6.6.3.2 *Role of fusion peptide in membrane fusion*

A fusion peptide is a typically short hydrophobic sequence of 20–25 amino acids that plays a crucial role in membrane fusion by modulating the organization and dynamics of lipid bilayers (Meher & Chakraborty, 2019). The fusion peptide gets inserted into the host membrane and induces local curvature

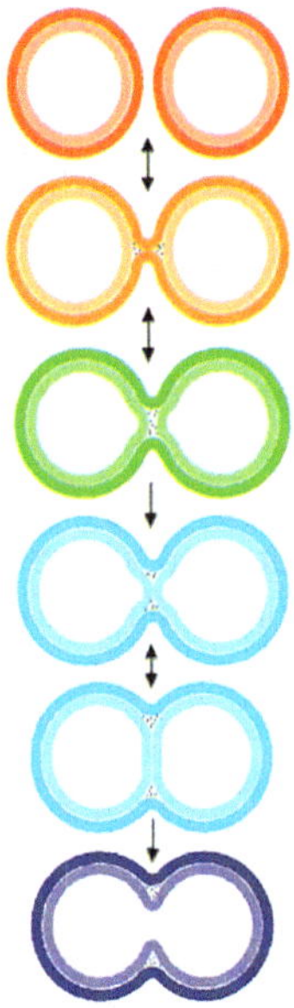

FIGURE 6.11

Schematic representation of the stepwise fusion process.

From Pattnaik, G. P., Meher, G., & Chakraborty, H. (2018). Exploring the mechanism of viral peptide-induced membrane fusion *(pp. 69–78). Springer. https://doi.org/10.1007/978-981-13-3065-0_6*

that catalyzes the formation of nonlamellar intermediates such as stalks and hemifusion diaphragms. The fusion peptide of gp41 from HIV switches between α-helical or β-stranded conformations depending on the lipid environment, underscoring their structural plasticity (Li & Tamm, 2007). Despite rigorous research on peptide-mediated membrane fusion, it remains unclear how peptides stabilize the nonbilayer intermediates during membrane fusion. It is assumed that the peptide allows water percolation in the acyl chain region of the membrane and reduces the activation barrier for the transition state formation, where lipidic tails are disposed to the aqueous environment (Meher et al., 2019). The X-ray scattering results demonstrated that the gp41 fusion peptide reduces the bending modulus of the membrane that supports nonbilayer intermediate formation (Tristram-Nagle & Nagle, 2007). Studies have revealed that fusion peptides of simian immunodeficiency virus, influenza, bovine leukemia virus, and Ebola demonstrated higher fusogenicity because of their oblique insertion in the membrane that reduces the bending energy of the membrane and supports fusion (Lins et al., 2008). Fusion peptides of gp41 (Haque et al., 2005; Joardar et al., 2022a), hemagglutinin (Haque et al., 2005), and the transmembrane domain of vesicular stomatitis virus (TMD-VSV) (Sengupta et al., 2014) are known to fill up the void space generated in the nonbilayer intermediates, thereby stabilizing the stalk state during fusion. The ability of fusion peptides to modulate the membrane curvature also plays an important role in regulating fusion. It has been speculated that the peptide that reduces the acyl chain ordering can promote fusion. Several depth-dependent fluorophores were used to assess the membrane's depth-dependent ordering, and the findings indicate that membrane fusion requires the ordering and disordering at a specific depth (Chakraborty et al., 2017).

6.6.3.3 Effect of lipid composition and curvature on membrane fusion

Lipid bilayers are not passive participants in the fusion process but actively influence the dynamics and efficiency of fusion through their composition and physical properties. Lipid composition plays a critical role in modulating the fusogenic properties of the membrane either by perturbing the organization and dynamics or altering the membrane curvature and conformation of peptides. The lipid composition is responsible for the generation of either positive or negative curvature in the membrane (Chernomordik & Kozlov, 2003). Different lipids demonstrate varying spontaneous curvature depending on their molecular shape and how they are packed in the membrane bilayer. Inverted cone-shaped lipids generate an intrinsic negative curvature in the membrane, whereas cone-shaped lipids generate an intrinsic positive curvature in the membrane because of their effective packing. A lipidic shape depends upon the ratio of headgroup to hydrophobic area of cross-section and interfacial hydration (Joardar et al., 2022b). According to the stalk hypothesis, curvature plays an important role in stabilizing the intermediates formed during the process. The fusion starts with the generation of a point-like protrusion in one of the apposed bilayers. The formation of a point-like protrusion generates high negative curvature in the membrane; therefore lipids with intrinsic negative curvature favor the formation of hemifusion formation. The hemifusion intermediate is of high energy because of the presence of structural voids in the bilayer, and it moves toward pore formation. As opposed to hemifusion formation, pore opening requires positive curvature. Therefore the plasticity of the fusion peptide in the lipidic environment is important as it supports the intrinsic negative curvature at the beginning of the process, whereas it generates positive curvature at the later part of the process. In addition to the plasticity of peptides the lipid distribution in two different leaflets plays an important role in driving the process to completion. PE and CH have an inverted cone shape because of their small headgroup size compared to the acyl chain region and generate intrinsic negative curvature. The intrinsic negative curvatures of DOPE and CH are -0.399 and -0.494 nm^{-1}, respectively (Chen & Rand, 1997; Fuller & Rand, 2001; Kollmitzer et al., 2013), and favor the formation of highly curved hemifusion intermediates (Liu & Boxer, 2020; Pattnaik and Chakraborty, 2021a). The PC/PE ratio is a critical determinant of membrane fusogenicity, as the otherwise lamellar PC membranes accumulate negative curvature in the presence of PE (Mishra & Chakraborty, 2023). On the other hand, lysophosphatidylcholine and oleic acid have intrinsic positive at physiological pH as they have a larger hydrated headgroup than their tail portion (Lähdesmäki et al., 2010). This positive curvature in the membrane raises the activation barrier for the intermediate formation and inhibits fusion (Joardar et al., 2021). It has been shown by the Lentz group that the presence of PE in the lipid mixture enhances the rate and extent of lipid mixing and content mixing, but still content leakage is controlled by other lipids in the membrane (Haque et al., 2001).

Negatively charged lipids such as PS and PG affect the fusion because of electrostatic repulsion between two negatively charged membrane surfaces (Tarafdar et al., 2012). In contrast, Ca^{2+}-mediated membrane fusion was promoted in the presence of PS in the membrane as it helps to bring two approaching vesicles closer (Tarafdar et al., 2012). PS and PIP regulate fusion by reorganizing the fusion machinery; that is, they affect the protein conformation and modulate the membrane domain (Tarafdar et al., 2015; van den Bogaart et al., 2011).

6.6.3.4 Role of cholesterol in membrane fusion

CH is the most dynamic constituent of the mammalian cell membrane. It is synthesized through a complex series of enzymatic steps by de novo synthesis in the endoplasmic reticulum (Nes, 2011). A huge amount of CH is present in the blood serum as lipoproteins and is taken up by cells via endocytosis. It has a rigid fused ring with a hydroxyl group that is the only polar part of the CH at one end and a branched aliphatic carbon tail at the other end. The only polar hydroxyl group of the CH orients close to the headgroup region of phospholipids, whereas the remaining nonpolar hydrocarbon tail aligns perpendicularly to the bilayer. The presence of a rigid fused ring of CH restricts the transgauche isomerism of the nearby lipid acyl chain, thereby rigidifying the membrane bilayer. The ordering effect of CH also affects the mechanical properties of the lipid bilayer, such as increasing the bending and compressibility moduli and restricting the lateral and diffusion of other lipids and embedded proteins in the membrane (Chiu et al., 2001). Because of the smaller headgroup to the hydrophobic cross-sectional area of the tail, CH induces intrinsic negative curvature to the membrane. The interaction of CH with sphingolipids forms the lipid domain that generates phase separation in the membrane. The CH-dependent modifications of membrane physical properties modulate viral infections by affecting the fusion between viral and host cell membranes. It plays an important role in the entry of viruses into the host cell via fusion, as it can induce negative curvature in the lipid bilayer (Pattnaik and Chakraborty, 2021b). It further affects the conformation, orientation, organization, and dynamics of fusion peptides in the membrane milieu, hence modulating the lipid–peptide interaction, which in turn affects the fusion process (Meher et al., 2019). CH promotes membrane fusion by stabilizing the nonlamellar stalk intermediates by modulating lipid packing and void space. It has been found that gp41 penetrates deeper into the membrane and alters the membrane ordering more than that in the absence of CH (Meher et al., 2019). Tamm group showed that gp41 fusion peptide assumes the α-helical conformation in lower CH-containing membrane, whereas it changes to β-sheet conformation when membrane CH concentration is increased (Lai et al., 2012). The fusion efficiency of HIV gp41 fusion peptide is further enhanced in CH-containing membranes (Aisenbrey et al., 2020; Yang et al., 2015). TG-23, a coronin 1-derived fusion inhibitory peptide, was inhibiting model membrane fusion of lipid composition of DOPC/DOPE/DOPG (60/30/10 mol%). However, this inhibitory efficiency was found to reduce when 10 mol% of CH was introduced in the lipid composition of DOPC/DOPE/DOPG (60/30/10 mol%). Interestingly, this inhibitory efficiency of TG-23 entirely vanished when CH concentration was increased to 20 mol%. Higher CH concentration in the membrane was altering the organization and dynamics of the membrane in such a way that it was hindering the fusion inhibitory effect of the peptide (Pattnaik & Chakraborty, 2018, 2019).

6.7 Conclusion

This chapter offers a detailed description of the membrane that provides a unique platform for the associated and integral membrane proteins to display their functions. The chemical and physical diversities of lipids generate heterogeneous surfaces and induce specificity of lipid–protein interaction. The intricate cross-talking between lipids and proteins is significantly important for biological processes such as protein trafficking and signaling. Moreover, the membrane is targeted for the successful development of antimicrobial drugs, as the lipid composition is widely different between the human cell and the microbes. We have further discussed how the lipid composition favors the entry of enveloped viruses by modulating the physical properties of membranes.

References

Ahmed, A., Siman-Tov, G., Hall, G., Bhalla, N., & Narayanan, A. (2019). Human antimicrobial peptides as therapeutics for viral infections. *Viruses, 11*, 704.

Aisenbrey, C., Rifi, O., & Bechinger, B. (2020). Structure, membrane topology and influence of cholesterol of the membrane proximal region: Transmembrane helical anchor sequence of gp41 from HIV. *Scientific Reports, 10*, 22278.

Al-Mualem, Z. A., Chen, X., Shafieenezhad, A., Senning, E. N., & Baiz, C. R. (2024). Binding-induced lipid domains: Peptide-membrane interactions with PIP(2) and PS. *Biophysical Journal, 123*, 2001–2011.

Aloia, R. C., Tian, H., & Jensen, F. C. (1993). Lipid composition and fluidity of the human immunodeficiency virus envelope and host cell plasma membranes. *Proceedings of the National Academy of Sciences of the United States of America, 90*, 5181–5185.

Ariotti, N., Fernández-Rojo, M. A., Zhou, Y., Hill, M. M., Rodkey, T. L., Inder, K. L., Tanner, L. B., Wenk, M. R., Hancock, J. F., & Parton, R. G. (2014). Caveolae regulate the nanoscale organization of the plasma membrane to remotely control Ras signaling. *Journal of Cell Biology, 204*, 777–792.

Bar, R. S., Deamer, D. W., & Cornwell, D. G. (1966). Surface area of human erythrocyte lipids: Reinvestigation of experiments on plasma membrane. *Science, 153*, 1010–1012.

Behnia, R., Panic, B., Whyte, J. R., & Munro, S. (2004). Targeting of the Arf-like GTPase Arl3p to the Golgi requires N-terminal acetylation and the membrane protein Sys1p. *Nature Cell Biology, 6*, 405–413.

Björkholm, P., Ernst, A. M., Hacke, M., Wieland, F., Brügger, B., & von Heijne, G. (2014). Identification of novel sphingolipid-binding motifs in mammalian membrane proteins. *Biochimica Et Biophysica Acta, 1838*, 2066–2070.

Blouin, C. M., Hamon, Y., Gonnord, P., Boularan, C., Kagan, J., Viaris de Lesegno, C., Ruez, R., Mailfert, S., Bertaux, N., Loew, D., Wunder, C., Johannes, L., Vogt, G., Contreras, F. X., Marguet, D., Casanova, J. L., Galès, C., He, H. T., & Lamaze, C. (2016). Glycosylation-dependent IFN-γR partitioning in lipid and actin nanodomains is critical for JAK activation. *Cell, 166*, 920–934.

Bretscher, M. S. (1973). Membrane structure: Some general principles. *Science, 181*, 622–629.

Brown, D. A., & Rose, J. K. (1992). Sorting of GPI-anchored proteins to glycolipid-enriched membrane subdomains during transport to the apical cell surface. *Cell, 68*, 533–544.

Budin, I., & Szostak, J. W. (2010). Expanding roles for diverse physical phenomena during the origin of life. *Annual Review of Biophysics, 39*, 245–263.

Byrne, E. F. X., Sircar, R., Miller, P. S., Hedger, G., Luchetti, G., Nachtergaele, S., Tully, M. D., Mydock-McGrane, L., Covey, D. F., Rambo, R. P., Sansom, M. S. P., Newstead, S., Rohatgi, R., & Siebold, C. (2016). Structural basis of Smoothened regulation by its extracellular domains. *Nature, 535*, 517–522.

Cail, R. C., & Drubin, D. G. (2023). Membrane curvature as a signal to ensure robustness of diverse cellular processes. *Trends in Cell Biology, 33*(5), 427–441.

Cannarozzo, C., Fred, S. M., Girych, M., Biojone, C., Enkavi, G., Róg, T., Vattulainen, I., Casarotto, P. C., & Castrén, E. (2021). Cholesterol-recognition motifs in the transmembrane domain of the tyrosine kinase receptor family: The case of TRKB. *European Journal of Neuroscience, 53*, 3311–3322.

Casares, D., Escribá, P. V., & Rosselló, C. A. (2019). Membrane lipid composition: Effect on membrane and organelle structure, function and compartmentalization and therapeutic avenues. *International Journal of Molecular Sciences, 20*, 2167.

Chakraborty, H., Lentz, B. R., Kombrabail, M., Krishnamoorthy, G., & Chattopadhyay, A. (2017). Depth-dependent membrane ordering by hemagglutinin fusion peptide promotes fusion. *Journal of Physical Chemistry B, 121*, 1640–1648.

Chapman, D. (1975). Phase transitions and fluidity characteristics of lipids and cell membranes. *Quarterly Reviews of Biophysics, 8*, 185–235.

Chattopadhyay, A. (2014). GPCRs: Lipid-dependent membrane receptors that act as drug targets. *Journal of Advances in Biology, 2014*, 143023.

Chaudhury, A., Swarnakar, S., Pattnaik, G. P., Varshney, G. K., Chakraborty, H., & Basu, J. K. (2023). Peptide-induced fusion of dynamic membrane nanodomains: Implications in a viral entry. *Langmuir, 39*, 17713–17722.

Chavent, M., Karia, D., Kalli, A. C., Domański, J., Duncan, A. L., Hedger, G., Stansfeld, P. J., Seiradake, E., Jones, E. Y., & Sansom, M. S. P. (2018). Interactions of the EphA2 kinase domain with PIPs in membranes: Implications for receptor function. *Structure (London, England), 26*, 1025–1034.

Chen, Z., & Rand, R. P. (1997). The influence of cholesterol on phospholipid membrane curvature and bending elasticity. *Biophysical Journal, 73*, 267–276.

Chernomordik, L. V., & Kozlov, M. M. (2003). Protein-lipid interplay in fusion and fission of biological membranes. *Annual Review of Biochemistry, 72*, 175–207.

Chernomordik, L. V., Melikyan, G. B., & Chizmadzhev, Y. A. (1987). Biomembrane fusion: A new concept derived from model studies using two interacting planar lipid bilayers. *Biochimica Et Biophysica Acta, 906*, 309–352.

Chiu, S. W., Jakobsson, E., & Scott, H. L. (2001). Combined Monte Carlo and molecular dynamics simulation of hydrated lipid-cholesterol lipid bilayers at low cholesterol concentration. *Biophysical Journal, 80*, 1104–1114.

Cho, W., & Stahelin, R. V. (2005). Membrane-protein interactions in cell signaling and membrane trafficking. *Annual Review of Biophysics and Biomolecular Structure, 34*, 119–151.

Clements, J. A. (1962). Surface phenomena in relation to pulmonary function. *The Physiologist, 5*, 11–28.

Cole, K. S. (1932). Surface forces of the Arbacia egg. *Journal of Cellular Physiology, 1*, 1–9.

Cole, L. K., Vance, J. E., & Vance, D. E. (2012). Phosphatidylcholine biosynthesis and lipoprotein metabolism. *Biochimica et Biophysica Acta (BBA)-Molecular Cell Research, 1821*, 754–761.

Corradi, V., Sejdiu, B. I., Mesa-Galloso, H., Abdizadeh, H., Noskov, S. Y., Marrink, S. J., & Tieleman, D. (2019). Emerging diversity in lipid–protein interactions. *Chemical Reviews, 119*, 5775–5848.

Couet, J., Sargiacomo, M., & Lisanti, M. P. (1997). Interaction of a receptor tyrosine kinase, EGF-R, with caveolins. Caveolin binding negatively regulates tyrosine and serine/threonine kinase activities. *Journal of Biological Chemistry, 272*, 30429–30438.

D'Alessio, A. (2023). Unraveling the cave: A seventy-year journey into the caveolar network, cellular signaling, and human disease. *Cells, 12*, 2680.

Danielli, J. F., & Davson, H. (1935). A contribution to the theory of permeability of thin films. *Journal of Cellular and Comparative Physiology, 5*, 495–508.

Deamer, D., Dworkin, J. P., Sandford, S. A., Bernstein, M. P., & Allamandola, L. J. (2002). The first cell membranes. *Astrobiology, 2*, 371–381.

Di Paolo, G., & De Camilli, P. (2006). Phosphoinositides in cell regulation and membrane dynamics. *Nature, 443*, 651–657.

Dumas, F., Tocanne, J. F., Leblanc, G., & Lebrun, M. C. (2000). Consequences of hydrophobic mismatch between lipids and melibiose permease on melibiose transport. *Biochemistry, 39*, 4846–4854.

Edidin, M. (2003). Lipids on the frontier: A century of cell-membrane bilayers. *Nature Reviews Molecular Cell Biology, 4*, 414–418.

Escribá, P. V., González-Ros, J. M., Goñi, F. M., Kinnunen, P. K., Vigh, L., Sánchez-Magraner, L., Fernández, A. M., Busquets, X., Horváth, I., & Barceló-Coblijn, G. (2008). Membranes: A meeting point for lipids, proteins and therapies. *Journal of Cellular and Molecular Medicine, 12*, 829–875.

Fattal, D. R., & Ben-Shaul, A. (1993). A molecular model for lipid-protein interaction in membranes: The role of hydrophobic mismatch. *Biophysical Journal, 65*, 1795–1809.

Fricke, H. (1925). The electric capacity of suspensions with special reference to blood. *Journal of General Physiology, 9*, 137–152.

Fuller, N., & Rand, R. P. (2001). The influence of lysolipids on the spontaneous curvature and bending elasticity of phospholipid membranes. *Biophysical Journal, 81*, 243–254.

Galbiati, F., Volonté, D., Engelman, J. A., Watanabe, G., Burk, R., Pestell, R. G., & Lisanti, M. P. (1998). Targeted downregulation of caveolin-1 is sufficient to drive cell transformation and hyperactivate the p42/44 MAP kinase cascade. *EMBO Journal, 17*, 6633–6648.

Garcia, P., Gupta, R., Shah, S., Morris, A. J., Rudge, S. A., Scarlata, S., Petrova, V., McLaughlin, S., & Rebecchi, M. J. (1995). The pleckstrin homology domain of phospholipase C-. delta. 1 binds with high affinity to phosphatidylinositol 4, 5-bisphosphate in bilayer membranes. *Biochemistry, 34*, 16228–16234.

Gong, H., Hu, X., Zhang, L., Fa, K., Liao, M., Liu, H., Fragneto, G., Campana, M., & Lu, J. R. (2023). How do antimicrobial peptides disrupt the lipopolysaccharide membrane leaflet of Gram-negative bacteria? *Journal of Colloid and Interface Science, 637*, 182–192.

Goodwin, J. S., Drake, K. R., Rogers, C., Wright, L., Lippincott-Schwartz, J., Philips, M. R., & Kenworthy, A. K. (2005). Depalmitoylated Ras traffics to and from the Golgi complex via a nonvesicular pathway. *Journal of Cell Biology, 170*, 261–272.

Gorter, E., & Grendel, F. (1925). On bimolecular layers of lipoids on the chromocytes of the blood. *Journal of Experimental Medicine, 41*, 439–443.

Grage, S. L., Afonin, S., Kara, S., Buth, G., & Ulrich, A. S. (2016). Membrane thinning and thickening induced by membrane-active amphipathic peptides. *Frontiers in Cell and Developmental Biology, 4*, 65.

Greaves, J., & Chamberlain, L. H. (2006). Dual role of the cysteine-string domain in membrane binding and palmitoylation-dependent sorting of the molecular chaperone cysteine-string protein. *Molecular Biology of the Cell, 17*, 4748–4759.

Greaves, J., & Chamberlain, L. H. (2007). Palmitoylation-dependent protein sorting. *Journal of Cell Biology, 176*, 249–254.

Groves, J. T., & Kuriyan, J. (2010). Molecular mechanisms in signal transduction at the membrane. *Natural Structural Biology, 17*, 659–665.

Guerrero-Valero, M., Marín-Vicente, C., Gómez-Fernández, J. C., & Corbalán-García, S. (2007). The C2 domains of classical PKCs are specific PtdIns (4, 5) P2-sensing domains with different affinities for membrane binding. *Journal of Molecular Biology, 371*, 608–621.

Han, J. M., Kim, Y., Lee, J. S., Lee, C. S., Lee, B. D., Ohba, M., Kuroki, T., Suh, P. G., & Ryu, S. H. (2002). Localization of phospholipase D1 to caveolin-enriched membrane via palmitoylation: Implications for epidermal growth factor signaling. *Molecular Biology of the Cell, 13*, 3976–3988.

Hanafusa, K., Hotta, T., & Iwabuchi, K. (2020). Glycolipids: Linchpins in the organization and function of membrane microdomains. *Frontiers in Cell and Developmental Biology, 8*, 589799.

Hanulová, M., & Weiss, M. (2012). Protein sorting and membrane-mediated interactions. *Biophysical Reviews, 4*, 117–124.

Haque, M. E., Koppaka, V., Axelsen, P. H., & Lentz, B. R. (2005). Properties and structures of the influenza and HIV fusion peptides on lipid membranes: Implications for a role in fusion. *Biophysical Journal, 89*, 3183–3194.

Haque, M. E., McIntosh, T. J., & Lentz, B. R. (2001). Influence of lipid composition on physical properties and peg-mediated fusion of curved and uncurved model membrane vesicles: "Nature's own" fusogenic lipid bilayer. *Biochemistry, 40*, 4340–4348.

Haucke, V., & Di Paolo, G. (2007). Lipids and lipid modifications in the regulation of membrane traffic. *Current Opinion in Cell Biology, 19*, 426–435.

Hayashi, T., Rumbaugh, G., & Huganir, R. L. (2005). Differential regulation of AMPA receptor subunit trafficking by palmitoylation of two distinct sites. *Neuron, 47*, 709–723.

Haydon, D. A. (1968). Properties of lipid bilayers at a water-water interface. *Journal of the American Oil Chemists' Society, 45*, 230–240.

Haydon, D. A., & Taylor, J. (1963). The stability and properties of bimolecular lipid leaflets in aqueous solutions. *Journal of Theoretical Biology, 4*, 281–296.

Huan, Y., Kong, Q., Mou, H., & Yi, H. (2020). Antimicrobial peptides: Classification, design, application and research progress in multiple fields. *Frontiers in Microbiolog, 11*, 582779.

Huang, P., Nedelcu, D., Watanabe, M., Jao, C., Kim, Y., Liu, J., & Salic, A. (2016). Cellular cholesterol directly activates smoothened in Hedgehog signaling. *Cell, 166*, 1176–1187.

Hwang, D. H., Kim, J.-A., & Lee, J. Y. (2016). Mechanisms for the activation of Toll-like receptor 2/4 by saturated fatty acids and inhibition by docosahexaenoic acid. *European Journal of Pharmacology, 785*, 24–35.

Ivanova, P. T., Myers, D. S., Milne, S. B., McClaren, J. L., Thomas, P. G., & Brown, H. A. (2015). Lipid composition of viral envelope of three strains of influenza virus—Not all viruses are created equal. *ACS Infectious Diseases, 1*, 399–452.

Jahn, R., Lang, T., & Südhof, T. C. (2003). Membrane fusion. *Cell, 112*, 519–533.

Javanainen, M., Enkavi, G., Guixà-Gonzaléz, R., Kulig, W., Martinez-Seara, H., Levental, I., & Vattulainen, I. (2019). Reduced level of docosahexaenoic acid shifts GPCR neuroreceptors to less ordered membrane regions. *PLoS Computational Biology, 15*, e1007033.

Jensen, D., & Schekman, R. (2011). COPII-mediated vesicle formation at a glance. *Journal of Cell Science, 124*, 1–4.

Joardar, A., Pattnaik, G. P., & Chakraborty, H. (2021). Effect of phosphatidylethanolamine and oleic acid on membrane fusion: Phosphatidylethanolamine circumvents the classical stalk model. *Journal of Physical Chemistry B, 125*, 13192–13202.

Joardar, A., Pattnaik, G. P., & Chakraborty, H. (2022a). Combination of oleic acid and the gp41 fusion peptide switches the phosphatidylethanolamine-induced membrane fusion mechanism from a nonclassical to a classical stalk model. *Journal of Physical Chemistry B, 126*, 3673–3684.

Joardar, A., Pattnaik, G. P., & Chakraborty, H. (2022b). Mechanism of membrane fusion: Interplay of lipid and peptide. *Journal of Membrane Biology, 255*, 211–224.

Johnson, D. H., Kou, O. H., Bouzos, N., & Zeno, W. F. (2024). Protein–membrane interactions: Sensing and generating curvature. *Trends in Biochemical Sciences, 49*, 401–416.

Juhola, H., Postila, P. A., Rissanen, S., Lolicato, F., Vattulainen, I., & Róg, T. (2018). Negatively charged gangliosides promote membrane association of amphipathic neurotransmitters. *Neuroscience, 384*, 214–223.

Kelkar, D. A., & Chattopadhyay, A. (2007a). The gramicidin ion channel: A model membrane protein. *Biochimica et Biophysica Acta (BBA)-Molecular Cell Research, 1768*, 2011–2025.

Kelkar, D. A., & Chattopadhyay, A. (2007b). Modulation of gramicidin channel conformation and organization by hydrophobic mismatch in saturated phosphatidylcholine bilayers. *Biochimica et Biophysica Acta (BBA)-Molecular Cell Researc, 1768*(5), 1103–1113.

Kirchhausen, T. (2000). Clathrin. *Annual Review of Biochemistry, 69*, 699–727.

Kleinschmidt, J. H., & Tamm, L. K. (2002). Secondary and tertiary structure formation of the beta-barrel membrane protein OmpA is synchronized and depends on membrane thickness. *Journal of Molecular Biology, 324*, 319–330.

Kollmitzer, B., Heftberger, P., Rappolt, M., & Pabst, G. (2013). Monolayer spontaneous curvature of raft-forming membrane lipids. *Soft Matter, 9*, 10877–10884.

Kooijman, E. E., & Burger, K. N. (2009). Biophysics and function of phosphatidic acid: A molecular perspective. *Biochimica et Biophysica Acta (BBA)-Molecular Cell Research, 1791*, 881–888.

Kozlov, M. M., & Markin, V. S. (1983). Possible mechanism of membrane fusion. *Biofizika, 28*, 242–247.

Lähdesmäki, K., Ollila, O. H., Koivuniemi, A., Kovanen, P. T., & Hyvönen, M. T. (2010). Membrane simulations mimicking acidic pH reveal increased thickness and negative curvature in a bilayer consisting of lysophosphatidylcholines and free fatty acids. *Biochimica Et Biophysica Acta, 1798*, 938–946.

Lai, A. L., Moorthy, A. E., Li, Y., & Tamm, L. K. (2012). Fusion activity of HIV gp41 fusion domain is related to its secondary structure and depth of membrane insertion in a cholesterol-dependent fashion. *Journal of Molecular Biology, 418*, 3–15.

Langmuir, I. (1916). The constitution and fundamental properties of solids and liquids. Part I. Solids. *Journal of the American Chemical Society, 38*, 2221–2295.
Langmuir, I. (1917). The constitution and fundamental properties of solids and liquids. *Journal of the Franklin Institute, 183*, 102–105.
Lee, K.-Y. (2024). Membrane-driven dimerization of the peripheral membrane protein KRAS: Implications for downstream signaling. *International Journal of Molecular Sciences, 25*, 2530.
Lemmon, M. A. (2008). Membrane recognition by phospholipid-binding domains. *Nature Reviews Molecular Cell Biology, 9*, 99–111.
Lemmon, M. A., Ferguson, K. M., & Abrams, C. S. (2002). Pleckstrin homology domains and the cytoskeleton. *FEBS Letters, 513*, 71–76.
Lentz, B. R. (2003). Exposure of platelet membrane phosphatidylserine regulates blood coagulation. *Progress in Lipid Research, 42*, 423–438.
Li, Y., & Tamm, L. K. (2007). Structure and plasticity of the human immunodeficiency virus gp41 fusion domain in lipid micelles and bilayers. *Biophysical Journal, 93*, 876–885.
Lin, B. J., Tsao, S. H., Chen, A., Hu, S. K., Chao, L., & Chao, P. G. (2017). Lipid rafts sense and direct electric field-induced migration. *Proceedings of the National Academy of Sciences of the United States of America, 114*, 8568–8573.
Lin, Q., & London, E. (2013). Altering hydrophobic sequence lengths shows that hydrophobic mismatch controls affinity for ordered lipid domains (rafts) in the multitransmembrane strand protein perfringolysin O. *Journal of Biological Chemistry, 288*, 1340–1352.
Lins, L., Decaffmeyer, M., Thomas, A., & Brasseur, R. (2008). Relationships between the orientation and the structural properties of peptides and their membrane interactions. *Biochimica Et Biophysica Acta, 1778*, 1537–1544.
Liu, K. N., & Boxer, S. G. (2020). Target membrane cholesterol modulates single influenza virus membrane fusion efficiency but not rate. *Biophysical Journal, 118*, 2426–2433.
Lopez, P. H., & Schnaar, R. L. (2009). Gangliosides in cell recognition and membrane protein regulation. *Current Opinion in Structural Biology, 19*, 549–557.
Luchetti, G., Sircar, R., Kong, J. H., Nachtergaele, S., Sagner, A., Byrne, E. F., Covey, D. F., Siebold, C., & Rohatgi, R. (2016). Cholesterol activates the G-protein coupled receptor Smoothened to promote Hedgehog signaling. *Elife, 5*, e20304.
Lyu, Y., & Narsimhan, G. (2023). Antimicrobial peptides-membrane interactions. In R. Nagarajan (Ed.). In *One hundred years of colloid symposia: Looking back and looking forward: 1457* (pp. 381–430). American Chemical Society.
Magee, T., & Seabra, M. C. (2005). Fatty acylation and prenylation of proteins: What's hot in fat. *Current Opinion in Cell Biology, 17*, 190–196.
Majumder, R., Quinn-Allen, M. A., Kane, W. H., & Lentz, B. R. (2008). A phosphatidylserine binding site in factor Va C1 domain regulates both assembly and activity of the prothrombinase complex. *Blood, 112*, 2795–2802.
Majumder, R., Weinreb, G., & Lentz, B. R. (2005). Efficient thrombin generation requires molecular phosphatidylserine, not a membrane surface. *Biochemistry, 44*, 16998–17006.
McLaughlin, S., & Murray, D. (2005). Plasma membrane phosphoinositide organization by protein electrostatics. *Nature, 438*, 605–611.
McMahon, H. T., & Boucrot, E. (2011). Molecular mechanism and physiological functions of clathrin-mediated endocytosis. *Nature Reviews Molecular Cell Biology, 12*, 517–533.
McMahon, H. T., & Boucrot, E. (2015). Membrane curvature at a glance. *Journal of Cell Science, 128*, 1065–1070.
McMahon, H. T., Kozlov, M. M., & Martens, S. (2010). Membrane curvature in synaptic vesicle fusion and beyond. *Cell, 140*, 601–605.
Meher, G., & Chakraborty, H. (2019). Membrane composition modulates fusion by altering membrane properties and fusion peptide structure. *Journal of Membrane Biology, 252*, 261–272.

Meher, G., Sinha, S., Pattnaik, G. P., Ghosh Dastidar, S., & Chakraborty, H. (2019). Cholesterol modulates membrane properties and the interaction of gp41 fusion peptide to promote membrane fusion. *Journal of Physical Chemistry B, 123*, 7113–7122.

Mirdha, L., Sengupta, T., & Chakraborty, H. (2022). Lipid composition dependent binding of apolipoprotein E signal peptide: Importance of membrane cholesterol in protein trafficking. *Biophysical Chemistry, 291*, 106907.

Mishra, S., & Chakraborty, H. (2023). Phosphatidylethanolamine and cholesterol promote hemifusion formation: A tug of war between membrane interfacial order and intrinsic negative curvature of lipids. *Journal of Physical Chemistry B, 127*, 7721–7729.

Mulgrew-Nesbitt, A., Diraviyam, K., Wang, J., Singh, S., Murray, P., Li, Z., Rogers, L., Mirkovic, N., & Murray, D. (2006). The role of electrostatics in protein–membrane interactions. *Biochimica et Biophysica Acta (BBA)-Molecular and Cell Biology of Lipids, 1761*, 812–826.

Nägeli, C. (1858). *Pflanzenphysiologische Untersuchungen*: 2. Schulthess.

Nakano, A. (2022). The Golgi apparatus and its next-door neighbors. *Frontiers in Cell and Developmental Biology, 10*, 884360.

Needham, D. (1995). Cohesion and permeability of lipid bilayer vesicles. In E. A. Disalvo, & S. A. Simon (Eds.), *Permeability and stability of lipid bilayers*. CRC Press.

Nes, W. D. (2011). Biosynthesis of cholesterol and other sterols. *Chemical Reviews, 111*, 6423–6451.

Oreopoulos, J., Epand, R. F., Epand, R. M., & Yip, C. M. (2010). Peptide-induced domain formation in supported lipid bilayers: Direct evidence by combined atomic force and polarized total internal reflection fluorescence microscopy. *Biophysical Journal, 98*, 815–823.

Overton, C. E. (1895). *Über die osmotischen Eigenschaften der lebenden Pflanzen-und Tierzelle*. Fäsi & Beer.

Paila, Y. D., Tiwari, S., Sengupta, D., & Chattopadhyay, A. (2011). Molecular modeling of the human serotonin$_{1A}$receptor: Role of membrane cholesterol in ligand binding of the receptor. *Molecular Biosystems, 7*, 224–234.

Pattnaik, G. P., & Chakraborty, H. (2018). Coronin 1 derived tryptophan-aspartic acid containing peptides inhibit membrane fusion. *Chemistry and Physics of Lipids, 217*, 35–42.

Pattnaik, G. P., & Chakraborty, H. (2019). Cholesterol alters the inhibitory efficiency of peptide-based membrane fusion inhibitor. *Biochimica Et Biophysica Acta, Biomembranes, 1861*, 183056.

Pattnaik, G. P., & Chakraborty, H. (2021a). Fusogenic effect of cholesterol prevails over the inhibitory effect of a peptide-based membrane fusion inhibitor. *Langmuir, 37*, 3477–3489.

Pattnaik, G. P., & Chakraborty, H. (2021b). Cholesterol: A key player in membrane fusion that modulates the efficacy of fusion inhibitor peptides. *Vitamins and Hormones, 117*, 133–155.

Pepinsky, R. B., Zeng, C., Wen, D., Rayhorn, P., Baker, D. P., Williams, K. P., Bixler, S. A., Ambrose, C. M., Garber, E. A., Miatkowski, K., Taylor, F. R., Wang, E. A., & Galdes, A. (1998). Identification of a palmitic acid-modified form of human Sonic Hedgehog. *Journal of Biological Chemistry, 273*, 14037–14045.

Peter, B. J., Kent, H. M., Mills, I. G., Vallis, Y., Butler, P. J., Evans, P. R., & McMahon, H. T. (2004). BAR domains as sensors of membrane curvature: The amphiphysin BAR structure. *Science, 303*, 495–499.

Pockels, A. (1891). Surface tension. *Nature, 43*, 437–439.

Prasanna, X., Chattopadhyay, A., & Sengupta, D. (2014). Cholesterol modulates the dimer interface of the β_2-adrenergic receptor via cholesterol occupancy sites. *Biophysical Journal, 106*, 1290–1300.

Prasanna, X., Chattopadhyay, A., & Sengupta, D. (2015). Role of lipid-mediated effects in β_2-adrenergic receptor dimerization. *Advances in Experimental Medicine and Biology, 842*, 247–261.

Prior, I. A., Muncke, C., Parton, R. G., & Hancock, J. F. (2003). Direct visualization of Ras proteins in spatially distinct cell surface microdomains. *Journal of Cell Biology, 160*, 165–170.

Rao, B. D., Shrivastava, S., & Chattopadhyay, A. (2017). *Hydrophobic mismatch in membranes: When the tail matters, Springer series in biophysics*: 20. Springer.

Rocks, O., Peyker, A., Kahms, M., Verveer, P. J., Koerner, C., Lumbierres, M., Kuhlmann, J., Waldmann, H., Wittinghofer, A., & Bastiaens, P. I. (2005). An acylation cycle regulates localization and activity of palmitoylated Ras isoforms. *Science, 307*, 1746–1752.

Rosenbaum, D. M., Rasmussen, S. G., & Kobilka, B. K. (2009). The structure and function of G-protein-coupled receptors. *Nature, 459*, 356–363.

Roth, A. F., Wan, J., Bailey, A. O., Sun, B., Kuchar, J. A., Green, W. N., Phinney, B. S., Yates, J. R., 3rd, & Davis, N. G. (2006). Global analysis of protein palmitoylation in yeast. *Cell, 125*, 1003–1013.

Roy, P., Tomassoni, D., Nittari, G., Traini, E., & Amenta, F. (2022). Effects of choline containing phospholipids on the neurovascular unit: A review. *Frontiers in Cellular Neuroscience, 16*, 988759.

Sani, M.-A., & Separovic, F. (2016). How membrane-active peptides get into lipid membranes. *Accounts of Chemical Research, 49*, 1130–1138.

Sengupta, D., & Chattopadhyay, A. (2015). Molecular dynamics simulations of GPCR-cholesterol interaction: An emerging paradigm. *Biochimica et Biophysica Acta (BBA)-Molecular Cell Research, 1848*, 1775–1782.

Sengupta, D., Prasanna, X., Mohole, M., & Chattopadhyay, A. (2018). Exploring GPCR–lipid interactions by molecular dynamics simulations: Excitements, challenges, and the way forward. *Journal of Physical Chemistry B, 122*, 5727–5737.

Sengupta, T., Chakraborty, H., & Lentz, B. R. (2014). The transmembrane domain peptide of vesicular stomatitis virus promotes both intermediate and pore formation during PEG-mediated vesicle fusion. *Biophysical Journal, 107*, 1318–1326.

Setty, S. R., Shin, M. E., Yoshino, A., Marks, M. S., & Burd, C. G. (2003). Golgi recruitment of GRIP domain proteins by Arf-like GTPase 1 is regulated by Arf-like GTPase 3. *Current Biology, 13*, 401–414.

Sezgin, E., & Schwille, P. (2011). Fluorescence techniques to study lipid dynamics. *Cold Spring Harbor Perspectives in Biology, 3*, a009803.

Shahane, G., Ding, W., Palaiokostas, M., & Orsi, M. (2019). Physical properties of model biological lipid bilayers: Insights from all-atom molecular dynamics simulations. *Journal of Molecular Modeling, 25*, 76.

Sharma, P., Varma, R., Sarasij, R. C., Ira, Gousset, K., Krishnamoorthy, G., Rao, M., & Mayor, S. (2004). Nanoscale organization of multiple GPI-anchored proteins in living cell membranes. *Cell, 116*, 577–589.

Short, S. A., & White, D. C. (1972). Biosynthesis of cardiolipin from phosphatidylglycerol in *Staphylococcus aureus*. *Journal of Bacteriology, 109*, 820–826.

Simons, K., & Sampaio, J. L. (2011). Membrane organization and lipid rafts. *Cold Spring Harbor Perspectives in Biology, 3*, a004697.

Singer, S. J., & Nicolson, G. L. (1972). The fluid mosaic model of the structure of cell membranes. *Science, 175*, 720–731.

Slotte, P. J. (2013). Molecular properties of various structurally defined sphingomyelins—Correlation of structure with function. *Progress in Lipid Research, 52*, 206–219.

Soubias, O., Teague, W. E., Jr., Hines, K. G., & Gawrisch, K. (2015). Rhodopsin/lipid hydrophobic matching-rhodopsin oligomerization and function. *Biophysical Journal, 108*, 1125–1132.

Stace, C. L., & Ktistakis, N. T. (2006). Phosphatidic acid- and phosphatidylserine-binding proteins. *Biochimica Et Biophysica Acta, 1761*, 913–926.

Stahelin, R. V. (2009). Lipid binding domains: More than simple lipid effectors. *Journal of Lipid Research, 50*, S299–S304.

Sunshine, H., & Iruela-Arispe, M. L. (2017). Membrane lipids and cell signaling. *Current Opinion in Lipidology, 28*, 408–413.

Sych, T., Levental, K. R., & Sezgin, E. (2022). Lipid-protein interactions in plasma membrane organization and function. *Annual Review of Biophysics, 51*, 135–156.

Tarafdar, P. K., Chakraborty, H., Bruno, M. J., & Lentz, B. R. (2015). Phosphatidylserine-dependent catalysis of stalk and pore formation by synaptobrevin JMR-TMD peptide. *Biophysical Journal, 109*, 1863–1872.

Tarafdar, P. K., Chakraborty, H., Dennison, S. M., & Lentz, B. R. (2012). Phosphatidylserine inhibits and calcium promotes model membrane fusion. *Biophysical Journal, 103*, 1880–1889.

Teixeira, V., Feio, M. J., & Bastos, M. (2012). Role of lipids in the interaction of antimicrobial peptides with membranes. *Progress in Lipid Research, 51*, 149–177.

Tristram-Nagle, S., & Nagle, J. F. (2007). HIV-1 fusion peptide decreases bending energy and promotes curved fusion intermediates. *Biophysical Journal, 93*, 2048–2255.

Tsai, M. J., & O'Malley, B. W. (1994). Molecular mechanisms of action of steroid/thyroid receptor superfamily members. *Annual Review of Biochemistry, 63*, 451–486.

Valdez-Taubas, J., & Pelham, H. (2005). Swf1-dependent palmitoylation of the SNARE Tlg1 prevents its ubiquitination and degradation. *EMBO Journal, 24*, 2524–2532.

Vance, J. E. (2008). Phosphatidylserine and phosphatidylethanolamine in mammalian cells: Two metabolically related aminophospholipids. *Journal of Lipid Research, 49*, 1377–1387.

van den Bogaart, G., Meyenberg, K., Risselada, H. J., Amin, H., Willig, K. I., Hubrich, B. E., Dier, M., Hell, S. W., Grubmüller, H., Diederichsen, U., & Jahn, R. (2011). Membrane protein sequestering by ionic protein-lipid interactions. *Nature, 479*, 552–555.

Vanderkooi, G., & Green, D. E. (1970). Biological membrane structure, I. The protein crystal model for membranes. *PNAS, 66*, 615–621.

Van Meer, G., Voelker, D. R., & Feigenson, G. W. (2008). Membrane lipids: Where they are and how they behave. *Nature Reviews Molecular Cell Biology, 9*, 112–124.

Vanni, S., Hirose, H., Barelli, H., Antonny, B., & Gautier, R. (2014). A sub-nanometre view of how membrane curvature and composition modulate lipid packing and protein recruitment. *Nature Communications, 5*, 4916.

Volovik, M. V., & Batishchev, O. V. (2024). Membrane activity of melittin and magainin-I at low peptide-to-lipid ratio: Different types of pores and translocation mechanisms. *Biomolecules, 14*, 1118.

von Blume, J., & Hausser, A. (2019). Lipid-dependent coupling of secretory cargo sorting and trafficking at the trans-Golgi network. *FEBS Letters, 593*, 2412–2427.

Watson, H. (2015). Biological membranes. *Essays in Biochemistry, 59*, 43–69.

Williamson, I. M., Alvis, S. J., East, J. M., & Lee, A. G. (2002). Interactions of phospholipids with the potassium channel KcsA. *Biophysical Journal, 83*, 2026–2038.

Wimley, W. C. (2010). Describing the mechanism of antimicrobial peptide action with the interfacial activity model. *ACS Chemical Biology, 5*, 905–917.

Xie, S., Naslavsky, N., & Caplan, S. (2015). Diacylglycerol kinases in membrane trafficking. *Cellular Logistics, 5*, e1078431.

Yang, S. T., Kiessling, V., Simmons, J. A., White, J. M., & Tamm, L. K. (2015). HIV gp41-mediated membrane fusion occurs at edges of cholesterol-rich lipid domains. *Nature Chemical Biology, 11*, 424–431.

Yang, S. T., Kiessling, V., & Tamm, L. K. (2016). Line tension at lipid phase boundaries as driving force for HIV fusion peptide-mediated fusion. *Nature Communications, 7*, 11401.

Zhao, Y. G., & Zhang, H. (2020). Phase separation in membrane biology: The interplay between membrane-bound organelles and membraneless condensates. *Developmental Cell, 55*, 30–44.

Zhukovsky, M. A., Filograna, A., Luini, A., Corda, D., & Valente, C. (2019). Phosphatidic acid in membrane rearrangements. *FEBS Letters, 593*, 2428–2451.

Zocher, M., Zhang, C., Rasmussen, S. G., Kobilka, B. K., & Müller, D. J. (2012). Cholesterol increases kinetic, energetic, and mechanical stability of the human β2-adrenergic receptor. *PNAS, 109*, E3463–E3472.

CHAPTER

7 Energy landscape and aggregation dynamics of TDP-43: Insights into proteins with intrinsically disordered regions in neurodegeneration

Meenakshi Pillai[a,b,c], Abhilasha A. Doke[a,b,d] and Santosh Kumar Jha[a,b,*]

[a]*Physical and Materials Chemistry Division, CSIR-National Chemical Laboratory, Pune, Maharashtra, India,* [b]*Academy of Scientific and Innovative Research (AcSIR), Ghaziabad, Uttar Pradesh, India,* [c]*Department of Biochemistry and Molecular Biology, Bloomberg School of Public Health, Johns Hopkins University, Baltimore, MD, United States,* [d]*Department of Biology, Indian Institute of Science Education and Research Pune, Pune, Maharashtra, India*

**Corresponding author. sk.jha@ncl.res.in*

7.1 Introduction

Neurons are unique, long-living cells in the human body that play critical roles in functions such as thinking, memory storage, and movement by communicating with other neurons in the brain. Despite their longevity the formation of new neurons in adults, known as adult neurogenesis, is limited to a few specific regions in the brain (Bergmann et al., 2015). As a result the loss of neurons in the body can have significant consequences, as even minor miscommunications between neurons can impair abilities such as thinking, recognition, and memory. These disruptions lead to conditions called neurodegenerative disorders, which result from the death of the neurons. As populations are growing and aging worldwide, the manifestation of neurodegenerative disorders increases continuously (Collaborators, 2016). The loss of neurons in the brain results in a wide spectrum of disorders, from movement impairment to dementia, where individuals lose memory or cognitive abilities. Each neurodegenerative disorder presents distinct symptoms and affects different brain regions (Fu et al., 2018). For instance, Parkinson's disease is characterized by neurodegeneration in the substantia nigra, whereas Huntington's disease primarily affects the basal ganglia (Fu et al., 2018).

Neurodegeneration is caused by a complex interplay among various factors such as mitochondrial dysfunction, ROS production, DNA damage, proteasomal dysfunction, and protein aggregation (Prasad et al., 2019). Among the many factors contributing to the complexity of the disease, one common phenomenon observed universally across neurodegenerative diseases is the process of protein aggregation. Protein aggregation resulting from disrupted proteostasis is a key factor in many neurodegenerative disorders, often driven by aging or chronic cellular stress (Balch et al., 2008). The mechanisms behind how cellular stress or aging disturbs protein homeostasis are not fully understood. It

Protein Dynamics: Experimental and Computational Approaches. DOI: https://doi.org/10.1016/B978-0-443-30042-4.00002-X

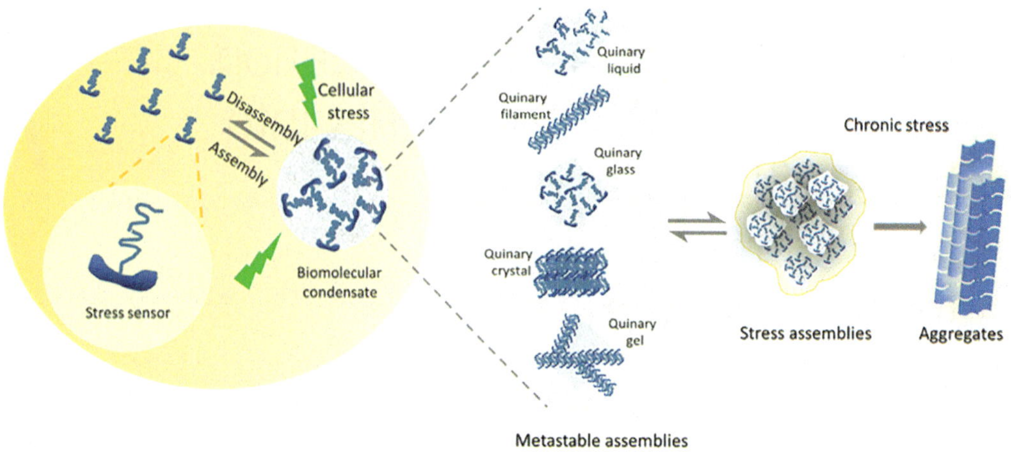

FIGURE 7.1

Schematic representation of the proposed mechanism for cellular stress sensing at the molecular level. Certain proteins are believed to function as biosensors within cells, undergoing site-specific modifications in response to subtle environmental changes, such as shifts in temperature, pH, or oxidative stress. Depending on the type of stress encountered, these modified proteins may assemble into metastable structures with different conformations. Under chronic or severe stress, however, these modifications can result in the formation of stable, irreversible aggregates.

is proposed that certain proteins act as biosensors within cells, undergoing site-specific modifications in response to subtle environmental changes (Fig. 7.1) such as variations in temperature, pH, or oxidative stress (Rabouille & Alberti, 2017).

Depending on the type of stress detected, these modified proteins can form metastable assemblies with a range of possible conformations (Fig. 7.1), allowing the cell to temporarily adapt to challenging conditions (Wheeler et al., 2016). However, when stress is chronic or overwhelming, these modifications may push the proteins toward forming stable, irreversible aggregates (Rabouille & Alberti, 2017) (Fig. 7.1). Over time, these aggregates accumulate, disrupting cellular function and contributing to the development of neurodegenerative diseases. Understanding how these biosensor proteins detect environmental changes and transition from adaptive conformations to harmful aggregates is crucial for developing strategies to prevent or treat these disorders.

Deposition of different proteins, such as Aβ, α-synuclein, TAR DNA-binding protein 43 (TDP-43), fused in sarcoma (FUS), superoxide dismutase (SOD1), and islet amyloid polypeptide precursor (IAPP), has been found in different diseases such as Alzheimer's, Parkinson's, amyotrophic lateral sclerosis (ALS), and frontotemporal lobar degeneration (FTLD) (Prasad et al., 2019; Zerovnik, 2002). These are termed conformational diseases because they involve a change in the conformation of a normally folded soluble polypeptide into predominantly β-sheet-rich insoluble aggregates (Uversky, 2010; Zerovnik, 2002). One such conformational disease is ALS, whereby 97% of diseased patients show deposition of TDP-43 (Ling et al., 2013). TDP-43 is a highly conserved multidomain protein involved in ALS and FTLD. It is found to be codeposited in many other diseases together termed "TDP-43 proteinopathies" (Arai et al., 2006; Gao et al., 2018; Neumann et al., 2006; Sun & Chakrabartty, 2017). Although recent studies have suggested that the protein also undergoes a passive diffusion between the nucleus and the cytoplasm to perform its function, it becomes primarily localized to the cytoplasm under diseased conditions (Ederle et al., 2018; Pinarbasi et al., 2018). TDP-43 performs a host of functions such as RNA binding, splicing, regulation of transcription, and translation (Sun & Chakrabartty, 2017). To perform such multiple functions in the cell, it binds to different RNA/DNA through its RNA recognition motif (RRM) domain and carries out protein–protein interactions through its C-terminal domain (CTD). Thus a multidomain protein like TDP-43 with diverse functions might have a complex energy landscape. To add to this complexity the protein also undergoes aggregation in many neurodegenerative diseases. Studying the different conformations that the protein can adapt thus becomes crucial to devise a strategy to overcome the aggregation process. However, because of the poor solubility and high aggregation propensity of the full-length protein, the structural and aggregation data are available mostly only for individual domains or modified full-length proteins. In this study, we review existing research on TDP-43 protein and reveal some of the possible conformations that the protein can adapt during misfolding and aggregation.

7.2 Full-length TDP-43: Environmental stress sensor

Full-length TDP-43 has a molecular weight of 43 kDa and is made up of 414 amino acids. TDP-43 protein consists of four domains (Fig. 7.2): an N-terminal domain (NTD) (1–102), two RRMs, that is, RRM1 (104–176) and RRM2 (192–262), and a CTD (274–414) (Kuo et al., 2009; Lukavsky et al., 2013; Qin et al., 2014).

Despite its inherent tendency to aggregate (Johnson et al., 2009), TDP-43 is crucial for cell function and survival under normal physiological conditions (Ratti & Buratti, 2016; Susnjar et al., 2022). The formation of TDP-43 aggregates in specific cells, particularly in neurodegenerative disorders such as ALS and FTLD (Fu et al., 2018), supports the idea that TDP-43 might act as an environmental stress sensor. Moreover, early research into the aggregation mechanism of TDP-43 has shown that cosolutes such as electrolytes and metal ions can influence its aggregation (Doke & Jha, 2024; Sun et al., 2019), impacting the equilibrium between its various structural states. TDP-43 exhibits a range of structural states, from soluble native forms such as monomers, dimers, and oligomers to larger self-assemblies such as aggregates and phase-separated structures (Doke & Jha, 2023; Pillai & Jha, 2024), adding complexity to the proteostasis network of the full-length protein in the cell. Among these various structural states, significant heterogeneity has been observed in terms of shape, size, and biochemical properties (Doke & Jha, 2023). This diversity hints at a potential role for TDP-43 as a cellular stress sensor. Although experimental evidence supporting the hypothesis that TDP-43 acts as an environmental stress-sensing protein is limited, recent studies have highlighted the protein's sensitivity to local environmental conditions (Doke & Jha, 2022, 2024).

The complete 3D structure and detailed structural information of the full-length TDP-43 protein are still largely unknown, primarily because of its strong tendency to aggregate. Although most structural insights come from studies on various fragments of the full-length protein (Fig. 7.2), a recent small-angle X-ray scattering (SAXS) study provided one of the first glimpses of the full-length protein's low-resolution structure (Wright et al., 2020). However, high-resolution data is still required to fully understand the arrangement of its domains and the protein as a whole. Along with the missing structural

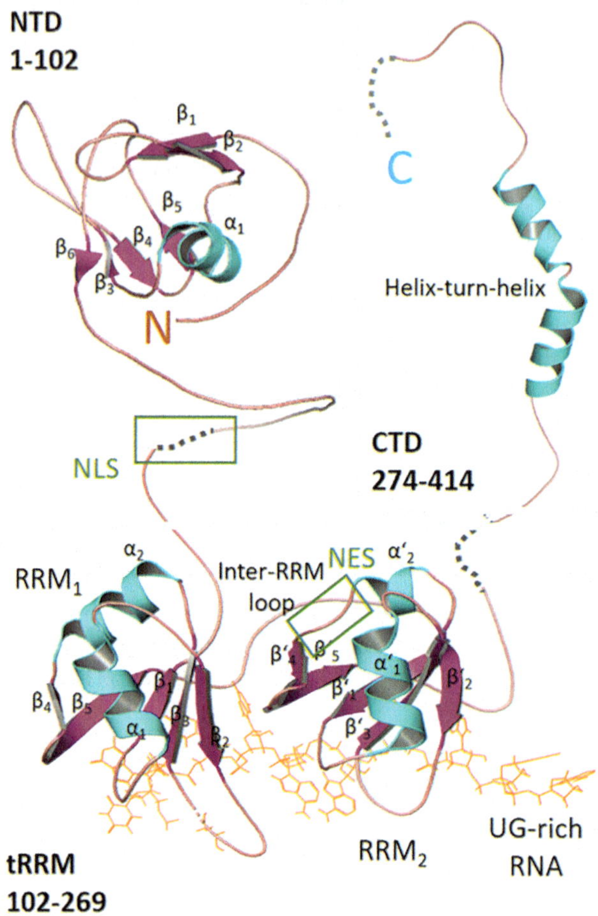

FIGURE 7.2

3D structure of TDP-43. Pictorial representation of the 3D structure of full-length TDP-43, which includes solution Nuclear Magnetic Resonance (NMR) structure of NTD (Protein Data Bank (PDB) file entry 5MRG), tRRM (PDB file entry 4BS2), and CTD (PDB file entry 2N3X). *CTD*, C-terminal domain; *NTD*, N-terminal domain; *RRM*, RNA recognition motif; *tRRM*, tethered RRM.

From Doke, A. A., & Jha, S. K. (2023). Shapeshifter TDP-43: Molecular mechanism of structural polymorphism, aggregation, phase separation and their modulators. Biophysical Chemistry, 295. *https://doi.org/10.1016/j.bpc.2023.106972.*

information the mechanism of aggregation of full-length protein is missing. Early studies on TDP-43 primarily focused on modified versions of the protein, using mutations or solubilization tags to manage its highly aggregation-prone nature (Sun & Chakrabartty, 2017; Wright et al., 2020). However, recent research has highlighted that TDP-43 can adopt various structural forms depending on environmental conditions. For instance, it has been observed in monomeric form under specific buffer conditions, with a chromatographic SAXS study revealing that a tryptophan-free variant of full-length TDP-43 predominantly exists as a soluble monomer (Wright et al., 2020). Other studies have shown that TDP-43 can exist as monomers and dimers and, under certain conditions, form larger assemblies (Doke & Jha, 2022; French et al., 2019; Johnson et al., 2009).

Research on unmodified full-length TDP-43 has been limited, but emerging studies (Doke & Jha, 2022; Staderini et al., 2022; Vega et al., 2019; Wright et al., 2020) suggest that the local environment plays a significant role in determining the protein's conformation and aggregation behavior. Understanding these influences is crucial for targeting TDP-43 aggregation, especially given its central role in several neurodegenerative disorders. Detailed structural and molecular insights are needed to fully grasp the mechanisms behind TDP-43 aggregation and its contribution to disease. In the absence of comprehensive studies on full-length TDP-43, researchers have often used fragments or individual domains to explore their roles in aggregation mechanisms and structural diversity (Kuo et al., 2009; Mompeán et al., 2014, 2016a). These studies have shown that certain domains can sense environmental changes and undergo structural modifications under stress-like conditions. The structure and function of each domain in TDP-43 aggregation are further detailed below.

7.2.1 Structure and role of the N-terminal domain in TDP-43 aggregation

The NTD of TDP-43 is made up of 102 amino acids (Fig. 7.2). The structure of the residues 1–77 has been studied by two independent groups with some contrasting results because of the different experimental conditions used. The high-resolution structure by Mompean et al. shows the residues adopting an axin-like fold consisting of $\beta1\beta2\alpha1\beta3\beta4\beta5\beta6$ topology, whereas the low-resolution structure showed $\beta4$ and $\beta5$ as a single β strand (Mompeán et al., 2014, 2016b; Pantoja-Uceda et al., 2021; Qin et al., 2014). The low-resolution structure described NTD as an ubiquitin-like fold present in equilibrium with an unfolding species (Qin et al., 2014). The remaining residues (78–102), however, are fairly disordered (Fig. 7.2) and rich in positively charged amino acids with the ability to bind nonspecifically to DNA at low pH (Mompeán et al., 2016a; Qin et al., 2014).

The NTD of TDP-43 undergoes dimer formation in the cell (Shiina et al., 2010). Indeed the functional state of TDP-43 is reported to be a homodimer mediated by NTD (Wang et al., 2013; Zhang et al., 2013). Multiple studies have shown that the NTD undergoes an oligomerization process both in vivo and in vitro (Afroz et al., 2017; Chang et al., 2012; Wang et al., 2018). NTD mediates the formation of homo-oligomers arranged in a head-to-tail fashion under physiological conditions in the cell and is independent of oxidation of cysteine residues and represents functional form of the protein (Jiang et al., 2017; Wang et al., 2018). The interaction between the domains occurs through the positively charged head (Arg52 and Arg55) and the negatively charged tail (Glu14, Glu17, and Glu21). This interaction interface is evolutionarily preserved, thus signifying a conserved mode of oligomerization for TDP-43 protein (Afroz et al., 2017). Formation of such TDP-43 polymers is vital for its splicing activity (Afroz et al., 2017; Jiang et al., 2017; Wang et al., 2018). Precisely the first 10 residues help in dimerization and recruitment of other RNA-binding proteins to carry out splicing activity (Zhang et al., 2013). But, however, this might also increase the local concentration of the protein, resulting in oligomerization of the NTD. This oligomerization of NTD is concentration-dependent (Chang et al., 2012) and also modulated by phosphorylation of the Ser48 residue (Wang et al., 2018). The formation of NTD-mediated oligomerization has been shown to be spatially separating the CTD and thus preventing formation of pathological aggregates (Afroz et al., 2017). An all-atom MD simulation of NTD at high temperature revealed presence of two metastable intermediates stabilized by nonnative interactions, one of which shows structural resemblance to a molten globule (MG) state. These intermediate ensembles may lead to formation of irreversible aggregates (Prakash et al., 2018).

7.2.2 Structure and role of the intrinsically disordered C-terminal domain in TDP-43 aggregation

The CTD of TDP-43 protein is intrinsically disordered (Fig. 7.2) in nature and has been shown to be similar to the prion domain of yeast prion proteins (Cushman et al., 2010; Lim et al., 2016; Mompeán et al., 2014). The CTD is majorly studied because of the many disease-associated mutations found within its intrinsically disordered region (IDR). The residues between 318 and 369 is known to be the amyloidogenic core of the protein because they contain a QN-rich segment known to form amyloid-like β-sheet structures. However, whether these regions form amyloid or nonamyloid fibrils remains highly debatable (Conicella et al., 2016; Jiang et al., 2013). Particularly the residue 320–343 within the amyloidogenic core forms a transient alpha-turn-alpha helix, which has been shown to undergo α-to-β transition and also participate in the formation of RNP granules (discussed later in the phase separation section) (Conicella et al., 2016; Jiang et al., 2013). Interestingly the CTD has been shown to self-assemble into dynamic oligomers, and the presence of mutations (A315E, Q331K, and M337V) simply accelerates the propensity of oligomerization (Lim et al., 2016). However, in the presence of nucleic acids the CTD forms a hydrogel, whereas the mutants undergo the formation of irreversible aggregates. This binding to nucleic acid might effectively increase the local concentration of the protein thereby increasing the propensity to aggregate (Lim et al., 2016). A study using different peptides of CTD suggested that the ability to form ordered fibrils or amorphous aggregates also varies with the type of peptide in study (Chen et al., 2010). A minimal segment/peptide required for TDP-43 aggregation was found when expressed in yeast and neuroblastoma cells (Johnson et al., 2008; Nonaka et al., 2009; Zhang et al., 2009). This segment consists of certain residues from RRM2 and the entire CTD domain. Another major reason why CTD is majorly studied is because of the presence of 35 and 25 kDa C-terminal fragments in the inclusions of the diseased patients generated by caspase cleavage (Zhang et al., 2007, 2009). But it is important to understand that the caspase cleavage site is located within the RRM domain, hinting that the early steps of aggregation involve the exposure of caspase cleavage site upon misfolding of the tethered RRM (tRRM) domain.

Although CTD has been considered to be the hallmark of disease pathology, it does not appear to be causal factor for disease onset (Berning & Walker, 2019). The CTD is extensively researched for its role in the aggregation process, primarily because of the large number of disease mutations present within this region. But it is interesting to understand that being IDR, the mutations can be better tolerated in the CTD in otherwise highly conserved TDP-43 protein (Buratti, 2015). It may be possible that mutations simply modulate the disease condition, as the same mutations have been shown to have different pharmacological/pathological effects in different patients (Buratti, 2015). Remarkably, around 30% of the proteins in humans that contain the IDP region also have a very well-structured RRM domain (Prasad et al., 2019). Many of these proteins, such as ataxin 1, ataxin 2, hnRNPA1, hnRNPA2/B1, and FUS/TLS, have been shown to be involved in various neurodegenerative diseases (Kim et al., 2013;

Millecamps et al., 2010; Tsai et al., 2011). Therefore it would be crucial to also focus on the RRM domain and study its role in the overall function and aggregation of TDP-43 protein.

7.2.3 Structure and role of the RNA-binding domain in TDP-43 aggregation

RRM domains or tRRM domains belong to a conserved set of amino acids that are known to bind to nucleic acids (Maris et al., 2005). Around 2% of the human protein coding gene codes for RRM domain (Maris et al., 2005). TDP-43 contains two RRM domains folded into five β-stranded sheets along with two α helices (Fig. 7.2). Unlike the typical four strands present in an RRM domain (Maris et al., 2005), there is an additional β strand present in TDP-43, which is believed to give stability to the protein (Shodai et al., 2013). The two RRM domains have the ability to bind UG- and TG-rich sequences (Kuo et al., 2009; Lukavsky et al., 2013). RRM1 binds to UG_6 with K_d of 65.2 nM, whereas RRM2 binds to UG_3 with K_d of 379 nM (Kuo et al., 2009). RRM1 contains a longer loop 3 region as compared to RRM2, which is believed to give it a higher affinity to bind to the targets. However, when both the RRM domains bind to UG_6, there is a synergistic binding effect that drastically reduces the K_d to 14.2 nM (Kuo et al., 2009). Interestingly the TDP-43 RRM domains bind to RNA in a reverse manner unlike the tandem RRM domains that usually bind in $3'$–$5'$ direction (Lukavsky et al., 2013). This reverse interaction allows the 15 residue linker present between the 2 RRM domains to participate more extensively with the nucleic acid targets (Lukavsky et al., 2013).

A recent Nuclear Magnetic Resonance (NMR) study coupled with X-ray fiber diffraction data points out a strong similarity in the structure of amyloid fibrils formed by full-length TDP-43 and TDP-35 (90–414 residues) (Shenoy et al., 2020). The C-terminal showed polymorphic structure upon forming fibrils alone or in combination with other domains. This points out that both the N-terminal and RRM domains are important and crucial in the amyloid transition of TDP-43 (Shenoy et al., 2020). In addition, other studies have also shown that the binding of nucleic acid to the protein prevents the aggregation, indirectly suggesting a role for the RRM domain in aggregation (Huang et al., 2013; Pillai & Jha, 2019; Sun et al., 2014). These domains have been studied individually (Zacco et al., 2018) and in tandem (Pillai & Jha, 2019, 2020; Zacco et al., 2018) and shown to undergo aggregation. Two regions within the RRM domain of TDP-43 have been identified to be playing an important role in the aggregation process, 166–173 in RRM1 and 246–255 in RRM2 (Shodai et al., 2012, 2013). The disease-linked mutation D169G in the RRM1 domain does not affect the aggregation process but induces a local conformational change because of the loss of hydrogen bond with T115 residue. It also increases the caspase cleavage activity, a process known to trigger aggregate formation (Chiang et al., 2016; Shodai et al., 2013). Interestingly, these regions are also shown to undergo oxidation and result in oligomerization of the protein (Chang et al., 2013). Cys173 of RRM1 domain gets preferentially oxidized, resulting in conformational change, and further oxidation of Cys175 results in the formation of the insoluble RRM1 dimer aggregate. Other cysteines present in the RRM2 and NTD are partially accessible, and hence, their oxidation does not affect the protein. However, oxidation of a buried cysteine residue results in the change of local electrostatics and results in exposure of hydrophobic patches because of destabilization of the hydrophobic interactions (Chang et al., 2013). The two cysteines, C173 and C175, make contacts with residues in the RRM domain, and it has been suggested that these contacts are vital and loss of these contacts because of oxidation can increase the rate of aggregation (Lukavsky et al., 2013). It has been suggested that the free thiols of C173 and C175 govern the conformation of the RRM1 domain, and the presence of intra- or intermolecular disulfide bond can cause the protein to undergo aggregation (Shodai et al., 2013).

Intriguingly, many RNA-binding proteins have been found to undergo aggregation in a number of different neurodegenerative diseases (King et al., 2012; Lukong et al., 2008). A recent study on RRM domains of multiple aggregation-prone proteins such as TDP-43, FUS, and RBM-45 was shown to form amyloid-like fibrils using X-ray diffraction and Fourier Transform Infrared (FT-IR) spectroscopy (Agrawal et al., 2019). Structural studies have revealed that RRM2 domain of TDP-43 populates an unfolding intermediate at 4 M GdmCl (Mackness et al., 2014). Multiple high energy microstates were characterized at 6 M urea concentration using RAM simulations coupled to chemical restraints from NMR (Tavella et al., 2018). Although these states were categorized into three regions depending on the type of structure present, a common feature among all these states showed exposure of hydrophobic regions on strands $\beta3$, $\beta4$, and $\beta5$ that are prone to aggregation in the solvent. These results suggest that RRM2 unfolding may drive the protein to aggregate formation (Tavella et al., 2018). In line with the above results, it was shown by an MD simulation study that RRM2 forms multiple metastable states governed by nonnative contact and hydrogen bond formation (Prakash et al., 2019). Moreover, it has also been seen experimentally that the partial destabilization of TDP-43 occurs because of binding of zinc to RRM domain. Such partial unfolding of the protein leads to formation of ThT-positive aggregates (Garnier et al., 2017). There are multiple factors that can bring about initiation of RRM misfolding, but we do not understand how these factors affect the protein at the molecular level. In an attempt to understand the stepwise change that the RRM undergoes during misfolding, a recent study showed that during the pH-induced stress, the ionizable residues in the RRM domains get protonated, which is coupled to a partial unfolding reaction (Pillai & Jha, 2020). This results in the formation of MG-like species, which are quite metastable in nature. Under salt stress, these molecules assemble through their hydrophobic patches to form ThT positive worm-like fibrils of around 20 nm in size (Pillai & Jha, 2020). The transition to fibril formation occurs in a multistep fashion involving an increase in the ellipticity, indicating the conversion of disordered regions into a more ordered structure, along with a growth in size (Pillai & Jha, 2023). Interestingly, upon binding to DNA, the formation of amyloid fibril is diminished, indicating the role of the DNA-binding region in the aggregation process (Pillai & Jha, 2019). Modulating electrostatics by adjusting the pH and ionic strength of the solution leads the protein to adopt different conformational changes, with the population and stability of these conformations being determined by the local environmental conditions (Pillai et al., 2023).

In spite of many studies focusing on the individual domains and their ability to undergo aggregation, comprehensive studies targeting the aggregation pathway and mechanism are lacking. Thus it will be highly interesting to understand the different number of steps and, consequently, the types of species formed during the transition from monomeric to oligomeric state of TDP-43. In general the aggregation process is quite complex; irrespective of the initial state of the protein, the final aggregated state is often rich in cross β-sheet structure. Hence studying the pathway will throw some light on the molecular process that generates the final irreversible fibrils.

7.3 Aggregation pathway in the context of energy landscape

The search of an unfolded polypeptide to form a stable native state is described by a widely accepted concept/unified framework of energy landscape (Fig. 7.2) (Leopold et al., 1992; Wolynes et al., 1995).

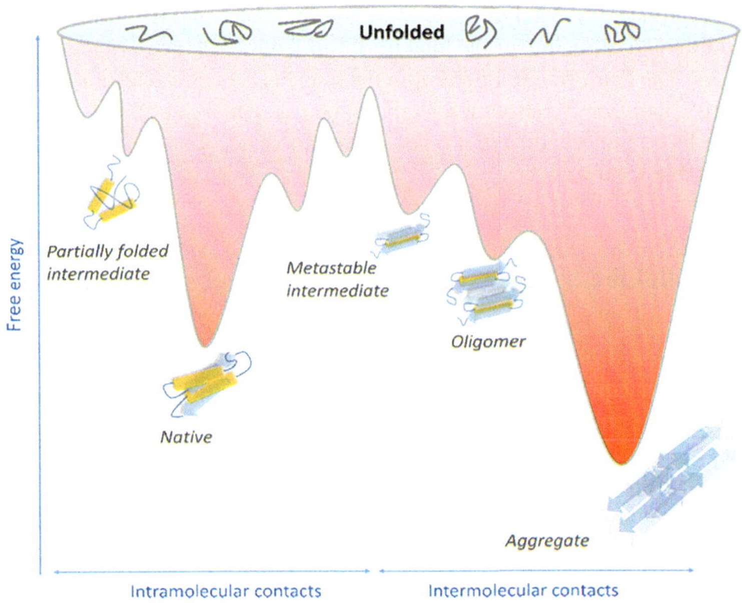

FIGURE 7.3

The folding funnel indicates the search of a polypeptide to reach the functional native state. The various energy minima indicate the stability of different types of intermediates along the folding and the aggregation energy landscape.

Features of an energy landscape for a peptide are highly dependent on the amino acid sequence, number of amino acids, solution condition, etc. The relative probabilities of multiple conformational states (thermodynamics) and the barriers (kinetics) between these states define the energy landscape (Jahn & Radford, 2005). Under unfolding condition the polypeptide has a high degree of freedom (entropy) and can achieve a number of conformations. But as the protein folds, the number of conformations that the polypeptide can achieve decreases, and this decrease in entropy determines the shape of the energy surface. However destabilizing solution conditions, mutations, oxidative stress, or posttranslational modifications might modulate the folding energy landscape, making the aggregation energy landscape easily accessible. Many native-like monomers or oligomers prone to aggregation have been trapped for different aggregation-prone proteins (Chiti & Dobson, 2009).

The process of aggregation is a complex phenomenon involving multiple steps. This is reflected in the aggregation energy landscape, which consists of many poorly defined energy minima representing the various species formed during the transition from soluble monomeric protein to highly stable ordered aggregates (Fig. 7.3).

Many different classes of oligomeric intermediates have been observed, and they have been classified according to their size, shape, and various biochemical and biophysical properties. Because of the transient nature of intermediates during the aggregation, it has become challenging to characterize them. However, the intermediates formed during the aggregation pathway have been found to be more toxic than the final fibrillar species (Guerrero-Muñoz et al., 2015; Kayed & Lasagna-Reeves, 2013;

Winner et al., 2011). Hence it is vital to study these intermediates and determine the characteristics of these states. In addition, the structure of the final fibril formed is often determined by the structure of the initial intermediates. The conformation present in the nucleus acts as a scaffold for the development of final fibril structure. Therefore the structural study of intermediates can offer a clue to the final fibrillar structure. We have listed out few different types of intermediates that the protein can undergo before forming the most stable aggregate.

7.3.1 Partially unfolded states

Studies have reported that partial unfolding reaction is the first major barrier for the aggregation of globular proteins (Dobson, 1999; Kelly, 1998; Uversky & Fink, 2004). Under physiological conditions, globular proteins are in equilibrium with partially unfolded molecules generated because of the conformational breathing process. However, the amount of such partially unfolded molecules is very low (Chamberlain et al., 1996; Uversky & Fink, 2004). This equilibrium, however, can be modulated in the presence of denaturant (Hamada & Dobson, 2002), mutation, and DNA binding (Nandi et al., 2002). Not many structural studies characterizing the partially unfolded states are available for TDP-43. However, few studies have revealed how partial unfolding under different stress (pH and denaturant) results in the formation of metastable states that are prone to undergo aggregation (Doke & Jha, 2024; Pillai & Jha, 2019, 2020, 2023). This has been discussed already in the RRM section. In brief, a partial destabilization reaction results in the exposure of hydrophobic patches that are otherwise buried in the core of the protein. This exposure of hydrophobic patches allows the protein molecules to come together and undergo an aggregation process.

7.3.2 Molten globule states

MGs are conformers that have an intact secondary structure and a compromised tertiary structure resulting in exposure of hydrophobic surfaces (Baldwin & Rose, 2013; Christensen & Pain, 1991; de Oliveira & Silva, 2017). Exposure of nonpolar groups and overall disruption in the packing can trigger the formation of higher order assemblies for many proteins. Dynamic MG competent states have been preferred by the proteins as a means to survive in the various compartments of the cell, which possess distinct chemical environments and pH ranges (de Oliveira & Silva, 2017). MG states of the protein are known to interact with the chaperones (Ellis, 1997). Indeed, one of the roles of chaperones inside the cell is to bind to MG-like misfolded conformations of the protein and help them to achieve a proper folded conformation. As an example, the enzyme glucosyltransferase recognizes the exposed hydrophobic patches on the MG-like conformers of the nascent glycoproteins in the cell (Caramelo et al., 2003). All these experimental studies prove the presence of MG-like conformations inside the cell. These MG-like conformers have been shown to be more prone to undergo aggregation for certain proteins. The protein p53 has been found to undergo preamyloidogenic MG-like conformers in the presence of denaturant and hydrostatic pressure (Pedrote et al., 2018). Formation of such species shifts the equilibrium toward aggregation, resulting in pathological implications. In the case of p53, MG states have also been found inside the cells (Ano Bom et al., 2010). Such presence of MG-like conformers has also been seen for the tRRM domain of TDP-43 (Pillai & Jha, 2019, 2020) and full-length protein (Doke & Jha, 2024). At low pH the protein has been shown to undergo a partial unfolding reaction, populating species with MG-like features (Fig. 7.4).

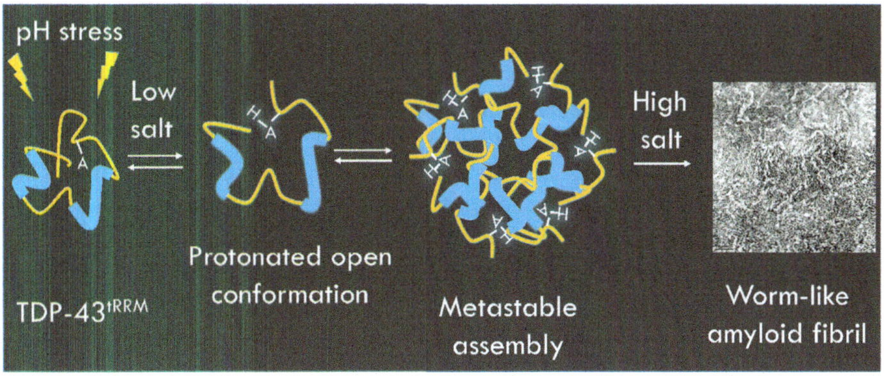

FIGURE 7.4

The illustration here represents how protonation–deprotonation reactions are linked to assembly–disassembly of TDP-43 tRRM. The partial opening of the protein causes the protein molecules to come together in an oligomeric complex termed "metastable assembly," which can undergo the formation of worm-like fibrils. *tRRM*, Tethered RRM.

Reproduced with permission from Pillai, M., & Jha, S. K. (2020). Early metastable assembly during the stress-induced formation of worm-like amyloid fibrils of nucleic acid binding domains of TDP-43. Biochemistry, 59*(3), 315–328. https://doi.org/10.1021/acs.biochem.9b00780. Copyright 2020 American Chemical Society.*

However, the exposed hydrophobic surfaces of the proteins in the MG conformation trigger it to undergo an oligomerization reaction to form amyloid fibrils (Doke & Jha, 2024; Pillai & Jha, 2019, 2020). The formation of MG-like states has been linked to the formation of aggregates for other proteins such as stefin B and β2-microglobulin (Čeru & Žerovnik, 2008; Skora et al., 2010).

7.3.3 Oligomeric intermediates

Full-length TDP-43 protein has been shown to form spherical oligomers of size 40–60 nm in diameter with exposed hydrophobic patches under in vitro and in vivo conditions (brains of FTLD-TDP patients and transgenic mice brains) (Fang et al., 2014). These oligomers are highly toxic and able to seed Aβ to form amyloid oligomers (Fang et al., 2014). Similar prion-like activity for TDP-43 has been shown recently in the diseased brain of patients (Nonaka et al., 2013). In line with these studies a spectrum of oligomers in the brain tissue samples has been identified and studied (Afroz et al., 2017; French et al., 2019; Guo et al., 2011). Fascinatingly, TDP-43 was shown to undergo the formation of amyloid-like oligomeric assemblies termed "myogranules" in the regenerating muscle cells; however, they lacked the typical cross β-sheet structure as seen by X-ray data (Vogler et al., 2018). These assemblies were functional and found to interact with sarcomeric mRNAs in the cell. The assemblies were able to act as seeds to initiate the amyloid-like fibril formation, establishing a link between the functional and the pathological TDP-43 aggregates (Vogler et al., 2018). The formation of physiological oligomeric assemblies may serve to increase the local concentration of protein, thereby facilitating the RNA-binding activity of the protein. However, the structural differences between the physiological and the pathological oligomers with respect to their size, shape, dynamics of side chains, affinity toward the nucleic acid, and ability to cross seed other proteins are not clearly distinguished. It will be important to understand the nature of the initial structural changes that consequently define the kind of oligomers that are formed. The type of interactions that dominate or drive the assembly formation often describes material property of the assemblies, and studying the characteristics of these oligomers will help us to target them.

For many aggregation-prone proteins a range of oligomeric species formed at different steps of the aggregation pathway has been studied (Cho et al., 2011; Frare et al., 2009; Jahn et al., 2006; Lee & Eisenberg, 2003; Saelices et al., 2015; Tanaka et al., 2003). Numerous studies have shown that the oligomers but not the final fibrils are toxic in nature (Guerrero-Muñoz et al., 2015; Kayed & Lasagna-Reeves, 2013; Winner et al., 2011). Some of the studied mechanisms through which oligomers act include disruption of intracellular and extracellular membranes, receptor-mediated toxicity by interacting with neuronal receptors, compromised conversion of autophagosomes to lysosomes because of oligomer accumulation, and impairment of P/Q-type calcium currents (Uversky, 2010). A similar study for TDP-43 will help us understand if the oligomers are toxic or nontoxic. It will also be interesting to understand the mechanism by which the oligomers operate to create toxicity in the cell.

7.3.4 Phase-separated assemblies

An emerging phenomenon of phase separation is being linked to a range of neurodegenerative diseases. Many RNA-binding proteins containing both RRM and an IDP region, such as FUS, hnRNPA1, hnRNPA2/B1, and TDP-43, have the ability to undergo phase separation (Prasad et al., 2019). There are many different types of assemblies that can be formed, such as liquid droplets, hydrogels, amyloid, or amorphous aggregates, depending upon the type of intermolecular interactions present within the assemblies. Liquid–liquid phase separation (LLPS) is majorly governed by weak transient intermolecular interactions such as cation–pi, pi–pi, hydrophobic, and electrostatic interactions (Shin & Brangwynne, 2017; Simon et al., 2017). The ability of the RNA-binding protein to phase separate is implicated in stress granule formation (Lin et al., 2015; Molliex et al., 2015). Stress granules are bimolecular condensates formed as a result of proteins experiencing stress and phase separating into membraneless condensates (Dewey et al., 2012). We do not yet understand clearly the mechanism of stress detection by proteins, but studies suggest that proteins respond to stress by undergoing phase separation.

Interestingly, more than 95% of the ALS cases are caused because of environmental stress (Sun & Chakrabartty, 2017). TDP-43 acts as a stress sensor and has been shown to codeposit with other stress granule proteins, indicating that stress is one of the major causal factors for ALS disease. Under in vitro conditions, TDP-43 is also shown to be recruited to stress granules in response to heat stress, oxidative stress, and chemical stressors (Colombrita et al., 2009; Dewey et al., 2011). In addition, TDP-43 is known to act as a regulator of stress granule formation (McDonald et al., 2011). The ability of LLPS formation of TDP-43 is highly sensitive, as supported by weak transient nature of interactions that govern the LLPS. It has been shown that a phosphomimetic substitution at S48 in NTD drastically reduces the LLPS formation of TDP-43 and decreases the polymerization of NTD (Wang et al., 2018). Although we do not understand the structural changes that occur in the protein undergoing LLPS, an NMR study on the residues 267–414 has shown that the liquid droplets are formed through a partially populated α-helix initiated by addition of RNA extract or salt (Conicella et al., 2016). Mutations within these residues (A321G, Q331K, and M337V) associated with the diseased state reduced the phase separation while aggregating the protein (Conicella et al., 2016).

A very interesting study on the axons of primary neurons revealed that the binding of RNA to TDP-43 is prerequisite in the formation of RNP granules, indicating the importance of RRM domain in TDP-43 (Gopal et al., 2017). Remarkably the biophysical characteristics of RNP transport granules varied depending on the location of the granule (proximal or mid-axonal region). These differences in the granule properties were observed to be dependent on the subcellular context and altered over time (Gopal et al., 2017). Certain disease-associated mutations (M337V and G298S) caused an increase in granule viscosity and disturbed the axonal transport functions, suggesting that altered viscoelastic properties of granules may cause disease pathogenesis (Gopal et al., 2017). Thus the formation of biomolecular condensates appears to be transient in nature, governed by weak interactions that can be modulated by the local environment. In general, these liquid droplets become less dynamic over time and age to undergo a liquid-to-solid transition to form irreversible aggregates (Bentmann et al., 2013; Li et al., 2013).

7.4 Amyloid fibrils or amorphous aggregates?

Although the native state of the protein is considered to be the most stable state under physiological conditions as compared to the unfolded states, recent studies have proved that the native states may only be metastable with respect to the amyloid fibrils because of the kinetic barrier associated with self-assembly (Baldwin et al., 2011). The barrier becomes even higher because of the cell quality control mechanisms or chaperones, which decrease the chances of aggregation in the normal life span of the cell. However, under certain conditions such as cellular stress, mutations, or aging, the barrier might reduce, and thus, the proteins can easily access the aggregation energy landscape in biologically relevant time scale, resulting in the formation of thermodynamically most stable amyloid fibrils (David et al., 2010; Narayanaswamy et al., 2009). The term "amyloid" refers to the formation of cross-β-sheet structure that has capacity to bind Congo red and Thioflavin T dye. Interestingly, many recent studies prove that the polypeptides in general have the ability to form amyloid fibrils under appropriate conditions (Dobson, 1999, 2001; Fändrich et al., 2001; Pertinhez et al., 2001). Early studies of TDP-43 deposition in cells were found to be nonamyloid because it did not bind to amyloid specific dyes such as Congo red and ThT (Johnson et al., 2009; Neumann et al., 2006). However, ThT S/ThT T staining amyloid-like aggregates have been reported from ALS cases recently (Bigio et al., 2013; Robinson et al., 2013). Thus it is not clear if the TDP-43 aggregates deposited in the cells are amyloid or nonamyloid (Prasad et al., 2019; Sun & Chakrabartty, 2017). In addition, different variants/peptides of TDP-43 protein have also been found to be binding to the amyloid dye differentially (Prasad et al., 2019; Saini & Chauhan, 2014). A segment of RRM2 domain of TDP-43 (residues 247–257) has been found to adopt different backbone conformations, resulting in the formation of amyloid polymorphs (Guenther et al., 2018). Such a polymorphic nature of the peptide often results in differences in the structure of the final fibril. Studies have shown that clinical presentations, morphology, and the neuroanatomical distribution of pathological TDP-43 vary, resulting in disease subtypes. This disease heterogeneity may be because of the different types of polymorphic structure the aggregated protein can adopt. These differences often stem from the differences in the nature of initial assemblies that are formed. It, thus, becomes crucial to understand the nature of the early steps involved in the transition of monomeric protein.

7.5 Kinetic models to describe the molecular mechanism of the aggregation pathway

To understand the nature of transition states and kinetic barriers between different conformational states on the energy landscape, kinetic studies have proved to be an elegant tool (Kumar & Udgaonkar, 2010; Zerovnik, 2002). The process of protein aggregation involving interactions between an ensemble of protein molecules can be quite complex. There is a multitude of different species that can be populated as the protein undergoes conformation and structural transition to form aggregates. Several models have been proposed (Kumar & Udgaonkar, 2010), which are built by monitoring the process of fibrillation using a range of spectroscopic, microscopic, and amyloid-specific dyes (Lomakin et al., 1997; Modler et al., 2003; Scheibel et al., 2004). Based on the polymerization theory, which is often used to evaluate the aggregation features, we can define two basic models (Fig. 7.5).

7.5.1 Linear (isodesmic) polymerization

An isodesmic polymerization mechanism does not show the presence of any lag phase (Cho et al., 2011; Frare et al., 2009; Saelices et al., 2015). It can be considered to be similar to the elongation step of a nucleation-dependent polymerization (NDP) reaction. In the case of the linear polymerization reaction, no critical concentration of protein is required. The rate of the aggregation reaction is fastest at the start when the concentration of monomers is highest; thereafter the reaction reaches equilibrium, and the rate subsequently decreases. It is important to note that NDP and linear polymerization are the two extreme cases that can be used to represent the aggregation mechanism. However, it is equally possible that a particular aggregation kinetic might involve both NDP and linear polymerization mechanisms (Frieden, 2007).

Based on the different kinetic models adopted by the protein during aggregation, few other mechanisms have been proposed. These models have been developed considering the structural features of the species undergoing aggregation and the rate-limiting step of the aggregation reaction. In the "templated assembly" (TA model) the nucleus can act as a template provided the monomers attached to the nucleus undergo a simultaneous conformational conversion such that the rate-limiting step of conversion is coincidental with that of the aggregate formation (Uratani et al., 1972).

In the case of the monomer-directed conversion model, conformational conversion acts as a rate-limiting step, but the process of conformational conversion of monomer occurs in the solution and not as a part of the growing aggregates (Prusiner, 1982). In the nucleated conformational conversion model, we observe the formation of oligomers (unlike the nucleation polymerization mechanism) that assemble to undergo the conformational conversion process (Fu et al., 2015; Serio et al., 2000).

7.5.2 Nucleation-dependent polymerization

An NDP reaction involves the formation of a nucleus, which is the rate-limiting step of the reaction. The formation of the nucleus involves high activation barrier and is thermodynamically unfavorable. The process of NDP reaction involves measuring both the forward and the reverse rate constants

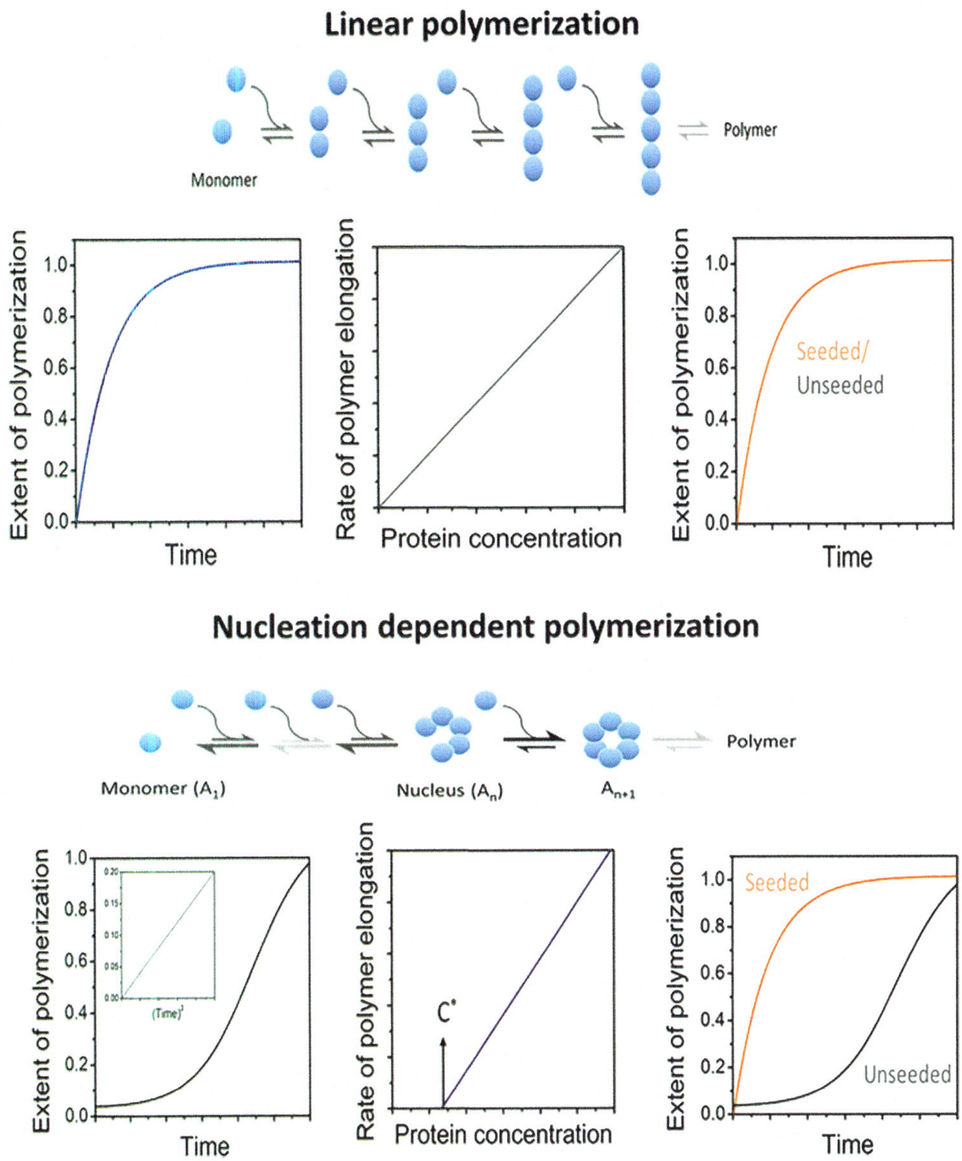

FIGURE 7.5

Schematic representation of the kinetic models describing the molecular mechanism of aggregation pathways.

(Ferrone, 1999; Kodaka, 2004). Once nucleation formation occurs, the subsequent steps in the aggregation reaction occur spontaneously. NDP reaction is associated with the following characteristics (Ferrone, 1999).

1. The kinetics of an NDP reaction shows a lag phase. The lag time is highly dependent on the protein concentration, size of the nucleus, and the association and dissociation rate constants (Kodaka, 2004). This weak lag phase is described by τ_2 function (Kumar & Udgaonkar, 2010).
2. The extent of the lag phase depends on the concentration of the protein; below a specific concentration, no nucleus formation occurs. This threshold concentration required for nucleus formation is called critical concentration, and it varies from polymer to polymer.
3. The addition of preformed nuclei (seeding) can abolish the lag phase of the reaction.

7.6 Modulation of the energy landscape by chaperones, small molecules, and nucleic acid

A single protein can adopt different conformations, such as MG, partially unfolded state, oligomers, LLPS or hydrogel formation, and aggregates. Proteins, which function as polyelectrolytes because of their charged amino acids, are highly sensitive to changes in local electrostatics. Even slight alterations, such as those caused by mutations, posttranslational modifications, or aging, can significantly impact protein conformation and behavior, ultimately disrupting healthy proteostasis (Chiti & Dobson, 2009; Dobson, 1999, 2001; Rabouille & Alberti, 2017). The type of conformation adopted by a protein, thus, depends upon the immediate environment around the protein molecule. In the view of an energy landscape, these species occupy different energy minima with respect to the native state. Some may occupy a very high energy state and, hence, may not populate under a given solution condition, whereas some are highly thermodynamically favorable, but because of a high energy barrier, their amount is negligible. Modulation of these energy barriers can be brought about by chaperones, nucleic acids, small molecules, local electrostatics, etc. For example, the interaction of nucleic acid can either prevent or accelerate the formation of protein assemblies (Maharana et al., 2018). According to the conformational selection model the binding of ligands to proteins preferentially selects a conformation from the preexisting distributions of conformations, thus shifting the equilibrium toward a particular state (Miller & Dill, 1997). Interestingly, it has been shown that the binding of nucleic acid can prevent TDP-43 from undergoing oligomerization and aggregate formation (Huang et al., 2013; Zacco et al., 2019). Some studies suggest that the presence of nucleic acid can increase the solubility of the protein (Sun et al., 2014). The type of response that the protein undergoes upon interaction depends on the number, length, and sequence of the nucleotides (Langdon et al., 2018). Intriguingly the ratio of RNA to protein in the cells acts to regulate the process of phase separation and assembly formation, having a strong implication toward the formation of stress granules and protein aggregation (Maharana et al., 2018). As a therapeutic strategy, small molecules can thus be designed to modulate these energy barriers and increase the solubility of protein to prevent it from undergoing aggregation. In line with this a recent study proposed that small molecules with planar moieties can efficiently modulate the recruitment of TDP-43 to stress granules and thus prevent persistent accumulation of the protein (Fang et al., 2019).

7.7 Conclusion

TDP-43 protein has emerged to be a vital protein for RNA metabolism in the cell. Any change in the protein because of mutations, solution condition, and aberrant posttranslational modification causes a disruption in RNA homeostasis and results in aggregation. The process of aggregation of TDP-43 seems to be very complex. Although many studies for TDP-43 protein have revealed the presence of oligomers under in vitro, in cell, and in brain tissues of patients, the transition from a soluble monomer to an irreversible insoluble aggregate is not clear. There are many questions that remain unanswered. How does the protein detect stress in the environment? How does the protein become destabilized? What are the initial molecular steps in stress detection, and how does the protein respond to the stress? What is the sequence of the steps and what forces govern the aggregation pathway? How heterogeneous is the process of aggregation? Can the process of aggregation be reversed at any early step? Are there any intermediates with common biophysical characteristics that can be targeted for therapeutics? It will, thus, be very interesting to study the aggregation pathway and obtain a global picture at the molecular level.

Studying the aggregation pathway will also help us understand the energy barriers and how it can be modulated to better design drugs and strategize to stop the aggregation process. For example, transthyretin amyloidosis is caused because of the dissociation of transthyretin (natively tetrameric) into monomers, which misfold and aggregate. An FDA-approved drug, tafamidis targets this rate-limiting tetramer dissociation step by binding and stabilizing the tetramer species (Bulawa et al., 2012). Studying the biochemical and biophysical properties of the various species involved in the aggregation process would be highly valuable for gaining a deeper understanding of the disease and developing more effective therapeutic targets.

Acknowledgment

This work was funded by a SERB-DST core research grant (project CRG/2023/002721) to S.K.J. M.P. is a recipient of a Senior Research Fellowship from the University Grants Commission, India. A.D. is a recipient of a Senior Research Fellowship from the Council of Scientific and Industrial Research, India.

References

Afroz, T., Hock, E. M., Ernst, P., Foglieni, C., Jambeau, M., Gilhespy, L. A. B., Laferriere, F., Maniecka, Z., Plückthun, A., Mittl, P., Paganetti, P., Allain, F. H. T., & Polymenidou, M. (2017). Functional and dynamic polymerization of the ALS-linked protein TDP-43 antagonizes its pathologic aggregation. *Nature Communications, 8*(1). doi:10.1038/s41467-017-00062-0.

Agrawal, S., Kuo, P. H., Chu, L. Y., Golzarroshan, B., Jain, M., & Yuan, H. S. (2019). RNA recognition motifs of disease-linked RNA-binding proteins contribute to amyloid formation. *Scientific Reports, 9*(1). doi:10.1038/s41598-019-42367-8.

Ano Bom, A. P. D., Freitas, M. S., Moreira, F. S., Ferraz, D., Sanches, D., Gomes, A. M. O., Valente, A. P., Cordeiro, Y., & Silva, J. L. (2010). The p53 core domain is a molten globule at low pH: Functional implications of a partially unfolded structure. *Journal of Biological Chemistry, 285*(4), 2857–2866. doi:10.1074/jbc.M109.075861.

Arai, T., Hasegawa, M., Akiyama, H., Ikeda, K., Nonaka, T., Mori, H., Mann, D., Tsuchiya, K., Yoshida, M., Hashizume, Y., & Oda, T. (2006). TDP-43 is a component of ubiquitin-positive tau-negative inclusions in frontotemporal lobar degeneration and amyotrophic lateral sclerosis. *Biochemical and Biophysical Research Communications, 351*(3), 602–611. doi:10.1016/j.bbrc.2006.10.093.

Balch, W. E., Morimoto, R. I., Dillin, A., & Kelly, J. W. (2008). Adapting proteostasis for disease intervention. *Science, 319*(5865), 916–919. doi:10.1126/science.1141448.

Baldwin, A. J., Knowles, T. P. J., Tartaglia, G. G., Fitzpatrick, A. W., Devlin, G. L., Shammas, S. L., Waudby, C. A., Mossuto, M. F., Meehan, S., Gras, S. L., Christodoulou, J., Anthony-Cahill, S. J., Barker, P. D., Vendruscolo, M., & Dobson, C. M. (2011). Metastability of native proteins and the phenomenon of amyloid formation. *Journal of the American Chemical Society, 133*(36), 14160–14163. doi:10.1021/ja2017703.

Baldwin, R. L., & Rose, G. D. (2013). Molten globules, entropy-driven conformational change and protein folding. *Current Opinion in Structural Biology, 23*(1), 4–10. doi:10.1016/j.sbi.2012.11.004.

Bentmann, E., Haass, C., & Dormann, D. (2013). Stress granules in neurodegeneration—Lessons learnt from TAR DNA binding protein of 43 kDa and fused in sarcoma. *FEBS Journal, 280*(18), 4348–4370. doi:10.1111/febs.12287.

Bergmann, O., Spalding, K. L., & Frisén, J. (2015). Adult neurogenesis in humans. *Cold Spring Harbor Perspectives in Biology, 7*(7), a018994. doi:10.1101/cshperspect.a018994.

Berning, B. A., & Walker, A. K. (2019). The pathobiology of TDP-43 C-terminal fragments in ALS and FTLD. *Frontiers in Neuroscience, 1*. doi:10.3389/fnins.2019.00335.

Bigio, E. H., Wu, J. Y., Deng, H. X., Bit-Ivan, E. N., Mao, Q., Ganti, R., Peterson, M., Siddique, N., Geula, C., Siddique, T., & Mesulam, M. (2013). Inclusions in frontotemporal lobar degeneration with TDP-43 proteinopathy (FTLD-TDP) and amyotrophic lateral sclerosis (ALS), but not FTLD with FUS proteinopathy (FTLD-FUS), have properties of amyloid. *Acta Neuropathologica, 125*(3), 463–465. doi:10.1007/s00401-013-1089-6.

Bulawa, C. E., Connelly, S., Devit, M., Wang, L., Weigel, C., Fleming, J. A., Packman, J., Powers, E. T., Wiseman, R. L., Foss, T. R., Wilson, I. A., Kelly, J. W., & Labaudinière, R. (2012). Tafamidis, a potent and selective transthyretin kinetic stabilizer that inhibits the amyloid cascade. *Proceedings of the National Academy of Sciences of the United States of America, 109*(24), 9629–9634. doi:10.1073/pnas.1121005109.

Buratti, E. (2015). Functional significance of TDP-43 mutations in disease. *Advances in Genetics, 9*, 1–53. doi:10.1016/bs.adgen.2015.07.001.

Caramelo, J. J., Castro, O. A., Alonso, L. G., De Prat-Gay, G., & Parodi, A. J. (2003). UDP-Glc:gLycoprotein glucosyltransferase recognizes structured and solvent accessible hydrophobic patches in molten globule-like folding intermediates. *Proceedings of the National Academy of Sciences of the United States of America, 100*(1), 86–91. doi:10.1073/pnas.262661199.

Čeru, S., & Žerovnik, E. (2008). Similar toxicity of the oligomeric molten globule state and the prefibrillar oligomers. *FEBS Letters, 582*(2), 203–209. doi:10.1016/j.febslet.2007.12.002.

Chamberlain, A. K., Handel, T. M., & Marqusee, S. (1996). Detection of rare partially folded molecules in equilibrium with the native conformation of RNaseH. *Nature Structural Biology, 3*(9), 782–787. doi:10.1038/nsb0996-782.

Chang, C. K., Chiang, M. H., Toh, E. K. W., Chang, C. F., & Huang, T. H. (2013). Molecular mechanism of oxidation-induced TDP-43 RRM1 aggregation and loss of function. *FEBS Letters, 587*(6), 575–582. doi:10.1016/j.febslet.2013.01.038.

Chang, C., Wu, T. H., Wu, C. Y., Chiang, M., Toh, E. K. W., Hsu, Y. C., Lin, K. F., Liao, Y., Huang, T., & Huang, J. J. T. (2012). The N-terminus of TDP-43 promotes its oligomerization and enhances DNA binding affinity. *Biochemical and Biophysical Research Communications, 425*(2), 219–224. doi:10.1016/j.bbrc.2012.07.071.

Chen, A. K. H., Lin, R. Y. Y., Hsieh, E. Z. J., Tu, P. H., Chen, R. P. Y., Liao, T. Y., Chen, W., Wang, C. H., & Huang, J. J. T. (2010). Induction of amyloid fibrils by the C-terminal fragments of TDP-43 in amyotrophic lateral sclerosis. *Journal of the American Chemical Society, 132*(4), 1186–1187. doi:10.1021/ja9066207.

Chiang, C. H., Grauffel, C., Wu, L. S., Kuo, P. H., Doudeva, L. G., Lim, C., Shen, C. K. J., & Yuan, H. S. (2016). Structural analysis of disease-related TDP-43 D169G mutation: Linking enhanced stability and caspase cleavage efficiency to protein accumulation. *Scientific Reports*. doi:10.1038/srep21581.

Chiti, F., & Dobson, C. M. (2009). Amyloid formation by globular proteins under native conditions. *Nature Chemical Biology, 5*(1), 15–22. doi:10.1038/nchembio.131.

Cho, K. R., Huang, Y., Yu, S., Yin, S., Plomp, M., Qiu, S. R., Lakshminarayanan, R., Moradian-Oldak, J., Sy, M. S., & De Yoreo, J. J. (2011). A multistage pathway for human prion protein aggregation in vitro: From multimeric seeds to β-oligomers and nonfibrillar structures. *Journal of the American Chemical Society, 133*(22), 8586–8593. doi:10.1021/ja1117446.

Christensen, H., & Pain, R. H. (1991). Molten globule intermediates and protein folding. *European Biophysics Journal, 19*(5), 221–229. doi:10.1007/BF00183530.

Collaborators, G. B. D. N. (2016). Global, regional, and national burden of neurological disorders, 1990–2016: A systematic analysis for the global burden of disease study. *Lancet Neurology, 1*, 459–480.

Colombrita, C., Zennaro, E., Fallini, C., Weber, M., Sommacal, A., Buratti, E., Silani, V., & Ratti, A. (2009). TDP-43 is recruited to stress granules in conditions of oxidative insult. *Journal of Neurochemistry, 111*(4), 1051–1061. doi:10.1111/j.1471-4159.2009.06383.x.

Conicella, A. E., Zerze, G. H., Mittal, J., & Fawzi, N. L. (2016). ALS mutations disrupt phase separation mediated by α-helical structure in the TDP-43 low-complexity C-terminal domain. *Structure, 24*(9), 1537–1549. doi:10.1016/j.str.2016.07.007.

Cushman, M., Johnson, B. S., King, O. D., Gitler, A. D., & Shorter, J. (2010). Prion-like disorders: Blurring the divide between transmissibility and infectivity. *Journal of Cell Science, 123*(8), 1191–1201. doi:10.1242/jcs.051672.

David, D. C., Ollikainen, N., Trinidad, J. C., Cary, M. P., Burlingame, A. L., & Kenyon, C. (2010). Widespread protein aggregation as an inherent part of aging in *C. elegans*. *PLoS Biology, 8*(8), 47–48. doi:10.1371/journal.pbio.1000450.

de Oliveira, G. A. P., & Silva, J. L. (2017). The push-and-pull hypothesis in protein unfolding, misfolding and aggregation. *Biophysical Chemistry, 23*, 20–26. doi:10.1016/j.bpc.2017.03.007.

Dewey, C. M., Cenik, B., Sephton, C. F., Johnson, B. A., Herz, J., & Yu, G. (2012). TDP-43 aggregation in neurodegeneration: Are stress granules the key? *Brain Research, 146*, 16–25. doi:10.1016/j.brainres.2012.02.032.

Dewey, C. M., Cenik, B., Sephton, C. F., Dries, D. R., Mayer, P., Good, S. K., Johnson, B. A., Herz, J., & Yu, G. (2011). TDP-43 is directed to stress granules by sorbitol, a novel physiological osmotic and oxidative stressor. *Molecular and Cellular Biology, 31*(5), 1098–1108. doi:10.1128/MCB.01279-10.

Dobson, C. M. (2001). Protein folding and its links with human disease. *Biochemical Society Symposium, 6*, 1–26. doi:10.1042/bss0680001.

Dobson, C. M. (1999). Protein misfolding, evolution and disease. *Trends in Biochemical Sciences, 24*(9), 329–332. doi:10.1016/S0968-0004(99)01445-0.

Doke, A. A., & Jha, S. K. (2022). Effect of in vitro solvation conditions on inter- and intramolecular assembly of full-length TDP-43. *Journal of Physical Chemistry B, 126*(26), 4799–4813. doi:10.1021/acs.jpcb.2c02203.

Doke, A. A., & Jha, S. K. (2024). Electrostatics choreographs the aggregation dynamics of full-length TDP-43 via a monomeric amyloid precursor. *Biochemistry, 63*(12), 1553–1568. doi:10.1021/acs.biochem.4c00060.

Doke, A. A., & Jha, S. K. (2023). Shapeshifter TDP-43: Molecular mechanism of structural polymorphism, aggregation, phase separation and their modulators. *Biophysical Chemistry, 29*. doi:10.1016/j.bpc.2023.106972.

Ederle, H., Funk, C., Abou-Ajram, C., Hutten, S., Funk, E. B. E., Kehlenbach, R. H., Bailer, S. M., & Dormann, D. (2018). Nuclear egress of TDP-43 and FUS occurs independently of Exportin-1/CRM1. *Scientific Reports, 8*(1). doi:10.1038/s41598-018-25007-5.

Ellis, R. J. (1997). Molecular chaperones: Avoiding the crowd. *Current Biology, 7*(9), R531. doi:10.1016/s0960-9822(06)00273-9.

Fändrich, M., Fletcher, M. A., & Dobson, C. M. (2001). Amyloid fibrils from muscle myoglobin. *Nature, 410*(6825), 165–166. doi:10.1038/35065514.

Fang, M. Y., Markmiller, S., Vu, A. Q., Javaherian, A., Dowdle, W. E., Jolivet, P., Bushway, P. J., Castello, N. A., Baral, A., Chan, M. Y., Linsley, J. W., Linsley, D., Mercola, M., Finkbeiner, S., Lecuyer, E., Lewcock, J. W., & Yeo, G. W. (2019). Small-molecule modulation of TDP-43 recruitment to stress granules prevents persistent TDP-43 accumulation in ALS/FTD. *Neuron, 103*(5), 802. doi:10.1016/j.neuron.2019.05.048.

Fang, Y. S., Tsai, K. J., Chang, Y. J., Kao, P., Woods, R., Kuo, P. H., Wu, C. C., Liao, J. Y., Chou, S. C., Lin, V., Jin, L. W., Yuan, H. S., Cheng, I. H., Tu, P. H., & Chen, Y. R. (2014). Full-length TDP-43 forms toxic amyloid oligomers that are present in frontotemporal lobar dementia-TDP patients. *Nature Communications*. doi:10.1038/ncomms5824.

Ferrone, F. (1999). Analysis of protein aggregation kinetics. *Methods in Enzymology, 30*, 256–274. doi:10.1016/S0076-6879(99)09019-9.

Frare, E., Mossuto, M. F., de Laureto, P. P., Tolin, S., Menzer, L., Dumoulin, M., Dobson, C. M., & Fontana, A. (2009). Characterization of oligomeric species on the aggregation pathway of human lysozyme. *Journal of Molecular Biology, 387*(1), 17–27. doi:10.1016/j.jmb.2009.01.049.

French, R. L., Grese, Z. R., Aligireddy, H., Dhavale, D. D., Reeb, A. N., Kedia, N., Kotzbauer, P. T., Bieschke, J., & Ayala, Y. M. (2019). Detection of TAR DNA-binding protein 43 (TDP-43) oligomers as initial intermediate species during aggregate formation. *Journal of Biological Chemistry, 294*(17), 6696–6709. doi:10.1074/jbc.RA118.005889.

Frieden, C. (2007). Protein aggregation processes: In search of the mechanism. *Protein Science, 16*(11), 2334–2344. doi:10.1110/ps.073164107.

Fu, H., Hardy, J., & Duff, K. E. (2018). Selective vulnerability in neurodegenerative diseases. *Nature Neuroscience, 21*(10), 1350–1358. doi:10.1038/s41593-018-0221-2.

Fu, Z., Aucoin, D., Davis, J., Van Nostrand, W. E., & Smith, S. O. (2015). Mechanism of nucleated conformational conversion of Aβ42. *Biochemistry, 54*(27), 4197–4207. doi:10.1021/acs.biochem.5b00467.

Gao, J., Wang, L., Huntley, M. L., Perry, G., & Wang, X. (2018). Pathomechanisms of TDP-43 in neurodegeneration. *Journal of Neurochemistry, 146*(1), 7–20. doi:10.1111/jnc.14327.

Garnier, C., Devred, F., Byrne, D., Puppo, R., Roman, A. Y., Malesinski, S., Golovin, A. V., Lebrun, R., Ninkina, N. N., & Tsvetkov, P. O. (2017). Zinc binding to RNA recognition motif of TDP-43 induces the formation of amyloid-like aggregates. *Scientific Reports, 7*(1). doi:10.1038/s41598-017-07215-7.

Gopal, P. P., Nirschl, J. J., Klinman, E., & Holzbaurb, E. L. F. (2017). Amyotrophic lateral sclerosis-linked mutations increase the viscosity of liquid-like TDP-43 RNP granules in neurons. *Proceedings of the National Academy of Sciences of the United States of America, 114*(12), E2466. doi:10.1073/pnas.1614462114.

Guenther, E. L., Ge, P., Trinh, H., Sawaya, M. R., Cascio, D., Boyer, D. R., Gonen, T., Zhou, Z. H., & Eisenberg, D. S. (2018). Atomic-level evidence for packing and positional amyloid polymorphism by segment from TDP-43 RRM2. *Nature Structural and Molecular Biology, 25*(4), 311–319. doi:10.1038/s41594-018-0045-5.

Guerrero-Muñoz, M. J., Gerson, J., & Castillo-Carranza, D. L. (2015). Tau oligomers: The toxic player at synapses in Alzheimer's disease. *Frontiers in Cellular Neuroscience*, 1–10. doi:10.3389/fncel.2015.00464.

Guo, W., Chen, Y., Zhou, X., Kar, A., Ray, P., Chen, X., Rao, E. J., Yang, M., Ye, H., Zhu, L., Liu, J., Xu, M., Yang, Y., Wang, C., Zhang, D., Bigio, E. H., Mesulam, M., Shen, Y., Xu, Q., … Wu, J. Y. (2011). An ALS-associated mutation affecting TDP-43 enhances protein aggregation, fibril formation and neurotoxicity. *Nature Structural and Molecular Biology, 18*(7), 822–831. doi:10.1038/nsmb.2053.

Hamada, D., & Dobson, C. M. (2002). A kinetic study of β-lactoglobulin amyloid fibril formation promoted by urea. *Protein Science, 11*(10), 2417–2426. doi:10.1110/ps.0217702.

Huang, Y. C., Lin, K. F., He, R. Y., Tu, P. H., Koubek, J., Hsu, Y. C., & Huang, J. J. T. (2013). Inhibition of TDP-43 aggregation by nucleic acid binding. *PLoS ONE, 8*(5). doi:10.1371/journal.pone.0064002.

Jahn, T. R., Parker, M. J., Homans, S. W., & Radford, S. E. (2006). Amyloid formation under physiological conditions proceeds via a native-like folding intermediate. *Nature Structural and Molecular Biology, 13*(3), 195–201. doi:10.1038/nsmb1058.

Jahn, T. R., & Radford, S. E. (2005). The Yin and Yang of protein folding. *FEBS Journal, 272*(23), 5962–5970. doi:10.1111/j.1742-4658.2005.05021.x.

Jiang, L. L., Che, M. X., Zhao, J., Zhou, C. J., Xie, M. Y., Li, H. Y., He, J. H., & Hu, H. Y. (2013). Structural transformation of the amyloidogenic core region of TDP-43 protein initiates its aggregation and cytoplasmic inclusion. *Journal of Biological Chemistry, 288*(27), 19614–19624. doi:10.1074/jbc.M113.463828.

Jiang, L. L., Xue, W., Hong, J. Y., Zhang, J. T., Li, M. J., Yu, S. N., He, J. H., & Hu, H. Y. (2017). The N-terminal dimerization is required for TDP-43 splicing activity. *Scientific Reports, 7*(1). doi:10.1038/s41598-017-06263-3.

Johnson, B. S., McCaffery, J. M., Lindquist, S., & Gitler, A. D. (2008). A yeast TDP-43 proteinopathy model: Exploring the molecular determinants of TDP-43 aggregation and cellular toxicity. *Proceedings of the National Academy of Sciences of the United States of America, 105*(17), 6439–6444. doi:10.1073/pnas.0802082105.

Johnson, B. S., Snead, D., Lee, J. J., McCaffery, J. M., Shorter, J., & Gitler, A. D. (2009). TDP-43 is intrinsically aggregation-prone, and amyotrophic lateral sclerosis-linked mutations accelerate aggregation and increase toxicity. *Journal of Biological Chemistry, 284*(30), 20329–20339. doi:10.1074/jbc.M109.010264.

Kayed, R., & Lasagna-Reeves, C. A. (2013). Molecular mechanisms of amyloid oligomers toxicity. In *Alzheimer's disease: Advances for a New Century* (pp. 67–78). IOS Pres. doi:10.3233/978-1-61499-154-0-67.

Kelly, J. W. (1998). The alternative conformations of amyloidogenic proteins and their multi-step assembly pathways. *Current Opinion in Structural Biology, 8*(1), 101–106. doi:10.1016/S0959-440X(98)80016-X.

Kim, H. J., Kim, N. C., Wang, Y. D., Scarborough, E. A., Moore, J., Diaz, Z., MacLea, K. S., Freibaum, B., Li, S., Molliex, A., Kanagaraj, A. P., Carter, R., Boylan, K. B., Wojtas, A. M., Rademakers, R., Pinkus, J. L., Greenberg, S. A., Trojanowski, J. Q., Traynor, B. J., ... Taylor, J. P. (2013). Mutations in prion-like domains in hnRNPA2B1 and hnRNPA1 cause multisystem proteinopathy and ALS. *Nature, 495*(7442), 467–473. doi:10.1038/nature11922.

King, O. D., Gitler, A. D., & Shorter, J. (2012). The tip of the iceberg: RNA-binding proteins with prion-like domains in neurodegenerative disease. *Brain Research, 146*, 61–80. doi:10.1016/j.brainres.2012.01.016.

Kodaka, M. (2004). Interpretation of concentration-dependence in aggregation kinetics. *Biophysical Chemistry, 109*(2), 325–332. doi:10.1016/j.bpc.2003.12.003.

Kumar, S., & Udgaonkar, J. B. (2010). Mechanisms of amyloid fibril formation by proteins. *Current Science, 98*(5), 639–656. http://www.ias.ac.in/currsci/10mar2010/639.pdf.

Kuo, P.-H., Doudeva, L. G., Wang, Y.-T., Shen, C.-K. J., & Yuan, H. S. (2009). Structural insights into TDP-43 in nucleic-acid binding and domain interactions. *Nucleic Acids Research, 37*(6), 1799–1808. doi:10.1093/nar/gkp013.

Langdon, E. M., Qiu, Y., Niaki, A. G., McLaughlin, G. A., Weidmann, C. A., Gerbich, T. M., Smith, J. A., Crutchley, J. M., Termini, C. M., Weeks, K. M., Myong, S., & Gladfelter, A. S. (2018). mRNA structure determines specificity of a polyQ-driven phase separation. *Science, 360*(6391), 922–927. doi:10.1126/science.aar7432.

Lee, S., & Eisenberg, D. (2003). Seeded conversion of recombinant prion protein to a disulfide-bonded oligomer by a reduction-oxidation process. *Nature Structural Biology, 10*(9), 725–730. doi:10.1038/nsb961.

Leopold, P. E., Montal, M., & Onuchic, J. N. (1992). Protein folding funnels: A kinetic approach to the sequence-structure relationship. *Proceedings of the National Academy of Sciences of the United States of America, 89*(18), 8721–8725. Www.pnas.org.

Li, Y. R., King, O. D., Shorter, J., & Gitler, A. D. (2013). Stress granules as crucibles of ALS pathogenesis. *Journal of Cell Biology, 201*(3), 361–372. doi:10.1083/jcb.201302044.

Lim, L., Wei, Y., Lu, Y., Song, J., & Petsko, G. A. (2016). ALS-causing mutations significantly perturb the self-assembly and interaction with nucleic acid of the intrinsically disordered prion-like domain of TDP-43. *PLoS Biology, 14*(1), e1002338. doi:10.1371/journal.pbio.1002338.

Lin, Y., Protter, D. S. W., Rosen, M. K., & Parker, R. (2015). Formation and maturation of phase-separated liquid droplets by RNA-binding proteins. *Molecular Cell, 60*(2), 208–219. doi:10.1016/j.molcel.2015.08.018.

Ling, S. C., Polymenidou, M., & Cleveland, D. W. (2013). Converging mechanisms in ALS and FTD: Disrupted RNA and protein homeostasis. *Neuron, 79*(3), 416–438. doi:10.1016/j.neuron.2013.07.033.

Lomakin, A., Teplow, D. B., Kirschner, D. A., & Benedeki, G. B. (1997). Kinetic theory of fibrillogenesis of amyloid β-protein. *Proceedings of the National Academy of Sciences of the United States of America, 94*(15), 7942–7947. doi:10.1073/pnas.94.15.7942.

Lukavsky, P. J., Daujotyte, D., Tollervey, J. R., Ule, J., Stuani, C., Buratti, E., Baralle, F. E., Damberger, F. F., & Allain, F. H. T. (2013). Molecular basis of UG-rich RNA recognition by the human splicing factor TDP-43. *Nature Structural and Molecular Biology, 20*(12), 1443–1449. doi:10.1038/nsmb.2698.

Lukong, K. E., Chang, K., Khandjian, E. W., & Richard, S. (2008). RNA-binding proteins in human genetic disease. *Trends in Genetics, 24*(8), 416–425. doi:10.1016/j.tig.2008.05.004.

Mackness, B. C., Tran, M. T., McClain, S. P., Matthews, C. R., & Zitzewitz, J. A. (2014). Folding of the RNA recognition motif (RRM) domains of the amyotrophic lateral sclerosis (ALS)-linked protein TDP-43 reveals an intermediate state. *Journal of Biological Chemistry, 289*(12), 8264–8276. doi:10.1074/jbc.M113.542779.

Maharana, S., Wang, J., Papadopoulos, D. K., Richter, D., Pozniakovsky, A., Poser, I., Bickle, M., Rizk, S., Guillén-Boixet, J., Franzmann, T. M., Jahnel, M., Marrone, L., Chang, Y. T., Sterneckert, J., Tomancak, P., Hyman, A. A., & Alberti, S. (2018). RNA buffers the phase separation behavior of prion-like RNA binding proteins. *Science, 360*(6391), 918–921. doi:10.1126/science.aar7366.

Maris, C., Dominguez, C., & Allain, F. H. T. (2005). The RNA recognition motif, a plastic RNA-binding platform to regulate post-transcriptional gene expression. *FEBS Journal, 272*(9), 2118–2131. doi:10.1111/j.1742-4658.2005.04653.x.

McDonald, K. K., Aulas, A., Destroismaisons, L., Pickles, S., Beleac, E., Camu, W., Rouleau, G. A., & Velde, C. V. (2011). TAR DNA-binding protein 43 (TDP-43) regulates stress granule dynamics via differential regulation of G3BP and TIA-1. *Human Molecular Genetics, 20*(7), 1400–1410. doi:10.1093/hmg/ddr021.

Millecamps, S., Salachas, F., Cazeneuve, C., Gordon, P., Bricka, B., Camuzat, A., Guillot-Noel, L., Russaouen, O., Bruneteau, G., Pradat, P.-F., Le Forestier, N., Vandenberghe, N., Danel-Brunaud, V., Guy, N., Thauvin-Robinet, C., Lacomblez, L., Couratier, P., Hannequin, D., Seilhean, D., … Meininger, V. (2010). SOD1, ANG, VAPB, TARDBP, and FUS mutations in familial amyotrophic lateral sclerosis: Genotype-phenotype correlations. *Journal of Medical Genetics, 47*(8), 554–560. doi:10.1136/jmg.2010.077180.

Miller, D. W., & Dill, K. A. (1997). Ligand binding to proteins: The binding landscape model. *Protein Science, 6*(10), 2166–2179. doi:10.1002/pro.5560061011.

Modler, A. J., Gast, K., Lutsch, G., & Damaschun, G. (2003). Assembly of amyloid protofibrils via critical oligomers—A novel pathway of amyloid formation. *Journal of Molecular Biology, 325*(1), 135–148. doi:10.1016/S0022-2836(02)01175-0.

Molliex, A., Temirov, J., Lee, J., Coughlin, M., Kanagaraj, A. P., Kim, H. J., Mittag, T., & Taylor, J. P. (2015). Phase separation by low complexity domains promotes stress granule assembly and drives pathological fibrillization. *Cell, 163*(1), 123–133. doi:10.1016/j.cell.2015.09.015.

Mompeán, M., Chakrabartty, A., Buratti, E., & Laurents, D. V. (2016a). Electrostatic repulsion governs TDP-43 C-terminal domain aggregation. *PLoS Biology, 14*(4). doi:10.1371/journal.pbio.1002447.

Mompeán, M., Buratti, E., Guarnaccia, C., Brito, R. M. M., Chakrabartty, A., Baralle, F. E., & Laurents, D. V. (2014). Structural characterization of the minimal segment of TDP-43 competent for aggregation. *Archives of Biochemistry and Biophysics, 54*, 53–62. doi:10.1016/j.abb.2014.01.007.

Mompeán, M., Romano, V., Pantoja-Uceda, D., Stuani, C., Baralle, F. E., Buratti, E., & Laurents, D. V. (2016b). The TDP-43 N-terminal domain structure at high resolution. *The FEBS Journal, 283*(7), 1242–1260. doi:10.1111/febs.13651.

Nandi, P. K., Leclerc, E., Nicole, J.-C., & Takahashi, M. (2002). DNA-induced partial unfolding of prion protein leads to its polymerisation to amyloid. *Journal of Molecular Biology, 322*(1), 153–161. doi:10.1016/s0022-2836(02)00750-7.

Narayanaswamy, R., Levy, M., Tsechansky, M., Stovall, G. M., O'Connell, J. D., Mirrielees, J., Ellington, A. D., & Marcotte, E. M. (2009). Widespread reorganization of metabolic enzymes into reversible assemblies upon nutrient starvation. *Proceedings of the National Academy of Sciences of the United States of America, 106*(25), 10147–10152. doi:10.1073/pnas.0812771106.

Neumann, M., Sampathu, D. M., Kwong, L. K., Truax, A. C., Micsenyi, M. C., Chou, T. T., Bruce, J., Schuck, T., Grossman, M., Clark, C. M., McCluskey, L. F., Miller, B. L., Masliah, E., Mackenzie, I. R., Feldman, H., Feiden, W., Kretzschmar, H. A., Trojanowski, J. Q., & Lee, V. M. Y. (2006). Ubiquitinated TDP-43 in frontotemporal lobar degeneration and amyotrophic lateral sclerosis. *Science, 314*(5796), 130–133. doi:10.1126/science.1134108.

Nonaka, T., Masuda-Suzukake, M., Arai, T., Hasegawa, Y., Akatsu, H., Obi, T., Yoshida, M., Murayama, S., Mann, D. M. A., Akiyama, H., & Hasegawa, M. (2013). Prion-like properties of pathological TDP-43 aggregates from diseased brains. *Cell Reports, 4*(1), 124–134. doi:10.1016/j.celrep.2013.06.007.

Nonaka, T., Kametani, F., Arai, T., Akiyama, H., & Hasegawa, M. (2009). Truncation and pathogenic mutations facilitate the formation of intracellular aggregates of TDP-43. *Human Molecular Genetics, 18*(18), 3353–3364. doi:10.1093/hmg/ddp275.

Pantoja-Uceda, D., Stuani, C., Laurents, D. V., McDermott, A. E., Buratti, E., & Mompean, M. (2021). NMR assignments for the C-terminal domain of human TDP-43. *Biomolecular NMR Assignments, 15*, 177–181.

Pedrote, M. M., De Oliveira, G. A. P., Felix, A. L., Mota, M. F., de A Marques, M., Soares, I. N., Iqbal, A., Norberto, D. R., Gomes, A. M. O., Gratton, E., Cino, E. A., & Silva, J. L. (2018). Aggregation-primed molten globule conformers of the p53 core domain provide potential tools for studying p53C aggregation in cancer. *Journal of Biological Chemistry, 293*(29), 11374–11387. doi:10.1074/jbc.RA118.003285.

Pertinhez, T. A., Bouchard, M., Tomlinson, E. J., Wain, R., Ferguson, S. J., Dobson, C. M., & Smith, L. J. (2001). Amyloid fibril formation by a helical cytochrome. *FEBS Letters, 495*(3), 184–186. doi:10.1016/S0014-5793(01)02384-5.

Pillai, M., Das, A., & Jha, S. K. (2023). Electrostatic modulation of intramolecular and intermolecular interactions during the formation of an amyloid-like assembly. *Biochemistry, 62*(12), 1890–1905. doi:10.1021/acs.biochem.3c00014.

Pillai, M., & Jha, S. K. (2020). Early metastable assembly during the stress-induced formation of worm-like amyloid fibrils of nucleic acid binding domains of TDP-43. *Biochemistry, 59*(3), 315–328. doi:10.1021/acs.biochem.9b00780.

Pillai, M., & Jha, S. K. (2023). Multistep molecular mechanism of amyloid-like aggregation of nucleic acid-binding domain of TDP-43. *Proteins: Structure, Function and Bioinformatics, 91*(5), 649–664. doi:10.1002/prot.26455.

Pillai, M., & Jha, S. K. (2019). The folding and aggregation energy landscapes of tethered RRM domains of human TDP-43 are coupled via a metastable molten globule-like oligomer. *Biochemistry, 58*(6), 608–620. doi:10.1021/acs.biochem.8b01013.

Pillai, M., & Jha, S. K. (2024). Conformational Enigma of TDP-43 misfolding in neurodegenerative disorders. *ACS Omega, 9*(39), 40286–40297. doi:10.1021/acsomega.4c04119.

Pinarbasi, E. S., Cağatay, T., Fung, H. Y. J., Li, Y. C., Chook, Y. M., & Thomas, P. J. (2018). Active nuclear import and passive nuclear export are the primary determinants of TDP-43 localization. *Scientific Reports, 8*(1). doi:10.1038/s41598-018-25008-4.

Prakash, A., Kumar, V., Meena, N. K., Hassan, M. I., & Lynn, A. M. (2019). Comparative analysis of thermal unfolding simulations of RNA recognition motifs (RRMs) of TAR DNA-binding protein 43 (TDP-43). *Journal of Biomolecular Structure and Dynamics, 37*(1), 178–194. doi:10.1080/07391102.2017.1422026.

Prakash, A., Kumar, V., Meena, N. K., & Lynn, A. M. (2018). Elucidation of the structural stability and dynamics of heterogeneous intermediate ensembles in unfolding pathway of the N-terminal domain of TDP-43. *RSC Advances, 8*(35), 19835–19845. doi:10.1039/c8ra03368d.

Prasad, A., Bharathi, V., Sivalingam, V., Girdhar, A., & Patel, B. K. (2019). Molecular mechanisms of TDP-43 misfolding and pathology in amyotrophic lateral sclerosis. *Frontiers in Molecular Neuroscience, 1*. doi:10.3389/fnmol.2019.00025.

Prusiner, S. B. (1982). Novel proteinaceous infectious particles cause scrapie. *Science, 216*(4542), 136–144. doi:10.1126/science.6801762.

Qin, H., Lim, L. Z., Wei, Y., & Song, J. (2014). TDP-43 N terminus encodes a novel ubiquitin-like fold and its unfolded form in equilibrium that can be shifted by binding to ssDNA. *Proceedings of the National Academy of Sciences of the United States of America, 111*(52), 18619–18624. doi:10.1073/pnas.1413994112.

Rabouille, C., & Alberti, S. (2017). Cell adaptation upon stress: The emerging role of membrane-less compartments. *Current Opinion in Cell Biology, 4*, 34–42. doi:10.1016/j.ceb.2017.02.006.

Ratti, A., & Buratti, E. (2016). Physiological functions and pathobiology of TDP-43 and FUS/TLS proteins. *Journal of Neurochemistry, 138*(S1), 95–111. doi:10.1111/jnc.13625.

Robinson, J. L., Geser, F., Stieber, A., Umoh, M., Kwong, L. K., Van Deerlin, V. M., Lee, V. M. Y., & Trojanowski, J. Q. (2013). TDP-43 skeins show properties of amyloid in a subset of ALS cases. *Acta Neuropathologica, 125*(1), 121–131. doi:10.1007/s00401-012-1055-8.

Saelices, L., Johnson, L. M., Liang, W. Y., Sawaya, M. R., Cascio, D., Ruchala, P., Whitelegge, J., Jiang, L., Riek, R., & Eisenberg, D. S. (2015). Uncovering the mechanism of aggregation of human transthyretin. *Journal of Biological Chemistry, 290*(48), 28932–28943. doi:10.1074/jbc.M115.659912.

Saini, A., & Chauhan, V. S. (2014). Self-assembling properties of peptides derived from TDP-43 C-terminal fragment. *Langmuir, 30*(13), 3845–3856. doi:10.1021/la404710w.

Scheibel, T., Bloom, J., & Lindquist, S. L. (2004). The elongation of yeast prion fibers involves separable steps of association and conversion. *Proceedings of the National Academy of Sciences of the United States of America, 101*(8), 2287–2292. doi:10.1073/pnas.0308754101.

Serio, T. R., Cashikar, A. G., Kowal, A. S., Sawicki, G. J., Moslehi, J. J., Serpell, L., Arnsdorf, M. F., & Lindquist, S. L. (2000). Nucleated conformational conversion and the replication of conformational information by a prion determinant. *Science, 289*(5483), 1317–1321. doi:10.1126/science.289.5483.1317.

Shenoy, J., El Mammeri, N., Dutour, A., Berbon, M., Saad, A., Lends, A., Morvan, E., Grélard, A., Lecomte, S., Kauffmann, B., Theillet, F. X., Habenstein, B., & Loquet, A. (2020). Structural dissection of amyloid aggregates of TDP-43 and its C-terminal fragments TDP-35 and TDP-16. *FEBS Journal, 287*(12), 2449–2467. doi:10.1111/febs.15159.

Shiina, Y., Arima, K., Tabunoki, H., & Satoh, J. I. (2010). TDP-43 dimerizes in human cells in culture. *Cellular and Molecular Neurobiology, 30*(4), 641–652. doi:10.1007/s10571-009-9489-9.

Shin, Y., & Brangwynne, C. P. (2017). Liquid phase condensation in cell physiology and disease. *Science, 357*(6357). doi:10.1126/science.aaf4382.

Shodai, A., Morimura, T., Ido, A., Uchida, T., Ayaki, T., Takahashi, R., Kitazawa, S., Suzuki, S., Shirouzu, M., Kigawa, T., Muto, Y., Yokoyama, S., Takahashi, R., Kitahara, R., Ito, H., Fujiwara, N., & Urushitani, M. (2013). Aberrant assembly of RNA recognition motif 1 links to pathogenic conversion of TAR DNA-binding protein of 43 kDa (TDP-43). *Journal of Biological Chemistry, 288*(21), 14886–14905. doi:10.1074/jbc.M113.451849.

Shodai, A., Ido, A., Fujiwara, N., Ayaki, T., Morimura, T., Oono, M., Uchida, T., Takahashi, R., Ito, H., Urushitani, M., & Buratti, E. (2012). Conserved acidic amino acid residues in a second RNA recognition motif regulate assembly and function of TDP-43. *PLoS ONE, 7*(12), e52776. doi:10.1371/journal.pone.0052776.

Simon, J. R., Carroll, N. J., Rubinstein, M., Chilkoti, A., & López, G. P. (2017). Programming molecular self-assembly of intrinsically disordered proteins containing sequences of low complexity. *Nature Chemistry, 9*(6), 509–515. doi:10.1038/nchem.2715.

Skora, L., Becker, S., & Zweckstetter, M. (2010). Molten globule precursor states are conformationally correlated to amyloid fibrils of human β-2-microglobulin. *Journal of the American Chemical Society, 132*(27), 9223–9225. doi:10.1021/ja100453e.

Staderini, T., Bigi, A., Mongiello, D., Cecchi, C., & Chiti, F. (2022). Biophysical characterization of full-length TAR DNA-binding protein (TDP-43) phase separation. *Protein Science, 31*(12). doi:10.1002/pro.4509.

Sun, Y., Arslan, P. E., Won, A., Yip, C. M., & Chakrabartty, A. (2014). Binding of TDP-43 to the 3′UTR of its cognate mRNA enhances its solubility. *Biochemistry, 53*(37), 5885–5894. doi:10.1021/bi500617x.

Sun, Y., Cruz, A. M., Hadley, K. C., Galant, N. J., Law, R., Vernon, R. M., Morris, V. K., Robertson, J., & Chakrabartty, A. (2019). Physiologically important electrolytes as regulators of TDP-43 aggregation and droplet-phase behavior. *Biochemistry, 58*(6), 590–607. doi:10.1021/acs.biochem.8b00842.

Sun, Y., & Chakrabartty, A. (2017). Phase to phase with TDP-43. *Biochemistry, 56*(6), 809–823. doi:10.1021/acs.biochem.6b01088.

Susnjar, U., Skrabar, N., Brown, A. L., Abbassi, Y., Phatnani, H., Consortium, N. A., Cortese, A., Cereda, C., Bugiardini, E., Cardani, R., Meola, G., Ripolone, M., Moggio, M., Romano, M., Secrier, M., Fratta, P., & Buratti, E. (2022). Cell environment shapes TDP-43 function with implications in neuronal and muscle disease. *Communications Biology, 5*, 314.

Tanaka, M., Machida, Y., Nishikawa, Y., Akagi, T., Hashikawa, T., Fujisawa, T., & Nukina, N. (2003). Expansion of polyglutamine induces the formation of quasi-aggregate in the early stage of protein fibrillization. *Journal of Biological Chemistry, 278*(36), 34717–34724. doi:10.1074/jbc.M209852200.

Tavella, D., Zitzewitz, J. A., & Massi, F. (2018). Characterization of TDP-43 RRM2 partially folded states and their significance to ALS pathogenesis. *Biophysical Journal, 115*(9), 1673–1680. doi:10.1016/j.bpj.2018.09.011.

Tsai, C. P., Soong, B. W., Lin, K. P., Tu, P. H., Lin, J. L., & Lee, Y. C. (2011). FUS, TARDBP, and SOD1 mutations in a Taiwanese cohort with familial ALS. *Neurobiology of Aging, 32*(3), 553. doi:10.1016/j.neurobiolaging.2010.04.009.

Uratani, Y., Asakura, S., & Imahori, K. (1972). A circular dichroism study of *Salmonella* flagellin: Evidence for conformational change on polymerization. *Journal of Molecular Biology, 67*(1), 85–98. doi:10.1016/0022-2836(72)90388-9.

Uversky, V. N. (2010). Mysterious oligomerization of the amyloidogenic proteins. *FEBS Journal, 277*(14), 2940–2953. doi:10.1111/j.1742-4658.2010.07721.x.

Uversky, V. N., & Fink, A. L. (2004). Conformational constraints for amyloid fibrillation: The importance of being unfolded. *Biochimica et Biophysica Acta—Proteins and Proteomics, 1698*(2), 131–153. doi:10.1016/j.bbapap.2003.12.008.

Vega, M. V., Nigro, A., Luti, S., Capitini, C., Fani, G., Gonnelli, L., Boscaro, F., & Chiti, F. (2019). Isolation and characterization of soluble human full-length TDP-43 associated with neurodegeneration. *FASEB Journal, 33*(10), 10780–10793. doi:10.1096/fj.201900474R.

Vogler, T. O., Wheeler, J. R., Nguyen, E. D., Hughes, M. P., Britson, K. A., Lester, E., Rao, B., Betta, N. D., Whitney, O. N., Ewachiw, T. E., Gomes, E., Shorter, J., Lloyd, T. E., Eisenberg, D. S., Taylor, J. P., Johnson, A. M., Olwin, B. B., & Parker, R. (2018). TDP-43 and RNA form amyloid-like myo-granules in regenerating muscle. *Nature, 563*(7732), 508–513. doi:10.1038/s41586-018-0665-2.

Wang, A., Conicella, A. E., Schmidt, H. B., Martin, E. W., Rhoads, S. N., Reeb, A. N., Nourse, A., Montero, D. R., Ryan, V. H., Rohatgi, R., Shewmaker, F., Naik, M. T., Mittag, T., Ayala, Y. M., & Fawzi, N. L. (2018). A single N-terminal phosphomimic disrupts TDP-43 polymerization, phase separation, and RNA splicing. *EMBO Journal, 37*(5). doi:10.15252/embj.201797452.

Wang, Y. T., Kuo, P. H., Chiang, C. H., Liang, J. R., Chen, Y. R., Wang, S., Shen, J. C. K., & Yuan, H. S. (2013). The truncated C-terminal RNA recognition motif of TDP-43 protein plays a key role in forming proteinaceous aggregates. *Journal of Biological Chemistry, 288*(13), 9049–9057. doi:10.1074/jbc.M112.438564.

Wheeler, J. R., Matheny, T., Jain, S., Abrisch, R., & Parker, R. (2016). Distinct stages in stress granule assembly and disassembly. *Elife*. doi:10.7554/eLife.18413.

Winner, B., Jappelli, R., Maji, S. K., Desplats, P. A., Boyer, L., Aigner, S., Hetzer, C., Loher, T., Vilar, M., Campioni, S., Tzitzilonis, C., Soragni, A., Jessberger, S., Mira, H., Consiglio, A., Pham, E., Masliah, E., Gage, F. H.,

& Riek, R. (2011). In vivo demonstration that α-synuclein oligomers are toxic. *Proceedings of the National Academy of Sciences of the United States of America, 108*(10), 4194–4199. doi:10.1073/pnas.1100976108.

Wolynes, P. G., Onuchic, J. N., & Thirumalai, D. (1995). Navigating the folding routes. *Science, 267*(5204), 1619–1620. doi:10.1126/science.7886447.

Wright, G. S. A., Watanabe, T. F., Amporndanai, K., Plotkin, S. S., Cashman, N. R., Antonyuk, S. V., & Hasnain, S. S. (2020). Purification and structural characterization of aggregation-prone human TDP-43 involved in neurodegenerative diseases. *iScience, 23*(6). doi:10.1016/j.isci.2020.101159.

Zacco, E., Graña-Montes, R., Martin, S. R., de Groot, N. S., Alfano, C., Tartaglia, G. G., & Pastore, A. (2019). RNA as a key factor in driving or preventing self-assembly of the TAR DNA-binding protein 43. *Journal of Molecular Biology, 431*(8), 1671–1688. doi:10.1016/j.jmb.2019.01.028.

Zacco, E., Martin, S. R., Thorogate, R., & Pastore, A. (2018). The RNA-recognition motifs of TAR DNA-binding protein 43 May play a role in the aberrant self-assembly of the protein. *Frontiers in Molecular Neuroscience, 1*. doi:10.3389/fnmol.2018.00372.

Zerovnik, E. (2002). Amyloid-fibril formation. Proposed mechanisms and relevance to conformational disease. *European Journal of Biochemistry, 26*, 3362–3371.

Zhang, Y. J., Xu, Y. F., Cook, C., Gendron, T. F., Roettges, P., Link, C. D., Lin, W. L., Tong, J., Castanedes-Casey, M., Ash, P., Gass, J., Rangachari, V., Buratti, E., Baralle, F., Golde, T. E., Dickson, D. W., & Petrucelli, L. (2009). Aberrant cleavage of TDP-43 enhances aggregation and cellular toxicity. *Proceedings of the National Academy of Sciences of the United States of America, 106*(18), 7607–7612. doi:10.1073/pnas.0900688106.

Zhang, Y. J., Xu, Y. F., Dickey, C. A., Buratti, E., Baralle, F., Bailey, R., Pickering-Brown, S., Dickson, D., & Petrucelli, L. (2007). Progranulin mediates caspase-dependent cleavage of TAR DNA binding protein-43. *Journal of Neuroscience, 27*(39), 10530–10534. doi:10.1523/JNEUROSCI.3421-07.2007.

Zhang, Y. J., Caulfield, T., Xu, Y. F., Gendron, T. F., Hubbard, J., Stetler, C., Sasaguri, H., Whitelaw, E. C., Cai, S., Lee, W. C., & Petrucelli, L. (2013). The dual functions of the extreme N-terminus of TDP-43 in regulating its biological activity and inclusion formation. *Human Molecular Genetics, 22*(15), 3112–3122. doi:10.1093/hmg/ddt166.

CHAPTER

8 Binding-induced folding of intrinsically disordered peptides

Matthias Heyden[a,*] and Sthitadhi Maiti[b]

[a]*School of Molecular Sciences, Arizona State University, Tempe, AZ, United States,* [b]*Department of Physics, University of Genoa, Genoa, Italy*

**Corresponding author: matthias.heyden@asu.edu*

8.1 Introduction

Intrinsically disordered proteins (IDPs) are characterized by their absence of a stable three-dimensional structure, primarily because of their unique amino acid composition, often rich in charged residues, glycines, and prolines (Uversky, 2002; Van Der Lee et al., 2014). Despite their structural variability, IDPs play crucial roles in vital physiological functions. They are involved in several fundamental cellular processes, including the regulation of transcription, signal transduction, the formation of multiprotein complexes, and the development of membraneless organelles (Babu, 2016; Wright & Dyson, 2015). Approximately 40% of the human proteome is believed to be disordered or to have lengthy disordered regions (exceeding 100 residues), with a more significant fraction containing disordered segments of up to 30 residues (Emenecker et al., 2021; Holehouse & Kragelund, 2024; Tunyasuvunakool et al., 2021). A key aspect of their functionality is their ability to interact with globular proteins, where IDPs can either adopt a specific conformation when bound or maintain their dynamic and disordered nature (Pancsa & Fuxreiter, 2012). These significant functional regions in IDPs are referred to as molecular recognition features (MoRFs) or molecular recognition elements (MoREs). MoRFs facilitate interactions with structured partner proteins and can experience transitions from disorder to order during interaction (Fatafta et al., 2021; Hadži et al., 2021; Sharma et al., 2019). MoRFs typically vary in size and can be up to 70 residues in length, often situated within larger intrinsically disordered regions (IDRs) (Cumberworth et al., 2013). Because of their flexible structure, MoRFs can accurately interact with their partners, making them crucial players in regulatory processes and signal transduction (Oldfield et al., 2005b).

Based on the specific folded structure they adopt after interacting with globular targets, MoRFs can be classified into four subtypes: α-MoRFs, β-MoRFs, ι-MoRFs, and complex-MoRFs (Cumberworth et al., 2013; Mohan et al., 2006; Van Der Lee et al., 2014). These subtypes are characterized by the formation of α-helices, β-strands, irregular secondary structures, and a combination of multiple secondary structures, respectively.

Among the various types of MoRFs the most prevalent and extensively studied class involves IDPs and IDRs that fold into α-helices upon binding to their targets (Oldfield et al., 2005a; Wright & Dyson, 2009; Yan et al., 2016). This type is known as an α-helix forming MoRE/feature α-MoRE/MoRF (Oldfield et al., 2005a) or simply as a helical binding motif (Carr et al., 1991). Usually,

Protein Dynamics: Experimental and Computational Approaches. DOI: https://doi.org/10.1016/B978-0-443-30042-4.00008-0

these sequences are composed of short amphipathic helices that transition from a disordered state to a helical conformation upon binding with their target partners (Dyson & Wright, 2002). These motifs can be predicted based on their sequence (Garner et al., 1999; Mészáros et al., 2009), and they are especially prevalent in eukaryotic transactivation domains (Dyson & Wright, 2016) and $\alpha\alpha$-hub proteins (Bugge et al., 2021).

Sections of proteins containing α-MoRFs were initially thought to be predominantly disordered when unbound and unrestrained. However, subsequent research has revealed that these regions can exhibit a considerable degree of residual helical structure (Arai et al., 2015). The partially formed helical configurations observed in α-MoRFs are often termed prefolded elements, prestructured motifs, or transient and residual helical structures (Kim & Han, 2018b). A previous bioinformatics study revealed a significant overlap between the residual structure and the structure of the IDP when bound to its target, suggesting that the residual structure predisposes the IDP to its target-bound conformation (Fuxreiter et al., 2004). Even with a limited number of systems available at that time the findings strongly suggested that prestructuring plays a crucial role in target binding. A contemporary review of experimentally analyzed IDPs in their free state documented more than 50 instances of α-MoRFs exhibiting considerable secondary structure, without exploring the correspondence between residual and target-bound conformations (Kim & Han, 2018a).

The transient, residual helical structures are believed to impact the binding affinity (thermodynamics), kinetics, specificity, and structural characteristics of the IDP-target ensemble. There is increasing interest in understanding how these structures influence these properties. An early hypothesis from Dunker et al. (2001) suggested that the disorder-to-order transition in α-MoRFs leads to a reduction in binding affinity because of the folding penalty associated with α-MoRFs. The significance of the folding penalty in IDPs, and particularly in α-MoRFs, is now well-established and has been thoroughly reviewed in the literature, as discussed in Skriver et al. (2023). The initial analysis of binding affinities from various studies revealed that the average affinity for associations involving folding upon binding is generally lower compared to that of globular complexes, indirectly suggesting an estimate of the folding penalty in such interactions (Lazar et al., 2022; Teilum et al., 2015). In a study conducted by Teilum et al. (2015), which examined a substantial number of protein complexes using experimental data, it was found that the less favorable entropic contribution in complexes involving ordered regions (helical in our case) and IDPs is primarily caused by a reduction in conformational entropy. This finding is consistent with the fundamental difference in the binding mechanisms of ordered and disordered ligands. A disordered polypeptide needs to undergo folding to create the final complex, resulting in a significant decrease in conformational entropy. It should be noted that equilibrium isothermal titration calorimetry data cannot definitively determine the pathway followed by the ligand's folding during the binding process. Therefore the contribution of the amount of folding free energy to the observed binding free energy is uncertain. It is likely that in some protein complexes, the IDP undergoes folding followed by binding in a conformational selection mechanism, whereas in others, the IDP folds upon binding in an induced fit mechanism. However, the variation in average Gibbs free energy ΔG° could still be explained by the necessary folding of the disordered ligand in the ordered-disordered complexes. The assessment of folding penalties has been conducted through a thermodynamic breakdown of entropic factors (Drobnak et al., 2013; Theisen et al., 2021) and the application of helix-coil models (Hadži & Lah, 2022). In addition, the alteration in the helicity of α-MoRFs upon binding to their targets can provide estimates of the folding penalty (Hadži & Lah, 2022). A thorough evaluation of 25 distinct α-MoRFs using the helix-coil approach has indicated that the average folding penalty stands at around 2.5 kcal/mol. Nonetheless, numerous instances of ultratight picomolar binding affinities in folding-upon-binding interactions have been documented (Zavrtanik et al., 2021), demonstrating that multiple factors such as posttranslational modifications, phosphorylation, oxidative stress, or initial binding to certain regions of the binding partners can effectively counteract the folding penalty to attain ultrahigh binding affinity (Flock et al., 2014). A key factor in overcoming the folding penalty is the presence of substantial amounts of residual helical structure, also referred to as prestructuring or prefolding. An examination of folding penalties in α-MoRFs revealed that prefolding can lower the penalty by an average of 25% (Hadži & Lah, 2022). Experimentally a reduction in the α-MoRF folding penalty can be seen when an enhancement in residual helicity (such as that induced by mutations) leads to an elevated binding affinity of the α-MoRF. Relationships between residual helicity and binding affinity have been documented across various systems, and they are in quantitative agreement with predictions from models (Hadži & Lah, 2022; Hadži et al., 2021), underscoring the significance of residual structure in influencing the affinities between IDPs and their targets. Nevertheless, some research has identified a less pronounced correlation or even a more intricate relationship between residual helicity and binding affinity (Poosapati et al., 2018; Rogers et al., 2014).

In multiple distinct systems exhibiting folding-upon-binding, experimental studies have shown that the residual helical structure plays a role in modulating both binding affinity and kinetics. However, further insights are needed into how the IDP sequence relates to the structural ensemble (both unbound and bound) and how this affects the associated binding parameters, including both kinetic and equilibrium aspects. Specifically the measurement of folding free energy using enhanced sampling techniques like umbrella sampling (US) has not been extensively documented because of challenges in defining an appropriate collective variable (CV) to describe the unfolding pathway. In addition, the quantitative effects of point mutations involving known helix stabilizers like alanine and helix disruptors such as proline and glycine on the overall folding free energy have yet to be reported in detail.

To address these unexplored pathways, we employed atomistic US to explore the folding free energies in disordered peptides that transition to an α-helix upon target binding, such as peptides containing α-MoRF, as well as fully disordered IDRs. After comparing the free energy minimum structures from simulation and from AlphaFold (Jumper et al., 2021) predictions, all the α-MoRF examples we studied exhibited folding free energies within a range of 10 kJ/mol (2.4 kcal/mol), which is consistent with the range predicted by Hadži and Lah (2022) in their study using a modified Lifson–Roig (LR) model (Chakrabartty et al., 1994; Hadži et al., 2021; Lifson & Roig, 1961). They considered different sequence lengths of IDPs, which is one of the factors contributing to the discrepancy between our results and theirs. The relatively low folding free energy penalty of these sequences can be implemented in designing peptide-based or small molecule drugs that stabilize the helix structure by inducing folding upon binding to α-MoRFs.

Furthermore, we calculated the binding free energies of two selected α-MoRFs from BH3 domain-binding proteins (PUMA and NOXA-A) to MCL-1 using coarse-grained (CG) simulations with the SIRAH force field. Our findings partially support our hypothesis that the experimental binding free energies reported by Dahal et al. (2018) are the sum of both the computed binding and folding free energies. This result offers valuable insights into the potential binding-induced folding pathway followed by our chosen set of α-MoRFs.

8.2 Methods

8.2.1 Simulation protocol for folding free energy calculations

We performed all simulations of folding free energies using GROMACS 2018.1 software package (Abraham et al., 2015) and the PLUMED library (Bonomi et al., 2009, 2019). For system preparation, we constructed fully α-helical peptide sequences. We generated a template Protein Data Bank (PDB) file of a helical polyalanine with the desired sequence length. The template file was then used to create a PDB file of the sequence with alanines replaced by other amino acids according to the sequence. We subsequently employed the SCWRL4 package (Krivov et al., 2009) to generate the side chains, thereby generating the initial helical state of the sequence. We adjusted the protonation state of each amino acid residue to correspond with a pH of 7, utilizing standard side chain pK_a values in solution.

Once the system was prepared, we carried out steered molecular dynamics (SMD) simulations (Lu & Schulten, 1999) on the helical peptide. This was performed to generate snapshots or starting configurations for further analysis. The peptide was placed in a cubic simulation box with an edge length of 80 A, and it was surrounded by explicit TIP3P water molecules to mimic an aqueous environment. To ensure the stability of the initial conformation of the peptide, we applied a force constant of 1000 kJ/mol to all atoms to maintain its initial configuration during equilibration. Next, we solvated the system with approximately 16,700 water molecules to ensure adequate hydration of the peptide. In addition, we introduced ionic species to the system to imitate a physiological salt concentration of 150 mM. We added close to 50 Na^+ and Cl^- ions. It is important to note that the exact numbers of these ions were adjusted based on the specific peptide sequence being studied, as different peptides may require different ionic concentrations to maintain charge neutrality and mimic physiological conditions. The presence of salt can affect the structural dynamics and phase separation tendencies of IDPs through mechanisms like electrostatic screening and the salting-out effect, as highlighted in a recent study by Wohl et al. (2021). This setup was aimed to create a realistic simulation environment for the helical peptide in dilute solution. Although it falls short of resembling the crowded biomolecular environment in vivo (Gnutt et al., 2015), it reflects the conditions of many experiments.

We implemented periodic boundary conditions across all dimensions to simulate an infinite system and avoid edge effects. For handling long-range electrostatic interactions, we employed the particle mesh Ewald algorithm (Darden et al., 1993), setting up a real-space grid with a resolution of 1.2 A. This method efficiently calculates electrostatic forces by dividing them into short-range and long-range components. For short-range electrostatic and Lennard–Jones interactions, we established a cutoff distance of 10 A. The linear constraint solver algorithm (Hess et al., 1997) was used to constrain high-frequency bond lengths such as NH, OH, and CH within the protein, allowing us to use a longer time step of 2 fs in our case when integrating the equations of motion. In addition, we employed the SETTLE algorithm (Miyamoto & Kollman, 1992) to preserve the geometry of water molecules in the system. We performed an energy minimization process using the steepest descent algorithm before starting the molecular dynamics (MD) simulations with all the selected force fields. This initial step, involving 2000 steps of minimization, helps in relieving any steric clashes or unfavorable conformations in the system, providing a stable starting point for the dynamic simulations.

We then conducted a two-step equilibration process for each system. Initially, we equilibrated the system in the isothermal–isochoric (NVT) ensemble, applying a position restraint of 1000 kJ/mol/nm^2

to each heavy atom to maintain structural stability. Subsequently, we transitioned to the isothermal–isobaric (NPT) ensemble, maintaining the system at a temperature of 300 K and a pressure of 1 bar for 1 ns. During this phase, we utilized a time step of 1 fs, a velocity rescaling thermostat with a time constant of 0.1 ps for temperature control, and a Berendsen weak-coupling barostat with a time constant of 2.0 ps to regulate pressure.

To represent the folding process of the helix during SMD, we selected a single CV, $\xi(R)$ that quantifies the number of hydrogen bonds between backbone atoms of residues that are four sites apart along the sequence (i.e., i, $i+4$ hydrogen bonds).

$$\xi(R) = \sum_{i=1}^{n-4} \frac{1 - (r_{\mathrm{i,i+4}}/d_0)^n}{1 - (r_{\mathrm{i,i+4}}/d_0)^m} \tag{8.1}$$

Here the variable R represents the microscopic coordinates of the system. This particular CV definition was previously introduced by Camilloni et al. (2008) in their study to effectively capture the folding dynamics of helical structures. In enhanced sampling simulations that involve dynamics, that is, integration of equations of motion based on forces, computing biasing forces requires CVs that are continuous and differentiable. The parameters in Eq. (8.1) were set to $n = 6$, $m = 10$, and $d_0 = 0.32$ nm. The variable $r_{i,i+4}$ represents the distance between the backbone oxygen of residue i and the backbone nitrogen of residue $i+4$. This CV does not account for angular contributions. Therefore conformations where an oxygen atom at site i is near a nitrogen atom at site $i+4$ will contribute to the CV even if their positions do not meet common angular criteria for a hydrogen bond (e.g., the angle $\theta_{\mathrm{N\text{–}H\text{–}O}}$ being less than 30 degrees). Incorporating an angle cutoff would, on one hand, enhance the accuracy with which the CV depicts the formation of helices. However, it would also introduce a region of the phase space with a zero gradient of the CV with respect to the microscopic atomic coordinates. Notably the expression in Eq. (8.1) does not require specific knowledge of the folded structure. For each system, we conducted 100 ns SMD simulations in the NPT ensemble at 300 K and 1 bar, using a time step of 2 fs. To maintain temperature and pressure, we employed a Nose–Hoover thermostat (Evans & Holian, 1985) and a Parrinello–Rahman barostat (Parrinello & Rahman, 1981), both with a time constant of 2.0 ps. These simulations were aimed at generating snapshots of the unfolding process, transitioning from a fully α-helical state with a nonzero CV value to an unfolded state with a ξ value close to zero. The PLUMED library was utilized to evaluate the CV and compute the biasing forces.

We generated 61 starting structures using SMD. Subsequently, we conducted US simulations for each of these structures. In the US method (Kästner, 2011) a biasing potential is applied at specific positions along the reaction coordinate (in this case the CV representing the hydrogen bond number, ξ) to enhance the sampling of states with high free energy barriers. The reaction coordinate was segmented into 61 windows, each corresponding to one of the starting structures. A harmonic potential, as detailed in Eq. (8.2), was added to the original potential (unbiased potential) in each window. This was done to facilitate the exploration of the CV space by the protein.

$$\omega_i(\xi) = \frac{1}{2}k[\xi(R) - \xi_i]^2 \tag{8.2}$$

In this expression, ξ_i serves as the reference coordinate for the ith US window. During the simulation, any deviation of the CV from its reference value results in an added bias potential, which in turn exerts a force to push the system back toward the reference position, ensuring enhanced sampling around this target coordinate. Here, k is the force constant, and R describes the microscopic coordinates of the system as in Eq. (8.1). We employed the weighted histogram analysis method (WHAM) (Kumar et al., 1992) to remove the bias from the restrained US simulations and to construct the potential of mean force (PMF) using 400 bins along the CV.

8.2.2 Simulation protocol for binding free energy calculations

As test systems, we chose two α-MoRF BH3-only domain proteins of PUMA (PDB ID: 2ROC) and NOXA-A (PDB ID: 2ROD), where each is bound in a helix conformation to the antiapoptotic protein MCL-1. BH3-only domain proteins act as proapoptotic agents in many organisms, including humans. Each test complex underwent identical preparation steps, starting with: (1) obtaining the experimental structures from the PDB website (http://www.rcsb.org); (2) utilizing PyMOL (Yuan et al., 2017) for editing and removing additional chains from the PDBs to create interacting protein–ligand systems; and (3) filling all the missing residues or atoms in the PDB file using MODELER v9.15 (Webb & Sali, 2016).

CG simulations for each of the three systems were carried out using the SIRAH force field (Darré et al., 2015). The SIRAH CG force field has been designed with the intent to overcome several limitations commonly found in CG force fields, such as the use of a uniform dielectric constant, the absence of long-range interactions, reliance on topological information for maintaining secondary structures, and limited consideration of ionic strength effects. Unlike the popular Martini force field (Marrink & Tieleman, 2013; Periole & Marrink, 2013), which typically maps four heavy atoms to one CG bead, SIRAH maintains a higher level of detail in the peptide bonds, retaining the positions of backbone nitrogen (N), α-carbon (C_α), and oxygen (O) atoms, whereas the side chains are represented more coarsely. The WT4 water model incorporated in SIRAH consists of four interconnected beads with partial charges, enabling the simulation of dielectric permittivity. Furthermore the CG electrolytes in SIRAH can replicate effects of ionic strength and osmotic pressure. The resulting CG Hamiltonian features only terms that are supported in most MD simulation packages.

Before using SIRAH for US, we generated 91 windows using Visual Molecular Dynamics (VMD) (Humphrey et al., 1996) software. As CV, we defined d_{COM}, the distance of separation of the two centers of masses (COM) between the MCL-1 and BH3-only domain proteins. With a window spacing of 0.05 nm, each receptor–ligand pair was separated to a maximum d_{COM} of 4.5 nm to allow sufficient separation, complete solvation, and avoid residual interaction at maximum separation. Next, we applied SIRAH CG simulations to perform US simulations for each window. We adhered to the protocol described in Darré et al. (2015). The mapping of coordinates and analysis were executed using the SIRAH tools (Machado & Pantano, 2016). Before transitioning to the SIRAH CG model, protonation states were determined assuming a neutral pH, utilizing the PDB2PQR server (Dolinsky et al., 2004) and selecting the AMBER (Hornak et al., 2006) naming scheme for the output. After converting from all-atom to CG representations the protein complexes were positioned in a cubic box of 8 nm filled with SIRAH WT4 water. To achieve a neutral charge, Na^+ and Cl^- ions were added at a concentration of 0.15 mol/L. The MCL-1 complexes with PUMA and NOXA-A were solvated with 4995 and 4991 SIRAH WT4 water molecules, respectively. In both cases a minimum distance of 0.3 nm was maintained between all water molecules and the closest protein CG bead. The corresponding periodic box dimensions were chosen to guarantee that the two proteins do not interact with their own periodic images at their maximum separation distance. Each system was initially minimized using the steepest descent method for 10,000 steps. To allow for the equilibration of water molecules around the protein complex, each system was then subjected to a 5 ns simulation with the positions of all CG protein atoms in the complex harmonically restrained. During these restrained simulations the temperature of the systems was maintained at 300 K and the pressure at 1 atm using the V-rescale thermostat (Bussi et al., 2007) and the Parrinello–Rahman barostat (Parrinello & Rahman, 1981) with isotropic pressure coupling, respectively. Following this, US simulations without position restraints were carried out for a duration of 15 ns for equilibration and 100 ns for sampling.

All simulations were conducted with a time step of 20 fs and updated neighbor lists every 10 steps. Electrostatic interactions were calculated using the particle mesh Ewald method (Darden et al., 1993) with a direct cutoff of 1.2 nm and a grid spacing of 0.2 nm. For van der Waals interactions a cutoff of 1.2 nm was employed. In each umbrella window a harmonic restraint potential with a force constant of 500 kJ/mol/nm^2 was applied to the center of mass distance between the proteins.

Prior to initiating the US simulations, we implemented distance restraints on the backbone atoms of MCL-1 and PUMA separately. This was done to preserve a stable CG structure while applying the biasing potential during US, as illustrated in Fig. 8.1. The elastic network model (ENM) restraints (Bahar et al., 1997; Haliloglu et al., 1997; Hinsen, 1998; Mishra & Jernigan, 2018; Tirion, 1996; Togashi & Flechsig, 2018) were applied between O–N, C–N, C–O, and C–C atom pairs within a 5.25 A radius. As a result the root-mean-square deviations of both MCL-1 and PUMA were maintained within a range of 5–6 A, ensuring the structural integrity of the proteins during the simulations. To remove the bias from the simulations, we utilized the WHAM and constructed the PMFs using 400 bins along the protein center-of-mass distance used to define the CV. The final one-dimensional WHAM profile was adjusted to account for the missing Jacobian contribution (Trzesniak et al., 2007) to obtain the accurate PMF (Chodera & Mobley, 2013; Herschbach et al., 1959; Hub et al., 2010; Khavrutskii et al., 2008; Qian, 2006).

$$\mathrm{PMF}(R) = \mathrm{PMF}_{\mathrm{WHAM}}(R) + k_{\mathrm{B}}T\,\ln\left(4\pi r^2\right) \tag{8.3}$$

This adjustment provided in Eq. (8.3) ensures a flat PMF at large distances by compensating for the entropic increase in the number of configurations on a sphere of a given radius (Hub et al., 2010). The binding free energy, $\Delta_{\mathrm{bind}}G$, for a protein complex was calculated as the difference in the PMF between the bound and unbound states.

$$\Delta G_{\mathrm{bind}} = -k_B T\,\ln\left[\frac{P(\xi_{\mathrm{bound}})}{P(\xi_{\mathrm{unbound}})}\right] \tag{8.4}$$

where $P(\xi_{\mathrm{bound}})$ and $P(\xi_{\mathrm{unbound}})$ represent the probabilities of states on the PMF associated with the bound and unbound states along the CV, that is, the center-of-mass distance.

8.3 Results and discussion

8.3.1 Folding free energy penalties

To confirm the accuracy of our method for quantitatively estimating the folding free energies of disordered regions transitioning from coiled to α-helical states, we conducted a set of analyses. We

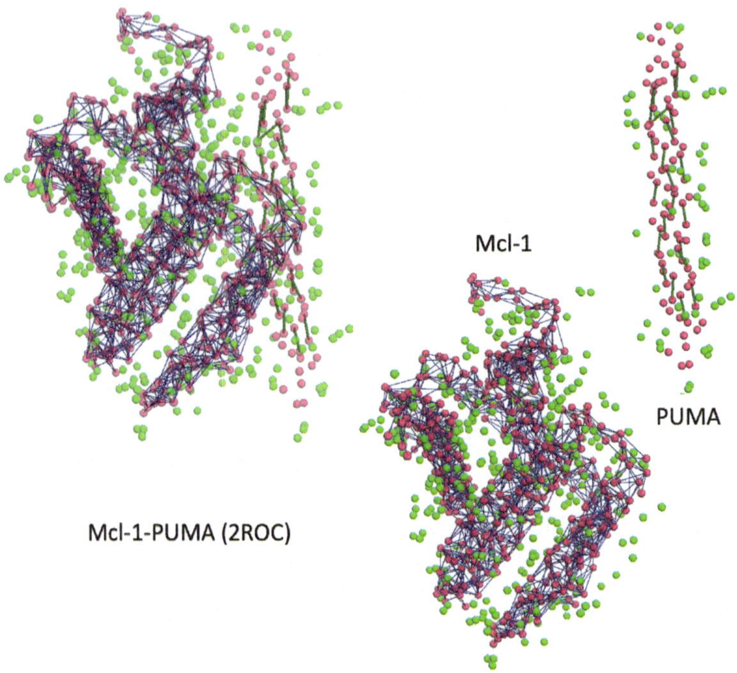

FIGURE 8.1

Elastic network models (ENMs): ENM applied to the backbone atoms of MCL-1 and PUMA individually to maintain a stable coarse-grained structure while applying biasing potential during umbrella sampling. The distance restraints are applied between O–N, C–N, C–O, and N–O atoms within 5.25 A radius. The *pink* and *green* spheres represent backbone and side chains, respectively.

focused initially on a 37-amino-acid segment of the c-Myc protein for this evaluation. This particular length was selected because it represents the longest disordered sequence identified as having the potential to fold into an α-helical structure. The c-Myc protein (Battey et al., 1983; Beaulieu et al., 2020; Dang, 2012; Jha et al., 2023; Madden et al., 2021; Meyer & Penn, 2008), with a molecular weight of 62 kDa and composed of 439 amino acid residues (https://www.uniprot.org/uniprotkb/P01106), contains multiple functional domains. These domains enable c-Myc to interact with a wide variety of binding partners, each contributing uniquely to the intricate network of interactions known as the c-Myc interactome. The transcription factor c-Myc is overexpressed in more than 70% of human cancers, both constitutively and abnormally. Studies have demonstrated that directly inhibiting c-Myc can lead to rapid tumor shrinkage in mice, accompanied by only mild and completely reversible side effects. This indicates that targeting c-Myc could be a feasible therapeutic approach. Although it does contain IDRs that contribute to its flexibility and interaction with various binding partners, c-Myc also has structured domains, such as the basic helix-loop-helix/leucine zipper domain, which is crucial for DNA binding and dimerization. The presence of both structured and disordered regions allows c-Myc to function as a versatile transcription factor involved in the regulation of numerous genes.

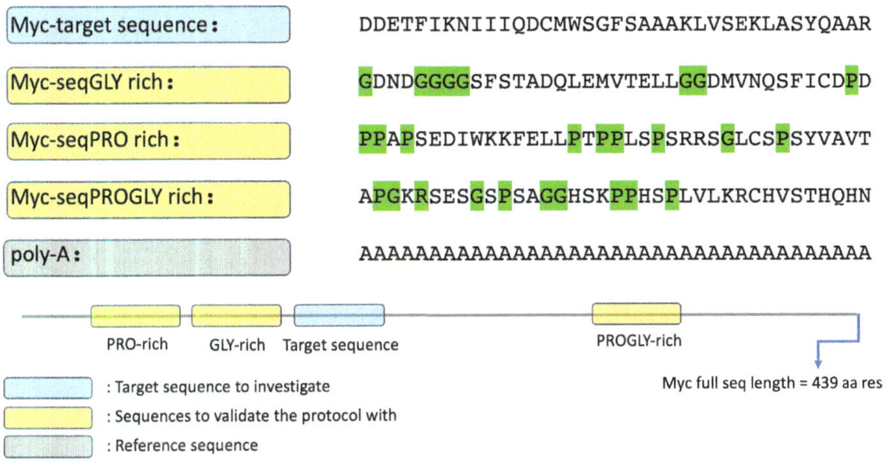

FIGURE 8.2

Sequences: Selected sequences to examine for validation of protocol. The 37-sequence polyalanine acts as a reference with a nearly intact α-helix structure.

We primarily chose four different sequences from the entire c-Myc protein sequence for our study. The first sequence was identified as a potential target because of the presence of several adjacent alanines and leucines known to stabilize helices. Moreover, AlphaFold's predicted structure for c-Myc (https://www.uniprot.org/uniprotkb/P01106) shows this region to be partially folded into an α-helix. The other three sequences were characterized by their high content of glycine, proline, or a combination of both. The residues of glycine and proline are known to destabilize α-helical structures in solution (Chou & Fasman, 1974; Lyu et al., 1990; O'Neil & DeGrado, 1990; Pace & Scholtz, 1998). For membrane proteins, however, they were found to stabilize the helical structures instead (Dong et al., 2012). As a reference sequence, we selected a polyalanine chain of 37 residues in length. This choice was based on its ability to maintain the highest α-helicity among all amino acids when simulated using the Amber03 force field (Best & Hummer, 2009; Duan et al., 2003) and the TIP3P water model (Jorgensen et al., 1983) in a 100 ns atomistic classical MD simulation. According to the literature, alanine is known to be the most stabilizing amino acid for α-helical structures, followed by leucine (Lyu et al., 1991; Pace & Scholtz, 1998; Rohl et al., 1999; Spek et al., 1999).

Fig. 8.2 presents the sequences we initially tested our protocol on. These sequences are located within the 439-residue chain of the c-Myc protein. The green highlighted residues are helix destabilizers present in the chosen sequences. These sequences were specifically chosen to deviate from the α-helical conformation. This deviation was intended to assist us in measuring the folding free energies with respect to a fully α-helical polyalanine reference.

Having determined the initial set of sequences for validating our protocol, we proceeded with the calculations of folding free energies. These calculations were carried out using US, as detailed in Section 8.2.

The resulting PMFs presented in Fig. 8.3 illustrate the unfolding process of a structure that begins in a fully α-helical conformation and transitions to an unfolded, coiled state. In this representation the

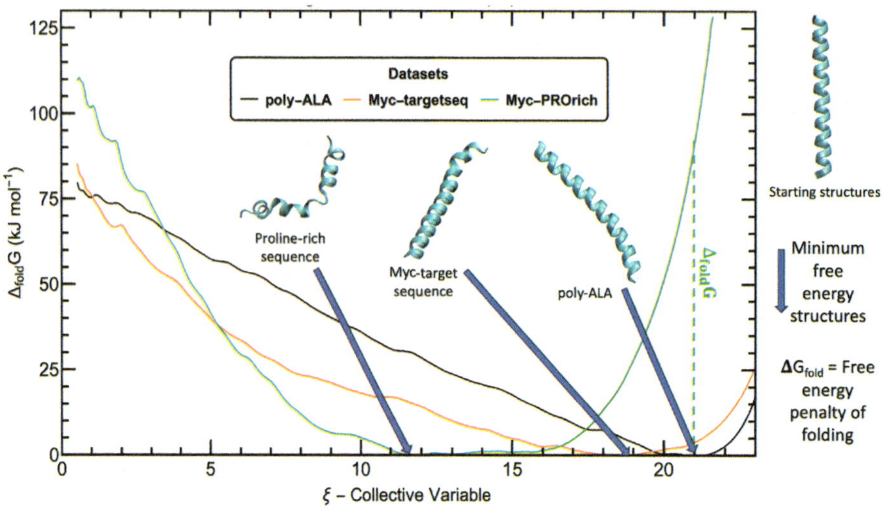

FIGURE 8.3

Potential of mean force (PMF): PMFs for the formation of α-helical secondary structures obtained via umbrella sampling simulations. Each structure shown as an inset corresponds to the corresponding free energy minimum (as indicated by *arrows*) along the folding pathway from coil to helix. The folding free energy penalty for a given sequence, $\Delta_{\text{fold}}G$, can be estimated by the PMF at the CV that corresponds to the free energy minimum of the polyalanine reference. *CV*, Collective variable.

reference structure of polyalanine is shown to maintain the closest resemblance to an α-helix. This is followed by our target sequence from the Myc protein and then by the sequence that is rich in prolines, which is expected to destabilize α-helices. The minimum of the PMF for polyalanine (all PMFs are shifted to yield a potential of zero for the respective minimum) represents a fully α-helical reference state at a value of ξ close to 21. The PMF of any other sequence at $\xi = 21$, $\Delta_{\text{fold}}G$ represents an estimate of the folding free energy penalty associated with transitioning from a disordered state to an α-helical state. This analysis helps us quantify the stability of α-helical structures and the energetic cost of folding for different sequences within the c-Myc protein.

In Fig. 8.4 the folding free energy penalties and the position of the folding free energy minimum along the CV are illustrated for different segments of the c-Myc protein, including our target fragment. The free energy penalty for folding our target c-Myc fragment into an α-helical structure is estimated to be around 5 kJ/mol. This value is significantly lower compared to the energy barrier of $\approx$98 kJ/mol observed for other segments rich in proline residues, which are known to destabilize α-helices. The relatively low folding free energy penalty for our target c-Myc fragment suggests that the folded conformation can be easily stabilized by favorable interactions upon binding. This characteristic makes it a promising candidate for drug development, as it implies that the segment could readily fold into an α-helix upon binding to a suitable ligand. This property is crucial for the binding-induced folding mechanism, where the interaction with a drug molecule triggers the folding of the target protein segment into a specific structure, in this case, an α-helix. Thus the favorable folding energetics of our target c-Myc fragment highlight its potential as a drug target sequence for

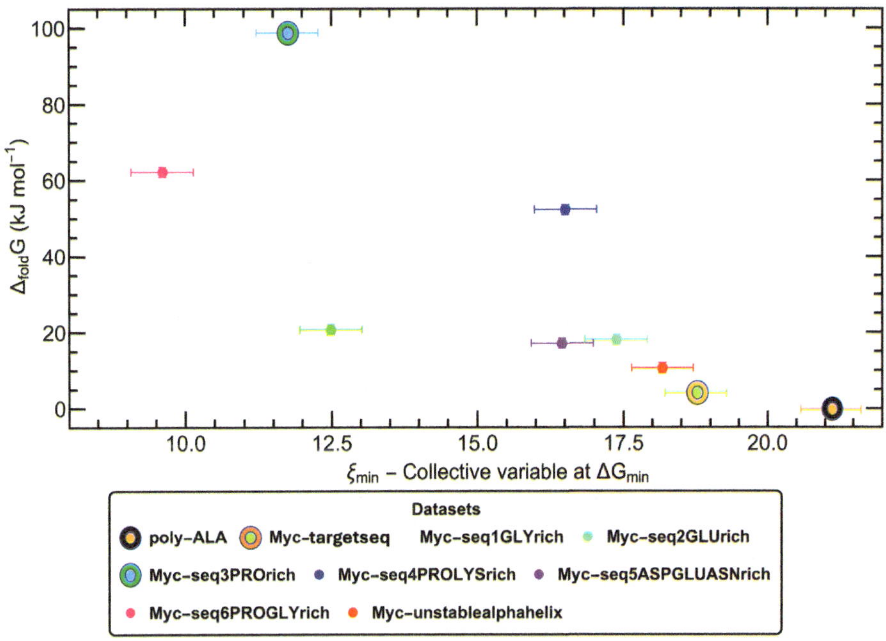

FIGURE 8.4

Folding free energies versus free energy minimum: Scatter plot of folding free energy penalties, $\Delta_{\text{fold}} G$, as a function of the position ξ_{min} of the free energy minimum along the coil-to-helix folding pathway for a collection of chosen fragments within the c-Myc sequence. In addition to the target sequence and the GLY and PRO-rich sequences illustrated in Fig. 8.2, we include data from c-Myc sequence fragments rich in other amino acids as indicated. As a reference, we include a sequence fragment of c-Myc that does not contain GLY or PRO residues but was predicted as unstructured by AlphaFold, that is, not to form a stable α-helix.

therapeutic interventions aimed at modulating the function of the c-Myc protein in cancer and other diseases.

Following our investigation of the target c-Myc fragment, we aimed to examine the impact of alanine point mutations on the stability of the α-helical conformations. Given that alanine is known to stabilize α-helices, substituting residues in a sequence with alanines should reduce the folding free energy penalty. To test this rationale, we introduced alanine mutations into sequences rich in glycine, proline, or both. The results of these mutations are presented in Fig. 8.5. We anticipated that replacing destabilizing residues with alanines would lead to a decrease in the folding free energy penalty, thereby increasing the stability of the α-helical structure. The observed trend in our results aligns with our hypothesis. We noticed a significant reduction in the free energy penalty for the proline-rich sequence upon mutation of proline residues to alanine. This reduction is expected because proline, with its rigid structure, impedes the formation of α-helices in solution by enthalpically destabilizing the helical structure. On the other hand the glycine-rich sequence, which is entropically unfavorable for α-helix formation, exhibited a smaller decrease in the folding free energy penalty upon mutation to alanine. This is because of the intrinsic flexibility of glycine, which contributes to a higher entropy in the unfolded state. The sequence

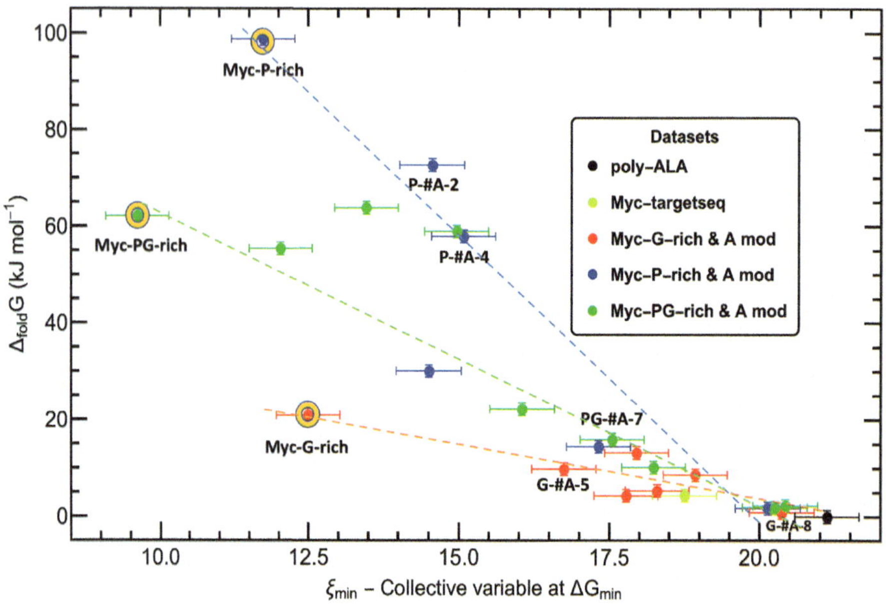

FIGURE 8.5

Helical folding propensities of alanine mutants: Scatter plot comparing the potential of mean force (PMF) features for polyalanine and the selected peptide sequences. A reduction in the folding free energy penalty, denoted as $\Delta_{\text{fold}}G$, is observed with increasing numbers of point mutations of destabilizing residues (PRO and GLY) in each sequence to ALA. Both the position of the free energy minimum along ξ_{min} and the folding free energy penalty gradually shifts toward the corresponding observations for polyalanine with increasing numbers of stabilizing mutations.

rich in both proline and glycine demonstrated an intermediate slope in the reduction of the folding free energy penalty upon mutation to alanine. This intermediate behavior reflects the combined effects of proline's enthalpic destabilization and glycine's (weaker) entropic destabilization of the α-helix.

To establish a general trend for the folding free energies, we conducted US on sequences of 37 residues in length for every naturally occurring polyamino acid peptide. Our results indicate that leucine, serine, and threonine have a high propensity for forming α-helical structures. In contrast, isoleucine, valine, tyrosine, and histidine demonstrate significant deviations from the α-helical form, primarily because of steric hindrance. The positively charged amino acids, lysine, and arginine, exhibit comparably low free energy penalties for α-helix formation. Among the negatively charged amino acids, glutamate shows a lower folding free energy penalty $\Delta_{\text{fold}}G$ compared to aspartate, which can be attributed to the reduced steric effects resulting from glutamate's longer side chain. These trends can be observed in Fig. 8.6.

Following the validation of our protocol using the selected sequences from the c-Myc protein, we expanded our research focus to include other naturally occurring IDRs and α-MoRFs. Our objective was to integrate the capability of our protocol to quantify the folding free energy penalty with the structural prediction prowess of AlphaFold (Jumper et al., 2021). By doing so, we aimed to accurately identify

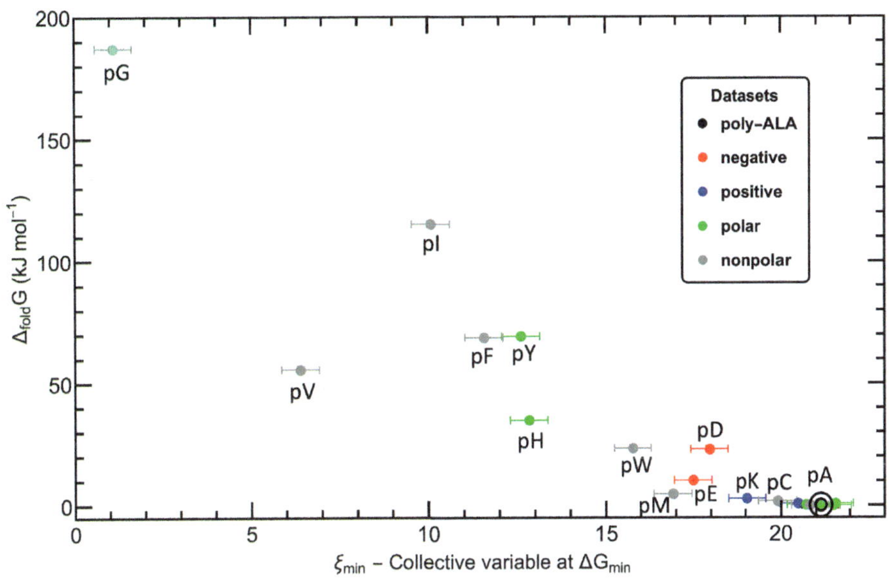

FIGURE 8.6

Helical folding propensities of polyamino acid peptides: Scatter plot comparing the folding free energy penalties $\Delta_{\text{fold}}G$ and position of free energy minima ξ_{min} for polyalanine and all other natural polyamino acid sequences.

or predict α-MoRFs and differentiate them from fully disordered IDRs. This approach would enhance our understanding of the structural dynamics of IDRs and their potential to transition into functionally relevant α-helical conformations upon interaction with binding partners. We selected two α-MoRFs: p53-CT (C-terminal domain of p53) and α-c-Myb-D (α-MoRF region of the c-Myb protein, where D stands for doubling the length of the single sequence for the purpose of our simulation) and three fully disordered sequences: dis-c-Myb (disordered fragment taken from c-Myb protein), dis2-p53-CT (disordered segment from p53 protein), and dis-Vc-HigA2-NT (disordered segment from the N-terminal region of Vc-HigA2 protein). We further include multiple other sequences from the same proteins with intermediate propensities for α-helical structure.

In Fig. 8.7, we observe a cluster of α-MoRF sequences (shown in *green*) and a cluster of fully disordered ones (shown in *red*). In evaluating the minimum free energy structures from our simulations against AlphaFold predictions, we observed that the folding free energies $\Delta_{\text{fold}}G$ of all the α-MoRF examples we investigated fell within the range of 0–10 kJ/mol (2.4 kcal/mol). This observation is in line with the findings of (Teilum et al., 2015), who reported complexes involving disordered proteins are, on average, less stable by approximately 2.5 kcal/mol or about 10.5 kJ/mol. This suggests that a certain amount of folding free energy might be associated with the binding-induced folding mechanism, where the binding event triggers the folding of the disordered protein into a more stable structure. This energy contribution is an important factor to consider when studying the dynamics and stability of protein complexes involving IDRs. Our finding also aligns with estimates provided by Hadži and Lah (2022), who utilized a modified LR model (Chakrabartty et al., 1994; Hadži et al., 2021; Lifson & Roig, 1961) to assess the folding free energies of IDPs (apart from discrepancies attributed to variations in IDP

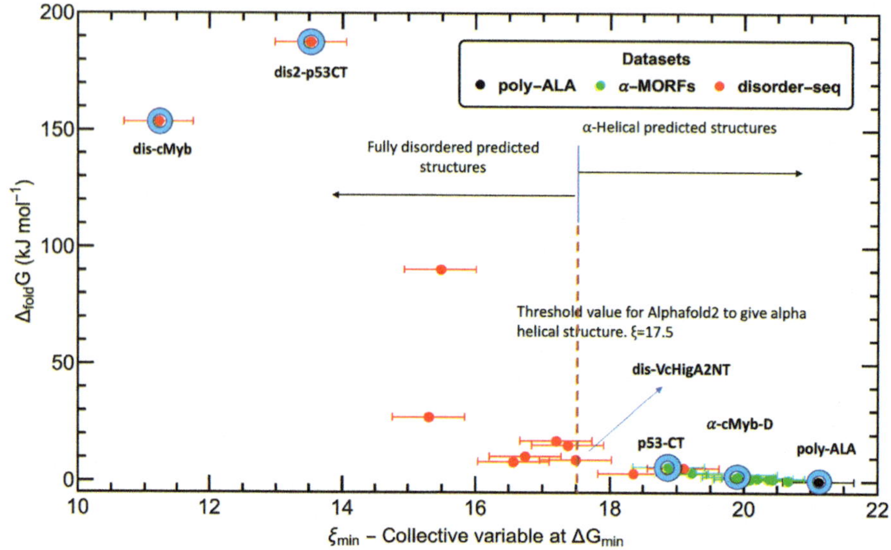

FIGURE 8.7

Helical folding propensities of naturally occurring IDRs: Folding free energy penalty $\Delta_{fold}G$ comparisons between IDRs from the c-Myb, p53, and Vc-HigA2 proteins with varying α-helical propensities. The threshold value of ξ_{min}, for which we consistently observed a prediction of α-helical structure by AlphaFold, is shown with a dotted line. *α-MoRFs*, α-Molecular recognition features; *IDRs*, intrinsically disordered regions.

sequence length). Low folding free energy penalties of a given sequence indicate its potential as a target, for example, in the development of peptide-based or small-molecule drugs. Such therapeutic agents have the potential to stabilize the helical structure of α-MoRFs by inducing folding upon binding, thereby offering a strategic approach for influencing protein–protein interactions in medical treatments. This approach could lead to the development of targeted therapies that specifically modulate the function of proteins involved in various diseases. We also identified a threshold value for the CV ξ_{min} that AlphaFold requires to predict α-helical structure. Specifically, for a 37-residue peptide, AlphaFold consistently predicts a fully disordered structure for a given sequence if we observe the position of the free energy minimum in our simulations below a value of 17.5. In addition, we observed that AlphaFold has a threshold for the folding free energy penalty, $\Delta_{fold}G$, of 0–10 kJ/mol. Consequently, all sequences with a folding free energy penalty below 10 kJ/mol were predicted to adopt an α-helical conformation. This finding is in alignment with the research conducted by Hadži and Lah (2022) and provides insights into the ability of the AlphaFold to distinguish preferences for secondary structure and disorder.

Finally, we compare the structures observed in unbiased simulations (representative structures observed in 100 ns MD trajectories starting from a helical conformation) with AlphaFold predictions. Specifically, we compare the simulated and predicted structures for the disordered sequences in c-Myb, p53, and VcHigA2 with the α-MoRF sequences in c-Myb and p53 with polyalanine as a reference in Fig. 8.8.

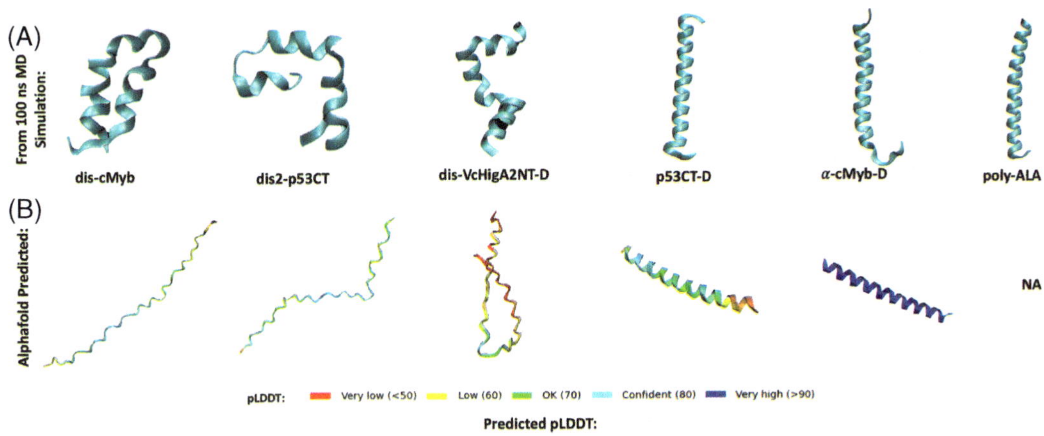

FIGURE 8.8

Comparison of simulated and predicted structures: Structure comparison of unbiased molecular dynamics simulations and AlphaFold predictions. (A) Representative conformations observed in 100 ns simulations starting from an α-helical conformation. (B) Structure predictions by AlphaFold with color-coded predicted local distance difference test (pLDDT) scores, which provide a confidence estimate and can be interpreted as a measure of disorder (low confidence = disorder).

We note that AlphaFold is proficient in predicting the helical structure of the two α-MoRFs, particularly in the case of α-c-Myb-D. In addition, AlphaFold is capable of accurately predicting disorder in sequences that yield a free energy minimum ξ_{min} below 17.5 in our folding free energy calculations for a 37-residue sequence. This threshold is exemplified by the structure of dis-VcHigA-2NT-D, a disordered protein that lies at the boundary of AlphaFold's ability to predict α-MoRFs. For values beyond this threshold, AlphaFold tends to predict that sequences will adopt an α-helical conformation. This observation underscores the ability of AlphaFold to recognize residual α-helical structure in α-MoRFs.

8.3.2 Binding free energy calculations of folded ensembles

Experimentally, characterizing the binding affinity of disordered α-MoRF peptides with their targets typically yields only the combined contributions of folding and binding (see Fig. 8.9). Separating these two contributions is not straightforward using conventional methods. Our goal here is to compare the sum of the folding and binding free energies derived from simulations with the experimentally determined binding free energies. Through this comparison, we aim to provide a more detailed understanding of the energetics involved in the interactions between IDPs and their binding partners.

To test our approach, we conducted CG US simulations of the binding process of two MCL-1-BH3 domain protein complexes. BH3 domains (Giam et al., 2008) are characteristic of the proapoptotic subset of Bcl-2 proteins (Youle & Strasser, 2008), which includes proteins like BIM, BAD, BID, PUMA, and NOXA-A. MCL-1 (Dahal et al., 2018; Garner et al., 2017; Senichkin et al., 2020; Youle & Strasser, 2008) is a key antiapoptotic protein that plays a critical role in cell survival and has

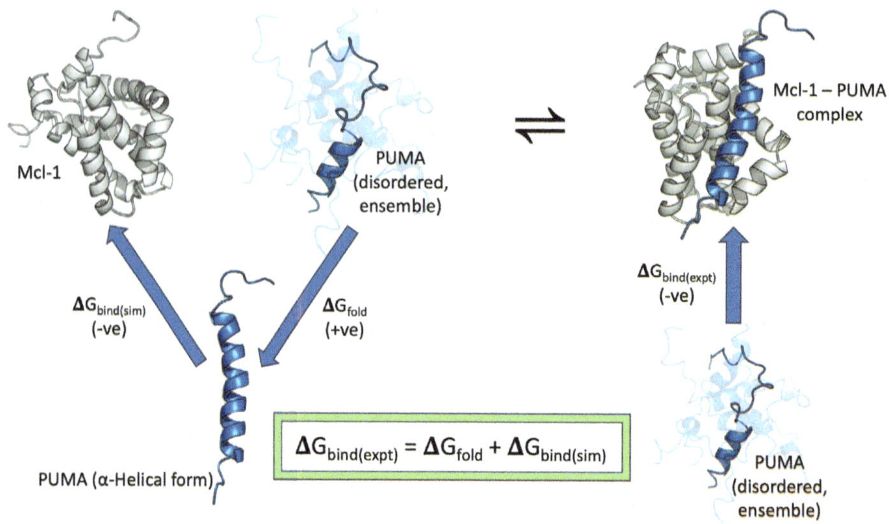

FIGURE 8.9

Binding as a two-step process: Expressing binding free energies as the sum of two processes in the binding-induced folding mechanism: folding and binding in the folded state. Adapted from Rogers et al. (2014).

been implicated in the development and progression of various cancers. Its ability to bind and inhibit proapoptotic proteins makes it a crucial regulator of the apoptotic pathway. As a result, MCL-1 has emerged as an attractive target for cancer therapy.

We note that both the structure of MCL-1 and peptide were stabilized by ENMs in our simulation as described in Section 8.2. This stabilized the native state of MCL-1 (which we observed to be necessary as in most other CG protein models despite the increased level of detail in SIRAH) as well as the helical state of the binding peptide. Although this limits the sampling problem in each distance-dependent umbrella window largely to relative positions and orientations of the protein, achieving adequate sampling can remain a challenge for unbound states in which one binding partner can freely explore the surface of a sphere around its target with a radius corresponding to the restrained center-of mass distance. As a consequence the PMF obtained from the simulation will, for sufficiently large distances and sampling, resemble $-k_BT\ln(4\pi r^2)$, which can be corrected as described in Eq. (8.3). As an alternative approach, cylindrical restraints can be used to restrict the separation path of the two binding partners. This approach restricts the ligand from exploring the entire space around the substrate in a spherical pattern, leading to faster convergence of the simulation with reduced sampling time and computational resources. Deriving the entropic correction term of the PMF for cylindrical restraints, akin to the one for Eq. (8.3), is an interesting analytical exercise. However, it is only noticeable for large cylindrical radii. Here, we verified that our PMFs resulting from simulations with and without cylindrical restraints (correcting the PMF using Eq. 8.3 in the latter case) yield equivalent binding free energies and PMFs.

The PMFs for the binding of PUMA and NOXA-A to MCL-1 are shown in Fig. 8.10. We determine the binding free energies, $\Delta_{\text{bind}}G_{\text{sim}}$, of both PUMA and NOXA-A in their prefolded helical states, from

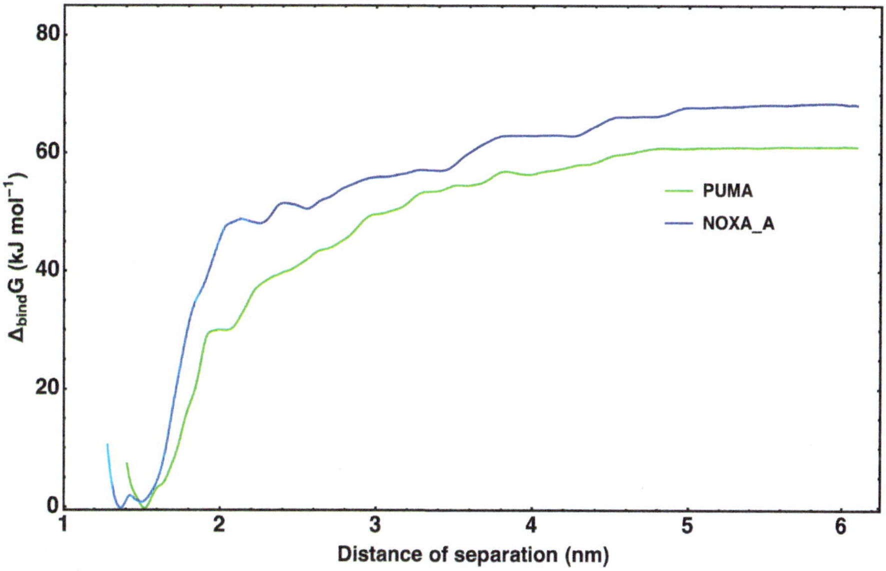

FIGURE 8.10

Binding free energy curves: Binding free energy curves for MCL-1 with the BH3-only domains of PUMA (*green*) and NOXA-A (*blue*).

the difference in free energy in the bound (free energy minimum) and unbound states (flat section of the PMF for large distances).

In Fig. 8.10 the calculated binding free energies $\Delta_{\text{bind}}G_{\text{sim}}$ for the protein complexes involving PUMA and NOXA-A were found to be −61.25 kJ/mol (−14.64 kcal/mol) and −68.58 kJ/mol (−16.39 kcal/mol), respectively. These values are noticeably lower than the experimental values $\Delta_{\text{bind}}G_{\text{exp}}$ reported by Dahal et al. (2018), which are −55.97 kJ/mol (−13.38 kcal/mol) for PUMA and −45.9 kJ/mol (−10.9 kcal/mol) for NOXA-A. This discrepancy suggests that, as expected, a positive folding free energy penalty, $\Delta_{\text{fold}}G$, needs to be accounted for to reconcile the excess binding free energy obtained from simulations of the helical peptides compared to experiments.

To calculate folding free energy penalties for both PUMA and NOXA-A peptides, we performed two sets of US simulations as already described for other peptides in the previous section. Fig. 8.11 illustrates that the folding free energy penalty for PUMA is 1.8 ± 1.3 kJ/mol, whereas for NOXA-A, it is approximately 0.5 ± 1.3 kJ/mol. Both are on the lower end of the 0–10 kJ/mol window expected for α-MoRFs, which indicates a significant population of the helical state even prior to binding. In both cases, the folding free energy penalty $\Delta_{\text{fold}}G$ does not fully capture the difference between $\Delta_{\text{bind}}G_{\text{exp}}$ and $\Delta_{\text{bind}}G_{\text{sim}}$, which amounts to 5.35 kJ/mol for PUMA and 22.68 kJ/mol for NOXA-A. However, for PUMA the total simulated binding free energy, obtained by combining the folding free energy penalty $\Delta_{\text{fold}}G$ of 1.8 ± 1.3 kJ/mol and the binding free energy $\Delta_{\text{bind}}G_{\text{sim}}$ of −61.25 kJ/mol from simulations of the helical peptides, amounts to −59.5 kJ/mol, which is within an acceptable error of 4 kJ/mol (less than 1 kcal/mol) from the experimental binding free energy of −55.97 kJ/mol.

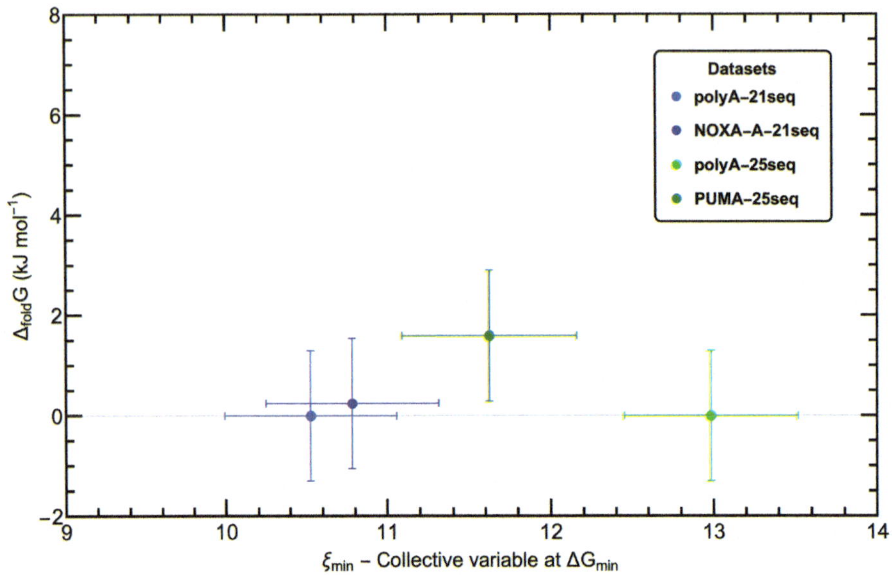

FIGURE 8.11

Helical folding propensity of PUMA and NOXA-A: Scattered plots depicting folding free energy penalties for MCL-1 with PUMA (*green*) and NOXA-A (*blue*).

On the other hand, our approach clearly overestimates binding for NOXA-A. Adding $\Delta_{\text{fold}}G$ and $\Delta_{\text{bind}}G_{\text{sim}}$ in that case yields −68.1 kJ/mol, which exceeds the affinity observed in the experiment (−45.9 kJ/mol) by 21 kJ/mol. Several factors may contribute to this discrepancy: (1) Sequence length: The discrepancy may arise from differences in the length of the sequences used in our simulations compared to those used in experiments. The length of the sequence can influence the folding dynamics and, consequently, the folding free energy penalty. (2) Mutations: The use of mutated versions of the proteins in the experiments described by Dahal et al. (2018), which differ from the structures in the PDB used as a starting point of the simulations, could lead to variations in the folding free energy. (3) Sequence composition: Our results for NOXA-A might have been more accurate if we had used a longer sequence that included proline and glycine residues. These residues are known to destabilize α-helices and could result in a higher folding free energy penalty, bringing our simulated values closer to the experimental ones. (4) Simulation model: Predictions of the folding free energy penalties of the peptides may be limited by the accuracy of the underlying all-atom force field potential. Potentially even more likely the CG protein–protein interactions described by the SIRAH model might not capture all types of interactions equally well, for example, protein–protein hydrogen bonds or hydrophobic interactions.

Addressing these factors in future simulations may reduce the discrepancy between the simulated and experimental folding free energy values for PUMA and NOXA-A. However, our hypothesis that the experimental binding free energy can be represented as the sum of the calculated folding and binding free energies is partially verified, and additional studies involving other sequences are necessary to gather more conclusive data. Expanding the scope of the simulations described here to a wider range of sequences will help to further enhance our understanding of the energetics involved in protein–protein interactions.

8.4 Conclusion

In this chapter, we delved into the intricate dynamics and thermodynamics of protein folding and binding processes, with a specific emphasis on IDRs and α-MoRFs present in the c-Myc protein and others. In addition, we examined the interactions between MCL-1 and BH3-only domains, which play pivotal roles in the regulation of apoptosis. Our primary aim was to quantify the folding free energy penalties that occur as proteins transition from disordered states to α-helical conformations. Furthermore, we sought to investigate the binding free energies of protein complexes, specifically those involving MCL-1 in combination with PUMA and NOXA-A, which are key players in apoptotic pathways. By understanding these energetics, we aimed to shed light on the molecular mechanisms underlying protein interactions and their implications for apoptotic regulation.

Through MD simulations and US, we first validated our protocol by analyzing sequences from the c-Myc protein, noting that the folding free energies fell within the anticipated range. This initial validation laid the groundwork for an extensive examination of naturally occurring IDRs and α-MoRFs. In this broader analysis, we harnessed the capabilities of our protocol in conjunction with structure predictions of AlphaFold. Our results demonstrated that AlphaFold was adept at predicting the α-helical structures of α-MoRFs, particularly in the case of α-c-Myb-D, and could identify sequences as fully disordered up to a specified threshold.

The computation of binding free energies as a sum of folding free energy penalties and binding free energies in the folded state unveiled partial agreement with experimental values. Specifically, PUMA binding to MCL-1 was reproduced quite accurately, whereas the binding free energy of NOXA-A was overestimated. In both cases the binding free energy penalty was comparably small and thus provided only a minor correction to the binding free energy. Overall, based on all sequences studied here, we typically expect a range from 0 to 10 kJ/mol for the positive penalty, which needs to be compensated by favorable interactions with the target to achieve binding.

In summary the decomposition of experimental binding free energies using simulation approaches as described here may reveal additional information for other systems. The intricate nature of protein folding and binding mechanisms underscores the necessity for additional investigations across a broader spectrum of sequences. This work enhances our comprehension of the energetic aspects of protein–protein interactions and has potential implications for the development of pharmaceuticals and the exploration of molecular-level biological phenomena.

AI disclosure

During the preparation of this work the author(s) used ChatGPT to do language editing. After using this tool/service the author(s) reviewed and edited the content as needed and take(s) full responsibility for the content of the publication.

References

Abraham, M. J., Murtola, T., Schulz, R., Páll, S., Smith, J. C., Hess, B., & Lindah, E. (2015). Gromacs: High performance molecular simulations through multi-level parallelism from laptops to supercomputers. *SoftwareX, 1–2*, 19–25. doi:10.1016/j.softx.2015.06.001.

Arai, M., Sugase, K., Dyson, H. J., & Wright, P. E. (2015). Conformational propensities of intrinsically disordered proteins influence the mechanism of binding and folding. *Proceedings of the National Academy of Sciences of the United States of America, 112*(31), 9614–9619. http://www.pnas.org/content/112/31/9614.full.pdf. doi:10.1073/pnas.1512799112.

Babu, M. M. (2016). The contribution of intrinsically disordered regions to protein function, cellular complexity, and human disease. *Biochemical Society Transactions, 44*(5), 1185–1200. http://www.biochemsoctrans.org/content/44/5/1185.full-text.pdf. doi:10.1042/BST20160172.

Bahar, I., Rana Atilgan, A., & Erman, B. (1997). Direct evaluation of thermal fluctuations in proteins using a single-parameter harmonic potential. *Folding and Design, 2*(3), 173–181. doi:10.1016/S1359-0278(97)00024-2.

Battey, J., Moulding, C., Taub, R., Murphy, W., Stewart, T., Potter, H., Lenoir, G., & Leder, P. (1983). The human c-myc oncogene: Structural consequences of translocation into the IgH locus in Burkitt lymphoma. *Cell, 34*(3), 779–787. doi:10.1016/0092-8674(83)90534-2.

Beaulieu, M.-E., Castillo, F., & Soucek, L. (2020). Structural and biophysical insights into the function of the intrinsically disordered myc oncoprotein. *Cells, 9*(4). doi:10.3390/cells9041038.

Best, R. B., & Hummer, G. (2009). Optimized molecular dynamics force fields applied to the helix-coil transition of polypeptides. *Journal of Physical Chemistry B, 113*(26), 9004–9015. doi:10.1021/jp901540t.

Bonomi, M., Branduardi, D., Bussi, G., Camilloni, C., Provasi, D., Raiteri, P., Donadio, D., Marinelli, F., Pietrucci, F., Broglia, R. A., & Parrinello, M. (2009). PLUMED: A portable plugin for free-energy calculations with molecular dynamics. *Computer Physics Communications, 180*(10), 1961–1972. doi:10.1016/j.cpc.2009.05.011.

Bonomi, M., Bussi, G., Camilloni, C., Tribello, G. A., Banáš, P., Barducci, A., Bernetti, M., Bolhuis, P. G., Bottaro, S., Branduardi, D., Capelli, R., Carloni, P., Ceriotti, M., Cesari, A., Chen, H., Chen, W., Colizzi, F., De, S., De La Pierre, M., … White, A. (2019). Promoting transparency and reproducibility in enhanced molecular simulations. *Nature Methods, 16*(8), 670–673. doi:10.1038/s41592-019-0506-8.

Bugge, K., Staby, L., Salladini, E., Falbe-Hansen, R. G., Kragelund, B. B., & Skriver, K. (2021). $\alpha\alpha$-Hub domains and intrinsically disordered proteins: A decisive combo. *Journal of Biological Chemistry, 296*, 100226. doi:10.1074/jbc.rev120.012928.

Bussi, G., Donadio, D., & Parrinello, M. (2007). Canonical sampling through velocity rescaling. *Journal of Chemical Physics, 126*(1). doi:10.1063/1.2408420.

Camilloni, C., Provasi, D., Tiana, G., & Broglia, R. A. (2008). Exploring the protein G helix free energy surface by solute tempering metadynamics. *Proteins: Structure, Function and Genetics, 71*(4), 1647–1654. doi:10.1002/prot.21852.

Carr, D. W., Stofko-Hahn, R. E., Fraser, I. D., Bishop, S. M., Acott, T. S., Brennan, R. G., & Scott, J. D. (1991). Interaction of the regulatory subunit (RII) of cAMP-dependent protein kinase with RII-anchoring proteins occurs through an amphipathic helix binding motif. *Journal of Biological Chemistry, 266*(22), 14188–14192. doi:10.1016/s0021-9258(18)98665-5.

Chakrabartty, A., Kortemme, T., & Baldwin, R. L. (1994). Helix propensities of the amino acids measured in alanine-based peptides without helix-stabilizing side-chain interactions. *Protein Science, 3*(5), 843–852. doi:10.1002/pro.5560030514.

Chodera, J. D., & Mobley, D. L. (2013). Entropy-enthalpy compensation: Role and ramifications in biomolecular ligand recognition and design. *Annual Review of Biophysics, 42*(1), 121–142. doi:10.1146/annurev-biophys-083012-130318.

Chou, P. Y., & Fasman, G. D. (1974). Conformational parameters for amino acids in helical, beta-sheet, and random coil regions calculated from proteins. *Biochemistry, 13*(2), 211–222.

Cumberworth, A., Lamour, G., Babu, M. M., & Gsponer, J. (2013). Promiscuity as a functional trait: Intrinsically disordered regions as central players of interactomes. *Biochemical Journal, 454*(3), 361–369. http://www.biochemj.org/bj/454/0361/4540361.pdf. doi:10.1042/BJ20130545Canada.

Dahal, L., Kwan, T. O. C., Hollins, J. J., & Clarke, J. (2018). Promiscuous and selective: How intrinsically disordered BH3 proteins interact with their pro-survival partner MCL-1. *Journal of Molecular Biology, 430*(16), 2468–2477. https://www.journals.elsevier.com/journal-of-molecular-biology. doi:10.1016/j.jmb.2018.04.004.

Dang, C. V. (2012). MYC on the path to cancer. *Cell, 149*(1), 22–35. doi:10.1016/j.cell.2012.03.003.

Darden, T., York, D., & Pedersen, L. (1993). Particle mesh Ewald: An N•log(N) method for Ewald sums in large systems. *The Journal of Chemical Physics, 98*(12), 10089–10092. doi:10.1063/1.464397.

Darré, L., Machado, M. R., Brandner, A. F., González, H. C., Ferreira, S., & Pantano, S. (2015). SIRAH: A structurally unbiased coarse-grained force field for proteins with aqueous solvation and long-range electrostatics. *Journal of Chemical Theory and Computation, 11*(2), 723–739. doi:10.1021/ct5007746.

Dolinsky, T. J., Nielsen, J. E., McCammon, J. A., & Baker, N. A. (2004). PDB2PQR: An automated pipeline for the setup of Poisson-Boltzmann electrostatics calculations. *Nucleic Acids Research, 32*(Web Server Issue). doi:10.1093/nar/gkh381.

Dong, H., Sharma, M., Zhou, H.-X., & Cross, T. A. (2012). Glycines: Role in α-helical membrane protein structures and a potential indicator of native conformation. *Biochemistry, 51*(24), 4779–4789. doi:10.1021/bi300090x.

Drobnak, I., De Jonge, N., Haesaerts, S., Vesnaver, G., Loris, R., & Lah, J. (2013). Energetic basis of uncoupling folding from binding for an intrinsically disordered protein. *Journal of the American Chemical Society, 135*(4), 1288–1294. doi:10.1021/ja305081b.

Duan, Y., Wu, C., Chowdhury, S., Lee, M. C., Xiong, G., Zhang, W., Yang, R., Cieplak, P., Luo, R., Lee, T., Caldwell, J., Wang, J., & Kollman, P. (2003). A point-charge force field for molecular mechanics simulations of proteins based on condensed-phase quantum mechanical calculations. *Journal of Computational Chemistry, 24*(16), 1999–2012. doi:10.1002/jcc.10349.

Dunker, A. K., Lawson, J. D., Brown, C. J., Williams, R. M., Romero, P., Oh, J. S., Oldfield, C. J., Campen, A. M., Ratliff, C. M., Hipps, K. W., Ausio, J., Nissen, M. S., Reeves, R., Kang, C. H., Kissinger, C. R., Bailey, R. W., Griswold, M. D., Chiu, W., Garner, E. C., & Obradovic, Z. (2001). Intrinsically disordered protein. *Journal of Molecular Graphics and Modelling, 19*(1), 26–59. doi:10.1016/S1093-3263(00)00138-8.

Dyson, H. J., & Wright, P. E. (2002). Coupling of folding and binding for unstructured proteins. *Current Opinion in Structural Biology, 12*(1), 54–60. http://www.elsevier.com/locate/sbi. doi:10.1016/S0959-440X(02)00289-0.

Dyson, H. J., & Wright, P. E. (2016). Role of intrinsic protein disorder in the function and interactions of the transcriptional coactivators CREB-binding protein (CBP) and p300. *Journal of Biological Chemistry, 291*(13), 6714–6722. http://www.jbc.org/content/291/13/6714.full.pdf+html. doi:10.1074/jbc.R115.692020.

Emenecker, R. J., Griffith, D., & Holehouse, A. S. (2021). Metapredict: A fast, accurate, and easy-to-use predictor of consensus disorder and structure. *Biophysical Journal, 120*(20), 4312–4319. http://www.sciencedirect.com/science/journal/00063495. doi:10.1016/j.bpj.2021.08.039.

Evans, D. J., & Holian, B. L. (1985). The Nose-Hoover thermostat. *The Journal of Chemical Physics, 83*(8), 4069–4074. doi:10.1063/1.449071.

Fatafta, H., Samantray, S., Sayyed-Ahmad, A., Coskuner-Weber, O., & Strodel, B. (2021). Molecular simulations of IDPs: From ensemble generation to IDP interactions leading to disorder-to-order transitions. *Progress in Molecular Biology and Translational Science, 183*, 135–185. http://www.elsevier.com/books/book-series/progress-in-molecular-biology-and-translational-science#. doi:10.1016/bs.pmbts.2021.06.003.

Flock, T., Weatheritt, R. J., Latysheva, N. S., & Babu, M. M. (2014). Controlling entropy to tune the functions of intrinsically disordered regions. *Current Opinion in Structural Biology, 26*(1), 62–72. http://www.elsevier.com/locate/sbi. doi:10.1016/j.sbi.2014.05.007.

Fuxreiter, M., Simon, I., Friedrich, P., & Tompa, P. (2004). Preformed structural elements feature in partner recognition by intrinsically unstructured proteins. *Journal of Molecular Biology, 338*(5), 1015–1026. doi:10.1016/j.jmb.2004.03.017.

Garner, E., Romero, P., Dunker, A. K., Brown, C., & Obradovic, Z. (1999). Predicting binding regions within disordered proteins. *Genome Informatics, 10*, 41–50.

Garner, T. P., Lopez, A., Reyna, D. E., Spitz, A. Z., & Gavathiotis, E. (2017). Progress in targeting the BCL-2 family of proteins. *Current Opinion in Chemical Biology, 39*, 133–142. doi:10.1016/j.cbpa.2017.06.014.

Giam, M., Huang, D. C. S., & Bouillet, P. (2008). BH3-only proteins and their roles in programmed cell death. *Oncogene, 27*. doi:10.1038/onc.2009.50.

Gnutt, D., Gao, M., Brylski, O., Heyden, M., & Ebbinghaus, S. (2015). Excluded-volume effects in living cells. *Angewandte Chemie—International Edition, 54*(8), 2548–2551. doi:10.1002/anie.201409847.

Hadži, S., & Lah, J. (2022). The free energy folding penalty accompanying binding of intrinsically disordered α-helical motifs. *Protein Science, 31*(7). doi:10.1002/pro.4370.

Hadži, S., Loris, R., & Lah, J. (2021). The sequence-ensemble relationship in fuzzy protein complexes. *Proceedings of the National Academy of Sciences of the United States of America, 118*(37). https://www.pnas.org/content/118/37/e2020562118. doi:10.1073/pnas.2020562118.

Haliloglu, T., Bahar, I., & Erman, B. (1997). Gaussian dynamics of folded proteins. *Physical Review Letters, 79*(16), 3090–3093. doi:10.1103/PhysRevLett.79.3090.

Herschbach, D. R., Johnston, H. S., & Rapp, D. (1959). Molecular partition functions in terms of local properties. *The Journal of Chemical Physics, 31*(6), 1652–1661. doi:10.1063/1.1730670.

Hess, B., Bekker, H., Berendsen, H. J. C., & Fraaije, J. G. E. M. (1997). LINCS: A linear constraint solver for molecular simulations. *Journal of Computational Chemistry, 18*(12), 1463–1472.

Hinsen, K. (1998). Analysis of domain motions by approximate normal mode calculations. *Proteins, 33*(3), 417–429.

Holehouse, A. S., & Kragelund, B. B. (2024). The molecular basis for cellular function of intrinsically disordered protein regions. *Nature Reviews Molecular Cell Biology, 25*(3), 187–211. https://www.nature.com/molcellbio. doi:10.1038/s41580-023-00673-0.

Hornak, V., Abel, R., Okur, A., Strockbine, B., Roitberg, A., & Simmerling, C. (2006). Comparison of multiple amber force fields and development of improved protein backbone parameters. *Proteins: Structure, Function and Genetics, 65*(3), 712–725. doi:10.1002/prot.21123.

Hub, J. S., de Groot, B. L., & van der Spoel, D. (2010). G-whams-A free weighted histogram analysis implementation including robust error and autocorrelation estimates. *Journal of Chemical Theory and Computation, 6*(12), 3713–3720. doi:10.1021/ct100494z.

Humphrey, W., Dalke, A., & Schulten, K. (1996). VMD: Visual molecular dynamics. *Journal of Molecular Graphics, 14*(1), 33–38. doi:10.1016/0263-7855(96)00018-5.

Jha, R. K., Kouzine, F., & Levens, D. (2023). MYC function and regulation in physiological perspective. *Frontiers in Cell and Developmental Biology, 11*. doi:10.3389/fcell.2023.1268275.

Jorgensen, W. L., Chandrasekhar, J., Madura, J. D., Impey, R. W., & Klein, M. L. (1983). Comparison of simple potential functions for simulating liquid water. *The Journal of Chemical Physics, 79*(2), 926–935. doi:10.1063/1.445869.

Jumper, J., Evans, R., Pritzel, A., Green, T., Figurnov, M., Ronneberger, O., Tunyasuvunakool, K., Bates, R., Žídek, A., Potapenko, A., Bridgland, A., Meyer, C., Kohl, S. A. A., Ballard, A. J., Cowie, A., Romera-Paredes, B., Nikolov, S., Jain, R., Adler, J., … Hassabis, D. (2021). Highly accurate protein structure prediction with AlphaFold. *Nature, 596*(7873), 583–589. doi:10.1038/s41586-021-03819-2.

Kästner, J. (2011). Umbrella sampling. *Wiley Interdisciplinary Reviews: Computational Molecular Science, 1*(6), 932–942. doi:10.1002/wcms.66.

Khavrutskii, I. V., Dzubiella, J., & McCammon, J. A. (2008). Computing accurate potentials of mean force in electrolyte solutions with the generalized gradient-augmented harmonic Fourier beads method. *Journal of Chemical Physics, 128*(4). doi:10.1063/1.2825620.

Kim, D. H., & Han, K. H. (2018a). PreSMo target-binding signatures in intrinsically disordered proteins. *Molecules and Cells, 41*(10), 889–899. http://www.molcells.org/main.html. doi:10.14348/molcells.2018.0192.

Kim, D. H., & Han, K. H. (2018b). Transient secondary structures as general target-binding motifs in intrinsically disordered proteins. *International Journal of Molecular Sciences, 19*(11). https://www.mdpi.com/1422-0067/19/11/3614/pdf. doi:10.3390/ijms19113614.

Krivov, G. G., Shapovalov, M. V., & Dunbrack, R. L., Jr. (2009). Improved prediction of protein side-chain conformations with SCWRL4. *Proteins: Structure, Function and Bioinformatics, 77*(4), 778–795. doi:10.1002/prot.22488.

Kumar, S., Rosenberg, J. M., Bouzida, D., Swendsen, R. H., & Kollman, P. A. (1992). THE weighted histogram analysis method for free-energy calculations on biomolecules. I. The method. *Journal of Computational Chemistry, 13*(8), 1011–1021. doi:10.1002/jcc.540130812.

Lazar, T., Tantos, A., Tompa, P., & Schad, E. (2022). Intrinsic protein disorder uncouples affinity from binding specificity. *Protein Science, 31*(11). doi:10.1002/pro.4455.

Lifson, S., & Roig, A. (1961). On the theory of helix—Coil transition in polypeptides. *The Journal of Chemical Physics, 34*(6), 1963–1974. doi:10.1063/1.1731802.

Lu, H., & Schulten, K. (1999). Steered molecular dynamics simulations of force-induced protein domain unfolding. *Proteins, 35*(4), 453–463.

Lyu, P. C., Liff, M. I., Marky, L. A., & Kallenbach, N. R. (1990). Side chain contributions to the stability of alpha-helical structure in peptides. *Science (New York, N.Y.), 250*(4981), 669–673.

Lyu, P. C., Sherman, J. C., Chen, A., & Kallenbach, N. R. (1991). Alpha-helix stabilization by natural and unnatural amino acids with alkyl side chains. *Proceedings of the National Academy of Sciences of the United States of America, 88*(12), 5317–5320.

Machado, M. R., & Pantano, S. (2016). SIRAH tools: Mapping, backmapping and visualization of coarse-grained models. *Bioinformatics, 32*(10), 1568–1570. doi:10.1093/bioinformatics/btw020.

Madden, S. K., Dantas de Araujo, A., Gerhardt, M., Fairlie, D. P., & Mason, J. M. (2021). Taking the Myc out of cancer: Toward therapeutic strategies to directly inhibit c-Myc. *Molecular Cancer, 20*(1). doi:10.1186/s12943-020-01291-6.

Marrink, S. J., & Tieleman, D. P. (2013). Perspective on the martini model. *Chemical Society Reviews, 42*(16), 6801–6822. doi:10.1039/C3CS60093A.

Mészáros, B., Simon, I., Dosztányi, Z., & Casadio, R. (2009). Prediction of protein binding regions in disordered proteins. *PLoS Computational Biology, 5*(5), e1000376. doi:10.1371/journal.pcbi.1000376.

Meyer, N., & Penn, L. Z. (2008). Reflecting on 25 years with MYC. *Nature Reviews Cancer, 8*(12), 976–990. doi:10.1038/nrc2231.

Mishra, S. K., & Jernigan, R. L. (2018). Protein dynamic communities from elastic network models align closely to the communities defined by molecular dynamics. *PLoS ONE, 13*(6). doi:10.1371/journal.pone.0199225.

Miyamoto, S., & Kollman, P. A. (1992). Settle: An analytical version of the SHAKE and RATTLE algorithm for rigid water models. *Journal of Computational Chemistry, 13*(8), 952–962. doi:10.1002/jcc.540130805.

Mohan, A., Oldfield, C. J., Radivojac, P., Vacic, V., Cortese, M. S., Dunker, A. K., & Uversky, V. N. (2006). Analysis of molecular recognition features (MoRFs). *Journal of Molecular Biology, 362*(5), 1043–1059. https://www.journals.elsevier.com/journal-of-molecular-biology. doi:10.1016/j.jmb.2006.07.087.

Oldfield, C. J., Cheng, Y., Cortese, M. S., Brown, C. J., Uversky, V. N., & Bunker, A. K. (2005a). Comparing and combining predictors of mostly disordered proteins. *Biochemistry, 44*(6), 1989–2000. doi:10.1021/bi047993o.

Oldfield, C. J., Cheng, Y., Cortese, M. S., Romero, P., Uversky, V. N., & Dunker, A. K. (2005b). Coupled folding and binding with α-helix-forming molecular recognition elements. *Biochemistry, 44*(37), 12454–12470. doi:10.1021/bi050736e.

O'Neil, K. T., & DeGrado, W. F. (1990). A thermodynamic scale for the helix-forming tendencies of the commonly occurring amino acids. *Science, 250*(4981), 646–651.

Pace, C. N., & Scholtz, J. M. (1998). A helix propensity scale based on experimental studies of peptides and proteins. *Biophysical Journal, 75*(1), 422–427. doi:10.1016/S0006-3495(98)77529-0.

Pancsa, R., & Fuxreiter, M. (2012). Interactions via intrinsically disordered regions: What kind of motifs? *IUBMB Life, 64*(6), 513–520. doi:10.1002/iub.1034.

Parrinello, M., & Rahman, A. (1981). Polymorphic transitions in single crystals: A new molecular dynamics method. *Journal of Applied Physics, 52*(12), 7182–7190. doi:10.1063/1.328693.

Periolse, X., & Marrink, S.-J. (2013). *The Martini coarse-grained force field* (pp. 533–565). Humana Press. doi:10.1007/978-1-62703-017-5_20.

Poosapati, A., Gregory, E., Borcherds, W. M., Chemes, L. B., & Daughdrill, G. W. (2018). Uncoupling the folding and binding of an intrinsically disordered protein. *Journal of Molecular Biology, 430*(16), 2389–2402. https://www.journals.elsevier.com/journal-of-molecular-biology. doi:10.1016/j.jmb.2018.05.045.

Qian, H. (2006). An asymptotic comparative analysis of the thermodynamics of non-covalent association. *Journal of Mathematical Biology, 52*(3), 277–289. doi:10.1007/s00285-005-0353-3.

Rogers, J. M., Wong, C. T., & Clarke, J. (2014). Coupled folding and binding of the disordered protein PUMA does not require particular residual structure. *Journal of the American Chemical Society, 136*(14), 5197–5200. http://pubs.acs.org/journal/jacsat. doi:10.1021/ja4125065.

Rohl, C. A., Fiori, W., & Baldwin, R. L. (1999). Alanine is helix-stabilizing in both template-nucleated and standard peptide helices. *Proceedings of the National Academy of Sciences of the United States of America, 96*(7), 3682–3687. doi:10.1073/pnas.96.7.3682.

Senichkin, V. V., Streletskaia, A. Y., Gorbunova, A. S., Zhivotovsky, B., & Kopeina, G. S. (2020). Saga of Mcl-1: Regulation from transcription to degradation. *Cell Death and Differentiation, 27*(2), 405–419. doi:10.1038/s41418-019-0486-3.

Sharma, R., Sharma, A., Patil, A., & Tsunoda, T. (2019). Discovering MoRFs by trisecting intrinsically disordered protein sequence into terminals and middle regions. *BMC Bioinformatics, 19*(S13). doi:10.1186/s12859-018-2396-7.

Skriver, K., Theisen, F. F., & Kragelund, B. B. (2023). Conformational entropy in molecular recognition of intrinsically disordered proteins. *Current Opinion in Structural Biology, 83*. http://www.elsevier.com/locate/sbi. doi:10.1016/j.sbi.2023.102697.

Spek, E. J., Olson, C. A., Shi, Z., & Kallenbach, N. R. (1999). Alanine is an intrinsic α-helix stabilizing amino acid. *Journal of the American Chemical Society, 121*(23), 5571–5572. doi:10.1021/ja990056x.

Teilum, K., Olsen, J. G., & Kragelund, B. B. (2015). Globular and disordered-the non-identical twins in protein-protein interactions. *Frontiers in Molecular Biosciences, 2*. https://www.frontiersin.org/articles/10.3389/fmolb.2015.00040/full. doi:10.3389/fmolb.2015.00040.

Theisen, F. F., Staby, L., Tidemand, F. G., O'Shea, C., Prestel, A., Willemoës, M., Kragelund, B. B., & Skriver, K. (2021). Quantification of conformational entropy unravels effect of disordered flanking region in coupled folding and binding. *Journal of the American Chemical Society, 143*(36), 14540–14550. http://pubs.acs.org/journal/jacsat. doi:10.1021/jacs.1c04214.

Tirion, M. M. (1996). Large amplitude elastic motions in proteins from a single-parameter, atomic analysis. *Physical Review Letters, 77*(9), 1905–1908. doi:10.1103/PhysRevLett.77.1905.

Togashi, Y., & Flechsig, H. (2018). Coarse-grained protein dynamics studies using elastic network models. *International Journal of Molecular Sciences, 19*(12). doi:10.3390/ijms19123899.

Trzesniak, D., Kunz, A.-P. E., & Van Gunsteren, W. F. (2007). A comparison of methods to compute the potential of mean force. *ChemPhysChem, 8*(1), 162–169. doi:10.1002/cphc.200600527.

Tunyasuvunakool, K., Adler, J., Wu, Z., Green, T., Zielinski, M., Žídek, A., Bridgland, A., Cowie, A., Meyer, C., Laydon, A., Velankar, S., Kleywegt, G. J., Bateman, A., Evans, R., Pritzel, A., Figurnov, M.,

Ronneberger, O., Bates, R., Kohl, S. A. A., ... Hassabis, D. (2021). Highly accurate protein structure prediction for the human proteome. *Nature, 596*(7873), 590–596. http://www.nature.com/nature/index.html. doi:10.1038/s41586-021-03828-1.

Uversky, V. N. (2002). What does it mean to be natively unfolded? *European Journal of Biochemistry, 269*(1), 2–12. doi:10.1046/j.0014-2956.2001.02649.x.

Van Der Lee, R., Buljan, M., Lang, B., Weatheritt, R. J., Daughdrill, G. W., Dunker, A. K., Fuxreiter, M., Gough, J., Gsponer, J., Jones, D. T., Kim, P. M., Kriwacki, R. W., Oldfield, C. J., Pappu, R. V., Tompa, P., Uversky, V. N., Wright, P. E., & Babu, M. M. (2014). Classification of intrinsically disordered regions and proteins. *Chemical Reviews, 114*(13), 6589–6631. http://pubs.acs.org/journal/chreay. doi:10.1021/cr400525m.

Webb, B., & Sali, A. (2016). Comparative protein structure modeling using MODELLER. *Current Protocols in Bioinformatics, 2016*. doi:10.1002/cpbi.3.

Wohl, S., Jakubowski, M., & Zheng, W. (2021). Salt-dependent conformational changes of intrinsically disordered proteins. *Journal of Physical Chemistry Letters, 12*(28), 6684–6691. doi:10.1021/acs.jpclett.1c01607.

Wright, P. E., & Dyson, H. J. (2015). Intrinsically disordered proteins in cellular signalling and regulation. *Nature Reviews Molecular Cell Biology, 16*(1), 18–29. http://www.nature.com/molcellbio. doi:10.1038/nrm3920.

Wright, P. E., & Dyson, H. J. (2009). Linking folding and binding. *Current Opinion in Structural Biology, 19*(1), 31–38. doi:10.1016/j.sbi.2008.12.003.

Yan, J., Dunker, A. K., Uversky, V. N., & Kurgan, L. (2016). Molecular recognition features (MoRFs) in three domains of life. *Molecular BioSystems, 12*(3), 697–710. http://pubs.rsc.org/en/journals/journal/mb. doi:10.1039/c5mb00640f.

Youle, R. J., & Strasser, A. (2008). The BCL-2 protein family: Opposing activities that mediate cell death. *Nature Reviews Molecular Cell Biology, 9*(1), 47–59. https://doi.org/10.1038/nrm2308.

Yuan, S., Chan, H. C. S., & Hu, Z. (2017). Using PyMOL as a platform for computational drug design. *Wiley Interdisciplinary Reviews: Computational Molecular Science, 7*(2). https://doi.org/10.1002/wcms.1298.

Zavrtanik, U., Hadži, S., & Lah, J. (2021). Unraveling the thermodynamics of ultra-tight binding of intrinsically disordered proteins. *Frontiers in Molecular Biosciences, 8*. https://doi.org/10.3389/fmolb.2021.726824.

CHAPTER

9

Probing protein structure and dynamics: Insights from THz spectroscopy and molecular modeling studies

Dan Florin Mihailescu[a,b,*] **and Maria Mernea**[b]

[a]*Faculty of Biology, Doctoral School of Biology, University of Bucharest, Bucharest, Romania,*
[b]*Department of Anatomy, Animal Physiology and Biophysics, Faculty of Biology, University of Bucharest, Bucharest, Romania*
**Corresponding author: dan.mihailescu@bio.unibuc.ro*

9.1 Introduction

The three-dimensional (3D) structures of proteins and their inherent dynamics are fundamental to their biological function (Hvidsten et al., 2009; Nam & Wolf-Watz, 2023). Knowledge on the 3D structures of proteins is essential for understanding their dynamic behavior and biological function (Nam & Wolf-Watz, 2023). The diversity of protein functions emerged through molecular evolution, a process in which new functions resulted from sequence variations while conserving the 3D structures (Campitelli et al., 2020). To date, the 3D structures of proteins are investigated by well-established methods that resulted in solving the more than 230,000 structures deposited in the Protein Data Bank (PDB) (Berman et al., 2000). According to the PDB statistics on data distribution by experimental method and molecular type (RCSB Protein Data Bank Statistics (n.d.)), most of these (>80% from the total) were determined by X-ray crystallography. A smaller number of structures were determined by cryo-electron microscopy (>10%) and nuclear magnetic resonance (>6%). The remaining structures were resolved using alternative techniques such as small-angle X-ray scattering or neutron diffraction.

Although static 3D structures provide valuable insights into protein architecture, they represent only snapshots of these highly dynamic molecular entities (Miller & Phillips, 2021). Proteins present intrinsic structural dynamics that allow them to sample functionally distinct conformations across their complex free energy landscape (Maria-Solano et al., 2018; Wingert et al., 2021; Woods & Pfeffer, 2020). The interconversion between the conformational substates corresponding to local or global energy minima depends on the height of the energy barriers and on particular environmental conditions such as solvent, temperature, or pressure (Maria-Solano et al., 2018). Conformational transitions play a central role in protein function, including switching between inactive and active states (Isu et al., 2024; Lee et al., 2021; Nam et al., 2023; Narayan et al., 2020); regulated unfolding associated with allostery (Latysheva et al., 2015); ligand binding (Ren et al., 2021; Stenström et al., 2024; Woods & Pfeffer, 2016;

Protein Dynamics: Experimental and Computational Approaches. DOI: https://doi.org/10.1016/B978-0-443-30042-4.00013-4

Yao & Hamelberg, 2022); catalyzing enzymatic reactions (Rivoire, 2024; Riziotis et al., 2022; Walker et al., 2024); mediating signal transduction (Banerjee et al., 2015; Woods & Pfeffer, 2016, 2020; Yao & Hamelberg, 2022); or responding to environmental changes (Domnauer et al., 2021).

Protein conformational transitions occur on various timescales, from subpicosecond bond vibrations and side chain rotations to nanosecond-scale loop fluctuations and up to microsecond or millisecond domain rearrangements (Kneller et al., 2022; Maria-Solano et al., 2018; Xu & Havenith, 2015). These motions are hierarchical and interdependent; fast thermal fluctuations can facilitate or trigger slower, large-scale conformational changes (Fitter et al., 1997; Hu et al., 2016, 2016; Woods, 2010; Zaccai et al., 2016). Across this temporal spectrum, the frequency of protein motions varies from high, reflecting local flexibility, to low, reflecting the global flexibility that accounts for proteins' functional dynamics(Kneller et al., 2022). Proteins exhibit collective vibrational motions primarily in the low-frequency range of approximately 5–50 cm^{-1} corresponding to timescales between 0.66 and 6.7 ps (Brooks & Karplus, 1983; Chou, 1988; Cusack et al., 1988; Kneller et al., 2022). Global "breathing" motions dominate the lower end of this range (below 15 cm^{-1}), involving large-scale, delocalized movements across the entire protein structure. Higher frequencies (~15–50 cm^{-1}) are associated with coordinated movements of secondary structural elements. Together, these modes define the mechanical flexibility or rigidity of a protein in this vibrational window (Brooks & Karplus, 1983; Chou, 1988; Cusack et al., 1988; Kneller et al., 2022).

The lowest frequency global modes tend to be conserved within protein families, whereas a subset of low-to-intermediate frequency motions differentiates between protein subfamilies and underlies the functional divergence (Orellana, 2023; Zhang et al., 2019). Long-range collective vibrations play a key role in enabling structural changes essential to protein function (Levantino et al., 2015; Niessen et al., 2017; Paciaroni et al., 2012). Any alteration of these motions induced by ligand binding or by mutations even at sites distant from the active site can modulate protein function (Niessen et al., 2017). For instance, inhibitor binding or higher catalytic rate mutations may minimally impact the vibrational energy distribution of a protein, but can markedly change the vibrational directionality (motion orientation) (Niessen et al., 2017). Moreover, Kneller et al. (2022) showed that substrate binding can increase protein rigidity, whereas inhibitor binding may induce a softening effect. This raises the possibility that mutations affecting vibrational stiffness, rather than direct interactions at the active site, might contribute to resistance mechanisms (Kneller et al., 2022).

Given the functional importance of low-frequency collective motions, significant effort is being devoted to probing these motions both experimentally and computationally. Terahertz (THz) spectroscopy has emerged as a powerful technique to detect these large-scale, low-energy vibrational modes, which are often difficult to detect using conventional spectroscopic methods (Balu et al., 2008; Heyden & Havenith, 2010; Mancini et al., 2022; Markelz et al., 2000; Whitmire et al., 2003). In parallel, computational approaches such as normal mode analysis (NMA) or molecular dynamics (MD) simulations offer detailed insights into these low-frequency modes and help interpret experimental THz spectra (Balu et al., 2008; Heyden & Havenith, 2010; Mancini et al., 2022; Markelz et al., 2000; Whitmire et al., 2003).

In the following sections, we will explore how THz spectroscopy, supported by computational modeling, provides unique insights into protein dynamics, including hydration effects, allosteric regulation, ligand binding, and posttranslational modifications. By integrating experimental observations with theoretical frameworks, we aim to highlight the role of collective protein motions in biological function and disease mechanisms.

9.2 Terahertz spectroscopy—Probing protein collective motions

9.2.1 Principles of terahertz spectroscopy

THz spectroscopy operates within the electromagnetic spectrum typically between 0.1 and 10 THz (corresponding to wavenumbers of approximately 3–333 cm^{-1} and wavelengths between 3000 and 30 μm), bridging the gap between microwave and infrared (IR) radiation (Ghann & Uddin, 2017). Historically, it is referred to as the "THz gap" because of the difficulty of generating and detecting radiation in this range. Advances in photonics and ultrafast laser technology have enabled its use in spectroscopic techniques with a wide range of applications, of particular interest being those in the biomedical field (Mancini et al., 2022; Plusquellic et al., 2007). THz radiation directly accesses collective vibrational, torsional, and conformational modes of biomolecules, as well as intermolecular interactions mediated by hydrogen bonding and van der Waals forces (Ghann & Uddin, 2017; Wei et al., 2018). These large-scale, low-frequency motions are essential for biological function and are often delocalized across secondary structure elements or entire protein domains.

Because of its long wavelength and low photon energy, THz radiation is nonionizing, minimally invasive, and sensitive to subtle structural dynamics. Its interaction with proteins is further modulated by the surrounding hydration shell, making it a valuable probe for solvent-coupled dynamics (Ghann & Uddin, 2017; Wei et al., 2018). The ultrashort, picosecond-scale pulses typically achieved by time-domain THz setups enhance temporal resolution while minimizing thermal effects and far-IR background interference (Rytik & Tuchin, 2025; Schroer et al., 2021; Wei et al., 2018). These characteristics position THz spectroscopy as a powerful tool for investigating protein flexibility and functionally relevant dynamics.

Two main experimental approaches are used in THz spectroscopy: THz time-domain spectroscopy (THz-TDS) and THz frequency-domain spectroscopy (THz-FDS). THz-TDS was the first method to be commercialized and used for the study of various materials (Shimura et al., 2024). In THz-TDS the temporal evolution of the THz electric field is directly measured, and the frequency spectrum is obtained through a Fourier transform of the time-domain signal. This method provides not only amplitude information, but also phase data, refractive index, and dielectric constants of the sample (Dinca et al., 2010). Along with THz-TDS systems, THz-FDS systems were developed but were commercialized and used to a smaller extent. THz-FDS measures the absorbance of a sample by scanning the THz output frequency and recording the corresponding intensity changes, yielding a direct frequency-dependent absorption spectrum (Shimura et al., 2024).

9.2.2 Measurements on dry samples

THz spectroscopy can be applied to different types of protein samples, including dry samples (lyophilized proteins or thin films), hydrated samples, and crystals. The first THz spectroscopy investigations were conducted on dry protein samples such as bovine serum albumin (BSA), type I collagen (Markelz et al., 2000), myoglobin, lysozyme (Markelz et al., 2002), and wild type and mutant bacteriorhodopsin (Whitmire et al., 2003). In the low frequency domain (up to 50–86 cm^{-1}), these studies showed that THz absorption spectra are typically broad and exhibit an almost linear increase of absorption with frequency. Even more, the absorbance of some proteins increases more rapidly with frequency, like in the case of type I collagen versus BSA (Markelz et al., 2000). As differences in absorption are also seen in the case of lysozyme versus myoglobin, it became apparent that THz spectroscopy can discriminate between protein species, but the spectra are not sufficiently distinct to allow protein identification (Markelz et al., 2002). The analysis of wild type and mutant bacteriorhodopsin in both Markelz et al. (2002) and Whitmire et al. (2003) showed that a point mutation of the protein alters the dipole moment of the protein, resulting in a decreased THz absorption.

The work performed on dry proteins demonstrated the sensitivity of THz spectroscopy to conformation (different protein species, wild type versus mutant proteins, and different oxidation states). As dehydration has a strong impact on protein structure, such results may not accurately reflect the *in vivo* conformations (Ebbinghaus et al., 2007; Markelz et al., 2000). Thus, subsequent studies have analyzed proteins under hydrated conditions, as we present in the following section.

9.2.3 Measurements on protein solutions

When analyzing aqueous protein solutions by THz spectroscopy, a major challenge arises from the strong absorption of THz radiation by polar water molecules (Ge et al., 2023). Nevertheless, with optimized experimental setups and short path lengths, THz spectra of aqueous solutions can be reliably and reproducibly measured. Variations in the THz absorbance of protein solutions compared to pure solvent have been attributed to the displacement of highly absorbing water molecules by proteins, resulting in a net decrease in THz absorption (Chen et al., 2007). This observation aligns with the concept of the "THz defect" (Leitner et al., 2008), which suggests that biomolecules, lacking some of the low-frequency collective motions present in bulk water, absorb less THz radiation over certain frequency ranges. For instance, in the study by Chen et al. (2007), solutions of chicken egg white lysozyme in buffer were measured in the 0.2–1.6-THz range. The protein solution exhibited slightly higher THz transmission than the pure buffer, confirming the THz defect. When the absorption of bulk water was suppressed by freezing, the trend reversed: the protein solution showed lower transmission than the buffer (Chen et al., 2007). In our previous work, we also measured the THz absorption of both diluted (10% w/w) and concentrated (50% w/w) BSA solutions, showing that absorption up to 1 THz increased linearly with frequency, with the more diluted solution showing higher absorption (Mernea et al., 2011).

Conversely the concept of "THz excess" (Leitner et al., 2008) describes cases where protein–water mixtures absorb more THz radiation than either the protein or bulk water alone. This excess absorption can be attributed to the presence of biological or hydration water, namely, the layer of water molecules surrounding the protein that exhibits distinct properties (e.g., altered density, relaxation rates, and reorientation dynamics) because of interactions with the protein surface (Leitner et al., 2008). The THz excess was reported by several studies that involved the analysis focused on different frequency domains (Heyden & Havenith, 2010; Sushko et al., 2015).

Heyden and Havenith (2010) observed an increase in THz absorbance around 2.4 THz at very low λ_{6-85}* (lambda repressor fragment 6–85) and ubiquitin concentrations (below 1 mM for λ_{6-85}* or below than 2 mM for ubiquitin). This concentration-dependent increase is absent in experiments on denatured proteins when the absorption follows the prediction of the simple water exclusion model (Heyden & Havenith, 2010). The contrast in hydration water behavior between folded and unfolded proteins makes THz spectroscopy a suitable tool for studying protein folding. Kinetic THz absorption experiments used to monitor the pseudo-wild type ubiquitin folding showed that THz pulse attenuation decreases in time, which supports the progressive organization of hydration shells. The full organization was estimated to occur after 100 ms, when the attenuation of THz pulses reaches a plateau (Heyden & Havenith, 2010).

Sushko et al. (2015) identified a THz excess at low concentrations of lysozyme, myoglobin, and BSA in the 0.220–0.325-THz range. In this frequency range, they identified a sharp increase of THz absorption with concentration up to the diffusion limit of proteins. The concentrations at which the absorbances of solutions plateau are 60, 50, and 55 mg/mL for lysozyme, myoglobin, and BSA. Wang et al. (2019) investigated the hydration of monomeric and dimeric insulin by analyzing the THz spectra of solutions across four frequency bands: 0.8–1.2, 1.3–1.7, 1.8–2.2, and 2.3–2.7 THz. In all bands, they observed an initial increase in THz absorption at low protein concentrations, peaking at approximately 11.19 ± 0.29 mg/mL for the monomer and 14.47 ± 0.24 mg/mL for the dimer. At higher concentrations, THz absorption decreased.

There are solutions for which the THz excess is minimal or absent, like in the case of BSA solutions. Bye et al. (2014) measured the THz absorption of BSA solutions with increasing concentrations up to 4 mM in the 0.3–3.3-THz frequency range. They analyzed the average absorption of solutions in the 0.9–1.1-, 1.9–2.1-, and 2.9–3.1-THz ranges. Their results show that 0–1 mM BSA solutions present a constant or slightly higher absorbance, showing the THz excess is minimal for these solutions. The absorbance decreases for solutions up to 2.5 mM and plateaus at higher concentrations.

THz spectroscopy measurements on protein solutions are essential for understanding the interplay between protein conformational flexibility and solvent interactions, which are central to biological function. The next section will explore measurements on protein crystals, which offer additional advantages in resolving directionality of vibrations and eliminating hydration-related complexity.

9.2.4 Measurements on protein crystals

A more recent line of work involves measuring the directionality of motions using anisotropic THz microscopy. Such experiments are performed on protein crystals, like in the case of chicken egg white lysozyme (Niessen et al., 2017). The method presents the advantage of suppressing the contribution of biological water and allows the discrimination of orientation-sensitive resonances from the background (Acbas, et al., 2012). It uses polarized THz radiation, and the protein crystal is rotated 360 degrees, starting with the 0-degree angle in which the crystal axis (Hvidsten et al., 2009) is parallel to the radiation polarization direction. The THz absorption of the aligned sample is measured as a function of the light's polarization direction relative to the sample's orientation. Thus the technique directly addresses the active protein intramolecular vibrations (Niessen et al., 2017). In their study on chicken egg white lysozyme in the absence and presence of a substrate, Niessen et al. (2017) showed that a change in vibrational directionality occurs because of ligand binding, whereas the vibrational energy distribution determined by inelastic neutron scattering (INS) shows modest changes.

9.2.5 Interpretation of terahertz spectra

Although THz spectroscopy allows direct access to collective protein motions, interpreting the broad and often overlapping spectral features of measured THz spectra remains a challenge. To bridge the gap between experimental data and molecular-level understanding, computational simulations play an essential role (Plusquellic et al., 2007). Methods such as MD and NMA provide detailed, atomistic views of protein motion that can be directly linked to the low-frequency vibrational modes observed in the THz range. These simulations allow the description of conformational transitions, hydration effects, and mutational impacts beyond experimental reach.

9.3 Methods of simulating terahertz spectra

9.3.1 Normal mode analysis

NMA is a computational technique used to explore the flexible conformational states available to a protein around its equilibrium structure (the minimum energy conformation). These accessible states are often closely linked to the molecule's functional mechanisms (Brooks & Karplus, 1983; Noguti & Gō, 1982). Among the available methods for investigating macromolecular dynamics, NMA stands out for its low computational cost, being suited for the analysis of most molecular systems, with the exception of the very largest complexes (Bauer et al., 2019).

Theoretical aspects of NMA and examples of applications to protein structure and flexibility investigations have been extensively reviewed in the literature (Bauer & Bauerová-Hlinková, 2021; Bauer et al., 2019). NMA relies on the harmonic approximation, in which the potential energy surface near the equilibrium structure is modeled as a multidimensional quadratic well (Brooks & Karplus, 1983; Hinsen, 2005). Although the approximation enables fast calculations, it limits the method's ability to account for anharmonic motions (Bauer et al., 2019).

In one of the first studies using THz spectroscopy accompanied by molecular modeling, NMA was applied to calculate the normal mode distributions of chicken egg white lysozyme and horse heart myoglobin (Markelz et al., 2002). Although the method did not consider the dipole strength of individual modes, the resulting normal mode distributions showed good agreement with experimentally measured THz absorptions, especially for spectra acquired at 77 K. This suggests that all low-frequency modes are optically active and contribute to THz absorption (Markelz et al., 2002).

Subsequent approaches improved this aspect by simulating THz spectra based on the intensities of normal modes, as described by Eq. (9.1) (Balu et al., 2008):

$$I_i \propto \left| < \upsilon' \left| \frac{\partial \mu_i}{\partial Q_i} \right| \upsilon'' > \right|^2 \cong \left| \sum_j^{3N} \frac{e_j q_{i,j} - e_j^0 q_{i,j}^0}{\Delta Q_i} \right|^2 \cong \left| \sum_j^{3N} \frac{e_j^0 \left(q_{i,j} - q_{i,j}^0 \right)}{\Delta Q_i} \right|^2. \tag{9.1}$$

In the equation, the intensity of the vibrational fundamental ($\upsilon' = \upsilon'' - 1$) for each normal mode Q_i is proportional to the square of the transition dipole moment derivative, $\partial \mu_i / \partial Q_i$. In the final approximation of the equation, the sum is calculated over the total 3N Cartesian coordinate displacements; $q_j - q_j^0$ are derived from the eigenvectors of the Hessian matrix by normalizing by square root of the atomic masses, e_j^0 represent the atomic partial charges assigned by the force field and ΔQ_i represents the root-mean-squared displacement of atoms in mode Q_i.

Eq. (9.1) allows the calculations of intensities for all normal modes. However, to obtain a spectrum comparable to experimental results, the intensity lines must be uniformly broadened using a Lorentzian (Balu et al., 2008) or Gaussian function (Globus et al., 2007). As Lorentzian broadening better describes the intrinsic damping of vibrational modes typically observed in biomolecular systems (Markelz et al., 2000), we chose to apply a Lorentzian function to fit the data in our previous work (Mernea et al., 2014), following the methodology presented by Balu et al. (2008). The Lorentzian function is given in Eq. (9.2):

$$I(\nu_i) = \frac{I_{max} \Gamma^2}{\pi \left[(\nu - \nu_i)^2 + \Gamma^2 \right]}. \tag{9.2}$$

The Lorentzian function is centered around the maximum intensity of each vibrational mode at its characteristic frequency v_i. Γ is the full width at half-maximum (FWHM). The choice of Γ significantly influences the calculated spectrum, as larger values lead to the loss of absorption features by averaging over neighboring frequencies (Globus et al., 2007). As different Γ values have been used in the literature, ranging from as small as 0.2 cm^{-1} (Thirumuruganandham et al., 2017) to as large as 20 cm^{-1} (Korter et al., 2006), we performed systematic calculations using various Γ values and found that 8 cm^{-1} was the smallest value that produces spectra in agreement with experimental data (Mernea et al., 2014).

The approach of simulating THz spectra based on NMA presents some limitations, as highlighted by Balu et al. (2008) in their study on rhodopsin. According to normal modes analysis theory the first six normal modes describe rigid-body motions of the protein, having zero energetic cost and eigenvalues of zero (Hinsen, 2005). Balu et al. (2008) noted that accurately minimizing the structure of rhodopsin to achieve a true local/global minimum was challenging, their calculation yielding a minimum nonzero frequency of 0.000029 cm^{-1}. In an intermediate (nonoptimal) minimum, they observed an imaginary frequency of −6.61 cm^{-1}. Although the frequency differences were minimal between these calculations, the mode motions and corresponding molecular interpretations varied significantly.

Another important limitation noted by Balu et al. (2008) is that the intensities calculated from normal modes rely on the harmonic approximation and therefore neglect anharmonic effects. In fact, experiments showed that the low-frequency normal modes are anharmonic, the anharmonic behavior resulting from substantial damping by the surrounding solvent and the complex, rugged topology of the protein's potential energy surface, which contains numerous local minima between the sampled conformations and the global energy minimum (Bauer et al., 2019). These limitations can be addressed by simulating THz spectra based on MD trajectories, which inherently capture anharmonicities and dynamic conformational sampling.

To address the limitations of performing NMA on a single structure, which does not capture conformational variability, structural snapshots from MD trajectories can be used as starting points for NMA. This approach enables an ensemble-averaged description of vibrational modes. We applied this method in our study aimed at validating a structural model of BSA using THz spectroscopy (Mernea et al., 2014). A 1-ns MD simulation was performed starting from a homology model of BSA, and 30 representative conformations were selected based on their root-mean-square deviations relative to the initial structure. NMA was carried out for each conformation, and the resulting spectra were compared with the THz spectrum measured from a lyophilized BSA sample. This comparison identified six BSA models whose computed spectra closely matched the experimental data. An average structure calculated from these conformations was proposed as the refined structural model of BSA (Mernea et al., 2014). Although this method still relies on the harmonic approximation for each individual snapshot, ensemble averaging offers a more realistic representation of the low-frequency vibrational landscape accessible to the protein.

9.3.2 Molecular dynamics simulations

MD simulations are a powerful tool to investigate protein flexibility and collective motions at atomic resolution. Based on classical Newtonian mechanics, MD integrates the equation of motion for all atoms in the system over discrete time steps, typically ranging from femtoseconds to microseconds

(Adak & Tripathi, 2019; Karplus & McCammon, 2002). These simulations allow the analysis of protein dynamics under near-native environments that include explicit solvent (such as water or lipid bilayers, depending on the protein type) and physiologically relevant ions (Adak & Tripathi, 2019; Mernea et al., 2014, 2020, 2022). The MD trajectories provide the basis for further computational analysis, including the simulation of vibrational spectra. Unlike NMA, which relies on the harmonic approximation, MD inherently captures anharmonic fluctuations and dynamic conformational sampling that are essential for a more realistic description of low-frequency vibrations (Choi & Cho, 2014; Wang, 2019).

MD trajectories are the input for extracting vibrational features from atomic motions that can be performed by two principal approaches. The first approach relies on calculating the autocorrelation function of the system's time-dependent properties [such as dipole moment (Thirumuruganandham et al., 2017) or velocity (Adak & Tripathi, 2019)] and applying a Fourier transform to obtain the vibrational spectrum. The second approach based on the quasiharmonic analysis involves deriving the vibrational modes from the covariance matrix of atomic displacements sampled during the MD simulation (He et al., 2011).

9.3.2.1 Vibrational density of states from velocity autocorrelation function

This method is based on the relationship between the vibrational motions and the time-dependent fluctuations of atomic velocities in MD simulations (Senske et al., 2018). Velocity autocorrelation function (VACF) measures how the velocity of an atom at a given time correlates with its velocity at a later time, as given by Eq. (9.3) (Adak & Tripathi, 2019; Kneller et al., 2022; Senske et al., 2018; Woods & Pfeffer, 2016):

$$C_v(\tau) = \frac{< v_i(\tau) \cdot v_i(0) >}{< v_i(0)^2 >} \tag{9.3}$$

where $v_i(\tau)$ is the velocity vector of atom i at time τ, and the angular brackets denote averaging over all atoms. The Fourier transform of VACF ($C_v(\tau)$) yields the vibrational density of states (VDOS) that reflects the distribution of kinetic energy across the frequency spectrum (ω). An example of calculation is given in Eq. (9.4) (Kneller et al., 2022):

$$\text{VDoS}(\omega) = \frac{1}{2\pi} \int \exp(-2i\pi\omega\tau) C_v(\tau) d\tau. \tag{9.4}$$

The VDOS encompasses both diffusive motions near zero frequency and vibrational contributions at higher, positive frequencies (Senske et al., 2018). This approach is particularly useful for comparing vibrational features across different environments, such as between atoms in bulk water and those in hydration shells or confined regions (Adak & Tripathi, 2019; Senske et al., 2018). However, because this method does not account for dipole moment fluctuations, it includes both IR-active and IR-inactive modes, and thus VDOS does not directly correspond to the THz absorption spectrum (Imoto et al., 2018).

9.3.2.2 Terahertz absorption spectra from dipole moment autocorrelation

One of the most direct ways to simulate THz absorption spectra from MD trajectories is through the calculation of the dipole moment autocorrelation function (DACF) (Woods, 2010). Most DACF calculations reported in the literature are based on the autocorrelation of the total dipole moment, as in Eq. (9.5) (Choi & Cho, 2014; Heyden & Havenith, 2010; Imoto et al., 2018):

$$C_{\mu\mu}(\tau) = < \mu(\tau) \cdot \mu(0) > \tag{9.5}$$

where $\mu(\tau)$ is the total dipole moment of the system at time τ, and the angular brackets denote averaging over all atoms. Fewer studies (Thirumuruganandham et al., 2017; Woods, 2010) calculate DAFCs over all atoms, using an equation similar to that used for VACF (Eq. 9.3). The THz absorption, $\alpha(\omega)$, is calculated as the Fourier transform of DACF, as shown in Eq. (9.6) (Choi & Cho, 2014; Heyden & Havenith, 2010; Imoto et al., 2018):

$$\alpha(\omega) = \frac{1}{4\pi\varepsilon_0 n(\omega)} \cdot \frac{2\pi}{3Vck_BT} \int \exp(-i\omega\tau)C_{\mu\mu}(\tau)d\tau \tag{9.6}$$

where $n(\omega)$ is the frequency-dependent refractive index, V is the system volume, c is the speed of light, k_B is the Boltzmann's constant, and T is the absolute temperature of the system. The $n(\omega)$ can be implicitly accounted for by reporting the absorption spectrum as $\alpha(\omega) \bullet n(\omega)$, which can further be compared to experiments (Heyden & Havenith, 2010).

As the dipole moment is dependent on atomic partial charges, the calculation is strongly influenced by the choice of force field. For example, the absorption spectrum of bulk water simulated using the simple point charge model showed notable discrepancies with the experimental results below 300 cm^{-1}, particularly missing the absorption band at 200 cm^{-1} (Heyden & Havenith, 2010). This deviation is attributed to the absence of polarization and charge transfer effects, which are especially important in hydrogen-bonding solvents like water. Although such effects are neglected in fixed-charge models, they are partially incorporated in polarizable force fields such as Atomic Multipole Optimized Energetics for Biomolecular Applications (AMOEBA) (Esser et al., 2017) and fully captured in *ab initio* MD (AIMD) simulations (Esser et al., 2017; Imoto et al., 2018). Because the electronic structure is computed dynamically in AIMD, these effects are inherently included according to the level of theory applied. AIMD simulations have been shown to accurately reproduce the far-IR hydrogen-bond network peak in pure water and allow a quantitative decomposition of IR spectra into solute, solvent, and solute–solvent interaction components (Imoto et al., 2018). Similarly, simulations using AMOEBA for zwitterionic amino acids in aqueous solution demonstrated good agreement with experimental THz spectra and enabled mode-specific spectral decomposition into intramolecular and solute–solvent contributions (Esser et al., 2017). Given the computational demands of AIMD simulations, especially for large or complex systems (Esser et al., 2017), fixed-charge models, such as all-atom CHARMM36m force field (Woods, 2023), and polarizable force fields, such as Drude-2013 force field (Pan et al., 2021), remain the most widely used in practice.

9.3.2.3 Quasiharmonic analysis of molecular dynamics trajectories

The quasiharmonic approximation (also referred to as principal component analysis) is applied to MD trajectories to derive effective vibrational modes from atomic fluctuations about an average structure. Unlike the classical NMA based on harmonic approximation, this approach captures anharmonic effects to some extent (Balog et al., 2004) and has been used to simulate THz absorption coefficients (Alijabbari et al., 2012; He et al., 2011). In this method, the covariance matrix of atomic positional fluctuations is computed and then inverted to approximate the force constant matrix. Diagonalizing this matrix yields the eigenvectors (mode shapes), eigenvalues (mode frequencies), and the dipole derivatives associated with each mode (Alijabbari et al., 2012; He et al., 2011).

To compute absorption coefficients, the integrated intensity values, Γ_k, for each vibrational mode of frequency ω_k is calculated using Eq. (9.7) (He et al., 2011):

$$\Gamma_k = \frac{N_0 \pi^2}{3c^2 \varepsilon_0 \omega_k} \left(\frac{\partial p}{\partial Q_k} \right)^2 \tag{9.7}$$

where N_0 is the Avogadro number, ε_0 is the permittivity of vacuum, c is the speed of light, and in parentheses is the dipole derivative of mode k. The integrated intensity is expressed in units of molar absorptivity.

As in NMA-based spectra, the absorption coefficient is calculated as a sum of Lorentzian functions fitting the integrated intensities of each mode. He et al. (2011) used a relaxation rate of 4 cm^{-1} that is consistent with the FWHM observed for THz vibrational resonances for molecular crystals.

However, the approach of calculating absorption coefficients using the quasiharmonic analysis presents some limitations (He et al., 2011). Notably, it cannot capture diffusive or large-amplitude motions, such as librations of side chains or rotational diffusion of solvent molecules, which are important in the low-frequency THz range. These contributions are better captured by computing the full dielectric response via the dipole autocorrelation function. Moreover, the accuracy of the absorption spectra from quasiharmonic analysis depends on the representation of dipole moments. As previously discussed, the use of fixed atomic partial charges can lead to discrepancies with experimental data. He et al. (2011) found that the quasiharmonic analysis-derived spectra of cytochrome *c* (Cytc) failed to reproduce the experimentally observed dependence of THz absorption on frequency, hydration, and oxidation state.

9.4 Protein conformational dynamics through terahertz spectroscopy and simulations

Building on the theoretical and methodological foundations outlined in the previous sections, we now explore how THz spectroscopy, in combination with computational modeling, has been applied to elucidate protein conformational dynamics across a variety of biological contexts. These studies highlight the sensitivity of THz spectroscopy to low-frequency motions associated with functionally relevant structural changes and emphasize the essential role of molecular simulations in interpreting experimental data.

9.4.1 Conformational dynamics and allosteric effects

THz spectroscopy, in combination with computational modeling, has emerged as a powerful tool for investigating protein conformational dynamics and allosteric regulation. The studies reviewed below reveal how low-frequency collective motions underpin functional transitions and long-range communication within proteins.

Rhodopsin and its mutations, especially P23H rhodopsin, were associated with autosomal dominant retinitis pigmentosa, a conformational disease. THz spectroscopy and molecular modeling were used to characterize the functional dynamics and conformational heterogeneity of P23H rhodopsin (Woods & Pfeffer, 2020). Experimentally recorded THz spectra show differences between wild type and

P23H rhodopsin, supporting that P23H rhodopsin is thermally unstable and activates via a distinct mechanism compared to the wild-type receptor. Simulations showed that this altered behavior stems primarily from mutation-induced disruption of the disulfide bond between Cys110 and Cys187 in the extracellular domain, leading to local misfolding near the retinal-binding site and instability of the Schiff base linkage. Importantly, P23H rhodopsin was shown to be nonfunctional as a monomer. Instead, it forms a specific homodimer (H4/H5–H4/H5 interface) that restores receptor function and introduces new activity. In the activated Meta-II state, this dimer exhibits reduced retinal affinity but supports a hyperactive conformation. Despite these changes the G-protein interaction interface remains similar to the WT, suggesting that P23H dimerization not only compensates for functional loss but may also alter downstream signaling dynamics (Woods & Pfeffer, 2020).

Woods and Pfeffer (2016) employed an integrative strategy combining THz spectroscopy, MD simulations, and evolutionary sequence analysis to investigate how long-range dynamic interactions contribute to functional divergence in the c-type lysozyme protein family. Their study focused on three structurally related lysozymes, namely, equine lysozyme, α-lactalbumin (LALBA), and conventional lysozyme, demonstrating that differences in intrinsic dynamics and conformational flexibility are closely linked to functional specialization. THz spectroscopy was used to probe protein dynamics across two spectral ranges: 20–100 cm^{-1}, capturing global, internal fluctuations related to intrinsic motions; and 100–250 cm^{-1}, highlighting localized interactions involved in allosteric signal propagation. The experimental results showed that both intrinsic and ligand-modulated dynamics vary significantly between family members and are associated with distinct functional roles and allosteric communication pathways. The computational component, particularly MD simulations in conjunction with sequence alignment and coevolutionary network analysis, provided detailed insights into how ligand binding reshapes the dynamic behavior and allosteric signaling. The findings suggest that shifts in the dynamic equilibrium of protein ensembles drive the emergence of new functional traits. These results support the broader view that protein evolution is closely linked to dynamic allosteric networks that modulate molecular interactions and enable functional diversification.

As in the previous studies, temperature-dependent THz spectroscopy, MD simulations, and sequence-based network analysis were used to characterize collective thermal fluctuations and their role in ligand-specific allosteric communication pathways in myoglobin (Mb) (Woods, 2014b). The work focuses on Mb complexed with either molecular oxygen (O_2) or carbon monoxide (CO), aiming to elucidate how low-frequency vibrational modes contribute to functional specificity. THz spectra were recorded for hydrated myoglobin films (97% relative humidity) in the 15–250-cm^{-1} range over a temperature interval from 83 to 303 K. The measurements captured spectral changes across the protein glass transition temperature (T_g), enabling identification of anharmonic, functionally relevant motions. Complementarily, equilibrium MD simulations were performed for both Mb–O_2 and Mb–CO complexes (Woods, 2014b).

In the 20–100 cm^{-1}-THz range (sensitive to global protein dynamics), Mb–O_2 exhibited well-defined anharmonic peaks (e.g., ~55 cm^{-1} above T_g), indicating highly collective motions that differ markedly from the more constrained, less temperature-responsive profile of Mb–CO. Above T_g the spectra diverged significantly, underscoring ligand-specific dynamic behavior. In the 100–250-cm^{-1} region, indicative of more localized motions, both complexes showed temperature-dependent vibrational features (e.g., peaks near 120, 144, and 165 cm^{-1}), but the amplitude and anharmonicity were consistently greater in Mb–O_2 (Woods, 2014b). The combined experimental and computational data reveal that myoglobin's collective low-frequency motions are distinctly modulated by ligand identity.

O_2 binding promotes more pronounced anharmonic fluctuations that reshape hydrophobic packing and enhance allosteric connectivity, consistent with an evolutionarily optimized mechanism for oxygen transport. In contrast, CO binding results in more localized, lower amplitude motions and a suboptimal allosteric response. These results emphasize the functional role of thermally excited collective motions in defining ligand specificity and allosteric efficiency in globular proteins (Woods, 2014b).

9.4.2 Protein stability

The stability of proteins under varying environmental conditions is a critical aspect of their functional integrity, especially in extreme environments such as high temperatures or dehydration. THz spectroscopy, combined with computational modeling techniques, has been instrumental in investigating how protein structure and low-frequency vibrational behavior are affected by hydration and temperature changes (Thirumuruganandham et al., 2017). Such a study investigated the thermal stability of three bacterial cold shock proteins (Csps) from mesophilic, thermophilic, and hyperthermophilic organisms (Thirumuruganandham et al., 2017). The VDOS and THz absorption spectra were calculated via NMA and MD simulations across a temperature range of 300–600 K. Normalized atomic fluctuations derived from normal modes, corroborated by IR spectral profiles, revealed clear distinctions in thermostability among the Csp homologues. Modes below 100 cm^{-1} were found to contribute significantly to internal protein fluctuations, with good agreement between NMA and MD-derived spectra. The evolution of THz spectra during dehydration (0–500 cm^{-1} range) further differentiated the thermal behavior of mesophilic versus thermophilic proteins, with noticeable spectral features, such as a double peak structure, linked to protein stability. Secondary structure–sensitive regions, such as amide-I and amide-II bands, also reflected conformational differences. During evaporation simulations, mesophilic Csp exhibited a more pronounced temperature drop, indicating faster structural destabilization compared to its thermophilic counterparts. This was attributed to the earlier loss of hydration shell and thermal contact with water clusters, especially under high-temperature conditions (400–600 K) (Thirumuruganandham et al., 2017).

9.4.3 Protein hydration and water–protein interactions

The hydration of proteins plays a critical role in modulating their structural dynamics, stability, and functional behavior. THz spectroscopy has emerged as a powerful technique to investigate the interplay between protein conformational flexibility and water dynamics, especially in the hydration shells. By probing low-frequency vibrational modes associated with both protein backbone fluctuations and hydrogen-bond network rearrangements in hydration shells, THz spectroscopy provides insights into how hydration affects protein function and how protein surfaces modulate water structure and mobility.

9.4.3.1 *Investigation of protein flexibility*

Woods (2010) investigated the effect of hydration on the low-frequency vibrational modes of chicken egg white lysozyme using THz spectroscopy in the 15–250-cm^{-1} frequency range. At low hydration (75% relative humidity) the spectra revealed three distinct absorption peaks at 28, 68, and 75 cm^{-1}. At intermediate hydration (85% relative humidity) a single peak appears near 36 cm^{-1}, whereas at high hydration (97% relative humidity) the spectrum presents a broad peak at 45 cm^{-1}. These spectral changes reflect the increasing influence of water–protein interactions. Under low hydration, the intramolecular interactions are the most prevalent, and protein dynamics are mainly localized, such as methyl group fluctuations. As hydration increases, water molecules promote more extensive protein–solvent interactions, activating collective backbone motions and anharmonic side-chain fluctuations. These solvent-mediated dynamics facilitate correlated motions across the protein, highlighting the critical role of water in modulating protein flexibility (Woods, 2010).

In another study, Woods (2014a) used THz spectroscopy and MD simulations to investigate the glass transition in crambin, a small plant protein that is highly hydrophobic and insoluble in water. The analysis involved two hydration states: low-hydration shell (LHS) and high-hydration shell (HHS), examined at three temperatures: 93, 200, and 303 K. Spectral data in the 20–100-cm^{-1} range revealed hydration-dependent differences in protein vibrational modes. At higher frequencies (100–170 cm^{-1}) the LHS condition was associated with the emergence of long-range vibrational coherence and localized intraprotein interactions, suggesting enhanced collective dynamics under low hydration. In contrast, HHS samples showed limited large-scale fluctuations, pointing to more restricted dynamics. Further analysis in the 170–240-cm^{-1} range focused on the vibrational behavior of water molecules. In highly hydrated samples, water exhibited self-associated, localized motions that interacted less effectively with protein dynamics, favoring water–water over water–protein interactions. Under LHS conditions, however, water adopted a more flexible, anharmonic motion within an open hydrogen-bonding network, which promoted extensive backbone fluctuations and vibrational delocalization across the protein. Overall the study underscores how hydration level critically shapes low-frequency protein dynamics. LHS conditions enhance anharmonic, collective motions that may underlie functional flexibility, whereas HHS conditions limit these motions to localized fluctuations, reducing solvent–protein coupling. These findings highlight the significant role of hydration in modulating both structural and functional properties of globular proteins such as crambin.

9.4.3.2 Characterization of protein hydration shells

THz spectroscopy has proven valuable for probing the dynamics of hydration water in protein solutions. This is possible because THz spectroscopy directly accesses the collective hydrogen-bond dynamics, which is sensitive to the coupling between protein and hydration shell motions. THz spectroscopy captures a broad range of dynamical processes, from nanosecond to subpicosecond timescales, including librational, translational, and collective intermolecular motions of hydration water, as well as large-amplitude conformational fluctuations of biomolecules (Xu & Havenith, 2015). THz absorption is sensitive to subtle changes in the dynamic orientation of water molecules, variations that are often averaged out in radial distribution functions obtained from scattering experiments. Moreover, it provides insights into distance scales that exceed the resolution limits of NMR spectroscopy (Leitner et al., 2008).

The THz spectrum of a protein solution arises from the combined contributions of the protein, its hydration shell, and bulk water. However, the three-component model that calculates THz absorption as the volume-weighted average of the solute, solvation water, and bulk water fails to account for the concentration-dependent behavior of THz absorption coefficients. This discrepancy suggests that protein–protein proximity and the overlap of hydration layers must also be considered (Ebbinghaus et al., 2007). To explore this, Ebbinghaus et al. (2007) analyzed the THz absorption of solutions with increased concentrations of $\lambda_{6\text{–}85}^{*}$, a five-helix bundle protein. Measurements were taken in the 2.24–2.55-THz range, which is particularly sensitive to water dynamics, capturing both intermolecular modes of the hydrogen-bond network and collective protein motions such as skeletal and breathing modes. Their findings revealed a nonmonotonic trend in absorption with respect to the water molar ratio, supporting the idea of overlapping hydration shells at higher protein concentrations. Based on experimental data and MD simulations, the dynamic hydration shell of λ_{6-85}* was estimated to extend beyond 20 Å (Ebbinghaus et al., 2007). Similar analyses were performed on ubiquitin, revealing an 18 Å hydration shell (Born et al., 2009).

Further studies across various proteins have extended these findings. Sushko et al. (2015) measured the THz absorption of increasingly concentrated solutions of lysozyme, myoglobin, and BSA and analyzed the results in the low-frequency range (0.220–0.325 THz). Although this low-frequency range is not dominated by hydration water (which absorbs less than bulk water here), the observed THz excess was attributed to enhanced protein dynamics in dilute solutions. Their analysis revealed hydration shell thicknesses of 16 Å for lysozyme, 19 Å for myoglobin, and 25 Å for BSA, which are larger than previous estimates, likely because of long-range dipolar interactions between water molecules and protein surfaces (Sushko et al., 2015).

Bye et al. (2014) also examined BSA solutions and found a nonlinear absorption–concentration relationship consistent with hydration shell overlap. Using a Beer-like fitting function (Born et al., 2009) to fit the average absorption coefficients for 1.0, 2.0, and 3.0 $\pm$ 0.1 THz, they estimated the BSA hydration layer to be 15 Å (Bye et al., 2014). The discrepancy with Sushko et al.'s (2015) result may arise from differences in sample preparation. Bye et al. (2014) used 10 mM NaCl as solvent, whereas Sushko et al.'s (2015) group used pure water. Ionic content is known to perturb the water hydrogen-bond network and affect THz absorption.

A more detailed frequency-resolved analysis was carried out by Wang et al. (2019), who studied monomeric and dimeric insulin across four frequency bands (0.8–1.2, 1.3–1.7, 1.8–2.2, and 2.3–2.7 THz). Using a random distribution model, they found the hydration shell of the dimer to be thicker (17.5 $\pm$ 0.5 Å) than that of the monomer (15.4 $\pm$ 0.4 Å), and the hydration number for the dimer was 1.64 times higher. By applying a cubic distribution model, the size of the hydration shell is 33.6 $\pm$ 0.4 Å for the monomer and 37.7 $\pm$ 0.3 Å for the dimeric insulin solution. This suggests that dimerization alters protein structure in a way that reduces the solvent-accessible surface of each monomer and influences hydration dynamics.

Novelli et al. (2017) investigated the hydration properties of native and fibrillar human lysozyme by analyzing average absorption coefficients in the 0.38–0.92-THz range. Their study based on the approach of Heyden et al. (2012) and employing a cubic distribution model under the assumption that bulk water is fully replaced by hydration water at 15 mg/mL native lysozyme revealed that the hydration layer of the native protein extends up to 42 Å from the protein surface. In contrast, fibrillar lysozyme at a concentration of 50 mg/mL exhibited THz absorption behavior similar to that of native lysozyme at less than 15 mg/mL. This suggests that fibrils bind significantly less hydration water, and in this case, bulk water is not fully replaced by hydration water. Consequently the hydration shell thickness for fibrils could not be determined using the same model. Furthermore, Novelli et al. (2017) observed that fibrils exhibit a larger induced dipole moment than native lysozyme. This may be attributed to the higher density of β-sheets in the fibrillar structure, which likely results in an increased surface charge density that perturbs the hydrogen-bonding network of surrounding water molecules.

9.4.3.3 *Characterization of hydration water near the active sites*

THz spectroscopy, in combination with molecular modeling, has revealed interesting details about hydration water dynamics in the vicinity of protein active sites. One key finding is the existence of a "hydration funnel," which refers to a gradient of increasingly slowed water dynamics close to the active site (Conti Nibali & Havenith, 2014). Water molecules within this funnel exhibit significantly retarded motion compared to those on the general protein surface. These highly ordered water molecules play a crucial role in molecular recognition processes, such as the detection of nanoscale ice crystals by antifreeze proteins or the selective substrate binding by human membrane-type 1 matrix metalloproteinase (MT1-MMP) (Conti Nibali & Havenith, 2014). This local slowdown of water dynamics reduces the entropic penalty typically associated with desolvation upon binding, thus favoring specific interactions over nonspecific ones (Xu & Havenith, 2015).

Beyond active sites, water also plays a central role in the formation and stabilization of biomolecular condensates via liquid–liquid phase separation (LLPS), a process increasingly recognized for its regulatory importance in cellular organization and its involvement in neurodegenerative diseases (Pakravan et al., 2021). Although earlier LLPS research focused primarily on protein–protein and crowding interactions, recent evidence points to the thermodynamic role of water as a key modulator of phase behavior (Mukherjee et al., 2024). A striking example comes from THz spectroscopy studies on γD-crystallin, a globular protein highly concentrated in the eye lens's cytoplasm. When γD-crystallin undergoes LLPS, two distinct hydration water populations are observed through THz absorption features and supported by MD simulations. The first population, termed "cavity water," corresponds to an absorption band near 150 cm^{-1}. As LLPS progresses, this band diminishes, indicating that these water molecules are released into the dilute phase. The second population, "bound water," is associated with the 400–650-cm^{-1} spectral regions. An increase in this band, reflected in positive spectral slopes, suggests a greater degree of water confinement within the condensates. This dual behavior reflects an entropic tug-of-war during LLPS, when water released into the surrounding medium increases the overall entropy and contributes favorably to phase separation, whereas retained water suffers entropic penalties because of its restricted mobility. The balance between these opposing forces is crucial in determining the stability and properties of the resulting condensates. These findings underscore the role of water entropy as a driving factor in LLPS, complementing traditional contributors such as protein–protein interactions and structural rearrangements (Mukherjee et al., 2024). Similar results were previously reported (Krevert et al., 2023).

9.4.3.4 The impact of hydration on protein-protein interactions

Woods (2023) investigated the impact of high dilution (HD) on protein dynamics and interactions using THz spectroscopy and MD simulations. The study focused on interferon gamma (IFN-γ) and antibodies to IFN-γ (antiIFN-γ and antiIFNGR1), prepared via a serial dilution protocol involving mechanical agitation and flotation-based separation. THz spectroscopy revealed distinct spectral signatures in HD protein solutions compared to pure water or HD-treated water. Notably, HD-IFN-γ exhibited prominent absorption peaks at ~140, 175, and 217 cm^{-1}, attributed to interprotein hydrogen bonding, bulk water–protein interactions, and fluctuations of hydrophobic side chains, respectively. In the <100 cm^{-1} range, HD samples showed shifts toward higher frequencies and distinct band patterns, suggesting altered collective dynamics. In contrast, standard (undiluted) protein solutions displayed typical solvent-induced global vibrational modes. Complementary MD simulations provided a molecular-level interpretation of these findings. In water-ethanol environments, IFN-γ dynamics was markedly altered, with increased rigidity and heterogeneous behavior, particularly in regions involved in receptor binding. Ethanol influenced hydration layers, promoting clathrate-like structures and enhancing side-chain fluctuations via modified hydrogen-bond networks. Simulations of the IFN-γ receptor complex (with IFNGR1/2) and antibody-receptor assemblies revealed enhanced conformational heterogeneity, superdiffusive motion, and long-range dynamic coupling, especially involving hydrophobic side chains. These effects were accompanied by shifts in methyl group mobility, heterogeneous relaxation times, and local structural rearrangements, all supporting a dynamic restructuring of protein–solvent interactions under HD conditions. Overall, the study demonstrates that HD preparation can significantly modulate protein–solvent dynamics through persistent nanoassociations and altered hydration structures, with implications for both the physical understanding of HD systems and the development of HD-based antibody formulations (Woods, 2023).

9.4.4 Protein–ligand interactions

The effect of ligand binding on the picosecond-scale structural dynamics of proteins was investigated in hen egg-white lysozyme (HEWL) when interacting with the trisaccharide inhibitor $(NAG)_3$ (Woods, 2014c). Experiments were performed on hydrated thin films of HEWL with and without ligand. The results showed that ligand binding alters collective vibrational modes in the 20–100-cm^{-1} range, shifting dominant peaks and reducing spectral intensity, as a consequence of changes in backbone motions and side-chain fluctuations. In the 100–200-cm^{-1} range, the free enzyme showed prominent solvent-mediated vibrational modes (e.g., at 120 and 143 cm^{-1}), which disappeared upon ligand binding. These suggest a marked dynamic reorganization and a key role for the hydration shell. A new mode at 153 cm^{-1} appeared in the complex at low temperatures, likely reflecting solvent-coupled stretching motions. MD simulations supported the experimental findings, revealing hinge-bending and twisting motions as dominant global modes. Ligand binding induced measurable changes in local compactness and vibrational frequencies, particularly at functionally important residues such as Phe38 and Gly102. Solvent structure and hydrogen bonding around these residues were reorganized upon binding, with specific low-frequency modes (e.g., 153 cm^{-1}) emerging as solvent-induced signatures of dynamic rearrangement (Woods, 2014c).

THz spectroscopy and molecular modeling were used to explore how the epithelial sodium channel (ENaC), a highly Na^+-selective ion channel expressed in water-absorbing epithelia, is inhibited by its specific blocker amiloride, a potassium-sparing diuretic (Mernea et al., 2022). The experimental work was performed on A6 cell suspensions, derived from *Xenopus laevis* kidney and overexpressing ENaC. THz measurements were conducted on both untreated cells and cells exposed to 100 μM amiloride. Spectra for the culture medium (GM) and for amiloride in GM served as controls. Spectral data, spanning 0.18–1 THz, showed a linear increase in absorption with frequency, without distinct peaks. Cells treated with amiloride exhibited significantly higher THz absorption than untreated ones. After subtracting solvent contributions, this increase was confirmed and attributed to changes in ENaC flexibility upon ligand binding. To support this hypothesis, molecular models of xENaC and its complex with amiloride were constructed by homology modeling and molecular docking. Simulated spectra agreed with experimental results, confirming greater THz absorption for the amiloride-bound state. Flexibility analysis via RMSF revealed reduced mobility in the TM domain upon amiloride binding, particularly in regions interacting with the ligand. Despite the increased rigidity, the protein–ligand complex exhibited more low-frequency vibrational modes, likely because of coupling between protein and ligand motions.

MD simulations also detailed amiloride's movement through the extracellular vestibule into the channel pore, identifying four quasistable states. The final binding site found deep within the pore near residues forming the selectivity filter is aligned with prior mutagenesis data (e.g., βG547 and γG539).

Intermediate positions likely represent steps in amiloride's permeation path from the extracellular space to its final site. This study demonstrates that THz spectroscopy, combined with molecular simulations, is a powerful tool for distinguishing between native and inhibited ENaC states and for uncovering detailed dynamics of ligand–channel interactions (Mernea et al., 2022).

INS and MD simulations were used to examine the vibrational dynamics of HIV-1 protease (PR) and how it is altered by binding the clinical inhibitor darunavir (DRV) (Kneller et al., 2022). The study focused on the low-frequency regime ($\sim$5–60 cm^{-1}), corresponding to collective motions on the picosecond timescale, which are highly relevant for THz spectroscopy. Collective vibrational modes involving α-helices, β-sheets, and loops ("breathing" motions) occur in this frequency domain and determine the mechanical softness or rigidity of the protein. HIV-1 PR bound to DRV exhibited an increased Boson peak intensity in the 5–60-cm^{-1} range, indicating enhanced vibrational amplitudes and a softening of dynamics compared to the apo form. MD simulations closely reproduced the experimental INS spectra. The PR/DRV complex showed elevated VDOS in the same frequency window, further supporting vibrational softening. VDOS analysis also revealed asymmetric dynamic alterations across structural elements. The flap and β-sheet regions of monomer A, along with an α-helix of monomer B, showed the largest increases in flexibility, suggesting that inhibitor binding induces localized softening. The study provides a clear example of how collective protein motions influence molecular recognition and binding energetics (Kneller et al., 2022).

Conti Nibali et al. (2017) used MD simulations to investigate how ligand binding affects fast protein dynamics in the THz frequency range (0.1–3 THz) for PDZ2 and PDZ3 domains, both in their ligand-free and ligand-bound states. By applying Fourier filtering to isolate high-frequency motions from short MD trajectories, they found that ligand binding induced a clear rigidification of the binding site, which was only evident at THz frequencies. This rigidification defined a "binding response nucleus" (BRN) and was accompanied by increased flexibility in surrounding regions. These localized changes were confirmed through VDOS analyses, which showed a blue shift in rigid regions and red shift in more flexible ones, whereas the total molecular VDOS remained largely unchanged between bound and unbound states. Importantly, these effects were absent in the low-frequency domain. Side-chain dynamics, particularly of methyl carbons, were more sensitive to ligand binding than backbone atoms, revealing distal effects and enhanced mapping of the BRN. Notably the THz-defined BRN overlapped with known allosteric regions, suggesting a potential role for fast, collective motions in allosteric regulation (Conti Nibali et al., 2017).

9.4.5 Posttranslational modification of proteins

The oxidation of bovine heart Cytc was investigated using THz spectroscopy, NMA, VDOS, and dipole–dipole correlation calculations derived from MD simulations (He et al., 2011). THz spectroscopy revealed that the oxidized form of Cytc exhibits a higher absorption coefficient compared to its reduced form. Regarding the hydration dependence, absorption increases with hydration up to approximately 20%–25%, beyond which a saturation-like behavior is observed in both redox states. Across all hydration levels and frequencies, the oxidized Cytc consistently shows higher absorption. Normal mode calculations failed to reproduce the experimentally observed oxidation and hydration dependence, indicating that this method does not adequately capture Cytc dynamics at room temperature. In contrast, dipole–dipole correlation functions successfully reproduced the oxidation dependence of the spectra.

MD simulations suggested that the spectral differences arise from changes in water dynamics near the heme group in the oxidized state. At low hydration levels, these internal water molecules dominate the spectral signal, whereas at higher hydration levels, their contribution diminishes, and the oxidation dependence becomes negligible (He et al., 2011).

In a previous study, we investigated the glycosylation state of the TRPM8 ion channel using THz spectroscopy combined with molecular modeling approaches (Mernea et al., 2020). THz absorption measurements were conducted on dense suspensions of human embryonic kidney (HEK) cells stably transfected with human TRPM8. Two experimental conditions were analyzed, namely, untreated control cells and cells treated with brefeldin A, a deglycosylation agent. The cell suspensions were normalized to contain the same number of cells, verified by cell counting, to ensure comparability between the samples. Control cell suspensions consistently exhibited higher THz absorption than their deglycosylated counterparts, suggesting a significant contribution of glycosylation to the THz spectral response. TRPM8 contains a single *N*-glycosylation site per subunit, occupied by complex glycans of approximately 2.5 kDa (Mernea et al., 2020). As the exact glycan structure was unknown, two models of glycosylated TRPM8 were constructed: one with shorter, branched glycans at the first sugar unit (glc1), and another with longer, unbranched glycans (glc2). These glycosylated models, along with the unglycosylated TRPM8 channel, were embedded in MD simulation systems that included a lipid bilayer, water, and ions. THz spectra were then calculated from the MD trajectories for each system. Only the simulated spectra corresponding to TRPM8 glycosylated with glc1-type glycans matched the experimental data, supporting this as the likely glycan structure in cells. These findings underscore the potential of THz spectroscopy, in conjunction with molecular modeling, as a sensitive technique for probing protein glycosylation states and potentially for assessing glycosylation in complex biological samples (Mernea et al., 2020).

9.5 Conclusion

Understanding the collective motions of proteins is essential for unraveling their functional mechanisms, particularly those involving large-scale conformational transitions and solvent-coupled dynamics. This chapter has demonstrated how THz spectroscopy, when integrated with computational modeling techniques such as NMA and MD simulations, provides a powerful framework for investigating these dynamic processes. Unlike traditional structural biology methods such as X-ray crystallography or NMR spectroscopy that primarily capture static snapshots or localized motions, THz spectroscopy directly probes low-frequency vibrational modes that span entire macromolecular structures and their surrounding hydration shells. These modes are increasingly recognized as central to key biological functions, including allosteric signaling, enzymatic catalysis, and ligand binding.

One of the principal advantages of THz spectroscopy lies in its sensitivity to both protein conformational states and their dynamical transitions, even when such transitions are not accompanied by large structural rearrangements. This sensitivity arises because THz frequencies correspond to collective vibrational modes, including hinge-bending, twisting, and domain motions, as well as picosecond-scale solvent fluctuations that are tightly coupled to protein surfaces. Moreover, the noninvasive nature of THz spectroscopy, requiring no labeling or immobilization, allows the investigation of proteins in solution under near-physiological conditions. Nevertheless, experimental THz spectra can be challenging to interpret because of the dense and overlapping nature of low-frequency modes and the strong contribution of water absorption. To address this, we reviewed the integration of simulation approaches (NMA and MD) that enable the assignment of spectral features to specific structural and dynamical components. These simulations not only help deconvolve the spectral contributions of internal protein motions versus solvent fluctuations but also allow hypothesis testing about mutation effects, ligand binding, or environmental changes.

The case studies presented throughout this chapter illustrate the power of combining THz spectroscopy with computational modeling to distinguish between different conformational substates, identify functionally relevant vibrational modes, and capture hydration-dependent dynamic signatures. Particularly interesting are findings that link changes in the THz spectra to collective vibrational softening upon ligand binding or to changes in hydration shell mobility in disease-associated mutations, phenomena that remain invisible to many other experimental techniques. The approach of using THz spectroscopy and modeling offers a unique vantage point on the protein energy landscape, emphasizing motions that bridge structure and function. As the field progresses, ongoing advances in experimental sensitivity (e.g., time-resolved and anisotropic THz spectroscopy) and modeling accuracy (e.g., enhanced force fields, polarizable water models, and machine-learned potential surfaces) are expected to further refine our capacity to explore protein dynamics at a fundamental level.

In conclusion, THz spectroscopy assisted by computational simulations emerges as a frontier method for probing the dynamic substrate of protein function. Its unique ability to access collective, solvent-coupled dynamics positions it as a critical tool for understanding biological processes in health and disease. As experimental sensitivity improves and modeling accuracy increases, the integration of THz spectroscopy with advanced computational tools will continue to open promising avenues for applications in drug design, diagnostics, and protein engineering. By bridging the gap between static structural information and dynamic functional behavior, THz spectroscopy contributes to a more complete picture of the protein energy landscape, one that emphasizes the interplay between structure, motion, and environment. This perspective is crucial for advancing our understanding of life at the molecular level and for developing novel therapeutic strategies that target protein dynamics rather than static conformations.

AI disclosure

During the preparation of this work the author(s) used OpenAI–ChatGPT. The tool was used to improve readability and language of the work, based on the text fully written by the authors. After using this tool/service the author(s) reviewed and edited the content as needed and take(s) full responsibility for the content of the publication.

References

Acbas, G., Snell, E., & Markelz, A. G. (2012). Orientation sensitive terahertz resonances observed in protein crystals. In *Proceedings of the 2012 37th International Conference on Infrared, Millimeter, and Terahertz Waves* (pp. 1–3). Australia: Wollongong, NSW. doi:10.1109/IRMMW-THz.2012.6380168.

Adak, S., & Tripathi, L. N. (2019). Nanoantenna enhanced terahertz interaction of biomolecules. *Analyst, 144*(21), 6172–6192. http://pubs.rsc.org/en/journals/journal/an. doi:10.1039/c9an00798a.

Alijabbari, N., Chen, Y., Sizov, I., Globus, T., & Gelmont, B. (2012). Molecular dynamics modeling of the sub-THz vibrational absorption of thioredoxin from *E. coli*. *Journal of Molecular Modeling, 18*(5), 2209–2218. doi:10.1007/s00894-011-1238-6.

Balog, E., Becker, T., Oettl, M., Lechner, R., Daniel, R., Finney, J., & Smith, J. C. (2004). Direct determination of vibrational density of states change on ligand binding to a protein. *Physical Review Letters, 93*(2), 1–28103. https://journals.aps.org/prl/issues. doi:10.1103/PhysRevLett.93.028103.

Bauer, J. A., & Bauerová-Hlinková, V. (2021). *Normal mode analysis: A tool for better understanding protein flexibility and dynamics with application to homology models*. IntechOpen. doi:10.5772/intechopen.94139.

Balu, R., Zhang, H., Zukowski, E., Chen, J.-Y., Markelz, A. G., & Gregurick, S. K. (2008). Terahertz spectroscopy of bacteriorhodopsin and rhodopsin: Similarities and differences. *Biophysical Journal, 94*(8), 3217–3226. doi:10.1529/biophysj.107.105163.

Banerjee, R., Yan, H., & Cukier, R. I. (2015). Conformational transition in signal transduction: Metastable states and transition pathways in the activation of a signaling protein. *Journal of Physical Chemistry B, 119*(22), 6591–6602. http://pubs.acs.org/journal/jpcbfk. doi:10.1021/acs.jpcb.5b02582.

Bauer, J. A., Pavlović, J., & Bauerová-Hlinková, V. (2019). Normal mode analysis as a routine part of a structural investigation. *Molecules, 24*(18), 3293. doi:10.3390/molecules24183293.

Berman, H. M., Westbrook, J., Feng, Z., Gilliland, G., Bhat, T. N., Weissig, H., Shindyalov, I. N., & Bourne, P. E. (2000). The Protein Data Bank. *Nucleic Acids Research, 28*(1), 235–242. https://academic.oup.com/nar/issue. doi:10.1093/nar/28.1.235.

Born, B., Kim, S. J., Ebbinghaus, S., Gruebele, M., & Havenith, M. (2009). The terahertz dance of water with the proteins: The effect of protein flexibility on the dynamical hydration shell of ubiquitin. *Faraday Discussions, 141*, 161–173. doi:10.1039/b804734k.

Brooks, B., & Karplus, M. (1983). Harmonic dynamics of proteins: Normal modes and fluctuations in bovine pancreatic trypsin inhibitor. *Proceedings of the National Academy of Sciences of the United States of America, 80*(21), 6571–6575. doi:10.1073/pnas.80.21.6571.

Bye, J. W., Meliga, S., Ferachou, D., Cinque, G., Zeitler, J. A., & Falconer, R. J. (2014). Analysis of the hydration water around bovine serum albumin using terahertz coherent synchrotron radiation. *Journal of Physical Chemistry A, 118*(1), 83–88. doi:10.1021/jp407410g.

Campitelli, P., Modi, T., Kumar, S., & Ozkan, S. B. (2020). The role of conformational dynamics and allostery in modulating protein evolution. *Annual Review of Biophysics, 49*(1), 267–288. doi:10.1146/annurev-biophys-052118-115517.

Chen, J.-Y., Knab, J. R., Ye, S., He, Y., & Markelz, A. G. (2007). Terahertz dielectric assay of solution phase protein binding. *Applied Physics Letters, 90*(24). doi:10.1063/1.2748852.

Choi, J. H., & Cho, M. (2014). Terahertz chiroptical spectroscopy of an α-helical polypeptide: A molecular dynamics simulation study. *Journal of Physical Chemistry B, 118*(45), 12837–12843. http://pubs.acs.org/journal/jpcbfk. doi:10.1021/jp508547y.

Chou, K. C. (1988). Low-frequency collective motion in biomacromolecules and its biological functions. *Biophysical Chemistry, 30*(1), 3–48. doi:10.1016/0301-4622(88)85002-6.

Conti Nibali, V., Morra, G., Havenith, M., & Colombo, G. (2017). Role of terahertz (THz) fluctuations in the allosteric properties of the PDZ domains. *The Journal of Physical Chemistry B, 121*(44), 10200–10208. doi:10.1021/acs.jpcb.7b06590.

Conti Nibali, V., & Havenith, M. (2014). New insights into the role of water in biological function: Studying solvated biomolecules using terahertz absorption spectroscopy in conjunction with molecular dynamics simulations. *Journal of the American Chemical Society, 136*(37), 12800–12807. doi:10.1021/ja504441h.

Cusack, S., Smith, J., Finney, J., Tidor, B., & Karplus, M. (1988). Inelastic neutron scattering analysis of picosecond internal protein dynamics. Comparison of harmonic theory with experiment. *Journal of Molecular Biology, 202*(4), 903–908. doi:10.1016/0022-2836(88)90566-9.

Dinca, M. P., Leca, A., Apostol, D., Mernea, M., Calborean, O., Mihailescu, D., & Dascalu, T. (2010). Transmission THz time domain system for biomolecules spectroscopy. *Journal of Optoelectronics and Advanced Materials, 12*(1), 110–114. http://inoe.inoe.ro/joam/download.php?idu=2274.

Domnauer, M., Zheng, F., Li, L., Zhang, Y., Chang, C. E., Unruh, J. R., Conkright-Fincham, J., McCroskey, S., Florens, L., Zhang, Y., Seidel, C., Fong, B., Schilling, B., Sharma, R., Ramanathan, A., Si, K., & Zhou, C. (2021). Proteome plasticity in response to persistent environmental change. *Molecular Cell, 81*(16), 3294. Www.molecule.org. doi:10.1016/j.molcel.2021.06.028.

Ebbinghaus, S., Seung, J. K., Heyden, M., Yu, X., Heugen, U., Gruebele, M., Leitner, D. M., & Havenith, M. (2007). An extended dynamical hydration shell around proteins. *Proceedings of the National Academy of Sciences of the United States of America, 104*(52), 20749–20752. http://www.pnas.org/cgi/reprint/104/52/20749. doi:10.1073/pnas.0709207104. Germany.

Esser, A., Belsare, S., Marx, D., & Head-Gordon, T. (2017). Mode specific THz spectra of solvated amino acids using the AMOEBA polarizable force field. *Physical Chemistry Chemical Physics, 19*(7), 5579–5590. doi:10.1039/C6CP07388C.

Fitter, J., Lechner, R. E., & Dencher, N. A. (1997). Picosecond molecular motions in bacteriorhodopsin from neutron scattering. *Biophysical Journal, 73*(4), 2126–2137. doi:10.1016/s0006-3495(97)78243-2.

Ge, H., Sun, Z., Jiang, Y., Wu, X., Jia, Z., Cui, G., & Zhang, Y. (2023). Recent advances in THz detection of water. *International Journal of Molecular Sciences, 24*(13), 10936. doi:10.3390/ijms241310936.

Globus, T., Bykhovski, A., Khromova, T., Gelmont, B., Tamm, L. K., & Salay, L. C. (2007). Low, terahertz spectroscopy of liquid water. *Proceedings SPIE 6772, Terahertz Physics, Devices, and Systems II*, 67720S. doi:10.1117/12.739859.

Ghann, W., & Uddin, J. (2017). *Terahertz spectroscopy—A Cutting edge technology*. IntechOpen.

He, Y., Chen, J.-Y., Knab, J. R., Zheng, W., & Markelz, A. G. (2011). Evidence of protein collective motions on the picosecond timescale. *Biophysical Journal, 100*(4), 1058–1065. doi:10.1016/j.bpj.2010.12.3731.

Heyden, M., Tobias, D. J., & Matyushov, D. V. (2012). Terahertz absorption of dilute aqueous solutions. *Journal of Chemical Physics, 137*(23). doi:10.1063/1.4772000.

Heyden, M., & Havenith, M. (2010). Combining THz spectroscopy and MD simulations to study protein-hydration coupling. *Methods, 52*(1), 74–83. doi:10.1016/j.ymeth.2010.05.007.

Hinsen, K. (2005). Normal mode theory and harmonic potential approximations. In *Normal mode analysis: Theory and applications to biological and chemical systems* (pp. 1–16). CRC Press. https://www.taylorfrancis.com/books/e/9781420035070. doi:10.1201/9781420035070.ch1.

Hu, X., Hong, L., Smith, M. D., Neusius, T., Cheng, X., & Smith, J. C. (2016). The dynamics of single protein molecules is non-equilibrium and self-similar over thirteen decades in time. *Nature Physics, 12*(2), 171–174. http://www.nature.com/nphys/index.html. doi:10.1038/nphys3553.

Hvidsten, T. R., Lægreid, A., Kryshtafovych, A., Andersson, G., Fidelis, K., Komorowski, J., & Sussman, J. L. (2009). A comprehensive analysis of the structure-function relationship in proteins based on local structure similarity. *PLoS One, 4*(7), e6266. doi:10.1371/journal.pone.0006266.

Imoto, S., Forbert, H., & Marx, D. (2018). Aqueous TMAO solutions as seen by theoretical THz spectroscopy: Hydrophilic versus hydrophobic water. *Physical Chemistry Chemical Physics, 20*(9), 6146–6158. doi:10.1039/C7CP07003A.

Isu, U. H., Polasa, A., & Moradi, M. (2024). Differential behavior of conformational dynamics in active and inactive states of cannabinoid receptor 1. *Journal of Physical Chemistry B, 128*(35), 8437–8447. http://pubs.acs.org/journal/jpcbfk. doi:10.1021/acs.jpcb.4c02828.

Karplus, M., & McCammon, J. A. (2002). Molecular dynamics simulations of biomolecules. *Nature Structural Biology, 9*(9), 646–652. doi:10.1038/nsb0902-646.

Kneller, D. W., Gerlits, O., Daemen, L. L., Pavlova, A., Gumbart, J. C., Cheng, Y., & Kovalevsky, A. (2022). Joint neutron/molecular dynamics vibrational spectroscopy reveals softening of HIV-1 protease upon binding of a tight inhibitor. *Physical Chemistry Chemical Physics, 24*(6), 3586–3597. http://pubs.rsc.org/en/journals/journal/cp. doi:10.1039/d1cp05487b.

Korter, T. M., Balu, R., Campbell, M. B., Beard, M. C., Gregurick, S. K., & Heilweil, E. J. (2006). Terahertz spectroscopy of solid serine and cysteine. *Chemical Physics Letters, 418*(1–3), 65–70. doi:10.1016/j.cplett.2005.10.097.

Krevert, C. S., Chavez, D., Chatterjee, S., Stelzl, L. S., Pütz, S., Roeters, S. J., Rudzinski, J. F., Fawzi, N. L., Girard, M., Parekh, S. H., & Hunger, J. (2023). Liquid-liquid phase separation of the intrinsically disordered domain of the fused in sarcoma protein results in substantial slowing of hydration dynamics. *Journal of Physical Chemistry Letters, 14*(49), 11224–11234. http://pubs.acs.org/journal/jpclcd. doi:10.1021/acs.jpclett.3c02790.

Latysheva, N. S., Flock, T., Weatheritt, R. J., Chavali, S., & Babu, M. M. (2015). How do disordered regions achieve comparable functions to structured domains? *Protein Science, 24*(6), 909–922. http://onlinelibrary.wiley.com/journal/10.1002/(ISSN)1469-896X. doi:10.1002/pro.2674.

Lee, B. H, Park, S. W., Jo, S., Kim, M. K, & Li, W. (2021). Protein conformational transitions explored by a morphing approach based on normal mode analysis in internal coordinates. *PLoS One, 16*(11), e0258818. doi:10.1371/journal.pone.0258818.

Leitner, D. M., Gruebele, M., & Havenith, M. (2008). Solvation dynamics of biomolecules: Modeling and terahertz experiments. *HFSP Journal, 2*(6), 314–323. doi:10.2976/1.2976661.

Levantino, M., Schirò, G., Lemke, H. T., Cottone, G., Glownia, J. M., Zhu, D., Chollet, M., Ihee, H., Cupane, A., & Cammarata, M. (2015). Ultrafast myoglobin structural dynamics observed with an X-ray free-electron laser. *Nature Communications, 6*. http://www.nature.com/ncomms/index.html. doi:10.1038/ncomms7772.

Mancini, T., Mosetti, R., Marcelli, A., Petrarca, M., Lupi, S., & D'Arco, A. (2022). Terahertz spectroscopic analysis in protein dynamics: Current status. *Radiation, 2*(1), 100–123. doi:10.3390/radiation2010008.

Maria-Solano, M. A., Serrano-Hervás, E., Romero-Rivera, A., Iglesias-Fernández, J., & Osuna, S. (2018). Role of conformational dynamics in the evolution of novel enzyme function. *Chemical Communications, 54*(50), 6622–6634. http://pubs.rsc.org/en/journals/journal/cc. doi:10.1039/c8cc02426j.

Markelz, A. G., Roitberg, A., & Heilweil, E. J. (2000). Pulsed terahertz spectroscopy of DNA, bovine serum albumin and collagen between 0.1 and 2.0 THz. *Chemical Physics Letters, 320*(1–2), 42–48. doi:10.1016/s0009-2614(00)00227-x.

Markelz, A., Whitmire, S., Hillebrecht, J., & Birge, R. (2002). THz time domain spectroscopy of biomolecular conformational modes. *Physics in Medicine and Biology, 47*(21), 3797–3805. doi:10.1088/0031-9155/47/21/318.

Mernea, M., Ulăreanu, R. Ş., Cucu, D., Al-Saedi, J. H., Pop, C. E., Fendrihan, S., Anghelescu, G. D. C., & Mihăilescu, D. F. (2022). Epithelial sodium channel inhibition by amiloride addressed with THz spectroscopy and molecular modeling. *Molecules, 27*(10). https://www.mdpi.com/1420-3049/27/10/3271/pdf?version=1653018974. doi:10.3390/molecules27103271.

Mernea, M., Calborean, O., Grigore, O., Dascalu, T., & Mihailescu, D. F. (2014). Validation of protein structural models using THz spectroscopy: A promising approach to solve three-dimensional structures. *Optical and Quantum Electronics, 46*(4), 505–514. doi:10.1007/s11082-013-9872-0.

Mernea, M., Ulăreanu, R., Călboreanu, O., Chiriţoiu, G., Cucu, D., & Mihăilescu, D. F. (2020). N-glycosylation state of TRPM8 protein revealed by terahertz spectroscopy and molecular modelling. *Biochimica et Biophysica Acta (BBA)—General Subjects, 1864*(7), 129580. doi:10.1016/j.bbagen.2020.129580.

Miller, M. D., & Phillips, G. N. (2021). Moving beyond static snapshots: Protein dynamics and the Protein Data Bank. *Journal of Biological Chemistry, 296*. https://www.jbc.org/article/S0021-9258(21)00539-1/fulltext. doi:10.1016/j.jbc.2021.100749.

Mukherjee, S., Ramos, S., Pezzotti, S., Kalarikkal, A., Prass, T. M., Galazzo, L., Gendreizig, D., Barbosa, N., Bordignon, E., Havenith, M., & Schäfer, L. V. (2024). Entropy tug-of-war determines solvent effects in the liquid-liquid phase separation of a globular protein. *Journal of Physical Chemistry Letters, 15*(15), 4047–4055. http://pubs.acs.org/journal/jpclcd. doi:10.1021/acs.jpclett.3c03421.

Nam, K., Tao, Y., & Ovchinnikov, V. (2023). Molecular simulations of conformational transitions within the insulin receptor kinase reveal consensus features in a multistep activation pathway. *The Journal of Physical Chemistry B, 127*(26), 5789–5798. doi:10.1021/acs.jpcb.3c01804.

Nam, K., & Wolf-Watz, M. (2023). Protein dynamics: The future is bright and complicated!. *Structural Dynamics, 10*(1), 014301. doi:10.1063/4.0000179.

Narayan, B., Fathizadeh, A., Templeton, C., He, P., Arasteh, S., Elber, R., Buchete, N. V., & Levy, R. M. (2020). The transition between active and inactive conformations of Abl kinase studied by rock climbing and Milestoning. *Biochimica et Biophysica Acta—General Subjects, 1864*(4). Www.elsevier.com/locate/bbagen. doi:10.1016/j.bbagen.2019.129508.

Niessen, K. A., Xu, M., Paciaroni, A., Orecchini, A., Snell, E. H., & Markelz, A. G. (2017). Moving in the right direction: Protein vibrational steering function. *Biophysical Journal, 112*(5), 933–942. http://www.sciencedirect.com/science/journal/00063495. doi:10.1016/j.bpj.2016.12.049.

Noguti, T., & Gō, N. (1982). Collective variable description of small-amplitude conformational fluctuations in a globular protein. *Nature, 296*(5859), 776–778. doi:10.1038/296776a0.

Novelli, F., Ostovar Pour, S., Tollerud, J., Roozbeh, A., Appadoo, D. R. T., Blanch, E. W., & Davis, J. A. (2017). Time-domain THz spectroscopy reveals coupled protein-hydration dielectric response in solutions of native and fibrils of human lysozyme. *Journal of Physical Chemistry B, 121*(18), 4810–4816. http://pubs.acs.org/journal/jpcbfk. doi:10.1021/acs.jpcb.7b02724.

Orellana, L. (2023). Are protein shape-encoded lowest-frequency motions a key phenotype selected by evolution? *Applied Sciences, 13*(11), 6756. doi:10.3390/app13116756.

Paciaroni, A., Orecchini, A., Haertlein, M., Moulin, M., Conti Nibali, V., De Francesco, A., Petrillo, C., & Sacchetti, F. (2012). Vibrational collective dynamics of dry proteins in the terahertz region. *The Journal of Physical Chemistry B, 116*(12), 3861–3865. doi:10.1021/jp211190q.

Pakravan, D., Michiels, E., Bratek-Skicki, A., De Decker, M., Van Lindt, J., Alsteens, D., Derclaye, S., Van Damme, P., Schymkowitz, J., Rousseau, F., Tompa, P., & Van Den Bosch, L. (2021). Liquid–liquid phase separation enhances TDP-43 LCD aggregation but delays seeded aggregation. *Biomolecules, 11*(4), 548. doi:10.3390/biom11040548.

Pan, Z., Huang, J., & Zhuang, W. (2021). Protein–ligand binding molecular details revealed by terahertz optical Kerr spectroscopy: A simulation study. *JACS Au, 1*(10), 1788–1797. doi:10.1021/jacsau.1c00356.

Plusquellic, D. F., Siegrist, K., Heilweil, E. J., & Esenturk, O. (2007). Applications of terahertz spectroscopy in biosystems. *ChemPhysChem, 8*(17), 2412–2431. http://onlinelibrary.wiley.com/journal/10.1002/(ISSN)1439-7641. doi:10.1002/cphc.200700332.

Mernea, M., Calborean, O., Petrescu, L., Zatreanu, D., Sandu, O., Dascalu, T., & Mihailescu, D. F. (2011). THz spectroscopy and molecular modeling of bovine serum albumin under various hydration conditions. *Proceedings SPIE 8306, Photonics, Devices, and Systems V*, 83060N. doi:10.1117/12.912460.

RCSB Protein Data Bank Statistics (n.d.). PDB Data Distribution by experimental method and molecular type. https://www.rcsb.org/stats/summary (Accessed Dec 23, 2024).

Ren, W., Dokainish, H. M., Shinobu, A., Oshima, H., & Sugita, Y. (2021). Unraveling the coupling between conformational changes and ligand binding in ribose binding protein using multiscale molecular dynamics and free-energy calculations. *Journal of Physical Chemistry B, 125*(11), 2898–2909. http://pubs.acs.org/journal/jpcbfk. doi:10.1021/acs.jpcb.0c11600.

Rivoire, O. (2024). A role for conformational changes in enzyme catalysis. *Biophysical Journal, 123*(12), 1563–1578. doi:10.1016/j.bpj.2024.04.030.

Riziotis, I. G., Ribeiro, A. J. M., Borkakoti, N., & Thornton, J. M. (2022). Conformational variation in enzyme catalysis: A structural study on catalytic residues. *Journal of Molecular Biology, 434*(7). https://www.journals.elsevier.com/journal-of-molecular-biology. doi:10.1016/j.jmb.2022.167517.

Rytik, A. P., & Tuchin, V. V. (2025). Effect of terahertz radiation on cells and cellular structures. *Frontiers of Optoelectronics, 18*(1). doi:10.1007/s12200-024-00146-y.

Schroer, M. A., Schewa, S., Gruzinov, A. Y., Rönnau, C., Lahey-Rudolph, J. M., Blanchet, C. E., Zickmantel, T., Song, Y. H., Svergun, D. I., & Roessle, M. (2021). Probing the existence of non-thermal terahertz radiation induced changes of the protein solution structure. *Scientific Reports, 11*(1). Www.nature.com/srep/index.html. doi:10.1038/s41598-021-01774-6.

Senske, M., Xu, Y., Bäumer, A., Schäfer, S., Wirtz, H., Savolainen, J., Weingärtner, H., & Havenith, M. (2018). Local chemistry of the surfactant's head groups determines protein stability in reverse micelles. *Physical Chemistry Chemical Physics, 20*(13), 8515–8522. doi:10.1039/c8cp00407b.

Shimura, K., Sasaki, T., Ono, T., Mohara, M., Aiko, K., & Sakamoto, T. (2024). Terahertz frequency-domain spectroscopy with a method for suppressing water vapor absorption peaks for analysis of pharmaceutical hydrate samples. *Journal of Infrared, Millimeter, and Terahertz Waves, 45*(9–10), 868–882. doi:10.1007/s10762-024-01004-w.

Stenström, O., Diehl, C., Modig, K., & Akke, M. (2024). Ligand-induced protein transition state stabilization switches the binding pathway from conformational selection to induced fit. *Proceedings of the National Academy of Sciences of the United States of America, 121*(14). doi:10.1073/pnas.2317747121.

Sushko, O., Dubrovka, R., & Donnan, R. S. (2015). Sub-terahertz spectroscopy reveals that proteins influence the properties of water at greater distances than previously detected. *Journal of Chemical Physics, 142*(5). http://scitation.aip.org/content/aip/journal/jcp. doi:10.1063/1.4907271.

Thirumuruganandham, S. P., Gomez, E. A., Lakshmanan, S., & Hamblin, M. R. (2017). Terahertz spectroscopy to determine cold shock protein stability upon solvation and evaporation-A molecular dynamics study. *IEEE Transactions on Terahertz Science and Technology, 7*(2), 131–141. http://www.mtt.org/publications/118-terahertz.html#thz. doi:10.1109/TTHZ.2016.2637380.

Walker, E. J., Hamill, C. J., Crean, R., Connolly, M. S., Warrender, A. K., Kraakman, K. L., Prentice, E. J., Steyn-Ross, A., Steyn-Ross, M., Pudney, C. R., van der Kamp, M. W., Schipper, L. A., Mulholland, A. J., & Arcus, V. L. (2024). Cooperative conformational transitions underpin the activation heat capacity in the temperature dependence of enzyme catalysis. *ACS Catalysis, 14*(7), 4379–4394. http://pubs.acs.org/page/accacs/about.html. doi:10.1021/acscatal.3c05584.

Wang, P., Wang, X., Liu, L., Zhao, H., Qi, W., & He, M. (2019). The hydration shell of monomeric and dimeric insulin studied by terahertz time-domain spectroscopy. *Biophysical Journal, 117*(3), 533–541. doi:10.1016/j.bpj.2019.06.028.

Wang, S. (2019). Efficiently calculating anharmonic frequencies of molecular vibration by molecular dynamics trajectory analysis. *ACS Omega, 4*(5), 9271–9283. doi:10.1021/acsomega.8b03364.

Wei, L., Yu, L., Jiaoqi, H., Guorong, H., Yang, Z., & Weiling, F. U. (2018). Application of terahertz spectroscopy in biomolecule detection. *Frontiers in Laboratory Medicine, 2*(4), 127–133. doi:10.1016/j.flm.2019.05.001.

Whitmire, S. E., Wolpert, D., Markelz, A. G., Hillebrecht, J. R., Galan, J., & Birge, R. R. (2003). Protein flexibility and conformational state: A comparison of collective vibrational modes of wild-type and D96N bacteriorhodopsin. *Biophysical Journal, 85*(2), 1269–1277. doi:10.1016/s0006-3495(03)74562-7.

Wingert, B., Krieger, J., Li, H., & Bahar, I. (2021). Adaptability and specificity: How do proteins balance opposing needs to achieve function? *Current Opinion in Structural Biology, 67*, 25–32. doi:10.1016/j.sbi.2020.08.009.

Woods, K. N. (2010). Solvent-induced backbone fluctuations and the collective librational dynamics of lysozyme studied by terahertz spectroscopy. *Physical Review E, 81*(3). doi:10.1103/physreve.81.031915.

Woods, K. N. (2014a). THz time scale structural rearrangements and binding modes in lysozyme-ligand interactions. *Journal of Biological Physics, 40*(2), 121–137. doi:10.1007/s10867-014-9341-4.

Woods, K. N. (2023). Modeling of protein hydration dynamics is supported by THz spectroscopy of highly diluted solutions. *Frontiers in Chemistry, 11*. http://journal.frontiersin.org/journal/chemistry. doi:10.3389/fchem.2023.1131935.

Woods, K. N. (2014b). The glassy state of crambin and the THz time scale protein-solvent fluctuations possibly related to protein function. *BMC Biophysics, 7*(1). http://www.biomedcentral.com/2046-1682/7/8. doi:10.1186/s13628-014-0008-0.

Woods, K. N. (2014c). Using THz time-scale infrared spectroscopy to examine the role of collective, thermal fluctuations in the formation of myoglobin allosteric communication pathways and ligand specificity. *Soft Matter, 10*(24), 4387–4402. doi:10.1039/c3sm53229a.

Woods, K. N., & Pfeffer, J. (2020). Conformational perturbation, allosteric modulation of cellular signaling pathways, and disease in P23H rhodopsin. *Scientific Reports, 10*(1). Www.nature.com/srep/index.html. doi:10.1038/s41598-020-59583-2.

Woods, K. N., & Pfeffer, J. (2016). Using THz spectroscopy, evolutionary network analysis methods, andmd simulation to map the evolution of allosteric communication pathways in c-type lysozymes. *Molecular Biology and Evolution, 33*(1), 40–61. http://mbe.oxfordjournals.org/. doi:10.1093/molbev/msv178.

Xu, Y., & Havenith, M. (2015). Perspective: Watching low-frequency vibrations of water in biomolecular recognition by THz spectroscopy. *The Journal of Chemical Physics, 143*(17). doi:10.1063/1.4934504.

Yao, X. Q., & Hamelberg, D. (2022). From distinct to differential conformational dynamics to map allosteric communication pathways in proteins. *Journal of Physical Chemistry B, 126*(14), 2612–2620. http://pubs.acs.org/journal/jpcbfk. doi:10.1021/acs.jpcb.2c00199.

Zaccai, G., Natali, F., Peters, J., Řihová, M., Zimmerman, E., Ollivier, J., Combet, J., Maurel, M.-C., Bashan, A., & Yonath, A. (2016). The fluctuating ribosome: Thermal molecular dynamics characterized by neutron scattering. *Scientific Reports, 6*(1). doi:10.1038/srep37138.

Zhang, S., Li, H., Krieger, J. M., Bahar, I., & Ozkan, B. (2019). Shared signature dynamics tempered by local fluctuations enables fold adaptability and specificity. *Molecular Biology and Evolution, 36*(9), 2053–2068. http://mbe.oxfordjournals.org/. doi:10.1093/molbev/msz102.

CHAPTER

Simulating the activation of enzymes

10

Ran Friedman*

Department of Chemistry and Biomedical Sciences, Linnaeus University, Kalmar, Sweden

**Corresponding author: ran.friedman@lnu.se*

10.1 Introduction

The environment in cells and tissues is tightly controlled to ensure the proper function of organisms. As part of this control mechanism, enzymes are typically not active all the time, but become activated when needed. The activation is a process through which the enzyme becomes ready for catalysis. This can happen through binding of another protein or cofactor, cleavage of a part of the enzyme that blocks the active site, or posttranslational modification. Regardless of how the process progresses, a structural modification takes place that makes the enzyme adopt an active state, ready for catalysis. Normal and pathological genetic variation can influence the activation process and might make enzymes more or less active than normal (or, in extreme cases, inactive or constitutively active).

Understanding how enzymes become active is important for many reasons. From a theoretical point of view, it is interesting to unravel how a single protein sequence leads to a protein that can adopt multiple states. From a more practical standpoint, any interference with the activation process (by genetic or other mechanisms) can have consequences such as cancer and genetically inherited diseases. Moreover, small molecular drugs that lead to enzyme activation can be used in the clinic (Matschinsky & Porte, 2010). It can be envisioned that knowledge of the activation process can also be used for biotechnological design of enzymes with higher activity.

Structural studies, using X-ray crystallography and nuclear magnetic resonance (NMR) spectroscopy, have been insightful for gaining knowledge of the end-states of enzymes (active and inactive states). Using mutants, inhibitors, or activators, it is sometimes possible to trap an enzyme also along the (in)activation pathway (e.g., Jeffrey et al., 1995; Venditti et al., 2015). However, such experimental studies remain difficult, and following the activation pathway of enzymes is mostly performed either by simulation methods or by a combination of simulation and experiment.

In this chapter, methods that have been used to follow large conformational transitions in biomolecules are introduced together with some applications. The aim is to provide the reader with a basic understanding of the theory and some information on the application. From a practical point of view such methods are always "advanced" in the sense that they can be tricky to master. Any scientist who wishes to use such methods is therefore advised to start by mastering the art of performing classical molecular dynamics (MD) simulations first.

MD simulations are routinely used to study biomolecules. Decades of research and applications have yielded reliable, fast, and relatively easy-to-use software and force fields that do an excellent job in

Protein Dynamics: Experimental and Computational Approaches. DOI: https://doi.org/10.1016/B978-0-443-30042-4.00005-5

modeling folded proteins. However, there are still limitations. Ligands or cofactors that involve metals can be difficult to study, as the metal ions' charges polarize the nearby environment, and including polarization in the force field is often not enough to remedy this (Ahlstrand et al., 2017); this limitation persists for all forcefield-based methods. When it comes to studying the activation of enzyme, we encounter a temporal problem. Enzyme activation is a slow process in molecular terms, extending beyond milliseconds (Volkman et al., 2001). MD simulations of proteins can be routinely run up to microseconds, but extending this by three orders of magnitude requires specialized hardware. Even if such long simulations can be carried out somehow, there is no guarantee that they end up showing the interesting transition. For this reason, there is a need to bias the simulation in such a way that the transition is reached and that we can gain some useful information from following it. Some methods to handle this are explained in this chapter.

There are many different methods that can be used to study enzyme activation. The methods that were chosen to be presented here are those that were deemed important as an introduction to the subject or those that had been successfully used to study enzyme activation. The focus was chosen to be on thermodynamics rather than kinetics. Some of the methods that use dynamics can be expanded to conclude on the kinetics of the process.

10.2 Methods to study large molecular transitions

Enzyme activation is in principle not different than any other large-scale molecular transition, and the methods that are described below can in principle be used for any molecular system as long as it can be described by the underlying potential. It is still important to consider what drives the reaction for accurate modeling, though—does this involve cleavage, binding of a cofactor or another protein, phosphorylation, or something else? Besides this, it is important to realize the pros and cons of the method of choice.

10.2.1 Chain of states methods

Chain-of-states methods seek to optimize the energy along a path by considering a chain of intermediate states $\mathbf{x}_1$, $\mathbf{x}_2$, ..., $\mathbf{x}_n$ between the initial configuration $\boldsymbol{R}_i$ and the final one $\boldsymbol{R}_f$. They normally yield a trajectory that progresses between two states (e.g., inactive and active enzyme) along the minimum-energy pathway (MEP). These optimization methods generally require an initial guess of points along the transition between the reactant and the product. Generating such a guess can be challenging. An interpolation between the active and inactive states is often used, but may yield highly distorted structures that, when subject to optimization, yield forces that are very large. This, in turn, leads to numerical instabilities. Thus, care should be exerted when considering the structures that form an input to such methods.

10.2.1.1 Steepest descent paths

Formally, identifying intermediates along an energy landscape that includes states of an enzyme requires that we consider the second derivatives of the energy function with respect to the position, which becomes formidable for systems that are large. To this end, Elber and Karplus have developed a method that relies on first derivatives. This was done in 1987 (Elber & Karplus, 1987), and many of the ideas presented in their seminal article are still useful today. The principle is as follows. Let us consider two conformations, $\boldsymbol{R}_i$ and $\boldsymbol{R}_f$. To approach $\boldsymbol{R}_f$, we start from $\boldsymbol{R}_i$ and continue along a curve that connects the values that correspond to $U(\boldsymbol{R}_i)$ and $U(\boldsymbol{R}_f)$ where U is an energy function that depends on the coordinates of all atoms in the system. This curve should be minimized; that is, the motion should be along the path where the energy difference should be minimal. Instead of minimizing the line integral that follows the curve, we replace it by a sum along M intermediate grid points. In total, we have $M + 2$ points where the first corresponds to $\boldsymbol{R}_i$ and the last to $\boldsymbol{R}_f$. The distances between each consecutive point (milestone) along this trajectory are constrained to be roughly equal and are calculated after removing any rotation or translation. The accuracy of the calculations depends on the number of points, the strength of the constraint on the distances (stronger constant leads to higher accuracy at higher cost), and if the chosen path is representative enough, because the generated path will be minimal on average but not at every point. The same considerations apply for many methods derived from this approach. One of the earliest applications of this method was a study of the activation of citrate synthase (El-Kettani & Durup, 1992). The method is implemented in the CHARMM package (Brooks et al., 2009) (TReK program).

10.2.1.2 Conjugate peak refinement

Conjugate peak refinement (CPR) is, in similarity with steepest descent paths (SDP), a method that relies on the energy and its gradient without any bias (Fischer & Karplus, 1992). CPR was developed specifically to identify saddle points along a complex free energy landscape, especially for systems with many different degrees of freedom, such as proteins. Unlike many other chain-of-states methods, CPR will typically generate and refine intermediate coordinates that are not equidistant from each other. It relies on the definition of conjugate vectors **s** as vectors that satisfy the condition:

$$\mathbf{s}_j^T \mathbf{H} \mathbf{s_j} < 0, \forall i \neq j \tag{10.1}$$

where H is the Hessian matrix. For each saddle point the energy along one of these vectors, s_0, has a local maximum, satisfying

$$\mathbf{s}_0^T \mathbf{H} \mathbf{s_0} < 0 \tag{10.2}$$

The energy along the remaining vectors will have local minima:

$$\mathbf{s}_j^T \mathbf{H} \mathbf{s_j} > 0, \ j > 0 \tag{10.3}$$

Given a path of M intermediate points, $M \geq 0$, the vector s_0 is identified between two minima, and a set of conjugate vectors is calculated. Thereafter the energy is maximized along s_0 and minimized along s_j to yield a saddle point. Intermediate points are added on each side of the saddle point, and the procedure is iteratively repeated.

The minimal input is to include only $\boldsymbol{R}_i$ and $\boldsymbol{R}_f$, whereas it is possible to include a guess of the trajectory (e.g., from SDP, an MD simulation starting with $\boldsymbol{R}_i$ and $\boldsymbol{R}_f$, or experimental structures that lie in-between these). The maximum number of intermediate points, that is, M, can be set and will affect the time it takes to run. The procedure is shown in Fig. 10.1. The larger the number of expected saddle points, the larger should M be. CPR was extended for use in QM/MM (Bondar et al., 2004) and QM simulations (Friedman et al., 2007). An example of the use of CPR that is relevant to enzymatic activation is its use to study the coupling between local and global conformational changes in the myosin motor, as summarized in (Kiani & Fischer, 2016). CPR is implemented in CHARMM (Brooks et al., 2009) (TReK program) and pyPCR (Gisdon et al., 2016).

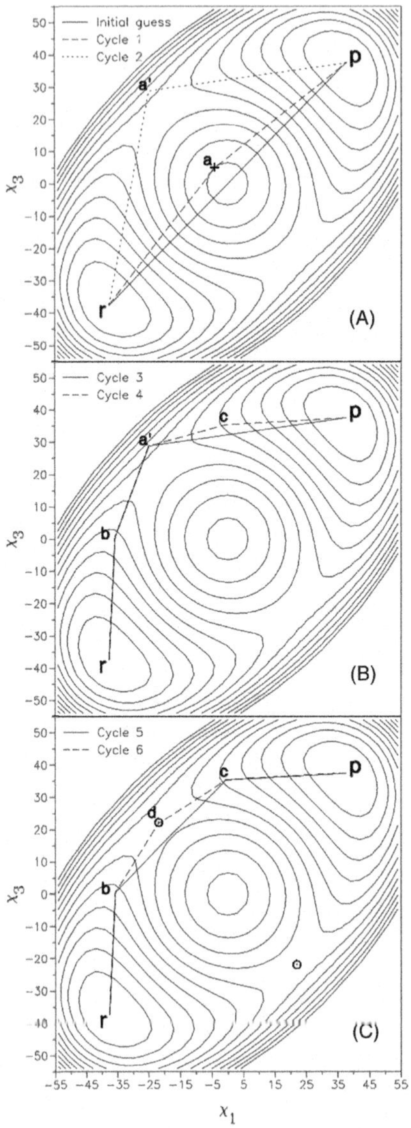

FIGURE 10.1

Conjugate peak refinement: Conjugate peak refinement of a multidimensional system, presented on a 2D surface: (A) the initial guess as a linear interpolation between the reactant **r** and product **p** (*solid line*). An intermediate point is introduced, and a search in all directions orthogonal to the initial path is being started from the maximum, identifying an intermediate point of high energy **a**. Minimization along the path yields **a′**. (B) The procedure is repeated between **r** and **a′** and between **p** and **a′**, yielding **b** and **c**. (C) An attempt to find a local energy maximum in any direction tangent to **a′** fails, which indicates that **a′** is not a true saddle point and it is removed. Further refinement between **b** and **c** identifies the saddle point, **d**.

Reproduced with permission from Fischer and Karplus (1992), see the original publication for additional details.

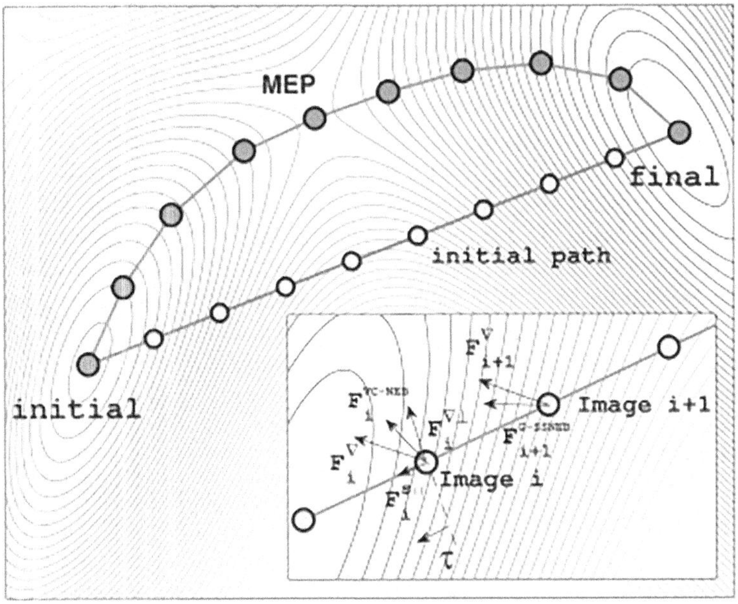

FIGURE 10.2

Nudged elastic band: A schematic of the NEB procedure. The optimization of the initial path allows it to converge to the MEP. $\mathbf{F}_i^{\nabla}$ is potential force in the gradient's direction, with $\mathbf{F}_i^{\nabla\perp}$ its transverse component. The spring force $\mathbf{F}_i^{s||}$ on each image keeps the distance between images constant during optimization of $\mathbf{F}_i^{\nabla\perp}$. *MEP*, Minimum-energy pathway; *NEB*, nudged elastic band.

Reproduced with permission from Qian et al. (2013).

10.2.1.3 ***Nudged elastic band***

Nudged elastic band (NEB) (Jónsson et al., 1998) is another algorithm that involves optimization of states leading from $\boldsymbol{R_i}$ to $\boldsymbol{R_f}$. Here, we start by generating configurations in-between $\boldsymbol{R_i}$ and $\boldsymbol{R_f}$ and not necessarily equally spaced at the beginning. Spring constraints are applied between each two consecutive configurations. In other words, configuration $\boldsymbol{x_i}$ is "bound" to configurations $\mathbf{x_{i-1}}$ and $\mathbf{x_{i+1}}$ with the end configurations held fixed. This prevents the system from converging into a single energy minimum. Optimization is carried out at each point ("image" in the NEB jargon), but keeping the string constraints that make the points relatively equidistant (depending on the strength of the spring) (Fig. 10.2). NEB can be combined with simulated annealing, which might help escape local energy minima along the potential. The NEB method was used to study large-scale conformational transitions in adenylate kinase (Arora & Brooks, 2007). NEB is implemented, for example, in CHARMM (Replica program) and AMBER (Case et al., 2005).

10.2.1.4 ***String method***

The string method can be seen as an extension of the NEB method, where the points are no longer connected by straight "springs," but instead they are connected by "arcs" (Weinan et al., 2007). The whole path is thereafter referred to as a "string." Instead of maintaining the distance between each consecutive pair of points along the trajectory fixed by applying a spring constant as in NEB, this is performed by a reparametrization of the string. During each iteration the configurations ("images" in the string method terminology) x_i are moved downhill along the gradient of the energy. To avoid collapsing into a single minimum state, the reparametrization ensures that the distances between the images are roughly equal. Reparametrization is performed by interpolating the positions of the images along the string and thereafter redistributing them at equal intervals. The string method had been suggested to improve the accuracy of finding the MEP in rough energy landscapes (Weinan et al., 2002). The arcs can be defined in the space of Cartesian coordinates or in another space that represents the trajectory (by so-called collective variables, CV). The MEP is then defined for this subspace, and choosing a different one will yield a different MEP. In the study of conformational changes during enzyme activation, it has been more common to use the string method in conjunction with MDs (see Section 10.2.3.1). The method is implemented in CHARMM (Brooks et al., 2009) (string program).

10.2.1.5 Rock climbing

The rock climbing method (Templeton et al., 2017) was developed to provide a solution to calculate the path between $\boldsymbol{R}_i$ and $\boldsymbol{R}_f$ without the need to generate an initial guess. The approach is similar to rock climbing in a sense that the trajectory progresses one point at a time instead of simultaneously optimizing the whole path, which is similar to a rock climber that has knowledge of the target and all previous steps and uses this knowledge to plan one step ahead. At first, we only know $\boldsymbol{R}_i$ and $\boldsymbol{R}_f$. A new configuration $\mathbf{x}_1$ is generated from $\boldsymbol{R}_i$ by a small displacement along the vector connecting $\boldsymbol{R}_i$ and $\boldsymbol{R}_f$. The path is then optimized, keeping the displacement constant. This procedure is repeated, but starting from $\boldsymbol{x}_1$, then $\boldsymbol{x}_2$, etc. This is referred to as the "Growth Stage." At the end, or every K steps, the path is relaxed without adding any new structures. The method has been used to study the activation of ABL1 kinase and identified a structural barrier for the transition (Narayan et al., 2020).

10.2.2 Enhanced sampling methods

This set of methods employs MD simulations where sampling is enhanced along the desired direction, $\boldsymbol{R}_i \rightarrow \boldsymbol{R}_f$, rather than being allowed to develop stochastically. The simulations may seem more realistic than mere energy optimizations, as they incorporate experimental conditions (temperature and pressure). Yet, the simulated system is not allowed to explore the complete energy landscape.

10.2.2.1 Targeted molecular dynamics

In targeted MD (TMD) (Schlitter et al., 1994) a time-dependent potential U_{TMD} is used to bias the trajectory, which starts at $\boldsymbol{R}_i$ toward $\boldsymbol{R}_f$:

$$U_{\mathrm{TMD}}(t) \propto \lambda(t) \cdot \mathrm{RMSD}(\mathbf{x}(\mathbf{t}), \mathbf{R}_i) + (1 - \lambda(t)) \cdot \mathrm{RMSD}\big(\mathbf{x}(\mathbf{t}), \mathbf{R}_f\big) \quad (10.4)$$

The bias is controlled by $\lambda(t) = 1 - \frac{t}{t_{\mathrm{TOT}}}$ where t_{TOT} is the total simulation time and controls the coupling as a function of root nean square deviation (RMSD) between the sampled conformation ($\mathbf{x}(\mathbf{t})$) and the initial and final ones. The implementation of Eq. (10.4) implies that the length of the run is crucial, with longer simulations leading to a smaller bias. Other implementations exist that enable control of the biasing constant's strength and desired RMSDs. For example, the implementation in the

NAMD code (Phillips et al., 2020) uses the following bias:

$$U_{\text{TMD}} = \frac{1}{2}\frac{k}{N}\left(\text{RMSD}(\mathbf{x}(\mathbf{t}), \mathbf{R}_f) - \text{RMSD}(\mathbf{x}(\mathbf{t}), \mathbf{R}_i)\right)^2 \tag{10.5}$$

where N is the number of targeted atoms and k is a coupling constant. In general the stronger is the biasing strength, the higher is the bias toward a trajectory that is a linear interpolation between the initial and final states.

TMD can be used to simulate an activation trajectory but not for calculating the energy barrier. Neither can it be used to directly identify intermediate structures. CHARMM (Brooks et al., 2009) (TMD program) and NAMD (Phillips et al., 2020) are two popular software packages in which TMD is implemented.

10.2.2.2 Essential dynamics sampling

The essential dynamics sampling (EDS) method was originally developed to sample the dynamics along a part of a protein's conformational space that represents the most significant motions (Amadei et al., 1993). Initially, an MD run is made without constraints, and the mean square deviation matrix is calculated, from which the principal components (PC) of the motions are extracted (by principal component analysis, PCA). When the sampled system does not undergo large conformational changes and is not dominated by random diffusion (Hess, 2000), the largest PC of the system shows the main component of the collective motion of the protein. Limiting the motions to some of the PCs (an "essential subspace") thereafter generates a trajectory that progresses only along this subspace.

The first use of EDS to study enzyme transitions could follow on the open-to-closed (or closed-to-open) transition in citrate synthase (Daidone et al., 2004). The procedure was as follows. First a simulation of several ns was run starting from the structure of origin, from which the PCs were calculated by PCA. Thereafter EDS was run with some of the most dominant PC. In each step the configuration was accepted if the RMSD to the target did not increase. Otherwise the next configuration was obtained by projecting the trajectory along the essential subspace subject to the constraint that the RMSD to the target should not increase. In a later study the essential subspace of the target was used as input, which led to rapid convergence toward the target structure, and the activation of an enzyme could be followed (Friedman & Caflisch, 2008). The limitations of EDS are that, similar to TMD, the method does not yield any estimation of the activation barriers, that any intermediate structures remain hidden (these are often deduced from structural information) and that, even if the RMSD to the target is guaranteed not to increase, the simulations can be stuck in a local minimum that is still quite far from the target. The latter limitation can be overcome by performing additional steps not subject to the bias toward the target between consecutive steps that are biased, or by running exploratory runs with EDS but without any targeting in the first place and starting from conformations that are observed in those. The stability and thermodynamics of selected structures along the trajectory can be examined with short unbiased MD simulations, which were used to obtain structural information on intermediate conformations along the activation pathway of the FLT3 kinase and to estimate the free energy barrier (Todde & Friedman, 2019). EDS is implemented in GROMACS (Abraham et al., 2015).

10.2.2.3 Metadynamics

In metadynamics an external bias potential is added to the Hamiltonian of a system that is simulated with MD (Barducci et al., 2011). An important underlying assumption is that the system can be described

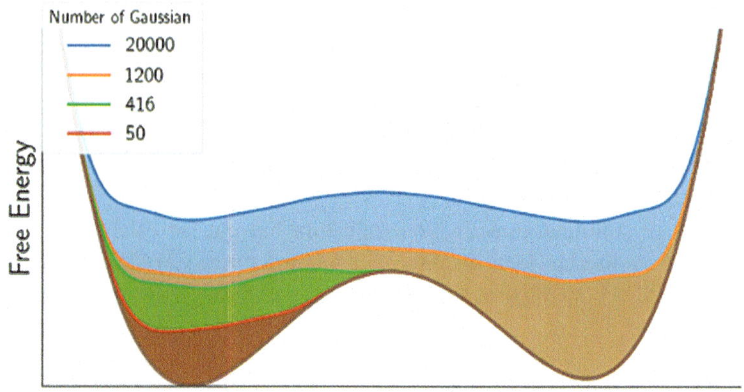

FIGURE 10.3

Metadynamics: A schematic of the principle of WT-Metadyn. Starting from the left, Gaussians are deposited and summed. The system does not resample already visited regions, but sampling of the energy basins is provided, and the barrier height can be calculated.

Reproduced from Hénin et al. (2022), CC BY 4.0 licence.

by a relatively small number of CVs. A time-dependent potential is added to the Hamiltonian of the system in the subspace of the chosen CVs. In the popular well-tempered metadynamics (WT-MetaDyn) (Barducci et al., 2008) the biasing potential along the set of CV **s** is

$$V(\mathbf{s}, t) = k_B \Delta T \, \mathrm{In}\left(1 + \frac{\omega N(\mathbf{s}, t)}{k_B \Delta T}\right) \tag{10.6}$$

where k_B is the Boltzmann constant, ΔT is an input variable with the same unit as the temperature, and $N\,(\mathrm{s},\,t) = \int_0^t \delta_{s,s(t')} dt\prime$ is a histogram of **s**. ω represents the initial rate (energy per time unit) of deposition of the bias. As the potential is a function of N, the system is effectively discouraged from sampling configurations that had already been visited (see Fig. 10.3 for an illustration). In practice $\delta_{s,s(t')}$ is replaced by a Gaussian function. The height of this function W is

$$W = \omega \tau_G e^{-\frac{V(s,t)}{k_B \Delta T}} \tag{10.7}$$

τ_G is the time interval after which a Gaussian function is added (deposition stride). In WT-MetaDyn the deposition stride is decreased as a reciprocal function of the time. Thus, as the simulation progresses, it approaches equilibrium sampling.

Running WT-MetaDyn simulations requires as input a set of CVs, the Gaussian height and σ to determine its width, the temperature and the initial stride, and a bias factor $\gamma = \frac{T+\Delta T}{T}$. The larger is the bias factor, the larger is the perturbation in the simulation.

Metadynamics is implemented in the PLUMED code (Tribello et al., 2014) that is interfaced with different MD packages and is very widely used. An example application to study conformational changes in enzymes is the study of casein kinase (Gouron et al., 2014). In this study, metadynamics was used to bridge between several distinct conformations of the enzyme. In a different study, the authors used metadynamics to follow the inactivation of the clinically important Aurora Kinase B, which suggested that a novel configuration is attained on-pathway (Lakkaniga et al., 2019). Other studies used metadynamics to study conformational transitions in various enzymes (e.g., Jiang et al., 2022; Li et al., 2015).

10.2.3 Combination of methods

Many of the methods that were discussed above require an initial input of intermediate structures, whereas some of the methods do not provide energy estimates, and hence no thermodynamic data (energy barriers and corresponding structures) can be deduced. A combination of several methods can be used to alleviate those limitations.

10.2.3.1 String method with swarm of trajectories

The string method with swarm of trajectories (SMwST) is an extension of the string method (Section 10.2.1.4) whereby a multitude of short MD runs are carried out starting from each conformation along the chain without bias (Chen et al., 2022; Pan et al., 2008). These are used to estimate the drift at each image, where the drift is estimated in directions that are orthogonal to the tangent of the string. This drift should become gradually smaller, which indicates that the path has converged. It is typical to consider a subset of CV for SMwST calculations. The method was used extensively to follow events related to enzyme activation, for example, in Src kinase (Fajer et al., 2016; Gan et al., 2009) and adenylate kinase (ligand-induced activation) (Matsunaga et al., 2012).

SMwST calculations are typically very demanding but scale well because of images can be separated. Moreover, each trajectory can be generated independently so that the loss of performance because of parallelization is reduced. For example, in a study of the activation of a multidomain kinase, the authors used 64 images and 32 swarms per image, leading to 2048 simulations at each step (Fajer et al., 2016). With a swarm length of 10 ps and 2000 iterations to reach convergence, this would require a total simulation time of about 40 μs for one system (of up to 148,000 atoms, Fajer et al., 2016).

10.2.3.2 Essential dynamics sampling and metadynamics

Although EDS can be used to study large conformational transitions, it does not provide information on the energies associated with the transitions and cannot be used in itself to identify intermediate states. Metadynamics can provide these, but it can be difficult to choose CVs to be used in the method. The two methods can be combined in the following way. First an EDS calculation is run, and the biased trajectory is used as an input to a WT-MetaDyn run (as a so-called pathCV). Thereafter a WT-MetaDyn simulation is used to refine the initial trajectory and provide a free energy profile (Oruganti & Friedman, 2021). The method was used successfully to study how the activation of the enzymes ABL1 (Oruganti & Friedman, 2021) and FLT3 (Freire et al., 2023) works. In the latter case, it was shown that the WT-MetaDyn simulations could refine the pathway, resulting in an activation barrier that was 3.7 kcal/mol lower than a previous estimate [from EDS followed by MD (Todde & Friedman, 2019)].

10.2.4 Conclusion: Choice of method and approach

Although the availability of many different methods means that there are plenty of opportunities to choose from, it also indicates that a particular method might not be optimal for a given study. Given that no systematic work was published that compares multiple methods for multiple and different systems, it is impractical to judge the accuracy of the methods. Rather, an estimation on the robustness of a certain approach should be based on experimental measurements if such are available and on comparison with values that are obtained for the system of interest or models that are similar to it. For example, if one is interested in the activation process for a kinase that has not been studied before, it is possible to consider the range of values that were obtained for calculations with other kinases. It is also important to examine intermediates. Do these make sense, that is, do the structures seem plausible? Are they similar to structures of related proteins? For example, in a study with EDS, an intermediate along the activation pathway of a proenzyme was found to be structurally similar to a related proenzyme (Friedman & Caflisch, 2008). Such findings increase the confidence in the obtained trajectory. Furthermore, in many cases proteins that bind to inhibitors or allosteric ligands obtain a distinctive structure that is somewhat different than those of the active and inactive states. If we come across an activation intermediate that mirrors this distinctive structure, it could be interpreted as a validation of the likelihood of the intermediate structure.

All the methods presented either do not fully account for the dynamics and only provide an MEP or bias the dynamics in some sense that may fail to include important contributions to the real activation pathway. This is a necessary consequence of our need for speed—running unbiased simulations and hoping to see the activation process is in most cases futile because the energy barrier is too large. Even if we could simulate the systems for seconds, finite difference inaccuracies in the MD algorithm would limit our ability to reach meaningful conclusions. What is important to consider, however, is whether the choice of CV or other biasing approach is representative enough for the problem at hand. In that context, it is important to realize that intermediates along the activation pathway that have the lowest energy do not need to have a lower RMSD to either $\boldsymbol{R}_i$ and $\boldsymbol{R}_f$ than structures that are higher in energy.

Methods that do not use MD simulations often require lower computational resources but yield an MEP that does not include entropic contributions. These can be added a posteriori (e.g., by running short MD trajectories and using ad hoc entropy calculations). Yet the underlying assumption is that the pathway is the same for the internal (ΔU) and free energy (ΔG or ΔA), which is not always the case.

Despite these considerations, all methods discussed here and many others are useful for studies involving enzyme activation and other large-scale transitions in proteins. These methods are built on solid ground and implemented in various software packages, making it possible to use them by researchers outside of the developers' groups. Such simulations, however, still require care and expertise in an even higher grade than necessary for running equilibrium MD simulations.

References

Abraham, M. J., Murtola, T., Schulz, R., Páll, S., Smith, J. C., Hess, B., & Lindahl, E. (2015). Gromacs: High performance molecular simulations through multi-level parallelism from laptops to supercomputers. *SoftwareX, 1–2*, 19–25. https://doi.org/10.1016/j.softx.2015.06.001.

Ahlstrand, E., Hermansson, K., & Friedman, R. (2017). Interaction energies in complexes of zn and amino acids: A comparison of ab initio and force field based calculations. *Journal of Physical Chemistry A, 121*(13), 2643–2654. https://doi.org/10.1021/acs.jpca.6b12969.

Amadei, A., Linssen, A. B. M., & Berendsen, H. J. C. (1993). Essential dynamics of proteins. *Proteins, 17*(4), 412–425. https://doi.org/10.1002/prot.340170408.

Arora, K., & Brooks, C. L. (2007). Large-scale allosteric conformational transitions of adenylate kinase appear to involve a population-shift mechanism. *Proceedings of the National Academy of Sciences of the United States of America, 104*(47), 18496–18501. https://doi.org/10.1073/pnas.0706443104.

Barducci, A., Bonomi, M., & Parrinello, M. (2011). Metadynamics. *WIREs Computational Molecular Science, 1*(5), 826–843. https://doi.org/10.1002/wcms.31.

Barducci, A., Bussi, G., & Parrinello, M. (2008). Well-tempered metadynamics: A smoothly converging and tunable free-energy method. *Physical Review Letter, 100*(2). https://doi.org/10.1103/physrevlett.100.020603.

Bondar, A.-N., Fischer, S., Smith, J. C., Elstner, M., & Suhai, S. (2004). Key role of electrostatic interac-tions in bacteriorhodopsin proton transfer. *Journal of the American Chemical Society, 126*(44), 14668–14677. https://doi.org/10.1021/ja047982i.

Brooks, B. R., Brooks, C. L., Mackerell, A. D., Nilsson, L., Petrella, R. J., Roux, B., Won, Y., Archontis, G., Bartels, C., Boresch, S., Caflisch, A., Caves, L., Cui, Q., Dinner, A. R., Feig, M., Fischer, S., Gao, J., Hodoscek, M., Im, W., … Karplus, M. (2009). Charmm: The biomolecular simulation program. *Journal of Computational Chemistry, 30*(10), 1545–1614. https://doi.org/10.1002/jcc.21287.

Case, D. A., Cheatham, T. E., Darden, T., Gohlke, H., Luo, R., Merz, K. M., Onufriev, A., Simmerling, C., Wang, B., & Woods, R. J. (2005). The amber biomolecular simulation programs. *Journal of Computational Chemistry, 26*(16), 1668–1688. https://doi.org/10.1002/jcc.20290.

Chen, H., Ogden, D., Pant, S., Cai, W., Tajkhorshid, E., Moradi, M., Roux, B., & Chipot, C. (2022). A companion guide to the string method with swarms of trajectories: Characterization, performance, and pitfalls. *Journal of Chemical Theory and Computation, 18*(3), 1406–1422. https://doi.org/10.1021/acs.jctc.1c01049.

Daidone, I., Roccatano, D., & Hayward, S. (2004). Investigating the accessibility of the closed domain conformation of citrate synthase using essential dynamics sampling. *Journal of Molecular Biology, 339*(3), 515–525. https://doi.org/10.1016/j.jmb.2004.04.007.

Elber, R., & Karplus, M. (1987). A method for determining reaction paths in large molecules: Application to myoglobin. *Chemical Physics Letters, 139*(5), 375–380. https://doi.org/10.1016/0009-2614(87)80576-6.

El-Kettani, M. A. E., & Durup, J. (1992). Theoretical determination of conformational paths in citrate synthase. *Biopolymers, 32*(5), 561–574. https://doi.org/10.1002/bip.360320512.

Fajer, M., Meng, Y., & Roux, B. (2016). The activation of c-Src tyrosine kinase: Conformational transition pathway and free energy landscape. *Journal of Physical Chemistry B, 121*(15), 3352–3363. https://doi.org/10.1021/acs.jpcb.6b08409.

Fischer, S., & Karplus, M. (1992). Conjugate peak refinement: An algorithm for finding reaction paths and accurate transition states in systems with many degrees of freedom. *Chemical Physics Letters, 194*(3), 252–261. https://doi.org/10.1016/0009-2614(92)85543-j.

Freire, T. S., Caracelli, I., Zukerman-Schpector, J., & Friedman, R. (2023). Resistance to a tyrosine kinase inhibitor mediated by changes to the conformation space of the kinase. *Physical Chemistry Chemical Physics, 25*(8), 6175–6183. https://doi.org/10.1039/d2cp05549j.

Friedman, R., & Caflisch, A. (2008). Pepsinogen-like activation intermediate of plasmepsin II revealed by molecular dynamics analysis. *Proteins, 73*(4), 814–827. https://doi.org/10.1002/prot.22105.

Friedman, R., Fischer, S., Nachliel, E., Scheiner, S., & Gutman, M. (2007). Minimum energy pathways for proton transfer between adjacent sites exposed to water. *Journal of Physical Chemistry B, 111*(21), 6059–6070. https://doi.org/10.1021/jp070781r.

Gan, W., Yang, S., & Roux, B. (2009). Atomistic view of the conformational activation of Src kinase using the string method with swarms-of-trajectories. *Biophysical Journal, 97*(4), L8–L10. https://doi.org/10.1016/j.bpj.2009.06.016.

Gisdon, F. J., Culka, M., & Ullmann, G. M. (2016). PyCPR – A python-based implementation of the conjugate peak refinement (CPR) algorithm for finding transition state structures. *Journal of Molecular Modeling, 22*(10). https://doi.org/10.1007/s00894-016-3116-8.

Gouron, A., Milet, A., & Jamet, H. (2014). Conformational flexibility of human casein kinase catalytic subunit explored by metadynamics. *Biophysical Journal, 106*(5), 1134–1141. https://doi.org/10.1016/j.bpj.2014.01.031.

Hénin, J., Lelièvre, T., Shirts, M. R., Valsson, O., & Delemotte, L. (2022). Enhanced sampling methods for molecular dynamics simulations [article v1.0]. *Living Journal of Computational Molecular Science, 4*(1), 1583. https://doi.org/10.33011/livecoms.4.1.1583.

Hess, B. (2000). Similarities between principal components of protein dynamics and random diffusion. *Physical Review E, 62*(6), 8438–8448. https://doi.org/10.1103/physreve.62.8438.

Jeffrey, P. D., Russo, A. A., Polyak, K., Gibbs, E., Hurwitz, J., Massagué, J., & Pavletich, N. P. (1995). Mechanism of CDK activation revealed by the structure of a cyclina-CDK2 complex. *Nature, 376*(6538), 313–320. https://doi.org/10.1038/376313a0.

Jiang, T., Liu, Z., Liu, W., Chen, J., Zheng, Z., & Duan, M. (2022). The conformational transition path-ways and hidden intermediates in DFG-flip process of c-Met kinase revealed by metadynamics simulations. *Journal of Chemical Information and Modeling, 62*(15), 3651–3663. https://doi.org/10.1021/acs.jcim.2c00770.

Jónsson, H., Mills, G., & Jacobsen, K. W. (1998). Nudged elastic band method for finding minimum energy paths of transitions. *Classical and quantum dynamics in condensed phase simulations* (pp. 385–404). World Scientific. https://doi.org/10.1142/97898128396640016.

Kiani, F. A., & Fischer, S. (2016). ATP-dependent interplay between local and global conformational changes in the myosin motor. *Cytoskeleton, 73*(11), 643–651. https://doi.org/10.1002/cm.21333.

Lakkaniga, N. R., Balasubramaniam, M., Zhang, S., Frett, B., & Li, H.-y. (2019). Structural characterization of the aurora kinase B "DFG-flip" using metadynamics. *The AAPS Journal [Electronic Resource], 22*(1). https://doi.org/10.1208/s12248-019-0399-6.

Li, D., Liu, M., & Ji, B. (2015). Mapping the dynamics landscape of conformational transitions in enzyme: The adenylate kinase case. *Biophysical Journal, 109*(3), 647–660. https://doi.org/10.1016/j.bpj.2015.06.059.

Matschinsky, F. M., & Porte, D. (2010). Glucokinase activators (GKAs) promise a new pharmacotherapy for diabetics. *F1000 Medicine Reports, 2*, 43.

Matsunaga, Y., Fujisaki, H., Terada, T., Furuta, T., Moritsugu, K., & Kidera, A. (2012). Minimum free energy path of ligand-induced transition in adenylate kinase. *PLoS Computational Biology, 8*(6), e1002555. https://doi.org/10.1371/journal.pcbi.1002555.

Narayan, B., Fathizadeh, A., Templeton, C., He, P., Arasteh, S., Elber, R., Buchete, N.-V., & Levy, R. M. (2020). The transition between active and inactive conformations of Abl kinase studied by rock climbing and Milestoning. *Biochimica et Biophysica Acta (BBA)—General Subjects, 1864*(4), 129508. https://doi.org/10.1016/j.bbagen.2019.129508.

Oruganti, B., & Friedman, R. (2021). Activation of Abl1 kinase explored using well-tempered metadynamics simulations on an essential dynamics sampled path. *Journal of Chemical Theory and Computation, 17*(11), 7260–7270. https://doi.org/10.1021/acs.jctc.1c00505.

Pan, A. C., Sezer, D., & Roux, B. (2008). Finding transition pathways using the string method with swarms of trajectories. *Journal of Physical Chemistry B, 112*(11), 3432–3440. https://doi.org/10.1021/jp0777059.

Phillips, J. C., Hardy, D. J., Maia, J. D. C., Stone, J. E., Ribeiro, J. V., Bernardi, R. C., Buch, R., Fiorin, G., Hénin, J., Jiang, W., McGreevy, R., Melo, M. C. R., Radak, B. K., Skeel, R. D., Singharoy, A., Wang, Y., Roux, B., Aksimentiev, A., Luthey-Schulten, Z., … Tajkhorshid, E. (2020). Scalable molecular dynamics on CPU and GPU architectures with NAMD. *Journal of Chemical Physics, 153*(4). https://doi.org/10.1063/5.0014475.

Qian, G.-R., Dong, X., Zhou, X.-F., Tian, Y., Oganov, A. R., & Wang, H.-T. (2013). Variable cell nudged elastic band method for studying solid–solid structural phase transitions. *Computer Physics Communications, 184*(9), 2111–2118. https://doi.org/10.1016/j.cpc.2013.04.004.

Schlitter, J., Engels, M., & Krüger, P. (1994). Targeted molecular dynamics: A new approach for searching pathways of conformational transitions. *Journal of Molecular Graphics, 12*(2), 84–89. https://doi.org/10.1016/0263-7855(94)80072-3.

Templeton, C., Chen, S.-H., Fathizadeh, A., & Elber, R. (2017). Rock climbing: A local-global algorithm to compute minimum energy and minimum free energy pathways. *Journal of Chemical Physics, 147*(15). https://doi.org/10.1063/1.4986298.

Todde, G., & Friedman, R. (2019). Activation and inactivation of the FLT3 kinase: Pathway intermediates and the free energy of transition. *Journal of Physical Chemistry B, 123*(26), 5385–5394. https://doi.org/10.1021/acs.jpcb.9b01567.

Tribello, G. A., Bonomi, M., Branduardi, D., Camilloni, C., & Bussi, G. (2014). Plumed 2: New feathers for an old bird. *Computer Physics Communications, 185*(2), 604–613. https://doi.org/10.1016/j.cpc.2013.09.018.

Venditti, V., Schwieters, C. D., Grishaev, A., & Clore, G. M. (2015). Dynamic equilibrium between closed and partially closed states of the bacterial Enzyme I unveiled by solution NMR and X-ray scattering. *Proceedings of the National Academy of Sciences of the United States of America, 112*(37), 11565–11570. https://doi.org/10.1073/pnas.1515366112.

Volkman, B. F., Lipson, D., Wemmer, D. E., & Kern, D. (2001). Two-state allosteric behavior in a single-domain signaling protein. *Science, 291*(5512), 2429–2433. https://doi.org/10.1126/science.291.5512.2429.

Weinan, E., Ren, W., & Vanden-Eijnden, E. (2007). Simplified and improved string method for computing the minimum energy paths in barrier-crossing events. *Journal of Chemical Physics, 126*(16). https://doi.org/10.1063/1.2720838.

Weinan, E., Ren, W., & Vanden-Eijnden, E. (2002). String method for the study of rare events. *Physical Review B, 66*(5). https://doi.org/10.1103/physrevb.66.052301.

CHAPTER

11

Simulating long-timescale conformational dynamics of proteins

Balint Dudas[a,b] and Erika Balog[a,*]

[a]*Department of Biophysics and Radiation Biology, Semmelweis University, Budapest, Hungary,*
[b]*Laboratory of Computational Biology, National Heart, Lung, and Blood Institute, National Institutes of Health, Bethesda, MD, United States*
[*]*Corresponding author: balog.erika@semmelweis.hu*

Biomolecular simulations are crucial tools for exploring detailed insights across various fields of biophysics, including drug discovery, protein structure and function, biomolecular interactions, enzyme mechanisms, and genetic diseases. These simulations aim to link molecular structure, energy landscapes, dynamics, and function. They offer a detailed, atomistic view of molecular systems, complementing experimental techniques. Experimental structures from X-ray crystallography, Nuclear Magnetic Resonance (NMR), or cryo-electron microscopy (cryo-EM) often serve as starting points for simulations, whereas experimental results are also used to refine and validate simulation models. However, experiments can lack fine-grained details, as they are often coarse-grained (CG) over time and/or space, leaving gaps in detailed molecular insights. Simulations bridge these gaps by capturing dynamic, atom-level details that are difficult to observe experimentally. This is especially useful for studying slow molecular motions, rare events, and critical transition states in biological mechanisms and revealing atomistic insights into rate-limiting transition states and corresponding mechanistic pathways, which are key elements in many biological functions but are often inaccessible through experiments alone. Integrating insights from experiments and simulations can often help answer questions that would otherwise be out of reach.

One of the most widely used techniques in biomolecular simulations is molecular dynamics (MD). This chapter introduces MD, discusses its limitations in conformational sampling, and reviews strategies to address these challenges for studying large-scale, cooperative conformational protein dynamics.

11.1 Molecular dynamics simulations

MD simulation was first introduced by Alder and Wainwright (1957) to study how a system of hard spheres with perfectly elastic collisions relaxes. Rahman and Stillinger (1971) used MD to study water, simulating a system of 216 rigid water molecules. McCammon et al. (1977) were the first to apply MD simulations to proteins, modeling the dynamics of a folded globular protein, the bovine pancreatic trypsin inhibitor. Since then, all-atom MD simulations of biomolecular systems have made remarkable

Protein Dynamics: Experimental and Computational Approaches. DOI: https://doi.org/10.1016/B978-0-443-30042-4.00006-7

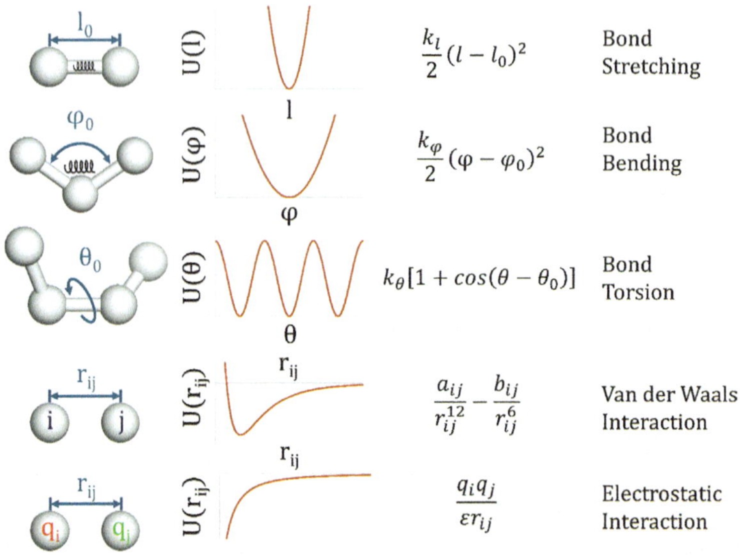

FIGURE 11.1

The potential energy function: The potential energy in classical force fields is composed of terms describing interactions between bonded and nonbonded atoms. Bonded interactions include bond stretching, bond bending, and bond torsion, whereas nonbonded interactions include van der Waals and electrostatic interactions between spatially close atoms. Typically, force fields are additive, meaning the total potential energy is obtained by summing contributions from individual atom pairs. In the equations, l_0, φ_0, and θ_0 represent equilibrium bond length, bond angle, and torsion angle, respectively, while the k's denote force constants. The Lennard–Jones (LJ) parameters include a_{ij}, the LJ repulsion coefficients, and b_{ij}, the attraction coefficients. The effective dielectric constant is represented by ε, and r_{ij} denotes interatomic distances.

progress. They have evolved from modeling small proteins in vacuum to simulating large protein complexes in solvated environments. This advancement has been made possible by improved computational power, advanced algorithms, and specialized hardware such as GPUs (Graphics Processing Units) and supercomputers.

In MD the motion of atoms in a system is determined by their interactions, including both bonded and nonbonded forces. These interactions are described by a force field (FF)—a set of equations and parameters that define the system's potential energy function, $U(r)$(Fig. 11.1). The simulations begin with an initial configuration of atoms, and their instantaneous positions and velocities are updated iteratively using Newton's equations of motion. This process generates a sequence of atomic configurations (conformations) over time, known as a trajectory. At each time step, Newton's equations of motion establish the relationship between atomic positions, the corresponding potential energy, and the acting forces:

$$\frac{d^2(m_i r_i)}{dt^2} = f_i = -\nabla_{r_i} U(r) \tag{11.1}$$

At each MD step the force acting on each atom is calculated by finding the gradient of the potential energy $U(r)$ with respect to the three components of the atom's position vector. This force directly relates to the atomic acceleration if the atomic mass m_i is constant. The negative sign indicates that atoms accelerate toward regions of lower potential energy. In MD, Newton's equations of motions are expressed in terms of finite differences, and the positions and velocities of atoms are updated by integrating over the previous timesteps. The timestep Δt must be shorter than the fastest motions in the system to ensure the numerical integrator's "small-step" assumption remains valid. If this condition is not met, unphysical energies and accelerations can result. In proteins the fastest motions are bond vibrations, so typical timesteps are set to around 1–2 femtoseconds. Common numerical integration methods in MD include the Verlet and leapfrog algorithms. Simulations usually begin with an energy-minimized 3D structure, and initial velocities are typically assigned randomly, following the probabilities defined by the Maxwell–Boltzmann distribution for a given simulation temperature T:

$$p(v_{x,i}) = \sqrt[2]{\frac{m_i}{2\pi RT}\exp\left(-\frac{m_i v_{x,i}^2}{2RT}\right)} \tag{11.2}$$

for each atom $i = 1 \ldots N$ of atomic mass m_i, and for each component x, y, and z, with R the universal gas constant.

The resulting set of conformations (trajectory) captures the system's dynamics, allowing for the observation of conformational changes and the extraction of statistical data. Although MD simulations are deterministic, their outcomes are highly sensitive to the initial conditions. To ensure consistency and reproducibility of average properties and dominant patterns, multiple replicas are typically run, each starting with different randomized initial velocities.

Today, MD simulations are widely used as powerful tools to provide valuable structural and energetic insights into the behavior of complex molecular systems. As the first MD simulation of proteins, which captured only 8.8 ps of protein dynamics (McCammon et al., 1977), it took about 20 years to achieve simulations on the microsecond scale for proteins in explicit solvent. Currently, molecular systems containing several hundred thousand atoms can be simulated for a few microseconds within a reasonable timeframe of days to weeks. This capability allows researchers to explore processes such as protein side-chain fluctuations, local conformational changes, and short binding or unbinding events. In addition, in the past decade, simulations reaching the millisecond scale have been reported. However, multiple runs of the same simulation are still necessary to ensure consistency and reproducibility of the results.

The sampling problem is a well-known challenge in MD simulations, arising from the difficulty of thoroughly exploring the vast conformational space of biomolecular systems within the limited timescales of simulations (Clarage et al., 1995; Bernardi et al., 2015). Biomolecular processes can occur on timescales of seconds or even minutes. Important transitions, such as ligand binding to proteins or large conformational changes (e.g., during the functional cycle of transmembrane transport proteins), are often rare events that happen infrequently during typical MD simulations. This makes the probability of capturing such events very unlikely.

In addition, transitions between metastable conformational states often require the system to overcome high energy barriers. In conventional MD simulations the system can stall in local energy minima, consuming significant computational resources and further restricting the timescales that can be simulated. Insufficient sampling can leave important transitions and their pathways unobserved, leading to an incomplete or inaccurate understanding of the underlying mechanisms.

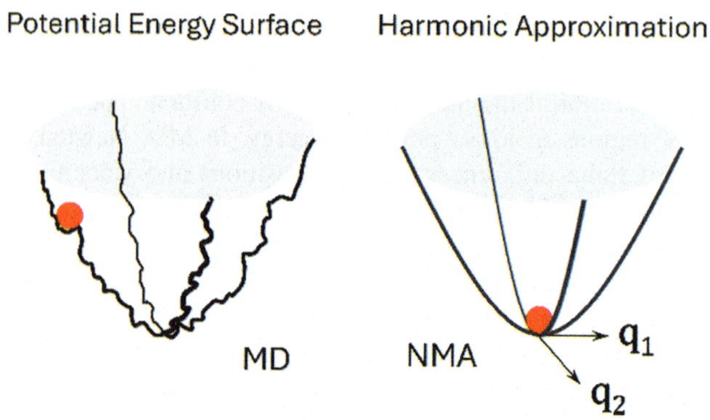

FIGURE 11.2

Schematic illustration of a two-dimensional classical force field potential and its harmonic approximation around a given local energy minimum (denoted by the *red* ball): The two normal modes, q_1 and q_2, represent orthogonal and independent oscillation directions, with their respective frequencies determining the steepness of the parabolic potential along each mode. The system's dynamics within this harmonic approximation can be described by linear combinations of oscillations along these modes. Although the normal mode approximation is strictly valid only near the local energy minimum, low-frequency normal modes often correspond well to large-scale collective motions that are functionally relevant in protein dynamics. *MD*, Molecular dynamics; NMA, normal mode analysis.

To overcome the limitations of conventional MD and gain insight into large-scale protein dynamics, various strategies exist. These include both approximative analytical methods, such as normal mode analysis (NMA), and enhanced simulation techniques. Enhanced simulation approaches aim to either broaden the exploration of conformational space or focus sampling on hard-to-reach regions, such as those near transition states.

One enhanced sampling strategy is to reduce relative energy barriers between metastable conformational states, accelerating sampling and promoting transitions. Another is to apply biases or restraints along specific collective variables (CVs), which guide the system toward or force it into desired conformational states. Coarse-graining strategies simplify molecular representations, reducing the system's degrees of freedom, allowing simulations to cover longer timescales, though at the expense of some atomic detail. Nonphysical, alchemical transformations are another strategy used to calculate free energy changes—a state function—by breaking down transitions between physical end states into smaller, nonphysical but more manageable steps. Finally, hybrid methods combine multiple of the above strategies, integrating their strengths while overcoming individual limitations.

11.2 Analytical analysis of protein dynamics—Normal modes

NMA is the study of harmonic potential wells by analytic means (Fig. 11.2). It is particularly useful for identifying and describing the slowest intrinsic motions of molecules, which in nature generally correspond to large, collective functional movements of the system. NMA relies on the harmonic approximation of the potential energy function around a local energy minimum. This inherently restricts the analysis to studying dynamics within a single potential well, neglecting anharmonic effects. However, the low-frequency modes identified by NMA appear to describe well the large conformational changes that are observed experimentally (Tama & Sanejouand, 2001), although without a certain guarantee of generality. In principle a low-energy motion at a large scale should also be low energy at a smaller scale.

The NMA theory relies on the analytic solution of Newton's equations of motion, assuming that the potential energy function $U(r)$ has a quadratic form. To perform NMA the reference conformation must correspond to a local energy minimum on the potential energy surface, so the analysis begins with an energy minimization. The potential energy can be expanded in a multivariate Taylor series around this local minimum:

$$U(r) \cong U(r_0) + \nabla U(r_0)^T (r - r_0) + \frac{1}{2}(r - r_0)^T H_0 (r - r_0) + \text{higher order terms} \tag{11.3}$$

Starting with a reference potential $U(r_0) = 0$ at a conformation r_0 corresponding to a local potential energy minimum where the gradient of the potential $\nabla U(r_0) = 0$, the harmonic approximation of the potential energy function around this minimum is obtained by neglecting all higher order terms. Then, Newton's equations of motion can then be written in matrix form as

$$M\frac{d^2 r(t)}{dt^2} = -\frac{1}{2}\nabla\left[(r - r_0)^T H_0 (r - r_0)\right] = -H_0(r - r_0) \tag{11.4}$$

where M is the mass matrix, which contains the atomic masses along its diagonal, and H_0 is the Hessian matrix. The elements of the Hessian H_0 are the second derivatives of the potential energy with respect to the atomic coordinates. To solve the $3N$ differential equations analytically, NMA involves diagonalizing the $3N \times 3N$ Hessian matrix of a system with N atoms. The dynamics are then expressed as linear combinations of $3N$ linearly independent harmonic oscillations, described by the $3N$ orthonormal vibration directions (the eigenvectors of the Hessian), each with a corresponding frequency (the eigenvalues of the Hessian).

These eigenvalues not only represent the frequencies of the vibrations but also relate to the energetic cost of displacing the system along each mode. This allows the analysis to classify directions into low-energy collective motions and high-energy local deformations. Typically, only a small number of low-frequency (low-energy) collective motions are used to describe directions that correspond to larger conformational transitions between metastable states.

11.3 Accelerated sampling methods—Sampling without collective variables

Accelerated sampling methods are useful when there is no prior knowledge of the dynamics of a system, meaning there is no information about its collective motions or metastable states. In such cases the conformational sampling is performed in a "blind" way, exploring all possible motions the system may encode. These methods accelerate conventional MD simulations by reducing the relative energy barriers that separate different metastable states. This can be achieved by raising potential energy wells to destabilize states where the simulation might get trapped, adding boost potentials, or increasing thermal fluctuations by simulating at higher temperatures. What both of these strategies have in common is that

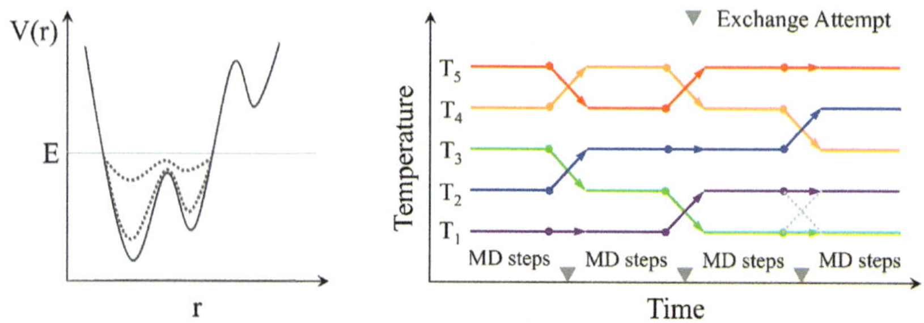

FIGURE 11.3

Two common accelerated sampling techniques used in MD simulations: *Left*: accelerated molecular dynamics (aMD)—A schematic representation of the original (unbiased) potential shown by a *solid curve* and the modified (biased) potential shown by *dashed curves*, where an additional boost energy is applied to smooth energy barriers and enhance conformational sampling. *Right*: replica exchange molecular dynamics (REMD)—Simulations are performed in parallel at different temperatures, with periodic attempts to exchange conformations between neighboring replicas (represented by different *colors*). If an exchange attempt is accepted, the corresponding conformations are swapped. This strategy improves sampling efficiency by facilitating transitions over energy barriers. *MD*, Molecular dynamics.

they do not require the careful selection of collective coordinates describing the system's conformational dynamics for the acceleration.

11.3.1 Accelerated molecular dynamics

Accelerated MD, introduced by Pal and Fichthorn (1999), was designed to accelerate conformational sampling in systems that may get stuck in potential energy minima with high free energy barriers for a large number of computational steps. Accelerated MD can be applied even when there is no prior knowledge about the locations of potential energy wells or saddle points of the system. It works by modifying the potential energy landscape such that the escape rates from these wells are enhanced. It uses a continuous, nonnegative boost potential $\Delta V(r)$. If the true potential $V(r)$ falls below a certain threshold E, the simulation switches to a modified potential $V^*(r) = V(r) + \Delta V(r)$. When $V(r)$ is above the threshold, the simulation returns to using the true potential (Fig. 11.3). Importantly, accurate thermodynamics and equilibrium properties can still be determined from accelerated MD simulations by taking corrected ensemble averages. In the corrected averages, each configuration is weighted by the Boltzmann factor of the bias energy, obtaining the equilibrium observable on the true potential.

Different definitions of the boost potential $\Delta V(r)$ exist, with the simplest creating a flat potential surface below the threshold E. In this case, $\Delta V(r) = E - V(r)$ and $V^*(r) = E$ for potentials below the threshold (Rahman & Tully, 2002). However, this definition causes derivation problems at the edges of the flat potential, when $V(r) = E$. Various choices for the boost potential exist that ensure that the derivative of $V^*(r)$ has no discontinuity (Hamelberg et al., 2004; Gong & Wilkins, 1999; Pal & Fichthorn, 1999).

11.3.2 Replica-exchange molecular dynamics

Replica-exchange MD (REMD, also known as parallel tempering) is a method used to improve the efficiency of sampling and address the sampling problem by exchanging between multiple noninteracting copies (replicas) of the system, each simulated at different temperatures (Fig. 11.3). The original replica exchange method was first developed for the Monte Carlo algorithm (Swendsen & Wang, 1986), and later, Sugita and Okamoto (1999) adapted it for use in MD simulations.

REMD involves using multiple replicas of the same system, each simulated at a different temperature. The method consists of two main alternating steps: (1) a simulation step and (2) a replica exchange step. In the simulation step the replicas do not interact with each other. Each replica is independently simulated at its assigned temperature within a canonical ensemble for a set number of MD steps, with exactly one replica at each temperature. In the replica exchange step, pairs of replicas at neighboring temperatures can be exchanged based on the Metropolis acceptance probability. To ensure the exchange process reaches an equilibrium distribution the detailed balance condition should be imposed (i.e., in equilibrium, each elementary process is in equilibrium with its reverse process).

The detailed balance condition for exchanging two replicas can be satisfied using the following Metropolis acceptance probability for the exchange:

$$p = \min\left(1, \exp\left(\left[\frac{1}{k_B T_i} - \frac{1}{k_b T_j}\right](V_i - V_j)\right)\right) \tag{11.5}$$

where k_B is the Boltzmann factor, T_i and V_i are the temperature and the potential energy of simulation i, whereas T_j and V_j correspond to simulation j. The probability p determines whether the temperatures of the two neighboring simulations i and j will be exchanged. If the exchange is accepted, the temperatures of the two systems are exchanged (exchanging the temperatures between replicas is equivalent to exchanging two configurations at adjacent temperatures).

The new temperatures are obtained by uniformly rescaling the velocities of all atoms in the system, multiplying by the square root of the ratio of the two temperatures. Allowing lower temperature simulations to explore higher temperature ranges helps the system avoid stalling in energy local minima for too long. The key parameters for REMD include the temperature distribution, the highest temperature, and the number of replicas.

11.4 Biased simulation approaches—Sampling along collective variables

If prior information about a system's dynamics is available, sampling along CVs can help guide the simulation to efficiently explore different states and connecting pathways. CVs are carefully chosen to describe the essential degrees of freedom relevant to the system's conformational transitions. Molecular systems exist in a high-dimensional configuration space, including the positions and velocities of all atoms. CVs simplify this complexity by reducing the system's dynamics to a low-dimensional space, where a few important variables capture the system's essential behavior.

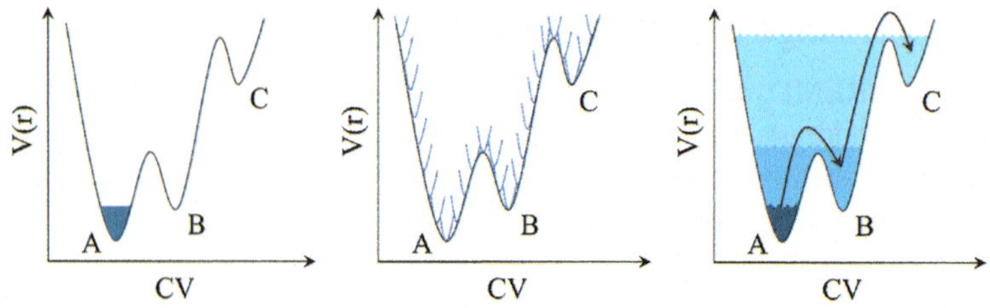

FIGURE 11.4

Examples of accelerated sampling techniques used in MD simulations: *Left*: Conformational flooding—The artificially applied flooding potential, derived from the initial conformation, is represented in *dark blue*. This method helps overcome energy barriers by flattening the energy landscape in predefined regions. *Middle*: Umbrella sampling—The harmonic restraining potentials, known as the "umbrella" potentials (shown in *blue*), bias the system toward specific regions of the conformational space, enabling efficient sampling along a reaction coordinate. *Right*: Metadynamics—As the system approaches an energy barrier separating two minima, Gaussian potentials are deposited to discourage revisiting the same states. Once the system reaches a barrier separating two minima, it evolves to a new state. *CV*, Collective variable.

Commonly used CVs include distances, angles, dihedral angles, and radii of gyration. To improve sampling efficiency, external restraining forces (biases) can be applied along the CVs, effectively guiding the simulation through transitions. In addition, free energy landscapes associated with conformational transitions are also typically expressed as functions of CVs. The enhancement strategies described below all rely on the use of CVs.

11.4.1 Conformational flooding

The conformational flooding method, introduced by Grubmüller (1995), helps overcome large energy barriers, relative to thermal energy, that would normally prevent conformational transitions within the timescales of conventional MD simulations. Conformational flooding applies an artificial external potential that destabilizes the initial conformation, effectively lowering the free energy barriers for structural transitions (Fig. 11.4). Importantly the flooding potential does not push the system toward any specific preselected states.

In conformational flooding, as with other biased simulation techniques, it is crucial to identify suitable CVs that capture the slow dynamics of proteins involved in conformational transitions. These CVs are usually derived from prior unbiased MD simulations using principal component analysis (PCA). The flooding potential is then constructed based on effective Hamiltonians, which provide a CG representation of the real potentials in the subspace defined by the CVs.

The flooding potential must meet two key requirements. First, it should be local, meaning it destabilizes only the potential well corresponding to the initial structure while leaving other regions, such as the transition region, unaffected. Second the flooding potential must be smooth and nonzero within the initial potential well. This ensures the system cannot "hide" in a niche within the well, avoiding the effects of the flooding potential. To satisfy these criteria a multivariate Gaussian bias can be used.

11.4.2 Essential dynamics sampling

Amadei et al. (1996) introduced an enhanced protein conformational sampling method, relying on "essential coordinates," better known as principal components (PCs), used as CVs. PCs are derived from PCA of the system's dynamics, where each PC represents a weighted combination of atomic coordinates from the MD simulation.

PCA is a technique used to reduce the dimensionality of complex data, increasing interpretability while retaining as much information as possible. This is achieved by performing an orthogonal linear transformation, converting the original Cartesian coordinates into a new space defined by uncorrelated CVs, the PCs. The PCs are ordered based on the amount of variation (or information) each one captures from the data.

The essential dynamics sampling method builds on the most significant PCs, which can be extracted from preceding unbiased MD simulations. Constraint forces are applied to guide the system within the subspace defined by these PCs (only the C_α atoms are included), whereas the remaining atomic coordinates move freely according to standard equations of motion.

Similarly to other CV-based biased sampling techniques, it should be noted that the essential dynamics sampling method cannot guarantee the identification of distinct conformational states of which no prior information is available.

11.4.3 Targeted molecular dynamics

The targeted MD method, introduced by Schlitter et al. (1993), induces conformational changes toward a known target structure at normal temperatures by imposing a time-dependent, purely geometrical constraint (Schlitter et al., 1994). Targeted MD allows for the characterization of potential pathways in a conformational transition, including the sequence of metastable intermediates and the energy barriers involved.

In targeted MD, one conformational state serves as the initial structure, whereas another state acts as the target and a directing restraint. The method generates pathways where the distance to the final configuration (measured as the mass–weighted Cartesian distance in the $3N$-dimensional space, where N is the number of atoms) decreases monotonously. Meanwhile the remaining $3N - 1$ degrees of freedom are free to move according to the FF, constrained to a hypersphere around the target structure.

The CV in targeted MD is the mass–weighted Cartesian distance from the target structure. At large initial distances, this constraint minimally affects internal motions. However, as the system approaches the target, dynamic fluctuations are gradually reduced until the structure becomes static. Using mass–weighted distances eliminates translation effects (provided the initial and target structures share the same center of mass), whereas rotational effects can be removed by additional constraints or stopping rotation periodically during the simulation.

11.4.4 Umbrella sampling

Umbrella sampling, introduced by Torrie and Valleau (1977), is a method used to calculate free-energy differences between a system of interest and a reference system using ensemble averages, both following the corresponding Boltzmann distributions. Umbrella sampling applies an external bias to overcome inherent energy barriers and improve sampling efficiency. This bias is applied based on a selected CV that effectively describes the transition between two metastable states.

A harmonic restraining potential, known as the "umbrella" potential, is used in umbrella sampling; the shape of the harmonic bias inspired the name "umbrella" (Fig. 11.4). A biasing potential is centered on different CV values in independent simulations, often referred to as "windows." Selecting the right CV describing the transition is critical, as it must allow for sufficient (and reasonably fast) equilibration along transverse degrees of freedom.

The windows, each corresponding to a different CV value that is used for centering the applied bias, are distributed to collectively span the entire reaction profile, and each simulation generates a biased distribution of the CV. The windows must overlap sufficiently to reconstruct the free-energy profile accurately. By unbiasing these distributions using the histogram of visited configurations and the applied bias in the CV space, the free-energy profile for each window can be reconstructed. These profiles are then combined to produce the full free-energy profile of the conformational transition (for methods reconstructing free energy profiles, see the end of this section).

11.4.5 Steered molecular dynamics

Steered MD, introduced by Grubmüller et al. (1996), is a computational method designed to estimate rupture forces during unbinding reactions. Their motivation was to provide atomic-level insights into binding kinetics, interpreting experimental data, which often lack such detail. Although techniques like free energy perturbation (FEP) simulations (see 'Alchemical Transformations below) can accurately calculate free energy differences between bound and unbound states using nonphysical reaction pathways, they cannot determine rupture forces. The rupture force is the maximum force experienced along the actual unbinding pathway, corresponding to the steepest slope in the free energy profile. Experimentally, it can be measured using techniques like atomic force microscopy (AFM).

Steered MD models AFM experiments by simulating the process of pulling a ligand away from its binding site and measuring the required force. In their original study the authors simulated the unbinding of biotin from streptavidin by pulling biotin's oxygen atom O2 with an external harmonic potential. The center of mass of streptavidin was fixed, and the pulling force was gradually increased, mimicking a constant spring elongation speed. The rupture force was defined as the maximum force observed during the unbinding process. Beyond estimating the rupture force the MD steer also reveals the detailed atomic-scale rupture mechanism.

11.4.6 Metadynamics

Metadynamics, introduced by Laio and Parrinello (2002), is a method for exploring the multidimensional free energy surfaces (FESs) of complex systems in the subspace of a few selected CVs (Barducci et al., 2011). It enhances sampling and reconstructs the FES by applying a history-dependent bias potential that is adaptively constructed in the CV space.

The bias potential is added to the system's Hamiltonian and is constructed as a sum of Gaussians placed along the trajectory in the CV space (Fig. 11.4). The purpose of the bias is to discourage the system from revisiting configurations it has already sampled. Initially the system is generally trapped in a local minimum, unable to cross energy barriers higher than thermal fluctuations. As

Gaussians are deposited on visited states, the bias grows, until eventually the system is pushed out of the starting basin. Once the system enters a new basin, the same process continues, with the bias compensating for the new potential basin. Eventually, all basins corresponding to relevant metastable states are compensated for, and the system evolution resembles a random walk on the now-flattened FES.

By tracking the applied bias during metadynamics the bias itself provides an unbiased estimate of the underlying free energy landscape. An advantage of metadynamics lies in its ability to perform multiple parallel runs to explore and reconstruct the same FES. However, determining when to stop the simulation can be challenging. A solution to this issue is provided by well-tempered metadynamics, where the bias deposition rate decreases over simulation time.

In the original publication the authors applied metadynamics to CG dynamics (see Section 11.5). They used it to study the dissociation of a NaCl ion pair in water (where the dissociated state is most stable, and the undissociated ion pair is a metastable local minimum) and the isomerization (conformational changes) of dialanine in solution.

11.4.7 String method

The string method, introduced by Weinan et al. (2002), is a technique for studying rare events and determining transition pathways, free energy barriers, and transition rates in complex systems with relatively smooth energy landscapes. The string method iteratively refines a "string"—a smooth curve with intrinsic parametrization—in the CVs space, ultimately converging toward and identifying the most probable transition path, known as the minimum energy path, between metastable states.

The original version of the string method operated on the potential energy landscape and did not account for thermal fluctuations or entropy, effectively assuming a zero-temperature environment. In Weinan et al. (2005) extended the method to finite temperatures, allowing it to account for temperature and entropic effects. In this finite-temperature version the string evolves within the free energy landscape, building on isoprobability surfaces. These surfaces represent points starting from where trajectories have equal probability of reaching one metastable state or the other. The isoprobability surface at 0.5 corresponds to transition states, from where trajectories are equally likely to reach either metastable state. These surfaces can also be weighted by the equilibrium Gibbs distribution on the surface, determining where transition trajectories are most likely to pass through. The string itself represents the system's equilibrium position on each isoprobability surface.

In practice an initial string connecting the two metastable states is discretized into "images." Biased simulations are run for each image, restricting motion to hyperplanes perpendicular to the string. Within these hyperplanes the system samples the equilibrium Gibbs distribution. Ensemble averages of these samples are used to update the string, which is then rediscretized. This iterative process continues until the string converges.

Once the string converges, it can be used to reconstruct the free energy profile, energy barriers, and transition rates along the path using methods such as umbrella sampling. The equilibrium distributions in the hyperplanes allow the construction of a "transition tube" with a specified probability cutoff, where the center of the tube corresponds to the converged string. The transition tube indicates where transitions will most probably proceed. The relative width of the tube reflects entropic effects; for example, a high potential energy barrier with a broad, flat energy profile (large tube diameter) may result in a lower overall free energy barrier compared to a lower potential energy barrier with a narrow, steep energy profile (smaller tube diameter).

Generally, mapping entire free energy landscapes becomes inefficient as the number of CVs increases. However, the string method focuses only on determining the minimum free energy path rather than the full energy landscape itself, making its computational cost scale linearly with the number of points on the discretized string, and importantly, its computational cost is independent of the dimensionality of the CV space.

11.4.8 Free energy analysis tools—Weighted histogram analysis method, multistate Bennett acceptance ratio, and dynamic histogram analysis method

Both biased and unbiased ensembles from molecular simulations can be used to study the energetics of a system. The above methods are often complemented with existing free energy analysis tools to accurately reconstruct the free energy profile along CVs. For stationary biasing, unbiasing is performed by calculating the ratio of averages between the biased and unbiased ensembles, taking the applied bias into account.

A widely used tool for reconstructing free energy profiles is the weighted histogram analysis method (**WHAM**) (Kumar et al., 1992). WHAM integrates information from all intermediate states simultaneously, without using time information, and assumes perfect equilibrium sampling. It relies on the principle of maximum-likelihood estimation, determining model parameters that describe a distribution that maximizes the probability of the observed data. This is achieved by iteratively solving a self-consistent equation pair. WHAM discretizes the CV into bins, creating a histogram that provides the relative probability of observing different states. The output is an estimated probability density along the CV, which can be converted into relative free energy values for each bin.

A related approach, the multistate Bennett acceptance ratio (**MBAR**) method, is equivalent to WHAM in the limit of infinitesimally small bin sizes (Shirts & Chodera, 2008). Unlike WHAM, MBAR does not introduce a bias from binning. In WHAM the use of sufficiently wide bins containing many samples is necessary to ensure sufficient sampling. MBAR also directly estimates uncertainties. However, MBAR is computationally more demanding than WHAM.

To address potentially significant errors arising from not fully equilibrated biasing windows, the dynamic histogram analysis method (**DHAM**) was introduced by Rosta and Hummer (2015). Errors can occur when simulations within windows are too short relative to the system's slow dynamics in directions orthogonal to the chosen CV, or when too-wide umbrella windows are used, resulting in global regions separated by energy barriers. DHAM accounts for the dynamics of the system and its time sequences; it relies on a global Markov state model to compute free energy along the CV.

Like WHAM, DHAM discretizes the biased CV into bins (or microstates). However, instead of simply estimating probabilities, it tracks transition counts between these microstates (from all umbrella sampling simulations) after unbiasing. The transition counts are then used to construct a global Markov matrix of transition probabilities building on the maximum-likelihood approach. The free energy profile is then derived as the stationary distribution of the resulting Markov matrix. By invoking the detailed balance condition, DHAM offers an approximate explicit solution for the transition matrix in the limit of short lag times, avoiding the need to iteratively solve coupled nonlinear equations.

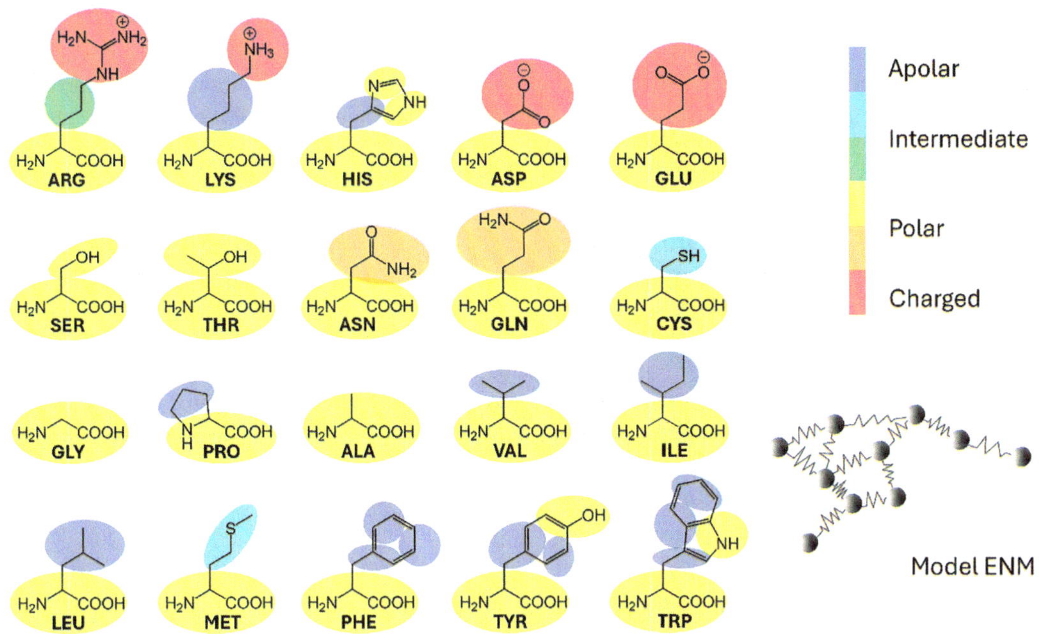

FIGURE 11.5

An example of atom grouping for coarse-grained representation: *Left*: The MARTINI coarse-grained force field amino acid representations for proteins. The shaded regions indicate atoms that are grouped together and represented by a single bead. MARTINI uses 18 different bead types, which vary based on their nature such as polarity (apolar, intermediate, polar, charged) and hydrogen bonding capabilities. Bottom right: schematic illustration of an elastic network model, where the beads represent C_α atoms in proteins, and the springs connecting them represent bonded and nonbonded interactions, capturing the essential flexibility and connectivity of the protein structure. *ENM*, Elastic network models.

11.5 Coarse-grained modeling

The methods discussed previously provide strategies to enhance simulations while retaining atomic-level detail. A significantly different strategy involves simplifying the representation of biomolecules, known as CG modeling, which sacrifices some atomic detail to enable the simulation of large-scale biomolecular processes over extended timescales that are inaccessible to conventional all-atom simulations. In the simplified CG models, molecules are represented not by individual atoms but by pseudo-atoms (also called united atoms), which group together a set of atoms, such as amino acid fragments or even entire amino acids (Fig. 11.5). This simplification significantly extends both the spatial and temporal scales of simulations.

There are two primary ways of coarse-graining. The first is a bottom-up approach, also referred to as structure- or physics-based modeling, where interaction parameters are derived from reference atomistic simulations. The second is a top-down approach, often called thermodynamic- or knowledge-based modeling, which focuses on reproducing key experimental data, particularly thermodynamic properties. In practice, many CG FFs combine these two routes. The reduced complexity of CG models with considerably fewer degrees of freedom and the elimination of high-frequency vibrations, which enables the use of larger integration timesteps and produces smoother energy landscapes, enable multiple magnitudes faster simulations compared to all-atom modeling.

Over the past decades, various CG models have been developed for different types of molecules, such as water, proteins, nucleic acids, lipids, and carbohydrates. For proteins, CG modeling dates back to the 1970s (Levitt & Warshel, 1975) and 1980s (Dill, 1985), when it was first introduced to enhance conformational sampling. Since then, numerous CG models have been proposed. In models that retain chemical specificity the level of coarse-graining strongly relates to the addressed problem. Specifically in protein modeling, the main chain might be represented by all heavy atoms or by just 1–2 united atoms per residue, whereas side chains are typically replaced by 1–2 united atoms.

Even though most CG models use a continuous representation of the geometry of molecular structures, some use lattice-based grids, which provide additional significant computational speedups. Beyond studying protein dynamics, coarse-graining is also widely applied in protein structure prediction, simulations of protein–protein interactions, and modeling membrane proteins.

11.5.1 Coarse-grained force fields

CG force fields (FFs) and corresponding biomolecule representations largely vary depending on how structurally realistic their model is. The high-resolution CG models are very close to all-atom representation, introducing only small unifications and simplifications. One example is the widely used **Rosetta** model developed by Baker et al. (Das & Baker, 2008) (various versions of the model exist) where coarse-graining and all-atom representations are combined. The model contains all heavy atoms of the backbone and, additionally, the C_β atom, while pseudoatoms represent amino acid side chains. The protein conformation is then defined in the dihedral space, with three degrees of freedom per amino acid backbone (phi, psi, and omega dihedrals corresponding to rotations around the N–C_α, C_α–C, and peptide C–N bonds, respectively) with possible chi dihedral angles (corresponding to rotations around the C_α–C_β bond) for side chains. The model is widespread in protein structure prediction, protein-folding simulations, protein-protein and protein–ligand docking, and for describing protein–DNA interactions and refinement of experimentally determined (NMR, X-ray, cryo-EM) structures.

Typical intermediate resolution models use one to three pseudoatoms to describe the geometry of amino acids. Such models enable quite realistic, although by no means exact, reconstruction of atomistic models. The **UNRES** (united residues) model (Liwo et al., 2014) is one well-known representative of this class; amino acids are represented by three pseudoatoms. The polypeptide chain is represented by a sequence of C_α atoms with additional pseudoatoms located halfway between the consecutive C_α atoms, and they are linked with virtual bonds. The third pseudoatom of an ellipsoid shape represents the different side chains attached to the C_α atoms linked with virtual bonds. The UNRES FF is derived from all-atom molecular mechanics (MM) interaction models. As Rosetta, it is used in protein structure prediction, macromolecular dockings, and simulating large-scale reorganizations of protein complexes.

Another intermediate resolution model is the **Martini** model, one of the most popular CG models in membrane protein modeling in membrane environment. It was originally developed for lipid molecules to simulate membranes (Marrink et al., 2004) and later extended to proteins (Monticelli et al., 2008).

In the Martini model each amino acid is represented by one backbone bead (placed at the center of mass of the amino acid backbone) and up to four side chain beads. To more accurately reproduce the chemical nature of the united atoms, different types of CG particles exist in the Martini model: polar, nonpolar, apolar, and charged, and these groups are further subdivided based on their hydrogen-bonding capabilities as well as their polarity. Interactions within the Martini FF are described by typical physical FF terms with parameters adjusted to reproduce experimental thermodynamics data, whereas parameters were tuned to satisfactory overlap with the distribution function resulting from all-atom MD simulations.

The **SICHO** (side chain only) model (Kolinski et al., 1998) has an even lower resolution using a single pseudoatom per amino acid residue. The explicitly simulated pseudoatoms are placed near the center of mass of each residue sidechain (or C_α in the case of glycine). SICHO is a lattice-based system, where the pseudoatoms are restricted to discrete positions on the lattice, further simplifying calculations. The SICHO FF defines potentials originating from statistical analysis of structural properties observed in known protein structures and extracts them to describe and guide distances between neighboring pseudoatoms, as well as planar and torsional angles. The SICHO resolution is about 2–3 Å and provides a crude although quite realistic representation of protein structure, including an acceptable picture of secondary structure (Kmiecik et al., 2016).

For an in-depth review of coarse-graining representations and FFs, we advise the reader to refer to the reviews of Kmiecik et al. (2016), Liwo et al. (2021), and Ingólfsson et al. (2014).

11.5.2 Rotations-translations of blocks

The rotations–translations of blocks (RTB) method (Durand et al., 1994) is quite different from the previously discussed CG FF models, as it does not change the all-atom FF; it only aims to reduce the degrees of freedom describing the dynamics to exploit the benefits of a simplified model. As discussed previously, NMA—which provides analytical solutions to Newton's equations of motion by relying on the harmonic approximation of the potential energy surface around local minima—can also be used to extract information on large-scale collective protein dynamics. However, in case of large systems, all-atom normal mode (NM) calculations with the corresponding Hessian matrix diagonalization (a $3N \times 3N$ matrix for a system of N atoms) may become unfeasible. Different coarse-graining strategies exist to efficiently retrieve low-frequency NMs of proteins, among which are the RTB method and elastic network models (ENMs) (see below).

The RTB method rests on the hypothesis that low-frequency NMs of proteins can be described as pure rigid-body (rotation–translation) motions of blocks made of one or a few consecutive amino-acid residues (Tama et al., 2000). The lowest frequency NMs of the protein are obtained as a linear combination of the rotations and translations of these blocks. In the RTB method the original Hessian is replaced by a projected Hessian expressed in a basis defined by the rotation and translation modes of the blocks (each block having six degrees of freedom). It is the projected Hessian that is diagonalized (a $6M \times 6M$ matrix for a system reduced to M blocks) and is transformed back to obtain the atomic displacements. A big advantage of the RTB method is that it can use the same potential energy function as in all-atom simulations, whereas other fast methods generally rely on highly simplified potential energy functions. Interestingly a later analysis has demonstrated that with six amino acids per block, the NMs are almost as accurate as with a single amino acid per block representation, enabling an even greater reduction in computational costs (Tama et al., 2000).

11.5.3 Elastic network models

ENMs treat molecular substructures as bead-like masses (representing groups of atoms) connected by springs to describe protein motions (Fig. 11.5). The first ENM used for describing protein dynamics was the Gaussian network model (**GNM**) introduced by Bahar et al. (1997) (Haliloglu et al., 1997). In GNM the protein is treated as a microscopically uniform elastic material at equilibrium. Each amino acid is represented by a single bead centered on its C_α, and the protein is viewed as a network, the connectivity between neighboring beads being established by a single-parameter harmonic potential, all the connecting springs having equal spring constants. In GNM, there is no distinction between the springs connecting bonded groups or nonbonded groups nearby in space—typically residues within a cutoff of 7 Å are connected by springs (Bahar & Jernigan, 1997). Importantly, in GNM, any deviation from the equilibrium interresidue distance vector between two connected residues in their instantaneous separation introduces an increase in the potential. This means that changes in the orientation are also considered—if the orientation is changed even with constant distance magnitude, the potential also changes.

As opposed to all-atomic NMs calculations, there is no requirement for an initial energy-minimization step. The initial structure is handled as a state of minimum energy by defining the equilibrium separations between beads used in the harmonic potentials as the internal distances found in the initial structure. The total elastic energy is computed by summing over all the springs (divided by two to correct for double counting the springs). With the help of GNM, residue mean-square fluctuations can be extracted and related to experimental B-factors after scaling. Importantly, to obtain the absolute magnitude of the motions, a universal rescaling of the size of all residue motions is required—this information is lost upon coarse-graining and can be obtained by, for example, parameter fitting to match average fluctuations of protein residues to their average B-factors observed in experiments. Furthermore, it can be used to predict cooperative actions in the protein.

Harmonic oscillations can be extracted from the GNM model, decomposing the structural dynamics into independent modes, each mode corresponding to a frequency and relative motion amplitudes of each protein residue for the corresponding mode. Using these modes, highly flexible and constrained regions can be identified. Typically, hinge regions and active sites correspond to regions without high flexibility, whereas, for example, substrate recognition loops tend to exhibit high flexibility.

Even though GNM provides information on the flexibility of the different protein regions as well as on the correlated behavior between subparts of the protein, it does not provide information about the directions of motions, which are critical to understanding biological mechanisms. Another ENM approach, the anisotropic network model (**ANM**) proposed by Atilgan et al. (2001), goes beyond GNM and explores also the directions of the motions. The representation of the protein is identical to GNM, with each residue represented as a single bead and spatially close beads (within a cutoff distance) connected by harmonic potentials with a uniform force constant. However, unlike in GNM where deviations from the equilibrium interresidue distance both in length and orientation are accounted for, in ANM only the interresidue distance amplitude change is considered, and changes in the orientation are neglected. On the ANM model NMs analysis can be performed, similar to all-atom NMA—obtaining both the frequency and direction for each mode. Note that the frequency has an arbitrary unit because of the loss of information upon coarse-graining. Similarly to GNM, no preceding energy minimization is required for the NMA using the ANM representation.

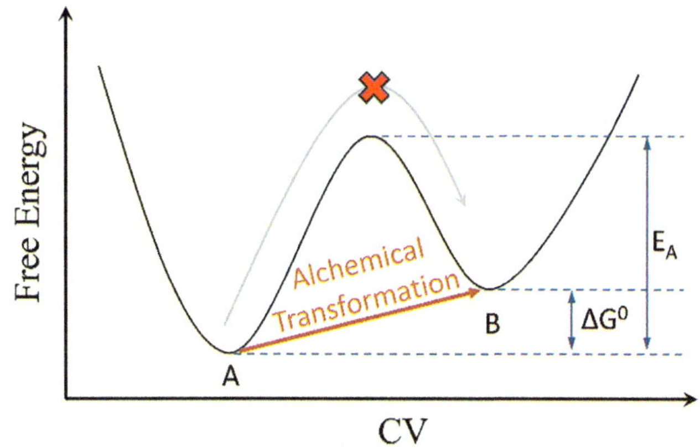

FIGURE 11.6

Schematic illustration of an alchemical transformation: The activation energy is denoted by E_A, and the free energy difference between states A and B by ΔG^0. Instead of simulating a physical path that passes over a large energy barrier, a nonphysical pathway can be sampled to compute the free energy difference between the two end states, A and B. Alchemical free energy calculations leverage the principle that free energy is a state function and, as such, it does not depend on the connecting path but exclusively on the two end states. (*CV* dnotes Collective Variable.)

11.6 Alchemical free energy calculations

In contrast to the formerly described strategies, alchemical transformations do not directly simulate large physical conformational changes. However, they can be used to evaluate the change in free energy associated with slow processes like ligand binding and conformational transitions between conformational states. During alchemical transformations two physical end states are connected via a pathway of nonphysical (alchemical) states (Fig. 11.6). These alchemical processes can often be computationally much more affordable than the physical process itself, yet they provide an estimate of the overall free energy difference that is a state function; it only depends on the initial and end states and is independent from the connecting paths.

Free Energy Perturbation (FEP) was first introduced by Zwanzig (1954). It is based on statistical mechanics for computing free energy differences using molecular simulations. The method relies on the Zwanzig equation for calculating the free energy difference between two states:

$$\Delta G_{A\to B} = G_B - G_A = -k_B T \ln \exp\left(-\frac{V_B - V_A}{k_B T}\right)_A \tag{11.6}$$

with k_B the Boltzmann constant, T the simulation temperature, and V_B and V_A the potential energies of states A and B, respectively. If the microstates corresponding to the canonical distribution of state A are of too low probability in the canonical distribution of state B, it results in a too large potential energy difference and consequent negligibly small exponential factor, and the free energy difference will converge very slowly in an actual simulation. For this reason, FEP can only efficiently handle situations where states A and B are not too different. To avoid slow convergence, intermediate states are typically introduced in FEP, and the full transformation is described as passing through these intermediate states and computing free energy differences between neighboring intermediate states. These differences are integrated to obtain the overall free energy difference between the two physical end states, A and B.

An alternative approach to FEP is *thermodynamic integration* (TI) (Kirkwood, 1935). It introduces a continuous series of intermediate states as a function of a variable $\lambda \in [0, 1]$connecting the end states, A and B. It also relies on free energy being a state function and that any path between the states yields the same free energy difference. In TI the system is slowly switched from the initial to the end state, fully relaxing the system at each intermediate point chosen along the path. The switching between the potential energy functions is described as

$$V(r, \lambda) = f(\lambda)V_A(r) + g(\lambda)V_B(r) \tag{11.7}$$

with $f(\lambda)$ and $g(\lambda)$ being the switching functions satisfying the criteria $V(r, 0) = V_A(r)$and $V(r, 1) = V_B(r)$. The free energy difference between the two states is then obtained as

$$\Delta G_{A\rightarrow B} = G_B - G_A = \int_0^1 \frac{\partial V(r, \lambda)}{\partial \lambda}_{\lambda} d\lambda \tag{11.8}$$

where the angle brackets denote average in the canonical ensemble with a fixed λ. A trivial choice for the switching functions is $f(\lambda) = 1 - \lambda$ and $g(\lambda) = \lambda$. In this case the free energy difference simplifies to

$$\Delta G_{A\rightarrow B} = \int_0^1 V_B - V_{A\,\lambda} d\lambda \tag{11.9}$$

In practice a discrete set of intermediate states is chosen for $\lambda \in [0, 1]$, where molecular simulations are performed to generate ensemble averages. The results are then numerically integrated to obtain the overall free energy difference between the physical states A and B.

A more general form of the previously discussed REMD (see Section 11.3) is the Hamiltonian REMD (H-REMD or HREX) (Fukunishi et al., 2002). Several H-REMD strategies have been developed, modifying parts of the Hamiltonian beyond temperature to enhance sampling. In H-REMD each replica corresponds to a different potential energy function rather than to different temperatures. It can be efficiently combined with both FEP and IT in order to improve convergence of free energy calculations (Meng et al., 2011).

11.7 Hybrid approaches

The previously discussed strategies all provide different toolkits to efficiently sample the conformational space and the possible transitions between functional metastable states. The different strategies were developed for specific problems; whereas they provide various advantages, they also come with different limitations and disadvantages. Since the appearance of the first enhanced sampling approaches, the combination of sampling methods has led to the creation of vast numbers of hybrid approaches with the goal of combining advantages and overcoming limitations of the specific tools. One such example is multiscale modeling where an optimal compromise is attempted in providing precise details while retaining efficiency, for which Warshel, Levitt, and Karplus won the Nobel Prize in Chemistry in 2013.

For example, the quantum mechanics/MM (**QM/MM**) approach (Warshel & Levitt, 1976) combines ab initio quantum calculations at regions of interest (e.g., at catalytic active sites) while more distant parts of the system are treated on the classical molecular mechanics level. These regions are coupled throughout the simulations, ensuring the QM and MM interactions are consistently integrated. A different level of multiscale modeling can be found in the previously discussed **Rosetta** model (Das & Baker, 2008) where coarse-graining and all-atom representations are combined in MD simulations.

Analytical approximations can also be integrated into MD simulations. NMA and ENM representations, for example, enable the extraction of the spectrum of modes uniquely accessible to supramolecular systems, providing mathematically exact and physically plausible information on cooperative events (see above), albeit neglecting anharmonicity and atomic details. Several hybrid methods that combine MD and NMA (using either ENMs or full atomic models) proved to be useful in recent years. These methods have demonstrated an accurate description of cooperative changes in structure, usually described by low-frequency NMs, while providing atomic details and incorporating local nonlinear effects from MD simulations that "recalibrate" these conformational changes (Kaynak et al., 2022).

One such example is **ClustENMD,** which performs iterative cycles of conformer generation using ENM for deformations along global modes, followed by clustering and short MD simulations (Kaynak et al., 2021). Another hybrid method, **CoMD** (Gur et al., 2013), provides a combination of ENM–NMA and targeted MD, coupled with energy minimization to adaptively generate a series of conformers. CoMD deforms the structure collectively along the modes predicted by ENM–NMA. It selects modes via a Monte Carlo/Metropolis algorithm corresponding to their probabilities and amplitudes based on their eigenvalues, such that the lowest frequency (energetically most favorable) global modes dominate the sampling. The target is then approached by using targeted MD. Finally an example for all-atom NMA-based hybrid methods is MDs with excited NMs (**MDeNM**) (Costa et al., 2015). The MDeNM method is a multi-replica protocol designed to enhance conformational exploration in a subspace defined by a set of low-frequency NMs, also including the couplings with localized motions occurring within the Cartesian space by MD. In this method, additional atomic velocities are introduced during MD simulations along different linear combinations of NMs. For a comparative analysis of these NMA-based hybrid approaches we refer to the reader to reference (Kaynak et al., 2022).

AI disclosure

During the preparation of this work the author(s) used ChatGPT to edit text, correct grammar, and improve English language. After using this tool/service the author(s) review(s) and edit(s) the content as needed and take(s) full responsibility for the content of the publication.

References

Alder, B. J., & Wainwright, T. E. (1957). Phase transition for a hard sphere system. *The Journal of Chemical Physics, 27*(5), 1208–1209. doi:10.1063/1.1743957.

Amadei, A., Linssen, A. B. M., de Groot, B. L., van Aalten, D. M. F., & Berendsen, H. J. C. (1996). An efficient method for sampling the essential subspace of proteins. *Journal of Biomolecular Structure and Dynamics, 13*(4), 615–625. doi:10.1080/07391102.1996.10508874.

Atilgan, A. R., Durell, S. R., Jernigan, R. L., Demirel, M. C., Keskin, O., & Bahar, I. (2001). Anisotropy of fluctuation dynamics of proteins with an elastic network model. *Biophysical Journal, 80*(1), 505–515. doi:10.1016/s0006-3495(01)76033-x.

Bahar, I., Atilgan, A. R., & Erman, B. (1997). Direct evaluation of thermal fluctuations in proteins using a single-parameter harmonic potential. *Folding and Design, 2*(3), 173–181. doi:10.1016/S1359-0278(97)00024-2.

Bahar, I., & Jernigan, R. L. (1997). Inter-residue potentials in globular proteins and the dominance of highly specific hydrophilic interactions at close separation. *Journal of Molecular Biology, 266*(1), 195–214. doi:10.1006/jmbi.1996.0758.

Barducci, A., Bonomi, M., & Parrinello, M. (2011). Metadynamics. *WIREs Computational Molecular Science., 1*(5), 826–843. doi:10.1002/wcms.31.

Bernardi, R. C., Melo, M. C. R., & Schulten, K. (2015). Enhanced sampling techniques in molecular dynamics simulations of biological systems. *Biochimica et Biophysica Acta—General Subjects, 1850*(5), 872–877. doi:10.1016/j.bbagen.2014.10.019.

Clarage, J. B., Romo, T., Andrews, B. K., Pettitt, B. M., & Phillips, G. N. (1995). A sampling problem in molecular dynamics simulations of macromolecules. *Proceedings of the National Academy of Sciences, 92*(8), 3288–3292. doi:10.1073/pnas.92.8.3288.

Costa, M. G. S., Batista, P. R., Bisch, P. M., & Perahia, D. (2015). Exploring free energy landscapes of large conformational changes: Molecular dynamics with excited normal modes. *Journal of Chemical Theory and Computation, 11*(6), 2755–2767. doi:10.1021/acs.jctc.5b00003.

Das, R., & Baker, D. (2008). Macromolecular modeling with Rosetta. *Annual Review of Biochemistry, 77*, 363–382. doi:10.1146/annurev.biochem.77.062906.171838.

Dill, K. A. (1985). Theory for the folding and stability of globular proteins. *Biochemistry, 24*(6), 1501–1509. doi:10.1021/bi00327a032.

Durand, P., Trinquier, G., & Sanejouand, Y. (1994). A new approach for determining low-frequency normal modes in macromolecules. *Biopolymers, 34*(6), 759–771. doi:10.1002/bip.360340608.

Fukunishi, H., Watanabe, O., & Takada, S. (2002). On the Hamiltonian replica exchange method for efficient sampling of biomolecular systems: Application to protein structure prediction. *Journal of Chemical Physics, 116*(20), 9058–9067. doi:10.1063/1.1472510.

Gong, X. G., & Wilkins, J. W. (1999). Hyper molecular dynamics with a local bias potential. *Physical Review B, 59*(1), 54–57. doi:10.1103/physrevb.59.54.

Grubmüller, H. (1995). Predicting slow structural transitions in macromolecular systems: Conformational flooding. *Physical Review E, 52*(3), 2893–2906. doi:10.1103/physreve.52.2893.

Grubmüller, H., Heymann, B., & Tavan, P. (1996). Ligand binding: Molecular mechanics calculation of the streptavidin-biotin rupture force. *Science, 271*(5251), 997–999. doi:10.1126/science.271.5251.997.

Gur, M., Madura, J. D., & Bahar, I. (2013). Global transitions of proteins explored by a multiscale hybrid methodology: Application to adenylate kinase. *Biophysical Journal, 105*(7), 1643–1652. doi:10.1016/j.bpj.2013.07.058.

Haliloglu, T., Bahar, I., & Erman, B. (1997). Gaussian dynamics of folded proteins. *Physical Review Letters, 79*(16), 3090–3093. doi:10.1103/PhysRevLett.79.3090.

Hamelberg, D., Mongan, J., & McCammon, J. A. (2004). Accelerated molecular dynamics: A promising and efficient simulation method for biomolecules. *Journal of Chemical Physics, 120*(24), 11919–11929. doi:10.1063/1.1755656.

Ingólfsson, H. I., Lopez, C. A., Uusitalo, J. J., de Jong, D. H., Gopal, S. M., Periole, X., & Marrink, S. J. (2014). The power of coarse graining in biomolecular simulations. *WIREs Computational Molecular Science, 4*(3), 225–248. doi:10.1002/wcms.1169.

Kaynak, B. T., Zhang, S., Bahar, I., & Doruker, P. (2021). ClustENMD: Efficient sampling of biomolecular conformational space at atomic resolution. *Bioinformatics, 37*(21), 3956–3958. doi:10.1093/bioinformatics/btab496.

Kaynak, B. T., Krieger, J. M., Dudas, B., Dahmani, Z. L., Costa, M. G. S., Balog, E., Scott, A. L., Doruker, P., Perahia, D., & Bahar, I. (2022). Sampling of protein conformational space using hybrid simulations: A critical assessment of recent methods. *Frontiers in Molecular Biosciences, 9*. doi:10.3389/fmolb.2022.832847.

Kirkwood, J. G. (1935). Statistical mechanics of fluid mixtures. *The Journal of Chemical Physics, 3*(5), 300–313. doi:10.1063/1.1749657.

Kmiecik, S., Gront, D., Kolinski, M., Wieteska, L., Dawid, A. E, & Kolinski, A. (2016). Coarse-grained protein models and their applications. *Chemical Reviews, 116*(14), 7898–7936. doi:10.1021/acs.chemrev.6b00163.

Kolinski, A., Jaroszewski, L., Rotkiewicz, P., & Skolnick, J. (1998). An efficient Monte Carlo model of protein chains. Modeling the short-range correlations between side group centers of mass. *The Journal of Physical Chemistry B, 102*(23), 4628–4637. doi:10.1021/jp973371j.

Kumar, S., Rosenberg, J. M., Bouzida, D., Swendsen, R. H., & Kollman, P. A. (1992). THE weighted histogram analysis method for free-energy calculations on biomolecules. I. The method. *Journal of Computational Chemistry, 13*(8), 1011–1021. doi:10.1002/jcc.540130812.

Laio, A., & Parrinello, M. (2002). Escaping free-energy minima. *Proceedings of the National Academy of Sciences of the United States of America, 99*(20), 12562–12566. doi:10.1073/pnas.202427399.

Levitt, M., & Warshel, A. (1975). Computer simulation of protein folding. *Nature, 253*(5494), 694–698. doi:10.1038/253694a0.

Liwo, A., Baranowski, M., Czaplewski, C., Gołaś, E., He, Y., Jagieła, D., Krupa, P., Maciejczyk, M., Makowski, M., Mozolewska, M. A., Niadzvedtski, A., Ołdziej, S., Scheraga, H. A., Sieradzan, A. K., Ślusarz, R., Wirecki, T., Yin, Y., & Zaborowski, B. (2014). A unified coarse-grained model of biological macromolecules based on mean-field multipole–multipole interactions. *Journal of Molecular Modeling, 20*(8). doi:10.1007/s00894-014-2306-5.

Liwo, A., Czaplewski, C., Sieradzan, A. K., Lipska, A. G., Samsonov, S. A., & Murarka, R. K. (2021). Theory and practice of coarse-grained molecular dynamics of biologically important systems. *Biomolecules, 11*(9), 1347. doi:10.3390/biom11091347.

Marrink, S. J., De Vries, A. H., & Mark, A. E. (2004). Coarse grained model for semiquantitative lipid simulations. *Journal of Physical Chemistry B, 108*(2), 750–760. doi:10.1021/jp036508g.

McCammon, J. A., Gelin, B. R., & Karplus, M. (1977). Dynamics of folded proteins. *Nature, 267*(5612), 585–590. doi:10.1038/267585a0.

Meng, Y., Sabri Dashti, D., & Roitberg, A. E. (2011). Computing alchemical free energy differences with Hamiltonian replica exchange molecular dynamics (H-REMD) simulations. *Journal of Chemical Theory and Computation, 7*(9), 2721–2727. doi:10.1021/ct200153u.

Monticelli, L., Kandasamy, S. K., Periole, X., Larson, R. G., Tieleman, D. P., & Marrink, S.-J. (2008). The MARTINI coarse-grained force field: Extension to proteins. *Journal of Chemical Theory and Computation, 4*(5), 819–834. doi:10.1021/ct700324x.

Pal, S., & Fichthorn, K. A. (1999). Accelerated molecular dynamics of infrequent events. *Chemical Engineering Journal, 74*(1–2), 77–83. doi:10.1016/s1385-8947(99)00055-8.

Rahman, A., & Stillinger, F. H. (1971). Molecular dynamics study of liquid water. *The Journal of Chemical Physics, 55*(7), 3336–3359. doi:10.1063/1.1676585.

Rahman, J. A., & Tully, J. C. (2002). Puddle-skimming: An efficient sampling of multidimensional configuration space. *Journal of Chemical Physics, 116*(20), 8750–8760. doi:10.1063/1.1469605.

Rosta, E., & Hummer, G. (2015). Free energies from dynamic weighted histogram analysis using unbiased Markov state model. *Journal of Chemical Theory and Computation, 11*(1), 276–285. doi:10.1021/ct500719p.

Schlitter, J., Engels, M., Krüger, P., Jacoby, E., & Wollmer, A. (1993). Targeted molecular dynamics simulation of conformational change-application to the T ↔ R transition in insulin. *Molecular Simulation, 10*(2-6), 291–308. doi:10.1080/08927029308022170.

Schlitter, J., Engels, M., & Krüger, P. (1994). Targeted molecular dynamics: A new approach for searching pathways of conformational transitions. *Journal of Molecular Graphics, 12*(2), 84–89. doi:10.1016/0263-7855(94)80072-3.

Shirts, M. R., & Chodera, J. D. (2008). Statistically optimal analysis of samples from multiple equilibrium states. *Journal of Chemical Physics, 129*(12). doi:10.1063/1.2978177.

Sugita, Y., & Okamoto, Y. (1999). Replica-exchange molecular dynamics method for protein folding. *Chemical Physics Letters, 314*(1–2), 141–151. doi:10.1016/S0009-2614(99)01123-9.

Swendsen, R. H., & Wang, J. S. (1986). Replica Monte Carlo simulation of spin-glasses. *Physical Review Letters, 57*(21), 2607–2609. doi:10.1103/PhysRevLett.57.2607.

Tama, F., & Sanejouand, Y.-H. (2001). Conformational change of proteins arising from normal mode calculations. *Protein Engineering, Design and Selection, 14*(1), 1–6. doi:10.1093/protein/14.1.1.

Tama, F., Xavier Gadea, F., Marques, O., & Sanejouand, Y.-H. (2000). Building-block approach for determining low-frequency normal modes of macromolecules. *Proteins: Structure, Function, and Genetics, 41*(1), 1–7. doi:10.1002/1097-0134(20001001)41:1⟨1::aid-prot10⟩3.0.co;2-p.

Torrie, G. M., & Valleau, J. P. (1977). Nonphysical sampling distributions in Monte Carlo free-energy estimation: Umbrella sampling. *Journal of Computational Physics, 23*(2), 187–199. doi:10.1016/0021-9991(77)90121-8.

Warshel, A., & Levitt, M. (1976). Theoretical studies of enzymic reactions: Dielectric, electrostatic and steric stabilization of the carbonium ion in the reaction of lysozyme. *Journal of Molecular Biology, 103*(2), 227–249. doi:10.1016/0022-2836(76)90311-9.

Weinan, E., Ren, W., & Vanden-Eijnden, E. (2005). Finite temperature string method for the study of rare events. *Journal of Physical Chemistry B, 109*(14), 6688–6693. doi:10.1021/jp0455430.

Weinan, E., Ren, W., & Vanden-Eijnden, E. (2002). String method for the study of rare events. *Physical Review B, 66*, 052301.

Zwanzig, R. W. (1954). High-temperature equation of state by a perturbation method. I. Nonpolar gases. *Journal of Chemical Physics, 22*(8), 1420–1426. doi:10.1063/1.1740409.

CHAPTER

12 Heparan-*N*-sulfatase and *N*-acetyl-α-glucosaminidase: Mechanisms, mutations, biophysical studies, and chaperone development

Lauren R. Dorsett[a,b], Neha S. Gandhi[a,b,c,*], Vito Ferro[d] and Luke Guddat[d]

[a]*School of Chemistry and Physics, Queensland University of Technology, Brisbane, QLD, Australia,* [b]*Centre for Genomics and Personalised Health, Queensland University of Technology, Brisbane, QLD, Australia,* [c]*Department of Biotechnology, Manipal Institute of Technology, Manipal Academy of Higher Education, Manipal, Karnataka, India,* [d]*School of Chemistry and Molecular Biosciences, The University of Queensland, Brisbane, QLD, Australia*

**Corresponding author: neha.gandhi@manipal.edu*

12.1 Introduction

Lysosomal storage disorders (LSDs) are a group of inherited diseases that affect approximately 1 in 5000–7000 people through improper functioning of the lysosome (Nagree et al., 2019). Mutations in lysosomal enzymes inhibit correct lysosomal function. As a result, toxic material is built up within the organelle, resulting in cell death. LSDs are commonly present as neurodegenerative diseases. Seventy diseases fall under the LSDs classification, and mucopolysaccharidosis (MPS) makes up the largest group of LSDs (Jurecka et al., 2015).

MPS is associated with the accumulation of polysaccharides known as glycosaminoglycans (GAGs). There are four main groups of GAGs: hyaluronan, heparan sulfate (HS)/heparin, keratan sulfate (KS), and chondroitin sulfate/dermatan sulfate (DS) (Fig. 12.1). These molecules are highly sulfated polysaccharides, except for hyaluronic acid (Bauer et al., 2021). GAGs comprise repeating disaccharide units of amino sugars (D-glucosamine or D-galactosamine) and uronic acids (D-glucuronic acid or L-iduronic acid). The GAGs associated with MPS are HS, DS, and KS. A build-up of one, two, or all three of HS, DS, and KS results in organ and tissue damage throughout the body, ultimately leading to a shorter lifespan (Bigger et al., 2018). The most common form of MPS is MPS III, which occurs at ~0.28–4.1 per 100,000 (Baehner et al., 2005; Valstar et al., 2008).

MPS III, also known as Sanfilippo syndrome, is an incurable and lethal form of MPS. This form of MPS is caused by the build-up of a single GAG, HS, within the central nervous system (CNS), which triggers the neurological symptoms characteristic of this disease. The accumulation of HS occurs because of mutations in one of the four key enzymes responsible for the breakdown of HS. These enzymes include heparan-*N*-sulfatase (SGSH) for type A, *N*-acetyl-α-glucosaminidase (NAGLU) for type B, Ac-CoA: α-glucosaminidase *N*-acetyltransferase (HGSNAT) for type C, and

Protein Dynamics: Experimental and Computational Approaches. DOI: https://doi.org/10.1016/B978-0-443-30042-4.00014-6

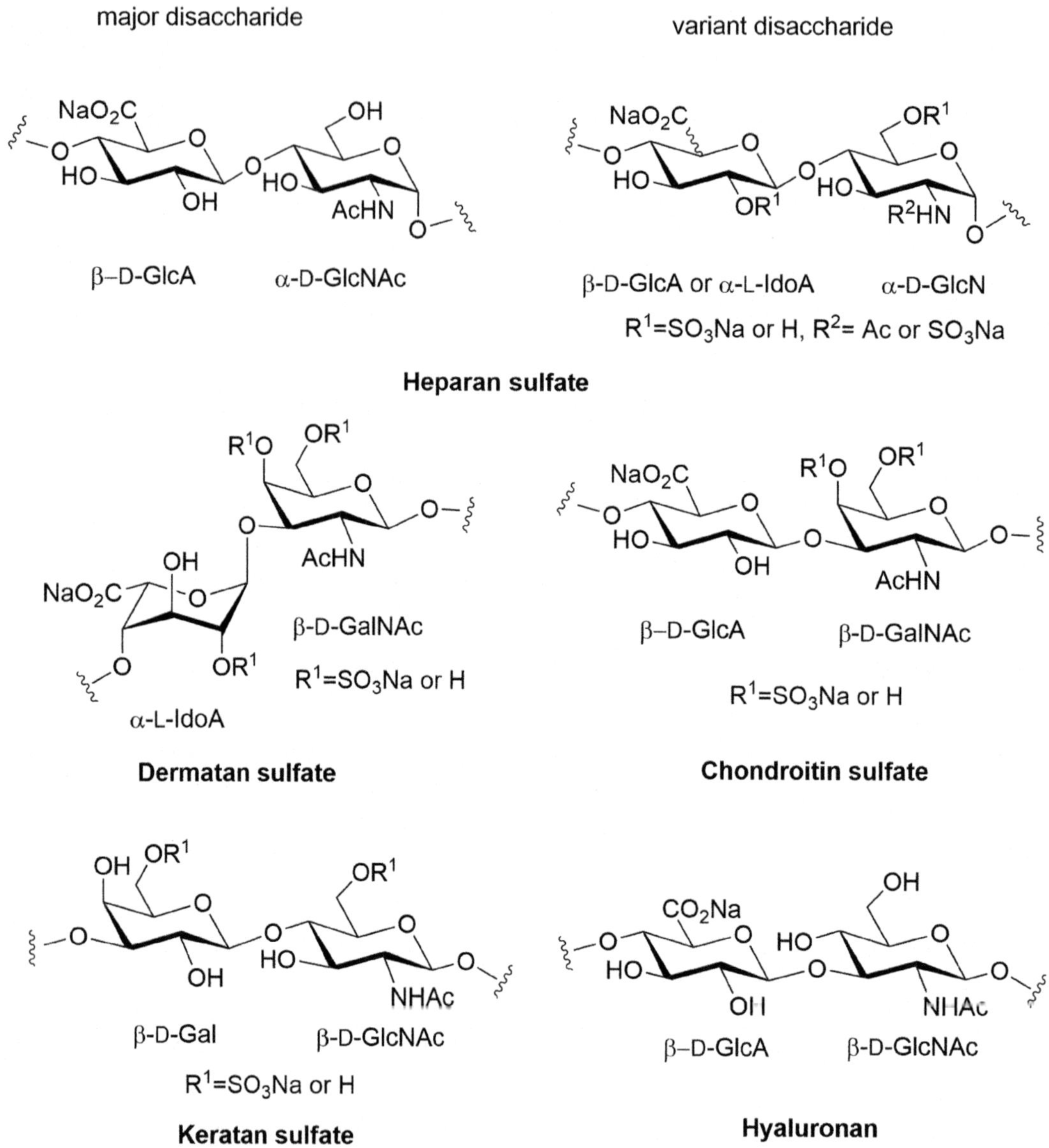

FIGURE 12.1

Structures of HS, CS, DS, KS, and HA disaccharide units. The top two compounds are the major and minor/variable disaccharide units of HS. The disaccharide units of CS, DS, KS, and HA are shown at the bottom of the figure. For the disaccharide units of HS, and DS, the location of the substitutions is denoted by X for the O-sulfo group substitution and Y for the substitution of a sulfo or acetyl group on the amino group. *CS*, Chondroitin sulfate; *DS*, dermatan sulfate; *HA*, hyaluronan; *HS*, heparan sulfate; *KS*, keratan sulfate.

Modified from Bauer et al., 2021.

N-acetylglucosamine 6-sulfatase (GNS) for type D (Andrade et al., 2015; Marcó et al., 2019; Valstar et al., 2008). The improper function of any of these enzymes results in the accumulation of HS within lysosomes, resulting in large quantities throughout the CNS, tissues, and organs. The buildup of HS, in turn, disrupts normal lysosomal function, resulting in the accumulation of secondary storage materials such as gangliosides. Without a way to break down, large amounts of HS and other storage products are excreted through the urinary system (Cleary & Wraith, 1993).

The progression of the disease is divided into three main stages, and the ages at which these occur vary depending on the subtype and patient. The antenatal and postnatal development of the baby is standard, with no significant signs. The first stage of the disease occurs at ages 1–3 with symptoms such as aggression, regression, and cognitive delay. The second stage occurs around 3–5 years, with increasing hyperactivity, sleep disturbances, and behavioral issues. Stage two of the disease can last several years until the child is in their mid to late teens. The final stage occurs when the patient is vegetative, losing motor function and intellectual processes. Stage three of the disease ultimately results in the death of the patient (Bodamer et al., 2014; Fedele, 2015).

The initial diagnosis of the disease occurs through a noninvasive urinary GAG assay. In the commonly used assay a mixture of urine with dimethyl methylene blue is made, and a reaction occurs when there are GAGs within the urine. A spectrophotometric assay can observe and measure any reaction (Bodamer et al., 2014; Farndale et al., 1982). Once there is a confirmed elevation in GAGs within the urine, mass spectrometry or electrophoretic methods are used to identify the specific GAG (Seker Yilmaz et al., 2021). An enzyme activity assay of leukocytes is then needed to determine the subtype (Bodamer et al., 2014; Seker Yilmaz et al., 2021). Diagnosis of the subtype will then help to develop strategies to ensure quality of life for the patient. Sanfilippo syndrome's most common subtypes are Sanfilippo syndrome types A and B (Garbuzova-Davis et al., 2013).

12.2 Sanfilippo types A and B

Sanfilippo syndrome type A is frequently the most aggressive form of Sanfilippo syndrome; however, but depending on the mutation that is causing the disease, it can present as a slow-progressing form. It is common for type A to have an early onset and a rapid progression (Garbuzova-Davis et al., 2013). Due to its rapid progression, sufferers also have a shorter life expectancy, with a mean age of 18 years when death occurs (Ugrinov et al., 2015). Sanfilippo syndrome type A is caused by a lack of the enzyme SGSH. SGSH belongs to a group of enzymes called *N*-sulfatases, a group under the sulfatase family. The *N*-sulfatases cleave nitrogen-linked sulfate groups (Daniele & Di Natale, 2001). Multiple 3D structures have been reported for human *N*-sulfatases, but they share only 25%–30% structural homology with SGSH. Five enzymes that are homologs of SGSH have been characterized to date. The five proteins all belong to the sulfatase family. They are arylsulfatase A (ASA), arylsulfatase B (ASB), arylsulfatase from *Pseudomonas aeruginosa* (PAS), galactosamine (*N*-acetyl)-6-sulfatase, and sulfatase/hydrolase from *Burkholderia caryophylli* PG2952 (BcPMH). Differences in protein folding result in distinct catalytic pocket architectures. However, commonalities between the catalytic residues help determine the SGSH active site structure (Sidhu et al., 2014).

Within this chain, roughly 115–130 mutations have been identified to contribute to the misfunction and removal of SGSH by endoplasmic reticulum-associated protein degradation. The mutations include missense, nonsense, insertions, and deletions (Andrade et al., 2015; Tanwar et al., 2019). Two missense

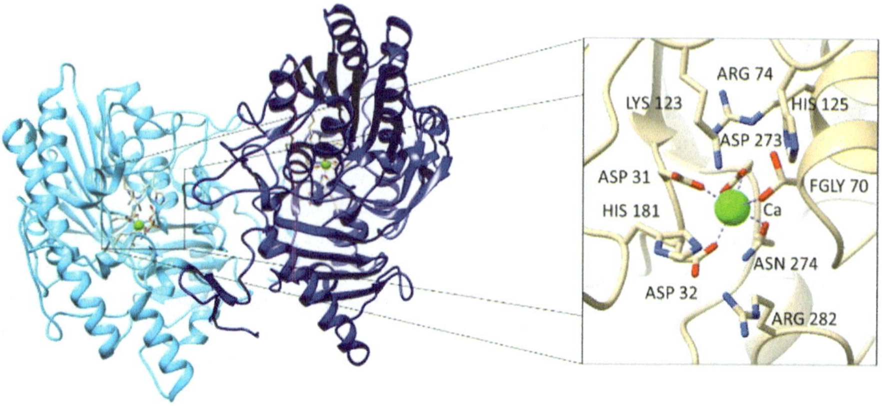

FIGURE 12.2

SGSH ribbon diagram and active site. The dimer structure of SGSH is shown in the ribbon representation shown on the left. The two subunits of the dimer are shown in *light blue* and *dark blue*. The calcium ion within the ribbon can be seen within each chain as a *green sphere*. On the right is the active site of the protein. The key residues are shown as stick models labeled with the three-letter code signifying the amino acid type and number. SGSH has a PDB ID 4MHX and is solved at 2.00 Å resolution. Images were modeled using UCSF ChimeraX.

From Pettersen, E. F., Goddard, T. D., Huang, C. C., Meng, E. C., Couch, G. S., Croll, T. I., Morris, J. H., & Ferrin, T. E. (2021). UCSF ChimeraX: Structure visualization for researchers, educators, and developers. Protein Science, 30*(1), 70–82. https://doi.org/10.1002/pro.3943.*

mutations have been linked to the aggressive presentation and rapid progression of the disease, which are R245H and S66W (Di Natale et al., 1998; Weber et al., 1998). The mutations can be located throughout the 502 amino acid sequence, resulting in an inability to form the quaternary protein structure of SGSH, a homodimer of two identical monomers that create a butterfly-shaped structure, which then fails to remove the *N*-sulfate groups from HS (Sidhu et al., 2014). Each monomer of SGSH interacts to create a binding site on either side of the protein. Within the protein the binding site is found at the bottom of the surface cleft near the end of the first *β*-strand and uses residues from both monomers to create the whole active site.

The protein's active site consists of 10 essential amino acids (Asp31, Asp32, FGly70, Arg74, Lys123, His125, His181, Asp273, Asn274, and Arg282) and a calcium ion (Fig. 12.2) (Sidhu et al., 2014). The active site contains a calcium ion coordinated with the Asp31, Asp32, FGly70, Asp273, and Asn274 oxygens, forming a distorted octahedral arrangement. The other residues then interact with the oxygens of FGly70, the critical catalytic residue of SGSH, which is derived from cysteine by a posttranslational modification. FGly is a critical component within the active site of all enzymes within the sulfatase family (Lübke & Damme, 2020). With FGly present within the active site of all sulfatases, it then allows for the prediction of how the mechanism will occur within SGSH. The desulfation of HS is proposed to occur from these residues first when the calcium ion enables the removal of hydrogen from a hydroxyl group of the hydrated form of FGly70, giving a negative charge on oxygen. The negatively charged FGly70 then attacks the *N*-sulfate group of the terminal glucosamine of the HS substrate, forming a covalent bond to the sulfate group and expelling the glucosamine-HS with protonation by His181.

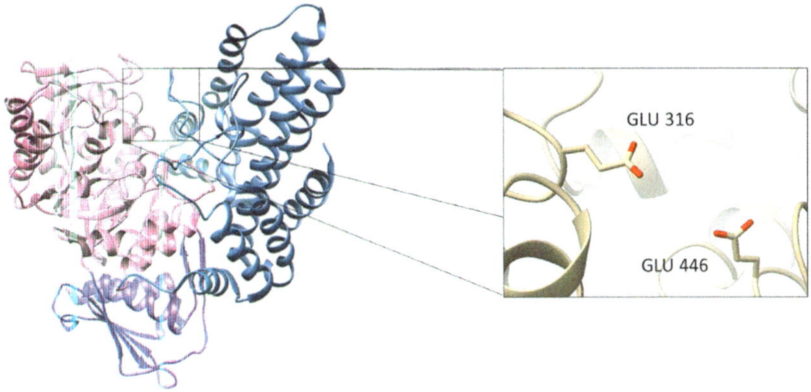

FIGURE 12.3

NAGLU ribbon diagram and active site. The monomer structure of NAGLU is shown in the ribbon representation on the left. The three domains of the monomer are shown in *blue, pink*, and *purple.* Domain one is *purple*, domain two is *pink*, and domain three is *blue.* On the right is the active site of the protein. The two key catalytic residues are shown as stick models labeled with the three-letter code signifying the amino acid type and number. NAGLU has a PDB ID 4XWH. The structure has been solved to 2.32 Å resolution. Images were modeled using UCSF ChimeraX. *NAGLU, N*-acetyl-*α*-glucosaminidase.

From Goddard, T. D., Huang, C. C., Meng, E. C., Pettersen, E. F., Couch, G. S., Morris, J. H., & Ferrin, T. E. (2018). UCSF ChimeraX: Meeting modern challenges in visualization and analysis. Protein Science, *27(1), 14–25. https://doi.org/10.1002/pro.3235.*

FGly70 is then desulfated via deprotonation of the other hydroxyl group by His125, which eliminates sulfate and regenerates FGly70 in the aldehyde form (Sidhu et al., 2014).

A shortage of functioning NAGLU results in Sanfilippo syndrome type B. Sanfilippo syndrome type B is frequently presented in the attenuated form, which means a slower progression of each stage and a longer life span than other subtypes. NAGLU belongs to one of the two largest groups of carbohydrate-active enzymes (CAZymes): glycoside hydrolases (GH), which contain 171 different families (Coines et al., 2022). The GH families are responsible for the cleavage of glycosidic bonds through hydrolysis (Davies & Henrissat, 1995). NAGLU is categorized into family GH89, also known as "*α*-*N*-acetylglucosaminidases." The mode of action of GH89 family enzymes is to cleave terminal *α*-linked *N*-acetylglucosamine residues from polysaccharides and glycoconjugates. Within the GH89 family, it is theorized that there are more than 1000 different enzymes. Out of all of the theorized enzymes, only four have been characterized, which are human *α*-*N*-acetylglucosaminidase (NAGLU) and bacterial *α-N-acetylglucosaminidases* from *Clostridium perfringens* ATCC 13124 (CpGH89), *C. perfringens* strain 13, and *Bifidobacterium bifidum* JCM1254 (Yang et al., 2021).

NAGLU and CpGH89 are homologues of each other and share a low sequence similarity of roughly 30%. Even with a low similarity in sequence the active site is highly conserved. Due to the conservation between the active site of CpGH89 and NAGLU, the two catalytic residues of NAGLU were identified as the glutamic acids at positions 316 and 446 (Fig. 12.3) (Ficko-Blean et al., 2008; Goddard et al., 2018; Yang et al., 2021). There are two main differentiating features between the two active sites. The first is that the asparagine residue within the active site of NAGLU has been substituted in CpGH89 for a histidine (Birrane et al., 2019; Ficko-Blean et al., 2008). Second, CpGH89 contains a seven amino acid loop, which is not included within NAGLU. The amino acid loop adds a tryptophan to the binding site. This tryptophan then closes off part of the entrance to the pocket, making the pocket more sock-shaped, whereas NAGLU, lacking this residue, has a wider opening into the pocket (Birrane et al., 2019). When the binding pockets for NAGLU and CpGH89 are compared, it is clear why similar small molecules would bind inside the pocket, but larger substrates would not fit similarly because of the width of the site's opening. The binding site being more open in NAGLU would then contribute to a difference in binding and why NAGLU recognizes GlcNAc-α(1 → 4)-GlcA or GlcNAc-α(1 → 4)IdoA, whereas CpGH89 does not and instead recognizes GlcNAc-α-(1 → 4)Gal found in synthetic oligosaccharides and class III mucin glycans (Yang et al., 2021). The knowledge of the similarities and differences between the two pockets is vital when virtual screening methods are conducted to screen for pharmacological chaperones (PC;see below).

12.3 Current therapeutic options

Treatment for MPS and LSD can be classified into four main types: substrate reduction therapy (SRT), enzyme replacement therapy (ERT), gene therapy (GT), and pharmacological chaperone therapy (PCT). SRT focuses on inhibiting enzymes involved in the substrate production that accumulates within the lysosome. SRT treatment is oral, which is an advantage over other forms. However, this method has had limited clinical success, with approved drugs only for Gaucher disease and Niemann–Pick type C disease (Boyd et al., 2013; Parenti, 2009).

ERT uses a recombinant form of the enzyme lacking in the disease. Human, animal, or plant cells can be used to produce the desired enzymes. Enzymes are administered intravenously to the patient, which, in turn, will reduce the quantity of the substrate within the cell. The disadvantage to this treatment method is that because of the enzyme size, they do not cross the blood–brain barrier (BBB), which means that this is not an effective treatment for all types of MPS (Concolino et al., 2018; Crawley et al., 1996).

GT treats a disease through the introduction of nucleic acid into affected cells. The nucleic acid can then encode the gene's correct code, allowing for proper function. Viruses are commonly used as transport in GT because they are designed to enter cells with ease. One of the key viruses - used in GT is adeno-associated virus (AAV). A limitation of GT is that some patients may already possess preexisting anti-AAV immunity, which would then mean that the treatment will not be effective. GT is also a very expensive form of treatment (Lukashev & Zamyatnin, 2016; Marcó et al., 2019; Mulligan, 1993).

Chaperones are a more viable treatment option for diseases involving the CNS, such as MPS III (Parenti, 2009). Chaperone therapies use small molecules to promote the correct function of proteins. This method has also shown promising results for MPS II and IV (Loo & Clarke, 2007). Chaperones may not have the same limitations of the BBB or immunity issues as the other therapies and are a less expensive form of treatment.

12.4 Chaperone therapy studies for Sanfilippo syndrome

The studies of chaperones for treating Sanfilippo syndrome type B are limited but have shown increased protein function. There have been no published works on Sanfilippo syndrome type A that have identified successful chaperones.

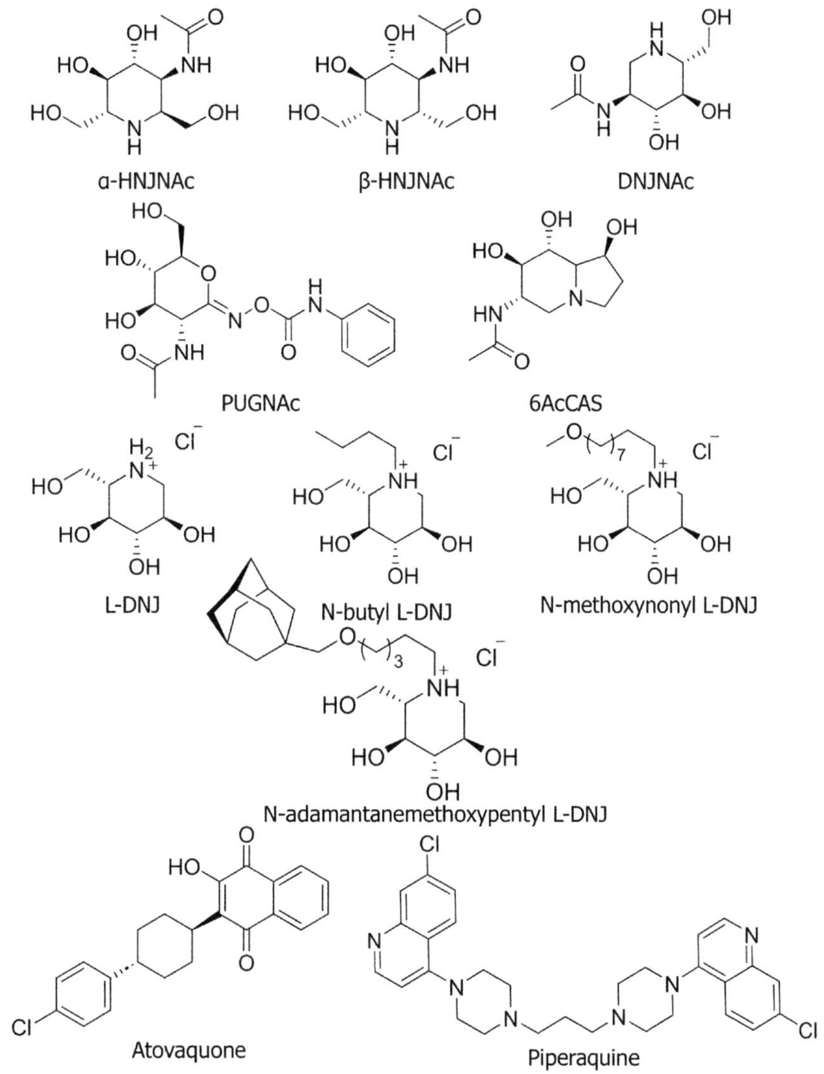

FIGURE 12.4

Structures of MPS III B potential chaperones. The structure of eleven compounds, which have been identified in the literature as potential chaperones for NAGLU.

Data from Muschol et al. (2011), Ficko-Blean et al., 2008, De Pasquale et al., 2023, Zhu et al., 2021, Losada et al., 2024.

A study by Muschol et al. (2011) tested five different compounds at varying concentrations against a mutant variant of SGSH. The five compounds tested were phenylbutyric acid, nojirimycin, *N*-butyl deoxynojirimycin, deoxygalactonojirimycin, and deoxynojirimycin. None of these five compounds showed chaperone capabilities against the SGSH variant (Muschol et al., 2011).

Multiple studies have been conducted that target the catalytic pocket of NAGLU or CpGH89. Ficko-Blean et al. tested the binding of three known inhibitors of other *N*-acetylglucosaminidases against

CpGH89. The compounds tested were 6-acetamido-6-deoxycastanospermine (6AcCAS), 2-acetamido-1,2-dideoxynojirimycin (DNJNAc), and *O*-(2-acetamido-2-deoxy-D-glucopyranosylidene) amino-*N*-phenylcarbamate (PUGNAc) (Fig. 12.4). Of the three, DNJNAc and 6AcCAS showed clear inhibition of the CpGH89, which was then identified as an initial lead compound for chaperone development for NAGLU (Ficko-Blean et al., 2008). However, these molecules are unlikely to useful because of their broad human hexosaminidase inhibition spectrum. Two significant papers detailing iminosugars were published in 2021 and 2023, which showed potential chaperone capabilities for NAGLU (De Pasquale et al., 2023; Zhu et al., 2021). Zhu et al. investigated two previously prepared iminosugar *C*-glycosides derived from GlcNAc, *α*-HNJNAc, and *β*-HNJNAc, along with a library of 15 new derivatives for potential chaperone capabilities (Fig. 12.4). Sixteen compounds were synthesized and tested as derivatives for *α*-HNJNAc and *β*-HNJNAc. Of the tested compounds the weak inhibitor *β*-HNJNAc showed the best potential chaperone capabilities, showing a 2.1-fold increase in NAGLU activity at a dose of 10 μM (Zhu et al., 2021). De Pasquale et al. tested seven different iminosugars with the L-configuration against different NAGLU cell lines. Of these seven different iminosugars, four showed potential as chaperones: L-Deoxynojirimycin (L-DNJ), *N*-butyl L-DNJ, *N*-adamantanemethoxypentyl L-DNJ, and *N*-methoxynonyl L-DNJ (De Pasquale et al., 2023). Finally, Losada et al. identified two potential chaperones atovaquone and piperaquine for NAGLU through virtual screening (Fig. 12.4). Once the two compounds were identified, their ability to increase the activity of NAGLU in MPS IIIB fibroblasts was examined. Atovaquone was able to increase the NAGLU activity 30–99-fold, and piperaquine was able to increase activity by 25–31-fold (Losada et al., 2024). Each of the abovementioned compounds has shown potential as a possible chaperone, but all are yet to be tested in vivo. These studies show that the prospect of different PCs as potential treatments for A and B subtypes of Sanfilippo syndrome is possible. The identification of possible chaperones can then lead to animal and human trials and result in a treatment for these Sanfilippo syndrome subtypes.

12.5 Conclusion

Sanfilippo syndrome is a rare recessive genetic disorder that currently has no cure. A wide variety of studies have been conducted on the identification of potential chaperones as a treatment option for Sanfilippo syndrome. For MPS IIIA no potential chaperones have been identified, but for MPS IIIB several potential chaperones have been identified through multiple studies. The identification of the chaperones then gives a possible treatment avenue for many patients and their families.

References

Andrade, F., Aldámiz-Echevarría, L., Llarena, M., & Couce, M. L. (2015). Sanfilippo syndrome: Overall review. *Pediatrics International, 57*, 331–338. https://doi.org/10.1111/ped.12636.

Baehner, F., Schmiedeskamp, C., Krummenauer, F., Miebach, E., Bajbouj, M., Whybra, C., Kohlschütter, A., Kampmann, C., & Beck, M. (2005). Cumulative incidence rates of the mucopolysaccharidoses in Germany. *Journal of Inherited Metabolic Disease, 28*, 1011–1017. https://doi.org/10.1007/s10545-005-0112-z.

Bauer, S., Zhang, F., & Linhardt, R. J. (2021). Implications of glycosaminoglycans on viral zoonotic diseases. *Diseases, 9*, 85. https://doi.org/10.3390/diseases9040085.

Bigger, B. W., Begley, D. J., Virgintino, D., & Pshezhetsky, A. V. (2018). Anatomical changes and pathophysiology of the brain in mucopolysaccharidosis disorders. *Molecular Genetics and Metabolism, 125*, 322–331. https://doi.org/10.1016/j.ymgme.2018.08.003.

Birrane, G., Dassier, A.-L., Romashko, A., Lundberg, D., Holmes, K., Cottle, T., Norton, A. W., Zhang, B., Concino, M. F., & Meiyappan, M. (2019). Structural characterization of the α-*N*-acetylglucosaminidase, a key enzyme in the pathogenesis of Sanfilippo syndrome B. *Journal of Structural Biology, 205*, 65–71. https://doi.org/10.1016/j.jsb.2019.02.005.

Bodamer, O. A., Giugliani, R., & Wood, T. (2014). The laboratory diagnosis of mucopolysaccharidosis III (Sanfilippo syndrome): A changing landscape. *Molecular Genetics and Metabolism, 113*, 34–41. https://doi.org/10.1016/j.ymgme.2014.07.013.

Boyd, R. E., Lee, G., Rybczynski, P., Benjamin, E. R., Khanna, R., Wustman, B. A., & Valenzano, K. J. (2013). Pharmacological chaperones as therapeutics for lysosomal storage diseases. *Journal of Medicinal Chemistry, 56*, 2705–2725. https://doi.org/10.1021/jm301557k.

Cleary, M. A., & Wraith, J. E. (1993). Management of mucopolysaccharidosis type III. *Archives of Disease in Childhood, 69*, 403–406. https://doi.org/10.1136/adc.69.3.403.

Coines, J., Cuxart, I., Teze, D., & Rovira, C. (2022). Computer simulation to rationalize "Rational" engineering of glycoside hydrolases and glycosyltransferases. *The Journal of Physical Chemistry B, 126*, 802–812. https://doi.org/10.1021/acs.jpcb.1c09536.

Concolino, D., Deodato, F., & Parini, R. (2018). Enzyme replacement therapy: Efficacy and limitations. *Italian Journal of Pediatrics, 44*, 120. https://doi.org/10.1186/s13052-018-0562-1.

Crawley, A. C., Brooks, D. A., Muller, V. J., Petersen, B. A., Isaac, E. L., Bielicki, J., King, B. M., Boulter, C. D., Moore, A. J., Fazzalari, N. L., Anson, D. S., Byers, S., & Hopwood, J. J. (1996). Enzyme replacement therapy in a feline model of Maroteaux-Lamy syndrome. *Journal of Clinical Investigation, 97*, 1864–1873. https://doi.org/10.1172/JCI118617.

Daniele, A., & Di Natale, P. (2001). Heparan *N*-sulfatase: Cysteine 70 plays a role in the enzyme catalysis and processing. *FEBS Letters, 505*, 445–448. https://doi.org/10.1016/s0014-5793(01)02867-8.

Davies, G., & Henrissat, B. (1995). Structures and mechanisms of glycosyl hydrolases. *Structure, 3*, 853–859. https://doi.org/10.1016/S0969-2126(01)00220-9.

De Pasquale, V., Esposito, A., Scerra, G., Scarcella, M., Ciampa, M., Luongo, A., D'Alonzo, D., Guaragna, A., D'Agostino, M., & Pavone, L. M. (2023). *N*-Substituted l-iminosugars for the treatment of Sanfilippo type B syndrome. *Journal of Medicinal Chemistry, 66*, 1790–1808. https://doi.org/10.1021/acs.jmedchem.2c01617.

Di Natale, P., Balzano, N., Esposito, S., & Villani, G. R. (1998). Identification of molecular defects in Italian Sanfilippo A patients including 13 novel mutations. *Human Mutation, 11*, 313–320. https://doi.org/10.1002/(SICI)1098-1004(1998)11:4⟨313::AID-HUMU9⟩3.0.CO;2-P.

Farndale, R. W., Sayers, C. A., & Barrett, A. J. (1982). A direct spectrophotometric microassay for sulfated glycosaminoglycans in cartilage cultures. *Connective Tissue Research, 9*, 247–248. https://doi.org/10.3109/03008208209160269.

Fedele, A. O. (2015). Sanfilippo syndrome: Causes, consequences, and treatments. *The Application of Clinical Genetics, 8*, 269–281. https://doi.org/10.2147/TACG.S57672.

Ficko-Blean, E., Stubbs, K. A., Nemirovsky, O., Vocadlo, D. J., & Boraston, A. B. (2008). Structural and mechanistic insight into the basis of mucopolysaccharidosis IIIB. *Proceedings of the National Academy of Sciences of the United States of America, 105*, 6560–6565. https://doi.org/10.1073/pnas.0711491105.

Garbuzova-Davis, S., Mirtyl, S., Sallot, S. A., Hernandez-Ontiveros, D. G., Haller, E., & Sanberg, P. R. (2013). Blood-brain barrier impairment in MPS III patients. *BMC Neurology, 13*, 174. https://doi.org/10.1186/1471-2377-13-174.

Goddard, T. D., Huang, C. C., Meng, E. C., Pettersen, E. F., Couch, G. S., Morris, J. H., & Ferrin, T. E. (2018). UCSF ChimeraX: Meeting modern challenges in visualization and analysis. *Protein Science, 27*, 14–25. https://doi.org/10.1002/pro.3235.

Jurecka, A., Ługowska, A., Golda, A., Czartoryska, B., & Tylki-Szymańska, A. (2015). Prevalence rates of mucopolysaccharidoses in Poland. *Journal of Applied Genetics, 56*, 205–210. https://doi.org/10.1007/s13353-014-0262-5.

Loo, T. W., & Clarke, D. M. (2007). Chemical and pharmacological chaperones as new therapeutic agents. *Expert Reviews in Molecular Medicine, 9*, 1–18. https://doi.org/10.1017/S1462399407000361.

Losada, J. C., Triana, H., Vanegas, E., Caro, A., Rodríguez-López, A., Espejo-Mojica, A. J., & Almécigа-Diaz, C. J. (2024). Identification of orthosteric and allosteric pharmacological chaperones for mucopolysaccharidosis type IIIB. *ChemBioChem, 25*, e202400081. https://doi.org/10.1002/cbic.202400081.

Lübke, T., & Damme, M. (2020). Lysosomal sulfatases: A growing family. *Biochemical Journal, 477*, 3963–3983. https://doi.org/10.1042/BCJ20200586.

Lukashev, A. N., & Zamyatnin, A. A. (2016). Viral vectors for gene therapy: Current state and clinical perspectives. *Biochemistry (Mosc), 81*, 700–708. https://doi.org/10.1134/S0006297916070063.

Marcó, S., Haurigot, V., & Bosch, F. (2019). In vivo gene therapy for mucopolysaccharidosis type III (Sanfilippo syndrome): A new treatment horizon. *Human Gene Therapy, 30*, 1211–1221. https://doi.org/10.1089/hum.2019.217.

Mulligan, R. C. (1993). The basic science of gene therapy. *Science, 260*, 926–932. https://doi.org/10.1126/science.8493530.

Muschol, N., Pohl, S., Meyer, A., Gal, A., Ullrich, K., & Braulke, T. (2011). Residual activity and proteasomal degradation of p.Ser298Pro sulfamidase identified in patients with a mild clinical phenotype of Sanfilippo A syndrome. *American Journal of Medical Genetics, A, 155A*, 1634–1639. https://doi.org/10.1002/ajmg.a.34053.

Nagree, M. S., Scalia, S., McKillop, W. M., & Medin, J. A. (2019). An update on gene therapy for lysosomal storage disorders. *Expert Opinion on Biological Therapy, 19*, 655–670. https://doi.org/10.1080/14712598.2019.1607837.

Parenti, G. (2009). Treating lysosomal storage diseases with pharmacological chaperones: From concept to clinics. *EMBO Molecular Medicine, 1*, 268–279. https://doi.org/10.1002/emmm.200900036.

Seker Yilmaz, B., Davison, J., Jones, S. A., & Baruteau, J. (2021). Novel therapies for mucopolysaccharidosis type III. *Journal of Inherited Metabolic Disease, 44*, 129–147. https://doi.org/10.1002/jimd.12316.

Sidhu, N. S., Schreiber, K., Pröpper, K., Becker, S., Usón, I., Sheldrick, G. M., Gärtner, J., Krätzner, R., & Steinfeld, R. (2014). Structure of sulfamidase provides insight into the molecular pathology of mucopolysaccharidosis IIIA. *Acta Crystallographica Section D Biological Crystallography, 70*, 1321–1335. https://doi.org/10.1107/S1399004714002739.

Tanwar, H., Kumar, D. T., Doss, C. G. P., & Zayed, H. (2019). Bioinformatics classification of mutations in patients with Mucopolysaccharidosis IIIA. *Metabolic Brain Disease, 34*, 1577–1594. https://doi.org/10.1007/s11011-019-00465-6.

Ugrinov, K. G., Freed, S. D., Thomas, C. L., & Lee, S. W. (2015). A multiparametric computational algorithm for comprehensive assessment of genetic mutations in mucopolysaccharidosis type IIIA (Sanfilippo syndrome). *PLoS One, 10*, e0121511. https://doi.org/10.1371/journal.pone.0121511.

Valstar, M. J., Ruijter, G. J. G., van Diggelen, O. P., Poorthuis, B. J., & Wijburg, F. A. (2008). Sanfilippo syndrome: A mini-review. *Journal of Inherited Metabolic Disease, 31*, 240–252. https://doi.org/10.1007/s10545-008-0838-5.

Weber, B., van de Kamp, J. J., Kleijer, W. J., Guo, X. H., Blanch, L., van Diggelen, O. P., Wevers, R., Poorthuis, B. J., & Hopwood, J. J. (1998). Identification of a common mutation (R245H) in Sanfilippo A patients from The Netherlands. *Journal of Inherited Metabolic Disease, 21*, 416–422. https://doi.org/10.1023/a:1005362826552.

Yang, X., Yang, X., Yu, H., Na, L., Ghosh, T., McArthur, J. B., Chou, T.-F., Dickson, P., & Chen, X. (2021). A GH89 human α-*N*-acetylglucosaminidase (hNAGLU) homologue from gut microbe Bacteroides thetaiotaomicron capable of hydrolyzing heparosan oligosaccharides. *AMB Express, 11*, 94. https://doi.org/10.1186/s13568-021-01253-1.

Zhu, S., Jagadeesh, Y., Tran, A. T., Imaeda, S., Boraston, A., Alonzi, D. S., Poveda, A., Zhang, Y., Désiré, J., Charollais-Thoenig, J., Demotz, S., Kato, A., Butters, T. D., Jiménez-Barbero, J., Sollogoub, M., & Blériot, Y. (2021). Iminosugar C-glycosides work as pharmacological chaperones of NAGLU, a glycosidase involved in MPS IIIB rare disease. *Chemistry, 27*, 11291–11297. https://doi.org/10.1002/chem.202101408.

CHAPTER

13 Membrane transporters

Magnus Andersson*
Department of Chemistry, Umeå University, Umeå, Sweden
**Corresponding author: magnus.p.andersson@umu.se*

13.1 Exposing proteins and their elusive intermediate states during membrane transport

Membrane structures consist of a lipid bilayer that defines the volume of cells and cellular compartments, such as the mitochondria, endoplasmic reticulum, and Golgi apparatus. The hydrophobic membrane core renders them essentially impenetrable to water-soluble molecules. Life processes—though—are dependent on transfer of ions and molecular compounds across the membrane boundaries. For instance, neuronal transmission, carbohydrate metabolism, and muscle control are directly dependent on such transfer across membranes. This membrane transfer needs to be selective and achieved in a controlled manner. Membrane protein transporters span biological membranes and have evolved specificity for a certain cargo and are also equipped with regulatory mechanisms to switch them on and off and are therefore ideally suited for the job.

In the alternating access model proposed by Jardetzky (1966), the transport protein—upon binding the transported cargo—rearranges its structure to open to the opposite side of the membrane and thereby facilitate cargo release (Fig. 13.1). In the simplest case the transported cargo follows the direction of the gradient, that is, from high to low concentration. This passive transport resembles that of membrane channels—with the distinction that a single cargo—and not multiple—is transported per opening event. Active transport, on the other hand, takes the direction against existing gradients and therefore requires energy input. In primary active transport the source of energy is hydrolysis of ATP. Secondary active transporters instead couple the uphill transport of the cargo with existing downhill gradients of other molecules (commonly ions). For symporters the cargo transport shares the same direction as the downhill gradient—while for antiporters—the directions are opposite.

Membrane transporters have evolved high selectivity for their cargos and are required for proper functioning of many physiological processes, including neuronal transmission, carbohydrate metabolism, and muscle control. Without efficient and controlled transfer across membranes, cells would be unable to maintain homeostasis, generate energy, or communicate with other cells. For example, in neuronal transmission, the movement of ions like sodium, potassium, and calcium across the cell membrane is essential for generating action potentials and propagating signals along nerves. Neurotransmitter sodium transporters (NSS) use sodium gradients for reuptake of neurotransmitters, such as dopamine and norepinephrine, from the synaptic cleft back into the neuron (Focke et al., 2013). Because NSS dysregulation is involved in, for example, attention deficit hyperactivity disorder,

Protein Dynamics: Experimental and Computational Approaches. DOI: https://doi.org/10.1016/B978-0-443-30042-4.00017-1

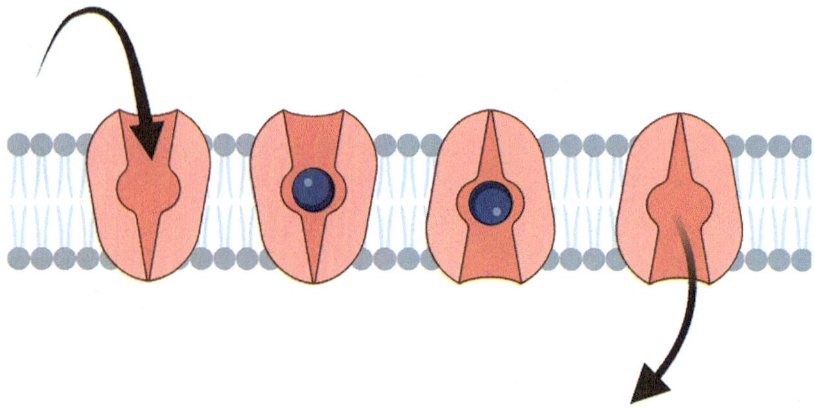

FIGURE 13.1

Schematic illustrating the alternating access model of membrane transport.

Created in https://help.biorender.com/hc/en-gb/articles/17605511350685-How-to-cite-your-BioRender-figure.

bipolar affective disorder, and autism (Kurian et al., 2011), these transporters are targets for developing psychostimulants and antidepressants (Navratna & Gouaux, 2019). Similarly, muscle cells rely on ion gradients and membrane transporters to control muscle contraction and relaxation. Relaxation of the heart muscle requires sarco/endoplasmic Ca^{2+} ATPase (SERCA) transporters for clearance of cytoplasmic calcium subsequent to contraction (Chen et al., 2020). Therefore therapeutic strategies are aimed at SERCA for treatment of cardiovascular disease, such as heart failure (Denniss et al., 2020). Transporters also play a critical role in movement of carbohydrates across membranes and have therefore been associated with metabolic diseases, such as diabetes and different forms of cancer (Nomura et al., 2015; Zambrano et al., 2019). In summary, because of their central role in physiological processes, disruptions in membrane protein transporters can lead to a wide range of medical conditions, making them important targets for diagnosing, understanding, and treating various diseases.

Transport parameters, such as affinity, selectivity, and transport-cycle turnover, are inherent to the amino acid sequence and three-dimensional (3D) fold of the membrane transporter protein. Therefore determination of membrane protein structure is critical to understanding the biological function (Piper et al., 2022). Technical developments in biophysical structural determination methods, such as cryo-electron microscopy (cryo-EM) and X-ray free electron lasers (XFELs), have transformed structural biology. In this context, it is important to realize that protein transporters are not static structures embedded in the membranes. On the contrary, they display both local and global structural rearrangements that are highly synchronized to perform the biological function. Therefore, after decades of determining single structures, the new frontier in membrane transporter biology is to understand the dynamics inherent to the structure. In the following, I will discuss biophysical methods aimed at capturing and revealing the structures of a selected number of membrane protein transporters.

13.2 Trapping intermediate states in membrane transport

Trapping intermediate states followed by structural determination with X-ray crystallography has resulted in considerable insight into transport dynamics. Light-sensitive transporters are well-suited for trapping intermediate states, and bacteriorhodopsin—known for its robustness under laboratory conditions—is a particularly attractive target protein. Bacteriorhodopsin is a proton transporter expressed in the membranes of halophilic archaebacteria—where it builds up a proton gradient in response to light. The structural rearrangements associated with proton transport originate from isomerization of the retinal, which is a cofactor bound inside the seven-helical fold (Fig. 13.2A). In response to retinal isomerization the protein steps through successive states of proton transfer by coordination of subtle dynamics, such as water displacement and side-chain reorientations and more large-scale helical movements. Each state from the ground state and intermediate states from K to O are defined by their unique absorption maxima and time scales (Fig. 13.2B).

Therefore different intermediate-state structures of bacteriorhodopsin can be accessed by illuminating crystals containing the protein by light of specific wavelengths at either cryogenic temperature or during thawing for a defined time and followed by quenching in liquid nitrogen. For instance an M intermediate of bacteriorhodopsin could be trapped by blocking the nitrogen cold stream to obtain a room-temperature crystal, followed by 514 nm illumination for 1 second and subsequently unblocking the cold stream to refreeze the crystal (Sass et al., 2000). Following similar trapping protocols, structures of the K- (Edman et al., 1999), L- (Royant et al., 2000), and M- (Luecke et al., 2000, 1999; Sass et al., 2000) intermediates were obtained. Although significantly advancing structural studies of intermediate states in membrane transport, there were concerns about the interpretation of the results. The major concerns were missing structural regions in the protein, merohedral twinning in the crystals, and heterogeneity of intermediate states within the obtained data (Wickstrand et al., 2015).

Although using illuminated crystals to trap intermediate states can be considered a special case, much ingenuity has been devoted to increase the general applicability of this approach. In this section, I will cover a set of major landmarks in trapping of membrane transporters in intermediate states—starting with transport fueled by existing gradients (i.e., secondary active transport). The major facilitator superfamily (MFS) is the largest known superfamily of secondary active transporters and displays high diversity and is consequently found in all living organisms (Pao et al., 1998). Much research efforts have generated >100 structures, including structural information on intermediates trapped in different states during the membrane transport reaction (Drew et al., 2021). A typical MFS-fold has emerged that consists of two six-transmembrane (TM) helix bundles connected by an extensive intracellular loop (Drew et al., 2021; Reddy et al., 2012) (Fig. 13.3A). The symport mechanism of lactose permease (LacY) that catalyzes H^+-facilitated transport of lactose into *Escherichia coli* was for an extended time the prototype for MFS transport—much owing to pioneering work by Kaback et al. (Kaback, 2021; Kaback & Guan, 2019). A first crystal structure trapped LacY in an inward-facing conformation by introduction of a cysteine-for-glycine mutation (Abramson et al., 2003), which was later followed by a wild-type structure trapped in a similar conformation by optimization of the phospholipid content in the crystallization trials (Guan et al., 2007). Several studies indicated destabilization of the inward-facing conformation upon replacement of two conserved glycine residues, which enabled tight helical packing in the wild-type protein (Jiang et al., 2016a; Smirnova et al., 2013). Indeed, two crystal structures of the LacY double mutant presented the protein bound with two different high-affinity lactose analogs—both

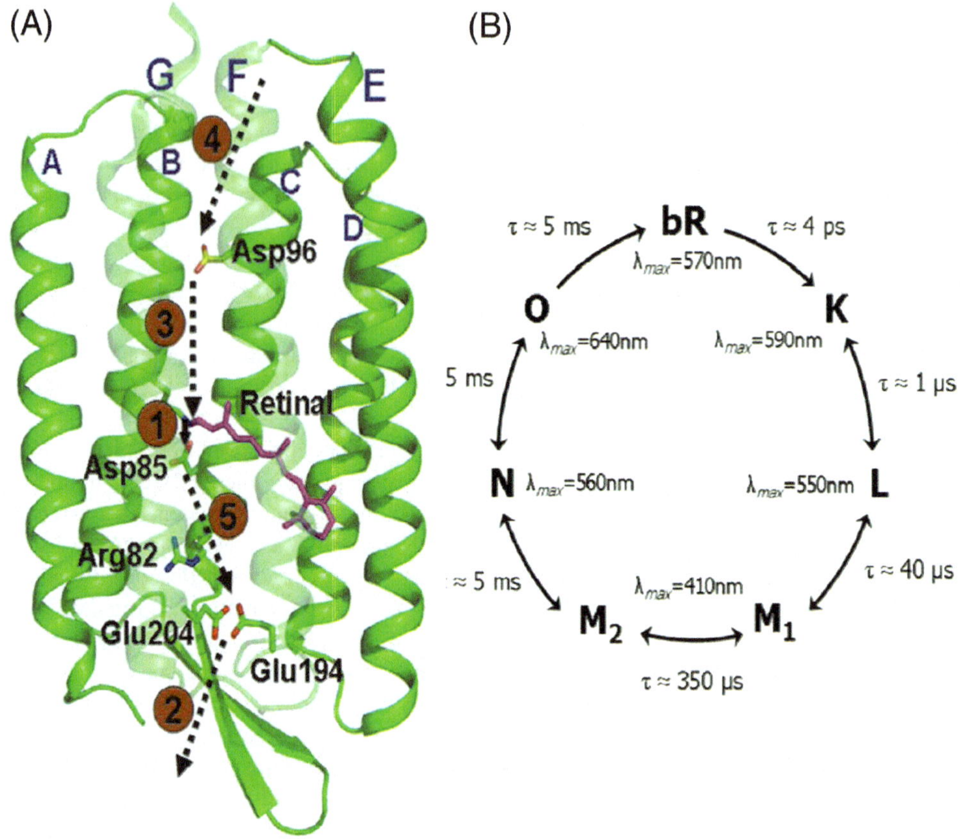

FIGURE 13.2

Bacteriorhodopsin structure and photocycle. (A) Principal proton transfer steps in bacteriorhodopsin depicted onto a ground-state crystal structure. The primary proton transfer from the retinal Schiff base to Asp85 (step 1) is followed by proton release (step 2) to the extracellular side. Reprotonation of the Schiff base from Asp96 (step 3) is then followed by reuptake of a proton from the cytoplasm (step 4). The final step is proton transfer from Asp85 to the proton release group (step 5). (B) The bacteriorhodopsin photocycle and absorption maxima and lifetimes of intermediate states at room temperature.

Adapted from Andersson, M., Malmerberg, E., Westenhoff, S., Katona, G., Cammarata, M., Wohri, A. B., Johansson, L. C., Ewald, F., Eklund, M., Wulff, M., Davidsson, J., & Neutze, R. (2009). Structural dynamics of light-driven proton pumps. Structure, *17(9), 1265–1275. https://doi.org/10.1016/j.str.2009.07.007.*

structures showing a closed seal toward the cytoplasmic side (Kumar et al., 2015, 2014). The opening toward the periplasm was narrow in both structures and they were, therefore, assigned as occluded states.

Existence of an occluded state is inherent to the alternating access model. To avoid dissipation of gradients, it is critical that the transporter is never simultaneously open to both sides of the membrane during the transport mechanism. For example, in the case of sugar import catalyzed by LacY, an

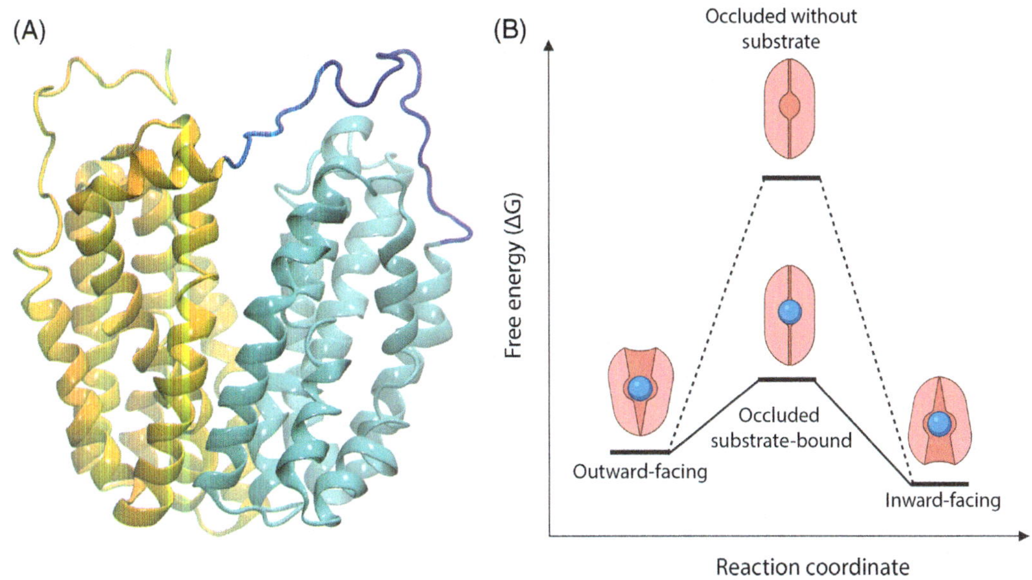

FIGURE 13.3

Major facilitator superfamily fold and reaction free energies: (A) inward-facing wild-type LacY crystal structure (PDB ID: 2V8N) (Guan et al., 2007) with six-helical N-terminal (*cyan*) and C-terminal (*orange*) domains interconnected by the intracellular loop (*blue*) and (B) schematic relative free energies of principal states of transport. The free energy barrier is lowered by substrate binding.

Created in https://help.biorender.com/hc/en-gb/articles/17605511350685-How-to-cite-your-BioRender-figure.

outward-facing state should bind lactose from the *E. coli* periplasm, followed by formation of an occluded state that coordinates lactose internally but is sealed to both sides of the membrane and subsequently open to the cytoplasm for lactose release. In analogy to enzymatic reactions the occluded state can be considered a transition state defined by its higher energy (Klingenberg, 2005). Binding of the transported substrate would lower the free energy barrier to induce the structural rearrangements necessary for opening to the opposite side of the membrane and release the substrate (Fig. 13.3B). As such, the occluded state is key to understand the coupling between substrate binding, utilization of an existing gradient as driving force, and the resulting structural rearrangements (Drew et al., 2021). Indeed, in the LacY occluded-state structures (Kumar et al., 2014, 2015), the amino acids coordinating the lactose analogs had earlier been identified as critical to the transport mechanism by biochemical experiments and computer simulation (Guan & Kaback, 2006; Jiang et al., 2014). Moreover, simulations identified a coupling between protonation states of an internal amino acid and formation of an occluded state (Andersson et al., 2012)—which is at the heart of proton gradient-driven transport.

To obtain a structural view of LacY in an outward-facing conformation, single-domain nanobodies (Muyldermans, 2013) were developed against the destabilizing double mutant. The nanobodies also stabilized outward-facing conformations of wild-type LacY (Smirnova et al., 2014, 2015), which was suggested as an entropic effect generated by asymmetric release of internal waters to each side of the membrane (Hariharan et al., 2016). Crystal structures of double mutant LacY in the presence of such nanobodies were obtained both with bound substrate (Kumar et al., 2018) and in the apo form (Jiang et al., 2016b). Moreover, in these structures the periplasmic opening was rather narrow—albeit with the apo structure hosting a somewhat more opened periplasmic part. Computer simulations based on one of the occluded crystal structures—made into an apo form—showed extensive dynamics on the periplasmic side that enabled entry of a lactose analog and subsequent binding (Kimanius et al., 2018). Hence the openings on each side of the membrane may not be symmetric. In fact, minimizing exposure to a potentially more hostile outside environment might constitute an evolutionary advantage.

Intermediate states in several other MFS subfamilies have been extensively studied by a range of different trapping strategies. For example, crystal structures of glucose transporters (GLUT) and related homologs were trapped in five different conformational states: outward open, outward occluded, fully occluded, inward occluded, and inward open. Here the outward-open conformation was stabilized either by an Fv antibody fragment targeting an intracellular helical bundle unique to GLUT proteins (Nomura et al., 2015) or by an inhibitor of sugar transport that binds from the outside (exofacial inhibitor) (Deng et al., 2015). Including the natural substrate seems to induce partially occluded conformations with limited openings to either the outside (Deng et al., 2015; Paulsen et al., 2019; Sun et al., 2012) or inside (Quistgaard et al., 2013), which are too narrow to allow substrate release. Luring out the inward-open conformation was achieved by inducing alternative crystal forms (Quistgaard et al., 2013; Wisedchaisri et al., 2014), a single missense mutation (Deng et al., 2014), or simply by targeting a GLUT member (GLUT 5), where this conformation likely is inherently more stable (Nomura et al., 2015). A fully occluded state was also obtained by cocrystallization of a hexose transporter from the malaria parasite *Plasmodium falciparum* (PfHT1) with glucose (Qureshi et al., 2020). As we have discussed, occluded structures of membrane transporters are key to understanding the transport mechanism, which in the case of PfHT1 would be coupling between sugar binding and gating. Indeed the crystal structure identified polar contacts located 15 Å from the bound sugar to be equally important for transport as the sugar–protein interactions. Enhanced molecular dynamic (MD) simulations generated a similar occluded conformation in GLUT5 and it was suggested that gating in GLUT proteins is triggered by allostery induced by the bound sugar rather than the magnitude of its affinity (McComas et al., 2023).

Other MFS transporters have been trapped in intermediate states by exploiting cysteine crosslinking (Coincon et al., 2016) and pH sensing (Winkelmann et al., 2022) in protein crystals. Using cryo-EM, different conformations of a mammalian Na^+/H^+ exchanger (NHA_2) were characterized by modification of the environment in detergent and nanodiscs (Matsuoka et al., 2022). The emerging view of MFS transport can be summarized by three fundamentally different ion-coupled alternating access mechanisms (Drew & Boudker, 2016, 2024) (Fig. 13.4). In the rocker-switch mechanism, bundles with high structural similarity rearrange symmetrically around a centrally located substrate-binding site. In the rocking-bundle mechanism, structurally dissimilar bundles rearrange asymmetrically around a centrally located substrate-binding site. Finally, in the elevator mechanism, the substrate is transported by movement in a single bundle against a structurally very different and more fixed bundle.

In primary active transport the fuel type is different. Here, ATP is the common energy source that accomplishes conformational change and alternating access. The specific mechanism by which ATP drives structural rearrangements depends on the type of primary active transporter. P-type AT-Pases belong to a highly diverse family that transports mostly ions but also lipids, polyamines, and helices (Palmgren, 2023; Sim & Park, 2023; Stock et al., 2023). The sarco/endoplasmic Ca^{2+} ATPase (SERCA1a) has become the prototype for P-type ATPase transport (Dyla et al., 2019a)—and like all other P-type ATPases, it consists of three soluble domains that extend into the cytoplasm: the actuator

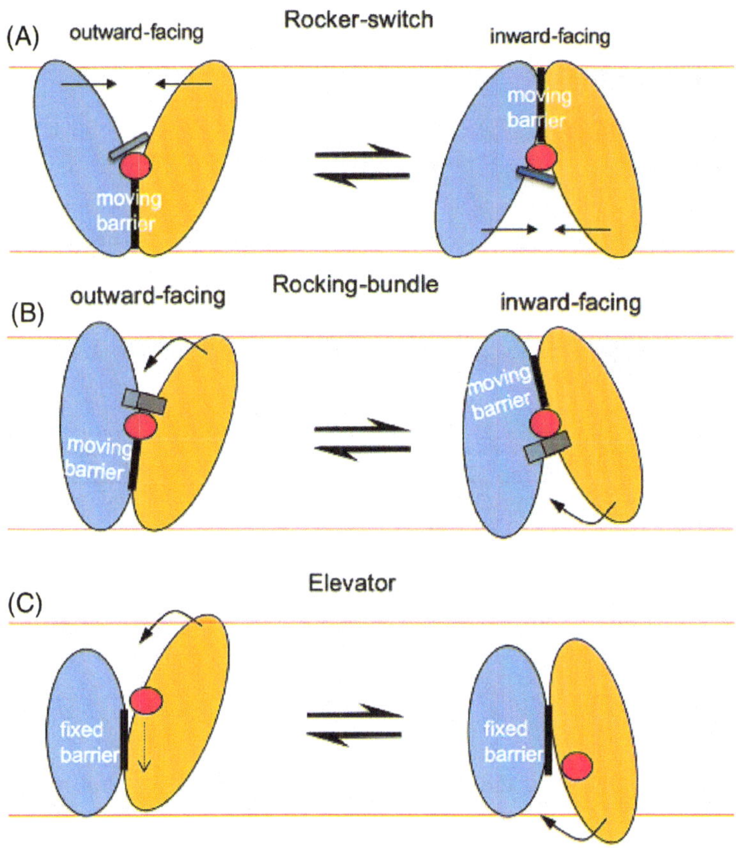

FIGURE 13.4

Transport mechanisms in the major facilitator superfamily: (A) the rocker-switch, (B) the rocking bundle, and (C) the elevator mechanisms.

Adapted from Drew, D., North, R. A., Nagarathinam, K., & Tanabe, M. (2021). Structures and general transport mechanisms by the major facilitator superfamily (MFS). Chemical Reviews, 121*(9), 5289–5335. https://doi.org/10.1021/acs.chemrev.0c00983.*

(A), nucleotide-binding (N), and phosphorylation (P) domains. In addition, there is a membrane-bound (M) domain and, in many cases, regulatory metal-binding domains that are attached to the N- or C-terminus (or both) (Fig. 13.5A). Structural determination of trapped intermediate SERCA1a states has been successfully carried out and has greatly contributed to understanding the transport mechanism. The trapping strategies have been diverse and include combining ion binding with ATP analogs, such as the nonhydrolyzable AMPPCP (Picard et al., 2007; Sörensen et al., 2004; Toyoshima & Mizutani, 2004) or the slowly hydrolyzable AMPPNP (Olesen et al., 2007), and structural analogs of phosphorylation (BeF_3^-), phosphoryl transfer (AlF_4^-), and dephosphorylation (MgF_x) (Karlsen & Bublitz, 2016). Such structural studies have resulted in a general transport mechanism (Dyla et al., 2019b) that starts with the protein facing the cytoplasm with accessible ion-binding sites in the M domain (E1) (Fig. 13.5B).

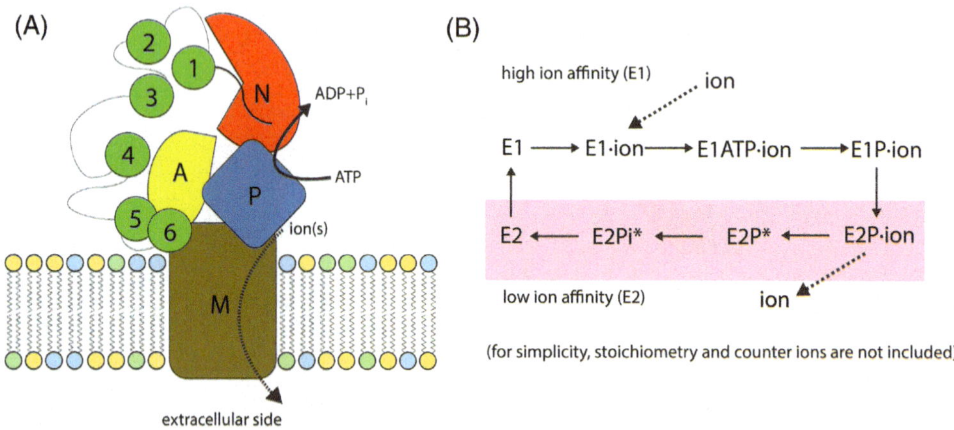

FIGURE 13.5

P-type ATPase structure and reaction cycle. (A) Schematic P-type ATPase structure showing cytoplasmic actuator (A), nucleotide-binding (N), and phosphorylation (P), and membrane (M) domains. As an example of regulatory domains, metal-binding domains 1–6 in the human Cu^{+}-transporter ATP7B are shown. (B) Principal intermediate states in a P-type ATPase reaction cycle. Here, ions represent the transported cargo, but others, e.g., lipids, polyamines, exist.

Binding of ions and ATP induces a closed conformation of the cytoplasmic domains and positions ATP bound to the N domain in proximity of a conserved Asp residue in the P domain (E1-ATP). The subsequent phosphorylation of the Asp residue (E1P) induces structural rearrangements in the cytoplasmic domains that drive opening of the membrane domain toward the extracytoplasmic side and Ca^{2+} release (E2P). H^{+} counterions can now bind, and dephosphorylation takes place (E2-P_i). The dephosphorylated protein (E2) now undertakes reorientations of the cytoplasmic domains to again open the M domain toward the cytoplasm (E1).

Calcium signaling is central to many physiological processes, and primary active transport of calcium is highly evolved across all kingdoms of life (Chen et al., 2020), also in prokaryotes (Dominguez et al., 2015). For example, although the SERCA1a transporter is expressed in fast-twitch skeletal muscle, other isoforms handle calcium transport in the cardiac and smooth muscle tissue (SERCA2a) or ubiquitously in the body (SERCA2b). Trapping studies of the SERCA2a isoform showed a high structural similarity to the corresponding states for SERCA1a (Inoue et al., 2019; Kabashima et al., 2020; Sitsel et al., 2019) and could not directly from the structural data explain the kinetic differences that result in heart muscle SERCA2a having a lower maximal turnover rate and higher calcium affinity (Clausen et al., 2012; Dode et al., 2002), although it was suggested that isoform-specific residues alter salt-bridge and hydrogen bond networks (Sitsel et al., 2019). Both SERCA1a and SERCA2a exhibit higher ATPase activity and lower calcium affinity compared to the SERCA2b isoform (Dode et al., 2003). Several trapped structures—using both X-ray crystallography (Inoue et al., 2019) and cryo-EM (Zhang et al., 2020, 2021, 2022)—identified roles for a unique C-terminal extension conveying isoform-specific conformational and functional regulation (Zhang & Inaba, 2022). As a final example of intermediate trapping in Ca^{2+} ATPase transport, a calcium transporter from *Listeria monocytogenes*

(LMCA1) has emerged as a prototype for prokaryotic primary active transport of calcium. LMCA1 has been structurally characterized by X-ray crystallography in E2 states (Hansen et al., 2021) and in a Ca^{2+}-bound E1 and E2 states using cryo-EM (Prabudiansyah et al., 2024). Recently, an intermediate Ca^{2+}-bound state of a four-glycine insertion LMCA1 mutant was captured by cryo-EM with the cytoplasmic domains in an E2P-like conformation, but the M domain configured as in an E1P-ADP-like state (Basse Hansen et al., 2025). The parts of the reaction mechanism elucidated so far point to unique features of prokaryotic calcium transport. For example, LMCA1 displayed a unique dephosphorylation mechanism in nanodiscs consisting of native-like lipids (Prabudiansyah et al., 2024), in agreement with amino acid partitioning into biological membranes depending on lipid chemistry (Andersson et al., 2011, 2013; Ulmschneider et al., 2011).

The sodium potassium transporter (Na^+/K^+-ATPase, NKA) converts ATP energy into a potential across the plasma membrane in all animal cells by exchange of three cytoplasmic Na^+ for two extracellular K^+ ions (Palmgren & Nissen, 2011). Several crystal structures have shown NKA trapped in E1P states with bound Na^+ (Kanai et al., 2013; Nyblom et al., 2013) or in E2-P_i states with bound K^+ (Morth et al., 2007; Shinoda et al., 2009). Cryo-EM efforts on human NKA have enabled characterization of E1, E1-ATP, E1P, E2P, E2-P_i, and E2 states (Guo et al., 2022; Nguyen et al., 2022). Of particular interest from the trapping strategy point of view was the use of the ATP analog adenosine 5′-O-(3-thio) triphosphate (ATPγS) that was proposed to trap an intermediate state preceding NKA phosphorylation (Guo et al., 2022). The increase in intermediate structures of human NKA is of high medical relevance given that mutations can cause, for example, rapid-onset dystonia parkinsonism, alternating hemiplegia of childhood, and therapy-resistant lethal epilepsy (Heinzen et al., 2014; Lazarov et al., 2020; Poulsen et al., 2010; Ygberg et al., 2021), with possible dependencies on the surrounding lipids (Mahato & Andersson, 2023).

As another example of intermediate trapping in P-type ATPases, one subtype within the family transports heavy metal ions, such as Cu^+, Zn^{2+}, Pb^{2+}, and Cd^{2+} (Smith et al., 2014). Although such ions are essential ingredients in, for example, metal-containing proteins, too high cellular concentrations become toxic. Heavy-metal transporting P-type ATPases are ion exporters that maintain this intricate cellular balance. Structural information resulted at first from studies on prokaryotic Cu^+, Zn^{2+}, and Co^{2+}/Zn^{2+} transporters, which were trapped in E2P and E2-P_i states (Andersson et al., 2014; Gourdon et al., 2011; Gronberg et al., 2021; Wang et al., 2014), and also a Cu^+ ATPase trapped in E1 states with and without bound ion (Salustros et al., 2022). Simulation work also showed ion uptake into the membrane domain (Gronberg et al., 2016). However, none of the determined structures showed electron density for C/N-terminal regulatory domains typical for heavy-metal transporting P-type ATPases (Sitsel et al., 2015; Smith et al., 2014). It was, therefore, a breakthrough when eukaryotic cryo-EM structures began to emerge. The ATP7B Cu^+ ATPase from frog is trapped in an E2-P_i state with two of the six N-terminal regulatory domains contacting the A- and P-domains (Bitter et al., 2022), with a similar A-domain interaction also observed in a Cu^+ ATPase from rice (Guo et al., 2024). In the human ATP7B structure trapped in an E1 state, the single resolved regulatory domain resided close to the ion-entry site (Yang et al., 2023). The structural data agree with computer simulations showing that metal ions binding to the regulatory domains reduce contacts to the cytoplasmic domains and induce an overall more extended domain configuration (Fig. 13.6) (Orädd et al., 2022). Thus, a model of ATP7B regulation has emerged where the regulatory domains extend from inhibitory interactions to the core protein upon binding copper (Dmitriev & Patry, 2024). Moreover, P-type ATPases that translocate phospholipids between lipid bilayer leaflets, so-called flippases (Duan & Li, 2024; Lyons et al., 2020;

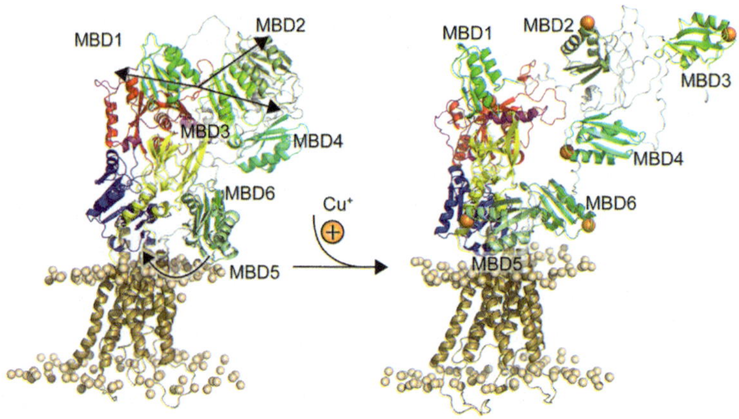

FIGURE 13.6

Simulated Cu^+-induced conformational dynamics in the human Cu^+-transporting P-type ATPase ATP7B. The apo state (left) forms an intermediate state with more extended metal-binding domains (MBD1-6) upon copper binding (right).

Modified from Orädd, F., Steffen, J. H., Gourdon, P., & Andersson, M. (2022). Copper binding leads to increased dynamics in the regulatory N-terminal domain of full-length human copper transporter ATP7B. PLoS Computational Biology, 18*(9), e1010074. https://doi.org/10.1371/journal.pcbi.1010074.*

Sai & Lee, 2024), and polyamines (Mu et al., 2023; Tomita et al., 2021), have been trapped in several different intermediate-state conformations with known P-type ATPase trapping protocols (Nakanishi et al., 2022; Xu et al., 2022).

Although trapping protein intermediates has provided incredibly valuable insights into their structure, function, and mechanisms, this approach is not without drawbacks. For example, trapping strategies require specific temperatures, inhibitors, analogs, or mutations that may induce nonphysiological conformations. It is also important to keep in mind that trapped intermediate states are snapshots in time and therefore have a static quality—rather than dynamic as is the case for the protein at work. Moreover, not all states will be amenable with trapping protocols, such as "high-energy" transition states. For instance the critical E1P-to-E2P transition associated with membrane transport, which is at the heart of the alternating access mechanism, has so far proven out of reach for trapping strategies.

13.2.1 Bringing order to structural heterogeneity with cryo-electron microscopy

Because structural dynamics are inherent to biological macromolecules, particle images collected by cryo-EM will display a certain degree of structural heterogeneity. Hence heterogeneous refinement in cryo-EM is a powerful approach for investigating intermediate states of proteins and other macromolecular complexes. Unlike traditional cryo-EM, which focuses on resolving a single, static protein conformation, heterogeneous refinement allows for capturing multiple structural conformations from the same dataset. The cryo-EM process typically involves imaging thousands to millions of individual particles, each in a unique orientation and potentially a different structural state. Advanced computational algorithms can achieve heterogeneous refinement and thereby segregate particles into distinct groups, each corresponding to a certain conformational state of the protein. Achieving heterogenic refinement with high robustness is challenging and constitutes an active research field (Kimanius & Schwab, 2024).

Heterogenic refinement has resulted in identification of several intermediate structures of the Ca^{2+}-transporting P-type ATPase SERCA2b (Zhang et al., 2021, 2022). Iterative 3D classification generated two major classes of conformations of a Ca^{2+}-bound E1 state—one conformation with a more open arrangement of the cytosolic domains, and the other conformation being more closed. The open SERCA2b conformation was consistent with crystal structures of the SERCA1a isoform in a similar Ca^{2+}-bound E1 state (Jensen et al., 2006; Toyoshima et al., 2000). This intermediate state has been assigned as a putative "off-cycle" state, in which the protein can reversibly become inactive (Dyla et al., 2019b), which can potentially be central to regulation of P-type ATPase activity. The more closed SERCA2b conformation of the Ca^{2+}-bound E1 state was proposed to correspond to a preformed state immediately prior to ATP binding (Zhang et al., 2021). Using a similar approach, three additional intermediate states of SERCA2b were identified that also displayed a high degree of structural heterogeneity (Zhang et al., 2022). The reaction steps covered a Ca^{2+}-bound phosphorylated (E1P) state and Ca^{2+}-free states at different stages of dephosphorylation (E2P and E2-P_i) (Fig. 13.7). An important observation in these experiments was that in the multiple substates observed for each intermediate, there existed also structures corresponding to the next state in the reaction cycle. It was suggested that these conformational overlaps provide insight into acceleration of catalytic cycles.

Calcium transport into the Golgi lumen is central to cellular calcium homeostasis, which is a necessity for protein synthesis (Krebs et al., 2015). Here the secretory pathway Ca^{2+} ATPase (SPCA1a) is a key player (Krebs et al., 2015; Pizzo et al., 2011). High-resolution cryo-EM structures of Ca^{2+} and Mn^{2+}-bound E1-ATP structures provided insight into ion specificity (Chen et al., 2023). The ion-bound states were trapped first by AMPPCP and a nanobody that displayed specific binding with high affinity to SPCA1a. This was followed by addition of a scaffold protein to the nanobody to create a so-called megabody (Uchanski et al., 2021), which enabled improvement of the cryo-EM map resolution. During this procedure, at least three subclass structures could be identified, where those with the clearest density maps for each bound ion type were chosen for structural analyses. In another structural study of SPCA1a, six intermediate states were trapped and characterized by cryo-EM (Wu et al., 2023). Here, heterogenic refinement was used to capture a Ca^{2+}-bound E2P state trapped with BeF_3^-, which was suggested to correspond to an "in-transition" intermediate subsequent to ADP release but before release of Ca^{2+}.

Heterogenic refinement has also advanced understanding of the flippase P-type ATPases by characterization of states associated with, for example, a transient phosphorylated (E1P) state subsequent to ADP release (Hiraizumi et al., 2019) and a lipid-bound state of dephosphorylation (E2-P_i) (Timcenko et al., 2021). Similar observations were made for a P-type ATPase polyamine transporter—showing coexistence of states at different stages of phosphorylation/dephosphorylation (Sim et al., 2021).

However, there are drawbacks and challenges inherent to the heterogenic refinement approach to identify and structurally characterize intermediate states. For example, a high structural flexibility of protein domains in intermediate states might prevent obtaining high-resolution structural data—despite additional rounds of 3D classification. Not being able to retrieve a high-resolution structure can also result from introducing damage to the particles during grid preparation and/or the data collection procedure (Zhang et al., 2021). In addition, many of the intermediate structures obtained by heterogenic refinement have so far been mimics of intermediate states achieved by various trapping protocols. Whether or not such trapped intermediates correspond to true structural rearrangements in the reaction

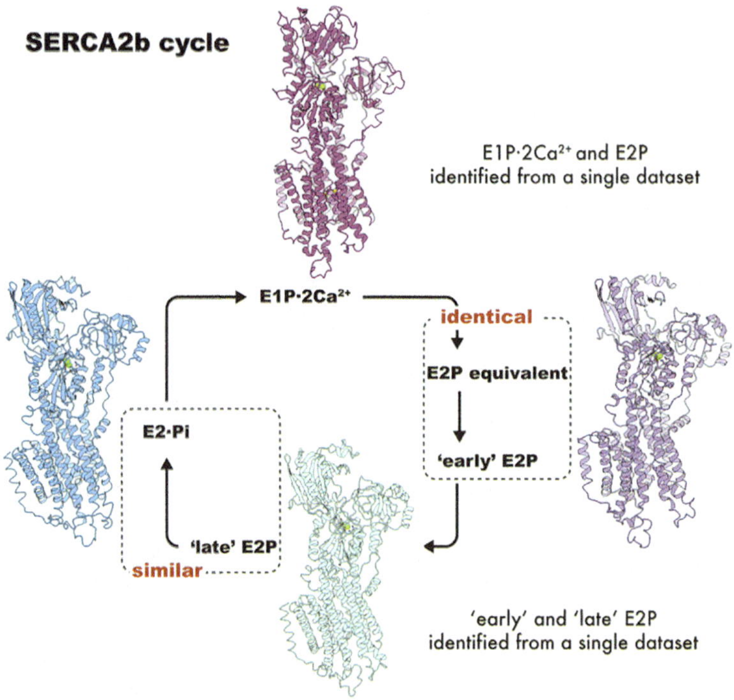

FIGURE 13.7

SERCA2b intermediates identified from cryo-EM heterogenic refinement. The intermediate states Ca^{2+}-bound E1P and E2P and also the early and late E2P states were structurally determined by cryo-EM data from single datasets. *cryo-EM*, Cryo-electron microscopy.

Adapted from Zhang, Y., Kobayashi, C., Cai, X., Watanabe, S., Tsutsumi, A., Kikkawa, M., Sugita, Y., & Inaba, K. (2022). Multiple sub-state structures of SERCA2b reveal conformational overlap at transition steps during the catalytic cycle. Cell Reports, *41(10), 111760. https://doi.org/10.1016/j.celrep.2022.111760.*

is not certain. Moreover, high-energy states might not be represented in enough particles to resolve the structure. Finally the obtained intermediate structures will not be linked in time; that is, it will not be possible to determine the order of the structural events.

13.3 Time-resolved cryo-electron microscopy

To overcome the limitation of conventional single-particle cryo-EM, such experiments can instead be carried out with a time-resolved design to obtain information on temporal scales and the relative order of the observed structural rearrangements. In such nonequilibrium, time-resolved cryo-EM (TR-cryo-EM) experiments, the distribution of conformational ensembles is shifted toward intermediate states. By employing methods that trigger the protein reaction followed by rapid freezing (flash-freezing), TR-cryo-EM can capture intermediate states in biological reactions (Amann et al., 2023). The resulting series of images provide a sequence of structural states, offering insights into the mechanisms and pathways of structural dynamics in real time.

Methods of triggering protein reactions for TR-cryo-EM studies include spraying droplets of one reactant onto a grid that contains the other reactant. Using this approach the structural rearrangements at 5–10 milliseconds associated with gating and channel opening were observed by spraying acetylcholine onto grids containing the acetylcholine receptor, albeit at low spatial resolution (6–9 Å) (Unwin, 1995; Unwin & Fujiyoshi, 2012). The disadvantage of this approach is that diffusion rates will be limiting the mixing of reactants. As an alternative the reactants could first be mixed, then sprayed onto the grid, followed by flash-freezing. A particularly appealing mixing device deposits samples onto the grid by piezo-assisted jets, which are comparable to inkjet printers (Jain et al., 2012). With this technique, it proved possible to observe, for example, binding of Ca^{2+} to a calcium-gated potassium channel MthK (Dandey et al., 2020). However, hydrodynamic and other physicochemical effects in the confined volumes of mixing devices pose challenges when setting the reference time point with respect to the subsequent time point (i.e. time zero). Despite this, recent development of mixing devices enabled impressive temporal control of ATP reacting with skeletal actomyosin (Klebl et al., 2021), with potential also to resolve submillisecond protein dynamics (Klebl et al., 2022).

Using a laser to trigger the reaction enables better control of the temporal development and assigning time tags to the observed intermediate structures. For example, a laser can activate a sample by illuminating the grid as it falls into liquid ethane during the flash-freezing step. By controlling the traveling distance subsequent to laser activation, different points in time of the reaction can be investigated. In this way, dissociation of the actin–myosin complex was observed by laser-induced activation of caged ATP (Menetret et al., 1991), which remains inert until a laser cleaves off the photo-protective group (McCray et al., 1980). The release time of ATP from the caged compound nitrophenylethyl ATP (NPE-caged ATP) depends greatly on conditions, such as pH and solvent, but is relatively slow, that is, on timescales of 1–10 milliseconds (Klan et al., 2013). Therefore the tracking actin–myosin dissociation was limited with a first time point at 20 milliseconds followed by 10-millisecond intervals (Menetret et al., 1991). Using a similar setup, transient states of the light-sensitive protein bacteriorhodopsin were characterized (Subramaniam & Henderson, 2000; Subramaniam et al., 1993). These time-resolved studies were enabled by use of a light-sensitive protein that formed two-dimensional crystalline samples, which must be considered a special case with limited general applicability. Moreover, using an infrared laser, it has been shown feasible to rapidly melt a local spot on the vitrified grid followed by fast revitrification (Voss et al., 2021). With this setup, disassembly of the GroEL protein in response to the electron beam was monitored on a microsecond timescale. However, to monitor protein functional dynamics, a second triggering laser would be required. This was realized by the same research group that induced a pH jump by releasing caged protons with the triggering laser, which enabled characterizing the structural response of a virus that mimics its entry into cells (Voss et al., 2021). Although this method poses technical challenges, it holds great potential to resolve intermediate structures of membrane protein transport on the microsecond timescale. Importantly the structural rearrangements would be observed in real time and hence enable determination of the order of the structural events.

13.4 Time-resolved serial femtosecond crystallography

Intermediate trapping within the traditional X-ray crystallography technique is restricted to conformational states that can be stabilized by various trapping agents and conditions and is also prone to result in structural heterogeneity (Wickstrand et al., 2019). In addition, X-ray-induced radiation damage can distort the intermediate structures with implications for the structural interpretation (Borshchevskiy et al., 2014). Time-resolved serial femtosecond crystallography (TR-SFX) enabled by XFELs bypasses these restrictions and concerns by real-time tracking in the absence of radiation damage (Hirata et al., 2014). Here, intense, femtosecond X-ray pulses enable collection of diffraction data before the sample has had time to respond to the energy deposition and break apart ("diffraction before destruction" Neutze et al., 2000). As a consequence of the XFEL pulse disrupting the crystals for each X-ray probing shot, one requirement for TR-SFX experiments is access to large quantities of microcrystals of the protein. These microcrystals will then be subjected to the activating laser, followed by the X-ray probe at different time intervals—all carried out at room temperature to better enable the biological reaction.

Several methods of delivering the randomly oriented microcrystals to the laser and X-ray pulses have been developed—all with their inherent pros and cons. The liquid microjet injector can deliver a continuous stream of microcrystals past the site of laser activation and then at a predefined point in time through the X-ray beam (Feld & Frank, 2014). A major challenge with liquid injectors is that only a small fraction of the sample will be probed by the X-rays—and as a consequence—large quantities of sample are required to collect a complete dataset. Injection of lipidic cubic phase with high viscosity slows down sample delivery and hence eases sample demands (Branden & Neutze, 2021) but is only amenable with a limited number of protein targets. Moreover, fixed-target methods alleviate sample demand by immobilizing the sample onto a solid support that is moved in the X-ray beam—which greatly increases the crystal hit rate (Schulz et al., 2022). 3D-printed microfluidic devices with multiple channels allow for mixing of crystals and substrate at predefined points in time before X-ray exposure (Monteiro et al., 2020) but can be less suitable for samples prone to clogging or adhering to surfaces. One particularly appealing sample delivery method is using a tape drive that moves individual crystal-containing droplets through the X-ray beam (Schulz et al., 2022). Droplets can then be activated by passing the tape by lasers or in different atmospheres.

In the following, I will briefly highlight landmark achievements in the development of TR-SFX with more elaborate descriptions of membrane protein transporter targets. An early important contribution was the determination of intermediate structures of a bacterial blue light receptor [photoactive yellow protein (PYP)] at 10 nanoseconds and 1 microsecond activated by a nanosecond laser (Tenboer et al., 2014). The difference electron density maps that display disappearing and appearing structural features showed isomerization of the buried cofactor followed by conformational changes originating proximal to the cofactor and spreading throughout the protein. By relying on microcrystals—rather than larger crystals used, for example, in time-resolved Laue diffraction (Wohri et al., 2010)—a large population of laser-excited PYP proteins could be achieved. This high occupancy of intermediate states enabled structural characterization of two intermediates at 10 microseconds. In principle, by instead using a femtosecond laser, it would open up for visualizing subnanosecond structural dynamics. However, such laser activation would reduce the quantum yield and hence the structural signal. Despite this, such experimental setups enabled tracking trans-to-cis chromophore isomerization in PYP from 100 femtoseconds to 3 picoseconds (Pande et al., 2016) and also further into the picosecond temporal regime (Pandey et al., 2020). Ultrafast structural rearrangement of a chromophore was also characterized in a photoswitchable fluorescent protein used in cellular imaging (Coquelle et al., 2018) and in the rhodopsin receptor (Gruhl et al., 2023), which transforms light into visual response (Hofmann et al., 2009). In addition, light-sensitive reactions have been monitored by TR-SFX in a bacterial photosynthetic reaction center (Dods et al., 2021), a phytochrome photosensor (Claesson et al., 2020), the prototypical B12 photoreceptor CarH (Rios-Santacruz et al., 2026), and electron transfer in a photolyase involved in DNA repair (Cellini et al., 2024).

Proteins that are not inherently light-sensitive have also been characterized with TR-SFX. For example, laser-induced dissociation of carbon monoxide bound to the hemoprotein myoglobin enabled tracking of heme dynamics and coupling to structural rearrangements in the neighboring secondary structure (Barends et al., 2015). Another study used a microfluidic device to mix a slurry of microcrystals containing cytochrome c oxidase in a reduced state with buffer saturated with oxygen to identify a catalytic intermediate (Ishigami et al., 2019). The intermediate structure offered insight into how oxygen reduction into water molecules is coupled to unidirectional proton translocation across the membrane channel. Opening of the cytochrome c oxidase proton channel was also characterized by TR-SFX after laser-induced release of carbon monoxide (Shimada et al., 2017). Here, nanosecond-to-microsecond dynamics identified a conformational relay system that prevented backflow of protons through the channel. Mixing devices have enabled monitoring of binding of inhibitors to lysozyme at 2 and 50 seconds (Beyerlein et al., 2017) and subsequently also at the microsecond timescale (Mehrabi et al., 2019). Inhibitor binding was also characterized in the millisecond-to-second temporal regime for β-lactamase, which is an antibiotic target prone to developing resistance (Kupitz et al., 2017; Olmos et al., 2018). Triggering by substrate mixing was also used to track dynamics on the second scale in the enzyme isocyanide hydratase (Dasgupta et al., 2019). One conclusion from TR-SFX experiments on nonlight-sensitive proteins is that while release of bound substrates enables capturing ultrafast dynamics, using mixing devices and caged compounds typically sets a limit to the temporal resolution.

Microbial rhodopsins are light-driven membrane protein ion transporters or sensors with a shared seven-helical fold containing a covalently attached retinal chromophore (Rozenberg et al., 2021). Proton transport by bacteriorhodopsin has been well studied using TR-SFX (Wickstrand et al., 2019). Two independent studies reported on subpicosecond trans-to-cis isomerization of the retinal that passed through a twisted geometry (Fig. 13.8) (Caramello & Royant, 2024; Nass Kovacs et al., 2019; Nogly et al., 2018). The isomerization was observed to induce collective motions of amino acid residues and water molecules proximal to the retinal. The structural dynamics in bacteriorhodopsin subsequent to retinal isomerization have also been characterized with TR-SFX on the nano-to-millisecond time scale. In this study, structural rearrangements were observed in the hydrogen-bond network between water molecules and charged amino acids, which enables the initial step in the proton transfer reaction (Nango et al., 2016) (Fig. 13.8). The observed dynamics ensure that protons are released to the extracellular side and prevent backflow into the cytoplasm. The slower millisecond dynamics of the bacteriorhodopsin photocycle were monitored using serial crystallography at a synchrotron (TR-SSX), which showed opening of the cytoplasmic part of the protein to enable proton uptake (Weinert et al., 2019) (Fig. 13.8).

Rhodopsins also regulate the internal concentration of ions other than protons. Although protons are not easily detected by X-rays, more strongly scattering ions can be directly observed from electron density maps, which can be exploited in TR-SFX (Nango & Iwata, 2023). In a TR-SFX study on the sodium-transporting rhodopsin KR2 a transient sodium-binding site was observed close to the retinal-binding pocket at 1 millisecond, which then faded and reappeared at 20 milliseconds at the extracellular exit site (Skopintsev et al., 2020). Ion translocation through a rhodopsin has also been observed in the inward-directed chloride pump NM-R3 found in a marine flavobacterium (Hosaka et al., 2022; Mous et al., 2022; Yun et al., 2021). Here a chloride ion was observed in the resting state and was displaced subsequent to isomerization and could be tracked along the ion pathway created by both

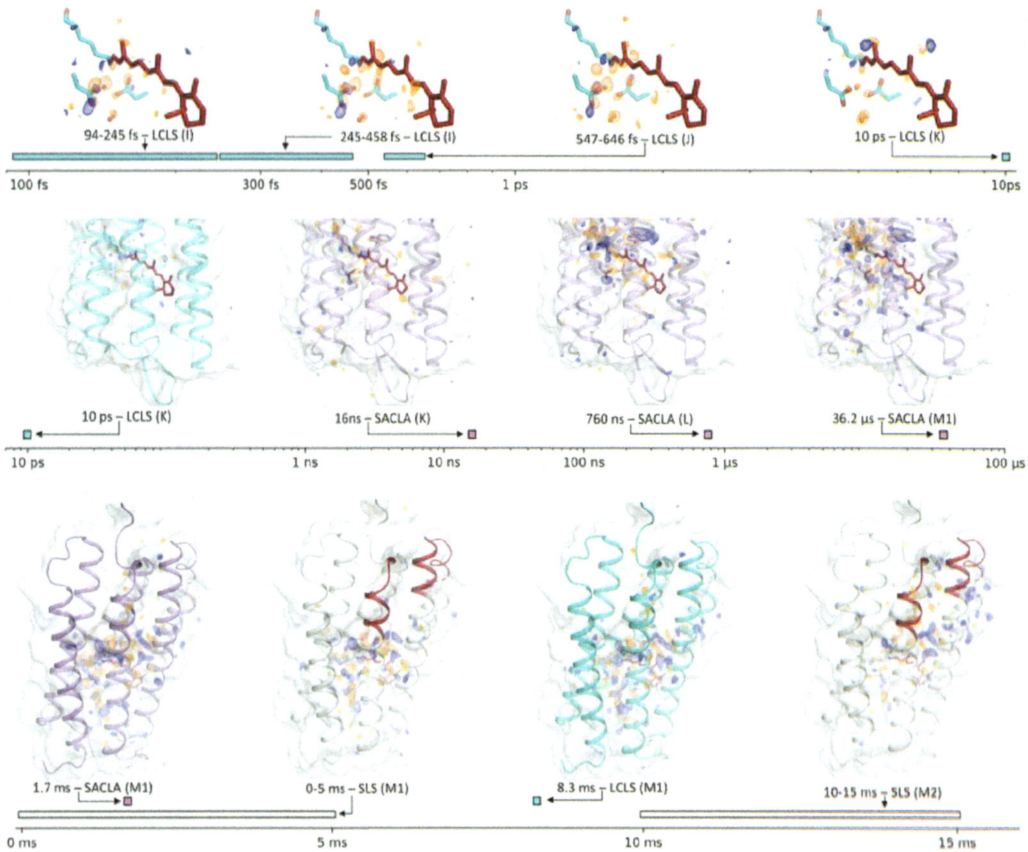

FIGURE 13.8

Bacteriorhodopsin intermediate states determined by TR-SFX and TR-SSX from femto-to-milliseconds. *TR*-SFX, Time-resolved serial femtosecond crystallography.

Adapted from Caramello, N., & Royant, A. (2024). From femtoseconds to minutes: Time-resolved macromolecular crystallography at XFELs and synchrotrons. Acta Crystalloaphica Section D: Structural Biology, 80*(Pt 2), 60–79. https://doi.org/10.1107/S2059798323011002.*

subtle side chain dynamics and helical rearrangements. One study also exploited the capability of NM-R3 to transport strongly scattering bromide and iodide ions (Hosaka et al., 2022), which constitutes an elegant approach to resolve transient ion-binding sites. Ultrafast dynamics were also monitored in a light-gated channel rhodopsin, which is a channel rather than a transporter, but consisting of the common rhodopsin structural fold. In this study, helical displacements near the retina were proposed to result in passive cation flow across the membrane (Oda et al., 2021). Hence these findings indicate that the shared rhodopsin fold works as a scaffold in which different reaction dynamics can accomplish diverse biological responses (Branden & Neutze, 2021).

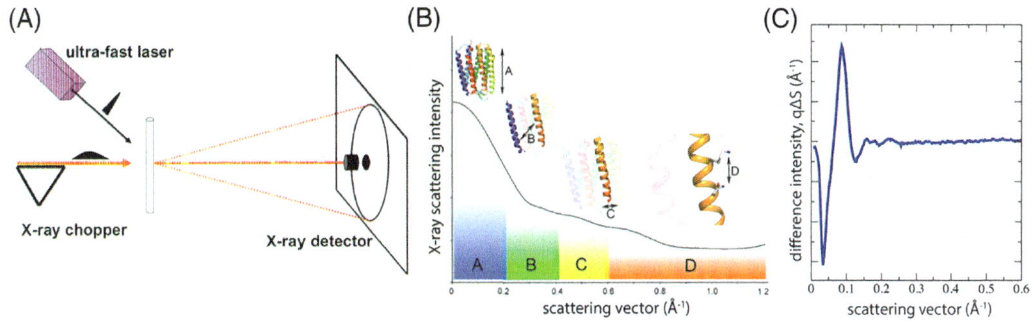

FIGURE 13.9

The time-resolved X-ray solution scattering (TR-XSS) methodology: (A) the synchrotron setup, (B) static (nontime-resolved) X-ray solution scattering as a function of scattering vector $q = 4\pi \sin(\theta)/\lambda = 4\pi/2d$, where 1/d is the crystallography spatial resolution, and (C) difference in X-ray scattering profile.

(B) Adapted from Abramson, J., & Chaptal, V. (2009). Keeping an eye on membrane transport by TR-WAXS. Structure, *17(9), 1153–1155. https://doi.org/10.1016/j.str.2009.08.003.*

Although TR-SFX has certainly developed our understanding of ultrafast dynamics, a major drawback is scarce accessibility to XFEL facilities (Orville, 2020). Another aspect to consider is the high sample requirements—in particular when using a continuous supply of microcrystals. Here, developments of alternative delivery methods are ongoing and will likely ease the demands on sample consumption. TR-SFX is best suited for systems where a distinct trigger (e.g., a laser pulse and temperature jump) can initiate the process. One concern that has been raised with regards to laser activation is that in the majority of TR-SFX studies, the laser energies were in a range that deposited several photons per chromophore (Barends et al., 2024; Miller et al., 2020). Such multiphoton absorption may induce nonphysiological structural responses in the proteins, which was shown by titrating laser fluences when inducing photodissociation of carbon monoxide from myoglobin (Barends et al., 2024). Indeed the structural dynamics in the heme group and against proximal residues were different in three experiments at increasing laser fluence. These results suggest that prior spectroscopic characterization can help in choosing adequate conditions for the laser triggering. In addition, systems lacking such inherent light-sensitive triggers are not directly amenable with TR-SFX characterization, but caged compounds are a possible way forward as an indirect trigger of protein dynamics. Finally, as with all crystallography, the crystal lattice can possibly impose constraints on protein conformational flexibility (Andersson et al., 2009).

13.5 Time-resolved X-ray solution scattering

To avoid possible artifacts of a crystal lattice or cryogenic temperatures, it would be desirable to study membrane transport intermediates in liquid phase at room or physiological temperatures. In time-resolved X-ray solution scattering (TR-XSS), structural dynamics are registered as a function of time by tracking the differences in distance between all atoms simultaneously (Fig. 13.9A) (Abramson & Chaptal, 2009; Orädd & Andersson, 2021). In this way, "high-energy" transition states can be visualized, which are difficult or impossible to detect with other structural biology techniques. The obtained structural information will be an average over the random orientations of the molecule under study—and therefore represent the low-resolution envelope of the structure (Fig. 13.9B). However, by subtracting a dark image from the laser-excited image at a certain time point, a difference scattering profile is obtained that only contains information on the structural rearrangements in the sample (Fig. 13.9C). Moreover, clever experimental designs can reveal structural and temporal differences under different conditions at impressive detail.

A requirement for time-resolved experiments is access to an efficient trigger that can initiate a large enough population of molecules in the sample to execute synchronized reaction dynamics that can be registered by X-rays. This setting of a time zero is often accomplished by a laser pulse. The TR-XSS methodology emerged by studying laser activation of small photoactive chemicals, such as I_2 (Neutze et al., 2001), CH_2I_2 (Davidsson et al., 2005), and $C_2H_4I_2$ (Ihee et al., 2005), and also solvent expansion (Georgiou et al., 2006) and resolved subtle differences of photoactivation dynamics in solvents of varying polarity (Vincent et al., 2009). Extending the method to monitoring large-scale conformational changes in proteins was predicted by MD simulations (Andersson et al., 2008). The proof-of-principle protein experiment showed structural rearrangements of hemoglobin relaxation following laser-induced release of carbon monoxide on the nano-to-millisecond timescale (Cammarata et al., 2008). No attempts to obtain a structural solution from the X-ray scattering data were made—but the different X-ray scattering profile at 100 microseconds corresponded well to the difference between static measurements of deoxyhemoglobin and carbon monoxide-liganded hemoglobin.

Because of the inherent trigger, light-sensitive proteins have been prime targets for TR-XSS characterization of protein structural dynamics (Lee et al., 2022b). For example, signal transduction mechanisms have been monitored in phytochrome proteins consisting of a photosensory domain and an output domain, typically a kinase that conveys signal transduction in bacteria and plants (Butler et al., 1959; Kehoe & Grossman, 1996). Light-induced conversion dynamics in bacterial phytochromes were observed on the millisecond timescale (Takala et al., 2014); identified the rate-limiting intermediate state (Björling et al., 2016); explored the effect of mutations at the interface between monomers (Takala et al., 2016); and explored different fundamental mechanistic dynamics (Lee et al., 2022a). It was also discovered that while interdomain communication was concerted in bacterial phytochromes, movements between the two domains in cyanobacteria occurred sequentially (Heyes et al., 2019).

The first membrane protein transporters to be characterized by TR-XSS were the proton-transporting archaeal rhodopsins—bacteriorhodopsin (bR) and proteorhodopsin (pR) (Andersson et al., 2009). Both transporters were extracted into detergent micelles, and their structural dynamics were monitored from nano- to milliseconds. For bR the structures of two intermediate states were determined by a refinement procedure consisting of rigid-body movements enabled by available intermediate crystal structures trapped in different states of the photocycle. The 22-microsecond intermediate showed an inward helical tilt nearby the retina, which explains the origin of the hydrogen-bond rearrangement required to achieve the initial transfer event of the proton (Fig. 13.10). The large-scale movements of helices to open the cytoplasmic end of the protein occurred on the millisecond timescale and were about twice the magnitude of the dynamics observed in crystal structures. Although the structural rearrangements were similar in pR, the kinetics showed faster appearance of the late intermediate state that also relaxed more slowly back to the resting state, which agreed with spectroscopic observations (Friedrich et al., 2002). Subsequent TR-XSS measurements showed faster kinetics in proteorhodopsin with an iodine-substituted retinal chromophore but associated with similar structural dynamics as for the wild-type

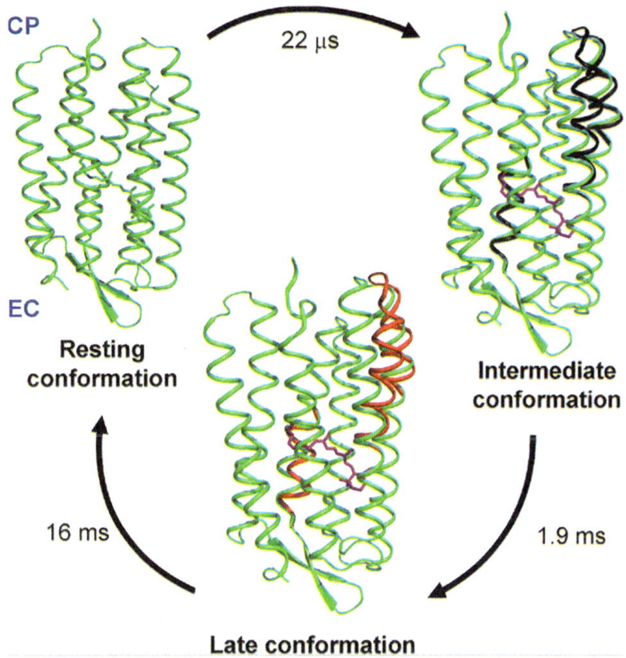

FIGURE 13.10

Bacteriorhodopsin intermediate states characterized by TR-XSS. The identified structural rearrangements for the intermediate (*black*) and late (*red*) conformations are displayed onto the resting (*green*) conformation. The retinal cofactor is depicted in *magenta*. *TR-XSS*, Time-resolved X-ray solution scattering.

Adapted from Andersson, M., Malmerberg, E., Westenhoff, S., Katona, G., Cammarata, M., Wohri, A. B., Johansson, L. C., Ewald, F., Eklund, M., Wulff, M., Davidsson, J., & Neutze, R. (2009). Structural dynamics of light-driven proton pumps. Structure, *17(9), 1265–1275. https://doi.org/10.1016/j.str.2009.07.007.*

protein (Malmerberg et al., 2011). This study explored the possibility of introducing heavy atoms that scatter X-rays more strongly as a means for resolving subtle structural rearrangements.

Ultrafast dynamics in light-dependent reactions in solution have also been tracked in XFEL experiments. Picosecond structural rearrangements were tracked from the light-sensitive cofactors in a bacterial photosynthetic reaction center and spread throughout the protein structure (Arnlund et al., 2014). These dynamics preceded heat propagation and were proposed to describe how a protein structure can dissipate strain introduced by excitation. Such structural responses have been coined protein quakes and are presumably encoded by evolution to prevent degradation of secondary structure upon exposure to excess energy. Ultrafast quake dynamics were also observed following laser-induced release of carbon monoxide in myoglobin (Levantino et al., 2015). Here a first increase in radius of gyration was followed by expansion of the global protein structure on the picosecond timescale. It was shown that the observed dynamics developed on a similar timescale as the acoustic speed of sound, which is in line with the protein quake hypothesis (Ansari et al., 1985). An alternative interpretation of the XFEL data was offered by MD simulations of myoglobin in explicit water, where the subpicosecond signal originated from protein dynamics but where the subsequent signals were attributed to wave propagation across the hydration layer and into the solvent environment (Brinkmann & Hub, 2016). Such hydration shell dynamics were observed in XFEL experiments on homodimeric hemoglobin, where the protein responded by <800 femtosecond helical rearrangements and subsequent expansion and contraction of the global structure on the picosecond timescale (Lee et al., 2021). The electron density in the hydration shell was observed to decrease up to 1.5 picoseconds and therefore lagged slightly behind the ultrafast protein response. Hence time-resolved XFEL measurements can resolve ultrafast protein dynamics, and further developments can be expected to advance our understanding of dynamic processes, such as protein quakes and hydration shell responses.

Because light-sensitive proteins only constitute a fraction of the proteome ($<0.5\%$Monteiro et al., 2021), finding alternative ways to trigger the protein reaction would drastically increase the number of targets for TR-XSS characterization. An obvious way to generalize triggering of a reaction is to simply mix the reactants and study the subsequent reaction. Such mixing-device setups can track folding/unfolding events on the micro-to-millisecond timescale (Akiyama et al., 2002; Pollack et al., 2001; Uzawa et al., 2006) but also slower conformational changes in, for example, $MgCl_2$-induced switches in DNA origami (Bruetzel et al., 2018) and ATP-driven dimerization of nucleotide-binding domains in ATP-binding cassette transporters (Josts et al., 2020) on the order of tens-to-hundreds of milliseconds. Faster dynamics can be accessible by mixer designs that enhance diffusion (Zielinski et al., 2023), which could pave way for TR-XSS characterization of a wide range of protein reactions.

Another—relatively general—way of triggering a protein reaction is by a sudden increase in temperature (T-jump)—induced by an infrared laser pulse—which can reveal how cooperative or allosteric effects influence protein function, mechanisms of folding/unfolding, or function-related conformational changes. Such T-jump triggering was used in combination with TR-XSS to study oligomerization dynamics in insulin (Rimmerman et al., 2017, 2018) and functional dynamics in human cyclophilin A (CypA), which is a proline isomerase involved in protein folding and modulation of signaling processes (Thompson et al., 2019). Although T-jump triggering can be considered general with wide applicability, care needs to be taken to achieve a uniform activation in the volume probed by the X-rays, and the structural refinement should separate the functional dynamics from dynamics associated with unfolding/folding.

A more specific way to trigger a reaction is to use protective groups (or cages) that can be removed by a laser pulse. The released compound can then activate the protein response of interest—that is, a form of indirect activation. Synthesis of photoprotective groups is an active research field, and many caged compounds exist, such as neurotransmitters, ions, small inorganic compounds, and acids (Klan et al., 2013). Several considerations and challenges arise upon designing time-resolved experiments with a caged-compound trigger. For example, the speed of release of the photoprotective group will limit which time scales are accessible, the extinction coefficient of the caged compound will determine the laser power required to achieve uniform activation, and the quantum yield will determine the fraction of release per absorbed photon (Monteiro et al., 2021). Therefore the use of caged compounds typically requires extensive optimization of release conditions (Ravishankar et al., 2018). In addition, unlike for samples containing light-sensitive proteins, a sample with caged compounds cannot be reused—and therefore demands on sample quantities typically are much higher.

Caged ATP was used to trigger structural changes in the *E. coli* adenylate kinase (AdK) enzyme following laser-induced activation of a protected ATP substrate (Orädd et al., 2021). AdK catalyzes the reversible transfer of a phosphate group, converting adenosine 5′-triphosphate (ATP) and adenosine

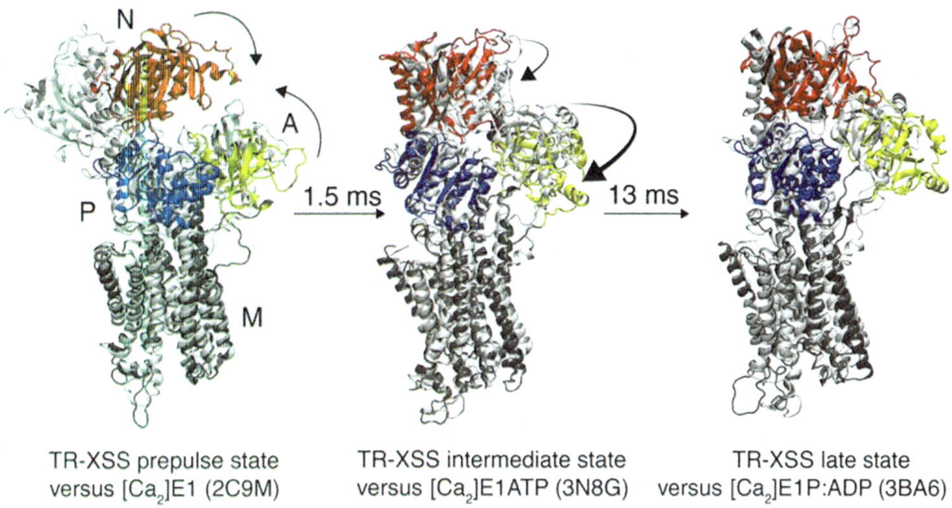

FIGURE 13.11

Structural intermediates identified by TR-XSS in Ca^{2+} transport by the SERCA1a transporter. The prepulse state (left), 1.5-millisecond intermediate (middle), and 13-millisecond intermediate (right) are superimposed onto the corresponding, closest, crystal structures (*white*). *TR-XSS*, Time-resolved X-ray solution scattering.

Modified from Ravishankar, H., Pedersen, M. N., Eklund, M., Sitsel, A., Li, C., Duelli, A., Levantino, M., Wulff, M., Barth, A., Olesen, C., Nissen, P., & Andersson, M. (2020). Tracking Ca(2+) ATPase intermediates in real time by x-ray solution scattering. Science Advances, *6(12), eaaz0981. https://doi.org/10.1126/sciadv.aaz0981.*

5′-monophosphate (AMP) into two molecules of adenosine 5′-diphosphate (ADP), which is a crucial reaction for maintaining cellular energy balance and homeostasis. In the TR-XSS study, a transient intermediate observed at 4.3 milliseconds revealed partial closure of both the ATP- and AMP-binding domains, suggesting a cooperative mechanism for domain closure. In addition, the ATP-binding domain exhibited localized unfolding, accompanied by the disruption of an Arg131–Asp146 salt bridge, which was further substantiated by nuclear magnetic resonance spectroscopy data of similar unfolding in an Arg131Ala AdK mutant, which subsequently refolded into a closed, substrate-bound conformation. These structural dynamics support the "cracking mechanism" hypothesis, which proposes a key role for such transitions in global structural transformations, such as allosteric regulation, in proteins. Recent work on AdK dynamics has developed methods accounting for heterogeneity in structural ensembles associated with caged-compound activation (Magkakis et al., 2025).

ATP-dependent Ca^{2+} transport by the sarco/endoplasmic reticulum Ca^{2+} ATPase (SERCA) transporter was also characterized by TR-XSS and caged ATP release (Ravishankar et al., 2020). The results revealed that the equilibrium [Ca^{2+}]E1 state before laser activation displayed a different domain arrangement compared to previously obtained crystal structures (Fig. 13.11). Following laser-induced release of caged ATP a 1.5-millisecond intermediate was observed that exhibited closure of the cytoplasmic domains, consistent with the E1 state in the presence of bound Ca^{2+} and ATP. A subsequent 13-millisecond transient state revealed a previously unresolved arrangement of the actuator (A) domain, which exposed the ADP-binding site following phosphorylation. These TR-XSS models provide new insights into the relative timing of domain rearrangements that were previously elusive, all within the context of the native membrane environment.

Although indirect triggering techniques can potentially increase TR-XSS characterization of membrane protein transporters, some challenges remain. First, dedicated time-resolved synchrotron beam stations are scarce, and allocations are highly competitive. A possible alternative is to exploit fast-readout detectors to achieve the temporal resolution (Herranz-Trillo et al., 2024; Westenhoff et al., 2010). In such experiments the sample will be exposed to a continuous X-ray beam—rather than a pulsed beam—with the risk of radiation damage effects. However, experimental design can minimize the effect of such X-ray damages (Magkakis et al., 2024). Another limiting factor for TR-XSS experiments is the lack of state-of-the-art analysis tools. As a result the community relies on custom scripts developed in individual laboratories, which are highly specific to each setup and lack standardization. Therefore, developing standardized, user-friendly software tailored to TR-XSS data analysis would greatly improve the consistency and accuracy of the results, facilitating more robust conclusions across different research environments. Efforts in this direction has begun (Sabzian-Molaei et al., 2026). Finally, structural interpretation of TR-XSS data is challenging and relies on existing structural information. MD-based refinement procedures have been developed but on a case-specific basis and might be limited by conformational sampling. Data-driven simulations might present a possible way forward—either by directly biasing the potential toward experimental data (Björling et al., 2015; Chen & Hub, 2015; Hsu et al., 2020), by exploring a free energy landscape based on collective variables set by the experimental data (Kimanius et al., 2015), or by using ensemble-optimization methods (Chan et al., 2023).

13.6 Future perspective

The next frontier in membrane transporter research lies in determining the structural dynamics associated with their biological function—and in particular, how such transient movements are regulated in, for example, cellular signaling and disease mechanisms. The integration of biophysical techniques such as cryo-EM, X-ray solution scattering, and crystallographic approaches is already now advancing the field. In addition, further developments of time-resolved methodologies to access real-time observations at high temporal resolution hold promise to further advance understanding of membrane protein transport and its regulation. Such advances in understanding of membrane transporters can be expected to have significant implications for drug development and disease treatment. For instance, dysfunctional transporters are implicated in a range of medical conditions, including heart disease and certain neurodegenerative diseases. By targeting the mechanisms that control transporter activity, new therapeutic strategies could emerge, offering more precise and effective treatments. In addition, this research may provide innovative approaches to bioengineering, with the potential to design custom transporters for applications in biotechnology and medicine.

References

Abramson, J., & Chaptal, V. (2009). Keeping an eye on membrane transport by TR-WAXS. *Structure, 17*(9), 1153–1155. https://doi.org/10.1016/j.str.2009.08.003.

Abramson, J., Smirnova, I., Kasho, V., Verner, G., Kaback, H. R., & Iwata, S. (2003). Structure and mechanism of the lactose permease of *Escherichia coli*. *Science, 301*(5633), 610–615. https://doi.org/10.1126/science.1088196.

Akiyama, S., Takahashi, S., Kimura, T., Ishimori, K., Morishima, I., Nishikawa, Y., & Fujisawa, T. (2002). Conformational landscape of cytochrome c folding studied by microsecond-resolved small-angle X-ray scattering. *Proceedings of the National Academy of Sciences of the United States of America, 99*(3), 1329–1334. https://doi.org/10.1073/pnas.012458999.

Amann, S. J., Keihsler, D., Bodrug, T., Brown, N. G., & Haselbach, D. (2023). Frozen in time: Analyzing molecular dynamics with time-resolved cryo-EM. *Structure, 31*(1), 4–19. https://doi.org/10.1016/j.str.2022.11.014.

Andersson, M., Bondar, A. N., Freites, J. A., Tobias, D. J., Kaback, H. R., & White, S. H. (2012). Proton-coupled dynamics in lactose permease. *Structure, 20*(11), 1893–1904. https://doi.org/10.1016/j.str.2012.08.021.

Andersson, M., Freites, J. A., Tobias, D. J., & White, S. H. (2011). Structural dynamics of the S4 voltage-sensor helix in lipid bilayers lacking phosphate groups. *Journal of Physical Chemistry B, 115*(27), 8732–8738. https://doi.org/10.1021/jp2001964.

Andersson, M., Malmerberg, E., Westenhoff, S., Katona, G., Cammarata, M., Wohri, A. B., Johansson, L. C., Ewald, F., Eklund, M., Wulff, M., Davidsson, J., & Neutze, R. (2009). Structural dynamics of light-driven proton pumps. *Structure, 17*(9), 1265–1275. https://doi.org/10.1016/j.str.2009.07.007.

Andersson, M., Mattle, D., Sitsel, O., Klymchuk, T., Nielsen, A. M., Moller, L. B., White, S. H., Nissen, P., & Gourdon, P. (2014). Copper-transporting P-type ATPases use a unique ion-release pathway. *Nature Structural & Molecular Biology, 21*(1), 43–48. https://doi.org/10.1038/nsmb.2721.

Andersson, M., Ulmschneider, J. P., Ulmschneider, M. B., & White, S. H. (2013). Conformational states of melittin at a bilayer interface. *Biophysical Journal, 104*(6), L12–L14. https://doi.org/10.1016/j.bpj.2013.02.006.

Andersson, M., Vincent, J., van der Spoel, D., Davidsson, J., & Neutze, R. (2008). A proposed time-resolved X-ray scattering approach to track local and global conformational changes in membrane transport proteins. *Structure, 16*(1), 21–28. https://doi.org/10.1016/j.str.2007.10.016.

Ansari, A., Berendzen, J., Bowne, S. F., Frauenfelder, H., Iben, I. E., Sauke, T. B., Shyamsunder, E., & Young, R. D. (1985). Protein states and proteinquakes. *Proceedings of the National Academy of Sciences of the United States of America, 82*(15), 5000–5004. https://doi.org/10.1073/pnas.82.15.5000.

Arnlund, D., Johansson, L. C., Wickstrand, C., Barty, A., Williams, G. J., Malmerberg, E., Davidsson, J., Milathianaki, D., DePonte, D. P., Shoeman, R. L., Wang, D., James, D., Katona, G., Westenhoff, S., White, T. A., Aquila, A., Bari, S., Berntsen, P., Bogan, M., … Neutze, R. (2014). Visualizing a protein quake with time-resolved X-ray scattering at a free-electron laser. *Nature Methods, 11*(9), 923–926. https://doi.org/10.1038/nmeth.3067.

Barends, T. R., Foucar, L., Ardevol, A., Nass, K., Aquila, A., Botha, S., Doak, R. B., Falahati, K., Hartmann, E., Hilpert, M., Heinz, M., Hoffmann, M. C., Kofinger, J., Koglin, J. E., Kovacsova, G., Liang, M., Milathianaki, D., Lemke, H. T., Reinstein, J., … Schlichting, I. (2015). Direct observation of ultrafast collective motions in CO myoglobin upon ligand dissociation. *Science, 350*(6259), 445–450. https://doi.org/10.1126/science.aac5492.

Barends, T. R. M., Gorel, A., Bhattacharyya, S., Schiro, G., Bacellar, C., Cirelli, C., Colletier, J. P., Foucar, L., Grunbein, M. L., Hartmann, E., Hilpert, M., Holton, J. M., Johnson, P. J. M., Kloos, M., Knopp, G., Marekha, B., Nass, K., Nass Kovacs, G., Ozerov, D., … Schlichting, I. (2024). Influence of pump laser fluence on ultrafast myoglobin structural dynamics. *Nature, 626*(8000), 905–911. https://doi.org/10.1038/s41586-024-07032-9.

Basse Hansen, S., Flygaard, R. K., Kjaergaard, M., & Nissen, P. (2025). Structure of the [Ca]E2P intermediate of Ca(2+)-ATPase 1 from Listeria monocytogenes. *EMBO Reports, 26*(7), 1709–1723. doi:10.1038/s44319-025-00392-x.

Beyerlein, K. R., Dierksmeyer, D., Mariani, V., Kuhn, M., Sarrou, I., Ottaviano, A., Awel, S., Knoska, J., Fuglerud, S., Jonsson, O., Stern, S., Wiedorn, M. O., Yefanov, O., Adriano, L., Bean, R., Burkhardt, A., Fischer, P., Heymann, M., Horke, D. A., … Oberthuer, D. (2017). Mix-and-diffuse serial synchrotron crystallography. *IUCrJ, 4*(Pt 6), 769–777. https://doi.org/10.1107/S2052252517013124.

Bitter, R. M., Oh, S., Deng, Z., Rahman, S., Hite, R. K., & Yuan, P. (2022). Structure of the Wilson disease copper transporter ATP7B. *Science Advances, 8*(9), eabl5508. https://doi.org/10.1126/sciadv.abl5508.

Björling, A., Berntsson, O., Lehtivuori, H., Takala, H., Hughes, A. J., Panman, M., Hoernke, M., Niebling, S., Henry, L., Henning, R., Kosheleva, I., Chukharev, V., Tkachenko, N. V., Menzel, A., Newby, G., Khakhulin, D., Wulff, M., Ihalainen, J. A., & Westenhoff, S. (2016). Structural photoactivation of a full-length bacterial phytochrome. *Science Advances, 2*(8), e1600920. https://doi.org/10.1126/sciadv.1600920.

Björling, A., Niebling, S., Marcellini, M., van der Spoel, D., & Westenhoff, S. (2015). Deciphering solution scattering data with experimentally guided molecular dynamics simulations. *Journal of Chemical Theory and Computation, 11*(2), 780–787. https://doi.org/10.1021/ct5009735.

Borshchevskiy, V., Round, E., Erofeev, I., Weik, M., Ishchenko, A., Gushchin, I., Mishin, A., Willbold, D., Buldt, G., & Gordeliy, V. (2014). Low-dose X-ray radiation induces structural alterations in proteins. *Acta Crystallographica. Section D, Biological Crystallography, 70*(Pt 10), 2675–2685. https://doi.org/10.1107/S1399004714017295.

Branden, G., & Neutze, R. (2021). Advances and challenges in time-resolved macromolecular crystallography. *Science, 373*(6558). https://doi.org/10.1126/science.aba0954.

Brinkmann, L. U., & Hub, J. S. (2016). Ultrafast anisotropic protein quake propagation after CO photodissociation in myoglobin. *Proceedings of the National Academy of Sciences of the United States of America, 113*(38), 10565–10570. https://doi.org/10.1073/pnas.1603539113.

Bruetzel, L. K., Walker, P. U., Gerling, T., Dietz, H., & Lipfert, J. (2018). Time-resolved small-angle X-ray scattering reveals millisecond transitions of a DNA origami switch. *Nano Letters, 18*(4), 2672–2676. https://doi.org/10.1021/acs.nanolett.8b00592.

Butler, W. L., Norris, K. H., Siegelman, H. W., & Hendricks, S. B. (1959). Detection, assay, and preliminary purification of the pigment controlling photoresponsive development of plants. *Proceedings of the National Academy of Sciences of the United States of America, 45*(12), 1703–1708. https://doi.org/10.1073/pnas.45.12.1703.

Cammarata, M., Levantino, M., Schotte, F., Anfinrud, P. A., Ewald, F., Choi, J., Cupane, A., Wulff, M., & Ihee, H. (2008). Tracking the structural dynamics of proteins in solution using time-resolved wide-angle X-ray scattering. *Nature Methods, 5*(10), 881–886. https://doi.org/10.1038/nmeth.1255.

Caramello, N., & Royant, A. (2024). From femtoseconds to minutes: Time-resolved macromolecular crystallography at XFELs and synchrotrons. *Acta Crystallographica Section D: Structural Biology, 80*(Pt 2), 60–79. https://doi.org/10.1107/S2059798323011002.

Cellini, A., Shankar, M. K., Nimmrich, A., Hunt, L. A., Monrroy, L., Mutisya, J., Furrer, A., Beale, E. V., Carrillo, M., Malla, T. N., Maj, P., Vrhovac, L., Dworkowski, F., Cirelli, C., Johnson, P. J. M., Ozerov, D., Stojkovic, E. A., Hammarstrom, L., Bacellar, C., … Westenhoff, S. (2024). Directed ultrafast conformational changes accompany electron transfer in a photolyase as resolved by serial crystallography. *Nature Chemistry, 16*(4), 624–632. https://doi.org/10.1038/s41557-023-01413-9.

Chan, A. M., Nijhawan, A. K., Hsu, D. J., Leshchev, D., Rimmerman, D., Kosheleva, I., Kohlstedt, K. L., & Chen, L. X. (2023). The role of transient intermediate structures in the unfolding of the Trp-cage fast-folding protein: Generating ensembles from time-resolved X-ray solution scattering with genetic algorithms. *The Journal of Physical Chemistry Letters, 14*(5), 1133–1139. https://doi.org/10.1021/acs.jpclett.2c03680.

Chen, J., Sitsel, A., Benoy, V., Sepulveda, M. R., & Vangheluwe, P. (2020). Primary active Ca(2+) transport systems in health and disease. *Cold Spring Harbor Perspectives in Biology, 12*(2). https://doi.org/10.1101/cshperspect.a035113.

Chen, P. C., & Hub, J. S. (2015). Interpretation of solution x-ray scattering by explicit-solvent molecular dynamics. *Biophysical Journal, 108*(10), 2573–2584. https://doi.org/10.1016/j.bpj.2015.03.062.

Chen, Z., Watanabe, S., Hashida, H., Inoue, M., Daigaku, Y., Kikkawa, M., & Inaba, K. (2023). Cryo-EM structures of human SPCA1a reveal the mechanism of Ca(2+)/Mn(2+) transport into the Golgi apparatus. *Science Advances, 9*(9), eadd9742. https://doi.org/10.1126/sciadv.add9742.

Claesson, E., Wahlgren, W. Y., Takala, H., Pandey, S., Castillon, L., Kuznetsova, V., Henry, L., Panman, M., Carrillo, M., Kubel, J., Nanekar, R., Isaksson, L., Nimmrich, A., Cellini, A., Morozov, D., Maj, M., Kurttila, M., Bosman, R., Nango, E., … Westenhoff, S. (2020). The primary structural photoresponse of phytochrome proteins captured by a femtosecond X-ray laser. *Elife, 9*. https://doi.org/10.7554/eLife.53514.

Clausen, J. D., Vandecaetsbeek, I., Wuytack, F., Vangheluwe, P., & Andersen, J. P. (2012). Distinct roles of the C-terminal 11th transmembrane helix and luminal extension in the partial reactions determining the high Ca2+ affinity of sarco(endo)plasmic reticulum Ca2+-ATPase isoform 2b (SERCA2b). *Journal of Biological Chemistry, 287*(47), 39460–39469. https://doi.org/10.1074/jbc.M112.397331.

Coincon, M., Uzdavinys, P., Nji, E., Dotson, D. L., Winkelmann, I., Abdul-Hussein, S., Cameron, A. D., Beckstein, O., & Drew, D. (2016). Crystal structures reveal the molecular basis of ion translocation in sodium/proton antiporters. *Nature Structural & Molecular Biology, 23*(3), 248–255. https://doi.org/10.1038/nsmb.3164.

Coquelle, N., Sliwa, M., Woodhouse, J., Schiro, G., Adam, V., Aquila, A., Barends, T. R. M., Boutet, S., Byrdin, M., Carbajo, S., De la Mora, E., Doak, R. B., Feliks, M., Fieschi, F., Foucar, L., Guillon, V., Hilpert, M., Hunter, M. S., Jakobs, S., … Weik, M. (2018). Chromophore twisting in the excited state of a photoswitchable fluorescent protein captured by time-resolved serial femtosecond crystallography. *Nature Chemistry, 10*(1), 31–37. https://doi.org/10.1038/nchem.2853.

Dandey, V. P., Budell, W. C., Wei, H., Bobe, D., Maruthi, K., Kopylov, M., Eng, E. T., Kahn, P. A., Hinshaw, J. E., Kundu, N., Nimigean, C. M., Fan, C., Sukomon, N., Darst, S. A., Saecker, R. M., Chen, J., Malone, B., Potter, C. S., & Carragher, B. (2020). Time-resolved cryo-EM using Spotiton. *Nature Methods, 17*(9), 897–900. https://doi.org/10.1038/s41592-020-0925-6.

Dasgupta, M., Budday, D., de Oliveira, S. H. P., Madzelan, P., Marchany-Rivera, D., Seravalli, J., Hayes, B., Sierra, R. G., Boutet, S., Hunter, M. S., Alonso-Mori, R., Batyuk, A., Wierman, J., Lyubimov, A., Brewster, A. S., Sauter, N. K., Applegate, G. A., Tiwari, V. K., Berkowitz, D. B., … Wilson, M. A. (2019). Mix-and-inject XFEL crystallography reveals gated conformational dynamics during enzyme catalysis. *Proceedings of the National Academy of Sciences of the United States of America, 116*(51), 25634–25640. https://doi.org/10.1073/pnas.1901864116.

Davidsson, J., Poulsen, J., Cammarata, M., Georgiou, P., Wouts, R., Katona, G., Jacobson, F., Plech, A., Wulff, M., Nyman, G., & Neutze, R. (2005). Structural determination of a transient isomer of CH2I2 by picosecond X-Ray diffraction. *Physical Review Letters, 94*, 245503.

Deng, D., Sun, P., Yan, C., Ke, M., Jiang, X., Xiong, L., Ren, W., Hirata, K., Yamamoto, M., Fan, S., & Yan, N. (2015). Molecular basis of ligand recognition and transport by glucose transporters. *Nature, 526*(7573), 391–396. https://doi.org/10.1038/nature14655.

Deng, D., Xu, C., Sun, P., Wu, J., Yan, C., Hu, M., & Yan, N. (2014). Crystal structure of the human glucose transporter GLUT1. *Nature, 510*(7503), 121–125. https://doi.org/10.1038/nature13306.

Denniss, A. L., Dashwood, A. M., Molenaar, P., & Beard, N. A. (2020). Sarcoplasmic reticulum calcium mishandling: Central tenet in heart failure? *Biophysical Reviews, 12*(4), 865–878. https://doi.org/10.1007/s12551-020-00736-y.

Dmitriev, O. Y., & Patry, J. (2024). Structure and mechanism of the human copper transporting ATPases: Fitting the pieces into a moving puzzle. *Biochimica et Biophysica Acta, Biomembranes, 1866*(4), 184306. https://doi.org/10.1016/j.bbamem.2024.184306.

Dode, L., Andersen, J. P., Leslie, N., Dhitavat, J., Vilsen, B., & Hovnanian, A. (2003). Dissection of the functional differences between sarco(endo)plasmic reticulum Ca2+-ATPase (SERCA) 1 and 2 isoforms and characterization of Darier disease (SERCA2) mutants by steady-state and transient kinetic analyses. *Journal of Biological Chemistry, 278*(48), 47877–47889. https://doi.org/10.1074/jbc.M306784200.

Dode, L., Vilsen, B., Van Baelen, K., Wuytack, F., Clausen, J. D., & Andersen, J. P. (2002). Dissection of the functional differences between sarco(endo)plasmic reticulum Ca^{2+}-ATPase (SERCA) 1 and 3 isoforms by steady-state and transient kinetic analyses. *Journal of Biological Chemistry, 277*(47), 45579–45591. https://doi.org/10.1074/jbc.M207778200.

Dods, R., Bath, P., Morozov, D., Gagner, V. A., Arnlund, D., Luk, H. L., Kubel, J., Maj, M., Vallejos, A., Wickstrand, C., Bosman, R., Beyerlein, K. R., Nelson, G., Liang, M., Milathianaki, D., Robinson, J., Harimoorthy, R., Berntsen, P., Malmerberg, E., … Neutze, R. (2021). Ultrafast structural changes within a photosynthetic reaction centre. *Nature, 589*(7841), 310–314. https://doi.org/10.1038/s41586-020-3000-7.

Dominguez, D. C., Guragain, M., & Patrauchan, M. (2015). Calcium binding proteins and calcium signaling in prokaryotes. *Cell Calcium, 57*(3), 151–165. https://doi.org/10.1016/j.ceca.2014.12.006.

Drew, D., & Boudker, O. (2016). Shared molecular mechanisms of membrane transporters. *Annual Review of Biochemistry, 85*, 543–572. https://doi.org/10.1146/annurev-biochem-060815-014520.

Drew, D., & Boudker, O. (2024). Ion and lipid orchestration of secondary active transport. *Nature, 626*(8001), 963–974. https://doi.org/10.1038/s41586-024-07062-3.

Drew, D., North, R. A., Nagarathinam, K., & Tanabe, M. (2021). Structures and general transport mechanisms by the major facilitator superfamily (MFS). *Chemical Reviews, 121*(9), 5289–5335. https://doi.org/10.1021/acs.chemrev.0c00983.

Duan, H. D., & Li, H. (2024). Consensus, controversies, and conundrums of P4-ATPases: The emerging face of eukaryotic lipid flippases. *Journal of Biological Chemistry, 300*(6), 107387. https://doi.org/10.1016/j.jbc.2024.107387.

Dyla, M., Basse Hansen, S., Nissen, P., & Kjaergaard, M. (2019a). Structural dynamics of P-type ATPase ion pumps. *Biochemical Society Transactions, 47*(5), 1247–1257. https://doi.org/10.1042/BST20190124.

Dyla, M., Kjaergaard, M., Poulsen, H., & Nissen, P. (2019b). Structure and mechanism of P-type ATPase ion pumps. *Annual Review of Biochemistry*. https://doi.org/10.1146/annurev-biochem-010611-112801.

Edman, K., Nollert, P., Royant, A., Belrhali, H., Pebay-Peyroula, E., Hajdu, J., Neutze, R., & Landau, E. M. (1999). High-resolution X-ray structure of an early intermediate in the bacteriorhodopsin photocycle. *Nature, 401*(6755), 822–826. https://doi.org/10.1038/44623.

Feld, G. K., & Frank, M. (2014). Enabling membrane protein structure and dynamics with X-ray free electron lasers. *Current Opinion in Structural Biology, 27*, 69–78. https://doi.org/10.1016/j.sbi.2014.05.002.

Focke, P. J., Wang, X., & Larsson, H. P. (2013). Neurotransmitter transporters: Structure meets function. *Structure, 21*(5), 694–705. https://doi.org/10.1016/j.str.2013.03.002.

Friedrich, T., Geibel, S., Kalmbach, R., Chizhov, I., Ataka, K., Heberle, J., Engelhard, M., & Bamberg, E. (2002). Proteorhodopsin is a light-driven proton pump with variable vectoriality. *Journal of Molecular Biology, 321*(5), 821–838. https://doi.org/10.1016/s0022-2836(02)00696-4.

Georgiou, P., Vincent, J., Andersson, M., Wohri, A. B., Gourdon, P., Poulsen, J., Davidsson, J., & Neutze, R. (2006). Picosecond calorimetry: Time-resolved X-ray diffraction studies of liquid CH_2Cl_2. *The Journal of Chemical Physics, 124*(23), 234507. https://doi.org/10.1063/1.2205365.

Gourdon, P., Liu, X. Y., Skjorringe, T., Morth, J. P., Moller, L. B., Pedersen, B. P., & Nissen, P. (2011). Crystal structure of a copper-transporting PIB-type ATPase. *Nature, 475*(7354), 59–64. https://doi.org/10.1038/nature10191.

Gronberg, C., Hu, Q., Mahato, D. R., Longhin, E., Salustros, N., Duelli, A., Lyu, P., Bagenholm, V., Eriksson, J., Rao, K. U., Henderson, D. I., Meloni, G., Andersson, M., Croll, T., Godaly, G., Wang, K., & Gourdon, P. (2021). Structure and ion-release mechanism of P(IB-4)-type ATPases. *Elife, 10*. https://doi.org/10.7554/eLife.73124.

Gronberg, C., Sitsel, O., Lindahl, E., Gourdon, P., & Andersson, M. (2016). Membrane anchoring and ion-entry dynamics in P-type ATPase copper transport. *Biophysical Journal, 111*(11), 2417–2429. https://doi.org/10.1016/j.bpj.2016.10.020.

Gruhl, T., Weinert, T., Rodrigues, M. J., Milne, C. J., Ortolani, G., Nass, K., Nango, E., Sen, S., Johnson, P. J. M., Cirelli, C., Furrer, A., Mous, S., Skopintsev, P., James, D., Dworkowski, F., Bath, P., Kekilli, D., Ozerov, D., Tanaka, R., … Panneels, V. (2023). Ultrafast structural changes direct the first molecular events of vision. *Nature, 615*(7954), 939–944. https://doi.org/10.1038/s41586-023-05863-6.

Guan, L., & Kaback, H. R. (2006). Lessons from lactose permease. *Annual Review of Biophysics and Biomolecular Structure, 35*, 67–91. https://doi.org/10.1146/annurev.biophys.35.040405.102005.

Guan, L., Mirza, O., Verner, G., Iwata, S., & Kaback, H. R. (2007). Structural determination of wild-type lactose permease. *Proceedings of the National Academy of Sciences of the United States of America, 104*(39), 15294–15298. https://doi.org/10.1073/pnas.0707688104.

Guo, Y., Zhang, Y., Yan, R., Huang, B., Ye, F., Wu, L., Chi, X., Shi, Y., & Zhou, Q. (2022). Cryo-EM structures of recombinant human sodium-potassium pump determined in three different states. *Nature Communications, 13*(1), 3957. https://doi.org/10.1038/s41467-022-31602-y.

Guo, Z., Oradd, F., Bagenholm, V., Gronberg, C., Ma, J. F., Ott, P., Wang, Y., Andersson, M., Pedersen, P. A., Wang, K., & Gourdon, P. (2024). Diverse roles of the metal binding domains and transport mechanism of copper transporting P-type ATPases. *Nature Communications, 15*(1), 2690. https://doi.org/10.1038/s41467-024-47001-4.

Hansen, S. B., Dyla, M., Neumann, C., Quistgaard, E. M. H., Andersen, J. L., Kjaergaard, M., & Nissen, P. (2021). The crystal structure of the Ca(2+)-ATPase 1 from *Listeria monocytogenes* reveals a pump primed for dephosphorylation. *Journal of Molecular Biology, 433*(16), 167015. https://doi.org/10.1016/j.jmb.2021.167015.

Hariharan, P., Andersson, M., Jiang, X., Pardon, E., Steyaert, J., Kaback, H. R., & Guan, L. (2016). Thermodynamics of nanobody binding to lactose permease. *Biochemistry, 55*, 5592–5917. https://doi.org/10.1021/acs.biochem.6b00826.

Heinzen, E. L., Arzimanoglou, A., Brashear, A., Clapcote, S. J., Gurrieri, F., Goldstein, D. B., Johannesson, S. H., Mikati, M. A., Neville, B., Nicole, S., Ozelius, L. J., Poulsen, H., Schyns, T., Sweadner, K. J., van den Maagdenberg, A., Vilsen, B., & Group, A. A. W. (2014). Distinct neurological disorders with ATP1A3 mutations. *Lancet Neurology, 13*(5), 503–514. https://doi.org/10.1016/S1474-4422(14)70011-0.

Herranz-Trillo, F., Sorensen, H. V., Dicko, C., Perez, J., Lenton, S., Fodera, V., Fornell, A., Skepo, M., Plivelic, T. S., Berntsson, O., Andersson, M., Magkakis, K., Oradd, F., Ahn, B., Appio, R., Da Silva, J., Da Silva, V., Lerato, M., & Terry, A. E. (2024). Time-resolved scattering methods for biological samples at the CoSAXS beamline, MAX IV Laboratory. *Methods in Enzymology, 709*, 245–296. https://doi.org/10.1016/bs.mie.2024.10.019.

Heyes, D. J., Hardman, S. J. O., Pedersen, M. N., Woodhouse, J., De La Mora, E., Wulff, M., Weik, M., Cammarata, M., Scrutton, N. S., & Schiro, G. (2019). Light-induced structural changes in a full-length cyanobacterial phytochrome probed by time-resolved X-ray scattering. *Communications Biology, 2*, 1. https://doi.org/10.1038/s42003-018-0242-0.

Hiraizumi, M., Yamashita, K., Nishizawa, T., & Nureki, O. (2019). Cryo-EM structures capture the transport cycle of the P4-ATPase flippase. *Science, 365*(6458), 1149–1155. https://doi.org/10.1126/science.aay3353.

Hirata, K., Shinzawa-Itoh, K., Yano, N., Takemura, S., Kato, K., Hatanaka, M., Muramoto, K., Kawahara, T., Tsukihara, T., Yamashita, E., Tono, K., Ueno, G., Hikima, T., Murakami, H., Inubushi, Y., Yabashi, M., Ishikawa, T., Yamamoto, M., Ogura, T., … Ago, H. (2014). Determination of damage-free crystal structure of an X-ray-sensitive protein using an XFEL. *Nature Methods, 11*(7), 734–736. https://doi.org/10.1038/nmeth.2962.

Hofmann, K. P., Scheerer, P., Hildebrand, P. W., Choe, H. W., Park, J. H., Heck, M., & Ernst, O. P. (2009). A G protein-coupled receptor at work: The rhodopsin model. *Trends in Biochemical Sciences, 34*(11), 540–552. https://doi.org/10.1016/j.tibs.2009.07.005.

Hosaka, T., Nomura, T., Kubo, M., Nakane, T., Fangjia, L., Sekine, S. I., Ito, T., Murayama, K., Ihara, K., Ehara, H., Kashiwagi, K., Katsura, K., Akasaka, R., Hisano, T., Tanaka, T., Tanaka, R., Arima, T., Yamashita, A., Sugahara, M., … Shirouzu, M. (2022). Conformational alterations in unidirectional ion transport of a light-driven chloride pump revealed using X-ray free electron lasers. *Proceedings of the National Academy of Sciences of the United States of America, 119*(9). https://doi.org/10.1073/pnas.2117433119.

Hsu, D. J., Leshchev, D., Kosheleva, I., Kohlstedt, K. L., & Chen, L. X. (2020). Integrating solvation shell structure in experimentally driven molecular dynamics using X-ray solution scattering data. *The Journal of Chemical Physics, 152*(20), 204115. https://doi.org/10.1063/5.0007158.

Ihee, H., Lorenc, M., Kim, T. K., Kong, Q. Y., Cammarata, M., Lee, J. H., Bratos, S., & Wulff, M. (2005). Ultrafast X-ray diffraction of transient molecular structures in solution. *Science, 309*(5738), 1223–1227. https://doi.org/10.1126/science.1114782.

Inoue, M., Sakuta, N., Watanabe, S., Zhang, Y., Yoshikaie, K., Tanaka, Y., Ushioda, R., Kato, Y., Takagi, J., Tsukazaki, T., Nagata, K., & Inaba, K. (2019). Structural basis of sarco/endoplasmic reticulum Ca(2+)-ATPase 2b regulation via transmembrane helix interplay. *Cell Reports, 27*(4), 1221–1230. https://doi.org/10.1016/j.celrep.2019.03.106. e1223.

Ishigami, I., Lewis-Ballester, A., Echelmeier, A., Brehm, G., Zatsepin, N. A., Grant, T. D., Coe, J. D., Lisova, S., Nelson, G., Zhang, S., Dobson, Z. F., Boutet, S., Sierra, R. G., Batyuk, A., Fromme, P., Fromme, R., Spence, J. C. H., Ros, A., Yeh, S. R., & Rousseau, D. L. (2019). Snapshot of an oxygen intermediate in the catalytic reaction of cytochrome c oxidase. *Proceedings of the National Academy of Sciences of the United States of America, 116*(9), 3572–3577. https://doi.org/10.1073/pnas.1814526116.

Jain, T., Sheehan, P., Crum, J., Carragher, B., & Potter, C. S. (2012). Spotiton: A prototype for an integrated inkjet dispense and vitrification system for cryo-TEM. *Journal of Structural Biology, 179*(1), 68–75. https://doi.org/10.1016/j.jsb.2012.04.020.

Jardetzky, O. (1966). Simple allosteric model for membrane pumps. *Nature, 211*(5052), 969–970. https://doi.org/10.1038/211969a0.

Jensen, A. M., Sorensen, T. L., Olesen, C., Moller, J. V., & Nissen, P. (2006). Modulatory and catalytic modes of ATP binding by the calcium pump. *The EMBO Journal, 25*(11), 2305–2314. https://doi.org/10.1038/sj.emboj.7601135.

Jiang, X., Andersson, M., Chau, B. T., Wong, L. Y., Villafuerte, M. K., & Kaback, H. R. (2016a). Role of conserved Gly-Gly pairs on the periplasmic side of LacY. *Biochemistry, 55*(31), 4326–4332. https://doi.org/10.1021/acs.biochem.6b00666.

Jiang, X., Smirnova, I., Kasho, V., Wu, J., Hirata, K., Ke, M., Pardon, E., Steyaert, J., Yan, N., & Kaback, H. R. (2016b). Crystal structure of a LacY-nanobody complex in a periplasmic-open conformation. *Proceedings of the National Academy of Sciences of the United States of America, 113*, 12420–12425. https://doi.org/10.1073/pnas.1615414113.

Jiang, X., Villafuerte, M. K., Andersson, M., White, S. H., & Kaback, H. R. (2014). Galactoside-binding site in LacY. *Biochemistry, 53*(9), 1536–1543. https://doi.org/10.1021/bi401716z.

Josts, I., Gao, Y., Monteiro, D. C. F., Niebling, S., Nitsche, J., Veith, K., Grawert, T. W., Blanchet, C. E., Schroer, M. A., Huse, N., Pearson, A. R., Svergun, D. I., & Tidow, H. (2020). Structural kinetics of MsbA investigated by stopped-flow time-resolved small-angle X-ray scattering. *Structure, 28*(3), 348–354. https://doi.org/10.1016/j.str.2019.12.001. e343.

Kaback, H. R. (2021). It's better to be lucky than smart. *Annual Review of Biochemistry, 90*, 1–29. https://doi.org/10.1146/annurev-biochem-011520-105008.

Kaback, H. R., & Guan, L. (2019). It takes two to tango: The dance of the permease. *Journal of General Physiology, 151*(7), 878–886. https://doi.org/10.1085/jgp.201912377.

Kabashima, Y., Ogawa, H., Nakajima, R., & Toyoshima, C. (2020). What ATP binding does to the Ca(2+) pump and how nonproductive phosphoryl transfer is prevented in the absence of Ca(2). *Proceedings of the National Academy of Sciences of the United States of America, 117*(31), 18448–18458. https://doi.org/10.1073/pnas.2006027117.

Kanai, R., Ogawa, H., Vilsen, B., Cornelius, F., & Toyoshima, C. (2013). Crystal structure of a Na+-bound Na+,K+-ATPase preceding the E1P state. *Nature, 502*(7470), 201–206. https://doi.org/10.1038/nature12578.

Karlsen, J. L., & Bublitz, M. (2016). How to compare, analyze, and morph between crystal structures of different conformations: The P-type ATPase example. *Methods in Molecular Biology, 1377*, 523–539. https://doi.org/10.1007/978-1-4939-3179-8_43.

Kehoe, D. M., & Grossman, A. R. (1996). Similarity of a chromatic adaptation sensor to phytochrome and ethylene receptors. *Science, 273*(5280), 1409–1412. https://doi.org/10.1126/science.273.5280.1409.

Kimanius, D., Lindahl, E., & Andersson, M. (2018). Uptake dynamics in the lactose permease (LacY) membrane protein transporter. *Scientific Reports, 8*(1), 14324. https://doi.org/10.1038/s41598-018-32624-7.

Kimanius, D., Pettersson, I., Schluckebier, G., Lindahl, E., & Andersson, M. (2015). SAXS-guided metadynamics. *Journal of Chemical Theory and Computation, 11*(7), 3491–3498. https://doi.org/10.1021/acs.jctc.5b00299.

Kimanius, D., & Schwab, J. (2024). Confronting heterogeneity in cryogenic electron microscopy data: Innovative strategies and future perspectives with data-driven methods. *Current Opinion in Structural Biology, 86*, 102815. https://doi.org/10.1016/j.sbi.2024.102815.

Klan, P., Solomek, T., Bochet, C. G., Blanc, A., Givens, R., Rubina, M., Popik, V., Kostikov, A., & Wirz, J. (2013). Photoremovable protecting groups in chemistry and biology: Reaction mechanisms and efficacy. *Chemical Reviews, 113*(1), 119–191. https://doi.org/10.1021/cr300177k.

Klebl, D. P., Kay, R. W., Sobott, F., Kapur, N., & Muench, S. P. (2022). Towards sub-millisecond cryo-EM grid preparation. *Faraday Discussions, 240*(0), 33–43. https://doi.org/10.1039/d2fd00079b.

Klebl, D. P., White, H. D., Sobott, F., & Muench, S. P. (2021). On-grid and in-flow mixing for time-resolved cryo-EM. *Acta Crystallographica Section D: Structural Biology, 77*(Pt 10), 1233–1240. https://doi.org/10.1107/S2059798321008810.

Klingenberg, M. (2005). Ligand-protein interaction in biomembrane carriers. The induced transition fit of transport catalysis. *Biochemistry, 44*(24), 8563–8570. https://doi.org/10.1021/bi050543r.

Krebs, J., Agellon, L. B., & Michalak, M. (2015). Ca(2+) homeostasis and endoplasmic reticulum (ER) stress: An integrated view of calcium signaling. *Biochemical and Biophysical Research Communications, 460*(1), 114–121. https://doi.org/10.1016/j.bbrc.2015.02.004.

Kumar, H., Finer-Moore, J. S., Jiang, X., Smirnova, I., Kasho, V., Pardon, E., Steyaert, J., Kaback, H. R., & Stroud, R. M. (2018). Crystal structure of a ligand-bound LacY-nanobody complex. *Proceedings of the National Academy of Sciences of the United States of America, 115*(35), 8769–8774. https://doi.org/10.1073/pnas.1801774115.

Kumar, H., Finer-Moore, J. S., Kaback, H. R., & Stroud, R. M. (2015). Structure of LacY with an alpha-substituted galactoside: Connecting the binding site to the protonation site. *Proceedings of the National Academy of Sciences of the United States of America, 112*(29), 9004–9009. https://doi.org/10.1073/pnas.1509854112.

Kumar, H., Kasho, V., Smirnova, I., Finer-Moore, J. S., Kaback, H. R., & Stroud, R. M. (2014). Structure of sugar-bound LacY. *Proceedings of the National Academy of Sciences of the United States of America, 111*(5), 1784–1788. https://doi.org/10.1073/pnas.1324141111.

Kupitz, C., Olmos, J. L., Jr., Holl, M., Tremblay, L., Pande, K., Pandey, S., Oberthur, D., Hunter, M., Liang, M., Aquila, A., Tenboer, J., Calvey, G., Katz, A., Chen, Y., Wiedorn, M. O., Knoska, J., Meents, A., Majriani, V., Norwood, T., … Schmidt, M. (2017). Structural enzymology using X-ray free electron lasers. *Structural Dynamics, 4*(4), 044003. https://doi.org/10.1063/1.4972069.

Kurian, M. A., Gissen, P., Smith, M., Heales, S., Jr., & Clayton, P. T. (2011). The monoamine neurotransmitter disorders: An expanding range of neurological syndromes. *Lancet Neurology, 10*(8), 721–733. https://doi.org/10.1016/S1474-4422(11)70141-7.

Lazarov, E., Hillebrand, M., Schroder, S., Ternka, K., Hofhuis, J., Ohlenbusch, A., Barrantes-Freer, A., Pardo, L. A., Fruergaard, M. U., Nissen, P., Brockmann, K., Gartner, J., & Rosewich, H. (2020). Comparative analysis of alternating hemiplegia of childhood and rapid-onset dystonia-parkinsonism ATP1A3 mutations reveals functional deficits, which do not correlate with disease severity. *Neurobiology of Disease, 143*, 105012. https://doi.org/10.1016/j.nbd.2020.105012.

Lee, S. J., Kim, T. W., Kim, J. G., Yang, C., Yun, S. R., Kim, C., Ren, Z., Kumarapperuma, I., Kuk, J., Moffat, K., Yang, X., & Ihee, H. (2022a). Light-induced protein structural dynamics in bacteriophytochrome revealed by time-resolved X-ray solution scattering. *Science Advances, 8*(21), eabm6278. https://doi.org/10.1126/sciadv.abm6278.

Lee, Y., Kim, J. G., Lee, S. J., Muniyappan, S., Kim, T. W., Ki, H., Kim, H., Jo, J., Yun, S. R., Lee, H., Lee, K. W., Kim, S. O., Cammarata, M., & Ihee, H. (2021). Ultrafast coherent motion and helix rearrangement of homodimeric hemoglobin visualized with femtosecond X-ray solution scattering. *Nature Communications, 12*(1), 3677. https://doi.org/10.1038/s41467-021-23947-7.

Lee, Y., Lee, H., & Ihee, H. (2022b). Structural dynamics of proteins explored via time-resolved X-ray liquidography. *Chemical Physics Reviews, 3*(4), 041304. https://doi.org/10.1063/5.0101155.

Levantino, M., Schiro, G., Lemke, H. T., Cottone, G., Glownia, J. M., Zhu, D., Chollet, M., Ihee, H., Cupane, A., & Cammarata, M. (2015). Ultrafast myoglobin structural dynamics observed with an X-ray free-electron laser. *Nature Communications, 6*, 6772. https://doi.org/10.1038/ncomms7772.

Luecke, H., Schobert, B., Cartailler, J. P., Richter, H. T., Rosengarth, A., Needleman, R., & Lanyi, J. K. (2000). Coupling photoisomerization of retinal to directional transport in bacteriorhodopsin. *Journal of Molecular Biology, 300*(5), 1237–1255. https://doi.org/10.1006/jmbi.2000.3884.

Luecke, H., Schobert, B., Richter, H. T., Cartailler, J. P., & Lanyi, J. K. (1999). Structural changes in bacteriorhodopsin during ion transport at 2 angstrom resolution. *Science, 286*(5438), 255–261. https://doi.org/10.1126/science.286.5438.255.

Lyons, J. A., Timcenko, M., Dieudonne, T., Lenoir, G., & Nissen, P. (2020). P4-ATPases: How an old dog learnt new tricks—Structure and mechanism of lipid flippases. *Current Opinion in Structural Biology, 63*, 65–73. https://doi.org/10.1016/j.sbi.2020.04.001.

Magkakis, K., Oradd, F., Ahn, B., Da Silva, V., Appio, R., Plivelic, T. S., & Andersson, M. (2024). Real-time structural characterization of protein response to a caged compound by fast detector readout and high-brilliance synchrotron radiation. *Structure, 32*(9), 1519–1527. https://doi.org/10.1016/j.str.2024.05.015. e1513.

Magkakis, K., Oradd, F., Pett, C., Lycksell, M., Pounot, K., Levantino, M., & Andersson, M. (2025). Addressing Structural Heterogeneity in Real-time Tracking of Protein Dynamics Triggered by Caged Compounds. *Journal of Molecular Biology, 437*(24), 169476. doi:10.1016/j.jmb.2025.169476.

Mahato, D. R., & Andersson, M. (2023). Dynamic lipid interactions in the plasma membrane Na(+),K(+)-ATPase. *Biochimica et Biophysica Acta - Molecular Cell Research, 1870*(7), 119545. https://doi.org/10.1016/j.bbamcr.2023.119545.

Malmerberg, E., Omran, Z., Hub, J. S., Li, X., Katona, G., Westenhoff, S., Johansson, L. C., Andersson, M., Cammarata, M., Wulff, M., van der Spoel, D., Davidsson, J., Specht, A., & Neutze, R. (2011). Time-resolved WAXS reveals accelerated conformational changes in iodoretinal-substituted proteorhodopsin. *Biophysical Journal, 101*(6), 1345–1353. https://doi.org/10.1016/j.bpj.2011.07.050.

Matsuoka, R., Fudim, R., Jung, S., Zhang, C., Bazzone, A., Chatzikyriakidou, Y., Robinson, C. V., Nomura, N., Iwata, S., Landreh, M., Orellana, L., Beckstein, O., & Drew, D. (2022). Structure, mechanism and lipid-mediated remodeling of the mammalian Na(+)/H(+) exchanger NHA2. *Nature Structural & Molecular Biology, 29*(2), 108–120. https://doi.org/10.1038/s41594-022-00738-2.

McComas, S. E., Reichenbach, T., Mitrovic, D., Alleva, C., Bonaccorsi, M., Delemotte, L., & Drew, D. (2023). Determinants of sugar-induced influx in the mammalian fructose transporter GLUT5. *Elife, 12*. https://doi.org/10.7554/eLife.84808.

McCray, J. A., Herbette, L., Kihara, T., & Trentham, D. R. (1980). A new approach to time-resolved studies of ATP-requiring biological systems; laser flash photolysis of caged ATP. *Proceedings of the National Academy of Sciences of the United States of America, 77*(12), 7237–7241. http://www.ncbi.nlm.nih.gov/entrez/query.fcgi?cmd=Retrieve&db=PubMed&dopt=Citation&list_uids=6938971.

Mehrabi, P., Schulz, E. C., Agthe, M., Horrell, S., Bourenkov, G., von Stetten, D., Leimkohl, J. P., Schikora, H., Schneider, T. R., Pearson, A. R., Tellkamp, F., & Miller, R. J. D. (2019). Liquid application method for time-resolved analyses by serial synchrotron crystallography. *Nature Methods, 16*(10), 979–982. https://doi.org/10.1038/s41592-019-0553-1.

Menetret, J. F., Hofmann, W., Schroder, R. R., Rapp, G., & Goody, R. S. (1991). Time-resolved cryo-electron microscopic study of the dissociation of actomyosin induced by photolysis of photolabile nucleotides. *Journal of Molecular Biology, 219*(2), 139–144. https://doi.org/10.1016/0022-2836(91)90554-j.

Miller, R. J. D., Pare-Labrosse, O., Sarracini, A., & Besaw, J. E. (2020). Three-dimensional view of ultrafast dynamics in photoexcited bacteriorhodopsin in the multiphoton regime and biological relevance. *Nature Communications, 11*(1), 1240. https://doi.org/10.1038/s41467-020-14971-0.

Monteiro, D. C. F., Amoah, E., Rogers, C., & Pearson, A. R. (2021). Using photocaging for fast time-resolved structural biology studies. *Acta Crystallographica Section D: Structural Biology, 77*(Pt 10), 1218–1232. https://doi.org/10.1107/S2059798321008809.

Monteiro, D. C. F., von Stetten, D., Stohrer, C., Sans, M., Pearson, A. R., Santoni, G., van der Linden, P., & Trebbin, M. (2020). 3D-MiXD: 3D-printed X-ray-compatible microfluidic devices for rapid, low-consumption serial synchrotron crystallography data collection in flow. *IUCrJ, 7*(Pt 2), 207–219. https://doi.org/10.1107/S2052252519016865.

Morth, J. P., Pedersen, B. P., Toustrup-Jensen, M. S., Sorensen, T. L., Petersen, J., Andersen, J. P., Vilsen, B., & Nissen, P. (2007). Crystal structure of the sodium-potassium pump. *Nature, 450*(7172), 1043–1049. https://doi.org/10.1038/nature06419.

Mous, S., Gotthard, G., Ehrenberg, D., Sen, S., Weinert, T., Johnson, P. J. M., James, D., Nass, K., Furrer, A., Kekilli, D., Ma, P., Brunle, S., Casadei, C. M., Martiel, I., Dworkowski, F., Gashi, D., Skopintsev, P., Wranik, M., Knopp, G., … Nogly, P. (2022). Dynamics and mechanism of a light-driven chloride pump. *Science, 375*(6583), 845–851. https://doi.org/10.1126/science.abj6663.

Mu, J., Xue, C., Fu, L., Yu, Z., Nie, M., Wu, M., Chen, X., Liu, K., Bu, R., Huang, Y., Yang, B., Han, J., Jiang, Q., Chan, K. C., Zhou, R., Li, H., Huang, A., Wang, Y., & Liu, Z. (2023). Conformational cycle of human polyamine transporter ATP13A2. *Nature Communications, 14*(1), 1978. https://doi.org/10.1038/s41467-023-37741-0.

Muyldermans, S. (2013). Nanobodies: Natural single-domain antibodies. *Annual Review of Biochemistry, 82*, 775–797. https://doi.org/10.1146/annurev-biochem-063011-092449.

Nakanishi, H., Hayashida, K., Nishizawa, T., Oshima, A., & Abe, K. (2022). Cryo-EM of the ATP11C flippase reconstituted in nanodiscs shows a distended phospholipid bilayer inner membrane around transmembrane helix 2. *Journal of Biological Chemistry, 298*(1), 101498. https://doi.org/10.1016/j.jbc.2021.101498.

Nango, E., & Iwata, S. (2023). Recent progress in membrane protein dynamics revealed by X-ray free electron lasers: Molecular movies of microbial rhodopsins. *Current Opinion in Structural Biology, 81*, 102629. https://doi.org/10.1016/j.sbi.2023.102629.

Nango, E., Royant, A., Kubo, M., Nakane, T., Wickstrand, C., Kimura, T., Tanaka, T., Tono, K., Song, C., Tanaka, R., Arima, T., Yamashita, A., Kobayashi, J., Hosaka, T., Mizohata, E., Nogly, P., Sugahara, M., Nam, D., Nomura, T., … Iwata, S. (2016). A three-dimensional movie of structural changes in bacteriorhodopsin. *Science, 354*(6319), 1552–1557. https://doi.org/10.1126/science.aah3497.

Nass Kovacs, G., Colletier, J. P., Grunbein, M. L., Yang, Y., Stensitzki, T., Batyuk, A., Carbajo, S., Doak, R. B., Ehrenberg, D., Foucar, L., Gasper, R., Gorel, A., Hilpert, M., Kloos, M., Koglin, J. E., Reinstein, J., Roome, C. M., Schlesinger, R., Seaberg, M., … Schlichting, I. (2019). Three-dimensional view of ultrafast dynamics in photoexcited bacteriorhodopsin. *Nature Communications, 10*(1), 3177. https://doi.org/10.1038/s41467-019-10758-0.

Navratna, V., & Gouaux, E. (2019). Insights into the mechanism and pharmacology of neurotransmitter sodium symporters. *Current Opinion in Structural Biology, 54*, 161–170. https://doi.org/10.1016/j.sbi.2019.03.011.

Neutze, R., Wouts, R., Techert, S., Davidsson, J., Kocsis, M., Kirrander, A., Schotte, F., & Wulff, M. (2001). Visualizing photochemical dynamics in solution through picosecond X-ray scattering. *Physical Review Letters, 87*(19), 195508. https://doi.org/10.1103/PhysRevLett.87.195508.

Neutze, R., Wouts, R., van der Spoel, D., Weckert, E., & Hajdu, J. (2000). Potential for biomolecular imaging with femtosecond X-ray pulses. *Nature, 406*(6797), 752–757. https://doi.org/10.1038/35021099.

Nguyen, P. T., Deisl, C., Fine, M., Tippetts, T. S., Uchikawa, E., Bai, X. C., & Levine, B. (2022). Structural basis for gating mechanism of the human sodium-potassium pump. *Nature Communications, 13*(1), 5293. https://doi.org/10.1038/s41467-022-32990-x.

Nogly, P., Weinert, T., James, D., Carbajo, S., Ozerov, D., Furrer, A., Gashi, D., Borin, V., Skopintsev, P., Jaeger, K., Nass, K., Bath, P., Bosman, R., Koglin, J., Seaberg, M., Lane, T., Kekilli, D., Brunle, S., Tanaka, T., … Standfuss, J. (2018). Retinal isomerization in bacteriorhodopsin captured by a femtosecond X-ray laser. *Science, 361*(6398). https://doi.org/10.1126/science.aat0094.

Nomura, N., Verdon, G., Kang, H. J., Shimamura, T., Nomura, Y., Sonoda, Y., Hussien, S. A., Qureshi, A. A., Coincon, M., Sato, Y., Abe, H., Nakada-Nakura, Y., Hino, T., Arakawa, T., Kusano-Arai, O., Iwanari, H., Murata, T., Kobayashi, T., Hamakubo, T., … Drew, D. (2015). Structure and mechanism of the mammalian fructose transporter GLUT5. *Nature, 526*(7573), 397–401. https://doi.org/10.1038/nature14909.

Nyblom, M., Poulsen, H., Gourdon, P., Reinhard, L., Andersson, M., Lindahl, E., Fedosova, N., & Nissen, P. (2013). Crystal structure of Na+, K(+)-ATPase in the Na(+)-bound state. *Science, 342*(6154), 123–127. https://doi.org/10.1126/science.1243352.

Oda, K., Nomura, T., Nakane, T., Yamashita, K., Inoue, K., Ito, S., Vierock, J., Hirata, K., Maturana, A. D., Katayama, K., Ikuta, T., Ishigami, I., Izume, T., Umeda, R., Eguma, R., Oishi, S., Kasuya, G., Kato, T., Kusakizako, T., … Nureki, O. (2021). Time-resolved serial femtosecond crystallography reveals early structural changes in channelrhodopsin. *Elife, 10*. https://doi.org/10.7554/eLife.62389.

Olesen, C., Picard, M., Winther, A. M., Gyrup, C., Morth, J. P., Oxvig, C., Moller, J. V., & Nissen, P. (2007). The structural basis of calcium transport by the calcium pump. *Nature, 450*(7172), 1036–1042. https://doi.org/10.1038/nature06418.

Olmos, J. L., Jr., Pandey, S., Martin-Garcia, J. M., Calvey, G., Katz, A., Knoska, J., Kupitz, C., Hunter, M. S., Liang, M., Oberthuer, D., Yefanov, O., Wiedorn, M., Heyman, M., Holl, M., Pande, K., Barty, A., Miller, M. D., Stern, S., Roy-Chowdhury, S., … Schmidt, M. (2018). Enzyme intermediates captured "on the fly" by mix-and-inject serial crystallography. *BMC Biology, 16*(1), 59. https://doi.org/10.1186/s12915-018-0524-5.

Orädd, F., & Andersson, M. (2021). Tracking membrane protein dynamics in real time. *Journal of Membrane Biology, 254*(1), 51–64. https://doi.org/10.1007/s00232-020-00165-8.

Orädd, F., Ravishankar, H., Goodman, J., Rogne, P., Backman, L., Duelli, A., Nors Pedersen, M., Levantino, M., Wulff, M., Wolf-Watz, M., & Andersson, M. (2021). Tracking the ATP-binding response in adenylate kinase in real time. *Science Advances, 7*(47), eabi5514. https://doi.org/10.1126/sciadv.abi5514.

Orädd, F., Steffen, J. H., Gourdon, P., & Andersson, M. (2022). Copper binding leads to increased dynamics in the regulatory N-terminal domain of full-length human copper transporter ATP7B. *PLoS Computational Biology, 18*(9), e1010074. https://doi.org/10.1371/journal.pcbi.1010074.

Orville, A. M. (2020). Recent results in time resolved serial femtosecond crystallography at XFELs. *Current Opinion in Structural Biology, 65*, 193–208. https://doi.org/10.1016/j.sbi.2020.08.011.

Palmgren, M. (2023). P-type ATPases: Many more enigmas left to solve. *Journal of Biological Chemistry, 299*(11), 105352. https://doi.org/10.1016/j.jbc.2023.105352.

Palmgren, M. G., & Nissen, P. (2011). P-type ATPases. *Annual Review of Biophysics, 40*, 243–266. https://doi.org/10.1146/annurev.biophys.093008.131331.

Pande, K., Hutchison, C. D., Groenhof, G., Aquila, A., Robinson, J. S., Tenboer, J., Basu, S., Boutet, S., DePonte, D. P., Liang, M., White, T. A., Zatsepin, N. A., Yefanov, O., Morozov, D., Oberthuer, D., Gati, C., Subramanian, G., James, D., Zhao, Y., … Schmidt, M. (2016). Femtosecond structural dynamics drives the trans/cis isomerization in photoactive yellow protein. *Science, 352*(6286), 725–729. https://doi.org/10.1126/science.aad5081.

Pandey, S., Bean, R., Sato, T., Poudyal, I., Bielecki, J., Villarreal, J. C., Yefanov, O., Mariani, V., White, T. A., Kupitz, C., Hunter, M., Abdellatif, M. H., Bajt, S., Bondar, V., Echelmeier, A., Doppler, D., Emons, M., Frank, M., Fromme, R., … Schmidt, M. (2020). Time-resolved serial femtosecond crystallography at the European XFEL. *Nature Methods, 17*(1), 73–78. https://doi.org/10.1038/s41592-019-0628-z.

Pao, S. S., Paulsen, I. T., & Saier, M. H., Jr. (1998). Major facilitator superfamily. *Microbiology and Molecular Biology Reviews, 62*(1), 1–34. https://doi.org/10.1128/MMBR.62.1.1-34.1998.

Paulsen, P. A., Custodio, T. F., & Pedersen, B. P. (2019). Crystal structure of the plant symporter STP10 illuminates sugar uptake mechanism in monosaccharide transporter superfamily. *Nature Communications, 10*(1), 407. https://doi.org/10.1038/s41467-018-08176-9.

Picard, M., Jensen, A. M., Sorensen, T. L., Champeil, P., Moller, J. V., & Nissen, P. (2007). Ca2+ versus Mg2+

coordination at the nucleotide-binding site of the sarcoplasmic reticulum Ca2+-ATPase. *Journal of Molecular Biology, 368*(1), 1–7. https://doi.org/10.1016/j.jmb.2007.01.082.

Piper, S. J., Johnson, R. M., Wootten, D., & Sexton, P. M. (2022). Membranes under the magnetic lens: A dive into the diverse world of membrane protein structures using Cryo-EM. *Chemical Reviews, 122*(17), 13989–14017. https://doi.org/10.1021/acs.chemrev.1c00837.

Pizzo, P., Lissandron, V., Capitanio, P., & Pozzan, T. (2011). Ca(2+) signalling in the Golgi apparatus. *Cell Calcium, 50*(2), 184–192. https://doi.org/10.1016/j.ceca.2011.01.006.

Pollack, L., Tate, M. W., Finnefrock, A. C., Kalidas, C., Trotter, S., Darnton, N. C., Lurio, L., Austin, R. H., Batt, C. A., Gruner, S. M., & Mochrie, S. G. (2001). Time resolved collapse of a folding protein observed with small angle X-ray scattering. *Physical Review Letters, 86*(21), 4962–4965. https://doi.org/10.1103/PhysRevLett.86.4962.

Poulsen, H., Khandelia, H., Morth, J. P., Bublitz, M., Mouritsen, O. G., Egebjerg, J., & Nissen, P. (2010). Neurological disease mutations compromise a C-terminal ion pathway in the Na(+)/K(+)-ATPase. *Nature, 467*(7311), 99–102. https://doi.org/10.1038/nature09309.

Prabudiansyah, I., Orädd, F., Magkakis, K., Pounot, K., Levantino, M., & Andersson, M. (2024). Dephosphorylation and ion binding in prokaryotic calcium transport. *Science Advances, 10*(41), eadp2916. https://doi.org/10.1126/sciadv.adp2916.

Quistgaard, E. M., Low, C., Moberg, P., Tresaugues, L., & Nordlund, P. (2013). Structural basis for substrate transport in the GLUT-homology family of monosaccharide transporters. *Nature Structural & Molecular Biology, 20*(6), 766–768. https://doi.org/10.1038/nsmb.2569.

Qureshi, A. A., Suades, A., Matsuoka, R., Brock, J., McComas, S. E., Nji, E., Orellana, L., Claesson, M., Delemotte, L., & Drew, D. (2020). The molecular basis for sugar import in malaria parasites. *Nature, 578*(7794), 321–325. https://doi.org/10.1038/s41586-020-1963-z.

Ravishankar, H., Barth, A., & Andersson, M. (2018). Probing the activity of a recombinant Zn(2+)-transporting P-type ATPase. *Biopolymers, 109*(2). https://doi.org/10.1002/bip.23087.

Ravishankar, H., Pedersen, M. N., Eklund, M., Sitsel, A., Li, C., Duelli, A., Levantino, M., Wulff, M., Barth, A., Olesen, C., Nissen, P., & Andersson, M. (2020). Tracking Ca(2+) ATPase intermediates in real time by X-ray solution scattering. *Science Advances, 6*(12), eaaz0981. https://doi.org/10.1126/sciadv.aaz0981.

Reddy, V. S., Shlykov, M. A., Castillo, R., Sun, E. I., & Saier, M. H., Jr. (2012). The major facilitator superfamily (MFS) revisited. *FEBS Journal, 279*(11), 2022–2035. https://doi.org/10.1111/j.1742-4658.2012.08588.x.

Rimmerman, D., Leshchev, D., Hsu, D. J., Hong, J., Abraham, B., Kosheleva, I., Henning, R., & Chen, L. X. (2018). Insulin hexamer dissociation dynamics revealed by photoinduced T-jumps and time-resolved X-ray solution scattering. *Photochemical & Photobiological Sciences, 17*(7), 874–882. https://doi.org/10.1039/c8pp00034d.

Rimmerman, D., Leshchev, D., Hsu, D. J., Hong, J., Kosheleva, I., & Chen, L. X. (2017). Direct observation of insulin association dynamics with time-resolved X-ray scattering. *The Journal of Physical Chemistry Letters, 8*(18), 4413–4418. https://doi.org/10.1021/acs.jpclett.7b01720.

Rios-Santacruz, R., Poddar, H., Pounot, K., Heyes, D. J., Coquelle, N., Mackintosh, M. J., … Schiro, G. (2026). Integrated structural dynamics uncover a new B(12) photoreceptor activation mode. *Nature, 650*(8103), 1045–1052. doi:10.1038/s41586-025-10074-2.

Royant, A., Edman, K., Ursby, T., Pebay-Peyroula, E., Landau, E. M., & Neutze, R. (2000). Helix deformation is coupled to vectorial proton transport in the photocycle of bacteriorhodopsin. *Nature, 406*(6796), 645–648. https://doi.org/10.1038/35020599.

Rozenberg, A., Inoue, K., Kandori, H., & Beja, O. (2021). Microbial rhodopsins: The last two decades. *Annual Review of Microbiology, 75*, 427–447. https://doi.org/10.1146/annurev-micro-031721-020452.

Sabzian-Molaei, F., Oradd, F., Magkakis, K., & Andersson, M. (2026). A light-triggered Time-Resolved X-ray Solution Scattering (TR-XSS) workflow with application to protein conformational dynamics. *FEBS Open Bio*. doi:10.1002/2211-5463.70200.

Sai, K. V., & Lee, J. E. (2024). Crossing the membrane-what does it take to flip a phospholipid? Structural and biochemical advances on P4-ATPase flippases. *Journal of Biological Chemistry, 300*(10), 107738. https://doi.org/10.1016/j.jbc.2024.107738.

Salustros, N., Gronberg, C., Abeyrathna, N. S., Lyu, P., Oradd, F., Wang, K., Andersson, M., Meloni, G., & Gourdon, P. (2022). Structural basis of ion uptake in copper-transporting P(1B)-type ATPases. *Nature Communications, 13*(1), 5121. https://doi.org/10.1038/s41467-022-32751-w.

Sass, H. J., Buldt, G., Gessenich, R., Hehn, D., Neff, D., Schlesinger, R., Berendzen, J., & Ormos, P. (2000). Structural alterations for proton translocation in the M state of wild-type bacteriorhodopsin. *Nature, 406*(6796), 649–653. https://doi.org/10.1038/35020607.

Schulz, E. C., Yorke, B. A., Pearson, A. R., & Mehrabi, P. (2022). Best practices for time-resolved serial synchrotron crystallography. *Acta Crystallographica Section D: Structural Biology, 78*(Pt 1), 14–29. https://doi.org/10.1107/S2059798321011621.

Shimada, A., Kubo, M., Baba, S., Yamashita, K., Hirata, K., Ueno, G., Nomura, T., Kimura, T., Shinzawa-Itoh, K., Baba, J., Hatano, K., Eto, Y., Miyamoto, A., Murakami, H., Kumasaka, T., Owada, S., Tono, K., Yabashi, M., Yamaguchi, Y., … Tsukihara, T. (2017). A nanosecond time-resolved XFEL analysis of structural changes associated with CO release from cytochrome c oxidase. *Science Advances, 3*(7), e1603042. https://doi.org/10.1126/sciadv.1603042.

Shinoda, T., Ogawa, H., Cornelius, F., & Toyoshima, C. (2009). Crystal structure of the sodium-potassium pump at 2.4 A resolution. *Nature, 459*(7245), 446–450. https://doi.org/10.1038/nature07939.

Sim, S. I., & Park, E. (2023). P5-ATPases: Structure, substrate specificities, and transport mechanisms. *Current Opinion in Structural Biology, 79*, 102531. https://doi.org/10.1016/j.sbi.2023.102531.

Sim, S. I., von Bulow, S., Hummer, G., & Park, E. (2021). Structural basis of polyamine transport by human ATP13A2 (PARK9). *Molecular Cell, 81*(22), 4635–4649. https://doi.org/10.1016/j.molcel.2021.08.017. e4638.

Sitsel, A., De Raeymaecker, J., Drachmann, N. D., Derua, R., Smaardijk, S., Andersen, J. L., Vandecaetsbeek, I., Chen, J., De Maeyer, M., Waelkens, E., Olesen, C., Vangheluwe, P., & Nissen, P. (2019). Structures of the heart specific SERCA2a Ca(2+)-ATPase. *The EMBO Journal, 38*(5). https://doi.org/10.15252/embj.2018100020.

Sitsel, O., Gronberg, C., Autzen, H. E., Wang, K., Meloni, G., Nissen, P., & Gourdon, P. (2015). Structure and function of Cu(I)- and Zn(II)-ATPases. *Biochemistry, 54*(37), 5673–5683. https://doi.org/10.1021/acs.biochem.5b00512.

Skopintsev, P., Ehrenberg, D., Weinert, T., James, D., Kar, R. K., Johnson, P. J. M., Ozerov, D., Furrer, A., Martiel, I., Dworkowski, F., Nass, K., Knopp, G., Cirelli, C., Arrell, C., Gashi, D., Mous, S., Wranik, M., Gruhl, T., Kekilli, D., … Standfuss, J. (2020). Femtosecond-to-millisecond structural changes in a light-driven sodium pump. *Nature, 583*(7815), 314–318. https://doi.org/10.1038/s41586-020-2307-8.

Smirnova, I., Kasho, V., Jiang, X., Pardon, E., Steyaert, J., & Kaback, H. R. (2014). Outward-facing conformers of LacY stabilized by nanobodies. *Proceedings of the National Academy of Sciences of the United States of America, 111*(52), 18548–18553. https://doi.org/10.1073/pnas.1422265112.

Smirnova, I., Kasho, V., Jiang, X., Pardon, E., Steyaert, J., & Kaback, H. R. (2015). Transient conformers of LacY are trapped by nanobodies. *Proceedings of the National Academy of Sciences of the United States of America, 112*(45), 13839–13844. https://doi.org/10.1073/pnas.1519485112.

Smirnova, I., Kasho, V., Sugihara, J., & Kaback, H. R. (2013). Trp replacements for tightly interacting Gly-Gly pairs in LacY stabilize an outward-facing conformation. *Proceedings of the National Academy of Sciences of the United States of America, 110*(22), 8876–8881. https://doi.org/10.1073/pnas.1306849110.

Smith, A. T., Smith, K. P., & Rosenzweig, A. C. (2014). Diversity of the metal-transporting P1B-type ATPases. *Journal of Biological Inorganic Chemistry, 19*(6), 947–960. https://doi.org/10.1007/s00775-014-1129-2.

Sörensen, T. L., Möller, J. V., & Nissen, P. (2004). Phosphoryl transfer and calcium ion occlusion in the calcium pump. *Science, 304*(5677), 1672–1675. https://doi.org/10.1126/science.1099366.

Stock, C., Heger, T., Basse Hansen, S., Thirup Larsen, S., Habeck, M., Dieudonne, T., Driller, R., & Nissen, P. (2023). Fast-forward on P-type ATPases: Recent advances on structure and function. *Biochemical Society Transactions, 51*(3), 1347–1360. https://doi.org/10.1042/BST20221543.

Subramaniam, S., Gerstein, M., Oesterhelt, D., & Henderson, R. (1993). Electron diffraction analysis of structural changes in the photocycle of bacteriorhodopsin. *The EMBO Journal, 12*(1), 1–8. https://doi.org/10.1002/j.1460-2075.1993.tb05625.x.

Subramaniam, S., & Henderson, R. (2000). Molecular mechanism of vectorial proton translocation by bacteriorhodopsin. *Nature, 406*(6796), 653–657. https://doi.org/10.1038/35020614.

Sun, L., Zeng, X., Yan, C., Sun, X., Gong, X., Rao, Y., & Yan, N. (2012). Crystal structure of a bacterial homologue of glucose transporters GLUT1-4. *Nature, 490*(7420), 361–366. https://doi.org/10.1038/nature11524.

Takala, H., Bjorling, A., Berntsson, O., Lehtivuori, H., Niebling, S., Hoernke, M., Kosheleva, I., Henning, R., Menzel, A., Ihalainen, J. A., & Westenhoff, S. (2014). Signal amplification and transduction in phytochrome photosensors. *Nature, 509*(7499), 245–248. https://doi.org/10.1038/nature13310.

Takala, H., Niebling, S., Berntsson, O., Bjorling, A., Lehtivuori, H., Hakkanen, H., Panman, M., Gustavsson, E., Hoernke, M., Newby, G., Zontone, F., Wulff, M., Menzel, A., Ihalainen, J. A., & Westenhoff, S. (2016). Light-induced structural changes in a monomeric bacteriophytochrome. *Structural Dynamics, 3*(5), 054701. https://doi.org/10.1063/1.4961911.

Tenboer, J., Basu, S., Zatsepin, N., Pande, K., Milathianaki, D., Frank, M., Hunter, M., Boutet, S., Williams, G. J., Koglin, J. E., Oberthuer, D., Heymann, M., Kupitz, C., Conrad, C., Coe, J., Roy-Chowdhury, S., Weierstall, U., James, D., Wang, D., … Schmidt, M. (2014). Time-resolved serial crystallography captures high-resolution intermediates of photoactive yellow protein. *Science, 346*(6214), 1242–1246. https://doi.org/10.1126/science.1259357.

Thompson, M. C., Barad, B. A., Wolff, A. M., Sun Cho, H., Schotte, F., Schwarz, D. M. C., Anfinrud, P., & Fraser, J. S. (2019). Temperature-jump solution X-ray scattering reveals distinct motions in a dynamic enzyme. *Nature Chemistry, 11*(11), 1058–1066. https://doi.org/10.1038/s41557-019-0329-3.

Timcenko, M., Dieudonne, T., Montigny, C., Boesen, T., Lyons, J. A., Lenoir, G., & Nissen, P. (2021). Structural basis of substrate-independent phosphorylation in a P4-ATPase lipid flippase. *Journal of Molecular Biology, 433*(16), 167062. https://doi.org/10.1016/j.jmb.2021.167062.

Tomita, A., Daiho, T., Kusakizako, T., Yamashita, K., Ogasawara, S., Murata, T., Nishizawa, T., & Nureki, O. (2021). Cryo-EM reveals mechanistic insights into lipid-facilitated polyamine export by human ATP13A2. *Molecular Cell, 81*(23), 4799–4809. https://doi.org/10.1016/j.molcel.2021.11.001. e4795.

Toyoshima, C., & Mizutani, T. (2004). Crystal structure of the calcium pump with a bound ATP analogue. *Nature, 430*(6999), 529–535. https://doi.org/10.1038/nature02680.

Toyoshima, C., Nakasako, M., Nomura, H., & Ogawa, H. (2000). Crystal structure of the calcium pump of sarcoplasmic reticulum at 2.6 A resolution. *Nature, 405*(6787), 647–655. https://doi.org/10.1038/35015017.

Uchanski, T., Masiulis, S., Fischer, B., Kalichuk, V., Lopez-Sanchez, U., Zarkadas, E., Weckener, M., Sente, A., Ward, P., Wohlkonig, A., Zogg, T., Remaut, H., Naismith, J. H., Nury, H., Vranken, W., Aricescu, A. R., Pardon, E., & Steyaert, J. (2021). Megabodies expand the nanobody toolkit for protein structure determination by single-particle cryo-EM. *Nature Methods, 18*(1), 60–68. https://doi.org/10.1038/s41592-020-01001-6.

Ulmschneider, J. P., Andersson, M., & Ulmschneider, M. B. (2011). Determining peptide partitioning properties via computer simulation. *Journal of Membrane Biology, 239*(1–2), 15–26. https://doi.org/10.1007/s00232-010-9324-8.

Unwin, N. (1995). Acetylcholine receptor channel imaged in the open state. *Nature, 373*(6509), 37–43. https://doi.org/10.1038/373037a0.

Unwin, N., & Fujiyoshi, Y. (2012). Gating movement of acetylcholine receptor caught by plunge-freezing. *Journal of Molecular Biology, 422*(5), 617–634. https://doi.org/10.1016/j.jmb.2012.07.010.

Uzawa, T., Kimura, T., Ishimori, K., Morishima, I., Matsui, T., Ikeda-Saito, M., Takahashi, S., Akiyama, S., & Fujisawa, T. (2006). Time-resolved small-angle X-ray scattering investigation of the folding dynamics of heme oxygenase: Implication of the scaling relationship for the submillisecond intermediates of protein folding. *Journal of Molecular Biology, 357*(3), 997–1008. https://doi.org/10.1016/j.jmb.2005.12.089.

Vincent, J., Andersson, M., Eklund, M., Wohri, A. B., Odelius, M., Malmerberg, E., Kong, Q., Wulff, M., Neutze, R., & Davidsson, J. (2009). Solvent dependent structural perturbations of chemical reaction intermediates visualized by time-resolved X-ray diffraction. *The Journal of Chemical Physics, 130*(15), 154502. https://doi.org/10.1063/1.3111401.

Voss, J. M., Harder, O. F., Olshin, P. K., Drabbels, M., & Lorenz, U. J. (2021). Rapid melting and revitrification as an approach to microsecond time-resolved cryo-electron microscopy. *Chemical Physics Letters, 778*, 138812. https://doi.org/10.1016/j.cplett.2021.138812.

Wang, K., Sitsel, O., Meloni, G., Autzen, H. E., Andersson, M., Klymchuk, T., Nielsen, A. M., Rees, D. C., Nissen, P., & Gourdon, P. (2014). Structure and mechanism of Zn2+-transporting P-type ATPases. *Nature, 514*(7523), 518–522. https://doi.org/10.1038/nature13618.

Weinert, T., Skopintsev, P., James, D., Dworkowski, F., Panepucci, E., Kekilli, D., Furrer, A., Brunle, S., Mous, S., Ozerov, D., Nogly, P., Wang, M., & Standfuss, J. (2019). Proton uptake mechanism in bacteriorhodopsin captured by serial synchrotron crystallography. *Science, 365*(6448), 61–65. https://doi.org/10.1126/science.aaw8634.

Westenhoff, S., Malmerberg, E., Arnlund, D., Johansson, L., Nazarenko, E., Cammarata, M., Davidsson, J., Chaptal, V., Abramson, J., Katona, G., Menzel, A., & Neutze, R. (2010). Rapid readout detector captures protein time-resolved WAXS. *Nature Methods, 7*(10), 775–776. https://doi.org/10.1038/nmeth1010-775c.

Wickstrand, C., Dods, R., Royant, A., & Neutze, R. (2015). Bacteriorhodopsin: Would the real structural intermediates please stand up? *Biochim Biophys Acta, 1850*(3), 536–553. https://doi.org/10.1016/j.bbagen.2014.05.021.

Wickstrand, C., Nogly, P., Nango, E., Iwata, S., Standfuss, J., & Neutze, R. (2019). Bacteriorhodopsin: Structural insights revealed using X-ray lasers and synchrotron radiation. *Annual Review of Biochemistry, 88*, 59–83. https://doi.org/10.1146/annurev-biochem-013118-111327.

Winkelmann, I., Uzdavinys, P., Kenney, I. M., Brock, J., Meier, P. F., Wagner, L. M., Gabriel, F., Jung, S., Matsuoka, R., von Ballmoos, C., Beckstein, O., & Drew, D. (2022). Crystal structure of the Na(+)/H(+) antiporter NhaA at active pH reveals the mechanistic basis for pH sensing. *Nature Communications, 13*(1), 6383. https://doi.org/10.1038/s41467-022-34120-z.

Wisedchaisri, G., Park, M. S., Iadanza, M. G., Zheng, H., & Gonen, T. (2014). Proton-coupled sugar transport in the prototypical major facilitator superfamily protein XylE. *Nature Communications, 5*, 4521. https://doi.org/10.1038/ncomms5521.

Wohri, A. B., Katona, G., Johansson, L. C., Fritz, E., Malmerberg, E., Andersson, M., Vincent, J., Eklund, M., Cammarata, M., Wulff, M., Davidsson, J., Groenhof, G., & Neutze, R. (2010). Light-induced structural changes in a photosynthetic reaction center caught by Laue diffraction. *Science, 328*(5978), 630–633. https://doi.org/10.1126/science.1186159.

Wu, M., Wu, C., Song, T., Pan, K., Wang, Y., & Liu, Z. (2023). Structure and transport mechanism of the human calcium pump SPCA1. *Cell Research, 33*(7), 533–545. https://doi.org/10.1038/s41422-023-00827-x.

Xu, J., He, Y., Wu, X., & Li, L. (2022). Conformational changes of a phosphatidylcholine flippase in lipid membranes. *Cell Reports, 38*(11), 110518. https://doi.org/10.1016/j.celrep.2022.110518.

Yang, G. M., Xu, L., Wang, R. M., Tao, X., Zheng, Z. W., Chang, S., Ma, D., Zhao, C., Dong, Y., Wu, S., Guo, J., & Wu, Z. Y. (2023). Structures of the human Wilson disease copper transporter ATP7B. *Cell Reports, 42*(5), 112417. https://doi.org/10.1016/j.celrep.2023.112417.

Ygberg, S., Akkuratov, E. E., Howard, R. J., Taylan, F., Jans, D. C., Mahato, D. R., Katz, A., Kinoshita, P. F., Portal, B., Nennesmo, I., Lindskog, M., Karlish, S. J. D., Andersson, M., Lindstrand, A., Brismar, H., & Aperia, A. (2021). A missense mutation converts the Na(+),K(+)-ATPase into an ion channel and causes therapy-resistant epilepsy. *Journal of Biological Chemistry, 297*(6), 101355. https://doi.org/10.1016/j.jbc.2021.101355.

Yun, J. H., Li, X., Yue, J., Park, J. H., Jin, Z., Li, C., Hu, H., Shi, Y., Pandey, S., Carbajo, S., Boutet, S., Hunter, M. S., Liang, M., Sierra, R. G., Lane, T. J., Zhou, L., Weierstall, U., Zatsepin, N. A., Ohki, M., … Liu, H. (2021). Early-stage dynamics of chloride ion-pumping rhodopsin revealed by a femtosecond X-ray laser. *Proceedings of the National Academy of Sciences of the United States of America, 118*(13). https://doi.org/10.1073/pnas.2020486118.

Zambrano, A., Molt, M., Uribe, E., & Salas, M. (2019). Glut 1 in cancer cells and the inhibitory action of resveratrol as a potential therapeutic strategy. *International Journal of Molecular Sciences, 20*(13). https://doi.org/10.3390/ijms20133374.

Zhang, Y., & Inaba, K. (2022). Structural basis of the conformational and functional regulation of human SERCA2b, the ubiquitous endoplasmic reticulum calcium pump. *Bioessays, 44*(7), e2200052. https://doi.org/10.1002/bies.202200052.

Zhang, Y., Inoue, M., Tsutsumi, A., Watanabe, S., Nishizawa, T., Nagata, K., Kikkawa, M., & Inaba, K. (2020). Cryo-EM structures of SERCA2b reveal the mechanism of regulation by the luminal extension tail. *Science Advances, 6*(33), eabb0147. https://doi.org/10.1126/sciadv.abb0147.

Zhang, Y., Kobayashi, C., Cai, X., Watanabe, S., Tsutsumi, A., Kikkawa, M., Sugita, Y., & Inaba, K. (2022). Multiple sub-state structures of SERCA2b reveal conformational overlap at transition steps during the catalytic cycle. *Cell Reports, 41*(10), 111760. https://doi.org/10.1016/j.celrep.2022.111760.

Zhang, Y., Watanabe, S., Tsutsumi, A., Kadokura, H., Kikkawa, M., & Inaba, K. (2021). Cryo-EM analysis provides new mechanistic insight into ATP binding to Ca(2+)-ATPase SERCA2b. *The EMBO Journal, 40*(19), e108482. https://doi.org/10.15252/embj.2021108482.

Zielinski, K. A., Katz, A. M., Calvey, G. D., Pabit, S. A., Milano, S. K., Aplin, C., San Emeterio, J., Cerione, R. A., & Pollack, L. (2023). Chaotic advection mixer for capturing transient states of diverse biological macromolecular systems with time-resolved small-angle X-ray scattering. *IUCrJ, 10*(Pt 3), 363–375. https://doi.org/10.1107/S2052252523003482.

CHAPTER

14 Dynamics of light-driven inward proton-pumping rhodopsins

Yoshitaka Kato[a] and Keiichi Inoue[a,b,*]

[a]*The Institute for Solid State Physics, The University of Tokyo, Kashiwa, Japan,* [b]*Graduate School of Frontier Sciences, The University of Tokyo, Kashiwa, Japan*

**Corresponding author: inoue@issp.u-tokyo.ac.jp*

14.1 Inward proton-pumping rhodopsins: A new class of ion-transporting rhodopsins

Rhodopsins are a superfamily of photoreceptive membrane proteins that utilize a retinal chromophore. They are classified into two distinct groups: animal and microbial rhodopsins. The formers function as light-dependent G protein-coupled receptors (GPCRs), mediating G protein signaling or acting as photoisomerases to convert all-*trans*- to 11-*cis*-retinal using light energy in animals (Koyanagi et al., 2014; Shichida et al., 2009). In contrast, microbial rhodopsins are found across a wide range of microorganisms, including bacteria, archaea, algae, fungi, protists, and even giant viruses (Nagata et al., 2021; Rozenberg et al., 2021). These proteins possess a seven transmembrane helical architecture (TM1–TM7, also referred to as helices A–G), in which an all-*trans*-retinal chromophore binds to a conserved lysine residue via a Schiff-base linkage. The functions of microbial rhodopsins are highly diverse: They include light-driven pumps, light-gated ion channels, photosensors, and light-regulated enzymes (Fig. 14.1).

The most abundant microbial rhodopsins in nature are outward proton-pumping rhodopsins, including the best characterized bacteriorhodopsin found in *Halobacterium salinarum* (*Hs*BR) and proteorhodopsins (PRs) from diverse bacteria inhabiting aquatic environments (Béjà et al., 2000; Ernst et al., 2014; Lanyi, 2004). Outward proton-pumping rhodopsins convert light energy to a proton motive force (pmf) that drives adenosine triphosphate (ATP) synthesis and proton-coupled substance transports. Recently a novel class of rhodopsins, inward proton-pumping rhodopsins, which unidirectionally transport H^+ from the outside to the inside of cells, has been discovered. These rhodopsins intriguingly eliminate the pmf via light-driven active proton transport. The inward proton transport by microbial rhodopsins was first observed in a mutant of *Anabaena* sensory rhodopsin (ASR), a nonion-transporting rhodopsin from the cyanobacterium *Anabaena* (*Nostoc*) sp. PCC7120. Subsequently, most members of xenorhodopsins (XeRs), a subfamily of microbial rhodopsins that includes ASR, have been revealed to function as inward proton-pumping rhodopsins. Furthermore, schizorhodopsins (SzRs), a newly identified subfamily of microbial rhodopsins phylogenetically distinct from XeRs, have been discovered as the second class of inward proton-pumping rhodopsins. There are considerable differences in the primary sequences between XeRs and SzRs (sequence identity and homology are ~15% and ~40%, respectively), suggesting their proton-transport mechanisms differ.

Protein Dynamics: Experimental and Computational Approaches. DOI: https://doi.org/10.1016/B978-0-443-30042-4.00001-8

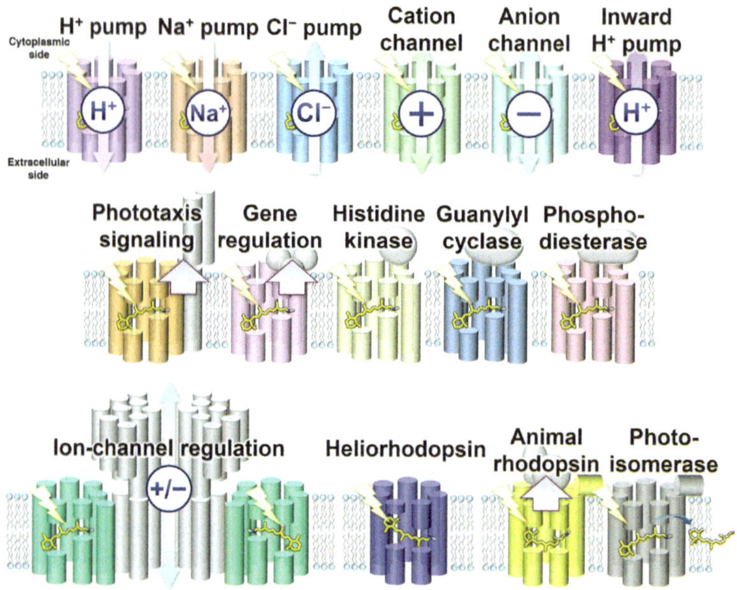

FIGURE 14.1

Diversity of the functions of rhodopsins. Microbial rhodopsins utilize an all-*trans*-retinal chromophore to exhibit a wide variety of functions, whereas animal rhodopsins mainly contain an 11-*cis*-ratinal and function as light-dependent G protein-coupled receptors or photoisomerases.

Although the mechanisms underlying inward proton-pumping rhodopsins are less well understood compared to outward proton-pumping rhodopsins, a mechanistic comparison between outward and inward proton-pumping rhodopsins is expected to provide valuable insights into the key determinants of ion transport directionality in membrane proteins. Recent studies have revealed essential structural and dynamic features of inward proton-pumping rhodopsins, whereas new applications for these proteins have also been proposed. In this chapter, we review more than 15 years of research on inward proton-pumping rhodopsins and discuss the unresolved questions that remain in this field.

14.2 Two families of light-driven inward proton-pumping rhodopsins: Xenorhodopsins and schizorhodopsins

The first member of XeR, ASR, was found in 2003 in the genome of a filamentous cyanobacteria, *Anabaena* (*Nostoc*) sp. PCC7120 (Jung et al., 2003), as the first sensory rhodopsin from the bacterial domain. ASR has a proline residue (P206) in the seventh transmembrane helix (TM7) at the position of D212 in *Hs*BR, which is highly conserved in most microbial rhodopsins as the counterion stabilizing the protonated retinal Schiff base (RSB) (Fig. 14.2) (Inoue et al., 2013; Jung et al., 2003). The ASR gene forms an operon with a 14-kDa protein. Although ASR does not show significant ion transport activity (Jung et al., 2003; Kawanabe et al., 2009), it interacts with the 14-kDa protein. This interaction

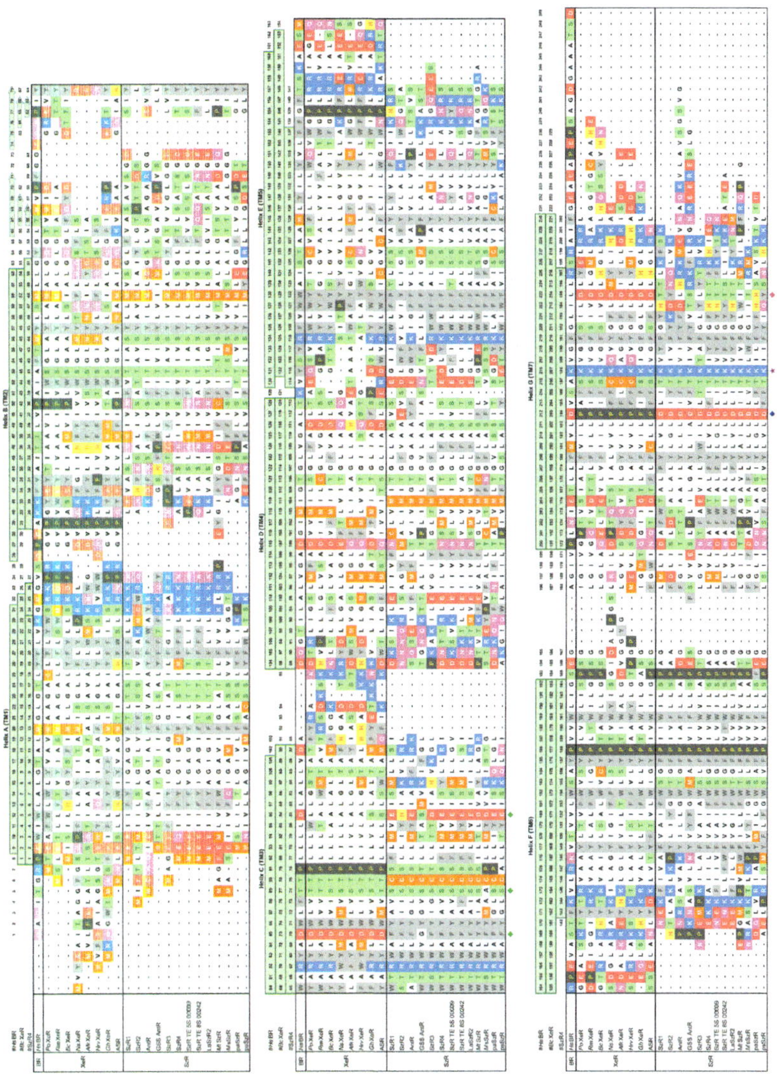

FIGURE 14.2

Amino acid sequence alignment of XeRs and SzRs. Amino acid sequence alignment of XeRs and SzRs with *Hs*BR. The positions of TMs and *β*-sheets, based on the X-ray crystallographic structures [PDB ID, *Hs*BR: 7Z09 (Borshchevskiy et al., 2022); *Bc*XeR: 7ZMY (Kovalev et al., 2023); SzR4: 7E4G (Higuchi et al., 2021)], were indicated by green rectangles and arrows, respectively. The positions of TM3 motif residues corresponding to *Hs*BR D85 (counterion), T89, and D96, as well as the typical TM7 counterion positions corresponding to *Hs*BR D212, the position of ASR D217, and the retinal-binding lysine in TM7, are indicated with *green diamonds*, a *blue diamond*, a *magenta diamond*, and a *purple star*, respectively. The amino acids are colored according to their chemical properties: hydrophobic aliphatic residues (*white*); hydroxyl group-bearing aliphatic residues (*green*); acidic residues (*red*); basic residues (*blue*); Asn and Gln (*pink*); sulfur-containing residues (*orange*); proline (*black*); and aromatic residues (*gray*). *ASR, Anabaena* sensory rhodopsin; *SzRs,* schizorhodopsins; *XeRs,* xenorhodopsins.

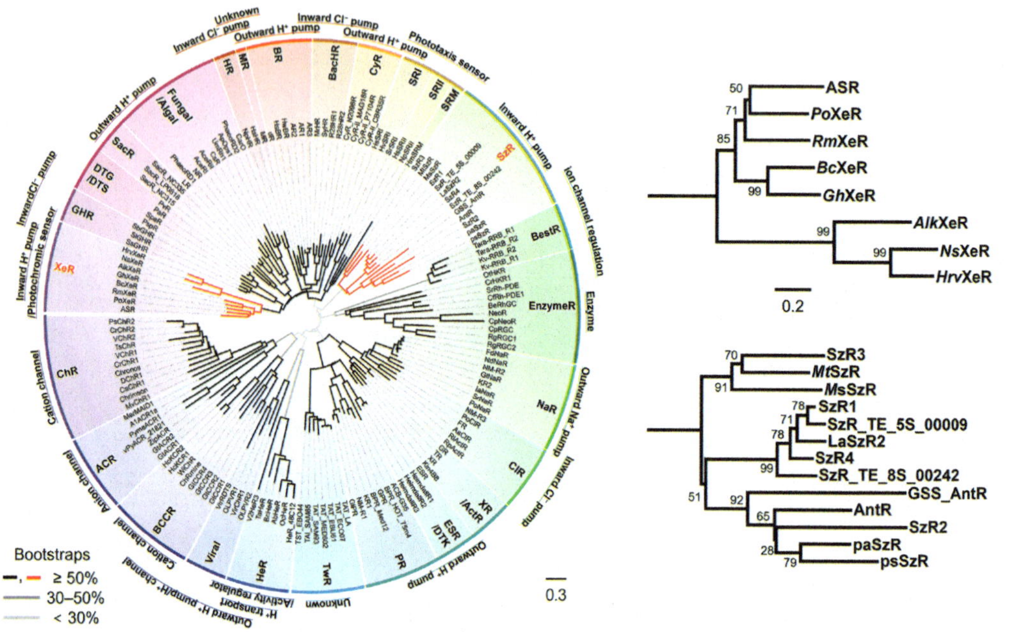

FIGURE 14.3

The phylogenetic tree of microbial rhodopsins and inward proton-pumping rhodopsins. The phylogenetic tree of microbial rhodopsins (left) and the branches of XRs and SzRs (right). The evolutionary history was inferred using the neighbor-joining method (Saitou et al., 1987). The tree is drawn to scale, with branch lengths in the same units as those of the evolutionary distances used to infer the phylogenetic tree. The evolutionary distances were computed using the Poisson correction method (Zuckerkandl et al., 1965) and are in the units of the number of amino acid substitutions per site. All ambiguous positions were removed for each sequence pair. Evolutionary analyses were conducted in MEGA X (Tamura et al., 2011). *SzRs*, schizorhodopsins; *XeRs*, xenorhodopsins.

was indicated by an accelerated photocycle (Jung et al., 2003), leading to the designation of the 14-kDa protein as ASR transducer (ASR_T). Because cyanobacteria exhibit chromatic adaptation in which the expression of phycobiliproteins, for example, blue-absorbing phycoerythrins and red-absorbing phycocyanins, is regulated by different colors of light (Kehoe et al., 1994; Kehoe, 2010), ASR was suggested to be a photosensor regulating the synthesis of biliproteins in a light-dependent manner (Jung et al., 2003). Interestingly, if D217 in the cytoplasmic side of TM7 is mutated to a glutamate residue (ASR D217E), inward proton-transporting activity is observed in *Escherichia coli* cells (Kawanabe et al. 2009, 2011). This is the first example of the artificial functional conversion from sensory rhodopsin to ion-transporting rhodopsin by amino acid substitution.

In 2011 several microbial rhodopsins closely related to ASR were identified through phylogenetic analysis (Ugalde et al., 2011). These rhodopsins conserve a proline residue identical to ASR P206, whereas they do not form operons with proteins homologous to ASR_T. These ASR-related rhodopsins were classified as XeRs (Fig. 14.3) (Ugalde et al., 2011). The absence of putative transducer genes for most XeR suggests their functions differ from that of ASR. In 2016 a member of the XeR family, *Po*XeR, from the deep-ocean α-proteobacterium *Parvularcula oceani*, was identified as the first natural light-driven proton-pumping rhodopsin, which actively transports H^+ from the outside to the inside of cells (Inoue et al., 2016). Subsequently, other XeRs, such as *Ns*XeR from *Nanosalina* sp., *Hrv*XeR, and *Alk*XeR from a crystallizer pond of a solar saltern in Spain and alkaline brines in Siberia, respectively (Shevchenko et al., 2017); *Rm*XeR from *Rubricoccus marinus* SG-29^T (Inoue et al., 2018b), *Bc*XeR from *Bacillus coahuilensis* (Alcaraz et al., 2008; Kovalev et al., 2023), and *Gh*XeR from *Guptibacillus hwajinpoensis* (Hattori et al., 2026) were also clarified to exhibit inward proton-pumping activity. These findings suggest that the inward proton pump is the primary function among XeRs lacking putative transducer genes. A recent bioinformatics study identified more than 70 XeR-like genes in nature, which have been classified into three distinct groups: bacterial XeRs, including *Po*XeR, *Rm*XeR, and *Bc*XeR; archaeal XeRs, including *Ns*XeR, *Hrv*XeR, and *Alk*XeR; and ASR-like sensors, including ASR (Kovalev et al., 2023).

In 2018 a large class of rhodopsin protein, heliorhodopsin (HeR), was discovered using functional metagenomics by observing pigment formation in *E. coli* cells carrying fosmids from a freshwater habitat in the presence of all-*trans*-retinal (Pushkarev et al., 2018). Although HeRs share $<15\%$ sequential identity with canonical microbial rhodopsins and are only distantly related, a new group of rhodopsins, SzRs (Fig. 14.3), was identified in the metagenome-assembled genomes (MAGs) of Asgard archaea (kingdom Promethearchaeati), phylogenetically positioned between microbial rhodopsins and HeRs (Bulzu et al., 2019). Interestingly the sequence similarity between SzRs, canonical microbial rhodopsins, and HeRs varies across their TMs: whereas, in TM3 of SzRs, there are aromatic residues homologous to those highly conserved among HeRs (W105 and Y108 in HeR 48C12, which is the first discovered HeR), the residues in TM6 and TM7 of SzRs are more similar to those in canonical microbial rhodopsins (e.g., *Hs*BR W182, Y185, P186, and D212 are conserved in most SzRs) (Fig. 14.2) (Inoue et al., 2020). These findings have made the classification of SzRs—whether they are part of canonical microbial rhodopsins or HeRs—ambiguous. Immunofluorescence imaging targeting artificial tags introduced at the N- and C-termini of SzRs clarified that their N- and C-termini are oriented toward the outside and inside of cells, respectively, which is the same as canonical microbial rhodopsins but differs from HeRs (Inoue et al. 2020). Furthermore, X-ray crystallography and high-speed atomic force microscopy (HS-AFM) showed that SzRs form a trimeric structure similar to *Hs*BR, distinct from the dimeric structure of HeRs (Higuchi et al., 2021; Inoue et al., 2020). Based on these structural similarities, SzRs were classified as canonical microbial rhodopsins. SzRs have also been identified in archaeal species other than Asgard archaea: in MAGs of Bathyarchaeota (Wong et al., 2020), in mesophilic *Methanoculleus taiwanensis* (*Mt*SzR), and in a MAG sampled at a hot spring, likely derived from *Methanoculleus* sp. (*M*sSzR) (Kawasaki et al., 2021). SzRs were even discovered in bacteria, specifically in MAGs of Parcubacteria, members of the candidate radiation phyla (CPR), or Parcubacteria (Wong et al., 2020). In Antarctic environments, two SzRs, paSzR from *Minisyncoccota* and psSzR from *Cyanobacteriota*, have also been identified (Marín, Konno, Rozenberg, Béjà, & Inoue, 2026). In addition, AntR, which is found in the metagenomic sequences sampled in the Antarctic lakes, was shown to have a highly similar amino acid sequence and forms a single clade with SzRs (Harris et al., 2020). As of today, over 90 SzR genes have been identified (Urui et al., 2024a). Phylogenetic analyses suggest that SzRs from archaea and bacteria form distinct groups (Marín, Konno, Rozenberg, Béjà, & Inoue, 2026).

Although the phenylalanine at the first position of the TM3 motif is highly conserved in the SzR family (Fig. 14.2), a new member, GSS AntR, identified in an Antarctic sample (Pinto et al., 2024), possesses a GSS motif. GSS AntR exhibits inward proton-pumping activity in *E. coli* cells, demonstrating that phenylalanine at the first position of the TM3 motif is not essential for the inward proton-pumping function in SzRs. The M formation and decay of GSS AntR are approximately three times faster than those of AntR, suggesting that it more efficiently transports H^+ compared to AntR. In addition, GSS AntR has a serine residue at the third position of the TM3 motif, where an acidic or histidine residue is typically found in other SzRs. As serine is generally not considered capable of accepting an H^+ from the RSB, unlike the proton-accepting acidic residues in most outward and inward proton-pumping rhodopsins, the robust activity of GSS AntR highlights the diverse functional roles of residues at this position. Another notable difference between GSS AntR and AntR is the absence of 13-*cis*-15-*syn* retinal accumulation at acidic pH, a feature observed in AntR but not in GSS AntR (Fig. 14.4).

14.3 Structures of inward proton-transporting rhodopsins

The three-dimensional structures at the atomic level of all five inward proton-transporting rhodopsins that have been resolved to date are shown in Fig. 14.5 and are compared with that of outward H^+ pumping *Hs*BR.

After the ASR wild-type (WT) structure, which was also the first XeR structure, was solved in 2004 (Vogeley et al., 2004), the structure of ASR D217E was determined using X-ray crystallography in 2016 (Fig. 14.5) (Dong et al., 2016). A significant change in the hydrogen-bonding network was observed around the mutated residue E217 (a detailed structural comparison was shown in the original work by Dr. Hartmut Luecke and his coworkers). In addition, the orientation of the main-chain carbonyl group of the retinal binding K210 is altered, forming a new hydrogen bond with a nearby cytoplasmic water molecule. This leads to the formation of a hydrogen-bonding network spanning >10 Å through K210, S214, E217, and two water molecules. In ASR D217E, E36, located at the cytoplasmic edge of TM2, is suggested to be protonated at pH 5 and partially at 7. This is based on a double hydrogen-bonding interaction with Q93 in TM3, as revealed by X-ray crystallography, and the C=O stretching vibration band of protonated carboxylic acid observed in FTIR spectroscopy, which was assigned to E36, on ASR WT (Kawanabe et al., 2008; Dong et al., 2016). The protonation of E36, located near the cytoplasmic surface, supports the hypothesis that E36 releases an H^+ to the cytoplasm and is subsequently protonated by E217 in ASR D217E (Dong et al., 2016).

As the first structure of natural inward proton-pumping rhodopsin the structure of *Ns*XeR was determined (Shevchenko et al., 2017). One of the most prominent features of XeR is the highly conserved proline in TM7 at the position corresponding to *Hs*BR D212, where the aspartic acid counterion is found in most microbial rhodopsins (Fig. 14.2). This difference results in the absence of a hydrogen bond between α-helical turns on the extracellular side in all XeR structures (Dong et al., 2016; Kovalev et al., 2023; Shevchenko et al., 2017) (Fig. 14.6).

On the cytoplasmic side of the RSB a hydrogen-bonding interaction is formed between H48 in TM2, which is highly conserved only among archaeal XeRs and replaced by alanine in bacterial XeRs and ASR-like sensors (Kovalev et al., 2023), and D220 in TM7, which is identical to D217 in ASR. This His–Asp interaction is reminiscent of the interaction between a histidine residue in TM2 and TM3

(A)

		λ_{max} (nm)	Retinal configuration (%)			pK_a	
			AT	13C	Others	Schiff base	Counter ion
ASR	dark	549	98	2	–	> 12.5	4.0 or much lower
	light	–	25	75	–		
*Po*XeR	dark	567	92	8	–	–	–
	light	560	50	50	–		
*Rm*XeR	dark	556	75	25	–	11.1	2.1
	light	550	45	55	–		
*Ns*XeR	dark	565	dominant	–	–	9.6	–
	light	~565	–	–	–		
*Hrv*XeR	dark	562	–	–	–	–	–
*Alk*XeR	dark	577	–	–	–	–	–
*Bc*XeR	dark	558	–	–	–	–	–
*Gh*XeR	dark	565	–	–	–	–	–

(B)

	λ_{max} (nm)	Retinal configuration (%)			pK_a	
		AT	13C	Others	Schiff base	Counter ion
SzR1	557	94.5	–	5.5	~13.5	< 2.3
SzR2	542	–	–	–	>13.5	< 2.5
SzR3	540	–	–	–	–	–
SzR4	557	–	–	–	12.5	–
LaSzR2	549	96.0	–	–	12.3	3.7
AntR	556	pH dependent		–	–	5.8
GSS AntR	~540	dominant	–	–	–	–
*Mt*SzR	540	95.1	2.9	1.9	11.1 13.5	2.4 4.0
*Ms*SzR	538	92.0	5.9	1.7	11.6 13.4	1.79 2.3
paSzR	550	96.0	2.1	1.9	12.9 13.6	2.1 4.3
psSzR	556	95.6	3.6	0.8	13.1	3.99

FIGURE 14.4

Molecular properties of XeRs and SzRs. (A and B) The absorption maximum wavelengths (λ_{max}), retinal compositions, and pK_a of XeRs (A) and SzRs (B). *13C*, 13-*cis; AT*: all-*trans; SzRs*, schizorhodopsins; *XeRs*, xenorhodopsins.

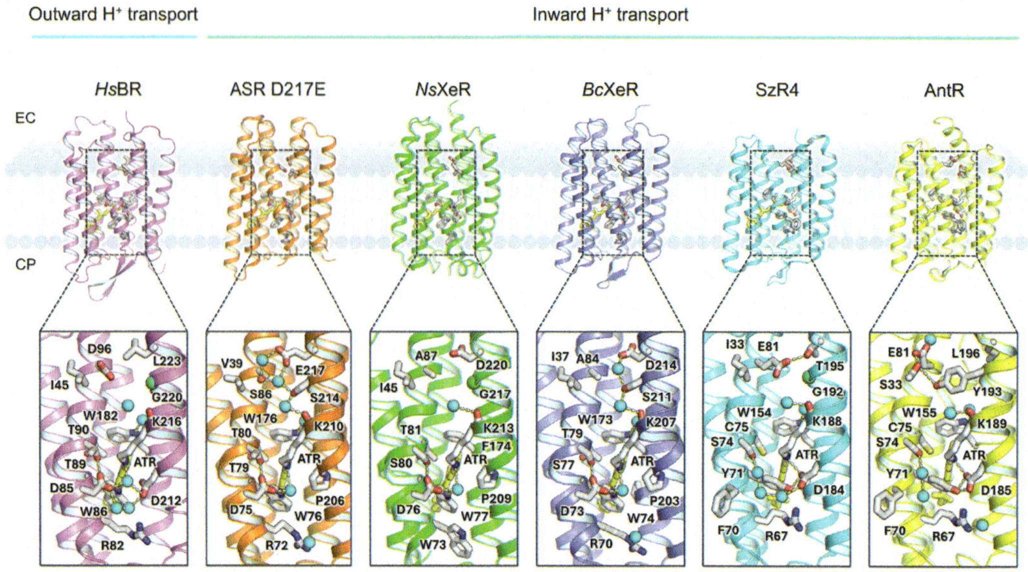

FIGURE 14.5

The three-dimensional structure of inward proton-pumping rhodopsins. The structures of ASR D217E (PDB ID: 4TL3) (Dong et al., 2016), *Ns*XeR (PDB ID: 6EYU) (Shevchenko et al., 2017), *Bc*XeR (PDB ID: 7ZMY) (Kovalev et al., 2023), SzR4 (PDB ID: 7E4G) (Higuchi et al., 2021), and AntR (PDB ID: 9ORK) (Besaw et al., 2025) are compared with that of *Hs*BR (PDB ID: 5ZIM) (Hasegawa et al., 2018). The retinal Schiff base, side chains of key residues, and the main-chain carbonyl group of the retinal-binding lysine are shown as stick models. Hydrogen bonds are represented by *yellow dashed lines*. Water molecules and the C_α atoms of glycine residues are depicted as *cyan* and *green spheres*, respectively. *ASR, Anabaena* sensory rhodopsin; *ATR*, all-*trans*-retinal; *SzRs*, schizorhodopsins; *XeRs*, xenorhodopsins.

counterion aspartate, corresponding to *Hs*BR D85, which is conserved in the extracellular side of the retinal in outward proton-pumping rhodopsins with a DTE motif such as PRs (Hirschi et al., 2021; Ran et al., 2013) and xanthorhodopsins (Chazan et al., 2023; Luecke et al., 2008; Morizumi et al., 2019; Tsukamoto et al., 2013). Because the extracellular His–Asp pair in outward proton-pumping rhodopsins acts as a proton acceptor from the RSB, H48 and D220 in *Ns*XeR have been suggested to receive an H^+ from the RSB. Another unique structural feature in *Ns*XeR is the hydrogen-bonding network between the counterion D76 in TM3, S55 in TM2, and Y20 in TM1 (Shevchenko et al., 2017). Although this interaction is conserved in most XeRs, including ASR, there is no hydrogen bonding interaction between *Hs*BR D85, which is identical to *Ns*XeR D76, and TM2 residues. The D–S–Y interaction conserved in XeRs is likely to be critical to forming functional proteins. In fact the S55A mutation of *Ns*XeR leads to the loss of colored protein formation (Shevchenko et al., 2017). Interestingly, *Ns*XeR and most archaeal XeRs have a tryptophan residue, W73, at the position corresponding to BR R82, which is one of the most conserved residues among microbial rhodopsins (Fig. 14.5) (Nagata et al., 2021; Shevchenko et al., 2017). W73 is located at the position, which is likely functionally critical, dividing extracellular cavities and the RSB region. The mutation of W73 to arginine or other amino acids inhibits the formation of

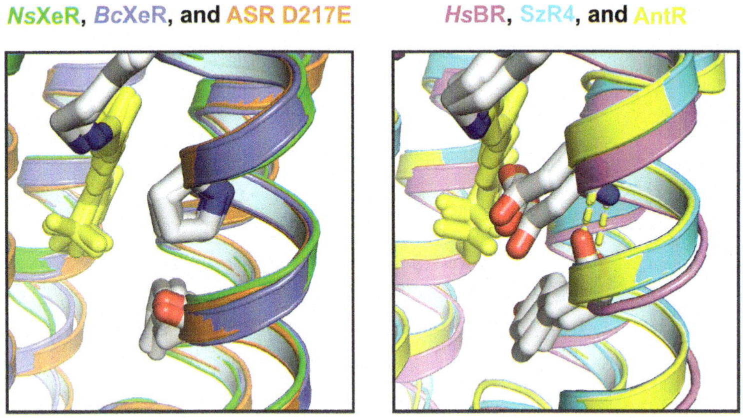

FIGURE 14.6

Disruption of hydrogen bonding interactions caused by a proline residue conserved in the XeR family. The substitution of the TM7 counterion, corresponding to *Hs*BR D212, SzR4 D184, and AntR D185, with a proline residue conserved in XeRs (left) disrupts hydrogen bonding interactions between α-helical turns in TM7 in microbial rhodopsins having the TM7 counterion (right). Side chains of key residues are shown as stick models. Hydrogen bonds are represented by *yellow dashed lines. SzRs,* schizorhodopsins; *XeRs,* xenorhodopsins.

colored protein or lowers the proton-pumping activity, respectively (Shevchenko et al., 2017). On the cytoplasmic surface of *Ns*XeR, E228 forms a hydrogen bond with H94. Although the H^+, transferred from the RSB to His48–D220, is proposed to further move to His94–E228 on the cytoplasmic surface, from where it is likely to be ultimately released into the cytoplasm, neither His94 nor E228 is conserved in ASR D217E and *Bc*XeR, suggesting that the proton-release process varies among different XeRs.

In 2023 the structures of the dark state and photointermediates, specifically the L and M, of *Bc*XeR at different pH values, were solved using cryo-trapping and time-resolved serial X-ray crystallography (Kovalev et al., 2023). Although an extensive hydrogen-binding network, including several water molecules, is present in the cytoplasmic side of *Bc*XeR, interestingly, an Na^+ binding site was observed on the cryo-trapped ground state and L state structures (Fig. 14.7). The Na^+ interacts with E34, D214, and T88. The orientation of E34 in *Bc*XeR differs from that of the corresponding E36 in ASR D217E, and in *Ns*XeR, the presence of a phenylalanine at the equivalent position prevents the Na^+ binding in a similar manner (Dong et al., 2016; Kovalev et al., 2023; Shevchenko et al., 2017). The Na^+ was not observed in the cryo-trapped structure of the subsequent M state. The E34–D214 pair region was designated as a Proton-Transit Group (PTG).

Notably, significant movements of residues, L81, L136, L169, W173, S206, and F210, on the cytoplasmic side of the retinal chromophore in *Bc*XeR, greater than those occurring in *Hs*BR, were observed in the L and M intermediates (Fig. 14.8) (Kovalev et al. 2023). This unique movement of the cytoplasmic residues is likely to be associated with the inward proton-pumping function by enabling the opposite vectoriality of H^+ transport compared to that of outward proton-pumping rhodopsins.

The flipping of L81 in the L intermediate creates a new space accommodating three water molecules that bridge the RSB and cytoplasmic hydrogen-bonding network, including the PTG (Fig. 14.8) (Kovalev et al., 2023). Consistently, low-temperature FTIR spectroscopy also observed an increase in the

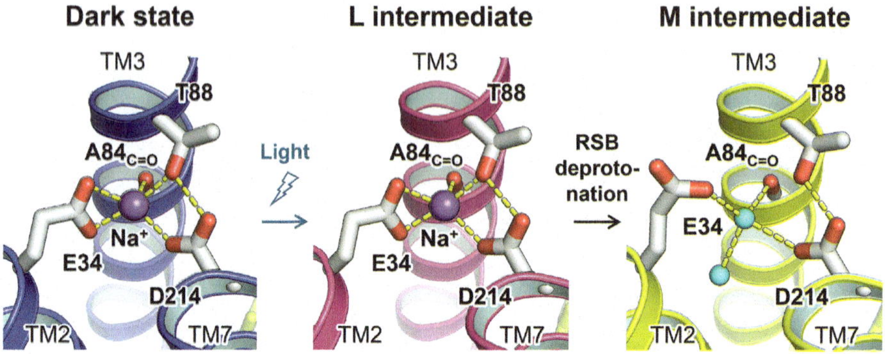

FIGURE 14.7

The cytoplasmic Na^+-binding site in *Bc*XeR. The cytoplasmic Na^+-binding site in *Bc*XeR in the dark (left, PDB ID: 7ZMY); L (middle, PDB ID: 7ZN3); and M (right, PDB ID: 7ZNO) intermediates (Kovalev et al. 2023). Side chains of key residues are shown as stick models. Hydrogen bonds are represented by *yellow dashed lines*. Water molecules and Na^+ are depicted as *cyan* and *purple spheres*, respectively. *RSB*, retinal Schiff base.

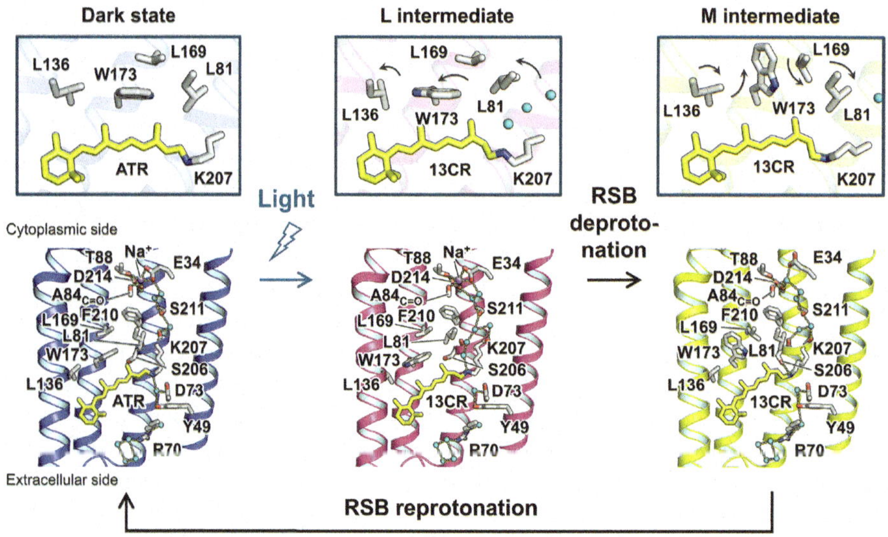

FIGURE 14.8

Structural change in the photocycle of *Bc*XeR. The structural change of the residues on the cytoplasmic side of the retinal chromophore (top) and the hydrogen-bonding network change (bottom) in *Bc*XeR. The structures in the dark (left, PDB ID: 7ZMY), L (middle, PDB ID: 7ZN3), and M (right, PDB ID: 7ZNO) states (Kovalev et al. 2023). The retinal Schiff base, side chains of key residues, and the main-chain carbonyl group of A84, S206, and K207 are shown as stick models. Hydrogen bonds are represented by *yellow dashed lines*. Water molecules and Na^+ are depicted as *cyan* and *purple spheres*, respectively. *13CR*, 13-*cis*-retinal; *ATR*, all-*trans*-retinal; *RSB*, retinal Schiff base.

internal water molecules in the L intermediate of *Bc*XeR and *Ns*XeR (Ito et al., 2026). It is assumed that the H^+ bound to the RSB is transferred to the cytoplasmic side during the L-to-M transition along with this hydrogen-bonding network. In the M intermediate of *Bc*XeR the RSB forms a hydrogen bond with S206, which is located one residue upstream of the retinal-binding K207 and is highly conserved among bacterial XeRs. This interaction switches the RSB interaction from one with the cytoplasmic side to one with the extracellular side. Furthermore the appearance of a new extracellular water molecule (Wat403) leads to the formation of a continuous hydrogen-bonding network from the RSB to Wat403, passing through Wat401, Y49, H6, R70, and Wat402 on the extracellular side, making the RSB accessible to an extracellular H^+. The flipping of W173 relieves steric repulsion between the isomerized retinal and the protein moiety, minimizing the helix movement on the cytoplasmic side to keep the cytoplasmic gate closed after the H^+ release and to avoid H^+ backflow during the M-to-dark-state transition.

Most XeRs have a proline residue at the same position as BR P186 in TM6. In many microbial rhodopsins, such as outward sodium-pumping rhodopsins, inward chloride-pumping rhodopsins, and anion channelrhodopsins, mutation of this proline induces a red shift in retinal absorption without significantly affecting their ion-transport activity (Inoue et al., 2019; Kojima et al., 2020). In contrast, although the mutation of P179T in *Po*XeR causes a red shift for 6 nm, the H^+ transport activity is considerably decreased (Inoue et al., 2019). In addition, mutation of the homologous P209 in *Ns*XeR to a glycine or aspartic acid residue results in considerably lowered pumping activity or a loss of the colored pigment formation, respectively (Shevchenko et al., 2017). These findings suggest that the proline residue in TM6 is more critical for XeRs compared to other microbial rhodopsins.

The first X-ray crystallographic structure of SzR was solved using SzR4, and this, along with an HS-AFM observation, revealed its trimetric assembly (Higuchi et al., 2021), which closely resembles haloarchaeal ion-pumping microbial rhodopsins than HeRs, which exclusively form dimers (Shihoya et al., 2019). Moreover, SzRs exhibit a protein orientation in which the N- and C-termini face the extracellular and cytoplasmic sides, respectively, which is the same as canonical microbial rhodopsins (Inoue et al., 2020). Based on these structural findings, SzRs are classified as members of microbial rhodopsins rather than HeRs.

Interestingly, F70, the first residue in the TM3 motif, is oriented toward the exterior of the helix bundle, whereas Y71 faces the RSB–counterion pocket and forms an interhelical hydrogen bond with D184 in TM7 (Fig. 14.5). A similar interhelical hydrogen bond is observed in bacteriorhodopsin-like cation channelrhodopsins, such as ChRmine, *Hc*KCR1, and *Hc*KCR2 (Kishi et al., 2022; Tajima et al., 2023). SzR4 Y71F mutant exhibits significantly reduced activity compared to the WT, whereas SzR4 Y71A completely loses its proton-transporting function, indicating that Y71 is essential for inward H^+ transport in SzR4 (Higuchi et al., 2021). In addition, a cysteine residue at the position corresponding to BR T90 in TM3 is fully conserved across the SzR family (Fig. 14.2). This cysteine is identical to the one conserved in channelrhodopsins, where it forms the so-called DC gate, as seen in many channelrhodopsins, including channelrhodopsin-2 from *Chlamydomonas reinhardtii* (*Cr*ChR2) (Nack et al., 2010). Among ion-pumping rhodopsins, only the SzR family possesses this cysteine. Mutation of this cysteine to Ala, Ser, or Thr in SzR1 (Inoue et al., 2020) and SzR4 (Higuchi et al., 2021) or to Ala in AntR (Harris et al., 2020) results in complete loss of function. Interestingly, although SzR4 has an unstructured short C-terminal tail, the structure of AntR, which is derived from cyanobacteria living in Antarctic environments (Marín, Konno, Rozenberg, Béjà, & Inoue, 2026), has a C-terminal helix (Fig. 14.5) (Besaw et al., 2025). This helix covers the extracellular side of the cavity, including cytoplasmic proton-transport pathway. Truncation of the C-terminal helix slightly lowers the thermal stability of AntR and also induces partial protein aggregation, indicating that it plays an important role in maintaining proper dispersibility of AntR under solubilized conditions, possibly contributing to its stability in aqueous environments (Besaw et al., 2025).

Notably, whereas the RSB in XeRs forms a hydrogen bond with a water molecule, as in *Hs*BR, the RSB in SzRs is hydrogen-bonded with a counterion residue in TM7 (SzR4 D184 and AntR D185). In contrast to XeRs, in which aspartate is conserved on the cytoplasmic side of TM7 at the position corresponding to ASR D217, most SzRs have Glu or Asp (E81 in SzR1, SzR4, and AntR) at the third position of the motif residues corresponding to BR D96 (Fig. 14.2). E81Q mutations in SzR1 and SzR4 result in a complete loss of inward proton-pumping activity, indicating that E81 plays a critical role in the proton-release process in these SzRs (Higuchi et al., 2021; Inoue et al., 2020). Consistently, recent $^{13}C/^{15}N$ solid-state NMR analyses, which assigned ^{13}C and ^{15}N chemical shifts for ~80% of the residues, showed that SzR1 E81 is deprotonated in the initial state, suggesting its potential role as a cytoplasmic proton acceptor (Kitaguchi et al., 2026). In contrast the AntR E81Q mutant retains activity similar to that of the WT (Harris et al., 2020), suggesting that the proton-release mechanism differs between SzR1/SzR4 and AntR. Consistently, in SzR4, E81 forms a hydrogen bond with T195, which is also located at the same position as ASR D217, in TM7. In contrast the corresponding residue in AntR, L196, cannot form a hydrogen bond with E81. Instead, Tyr193, which is located one α-helical turn toward the extracellular side from L196, is hydrogen-bonded with S33 and subsequent E81 (Fig. 14.5) (Besaw et al., 2025). This Y193–S33–E81 network is suggested to function as a proton-release pathway in AntR, and differences in the hydrogen-bonding network involving E81 may underlie the distinct ion-transport mechanisms between SzR1/SzR4 and AntR. Although bacterial SzRs (AntR, SzR2, GSS AntR, paSzR, and psSzR) possess a tyrosine residue homologous to AntR Tyr193, this position is occupied by glycine in archaeal SzRs (SzR1, SzR3, SzR4, SzR TE_5S_00009, SzR TE_8S_00242, LaSzR2, *Mt*SzR, and *Ms*SzR) (Fig. 14.2), suggesting that distinct proton-release mechanisms have evolved in these two SzR groups. Interestingly, SzR2 possesses a histidine residue at the position corresponding to E81 in SzR4/AntR. Although histidine can contribute to H^+ transfer, SzR2 H82F retains 66% of the activity of the WT (Inoue et al., 2020), suggesting that the functional importance of the third motif-residue position varies among SzRs.

Solid-state NMR analysis of the retinal chromophore in GSS AntR and molecular dynamics simulation on the structure-modeled GSS AntR suggested that the environment around the retinal, especially on the cytoplasmic side of the middle portion of the polyene chain and the Schiff base side, is more hydrophilic compared to that in outward proton-pumping rhodopsins such as *Hs*BR and green-absorbing PR (Pinto et al., 2024). Together with the X-ray crystallographic structures of other inward proton-pumping rhodopsins, which revealed a high number of water molecules on the cytoplasmic side of the proteins, the hydrophilic cytoplasmic environment and extensive hydrogen-bonding network are considered to be critical elements for reversing the vectoriality of the proton transport relative to outward proton-pumping rhodopsins.

14.4 Photocycle of inward proton-transporting rhodopsins

The photocycles of several ASR/XeRs and SzRs are shown in Fig. 14.9. They consist of photo-intermediates similar to those that appear in the photocycle of *Hs*BR: the K and the O with red-shifted absorption compared to that of the initial state; the L, whose λ_{max} is close to that of the initial state; and

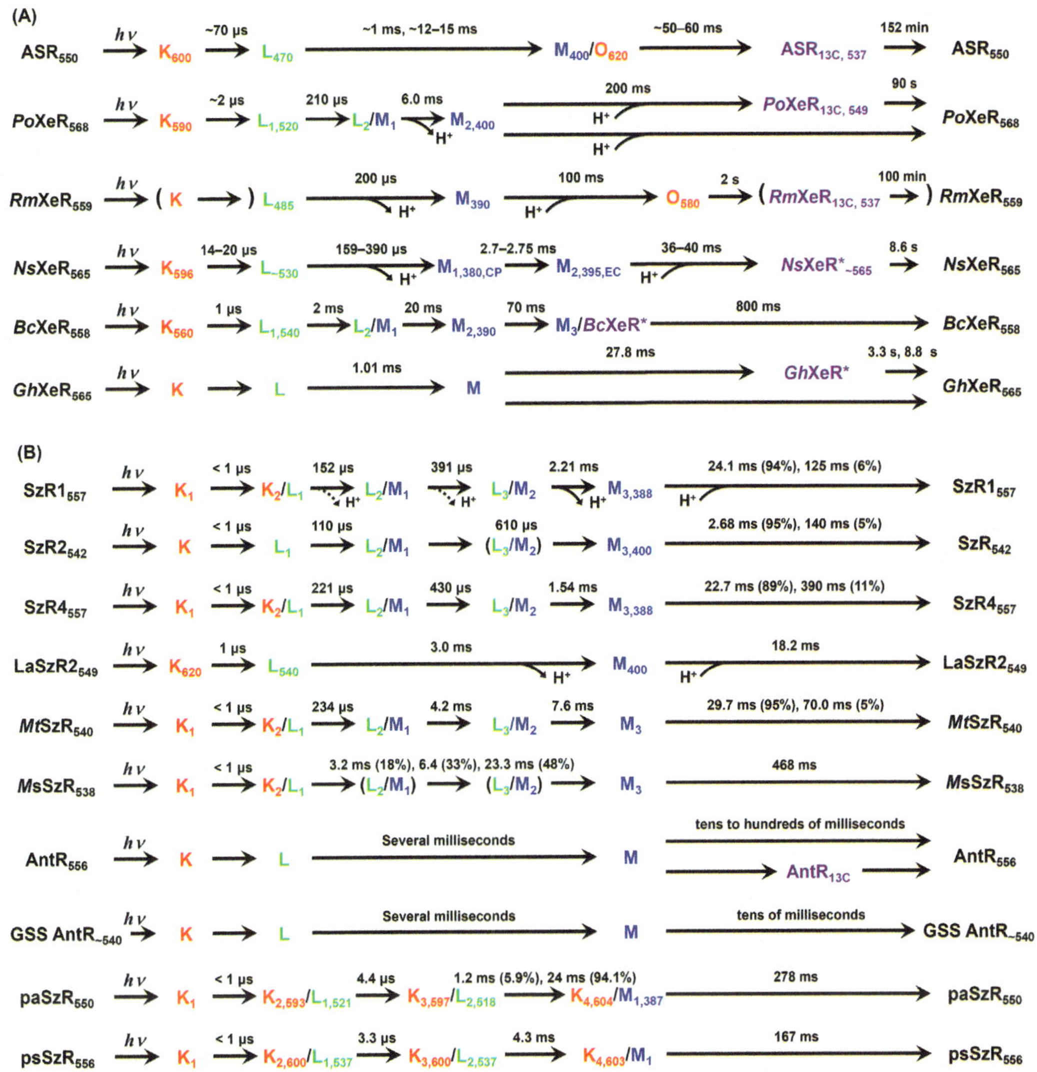

FIGURE 14.9

Photocycles of XeRs and SzRs. (A and B) Photocycles of XeRs (A) and SzRs (B). The λ_{max} values are indicated as subscripts. For the initial states of ASR, *Po*XeR, and *Rm*XeR, the λ_{max} values of the all-trans form are shown. *ASR, Anabaena* sensory rhodopsin; *SzRs*, schizorhodopsins; *XeRs*, xenorhodopsins.

the highly blue-shifted M, which represents the deprotonated RSB. However, the intermediate similar to the N of *Hs*BR, whose λ_{max} is similar to that of the initial state, is not observed in the photocycles of all XeRs and SzRs.

By contrast the photocycles of SzRs tend to be faster than those of XeRs. In addition, any long-lived intermediate having 13-*cis*-15-*syn*-retinal was not observed for SzRs. Consistent with their fast photocycles, SzRs show fast off-kinetics in electrophysiological measurements and high H^+ transport activity in mammalian and *E. coli* cells (Inoue et al., 2020; Okuyama et al., 2024). Interestingly, AntR possesses all-*trans*-15-*anti* and 13-*cis*-15-*syn*-retinal in the dark state, and the isomeric composition changes from being all-*trans*-retinal dominant at alkaline pH to predominantly 13-*cis*-retinal at acidic pH, in both the dark-adapted and light-adapted states. In addition, a branched photoreaction model was suggested (Fig. 14.9A) (Harris et al. 2020). In paSzR and psSzR the overall photocycle turnover is notably slower than that of SzR1 and SzR4, and the accumulation of the M intermediate is considerably smaller than that in other characterized SzRs, indicating reduced deprotonation of the RSB. Consistent with these photoreaction characteristics, the proton-pumping activities of paSzR and psSzR are relatively lower than those of others (Marín, Konno, Rozenberg, Béjà, & Inoue, 2026).

The primary photoreaction dynamics of *Po*XeR, *Ns*XeR, SzR1, and AntR were studied by femtosecond time-resolved absorption spectroscopy (Chang et al., 2023; Inoue et al., 2018a; Malakar et al., 2022; Weissbecker et al., 2021). The photoisomerization in these distinct inward H^+ pumps proceeds slower than typical outward proton-pumping rhodopsins: *Po*XeR: $\tau = 1.2 \pm 0.1$ ps; *Ns*XeR: $\tau = 0.65$ ps; SzR: $\tau = 0.99$ ps; AntR: $\tau = 1$ ps versus *Hs*BR: $\tau = 0.56 \pm 0.01$ ps and PR: $\tau = 0.25 \pm 0.01$ ps. The relative quantum yields of both inward H^+ pumps, *Po*XeR and SzR1, estimated based on their transient absorption spectra, are significantly lower than those of the outward H^+ pumps (Chang et al., 2023; Inoue et al., 2018a). On the other hand the quantum yield of AntR is high ($\phi_{at} = 0.85 \pm 0.15$) (Malakar et al., 2022).

Notably the mutation of C75, which is completely conserved in the SzR family, to a serine in SzR1 further slowed down the isomerization rate. Impulsive stimulated Raman spectroscopy suggested that the retinal in SzR4 C75S is significantly twisted compared to the chromophore in the WT, indicating that C75 in SzR4 flattens the retinal chromophore and plays an essential role in facilitating the all-*trans*-to-13-*cis* isomerization (Chang et al., 2023).

In *Po*XeR, H^+ is translocated from the Schiff base to a cytoplasmic aspartic acid, D216, during the L-to-L/M_1 transition. Subsequently, it is released from D216 to balk solution during L/M_1-to-M_2. Finally, H^+ uptake occurs during the M_2 decay. In outward H^+ pumps the mutation of the internal H^+ donor for deprotonated Schiff base makes the rate of Schiff base reprotonation process pH-dependent (Otto et al., 1989). In contrast, none of *Po*XeR mutants of the extracellular amino acids show such behavior. Furthermore the kinetic isotope effect (KIE) of the M_2 decay (1.69 ± 0.04) is significantly smaller than those for the L and L/M_1 (8.58 ± 0.05, 7.89 ± 0.06), where deprotonation of the RSB occurs. The smaller KIE of the M_2 decay, which involves the reprotonation of the RSB, suggests a mechanistic difference in proton transfer from the RSB deprotonation. The KIE was also investigated for *Ns*XeR: K-to-L: 1.1; L-to-M_1: 4.2; M_1-to-M_2: 1.8; M_2-to-*Ns*XeR*: 3.6 (Schubert et al., 2023). Although *Ns*XeR also exhibits a large KIE during the deprotonation of the RSB as *Po*XeR, the KIE of the RSB reprotonation is much larger than that of *Po*XeR. In addition, thermal reisomerization of the retinal was suggested to be the rate-limiting process in proton uptake (Inoue et al., 2016, 2018a). For the outward H^+ pump, *Hs*BR, the M intermediate possesses deprotonated 13-*cis*-15-*anti*-retinal. In this state the lone pair on the nitrogen atom of the RSB remains oriented toward the cytoplasmic side and accepts a proton from the cytoplasmic proton donor, D96, during the subsequent M-to-N conversion. For the inward H^+ pump, *Po*XeR, the N intermediate is not observed. The direction of the lone pair of deprotonated Schiff base switches from the cytoplasmic side toward the extracellular side during the M_2-to-*Po*XeR$_{13C}$ or M_2-to-*Po*XeR$_{AT}$ conversions. These conversions should elevate the pK_a of the RSB, and the RSB receives an H^+ from the extracellular side. At the same time an H^+ is taken up from bulk solution. Cryo-trapped FTIR showed that *Po*XeR contains a strongly hydrogen-bonded water molecule in the vicinity of the RSB, as observed for outward proton pumps (Ito et al., 2017). Because strongly hydrogen-bonded waters are considered to be essential for outward H^+ pumping rhodopsins (Kandori, 2020) and are absent in ASR, which functions as a sensor (Furutani et al., 2005), the strongly hydrogen-bonded water in *Po*XeR has also been thought to be critical for its inward proton-pumping function. In contrast, no strongly hydrogen-bonded water molecules were observed in *Ns*XeR and *Gh*XeR, indicating that their hydrogen-bonding networks are distingt from that in *Po*XeR (Ito et al., 2025; Hattori et al., 2026). In *Rm*XeR, although proton movement inside the protein is not clear, H^+ release and uptake associate with the M accumulation and decay, respectively (Inoue et al., 2018b).

One of the most unique points of ASR/XeR's photocycles is the 15-*anti*-to-*syn* isomerization generating long-lived intermediates having 13-*cis*-15-*syn* retinal (ASR_{13C}, $PoXeR_{13C}$, $RmXeR_{13C}$, and *Gh*XeR*) (Inoue et al., 2016,2018a; Sineshchekov et al. 2005; Hattori et al. 2026). Whereas the long-lived intermediate has not been reported for *Ns*XeR at first report, later several studies have suggested the existence of an additional intermediate with absorption characteristics similar to that of the dark state (*Ns*XeR*) (Schubert et al., 2023; Weissbecker et al., 2021). The decay time constant of *Ns*XeR* is 8.6 s (Weissbecker et al., 2021). FTIR studies indicated that *Ns*XeR* corresponds to the O intermediate or has a 13-*cis*-15-*syn* retinal (Asido et al., 2024; Schubert et al., 2023; Shevchenko et al., 2017; Weissbecker et al., 2021). The recovery time constants of the all-*trans*-15-*anti* form from the 13-*cis*-15-*syn* form are diverse, and the values range from 800 ms (*Bc*XeR) to 130 min (ASR) (Kawanabe et al., 2007; Kovalev et al., 2023). The *anti*-to-*syn* thermal isomerization at the C15=N position can compete with *cis–trans* thermal (re)isomerization at the C13=C14 position. Although the efficiency of 15-*anti*-to-*syn* thermal isomerization is nearly 100% on ASR (Kawanabe et al., 2007), the photoreaction of *Po*XeR branches during the M decay is 57.2% and 43.8% of molecules proceed to $PoXeR_{13C}$, where the retinal adopts 13-*cis*-15-*syn* configuration, or return to the initial state with all-*trans*-15-*anti* retinal (Inoue et al., 2018b).

The mechanism of *Ns*XeR exhibits significant differences from that of *Po*XeR. In *Ns*XeR, D220, which corresponds to D216 in *Po*XeR, is protonated in the dark state (Asido et al., 2024; Ito et al., 2026; Schubert et al., 2023). Upon light illumination a red-shifted K intermediate is produced. Notably the low intensity of the hydrogen out-of-plane (HOOP) modes in low-temperature FTIR spectra suggests that the retinal polyene chain in *Ns*XeR is less twisted than in other XeRs (Ito et al., 2025). The *Ns*XeR P205G and P205A mutants exhibited larger twisting of the retinal polyene chain in the K-intermediate and reduced proton-pumping activity, indicating that the less-twisted retinal chromophore in the K-intermediate is important for the higher proton-pumping activity of *Ns*XeR compared to other XeRs. In the L intermediate the N–D group of the RSB in *Ns*XeR is strongly hydrogen-bonded, which may facilitate subsequent proton release toward the cytoplasmic side (Ito et al., 2025). Subsequently an H^+ at D220 is released to the extracellular side (translocated to the extracellular membrane/micelle surface) during the L-to-M_1 transition. At the same time the H^+ at the Schiff base is transferred to an unidentified primary H^+ acceptor, which should exist between the RSB and D220. During the M_1-to-M_2 transition the H^+ at the unidentified primary proton donor moves to D220. Protein structural changes are coupled with the protonation state of D220, which may facilitate the proton transfer steps between the Schiff base and D220. In patch-clamp experiments, 355-nm laser nanosecond flash stimulation under yellow background illumination, which accumulates the M intermediate, induces a distinct voltage dependence and an inversion of the current direction, suggesting that the Schiff base reprotonation might occur from both sides with the two M_{EC} and M_{CP} intermediates (Weissbecker et al., 2021). The structural alterations during the M_1-to-M_2 transition may switch the accessibility of the RSB from the cytoplasmic side ($M_{1,CP}$) to the extracellular side ($M_{2,EC}$) as suggested by the isomerization-switch-transfer model, which may correspond to the direction switching mechanism of the Schiff base nitrogen lone pair in *Po*XeR and SzRs (described later) (Asido et al., 2024; Haupts et al., 1997; Weissbecker et al., 2021). It is also suggested that Y216, which is located between the RSB and D220, is crucial for the arrangement of water molecules in the cytoplasmic half channel and proton transfer from the Schiff base to D220 (Schubert et al., 2023).

Time-resolved resonance Raman spectroscopy was conducted to characterize the structure of the retinal chromophore in the unphotolyzed state and the photointermediates of SzRs (Urui et al., 2024b). In unphotolyzed SzRs (SzR1, SzR2, SzR3, and SzR4), the all-*trans*-15-*anti*-retinal chromophore has a planar polyene chain at its center but is twisted near the RSB (Shionoya et al., 2021). These protonated RSBs form a stronger hydrogen bond than that in outward proton-pumping rhodopsins, indicating that it forms a hydrogen bond directly with the counterion aspartate on TM7 without intervening water molecules (Shionoya et al., 2021), as also supported by X-ray crystallographic analyses (Fig. 14.5) (Besaw et al., 2025; Higuchi et al., 2021). Further analysis revealed that the degree of the polyene chain twist decreases as the hydrogen bond of the RSB becomes stronger, suggesting that the twist is caused by tight atomic contacts between the chromophore and nearby residues, probably including C75 (vide supra) (Chang et al., 2023). In addition, the hydrogen bond strength of RSB is associated with the color-tuning mechanism of SzRs (Urui et al., 2024c).

For the K intermediate of SzR4 the twist of the Schiff base side of the retinal chromophore is smaller than that of the outward proton pump, *Gloeobacter* rhodopsin. In the L intermediate of SzR4 the RSB forms strong hydrogen bonds with the water molecule and amino acid residue, which is proposed to be S74, and the twist of the polyene chain is relaxed (Urui et al., 2024a). One of the key steps for inward proton transport of SzR4 occurs during the multiple M intermediates. In SzR4, early M, corresponding to the M_2, possesses deprotonated 13-*cis*-15-*anti*-retinal, similar to the M of *Hs*BR. However, the late M, corresponding to the M_3, features deprotonated all-*trans*-retinal, which differs from the M of *Hs*BR (Hayashi et al., 2022). The reisomerization from the 13-*cis* to all-*trans* form reorients the nitrogen lone pair of the RSB from the cytoplasmic to the extracellular side, facilitating the deprotonated Schiff base to take up a proton from the extracellular medium via a water-mediated transport network (Hayashi et al., 2022). Identical reaction processes were also observed in SzR1, SzR2, SzR3, *Mt*SzR, *Ms*SzR, and AntR, indicating that *cis–trans* thermal reisomerization preceding reprotonation of the RSB is a universal mechanism across the SzR family (Urui et al., 2024a).

14.5 The similarities and differences in the proton-pumping mechanism between xenorhodopsins and schizorhodopsins

One of the most prominent common features between the proton-transporting mechanisms of SzRs and XeRs is their single counterion architecture because of the substitution of the typical TM3 and TM7 counterions, corresponding to D85 and D212 in *Hs*BR, with mainly a phenylalanine and a proline in SzRs and XeRs, respectively. The decrease in the number of counterions on the extracellular side of the RSB lowers the electrostatic force attracting the RSB proton toward the extracellular side, and the H^+ can be more preferentially moved to the cytoplasmic side. Moreover, the loss of counterion in TM3 in SzRs removes the proton acceptor (Braiman et al., 1988), thereby completely inhibiting the proton transfer to the extracellular side. In contrast, whereas E217 in ASR D217E and D216 in *Po*XeR serve as proton acceptors from the RSB (Inoue et al., 2016; Kawanabe et al., 2009), homologous residues D220 in *Ns*XeR and D214 in *Bc*XeR form a hydrogen bonding cluster with E34, which is suggested to form the proton transfer pathway that connects the RSB and the cytoplasmic milieu. Hence the cytoplasmic TM7 aspartate is essential for the inward proton-pumping mechanism of XeRs. By contrast, although the aspartate in TM7 is not conserved in SzRs, the glutamate in TM3—E81 in SzR1 and SzR4—is essential for the inward proton-pumping function. Therefore although single-counterion architecture and the importance of the cytoplasmic acidic residues are common to both XeRs and SzRs, their positions are swapped: in XeRs, the counterion and the conserved acidic residue are in TM3 and TM7, whereas in SzRs, those are in TM7 and TM3, respectively. Notably the cytoplasmic glutamate is not essential for the proton transport in AntR, and it is substituted with a histidine in some SzRs, suggesting that the proton release mechanism differs among SzRs.

Although the extracellular proton donor has not been identified, the reorientation of the lone pair on the Schiff base nitrogen atom through the thermal isomerization of the retinal chromophore has been proposed as playing an essential role in the reprotonation of the RSB in both XeR and SzR. This process is achieved through a 13-*cis*-15-*anti*-to-all-*trans*-15-*anti* or 13-*cis*-15-*anti*-to-all-*trans*-15-*syn* isomerization in XeRs and exclusively through a 13-*cis*-15-*anti*-to-all-*trans*-15-*anti* isomerization in SzRs (Hayashi et al., 2022; Inoue et al., 2016, 2018a; Urui et al. 2024a,2024b). These isomerizations occur concurrently with or prior to the reprotonation of the RSB in XeR and SzR, respectively, suggesting the difference in the accessibility of the nitrogen lone pair to H^+ from the extracellular side. In contrast to the mechanisms of inward proton-pumping rhodopsins, the RSB in outward proton-pumping rhodopsins is reprotonated by the cytoplasmic proton donor. This process involves a conformational change in the protein moiety, which forms a water chain between the proton donor residue and the RSB. Subsequently the protonated RSB undergoes isomerization to the all-*trans*-15-*anti* form (Ernst et al., 2014; Schobert et al., 2003). This distinction in the order of isomerization and RSB reprotonation is also a critical difference between outward and inward proton-pumping rhodopsins. In addition, it accounts for the absence of the N intermediate in the photocycles of XeR and SzR, which is formed through the reprotonation of the Schiff base of 13-*cis*-15-*anti* retinal in outward proton-pumping rhodopsins.

Three shared mechanistic features—the single counterion architecture, cytoplasmic proton-accepting residues, and unique thermal isomerization of the retinal chromophore—suggest that, to achieve the same inward proton-pumping function, convergent molecular evolution occurred between XeRs and SzRs, which are phylogenetically highly distinct.

14.6 Functional conversion of inward proton-transporting rhodopsins

Before the discovery of light-driven inward proton pumps in natural environments, an inward proton-transporting rhodopsin was engineered from ASR, a photochromic sensor, through a single amino acid substitution of ASR D217 to a glutamate residue (Kawanabe et al., 2009). Although ASR WT does not transport H^+, ASR D217E exhibited light-induced inward H^+ transport when expressed in *E. coli* cells. FTIR analysis showed an increased proton affinity for E217 in the M intermediate. However, in the original work, it was not possible to determine whether the mutant functioned as a "pump" or a "channel," as the membrane potential of *E. coli* cells could not be controlled.

To elucidate the essential elements determining the direction of H^+ transport between outward and inward H^+ pumps, Marín et al. attempted to convert the outward proton-pumping rhodopsin *Psp*R from *Pseudomonas putida* by swapping five residues that are conserved differently between outward proton-pumping rhodopsins, such as *Psp*R and SzRs. They ultimately found that converting *Psp*R into an inward proton-pumping rhodopsin requires mutation in a minimum of three residues—D73, T77, and T78 (corresponding to *Hs*BR D85, T89, and T90)—to phenylalanine, serine, and cysteine, respectively, mimicking F70, S74, and C75 in SzR4 (Marín et al., 2023). These three mutations are considered to play different roles: the D73F mutation removes the proton acceptor from the RSB, thereby inhibiting the proton transfer to the extracellular side while facilitating transfer to the cytoplasmic side (Urui et al., 2024c). In contrast the T77S and T78C mutations elevate the pK_a of the RSB, which is lowered by the loss of the counterion D73, thereby restoring the visible absorption of the RSB. This study has also suggested that the double and single counterion architectures are essential for outward and inward proton-pumping functions, respectively.

14.7 Application of inward proton-transporting rhodopsins

The unique active inward proton-transporting function of XeRs and SzRs forms a new basis for molecular tool development. Among them, because of its high ion-transport activity, *Ns*XeR is capable of inducing neuron firing as an excitatory optogenetic tool (Shevchenko et al., 2017). Notably the fast photocycle turnover rate of *Ns*XeR enables high-frequency neuron firing at up to 40 Hz (Jahnke et al., 2021). Furthermore, *Ns*XeR can induce a significant decrease in cytosolic pH (Vlasova et al., 2023). It provides a new methodology for investigating the effect of cytosolic acidification on the cell fate and proliferation without applying mechanical and chemical cellular stress. In addition, light-induced inward H^+ transport by *Rm*XeR can decelerate the alkalization-induced shrinkage of human HeLa cells cultured at pH 9.0 (Nakao et al., 2022). Furthermore, *Rm*XeR has been proposed as a tool for assaying electrogenic transporters, such as oxalate transporter (OxlT), by expressing them in *E. coli* cells and monitoring pH changes (ΔpH) in the outer solvent upon light illumination, which reflects H^+ uptake into the cells. The magnitude of the ΔpH changes upon addition of the substrate of the electrogenic transporters allows the estimation of the transporter activity by comparing the ΔpH values in the presence and absence of the substrate (Hayashi et al., 2021).

The inward H^+ transport by inward proton-pumping rhodopsins can acidify the interior of liposomes by incorporating the proteins into lipid bilayers. The acidification elicited by *Rm*XeR can disrupt pH-sensitive liposomes, such as those composed of 1,2-dioleoyl-*sn*-glycero-3-phosphoethanolamine and cholesteryl hemisuccinate, leading to the release of encapsulated chemicals in a light-dependent manner (Tsuneishi et al., 2023). These proteoliposomes, termed light-induced disruptive liposomes (LiDLs), are internalized by HeLa cells through endocytosis. Subsequently, LiDLs incorporated in the cells can be disrupted by green light, releasing various encapsulated chemicals into the cytoplasm. This approach holds promise as a novel intracellular delivery system at high spatiotemporal resolution.

On the other hand, inward proton-pumping rhodopsins can raise the pH in endosomes and lysosomes by localizing proteins in these cellular organelles. ASR D217E can be targeted to early and late endosomes by fusing with the C-terminal cytoplasmic region of chloride channel protein 5 (ClC5) (Kakegawa et al., 2018; Vlasova et al., 2024). This construct, termed PhotonSABER (photon-sensitive ASR-based endocytosis regulator), is capable of increasing the pH within endosomes in dendrites of hippocampal neurons by more than 1.5 units during a 30-min illumination. The elevation of endosomal pH inhibits endocytosis of AMPA receptors, enabling the causal relationship between AMPA receptor endocytosis and neuroscientific phenotypes such as long-term depression (LTD). The strong inward proton-pumping activity of *Ns*XeR is also useful for the optogenetic alkalization of intracellular vesicles. To this aim, lyso-*Ns*XeR, in which *Ns*XeR is fused with the N- and C-terminal parts of cluster of differentiation 63 (CD63), superecliptic pHluorin emitting pH-dependent fluorescence, red-fluorescent mKate2, and β-subunit of H^+, K^+-ATPase (βHK) as lyso-Arch3, originally named lyso-pHoenix (Rost et al., 2015), was designed to localize *Ns*XeR in lysosomes (Ilyinsky et al., 2023; Vlasova et al., 2024). Upon 590-nm illumination, lyso-*Ns*XeR increases the pH within lysosomes up to 1 unit. It suppresses the protease activity of cathepsin B in lysosomes, suggesting the usefulness of lyso-*Ns*XeR as a new molecular tool to elucidate the biological role of lysosomal acidic pH. Moreover, using lyso-*Ns*XeR in conjunction with lyso-Arch3, lysosomal pH can be regulated between pH 5 to 6.5. For a different type of optogenetics, *Po*XeR (Schmidlechner, 2024) and *Rm*XeR (Nakanishi et al., 2024) have been proposed as tools to deplete the pmf of bacterial cells, thereby facilitating investigations into cellular activity or metabolism under reduced pmf conditions. Notably the torque of the flagellar motor is decreased to nearly zero by diminishing the pfm by *Rm*XeR (Nakanishi et al., 2024).

14.8 Perspective

Spectroscopic and structural studies on XeRs and SzRs have elucidated several key mechanisms underlying their inward proton-pumping functions. However, many critical questions remain unsolved. One of such unresolved issues is the identity of the extracellular donor residue, which remains unclear in both XeR and SzR. The decay rate of the M-intermediates is independent of pH for *Po*XeR (Inoue et al., 2016) and LaSzR2 (Kojima et al., 2020), or only slightly accelerated at higher pH in *Bc*XeR (Kovalev et al., 2023). This pH independence of the rate of the M decay is similar to that observed for outward proton-pumping rhodopsins, such as *Hs*BR, where internal H^+ donor residue (e.g., *Hs*BR D96) transfers H^+ to the RSB during the M decay (Bousch et al., 1991; Otto et al., 1989). By analogy, inward proton-pumping rhodopsins are also hypothesized to possess H^+ donor residues on the extracellular side. In addition, in SzRs, the retinal chromophore isomerizes from the 13-*cis* to the all-*trans* prior to the RSB protonation (Hayashi et al., 2022; Urui et al., 2024a). Interestingly the rate of this isomerization process in SzR1 exhibits a large KIE value (5.4 ± 0.2) without a change in the protonation state of the RSB (Kawasaki et al., 2022), indicating the 13-*cis*-to-all-*trans* isomerization in SzRs is rate-limited by proton transfer occurring at sites in the protein other than RSB. However, the residues responsible for both the proposed H^+ donation and the rate-limiting proton transfer have not been identified yet so far, and further studies employing advanced structural and spectroscopic approaches are needed in the future.

In addition to unresolved molecular mechanisms, the biological significance of inward proton-pumping rhodopsins in their native organisms remains largely unknown, despite their molecular mechanisms being understood in detail, as reviewed in this chapter. This is primarily due to the difficulty of cultivating and genetically manipulating their host organisms. However, *Gh*XeR was recently report to be the only rhodopsin present in the cultivatable bacterium *G. hwajinpoensis* (Hattori et al., 2026), and the physiological role of *Gh*XeR may be elucidated by investigating the effect of light on the growth and/or other physiological responces *G. hwajinpoensis*. Although inward proton-pumping rhodopsins have been hypothesized to help mitigate rising intracellular pH in alkaline environments or regulate the activity of other membrane proteins by lowering pH near the cytoplasmic surface of the cell membrane, their ecological and physiological roles could be elucidated in the future through studies on the effect of light on the physiological responses of these organisms, combined with emerging genetic tools.

Lastly, molecular tools based on inward proton-pumping rhodopsins are proposed to be capable of regulating cytosolic and intravascular pH or of eliciting neural firing. However, more optimization of their cellular localization and achieving higher proton-transport activity are required to utilize them in practical medical treatments and industrial applications. While XeRs have been primarily utilized in application studies, the rapid photocycles of SzRs also suggest their promise as highly efficient molecular tools.

In summary, based on their novel functionality and mechanism, inward proton-pumping rhodopsins provide a foundation for gaining new insights into the universal mechanism of ion transport by membrane proteins and hold significant potential for developing next-generation optogenetic tools. Future interdisciplinary research will undoubtedly expand our understanding of these fascinating proteins and unlock their full application potential.

Acknowledgments

This work was supported by JSPS KAKENHI Grants-in-Aid (grants JP24K23073 to Y.K.; JP23K21092, JP23K18090, JP23H04404, JP24H02268, JP25K24717 to K.I.); JST CREST (grant JPMJCR22N2 to K.I.); and MEXT Promotion of Development of a Joint Usage/Research System Project: Coalition of Universities for Research Excellence Program (CURE) (grant JPMXP1323015482 to K.I.).

References

Alcaraz, L. D., Olmedo, G., Bonilla, G., Cerritos, R., Hernández, G., Cruz, A., Ramírez, E., Putonti, C., Jiménez, B., Martínez, E., López, V., Arvizu, J. L., Ayala, F., Razo, F., Caballero, J., Siefert, J., Eguiarte, L., Vielle, J. P., Martínez, O, … Herrera-Estrella, L. (2008). The genome of *Bacillus coahuilensis* reveals adaptations essential for survival in the relic of an ancient marine environment. *Proceedings of the National Academy of Sciences of the United States of America, 105*(15), 5803–5808.

Asido, M., Boumrifak, C., Weissbecker, J., Bamberg, E., & Wachtveitl, J. (2024). Vibrational study of the inward proton pump xenorhodopsin *Ns*XeR: Switch order determines vectoriality. *Journal of Molecular Biology, 436*(5), 168447.

Béjà, O., Aravind, L., Koonin, E. V., Suzuki, M. T., Hadd, A., Nguyen, L. P., Jovanovich, S. B., Gates, C. M., Feldman, R. A., Spudich, J. L., Spudich, E. N., & DeLong, E. F. (2000). Bacterial rhodopsin: Evidence for a new type of phototrophy in the sea. *Science, 289*(5486), 1902–1906.

Besaw, J. E., Peng, S., Kuo, A., Reichenwallner, J., Miller, R. J. D., Brown, L. S., & Ernst, O. P. (2025). Structural insights into an inward proton-pumping rhodopsin. *Biophysical Journal*. doi:10.1016/j.bpj.2025.1011.1011.

Borshchevskiy, V., Kovalev, K., Round, E., Efremov, R., Astashkin, R., Bourenkov, G., Bratanov, D., Balandin, T., Chizhov, I., Baeken, C., Gushchin, I., Kuzmin, A., Alekseev, A., Rogachev, A., Willbold, D., Engelhard, M., Bamberg, E., Büldt, G., & Gordeliy, V. (2022). True-atomic-resolution insights into the structure and functional role of linear chains and low-barrier hydrogen bonds in proteins. *Nature Structural & Molecular Biology, 29*(5), 440–450.

Bousché, O., Braiman, M., He, Y. W., Marti, T., Khorana, H. G., & Rothschild, K. J. (1991). Vibrational spectroscopy of bacteriorhodopsin mutants. Evidence that ASP-96 deprotonates during the M → N transition. *Journal of Biological Chemistry, 266*(17), 11063–11067.

Braiman, M. S., Mogi, T., Marti, T., Stern, L. J., Khorana, H. G., & Rothschild, K. J. (1988). Vibrational spectroscopy of bacteriorhodopsin mutants: Light-driven proton transport involves protonation changes of aspartic acid residues 85, 96, and 212. *Biochemistry, 27*(23), 8516–8520.

Bulzu, P. A., Andrei, A. S., Salcher, M. M., Mehrshad, M., Inoue, K., Kandori, H., Béjà, O., Ghai, R., & Banciu, H. L. (2019). Casting light on Asgardarchaeota metabolism in a sunlit microoxic niche. *Nature Microbiology, 4*(7), 1129–1137.

Chang, C. F., Konno, M., Inoue, K., & Tahara, T. (2023). Effects of the unique chromophore-protein interactions on the primary photoreaction of schizorhodopsin. *Journal of Physical Chemistry Letters, 14*(31), 7083–7091.

Chazan, A., Das, I., Fujiwara, T., Murakoshi, S., Rozenberg, A., Molina-Márquez, A., Sano, F. K., Tanaka, T., Gómez-Villegas, P., Larom, S., Pushkarev, A., Malakar, P., Hasegawa, M., Tsukamoto, Y., Ishizuka, T., Konno, M., Nagata, T., Mizuno, Y., Katayama, K., ... Béjà, O. (2023). Phototrophy by antenna-containing rhodopsin pumps in aquatic environments. *Nature, 615*(7952), 535–540.

Dong, B., Sanchez-Magraner, L., & Luecke, H. (2016). Structure of an inward proton-transporting *Anabaena* sensory rhodopsin mutant: Mechanistic insights. *Biophysical Journal, 111*(5), 963–972.

Ernst, O. P., Lodowski, D. T., Elstner, M., Hegemann, P., Brown, L. S., & Kandori, H. (2014). Microbial and animal rhodopsins: Structures, functions, and molecular mechanisms. *Chemical Reviews, 114*(1), 126–163.

Furutani, Y., Kawanabe, A., Jung, K. H., & Kandori, H. (2005). FTIR spectroscopy of the all-*trans* form of *Anabaena* sensory rhodopsin at 77 K: Hydrogen bond of a water between the Schiff base and Asp75. *Biochemistry, 44*(37), 12287–12296.

Harris, A., Lazaratos, M., Siemers, M., Watt, E., Hoang, A., Tomida, S., Schubert, L., Saita, M., Heberle, J., Furutani, Y., Kandori, H., Bondar, A. N., & Brown, L. S. (2020). Mechanism of inward proton transport in an antarctic microbial rhodopsin. *Journal of Physical Chemistry B, 124*(24), 4851–4872.

Hasegawa, N., Jonotsuka, H., Miki, K., & Takeda, K. (2018). X-ray structure analysis of bacteriorhodopsin at 1.3 Å resolution. *Scientific Reports, 8*(1), 13123.

Hattori, N., Ito, Y., Furutani, Y., Nishikino, T., & Kandori, H. (2026). Molecular properties of a novel inward proton pump rhodopsin, GhXeR. *Journal of Physical Chemistry B.* https://doi.org/10.1021/acs.jpcb.1026c01199.

Haupts, U., Tittor, J., Bamberg, E., & Oesterhelt, D. (1997). General concept for ion translocation by halobacterial retinal proteins: The isomerization/switch/transfer (IST) model. *Biochemistry, 36*(1), 2–7.

Hayashi, K., Mizuno, M., Kandori, H., & Mizutani, Y. (2022). *Cis–trans* reisomerization precedes reprotonation of the retinal chromophore in the photocycle of schizorhodopsin 4. *Angewandte Chemie International Edition, 61*(33), e202203149.

Hayashi, M., Kojima, K., Sudo, Y., & Yamashita, A. (2021). An optogenetic assay method for electrogenic transporters using *Escherichia coli* co-expressing light-driven proton pump. *Protein Science, 30*(10), 2161–2169.

Higuchi, A., Shihoya, W., Konno, M., Ikuta, T., Kandori, H., Inoue, K., & Nureki, O. (2021). Crystal structure of schizorhodopsin reveals mechanism of inward proton pumping. *Proceedings of the National Academy of Sciences of the United States of America, 118*(14), e2016328118.

Hirschi, S., Kalbermatter, D., Ucurum, Z., Lemmin, T., & Fotiadis, D. (2021). Cryo-EM structure and dynamics of the green-light absorbing proteorhodopsin. *Nature Communications, 12*(1), 4107.

Ilyinsky, N. S., Bukhalovich, S. M., Bagaeva, D. F., Alekseev, A. A., Nesterov, S. V., Tsybrov, F. M., ... Gordeliy, V. (2023). *Optogenetic control of lysosome function. bioRxiv* https://doi.org/10.1101/2025.10.31.685931.

Inoue, K., Del Carmen Marin, M., Tomida, S., Nakamura, R., Nakajima, Y., Olivucci, M., & Kandori, H. (2019). Red-shifting mutation of light-driven sodium-pump rhodopsin. *Nature Communications, 10*(1), 1993.

Inoue, K., Ito, S., Kato, Y., Nomura, Y., Shibata, M., Uchihashi, T., Tsunoda, S. P., & Kandori, H. (2016). A natural light-driven inward proton pump. *Nature Communications, 7*, 13415.

Inoue, K., Tahara, S., Kato, Y., Takeuchi, S., Tahara, T., & Kandori, H. (2018a). Spectroscopic study of proton-transfer mechanism of inward proton-pump rhodopsin, *Parvularcula oceani* xenorhodopsin. *Journal of Physical Chemistry B, 122*(25), 6453–6461.

Inoue, K., Tsukamoto, T., & Sudo, Y. (2013). Molecular and evolutionary aspects of microbial sensory rhodopsins. *Biochimica et Biophysica Acta, 1837*(5), 562–577.

Inoue, K., Tsunoda, S. P., Singh, M., Tomida, S., Hososhima, S., Konno, M., Nakamura, R., Watanabe, H., Bulzu, P. A., Banciu, H. L., Andrei, A., Uchihashi, T., Ghai, R., Béjà, O., & Kandori, H. (2020). Schizorhodopsins: A family of rhodopsins from Asgard archaea that function as light-driven inward H^+ pumps. *Science Advances, 6*(15), eaaz2441.

Inoue, S., Yoshizawa, S., Nakajima, Y., Kojima, K., Tsukamoto, T., Kikukawa, T., & Sudo, Y. (2018b). Spectroscopic characteristics of *Rubricoccus marinus* xenorhodopsin (*Rm*XeR) and a putative model for its inward H^+ transport mechanism. *Physical Chemistry Chemical Physics, 20*(5), 3172–3183.

Ito, S., Sugita, S., Inoue, K., & Kandori, H. (2017). FTIR analysis of a light-driven inward proton-pumping rhodopsin at 77 K. *Photochemistry and Photobiology, 93*(6), 1381–1387.

Ito, Y., Kovalev, K., Nishikino, T., Kandori, H., & Furutani, Y. (2026). Formation of a proton-conducting hydrogen-bond network during the L/M transition of *Ns*XeR uncovered by light-induced FTIR spectroscopy. *Journal of Physical Chemistry Letters, 17*(15), 4399–4407.

Ito, Y., Nishikino, T., Kandori, H., & Furutani, Y. (2025). The role of retinal chromophore photoisomerization in enhanced inward proton-pumping activity of xenorhodopsin from *Nanosalina*. *Biochimica et Biophysica Acta—Bioenergetics, 1866*(3), 149556.

Jahnke, K., Ritzmann, N., Fichtler, J., Nitschke, A., Dreher, Y., Abele, T., Hofhaus, G., Platzman, I., Schröder, R. R., Müller, D. J., Spatz, J. P., & Göpfrich, K. (2021). Proton gradients from light-harvesting *E. coli* control DNA assemblies for synthetic cells. *Nature Communications, 12*(1), 3967.

Jung, K. H., Trivedi, V. D., & Spudich, J. L. (2003). Demonstration of a sensory rhodopsin in eubacteria. *Molecular Microbiology, 47*(6), 1513–1522.

Kakegawa, W., Katoh, A., Narumi, S., Miura, E., Motohashi, J., Takahashi, A., Kohda, K., Fukazawa, Y., Yuzaki, M., & Matsuda, S. (2018). Optogenetic control of synaptic AMPA receptor endocytosis reveals roles of LTD in motor learning. *Neuron, 99*(5), 985–998.e986.

Kandori, H. (2020). Structure/function study of photoreceptive proteins by FTIR spectroscopy. *Bulletin of the Chemical Society of Japan, 93*(7), 904–926.

Kawanabe, A., Furutani, Y., Jung, K. H., & Kandori, H. (2007). Photochromism of *Anabaena* sensory rhodopsin. *Journal of the American Chemical Society, 129*(27), 8644–8649.

Kawanabe, A., Furutani, Y., Jung, K. H., & Kandori, H. (2009). Engineering an inward proton transport from a bacterial sensor rhodopsin. *Journal of the American Chemical Society, 131*(45), 16439–16444.

Kawanabe, A., Furutani, Y., Jung, K. H., & Kandori, H. (2011). An inward proton transport using *Anabaena* sensory rhodopsin. *Journal of Microbiology, 49*(1), 1–6.

Kawanabe, A., Furutani, Y., Yoon, S. R., Jung, K. H., & Kandori, H. (2008). FTIR study of the L intermediate of *Anabaena* sensory rhodopsin: Structural changes in the cytoplasmic region. *Biochemistry, 47*(38), 10033–10040.

Kawasaki, Y., Konno, M., & Inoue, K. (2021). Thermostable light-driven inward proton pump rhodopsins. *Chemical Physics Letters, 779*, 138868.

Kawasaki, Y., Konno, M., & Inoue, K. (2022). Kinetic study on the molecular mechanism of light-driven inward proton transport by schizorhodopsins. *Biochimica et Biophysica Acta—Biomembranes, 1864*(11), 184016.

Kehoe, D. M. (2010). Chromatic adaptation and the evolution of light color sensing in cyanobacteria. *Proceedings of the National Academy of Sciences of the United States of America, 107*(20), 9029–9030.

Kehoe, D. M., & Grossman, A. R. (1994). Complementary chromatic adaptation: Photoperception to gene regulation. *Seminars in Cell Biology, 5*(5), 303–313.

Kishi, K. E., Kim, Y. S., Fukuda, M., Inoue, M., Kusakizako, T., Wang, P. Y., Ramakrishnan, C., Byrne, E. F. X., Thadhani, E., Paggi, J. M., Matsui, T. E., Yamashita, K., Nagata, T., Konno, M., Quirin, S., Lo, M., Benster, T., Uemura, T., Liu, K., … Inoue, K. (2022). Structural basis for channel conduction in the pump-like channelrhodopsin ChRmine. *Cell, 185*(4), 672–689.e23.

Kitaguchi, A., Kanazawa, T., Nagashima, T., Yamazaki, T., Okitsu, T., Inoue, K., & Kawamura, I. (2026). Site-resolved $^{13}C/^{15}N$ solid-state nuclear magnetic resonance analysis of schizorhodopsin 1 in lipid bilayers. *Biophysical Chemistry, 335*, 107635.

Kojima, K., Miyoshi, N., Shibukawa, A., Chowdhury, S., Tsujimura, M., Noji, T., Ishikita, H., Yamanaka, A., & Sudo, Y. (2020). Green-sensitive, long-lived, step-functional anion channelrhodopsin-2 variant as a high-potential neural silencing tool. *Journal of Physical Chemistry Letters, 11*(15), 6214–6218.

Kojima, K., Yoshizawa, S., Hasegawa, M., Nakama, M., Kurihara, M., Kikukawa, T., & Sudo, Y. (2020). Lokiarchaeota archaeon schizorhodopsin-2 (LaSzR2) is an inward proton pump displaying a characteristic feature of acid-induced spectral blue-shift. *Scientific Reports, 10*(1), 20857.

Kovalev, K., Tsybrov, F., Alekseev, A., Shevchenko, V., Soloviov, D., Siletsky, S., Bourenkov, G., Agthe, M., Nikolova, M., von Stetten, D., Astashkin, R., Bukhdruker, S., Chizhov, I., Royant, A., Kuzmin, A., Gushchin, I., Rosselli, R., Rodriguez-Valera, F., Ilyinskiy, N., … Gordeliy, V. (2023). Mechanisms of inward transmembrane proton translocation. *Nature Structural & Molecular Biology, 30*(7), 970–979.

Koyanagi, M., & Terakita, A. (2014). Diversity of animal opsin-based pigments and their optogenetic potential. *Biochimica et Biophysica Acta, 1837*(5), 710–716.

Lanyi, J. K. (2004). Bacteriorhodopsin. *Annual Review of Physiology, 66*, 665–688.

Luecke, H., Schobert, B., Stagno, J., Imasheva, E. S., Wang, J. M., Balashov, S. P., & Lanyi, J. K. (2008). Crystallographic structure of xanthorhodopsin, the light-driven proton pump with a dual chromophore. *Proceedings of the National Academy of Sciences of the United States of America, 105*(43), 16561–16565.

Malakar, P., Das, I., Bhattacharya, S., Harris, A., Sheves, M., Brown, L. S., & Ruhman, S. (2022). Bidirectional photochemistry of antarctic microbial rhodopsin: Emerging trend of ballistic photoisomerization from the 13-*cis* resting state. *Journal of Physical Chemistry Letters, 13*(34), 8134–8140.

Marín, M. D. C., Konno, M., Yawo, H., & Inoue, K. (2023). Converting a natural-light-driven outward proton pump rhodopsin into an artificial inward proton pump. *Journal of the American Chemical Society, 145*(20), 10938–10942.

Morizumi, T., Ou, W. L., Van Eps, N., Inoue, K., Kandori, H., Brown, L. S., & Ernst, O. P. (2019). X-ray crystallographic structure and oligomerization of *Gloeobacter* rhodopsin. *Scientific Reports, 9*(1), 11283.

Nack, M., Radu, I., Gossing, M., Bamann, C., Bamberg, E., von Mollard, G. F., & Heberle, J. (2010). The DC gate in channelrhodopsin-2: Crucial hydrogen bonding interaction between C128 and D156. *Photochemical & Photobiological Sciences, 9*(2), 194–198.

Nagata, T., & Inoue, K. (2021). Rhodopsins at a glance. *Journal of Cell Science, 134*(22), jcs258989.

Nakanishi, K., Kojima, K., Sowa, Y., & Sudo, Y. (2024). Bidirectional optical control of proton motive force in *Escherichia coli* using microbial rhodopsins. *Journal of Physical Chemistry B, 128*(27), 6509–6517.

Nakao, S., Kojima, K., & Sudo, Y. (2022). Phototriggered apoptotic cell death (PTA) using the light-driven outward proton pump rhodopsin archaerhodopsin-3. *Journal of the American Chemical Society, 144*(9), 3771–3775.

Okuyama, A., Hososhima, S., Kandori, H., & Tsunoda, S. P. (2024). Driving forces of proton-pumping rhodopsins. *Biophysical Journal, 123*(24), 4274–4284.

Otto, H., Marti, T., Holz, M., Mogi, T., Lindau, M., Khorana, H. G., & Heyn, M. P. (1989). Aspartic acid-96 is the internal proton donor in the reprotonation of the Schiff base of bacteriorhodopsin. *Proceedings of the National Academy of Sciences of the United States of America, 86*(23), 9228–9232.

Pinto, M., Saliminasab, M., Harris, A., Lazaratos, M., Bondar, A. N., Ladizhansky, V., & Brown, L. S. (2024). The retinal chromophore environment in an inward light-driven proton pump studied by solid-state NMR and hydrogen-bond network analysis. *Physical Chemistry Chemical Physics, 26*(36), 24090–24108.

Pushkarev, A., Inoue, K., Larom, S., Flores-Uribe, J., Singh, M., Konno, M., Tomida, S., Ito, S., Nakamura, R., Tsunoda, S. P., Philosof, A., Sharon, I., Yutin, N., Koonin, E. V., Kandori, H., & Béjà, O. (2018). A distinct abundant group of microbial rhodopsins discovered using functional metagenomics. *Nature, 558*(7711), 595–599.

Ran, T., Ozorowski, G., Gao, Y., Sineshchekov, O. A., Wang, W., Spudich, J. L., & Luecke, H. (2013). Cross-protomer interaction with the photoactive site in oligomeric proteorhodopsin complexes. *Acta Crystallographica Section D: Biological Crystallography, 69*(Pt 10), 1965–1980.

Rost, B. R., Schneider, F., Grauel, M. K., Wozny, C., Bentz, C. G., Blessing, A., Rosenmund, T., Jentsch, T. J., Schmitz, D., Hegemann, P., & Rosenmund, C. (2015). Optogenetic acidification of synaptic vesicles and lysosomes. *Nature Neuroscience, 18*(12), 1845–1852.

Rozenberg, A., Inoue, K., Kandori, H., & Béjà, O. (2021). Microbial rhodopsins: The last two decades. *Annual Review of Microbiology, 75*, 427–447.

Saitou, N., & Nei, M. (1987). The neighbor-joining method: A new method for reconstructing phylogenetic trees. *Molecular Biology and Evolution, 4*(4), 406–425.

Schmidlechner, T. (2024). Optogenetic manipulation of cellular energetics in *Escherichia coli*. The University of Edinburgh.

Schobert, B., Brown, L. S., & Lanyi, J. K. (2003). Crystallographic structures of the M and N intermediates of bacteriorhodopsin: Assembly of a hydrogen-bonded chain of water molecules between Asp-96 and the retinal Schiff base. *Journal of Molecular Biology, 330*(3), 553–570.

Schubert, L., Chen, J. L., Fritz, T., Marxer, F., Langner, P., Hoffmann, K., Gamiz-Hernandez, A. P., Kaila, V. R. I., Schlesinger, R., & Heberle, J. (2023). Proton release reactions in the inward H^+ pump *Ns*XeR. *Journal of Physical Chemistry B, 127*(39), 8358–8369.

Shevchenko, V., Mager, T., Kovalev, K., Polovinkin, V., Alekseev, A., Juettner, J., Chizhov, I., Bamann, C., Vavourakis, C., Ghai, R., Gushchin, I., Borshchevskiy, V., Rogachev, A., Melnikov, I., Popov, A., Balandin, T., Rodriguez-Valera, F., Manstein, D. J., Bueldt, G., … Gordeliy, V. (2017). Inward H^+ pump xenorhodopsin: Mechanism and alternative optogenetic approach. *Science Advances, 3*(9), e1603187.

Shichida, Y., & Matsuyama, T. (2009). Evolution of opsins and phototransduction. *Philosophical Transactions of the Royal Society of London, Series B: Biological Sciences, 364*(1531), 2881–2895.

Shihoya, W., Inoue, K., Singh, M., Konno, M., Hososhima, S., Yamashita, K., Ikeda, K., Higuchi, A., Izume, T., Okazaki, S., Hashimoto, M., Mizutori, R., Tomida, S., Yamauchi, Y., Abe-Yoshizumi, R., Katayama, K., Tsunoda, S. P., Shibata, M., Furutani, Y., … Nureki, O. (2019). Crystal structure of heliorhodopsin. *Nature, 574*(7776), 132–136.

Shionoya, T., Singh, M., Mizuno, M., Kandori, H., & Mizutani, Y. (2021). Strongly hydrogen-bonded Schiff base and adjoining polyene twisting in the retinal chromophore of schizorhodopsins. *Biochemistry, 60*(41), 3050–3057.

Sineshchekov, O. A., Trivedi, V. D., Sasaki, J., & Spudich, J. L. (2005). Photochromicity of *Anabaena* sensory rhodopsin, an atypical microbial receptor with a *cis*-retinal light-adapted form. *Journal of Biological Chemistry, 280*(15), 14663–14668.

Tajima, S., Kim, Y. S., Fukuda, M., Jo, Y., Wang, P. Y., Paggi, J. M., Inoue, M., Byrne, E. F. X., Kishi, K. E., Nakamura, S., Ramakrishnan, C., Takaramoto, S., Nagata, T., Konno, M., Sugiura, M., Katayama, K., Matsui, T. E., Yamashita, K., Kim, S., … Kato, H. E. (2023). Structural basis for ion selectivity in potassium-selective channelrhodopsins. *Cell, 186*(20), 4325–4344.e26.

Tamura, K., Peterson, D., Peterson, N., Stecher, G., Nei, M., & Kumar, S. (2011). MEGA5: Molecular evolutionary genetics analysis using maximum likelihood, evolutionary distance, and maximum parsimony methods. *Molecular Biology and Evolution, 28*(10), 2731–2739.

Tsukamoto, T., Kikukawa, T., Kurata, T., Jung, K. H., Kamo, N., & Demura, M. (2013). Salt bridge in the conserved His–Asp cluster in *Gloeobacter* rhodopsin contributes to trimer formation. *FEBS Letters, 587*(4), 322–327.

Tsuneishi, T., Kojima, K., Kubota, F., Harashima, H., Yamada, Y., & Sudo, Y. (2023). Development of light-induced disruptive liposomes (LiDL) as a photoswitchable carrier for intracellular substance delivery. *Chemical Communications, 59*(49), 7591–7594.

Ugalde, J. A., Podell, S., Narasingarao, P., & Allen, E. E. (2011). Xenorhodopsins, an enigmatic new class of microbial rhodopsins horizontally transferred between archaea and bacteria. *Biology Direct, 6*, 52.

Urui, T., Hayashi, K., Mizuno, M., Inoue, K., Kandori, H., & Mizutani, Y. (2024). *Cis–trans* reisomerization preceding reprotonation of the retinal chromophore is common to the schizorhodopsin family: A simple and rational mechanism for inward proton pumping. *Journal of Physical Chemistry B, 128*(3), 744–754.

Urui, T., Mizuno, M., Abe-Yoshizumi, R., Kandori, H., & Mizutani, Y. (2024). Structural evolution of retinal chromophore in early intermediates of inward and outward proton-pumping rhodopsins. *Journal of Physical Chemistry B,, 129*(1), 41–51.

Urui, T., & Mizutani, Y. (2024). Origin of the difference in proton transport direction between inward and outward proton-pumping rhodopsins. *Accounts of Chemical Research,, 57*(22), 3292–3302.

Urui, T., Shionoya, T., Mizuno, M., Inoue, K., Kandori, H., & Mizutani, Y. (2024). Chromophore–protein interactions affecting the polyene twist and π–π^* energy gap of the retinal chromophore in schizorhodopsins. *Journal of Physical Chemistry B, 128*(10), 2389–2397.

Vlasova, A., Polyakova, A., Gromova, A., Dolotova, S., Bukhalovich, S., Bagaeva, D., Bondarev, N., Tsybrov, F., Kovalev, K., Mikhailov, A., Sidorov, D., Bogorodskiy, A., Ilyinsky, N., Kuklin, A., Vlasov, A., Borshchevskiy, V., & Ivanovich, V. (2023). Optogenetic cytosol acidification of mammalian cells using an inward proton-pumping rhodopsin. *International Journal of Biological Macromolecules, 242*(Pt 3), 124949.

Vlasova, A. D., Bukhalovich, S. M., Bagaeva, D. F., Polyakova, A. P., Ilyinsky, N. S., Nesterov, S. V., Tsybrov, F. M., Bogorodskiy, A. O., Zinovev, E. V., Mikhailov, A. E., Vlasov, A. V., Kuklin, A. I., Borshchevskiy, V. I., Bamberg, E., Uversky, V. N., & Gordeliy, V. I. (2024). Intracellular microbial rhodopsin-based optogenetics to control metabolism and cell signaling. *Chemical Society Reviews, 53*(7), 3327–3349.

Vogeley, L., Sineshchekov, O. A., Trivedi, V. D., Sasaki, J., Spudich, J. L., & Luecke, H. (2004). *Anabaena* sensory rhodopsin: A photochromic color sensor at 2.0 Å. *Science, 306*(5700), 1390–1393.

Weissbecker, J., Boumrifak, C., Breyer, M., Wießalla, T., Shevchenko, V., Mager, T., Slavov, C., Alekseev, A., Kovalev, K., Gordeliy, V., Bamberg, E., & Wachtveitl, J. (2021). The voltage dependent sidedness of the reprotonation of the retinal Schiff base determines the unique inward pumping of xenorhodopsin. *Angewandte Chemie International Edition, 60*(42), 23010–23017.

Wong, H. L., MacLeod, F. I., White, R. A., 3rd, Visscher, P. T., & Burns, B. P. (2020). Microbial dark matter filling the niche in hypersaline microbial mats. *Microbiome, 8*(1), 135.

Zuckerkandl, E., & Pauling, L. (1965). Evolutionary divergence and convergence in proteins. In V. Bryson, & H. J. Vogel (Eds.), *Evolving genes and proteins* (pp. 97–166). Academic Press.

Marín, M. D. C., Konno, M., Rozenberg, A., Béjà, O., & Inoue, K. (2026). Novel light-driven schizorhodopsins from antarctic *Minisyncoccota* (Patescibacteria) and cyanobacteria. *Biophysical Journal, 125*(10), 2391–2404.

CHAPTER

Computational approaches to understanding GPCR function

15

Shun Yokoi[a,b,c] and Ayori Mitsutake[a,*]

[a]*Department of Physics, School of Science and Technology, Meiji University, Kawasaki, Kanagawa, Japan,* [b]*International Institute for Integrative Sleep Medicine (WPI-IIIS), Tsukuba Institute for Advanced Research (TIAR), University of Tsukuba, Tsukuba, Ibaraki, Japan,* [c]*Department of Structural Biology, School of Medicine, Stanford University, Stanford, CA, United States*

Corresponding author: ayori@meiji.ac.jp

15.1 Introduction

G protein-coupled receptors (GPCRs) are a superfamily of membrane proteins characterized by seven transmembrane (TM) domains. GPCRs play crucial roles in various physiological functions, including sensory perception, immune response, and metabolic regulation (see reviews Rosenbaum et al., 2009; Zhang et al., 2024). Owing to their significant involvement in cellular signaling, over 30% of commercially available small-molecule drugs target GPCRs (Santos et al., 2016). GPCR activation is triggered by the extracellular binding of neurotransmitters, hormones, or other ligands, leading to conformational changes that enable interactions with intracellular signaling molecules, such as G proteins and arrestins. During GPCR activation the intracellular region of the sixth TM helix (TM6) moves outward, whereas that of the seventh TM helix (TM7) shifts inward. These structural rearrangements play a critical role in GPCR-mediated signal transduction by modulating the binding sites of other biomolecules (Hilger et al., 2018). GPCR signaling occurs through two distinct pathways: the G protein-dependent and G protein-independent pathways. GPCRs have diverse ligand specificities and pharmacological properties. There are two main types of ligands: agonists that mimic endogenous ligands to activate the receptor and antagonists that inhibit receptor activation. Moreover, a particularly important concept in GPCR pharmacology is biased signaling, wherein GPCRs selectively activate specific signaling pathways depending on their ligands. Biased signaling reflects the dynamic conformational state of the receptor and contributes to signal selectivity and functional diversity. This phenomenon has significant implications for drug development.

Recent advancements in structural biology, including X-ray crystallography (X-ray) and cryo-electron microscopy single-particle analysis (cryo-EM SPA), have significantly contributed to elucidating the three-dimensional structures of various GPCRs (Rasmussen et al., 2011; Venkatakrishnan et al., 2013). The antagonist- and agonist-bound structures correspond to the inactive and active structures, respectively. The large difference is in the arrangement of TM6 and TM7, which was mentioned earlier. Usually the active structure is solved with the G protein using cryo-EM SPA. Various pieces of information on GPCRs are organized in the GPCRdb (Isberg et al., 2016; Kolakowski, 1994; Kooistra et al., 2021; Munk et al., 2016). These structural insights have greatly advanced our understanding of GPCR functions and facilitated structure-based drug design (SBDD). Class A GPCRs contain the CWxP, PIF,

Protein Dynamics: Experimental and Computational Approaches. DOI: https://doi.org/10.1016/B978-0-443-30042-4.00010-9

Na^+ pocket, NPxxY, and DRY motifs (for recent reviews, see Conflitti et al., 2025; Zhou et al., 2019). The differences of these motifs between inactive and active states were discussed using the results of X-ray and cryo-EM experiments. These motifs are directly linked from the bottom of the ligand-binding pocket to the G-protein-coupling region during activation.

Although significant advances in experimental structural biology have provided static structural information on GPCRs, there is still a need for more dynamic structural information (Conflitti et al., 2025; Zhou et al., 2019). A significant understanding of the structural dynamics of GPCRs will facilitate SBDD and drug development. Molecular dynamics (MD) simulations are powerful tools for investigating the dynamic properties of GPCRs. MD simulations provide insights into dynamic conformational fluctuations and thermal perturbations, making them invaluable for studying GPCR activation dynamics (Gusach et al., 2020). GPCRmd, a database for MD simulation data on GPCRs, was constructed (Rodríguez-Espigares et al., 2020), and research on GPCRs using MD simulations has progressed (Dror et al., 2011; Latorraca et al., 2017). Given that the GPCR superfamily consists of over 800 distinct membrane proteins, comparative analyses of GPCR dynamics are essential to identify common activation mechanisms.

In this study, we focused on the orexin 2 receptor (OX2R) and explored its activation mechanism through large-scale MD simulations and a dynamic analysis method, relaxation mode (RM) analysis (RMA). OX2R is a crucial GPCR involved in the regulation of sleep-wake cycles, feeding behavior, and energy homeostasis in the central nervous system (de Lecea et al., 1998; Sakurai et al., 1998). OX2R belongs to the orexin system, which consists of two endogenous peptide ligands, orexin A and orexin B, and two GPCRs, orexin 1 receptor and OX2R. Among the disorders associated with the orexin system, narcolepsy is a chronic neurological disorder characterized by the inability to maintain stable sleep-wake patterns, leading to excessive daytime sleepiness and cataplexy (emotionally triggered muscle weakness) (Mahoney et al., 2019). In Japan the prevalence of narcolepsy is estimated to be approximately 1 in 600 individuals, highlighting the importance of understanding the orexin system for the development of diagnostic tools and effective treatments.

Recently, X-ray and cryo-EM have determined inactive and active structures (Heifetz et al., 2013; Hong et al., 2021; Rappas et al., 2020; Suno et al., 2018; Yin et al., 2015, 2022). The inactive structures of OX2R with small-molecule antagonists such as suvorexant or EMPA [Protein Data Bank (PDB) ID: 4S0V, 5WS3] were revealed by X-rays. The active structures of OX2R with agonists such as XGD, a few C-terminal amino acid residues of orexin B, or TAK-925 (PDB ID: 7L1V, 7L1U, and 7SQO) were determined using cryo-EM (Hong et al., 2021; Yin et al., 2022). Based on the difference between the inactive and active structures, important amino acid residues for activation are discussed in Yin et al. (2022).

For OX2R, we performed microsecond-scale MD simulations of the wild type (WT) and mutants in the ligand-free apo state, including the F(6.44)R mutant, which stabilizes the inactive state, and the L(3.43)Q, L(3.43)N, L(3.43)A, L(3.43)R, V(6.40)Y, and F(6.44)A mutants, which are constitutively active in other class A GPCRs. As similar dynamics are expected for OX2R, which belongs to the class A GPCRs, we expected that in constitutively active mutants (CAMs), we could capture the structural changes in TM6 outward movement and TM7 inward movement related to GPCR activation. Next, we performed extensive microsecond-scale MD simulations of the inactive states with and without an antagonist (EMPA) and active states with and without an agonist (XGD). Furthermore, we used simulations of the V(6.40)Y mutant for the analysis. To elucidate the atomic level dynamics and intermediate states in OX2R activation, we performed MD simulations for the six different systems, over 36 μs in total. We investigated the collective modes of TM1-7 and the structural flexibility related to their activation. In the MD simulations of both systems, data analysis was performed using RMA. Consequently, we successfully extracted the modes related to activation and investigated the structural dynamics using these results.

In this book, we review our results of microsecond-scale MD simulations of OX2R (Yokoi & Mitsutake, 2021, 2022; Yokoi et al., 2024) and reconsider the relationship between the collective modes obtained by RMA and the characteristic distances between atoms or the motifs related to the activation mechanism of OX2R.

15.2 Characteristic amino acid residues

15.2.1 Characteristic motifs

Here, we summarize the common key motifs of class A GPCRs and important amino acid residues during the activation of OX2R. Class A GPCRs contain CWxP, PIF, Na^+ pockets, NPxxY, and DRY motifs (for reviews see Conflitti et al., 2025; Zhou et al., 2019), as shown in Fig. 15.1. These motifs are directly linked from the bottom of the ligand-binding pocket to the G-protein-coupling region during activation. The binding of agonists triggers the rotamer switch W(6.48) (Y(6.48) in OX2R), which is a highly conserved amino acid residue in the CWxP motif that links the rotation of F(6.44). Conformational rearrangements occurred in the PIF [P(5.50), I(3.40), and F(6.44)] (PVF in OX2R) and the Na^+ pocket [D(2.50), S(3.39), N(7.45), and N(7.49)]. These rearrangements triggered the outward movement of TM6. Repacking of amino acid residues in the Na^+ pocket causes TM7 to move toward TM3. Y(7.53) in the NPxxY motif rotates toward TM3; as a result, new contacts between Y(7.53) and amino acid residues in TM3 [L(3.43), I(3.46), and R(3.50)] are generated. Finally, the DRY motif, one of the most conserved motifs, is located near the G protein-binding site and forms a salt bridge between D/E(3.49) and R(3.50). R(3.50) formed an additional salt bridge with D(6.30). D(6.30) is known as the ionic lock that connects the intracellular ends of TM3 and TM6 to stabilize receptors in an inactive state. These contacts are eliminated after agonist binding, and R(3.50) is released to interact with other amino acid residues and facilitate G protein coupling. Notably an acidic residue at position 6.30 is less conserved in 30% of class A receptors. Alternatively, R(3.50) may form polar interactions with other polar residues in TM6 [i.e., T(6.34) in κ-OR and μ-OR] to mediate the activation. Collectively, these rearrangements and reorganizations of conserved motifs are critical to the activation of class A GPCRs.

In Fig. 15.1, we show the inactive (A and D) and active (C and E) structures of OX2R, with amino acid residues related to these motifs. In OX2R, for CWxP motif, W is Y [C(6.47), Y(6.48), L(6.49), and P(6.50)]; for PIF motif, I is V [P(5.50), V(3.40), and F(6.44)]; and DRY motif is DRWY motif [D(3.49), R(3.50), W(3.51), and Y(3.52)]. The amino acid residues of the CYxP, PVF, Na^+ pocket, NPxxY, and DRWY motifs are colored *green, yellow, red, yellow green*, and *blue*, respectively. The heavy atoms of Y(6.48) in the CYxP motif, Y(7.53) in the NPxxY motif, and D(3.49) and R(3.50) in the DRWY motif are shown in Fig. 15.1. Here the heavy atoms of Q(3.32) are also shown in orange because Q(3.32) is important for ligand binding. As shown in the differences between the inactive and active structures of Q(3.32), F(6.44), Y(7.53), and R(3.50) were large. The side chain of Q(3.32) changes its structure from downward to upward during activation (*orange*). In the structure of agonist-bound OX2R the side chain

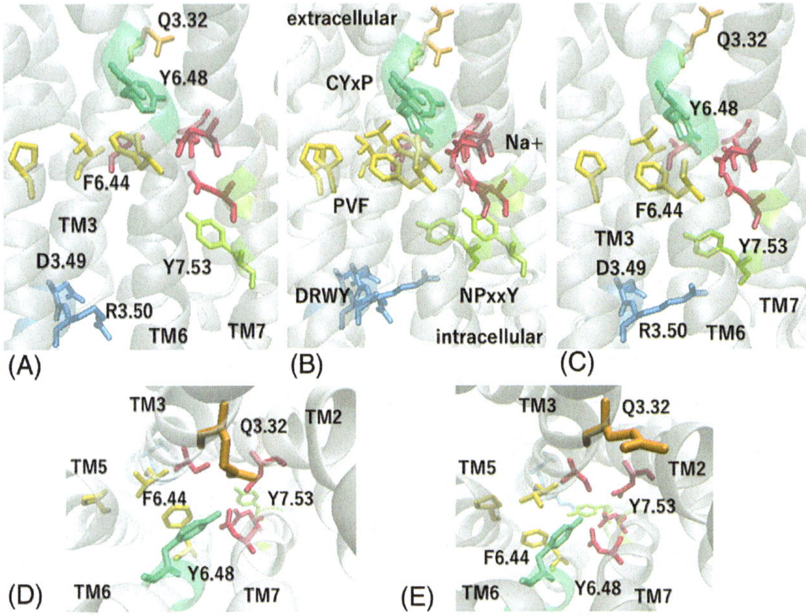

FIGURE 15.1

Inactive (A and D) and active (C and E) structures and overlay of two structures (B) of OX2R. Top images are a side view relative to the membrane, and the bottom images are a view from the extracellular side toward the intracellular direction. Amino acid residues of CWxP, PVF, Na^+ pocket, NPxxY, or DRWY motif were colored in *green, yellow, red, yellow green*, or *blue*, respectively. Especially the heavy atoms of Y(6.48) in CWxP motif, Y(7.53) in NPxxY motif, D(3.49), R(3.50) in DRWY motif, and Q(3.32) are shown. All figures of structures are drawn using VMD (Humphrey et al., 1996). *OX2R*, Orexin 2 receptor.

of Q(3.32) is extended and projects upward, and the upward displacement of TM3 is discussed. The side chains of Y(6.48) and F(6.44) appeared to push other amino acid residues downward (*green* and *yellow*). Y(7.53) in NPxxY motif (*yellow green*) moves inward by about 4 Å and moves toward TM3 (Yin et al., 2022). The side chain of R(3.50) in the DRWY motif changes its structure from downward to upward during activation (*blue*). In the structure with the antagonist, suvorexant D(3.49) and R(3.50) formed a salt bridge, whereas R(3.50) and W(3.51) contacted the cytoplasmic ends of TM5 and TM6 (Q(5.60), I(5.61), and L(6.37). These structural differences are similar in Class A GPCRs. We focused on the relationship between our results and the characteristic amino acid residues in these motifs.

15.2.2 Characteristic distances

Recently, in GPCRmd, the distance of the Cα atoms between x(2.46) and x(6.37) is used to distinguish between inactive and active states in Class A GPCR. The average distances of inactive, intermediate, and active states are 12.2, 13.6, and 16.4 Å, respectively. The distance was used for the classification of closed, intermediate, and open states. Closed and open states correspond to inactive and active states, respectively. GPCRdb also used the distance to distinguish between inactive and active states (https://docs.gpcrdb.org/structures.html).

15.3 Method

15.3.1 Relaxation mode analysis

We briefly explain the dynamic analysis method and RMA. MD simulations are powerful tools for investigating the motions of complex molecular systems. As proteins exhibit highly intricate atomic-level motions, it is essential to extract functionally relevant degrees of freedom from these dynamics. When experimental data provide characteristic interatomic distances or reference structures, it is possible to analyze structural changes using these values. However, analyses limited to predefined information are inherently constrained. Thus automated approaches are required to identify functionally important degrees of freedom. To address this challenge, mode decomposition methods have been applied to protein systems.

In this study, we introduced and developed a dynamic analysis technique, RMA (Hirao et al., 1997; Koseki et al., 1997; Takano & Miyashita, 1995), which can automatically extract characteristic motions and analyze simulation data. RMA was developed to investigate the "dynamic" properties of spin systems and homopolymer systems for Monte Carlo and MD simulations (Hirao et al., 1997; Koseki et al., 1997) and has been applied to various polymer systems. RMA can extract slow RMs. RMA has also been applied to biomolecular systems (Mitsutake & Takano, 2018; Mitsutake et al., 2011), and several new methods related to RMA have been developed, such as principal component (PC) RMA (PCRMA) (Nagai et al., 2013), two-step RMA with multiple evolution times (two-step RMA) (Karasawa et al., 2017; Natori & Takano, 2017), Markov-state RMA (MSRMA) (Mitsutake & Takano, 2015), and positive-definite RMA (PDRMA) (Karasawa et al., 2019). A comparison between RMA and diffusion map methods was presented in Fujisaki et al. (2018). RMA has also been applied to protein folding simulations (Mitsutake & Takano, 2015, 2019; Yamamoto et al., 2021). A theoretical construction for comparison with the neutron scattering experiments is introduced in Karasawa et al. (2024).

Previously, we investigated the dynamics of OX2R to examine collective modes using RMA (Yokoi & Mitsutake, 2021, 2022; Yokoi et al., 2024). Principal component analysis (PCA) is a broadly used mode decomposition method. Whereas PCA extracts modes with large variances from the average structure, RMA extracts modes with slow relaxation and then extracts rare events wherein structural fluctuations are relatively small, such as conformational changes in the side chains. In PCA, structural changes with large fluctuations can be detected to monitor PC mode during a simulation. On the other hand, in RMA, to monitor RM, the structural change for slow motion or rare events can be detected. Similar to PCA, RMA can also be used to calculate structural fluctuations along the RM. The collective motion can be obtained.

Here, we consider the 3N-dimensional column vector $\boldsymbol{R}$ composed of atomic coordinates relative to their average coordinates,

$$\boldsymbol{R} = \left(r'^{T}_{1}, \boldsymbol{r}'^{T}_{1}, \cdots, \boldsymbol{r}'^{T}_{N}\right) = \left(x'_{1}, y'_{1}, z'_{1}, \cdots, x'_{N}, y'_{N}, z'_{N}\right)$$

with $\boldsymbol{r'}_i = \boldsymbol{r}_i - \boldsymbol{r}_i$, where $\boldsymbol{r}_i$ is the coordinate of the ith atom, $\boldsymbol{r}_i$ is its average coordinate after removing the translational and rotational degrees of freedom, and N is the number of atoms in the protein. The RMA approximately estimates the slow RMs and their relaxation rates by solving the generalized eigenvalue problem of the correlation matrices of the coordinates,

$$\sum_{j=1}^{3N} C_{i,j}(t_0+\tau)f_{p,j} = e^{-\lambda_p\tau}\sum_{j=1}^{3N} C_{i,j}(t_0)f_{p,j},$$

where $C_{i,j}(t)$ is the component of the $3N \times 3N$ symmetric matrix $C_{i,j}(t)$ defined by $C_{i,j}(t) = \langle R_i(t)R_j(0)\rangle$.

Here, t_0 is the evolution time, τ is a time interval, λ_p is the relaxation state of the estimated RMs, $f_{p,j}$, and $\langle\cdots\rangle$ is the ensemble average. The parameter t_0 is introduced to reduce the relative weight of the faster modes contained in $\boldsymbol{R}$, and better estimation of the slow RMs is expected with sufficiently large values of t_0. In the RMA, $3N$–6 RMs are obtained because the translational and rotational degrees of freedom are removed from $\boldsymbol{R}$. By multiplying f_p^T the RM X_p is given by

$$X_p \approx \sum_{i=1}^{3N} \tilde{f}_{p,i}R_i$$

Here, $\tilde{f}_{p,i} = e^{-\lambda_p\tau}f_{p,i}$ and $0 < \lambda_1 < \lambda_2 < \cdots$. The mode with the slowest relaxation time ($\sim 1/\lambda_1$) corresponds to X_1. Below, we refer to X_p as RMp.

The $\boldsymbol{R}$ can be reconstructed by

$$R_i \approx \sum_{i=1}^{3N-6} \tilde{g}_{i,p}X_p$$

Here, $g_{i,p} = \sum_{i=1}^{3N} C_{i,j}(t_0)f_{p,j}$ and $\tilde{g}_{i,p} = e^{\lambda_p t_0/2}g_{i,p}$.

The direction of structural fluctuation along RMp is proportional to $\tilde{g}_{i,p}$.

In the present work the coordinates of the Cα atoms (in some cases the Cα and Cβ atoms) in the TM domain (TM1: residues 53–81, TM2: residues 88–117, TM3: residues 128–157, TM4: residues 165–190, TM5: residues 217–251, TM6: residues 297–330, and TM7: residues 339–367) were used to perform RMA for clarifying the structural differences between the TM domains. The evolution time, t_0, and time interval, τ, are set to 0 and 100 ps, respectively.

15.3.2 Molecular dynamics simulations of an inactive structure of wild type and mutants without ligand

First, we performed microsecond-scale MD simulations of an inactive structure without a ligand, introducing mutations known to stabilize the inactive state and constitutively active mutations in other Class A GPCRs. For systems with constitutively active mutations, we hypothesized that structural transitions from the inactive to active conformation could be observed even in the absence of a ligand. Using the inactive structure obtained by X-ray (PDB ID: 5WS3) as the WT structure, we created an F(6.44)R mutant that stabilizes the inactive state in the same class of GPCRs. In addition, we generated six CAMs of the same class of GPCRs: L(3.43)Q, L(3.43)N, L(3.43)A, L(3.43)R, V(6.40)Y, and F(6.44)A

(Zhou et al., 2019). As mentioned in Feng et al. (2014) the x(3.43)Q/N/A/R and x(6.40)Y mutants loosened the hydrophobic lock, thereby weakening contact between TM3 and TM6. Consequently the outward movement of TM6 was promoted, eventually rendering the receptor constitutively active. The x(6.44)A mutants also weakened the contact between TM3 and TM6, which reduced the energy barrier for the release of TM6 and facilitated the outward movement of TM6. F(6.44) in OX2R was located in the PIF motif. The V(6.40)Y mutant was also introduced for the structural analysis of its active state. As similar dynamics are expected for OX2R, which belongs to class A GPCRs, we investigated the differences in dynamics between the WT and mutants. Here, the missing structure, which consisted of 45 amino acid residues on the intracellular side of TM5, TM6, and ICL3 in GPCRdb, was modeled with 23 amino acid residues. MD simulations were performed for approximately 500–1200 ns for each. Further details are provided elsewhere (Yokoi & Mitsutake, 2021).

15.3.3 Molecular dynamics simulations of the inactive and active structures without/with ligands

Based on the inactive and active structures (PDB ID: 5WS3, 7L1V), to elucidate the atomic-level dynamics and intermediate state of OX2R activation, we performed MD simulations of five different systems: inactive- and active-state WT OX2R monomer, complexes with an antagonist (EMPA) or an agonist (XGD), and the V(6.40)Y mutant, which is known as a CAM in other Class A GPCRs. The antagonist- and agonist-bound structures correspond to inactive and active states, respectively. Recently an antagonist (EMPA)-bound structure was solved using X-ray crystallography, and the agonist (XGD)-bound structure was solved using cryo-EM SPA. The G protein bound to the active structure was then removed from the system. For the V(6.40)Y mutant, previous simulation data were used here, because TM6 outward and TM7 inward movements, which are structural changes during activation, were observed. The structural shift from the inactive to the active state was not a complete active-state structure; however, it was similar. Thus this structure is referred to as an active-like structure. MD simulations were performed for approximately 1100–2000 ns for each. Further details are provided elsewhere (Yokoi et al., 2024).

15.4 Results and discussion

15.4.1 Molecular dynamics simulations of an inactive structure of wild type and mutants without ligand

Here, we review the results of previous MD simulations of ligand-free inactive structures and discuss the relationship between the obtained modes and the characteristic distances relevant to receptor activation. For further details, please refer to Yokoi and Mitsutake (2021). We also conducted simulations with amino acid substitutions known to cause constitutive activation in other Class A GPCRs, aiming to capture the structural transitions associated with the outward movement of TM6 and the inward movement of TM7 during activation. RMA was applied to all 20 MD simulations of the WT and mutant receptors to extract the characteristic modes of structural change.

Fig. 15.2 shows the time series of the first RMA mode (RM1) for both WT and mutant simulations. The V(6.40)Y mutant exhibits a different tendency of RM1 from the other simulations. To further

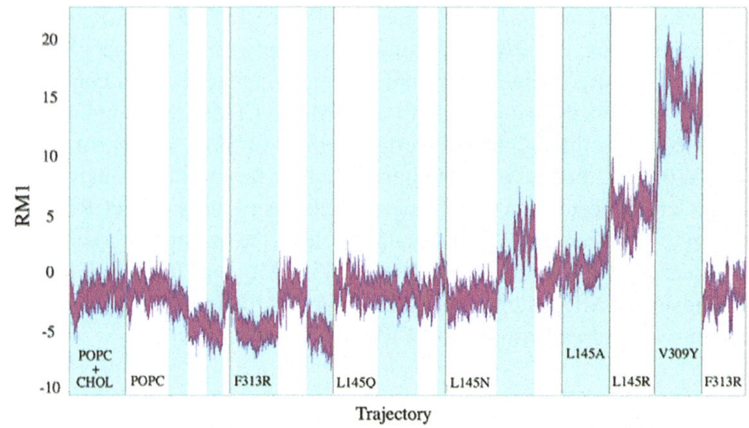

FIGURE 15.2

Time series of RM1 for all the simulations (20 simulations).

Reprinted with permission from Yokoi, S., & Mitsutake, A. (2021). Molecular dynamics simulations for the determination of the characteristic structural differences between inactive and active states of wild type and mutants of the orexin2 receptor. The Journal of Physical Chemistry B, 125*(17), 4286–4298. https://doi.org/10.1021/acs.jpcb.0c10985. Copyright 2021 American Chemical Society.*

investigate the characteristic structural dynamics, RMA was applied to a 900 ns MD simulation of the V(6.40)Y mutant. As shown in Fig. 15.3 a large shift in RM1 occurs at approximately 200 ns. To characterize the shift, we analyzed the structural fluctuations along RM1, similar to the PCA. The resulting fluctuation patterns of the Cα and Cβ atoms in the TM domains along RM1 revealed that these changes corresponded to an outward movement of TM6 and an inward movement of TM7 on the intracellular side. We observed the characteristic structural change from an inactive to an active-like structure in V(6.40)Y mutant. Focusing on the 0–200 ns window a swapping event between M(6.36) and L(7.56) was detected, as shown in Fig. 15.4. This rearrangement was influenced by interaction between Y(6.40) and Y(7.53), located at the mutation site and within the NPxxY motif, respectively. The conformational change involved the amino acid residues rearranging from their initial positions (Fig. 15.4A) to swapped positions (Fig. 15.4B), coinciding with the characteristic movements of TM6 and TM7 observed around 200 ns.

We quantitatively evaluated the interatomic distances corresponding to the degrees of freedom of movement of TM6 and TM7 by calculating the interatomic distances between various amino acid residues and investigating the interatomic distances that showed a similar tendency to that of the time series of RM1. We identified that the distance of the Cα atoms between I(3.46) and Y(7.53) is a good index to quantify these movements. The amino acid residue at x(3.46) is not conserved in all class A GPCRs but corresponds to a hydrophobic residue, such as I, L, or M, in many class A GPCRs. The amino acid residue at Y(7.53) is a part of the NPxxY motif. To validate the relevance of the distance as an activation index, we analyzed the distance for the active and inactive structures of other class A GPCRs using the structural database (PDB). In the inactive state the distance is typically 14–15 Å, whereas in the active state, it is significantly reduced to 3–5 Å. In the V(6.40)Y mutant simulations the distance was

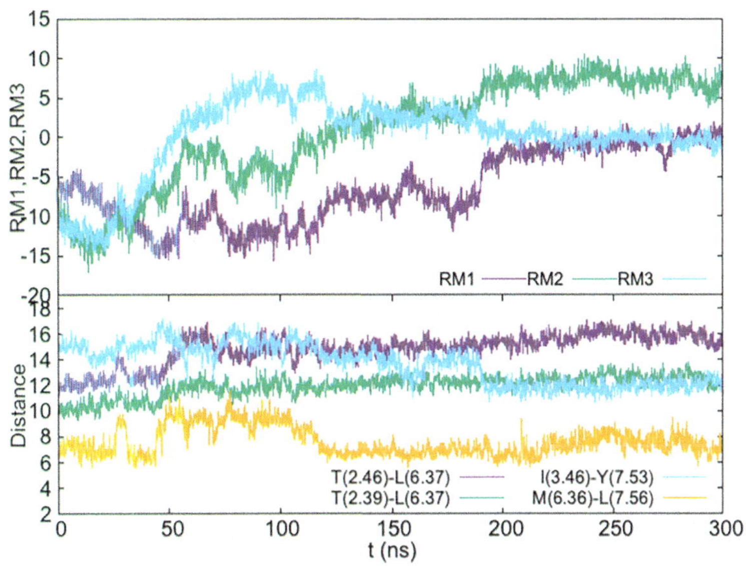

FIGURE 15.3

Time series of RM1–RM3, the distance between M(6.36) and L(7.56), the distance between I(3.46) and Y(7.53), the distance between T(2.39) and L(3.67), and the distance between T(2.46) and L(6.37) for the 0–300 ns MD simulation of V(6.40)Y.

initially 14–15 Å and decreased by 1.5–4 Å after TM6 and TM7 movements, indicating a shift toward an active-like state. In addition, we identified the T(2.39)–L(6.37) Cα distance as another useful metric for monitoring TM6 outward movement. The distances between I(3.46) and Y(7.53) and between T(2.39) and L(6.37) are highly correlated with activation (Fig. 15.5 in Yokoi & Mitsutake, 2021).

Fig. 15.3 presents the time series of RM1–RM3, distances M(6.36) and L(7.56), and distances I(3.46)–Y(7.53), T(2.39)–L(6.37), and T(2.46)–L(6.37) during the first 300 ns of the V(6.40)Y simulation. Snapshots of structures with these residues are illustrated in Fig. 15.4. The highly correlated changes in the T(2.39)–L(6.37) and T(2.46)–L(6.37) distances reflect the distance between TM2 and TM6. The maximum M(6.36)–L(7.56) distance indicates the precise onset of the swapping event. Swapping events for TM6 and TM7 as illustrated in Fig. 15.4A and B began at approximately 50 ns. Owing to this event the T(2.46)–L(6.37) distance increases from 12 to 16 Å, as shown in Fig. 15.3. RM3 also changed near 50 ns and detected the structural change that corresponded to the swapping event. Subsequently the I(3.46)–Y(7.53) distance decreased from 15 to 11 Å near 200 ns. RM1 detects structural changes related to the distance between TM3 and TM7. Both final distances correspond to one for an active-like structure. These analyses suggest that the swapping of M(6.36) and L(7.56) triggers the structural rearrangements of TM6 and TM7. Once this event occurred, TM7 appeared to move more readily toward TM3. RMA automatically extracts structural changes related to GPCR activation from complex protein motions. Based on the RMA results, in combination with available structural data from X-ray crystallography and cryo-EM, we propose a detailed mechanism for the structural transition of OX2R from an inactive to an active-like state.

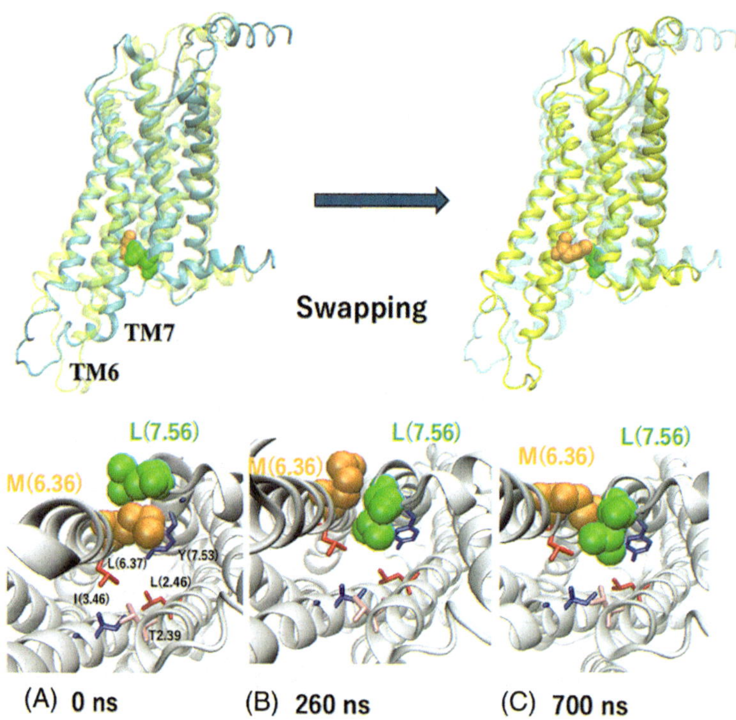

FIGURE 15.4

Inactive (*cyan*) and active-like (*yellow*) structures with M(6.36) (*orange*) and L(7.56) (*green*). The structures at 0 ns (A), 260 ns (B), and 700 ns (C) are also shown. The heavy atoms of I(3.46) and Y(7.53) are colored in *blue*, those of T(2.46) and L(6.37) are colored in *red*, and those of T(2.39) are colored in *pink*. The distances related to these residues are shown in Fig. 15.3.

15.4.2 Molecular dynamics simulations for the inactive or active structures without/with ligand

We investigated structural dynamics and intermediate states of OX2R by analyzing 30 μs scale MD simulations under four conditions: inactive structure without/with antagonist (EMPA) and active structure without/with agonist (XGD). In addition, simulations of the V(6.40)Y mutant were performed to obtain further insights into the receptor activation pathway. To explore the flexibility of OX2R, we focused on local fluctuations in both inactive and active states. Furthermore, the collective motion of the seven TM helices was analyzed in detail. We further elaborated on these findings, focusing on the relationship between the obtained modes and known structural motifs. See Yokoi et al. (2024) for further details.

Fig. 15.5 shows the results of the RMA for all the simulation data. RM1 exhibited negative values for the inactive state and positive values for the active state, effectively distinguishing between the two states. In the V(6.40)Y mutant a shift from an inactive to an active-like structure was observed. Notably

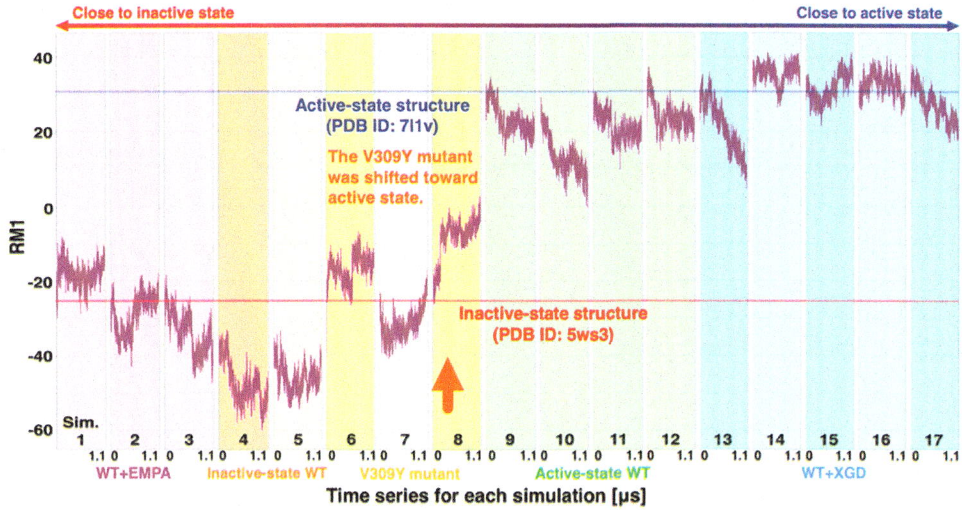

FIGURE 15.5

Time series of RM1 for all the simulations. *PDB*, Protein Data Bank.

Reprinted with permission from Yokoi, S., Suno, R., & Mitsutake, A. (2024). Structural and computational insights into dynamics and intermediate states of orexin 2 receptor signaling. The Journal of Physical Chemistry B, 128*(25), 6082–6096. https://doi.org/10.1021/acs.jpcb.4c00730. Copyright 2024 American Chemical Society.*

the active structure displayed partial reversion to an inactive state in the absence of an agonist. These observations confirm that RM1 captures the essential structural transitions associated with activation.

Next, we examined the structural fluctuations along RM1 and identified the key amino acid residues related to the characteristic structural changes between the inactive and active states, as shown in Fig. 15.6. In our previous study, we focused on three amino acid residues in TM3-Q(3.32), V(3.40), and R(3.50) that showed large fluctuations. These amino acid residues correspond to well-known motifs in class A GPCRs. Q(3.32) is implicated in ligand binding, V(3.40) is part of the PVF motif, and R(3.50) is part of the conserved DRWY motif. Further analysis highlighted the correlated motions across different TM helices. As shown in Fig. 15.6, large fluctuations were observed in the intracellular regions of the TM1, TM6, and TM7. Specifically, in TM6, increased fluctuations occurred toward the cytoplasmic end from amino acid residue Y(6.48). RM1 captured the "toggle switch" behavior involving amino acid residues Y(6.48) and F(6.44). Y(7.53) in the NPxxY motif also has large structural fluctuations. These differences consist of the results of the structural differences of the motifs.

To visualize the collectivity of the seven TM helices, we calculated the direction of the structural fluctuation along RM1 in three-dimensional space. Although the dynamics are best appreciated through animation, in this study, we illustrate the structural changes along RM1 by displaying structures corresponding to negative and positive RM1 values, as shown in Fig. 15.7(A and C) and (B and D), respectively. The structures in Fig. 15.7A and C, which have negative RM1 values, correspond to the inactive state, whereas those in Fig. 15.7B and D with positive RM1 values represent the active state. The motion from (A) to (B) along RM1 captures the structural transition associated with the activation.

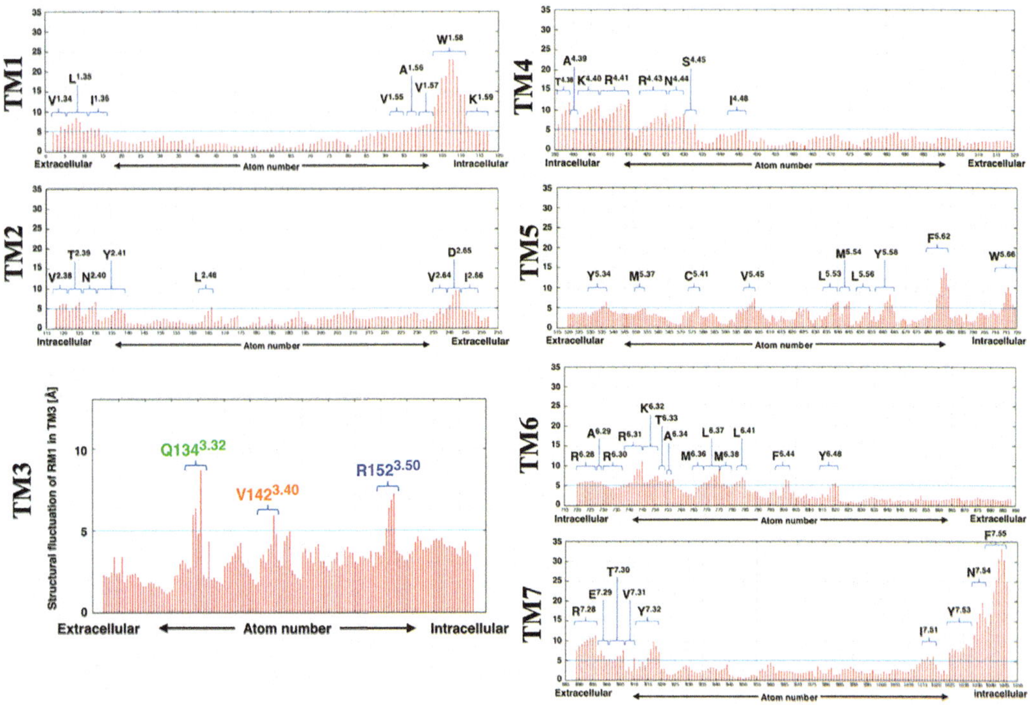

FIGURE 15.6

Structural fluctuations along RM1 on seven TM helices. *TM*, Transmembrane

Modified from Yokoi, S., Suno, R., & Mitsutake, A. (2024). Structural and computational insights into dynamics and intermediate states of orexin 2 receptor signaling. The Journal of Physical Chemistry B, 128*(25), 6082–6096. https://doi.org/10.1021/acs.jpcb.4c00730. Copyright 2024 American Chemical Society.*

This collective model allowed us to simplify and interpret the structural changes involved in GPCR activation.

We examined how the collective mode obtained by RMA was related to known activation motifs. Fig. 15.7 compares the conformational changes of key amino acid residues, such as R(2.50) (*blue*), which shifted from a downward to a lateral orientation, and Q(3.32) (*orange*), which transitioned from a downward to an upward orientation during activation. In addition, the packing of the PVF motif (*yellow*) and Na^+ pocket (*red*) increased upon activation, which is consistent with findings from crystallographic studies. Thus RM1 effectively captures these structural differences. Furthermore, Fig. 15.7E illustrates the directionality of the atomic motion along RM1 relative to the membrane plane. Amino acid residues on TM3 have values close to one, indicating that TM3 undergoes an upward movement toward the extracellular side. In contrast, TM6 and TM7 exhibited downward motion toward the cytoplasmic side. Overall, RM1 represents the structural differences between the inactive and active states. Upon agonist binding, TM3 shifts upward as Q(3.32) adopts an upward-facing conformation in its active state. Consequently, TM6 and TM7 shifted downward, suggesting a swapping-like movement.

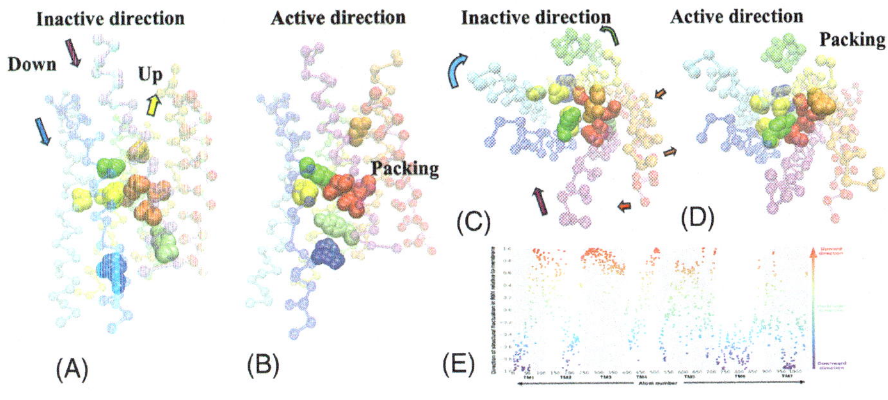

FIGURE 15.7

Structures corresponding to negative and positive RM1 values. Structures in (A) and (C), which have negative RM1 values, correspond to the inactive state, whereas those in (B) and (D) with positive RM1 values represent the active state. The inner product between the direction of residue motion along RM1 and the vector directed from the membrane interior to exterior (E) are shown. Structures of (C) and (D) are views from the extracellular side toward the intracellular direction. Structures are generated with a threefold magnification to clarify the structural differences. Cα atoms and bonds connecting them are illustrated for TM1 (*red*), TM2 (*orange*), TM3 (*yellow*), TM4 (*green*), TM5 (*cyan*), TM6 (*blue*), and TM7 (*purple*). The heavy atoms for PVF, Na^+ pocket, and DRWY motif were colored in *yellow, red*, and *blue*, respectively. The heavy atoms of Q(3.32) (*orange*), Y(6.48) (*green*), and Y(7.53) (*light green*) are also illustrated. *TM*, Transmembrane.

Reprinted with permission from Yokoi, S., Suno, R., & Mitsutake, A. (2024). Structural and computational insights into dynamics and intermediate states of orexin 2 receptor signaling. The Journal of Physical Chemistry B, 128*(25), 6082–6096. https://doi.org/10.1021/acs.jpcb.4c00730. Copyright 2024 American Chemical Society.*

These observations are consistent with previous structural biology studies, including X-ray crystallography and cryo-EM analyses, and our results reinforce these findings by reflecting the structural transitions between inactive and active structures. RMA is a powerful tool that can automatically extract the degrees of freedom to distinguish inactive from active structures, enabling a detailed investigation of the collective modes of the seven TM helices.

Previously, we analyzed the specific behavior of three key amino acid residues in TM3 during activation. Based on structural fluctuations in these characteristic amino acid residues, we evaluated the flexibility of OX2R as follows: Fig. 15.8 summarizes the conformational variability under antagonist- and agonist-bound conditions (five different states were compared in Yokoi et al., 2024).

Q(3.32), which plays a crucial role in ligand binding, exhibited substantial fluctuations in antagonist-bound simulations, deviating from the crystal structures. In contrast, in agonist-bound simulations, Q(3.32) remained stably fixed in an upward conformation. This observation is consistent with previous crystallographic findings and is further supported by the MD simulation data. In the inactive state the extracellular domain remains relatively flexible, whereas in the active state, it is tightly packed. The PVF motif displays high conformational variability in the inactive state but forms a stable pocket upon activation. Intracellularly, R(3.50) is fixed in a downward conformation in inactive structures, limiting accessibility for G protein binding. This rigidity was corroborated by MD simulations, which

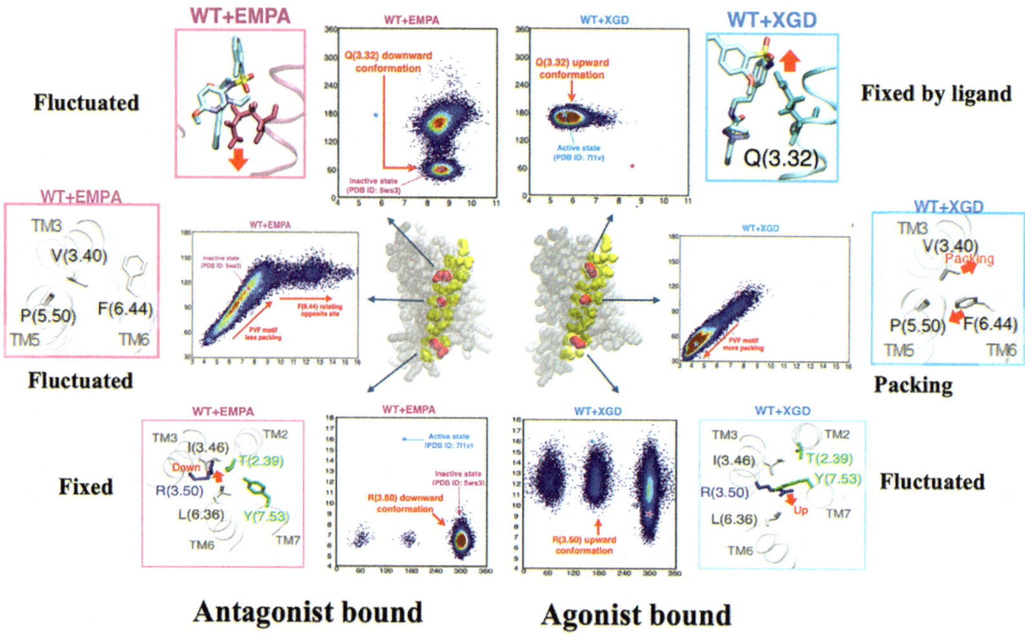

FIGURE 15.8

Summary of dynamics of OX2R in two systems: inactive-state WT with an antagonist (EMPA) and active-state WT with an agonist (XGD) obtained by our simulations. *OX2R*, Orexin 2 receptor; *WT*, wild type.

Modified from Yokoi, S., Suno, R., & Mitsutake, A. (2024). Structural and computational insights into dynamics and intermediate states of orexin 2 receptor signaling. The Journal of Physical Chemistry B, 128*(25), 6082–6096. https://doi.org/10.1021/acs.jpcb.4c00730. Copyright 2024 American Chemical Society.*

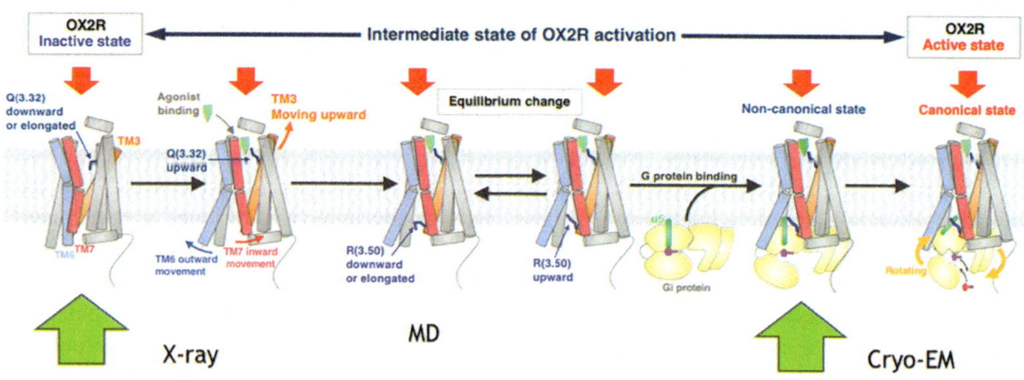

FIGURE 15.9

The insights into the activation mechanism obtained from our simulation results.

Modified from Yokoi, S., Suno, R., & Mitsutake, A. (2024). Structural and computational insights into dynamics and intermediate states of orexin 2 receptor signaling. The Journal of Physical Chemistry B, 128*(25), 6082–6096. https://doi.org/10.1021/acs.jpcb.4c00730. Copyright 2024 American Chemical Society.*

revealed minimal fluctuations in the ligand-bound inactive states. In contrast, in the active structure, R(3.50) adopts multiple conformations, indicating increased flexibility and reduced steric hindrance for G protein coupling because of the rearrangements of TM6 and TM7.

In summary, in the inactive structure, the extracellular side and the central region (including the PVF motif) are loosely packed, whereas the intracellular side is compact, immobilizing R(3.50). In the active structure, Q(3.32) was firmly bound and fixed by the ligand, and the PVF motif was well-packed, whereas the intracellular region became flexible, facilitating G protein binding. By examining the structural fluctuations of key amino acid residues through MD simulations, we gained insight into the intrinsic flexibility of the protein. The structural changes observed along RM1, as discussed previously, are in good agreement with these findings. Using RMA-derived models, we identified the critical amino acid residues and assessed their local flexibility to provide valuable insights into the relationship between structural fluctuations and function. The correlation between the modes and known functional motifs further supported the effectiveness of the RMA approach.

15.5 Conclusion

Based on the calculations in Yokoi and Mitsutake (2021, 2022) and Yokoi et al. (2024), the activation mechanism was elucidated, as shown in Fig. 15.9. These results are consistent with the static structural information obtained using techniques such as X-ray crystallography and cryo-EM, as indicated by the arrows in Fig. 15.9. Furthermore, we analyzed the results of the MD simulations under conditions that are experimentally difficult to achieve, such as in agonist-bound systems without G protein binding. We investigated structural fluctuations in key amino acid residues across various systems, as shown in Fig. 15.9. This analysis enabled us to complement the static structures and reinforce the previously reported findings. Moreover, combining a detailed RMA analysis with experimental data provided new insights into the flexibility of well-characterized structural motifs in each conformational state and the collective motion between TM helices. These findings, accumulated from various systems, are expected to contribute to a clearer understanding of ligand-dependent inactivation and activation mechanisms.

Here, we introduce the results that focus on the relationship between the modes obtained from RMA and the characteristic interatomic distances or structural motifs. In addition, previous simulations, which served as the basis for these data, provided the following insights: For example, in Yokoi and Mitsutake (2021), we discussed the results of MD simulations of the receptor-G protein complex, demonstrated that the active-like structure of the V(3.40)Y mutant was favorable for G protein binding, and identified key amino acid residues involved in the interaction with the G protein. In Yokoi et al. (2024), we performed MD simulations of systems with G protein binding and discussed the differences in the state of R(3.50) between systems with and without G protein binding, along with detailed findings regarding the interface amino acid residues. Notably, we also reported that OX2R might adopt a noncanonical intermediate state, which has previously been proposed to activate neurotensin receptor 1 (Kato et al., 2019).

Recently, improvements in software for protein simulation (Abraham et al., 2015; Case et al., 2005; Hwang et al., 2024; Jung et al., 2024), the accuracy of force fields, and useful tools such as CHARMM-GUI (Jo et al., 2007; Wu et al., 2014) have significantly simplified membrane protein simulations. Analytical methods and enhanced sampling techniques have simultaneously advanced. From the perspective of computational chemistry, it has become possible to perform ligand-binding

MD simulations using various approaches and investigate their dynamics in detail. For example, it is now feasible to combine enhanced sampling simulations of ligand binding to the RMA.

In the future, with advances in cryo-EM technology, the static structural information of G protein-bound active states is expected to become more accessible, and we can challenge the MD simulations of more complex systems. MD simulations can be conducted under conditions that are difficult to assess experimentally, allowing the investigation of structural dynamics at the atomic level and contributing to the elucidation of ligand selectivity. These challenges highlight the increasing importance of collaborative research in computational sciences.

References

Abraham, M. J., Murtola, T., Schulz, R., Páll, S., Smith, J. C., Hess, B., & Lindah, E. (2015). Gromacs: High performance molecular simulations through multi-level parallelism from laptops to supercomputers. *SoftwareX, 1–2*, 19–25. doi:10.1016/j.softx.2015.06.001.

Case, D. A., Cheatham, T. E., Darden, T., Gohlke, H., Luo, R., Merz, K. M., Onufriev, A., Simmerling, C., Wang, B., & Woods, R. J. (2005). The Amber biomolecular simulation programs. *Journal of Computational Chemistry, 26*(16), 1668–1688. doi:10.1002/jcc.20290.

Conflitti, P., Lyman, E., Sansom, M. S. P., Hildebrand, P. W., Gutiérrez-de-Terán, H., Carloni, P., Ansell, T. B., Yuan, S., Barth, P., Robinson, A. S., Tate, C. G., Gloriam, D., Grzesiek, S., Eddy, M. T., Prosser, S., & Limongelli, V. (2025). Functional dynamics of G protein-coupled receptors reveal new routes for drug discovery. *Nature Reviews Drug Discovery, 24*(4), 251–275. doi:10.1038/s41573-024-01083-3.

de Lecea, L., Kilduff, T. S., Peyron, C., Gao, X.-B., Foye, P. E., Danielson, P. E., Fukuhara, C., Battenberg, E. L. F., Gautvik, V. T., Bartlett, F. S., Frankel, W. N., van den Pol, A. N., Bloom, F. E., Gautvik, K. M., & Sutcliffe, J. G. (1998). The hypocretins: Hypothalamus-specific peptides with neuroexcitatory activity. *Proceedings of the National Academy of Sciences, 95*(1), 322–327. doi:10.1073/pnas.95.1.322.

Dror, R. O., Arlow, D. H., Maragakis, P., Mildorf, T. J., Pan, A. C., Xu, H., Borhani, D. W., & Shaw, D. E. (2011). Activation mechanism of the β 2-adrenergic receptor. *Proceedings of the National Academy of Sciences of the United States of America, 108*(46), 18684–18689. doi:10.1073/pnas.1110499108.

Feng, Z., Alqarni, M. H., Yang, P., Tong, Q., Chowdhury, A., Wang, L., & Xie, X. Q. (2014). Modeling, molecular dynamics simulation, and mutation validation for structure of cannabinoid receptor 2 based on known crystal structures of GPCRs. *Journal of Chemical Information and Modeling, 54*(9), 2483–2499. doi:10.1021/ci5002718.

Fujisaki, H., Moritsugu, K., Mitsutake, A., & Suetani, H. (2018). Conformational change of a biomolecule studied by the weighted ensemble method: Use of the diffusion map method to extract reaction coordinates. *The Journal of Chemical Physics, 149*(13). doi:10.1063/1.5049420.

Gusach, A., Maslov, I., Luginina, A., Borshchevskiy, V., Mishin, A., & Cherezov, V. (2020). Beyond structure: Emerging approaches to study GPCR dynamics. *Current Opinion in Structural Biology, 6*, 18–25. doi:10.1016/j.sbi.2020.03.004.

Heifetz, A., Barker, O., Benjamin Morris, G., Law, R. J., Slack, M., & Biggin, P. C. (2013). Toward an understanding of agonist binding to human orexin-1 and orexin-2 receptors with G-protein-coupled receptor modeling and site-directed mutagenesis. *Biochemistry, 52*(46), 8246–8260. doi:10.1021/bi401119m.

Hilger, D., Masureel, M., & Kobilka, B. K. (2018). Structure and dynamics of GPCR signaling complexes. *Nature Structural and Molecular Biology, 25*(1), 4–12. doi:10.1038/s41594-017-0011-7.

Hirao, H., Koseki, S., & Takano, H. (1997). Molecular dynamics study of relaxation modes of a single polymer chain. *Journal of the Physical Society of Japan, 66*(11), 3399–3405. doi:10.1143/jpsj.66.3399.

Hong, C., Byrne, N. J., Zamlynny, B., Tummala, S., Xiao, L., Shipman, J. M., Partridge, A. T., Minnick, C., Breslin, M. J., Rudd, M. T., Stachel, S. J., Rada, V. L., Kern, J. C., Armacost, K. A., Hollingsworth, S. A.,

O'Brien, J. A., Hall, D. L., McDonald, T. P., Strickland, C., … Hollenstein, K. (2021). Structures of active-state orexin receptor 2 rationalize peptide and small-molecule agonist recognition and receptor activation. *Nature Communications, 12*(1). doi:10.1038/s41467-021-21087-6.

Humphrey, W., Dalke, A., & Schulten, K. (1996). VMD: Visual molecular dynamics. *Journal of Molecular Graphics, 14*(1), 33–38. doi:10.1016/0263-7855(96)00018-5.

Hwang, W., Austin, S. L., Blondel, A., Boittier, E. D., Boresch, S., Buck, M., Buckner, J., Caflisch, A., Chang, H. T., Cheng, X., Choi, Y. K., Chu, J. W., Crowley, M. F., Cui, Q., Damjanovic, A., Deng, Y., Devereux, M., Ding, X., Feig, M. F., … Karplus, M. (2024). CHARMM at 45: Enhancements in accessibility, functionality, and speed. *Journal of Physical Chemistry B*. doi:10.1021/acs.jpcb.4c04100.

Isberg, V., Mordalski, S., Munk, C., Rataj, K., Harpsøe, K., Hauser, A. S., Vroling, B., Bojarski, A. J., Vriend, G., & Gloriam, D. E. (2016). GPCRdb: An information system for G protein-coupled receptors. *Nucleic Acids Research, 44*(1), D356. doi:10.1093/nar/gkv1178.

Jo, S., Kim, T., & Im, W. (2007). Automated builder and database of protein/membrane complexes for molecular dynamics simulations. *PLoS ONE, 2*(9). doi:10.1371/journal.pone.0000880.

Jung, J., Yagi, K., Tan, C., Oshima, H., Mori, T., Yu, I., Matsunaga, Y., Kobayashi, C., Ito, S., La Torre, D. U., & Sugita, Y. (2024). GENESIS 2.1: High-performance molecular dynamics software for enhanced sampling and free-energy calculations for atomistic, coarse-grained, and quantum mechanics/molecular mechanics models. *The Journal of Physical Chemistry B, 128*(25), 6028–6048. doi:10.1021/acs.jpcb.4c02096.

Karasawa, N., Mitsutake, A., & Takano, H. (2019). Identification of slow relaxation modes in a protein trimer via positive definite relaxation mode analysis. *Journal of Chemical Physics, 150*(8). doi:10.1063/1.5083891.

Karasawa, N., Mitsutake, A., & Takano, H. (2017). Two-step relaxation mode analysis with multiple evolution times applied to all-atom molecular dynamics protein simulation. *Physical Review E, 96*(6). doi:10.1103/physreve.96.062408.

Karasawa, N., Mitsutake, A., & Takano, H. (2024). Intermediate scattering function for polymer molecules: An approach based on relaxation mode analysis. *The Journal of Chemical Physics, 161*(2). doi:10.1063/5.0211504.

Kato, H. E., Zhang, Y., Hu, H., Suomivuori, C. M., Kadji, F. M. N., Aoki, J., Kumar, K. K., Fonseca, R., Hilger, D., Huang, W., Latorraca, N. R., Inoue, A., Dror, R. O., Kobilka, B. K., & Skiniotis, G. (2019). Conformational transitions of a neurotensin receptor 1–Gi1 complex. *Nature, 572*(7767), 80–85. doi:10.1038/s41586-019-1337-6.

Kolakowski, L. F. (1994). GCRDb: A G-protein-coupled receptor database. *Receptors and Channels, 2*(1), 1–7.

Kooistra, A. J., Mordalski, S., Pándy-Szekeres, G., Esguerra, M., Mamyrbekov, A., Munk, C., Keserű, G. M., & Gloriam, D. E. (2021). GPCRdb in 2021: Integrating GPCR sequence, structure and function. *Nucleic Acids Research, 49*(1), D335. doi:10.1093/nar/gkaa1080.

Koseki, S., Hirao, H., & Takano, H. (1997). Monte Carlo study of relaxation modes of a single polymer chain. *Journal of the Physical Society of Japan, 66*(6), 1631–1637. doi:10.1143/jpsj.66.1631.

Latorraca, N. R., Venkatakrishnan, A. J., & Dror, R. O. (2017). GPCR dynamics: Structures in motion. *Chemical Reviews., 117*(1), 139–155. doi:10.1021/acs.chemrev.6b00177.

Mahoney, C. E., Cogswell, A., Koralnik, I. J., & Scammell, T. E. (2019). The neurobiological basis of narcolepsy. *Nature Reviews Neuroscience, 20*(2), 83–93. doi:10.1038/s41583-018-0097-x.

Mitsutake, A., Iijima, H., & Takano, H. (2011). Relaxation mode analysis of a peptide system: Comparison with principal component analysis. *The Journal of Chemical Physics, 135*(16). doi:10.1063/1.3652959.

Mitsutake, A., & Takano, H. (2019). Folding pathways of NuG2—a designed mutant of protein G—using relaxation mode analysis. *The Journal of Chemical Physics, 151*(4). doi:10.1063/1.5097708.

Mitsutake, A., & Takano, H. (2015). Relaxation mode analysis and Markov state relaxation mode analysis for chignolin in aqueous solution near a transition temperature. *The Journal of Chemical Physics, 143*(12). doi:10.1063/1.4931813.

Mitsutake, A., & Takano, H. (2018). Relaxation mode analysis for molecular dynamics simulations of proteins. *Biophysical Reviews, 10*(2), 375–389. doi:10.1007/s12551-018-0406-7.

Munk, C., Isberg, V., Mordalski, S., Harpsøe, K., Rataj, K., Hauser, A. S., Kolb, P., Bojarski, A. J., Vriend, G., & Gloriam, D. E. (2016). GPCRdb: The G protein-coupled receptor database—An introduction. *British Journal of Pharmacology*, 2195–2207. doi:10.1111/bph.13509.

Nagai, T., Mitsutake, A., & Takano, H. (2013). Principal component relaxation mode analysis of an all-atom molecular dynamics simulation of human lysozyme. *Journal of the Physical Society of Japan, 82*(2), 023803. doi:10.7566/jpsj.82.023803.

Natori, S., & Takano, H. (2017). Two-step relaxation mode analysis with multiple evolution times: Application to a single [n]polycatenane. *Journal of the Physical Society of Japan, 86*(4), 043003. doi:10.7566/jpsj.86.043003.

Rappas, M., Ali, A. A. E., Bennett, K. A., Brown, J. D., Bucknell, S. J., Congreve, M., Cooke, R. M., Cseke, G., De Graaf, C., Doré, A. S., Errey, J. C., Jazayeri, A., Marshall, F. H., Mason, J. S., Mould, R., Patel, J. C., Tehan, B. G., Weir, M., & Christopher, J. A. (2020). Comparison of orexin 1 and orexin 2 ligand binding modes using X-ray crystallography and computational analysis. *Journal of Medicinal Chemistry, 63*(4), 1528–1543. doi:10.1021/acs.jmedchem.9b01787.

Rasmussen, S. G. F., DeVree, B. T., Zou, Y., Kruse, A. C., Chung, K. Y., Kobilka, T. S., Thian, F. S., Chae, P. S., Pardon, E., Calinski, D., Mathiesen, J. M., Shah, S. T. A., Lyons, J. A., Caffrey, M., Gellman, S. H., Steyaert, J., Skiniotis, G., Weis, W. I., Sunahara, R. K., & Kobilka, B. K. (2011). Crystal structure of the $\beta 2$ adrenergic receptor–Gs protein complex. *Nature, 477*(7366), 549–555. doi:10.1038/nature10361.

Rodríguez-Espigares, I., Torrens-Fontanals, M., Tiemann, J. K. S, et al. (2020). GPCRmd uncovers the dynamics of the 3D-GPCRome. *Nature Methods, 17*(8), 777–787. doi:10.1038/s41592-020-0884-y.

Rosenbaum, D. M., Rasmussen, S. G. F., & Kobilka, B. K. (2009). The structure and function of G-protein-coupled receptors. *Nature, 459*(7245), 356–363. doi:10.1038/nature08144.

Sakurai, T., Amemiya, A., Ishii, M., Matsuzaki, I., Chemelli, R. M., Tanaka, H., Williams, S. C., Richardson, J. A., Kozlowski, G. P., Wilson, S., Arch, J. R. S., Buckingham, R. E., Haynes, A. C., Carr, S. A., Annan, R. S., McNulty, D. E., Liu, W. S., Terrett, J. A., Elshourbagy, N. A., … Yanagisawa, M. (1998). Orexins and orexin receptors: A family of hypothalamic neuropeptides and G protein-coupled receptors that regulate feeding behavior. *Cell, 92*(4), 573–585. doi:10.1016/S0092-8674(00)80949-6.

Santos, R., Ursu, O., Gaulton, A., Bento, A. P., Donadi, R. S., Bologa, C. G., Karlsson, A., Al-Lazikani, B., Hersey, A., Oprea, T. I., & Overington, J. P. (2016). A comprehensive map of molecular drug targets. *Nature Reviews Drug Discovery, 16*(1), 19–34. doi:10.1038/nrd.2016.230.

Suno, R., Kimura, K. T., Nakane, T., Yamashita, K., Wang, J., Fujiwara, T., Yamanaka, Y., Im, D., Horita, S., Tsujimoto, H., Tawaramoto, M. S., Hirokawa, T., Nango, E., Tono, K., Kameshima, T., Hatsui, T., Joti, Y., Yabashi, M., Shimamoto, K., & Kobayashi, T. (2018). Crystal structures of human orexin 2 receptor bound to the subtype-selective antagonist EMPA. *Structure, 26*(1), 7. doi:10.1016/j.str.2017.11.005.

Takano, H., & Miyashita, S. (1995). Relaxation modes in random spin systems. *Journal of the Physical Society of Japan, 64*(10), 3688–3698. doi:10.1143/jpsj.64.3688.

Venkatakrishnan, A. J., Deupi, X., Lebon, G., Tate, C. G., Schertler, G. F., & Babu, M. M. (2013). Molecular signatures of G-protein-coupled receptors. *Nature, 494*(7436), 185–194. doi:10.1038/nature11896.

Wu, E. L., Cheng, X., Jo, S., Rui, H., Song, K. C., Dávila-Contreras, E. M., Qi, Y., Lee, J., Monje-Galvan, V., Venable, R. M., Klauda, J. B., & Im, W. (2014). CHARMM-GUI membrane builder toward realistic biological membrane simulations. *Journal of Computational Chemistry, 35*(27), 1997–2004. doi:10.1002/jcc.23702.

Yamamoto, E., Akimoto, T., Mitsutake, A., & Metzler, R. (2021). Universal relation between instantaneous diffusivity and radius of gyration of proteins in aqueous solution. *Physical Review Letters, 126*(12). doi:10.1103/physrevlett.126.128101.

Yin, J., Kang, Y., McGrath, A. P., Chapman, K., Sjodt, M., Kimura, E., Okabe, A., Koike, T., Miyanohana, Y., Shimizu, Y., Rallabandi, R., Lian, P., Bai, X., Flinspach, M., De Brabander, J. K., & Rosenbaum, D. M. (2022). Molecular mechanism of the wake-promoting agent TAK-925. *Nature Communications, 13*(1). doi:10.1038/s41467-022-30601-3.

Yin, J., Mobarec, J. C., Kolb, P., & Rosenbaum, D. M. (2015). Crystal structure of the human OX2 orexin receptor bound to the insomnia drug suvorexant. *Nature, 519*(7542), 247–250. doi:10.1038/nature14035.

Yokoi, S., & Mitsutake, A. (2022). Characteristic structural difference between inactive and active states of orexin 2 receptor determined using molecular dynamics simulations. *Biophysical Reviews, 14*(1), 221–231. doi:10.1007/s12551-021-00862-1.

Yokoi, S., & Mitsutake, A. (2021). Molecular dynamics simulations for the determination of the characteristic structural differences between inactive and active states of wild type and mutants of the orexin2 receptor. *The Journal of Physical Chemistry B, 125*(17), 4286–4298. doi:10.1021/acs.jpcb.0c10985.

Yokoi, S., Suno, R., & Mitsutake, A. (2024). Structural and computational insights into dynamics and intermediate states of orexin 2 receptor signaling. *The Journal of Physical Chemistry B, 128*(25), 6082–6096. doi:10.1021/acs.jpcb.4c00730.

Zhang, M., Chen, T., Lu, X., Lan, X., Chen, Z., & Lu, S. (2024). G protein-coupled receptors (GPCRs): Advances in structures, mechanisms and drug discovery. *Signal Transduction and Targeted Therapy, 9*(1). doi:10.1038/s41392-024-01803-6.

Zhou, Q., Yang, D., Wu, M., Guo, Y., Guo, W., Zhong, L., Cai, X., Dai, A., Jang, W., Shakhnovich, E., Liu, Z. J., Stevens, R. C., Lambert, N. A., Babu, M. M., Wang, M. W., & Zhao, S. (2019). Common activation mechanism of class a GPCRs. *eLife*. doi:10.7554/eLife.50279.

CHAPTER 16

FTIR study of light-sensitive and nonlight-sensitive G protein-coupled receptors

Kota Katayama*

Department of Life Science and Applied Chemistry, Nagoya Institute of Technology, OptoBioTechnology Research Center, Nagoya, Japan

**Corresponding author: katayama.kota@nitech.ac.jp*

16.1 Introduction

Biomolecules utilize temporal and spatial hierarchical structures with remarkable efficiency to perform diverse biological functions. Among these, photoreceptive proteins, which play critical roles in vision and photosynthesis, are particularly fascinating because their functional processes can be precisely and comprehensively studied across multiple timescales. For example, the initial events in electronically excited states, where quantum chemistry plays a pivotal role, occur on ultrafast femtosecond (10^{-15} s) timescales. In contrast the evolutionary processes that have optimized protein structures operate over petasecond (10^{15} s) timescales. This vast temporal spectrum, spanning 30 orders of magnitude, encompasses an extensive range of phenomena that can be explored to understand the dynamic nature of biomolecular systems.

Photoreceptive proteins containing a retinal molecule are collectively referred to as rhodopsins (Ernst et al., 2014). In humans, these include the visual pigment rhodopsin, which mediates scotopic (dim-light) vision, and three types of cone pigments (red, green, and blue), which enable color vision. A major unresolved question in visual research is how these pigments utilize the same chromophore molecule, 11-*cis*-retinal, to perform distinct functions such as scotopic vision and color discrimination. This functional diversity highlights the ability of the protein environment to precisely modulate the electronic states of the retinal chromophore, providing a finely tuned mechanism for spectral and functional differentiation.

To address these questions, we have employed advanced spectroscopic techniques to elucidate the molecular mechanisms underlying spectral tuning and signal transduction in cone pigments. In particular, we have extended light-induced Fourier-transform infrared (FTIR) difference spectroscopy to vibrational frequency regions sensitive to hydrogen bond donors. This approach has allowed us to monitor hydrogen bond changes during the functional activation of rhodopsins, including cone pigments. More recently, we have applied FTIR difference spectroscopy to investigate structural changes induced not only by light but also by ligand or drug binding. These studies have demonstrated the potential of FTIR as a powerful tool for probing the structural dynamics of photoreceptive proteins and G protein-coupled receptors (GPCRs), a major class of drug discovery targets.

Protein Dynamics: Experimental and Computational Approaches. DOI: https://doi.org/10.1016/B978-0-443-30042-4.00009-2

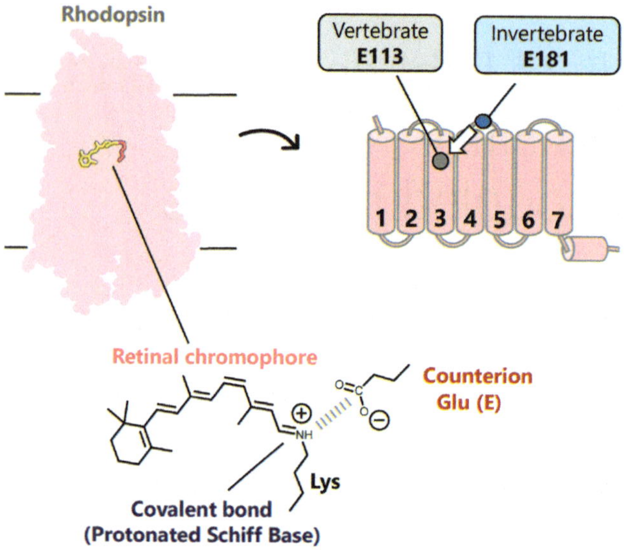

FIGURE 16.1

X-ray crystal structure and schematic representation of bovine rhodopsin (PDB ID: 1U19) (19). The chromophore 11-*cis*-retinal is covalently bound within the protein moiety via a protonated Schiff base linkage to a lysine residue. An acidic amino acid residue, serving as the counterion, is positioned in close proximity to the retinal. This counterion is positioned at residue 113 in vertebrates and at residue 181 in invertebrates.

This article presents the insights gained from these spectroscopic approaches, highlighting their application in elucidating the molecular mechanisms of photoreceptive protein function and their broader implications for GPCR research.

16.2 Spectral tuning mechanisms in visual pigments

16.2.1 Basic structure of visual pigments and diversity in absorption wavelengths

Rhodopsin is a photoreceptive protein located in the lipid bilayer of the outer segment of photoreceptor cells. It consists of an apoprotein, opsin, and a chromophore, 11-*cis*-retinal. Opsin is composed of approximately 350 amino acids and contains seven transmembrane α-helices (designated helices I–VII from the N-terminus). The chromophore, retinal, consists of a β-ionone ring and a polyene chain (Fig. 16.1). Retinal is covalently linked to a lysine (Lys) residue in helix VII (Lys296 in bovine rhodopsin) via a protonated Schiff base (PSB) and is embedded in a transmembrane helical structure. In addition, a glutamate (Glu) residue in helix III is deprotonated and functions as a counterion for the PSB (Ernst et al., 2014; Hofmann & Lamb, 2023; Palczewski, 2012).

Fig. 16.1 X-ray crystal structure and schematic representation of bovine rhodopsin (PDB ID: 1U19) (19). The chromophore 11-*cis*-retinal is covalently bound within the protein moiety via a PSB linkage

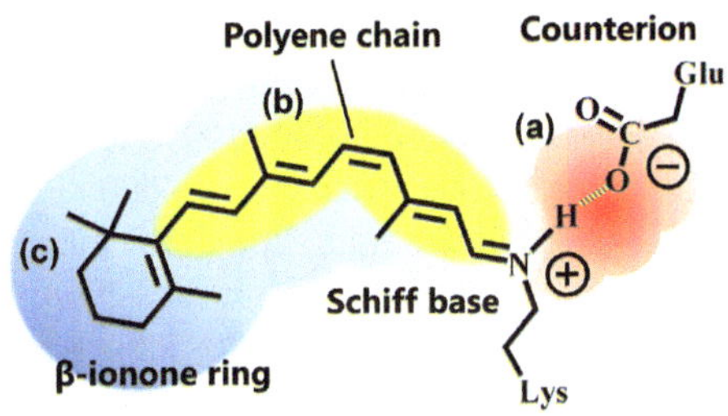

FIGURE 16.2

Schematic drawing of the spectral tuning mechanism in visual pigments.

to a lysine residue. An acidic amino acid residue, serving as the counterion, is positioned in close proximity to the retinal. This counterion is positioned at residue 113 in vertebrates and at residue 181 in invertebrates.

In the human eye, cone photoreceptor pigments (cone pigments) are present in addition to rhodopsin. These pigments are sensitive to red, green, and blue wavelengths, forming the molecular basis of trichromatic color vision (Imamoto & Shichida, 2014; Shichida & Imai, 1998). Many vertebrates, ranging from birds to fish, possess four types of cone pigments, enabling their color vision. However, most mammals, except for some primates, have only two types of cone pigments. This suggests that color discrimination ability regressed during mammalian evolution. Subsequently, in the evolutionary process of primates, gene duplication led to the acquisition of a new cone pigment, resulting in the development of trichromatic color vision.

Visual pigments exhibit diversity in their λ_{max}, which corresponds to the variability in the amino acid sequences of opsins. Based on amino acid sequence homology, cone pigments are classified into four groups. The pigments within each group exhibit similar λ_{max} and are designated, in order of increasing wavelength, as Group S (violet), M1 (blue), M2 (green), and L (red) (Okano et al., 1992). The similarity in λ_{max} among pigments with high amino acid sequence homology suggests that conserved amino acid residues within each group may play a critical role in spectral tuning. What are the molecular mechanisms underlying the control of λ_{max}?

16.2.2 Spectral tuning mechanisms in rhodopsin and cone pigments

In an organic solvent, 11-*cis*-retinal (λ_{max} ~380 nm) forms a Schiff base with butylamine (a derivative of a lysine residue), resulting in a λ_{max} at approximately 360 nm. When hydrochloric acid is added to the solution to protonate the Schiff base, forming a PSB, the λ_{max} shifts to approximately 440 nm. Based on this observation, numerous studies have investigated mechanisms to explain the differences in λ_{max} between rhodopsin and cone pigments, termed the "opsin shift," using the PSB in an organic solvent as a reference (Bravaya et al., 2007; Coto et al., 2006; Ernst et al., 2014; Pedraza-González et al., 2022; Rajamani et al., 2011; Wanko et al., 2008).

Quantum chemical calculations have proposed three major mechanisms underlying the opsin shift (Fig. 16.2) (Argade et al., 1981; Asenjo et al., 1994; Baasov et al., 1987; Barth, 2007; Bravaya et al., 2007; Coto et al., 2006; Doig et al., 1991; Doki et al., 2013; Ernst et al., 2014; Eyring & Mathies, 1979; Furutani et al., 2011; Haga et al., 2012; Haga, 2013; Hirano et al., 2006, 2001;

Hofmann & Lamb, 2023; Imai et al., 2000; Imamoto & Shichida, 2014; Inoue et al., 2019; Jiang et al., 2008; Kandori & Katayama, 2020; Katayama et al., 2010, 2012, 2015, 2017, 2018, 2019a, 2019b, 2021; Kitade et al., 2009; Kochendoerfer et al., 1999, 1997; Kruse et al., 2013; Lin et al., 1998; Neitz et al., 1991; Okano et al., 1992; Palczewski et al., 2000; Palczewski, 2012; Palings et al., 1989; Pedraza-González et al., 2022; Rajamani et al., 2011; Shichida & Imai, 1998; Stenkamp et al., 2002; Sugiura et al., 2025; Suzuki et al., 2021; Wanko et al., 2008; Xu et al., 2023; Yokoyama & Yokoyama, 1990; Yoshizawa & Kitô, 1958; Zhang & Fu, 2013):

1. Electrostatic interaction between the PSB proton and the counterion.
2. Steric interactions between the chromophore and nearby side chains, causing distortion of the polyene chain.
3. Electrostatic interactions with nearby side chains having high polarity or polarizability.

These factors alter the conjugated π-electron state of the chromophore, modulating the energy gap between the ground state and the excited state, thereby controlling the absorption wavelength. To elucidate how these spectral tuning mechanisms operate within the protein environment, experimental investigations of the chromophore structure, the surrounding protein environment, and their electronic states are required.

16.3 Vibrational spectroscopy of visual pigments to elucidate spectral tuning mechanisms

16.3.1 Vibrational spectroscopy identifying factors in visual pigment spectral tuning

Vibrational spectroscopy does not directly provide atomic positional information and is, therefore, not generally recognized as a primary tool in structural biology. However, molecular vibrations are highly sensitive to environmental changes, making vibrational spectroscopy a powerful method for detecting fine structures and structural changes in biomolecules. In fact, infrared absorption arises from changes in the dipole moment associated with molecular vibrations. Consequently the mid-infrared region (4000–800 cm^{-1}) contains a wealth of structural information, including data on protein secondary structures and hydrogen bonding in amino acid side chains (Fig. 16.3A) (Barth, 2007; Kandori & Katayama, 2020).

Fig. 16.3B shows the FTIR absolute absorption spectrum of a protein as an example, using a hydrated film sample of cone pigment reconstituted in a lipid membrane (Kandori & Katayama, 2020). Vibrational modes of the protein backbone include the amide-I band (~1655 cm^{-1}), derived from C=O stretching vibrations; the amide-II band (~1540 cm^{-1}), from N–H bending vibrations; and the amide-A band (~3290 cm^{-1}), from N–H stretching vibrations. These backbone vibrations are highly sensitive to protein secondary structure. For instance, in the amide-I region, α-helices typically exhibit vibrational frequencies between 1660 and 1650 cm^{-1}, whereas β-sheets are characterized by frequencies in the range of 1640–1620 cm^{-1}.

In addition, the C–H stretching vibrations of lipids appear at 2925 cm^{-1}, whereas lipid C=O stretching vibrations are observed at 1740 cm^{-1}. The latter region also includes characteristic vibrations

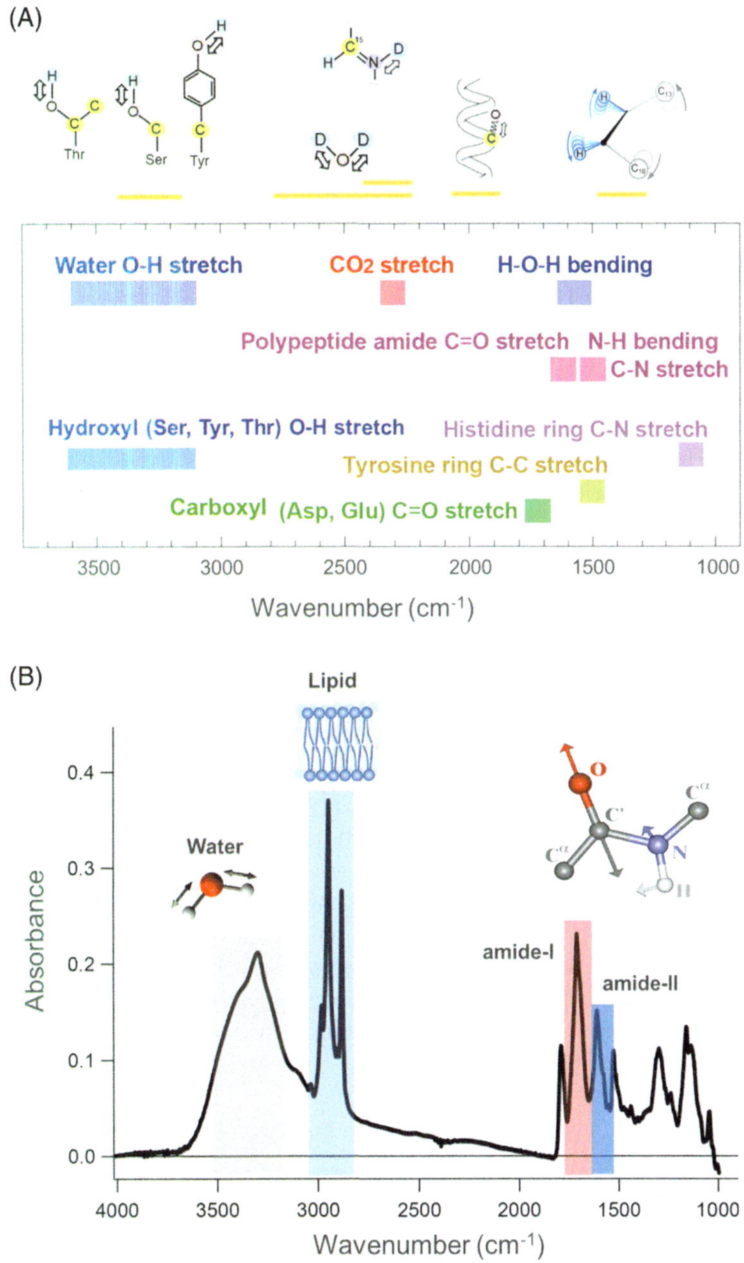

FIGURE 16.3

(A) Overview of the infrared absorption bands corresponding to protein secondary structure elements, amino acid side chains, and water molecules. In protein structural analysis, characteristic absorptions arising from amino bonds and water are important spectral indicators. (B) Representative absolute infrared absorption from protonated carboxylic acids (COOH). Moreover, the broad absorption band observed in the 3700–2800 cm^{-1} region corresponds to the symmetric and asymmetric O–H stretching vibrations of water molecules.

Infrared spectroscopy provides valuable information, but how can it be applied to elucidate the mechanisms of spectral tuning in visual pigments? As shown in b, the regulation of the λ_{max} in visual pigments is thought to arise from three main factors: (1) interactions between the PSB and its counterion, (2) retinal distortion, and (3) interactions with amino acids or water molecules possessing dipole moments.

In the analysis of factor (a) the stretching vibrations of the N–H bond between the PSB proton and the nitrogen atom to which it is attached, or the corresponding N–D stretching vibrations under deuterated conditions, are measured. The strength of these interactions can be directly assessed based on the frequency of these vibrations (Kandori & Katayama, 2020). In addition, in the C=NH group of the PSB, the C=N stretching vibration and the N–H bending vibration have similar frequencies, leading to coupled vibrations that deviate from the intrinsic C=N stretching vibration (a phenomenon referred to as "coupling"). However, under deuterated conditions, replacing the hydrogen in the C=NH group with deuterium significantly shifts the N–D bending vibration to a lower frequency. As a result the N–D bending vibration no longer couples with the C=N stretching vibration, allowing the intrinsic C=N stretching vibration band to appear prominently. The magnitude of the observed band shift indicates the strength of the interaction between the N–H region of the PSB and the counterion. This characteristic can be used as a quantitative measure of the interaction strength (Baasov et al., 1987; Kandori & Katayama, 2020).

In the analysis of factor (b) the fingerprint region of the vibrational spectrum can serve as an indicator. The fingerprint region generally refers to the 1300–1100 cm^{-1} range, which is widely used in structural analysis because it reflects changes in molecular structure, such as the geometric arrangement of atoms. This region primarily includes vibrational modes of retinal, such as C=C stretching vibrations (1550–1500 cm^{-1}), C–C stretching vibrations (1250–1180 cm^{-1}), and out-of-plane hydrogen bending vibrations associated with the polyene chain (HOOP, 1000–800 cm^{-1}). Notably the HOOP band is known to increase in intensity as retinal torsion increases, making it a critical vibrational mode that reflects retinal twisting (Doig et al., 1991; Eyring & Mathies, 1979; Palings et al., 1989).

In the analysis of factor (c), interactions with polar amino acids or water molecules can be examined in detail by measuring O–H stretching vibrations, N–H stretching vibrations, or their deuterated counterparts, O–D stretching vibrations and N–D stretching vibrations, under deuterated conditions. These measurements provide insights into the interactions between dipole-moment-bearing amino acids or water molecules and the pigment.

spectrum of a membrane protein. The amide I band (approximately 1600–1700 cm^{-1}), originating from the C=O stretching vibrations of the peptide backbone, and the amide II band (approximately 1500–1600 cm^{-1}), corresponding to N–H bending vibrations, are prominently observed. In addition, C–H stretching bands derived from lipid bilayers (approximately 2800–3000 cm^{-1}) and O–H stretching bands associated with surrounding water molecules (approximately 3200–3600 cm^{-1}) are clearly detected. These spectral features reflect the membrane environment and hydration state, and their analysis enables insights into the structural characteristics and dynamics of membrane proteins.

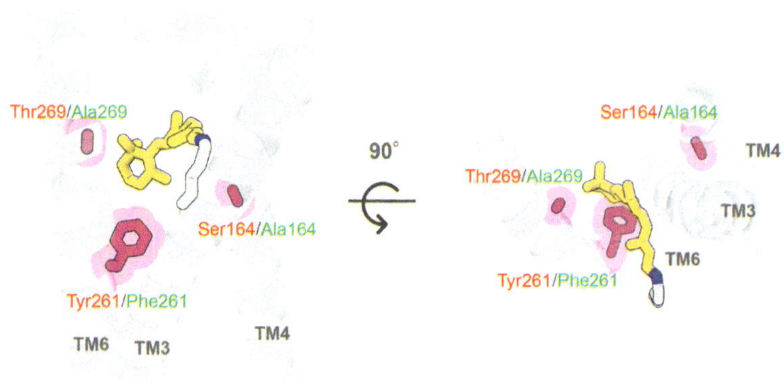

FIGURE 16.4

Left: X-ray crystal structure of the retinal binding site in bovine rhodopsin (PDB ID: 1U19), viewed from the helix VII side. The three amino acid residues highlighted in the figure are thought to play critical roles in the spectral tuning mechanisms that differentiate the absorption maxima of *red* and *green* cone pigments. Right: the same structure rotated 90 degrees, viewed from the extracellular side. The spatial arrangement of the retinal chromophore and surrounding amino acid residues is more clearly visualized, providing insights into how these residues contribute to the molecular basis of spectral tuning.

The following sections introduce examples of studies on spectral tuning mechanisms using cone pigments.

16.3.2 Raman spectroscopy of cone pigments

In primates the red cone pigment (λ_{max} = 560 nm) and the green cone pigment (λ_{max} = 530 nm) both belong to the L-group and are thought to have originated from a gene duplication approximately 35 million years ago. Despite their close evolutionary relationship, their amino acid sequences differ at only 15 out of 365 residues. Nevertheless, these minimal sequence differences result in a substantial spectral shift of approximately 30 nm in their maximal absorption wavelengths (λ_{max}). This spectral shift plays a critical role in primates' ability to discriminate colors, highlighting the functional significance of these two visual pigments in trichromatic vision.

Asenjo et al. systematically introduced site-directed mutations at these 15 residues and demonstrated that seven of them contribute significantly to the shift in λ_{max}. Among these, six involve changes in polarity, specifically the presence or absence of a hydroxyl group. Notably, substitutions at positions 164, 261, and 269 (numbering based on bovine rhodopsin) were found to almost entirely account for the λ_{max} difference between the two pigments (Asenjo et al., 1994). Furthermore, structural modeling based on bovine rhodopsin (Palczewski et al., 2000) indicated that these three residues are localized near the β-ionone ring of the retinal chromophore (Fig. 16.4) (Stenkamp et al., 2002). These findings support the involvement of factor (c) in wavelength regulation.

Furthermore, resonance Raman spectroscopy was utilized to investigate the mechanism underlying the spectral shift in λ_{max} associated with the three aforementioned residues (Kochendoerfer et al., 1997, 1999). As discussed in Section 16.3.1, vibrational modes in the fingerprint region are generally sensitive to local perturbations, such as backbone fluctuations and the presence of point charges. However, no significant differences in these vibrational modes were detected between the red and green cone pigments. In addition, the stretching vibration of the Schiff base (C=NH) showed no notable differences, suggesting that the interaction between the PSB and its counterion remains similar in both pigments (Kochendoerfer et al., 1997, 1999).

Based on these results the contributions of factors (a) and (b) to the spectral tuning mechanism can be excluded. Instead the λ_{max} difference between the two pigments is most likely attributable to hydroxyl groups uniquely present in the red cone pigment, which acts as dipoles and mediates long-range interactions with the retinal chromophore.

16.3.3 Fourier-transform infrared spectroscopy of cone pigments

Raman spectroscopic measurements performed by the Mathies research group provided the first experimental structural insights into the spectral tuning mechanism of cone pigments (Kochendoerfer et al., 1997, 1999). Their findings suggest that the dipole moment surrounding the retinal chromophore plays a crucial role in this process. However, the molecular details underlying this mechanism remain to be fully elucidated. For instance the C=NH stretching vibration band was analyzed to investigate the interaction between the PSB and its counterion. The results revealed no significant differences between the red and green cone pigments in this interaction. However, this assessment remains inherently indirect; to reach a more definitive conclusion, direct detection of the N–H or N–D stretching vibration bands of the PSB is essential.

Furthermore, UV–visible absorption spectroscopy (Asenjo et al., 1994; Neitz et al., 1991; Yokoyama & Yokoyama, 1990) and theoretical calculations on mutant visual pigments have suggested that three factors, (a–c), proposed by Section 16.2.2, may contribute to the spectral tuning mechanism. However, the possibility that other molecular mechanisms also play a role cannot be excluded. To address these challenges, we employed FTIR spectroscopy, a vibrational spectroscopic technique complementary to Raman spectroscopy, to further investigate the structural properties of visual pigments.

16.3.3.1 *Fourier-transform infrared spectroscopy of primate red and green cone pigments*

Previous studies have reported FTIR spectroscopy of cone pigments extracted from native chicken retina (Hirano et al., 2001, 2006). However, because of the use of native samples, the assignment of vibrational bands was challenging. In this study, we thus expressed primate red and green cone pigments in HEK293 cells and conducted FTIR spectroscopic measurements. In general the low expression levels of cone pigments compared to rhodopsin have posed a major challenge for structural studies (Imai et al., 2000). To overcome this, we utilized culture plates at a scale tens of times larger than those used for rhodopsin, successfully producing the required amount of cone pigments for FTIR measurements. The expressed pigments were subsequently solubilized, purified, and reconstituted into lipid liposomes. Low-temperature (77 K) light-induced FTIR difference spectroscopy was then performed.

Fig. 16.5 presents the light-induced FTIR difference spectra of the red and green cone pigments measured at 77 K in D_2O (Katayama et al., 2010). The *black line* represents the spectrum of rhodopsin for comparison. The positive and negative peaks correspond to the initial photoreaction intermediate, batho, and the initial state, respectively. Batho intermediates are known to exhibit red-shifted absorption

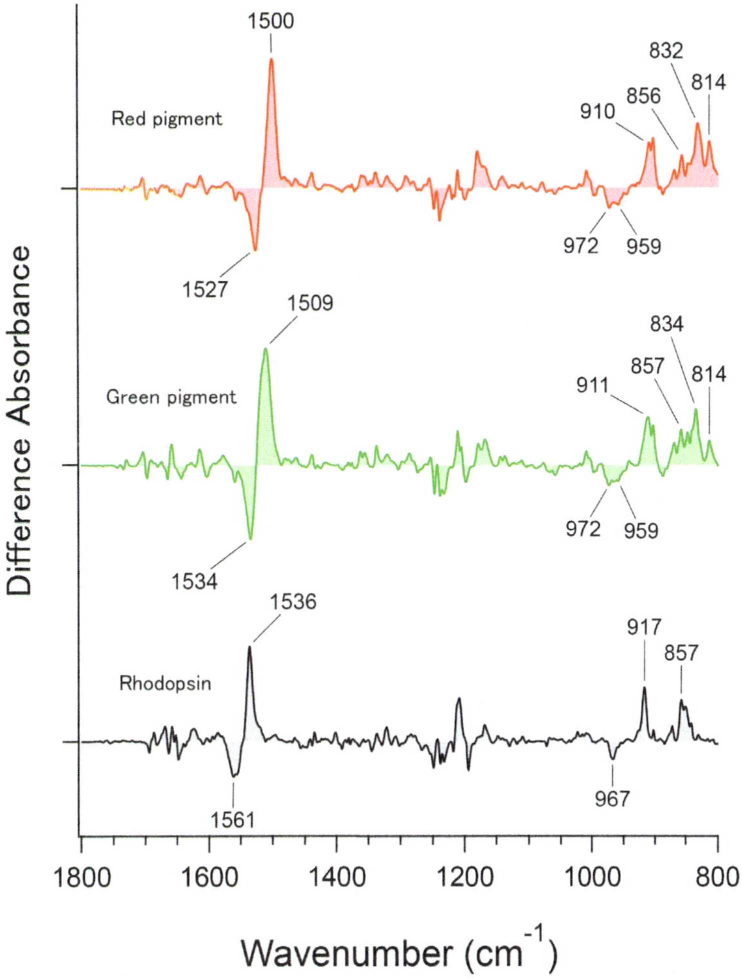

FIGURE 16.5

Light-induced FTIR difference spectra of *red* (top panel), *green* (middle panel) cone pigments, and rhodopsin (bottom panel) in the 1800–800 cm^{-1} region. Spectra were measured at 77 K in D_2O. Positive and negative bands correspond to the photointermediate (batho) and the initial (dark) state, respectively. One division on the *y*-axis represents 0.003 absorbance units. *FTIR*, Fourier-transform infrared.

(Yoshizawa & Kitô, 1958), and in this study, a downshift of the retinal C=C stretching vibration band was observed. Specifically, band pairs at 1527 (−)/1500 (+) cm^{-1} for red cone pigment, 1534 (−)/1509 (+) cm^{-1} for green cone pigment, and 1561 (−)/1536 (+) cm^{-1} for rhodopsin were detected, confirming the proper formation of batho intermediates. Based on these results, we further investigate the factors underlying the 30 nm spectral shift in λ_{max}.

16.3.3.2 Vibrational properties of retinal and structural changes associated with its isomerization

In the vibrational frequency region of 1300–1100 cm^{-1}, C–C stretching vibration bands of retinal are observed. These bands correspond to the vibrational modes of retinal associated with its isomerization from 11-*cis*-retinal to all-*trans*-retinal. As shown in Fig. 16.5 the spectral features of red and green cone pigments closely resemble that of rhodopsin. This similarity suggests that the mode of retinal isomerization and its overall chemical structure are shared among these pigments. On the other hand, in the vibrational frequency region of 1000–800 cm^{-1}, which corresponds to the HOOP bands reflecting retinal distortion, spectral features distinct from those of rhodopsin were observed. The negative 967 cm^{-1} band in rhodopsin appeared as two distinct peaks at 972 and 959 cm^{-1} in the red and green cone pigments. In addition, in the positive region, extra bands at 834 and 814 cm^{-1} were observed, in addition to the 917 and 857 cm^{-1} bands seen in rhodopsin. These findings suggest that retinal is more strongly distorted in red and green cone pigments compared to rhodopsin.

Meanwhile the number and frequencies of the HOOP bands were remarkably similar between the red and green cone pigments. This indicates that there is no significant difference in the retinal distorting state between these two pigments, which is consistent with previous Raman spectroscopic measurements (Kochendoerfer et al., 1997, 1999). Therefore it was revealed that factor (b) does not contribute to the spectral tuning of the red and green cone pigments.

16.3.3.3 Investigation of protonated Schiff base-counterion interactions and assignment of N–D stretching vibrations

As previously mentioned, the N–H stretching vibration of the PSB serves as the most direct probe for investigating the interaction between PSB and the counterion. Kandori and Katayama (2020) have systematically examined various microbial rhodopsins and determined the vibrational frequencies of the N–D stretching band in D_2O. In these studies the assignment of the vibrational band was successfully achieved by incorporating ^{15}N labeling into the nitrogen atom of the lysine residue that forms the Schiff base linkage during protein expression in *Escherichia coli* (Argade et al., 1981). However, when expressing cone pigments using a mammalian cell expression system, isotope labeling of specific amino acids is extremely challenging. Moreover, the expression levels are drastically reduced, making this approach infeasible.

To overcome this limitation, we synthesized deuterated retinal, in which the carbon at position 15 (C_{15}) adjacent to the nitrogen involved in Schiff base formation was selectively deuterated. We then incorporated this retinal into green cone pigment and performed FTIR spectroscopy in D_2O. As a result, isotope shifts induced by the coupling between C_{15}-D stretching and N–D stretching vibrations were observed, leading to the identification of a characteristic pair of bands at 2099 (−)/2083 (+) cm^{-1} (Fig. 16.6). This enabled us to successfully assign the N–D stretching vibration band of PSB in the green cone pigment. In addition, a corresponding pair of bands at 2099 (−)/2086 (+) cm^{-1} was observed in the red cone pigment (Katayama et al., 2015).

Notably the vibrational frequency of the negative band, corresponding to the initial state, was identical between the red and green cone pigments. This finding indicates that the hydrogen bonding strength of the N–D stretching vibration is the same in both pigments. Consequently, our results demonstrate that the interaction strength between PSB and the counterion is equivalent in red and green

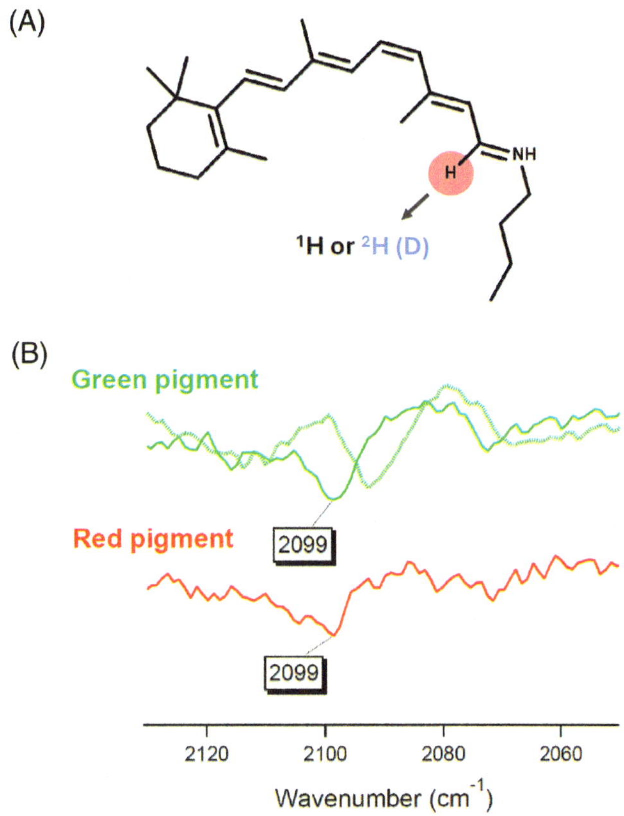

FIGURE 16.6

(A) Chemical structure of the C15-H and C15-D retinal chromophores. (B) Light-induced FTIR difference spectra of the *green* cone pigment in the 2130–2050 cm^{-1} region, measured at 77 K. Solid and *dotted lines* represent spectra with unlabeled (C15-H) and deuterium-labeled (C15-D) retinal, respectively. The bottom spectrum corresponds to the *red* cone pigment for comparison. *FTIR*, Fourier-transform infrared.

cone pigments, providing direct evidence that factor (a) does not contribute to the 30 nm spectral shift in λ_{max}.

16.3.3.4 Detection of polar amino acids and identification of a unique vibrational band in red cone pigment by Fourier-transform infrared spectroscopy

One of the strengths of FTIR spectroscopy is its ability to detect infrared signals from polar amino acids possessing dipole moments. Furthermore, measurements in D_2O allow for the separation and analysis of signals from polar amino acids in hydrophilic environments, which are susceptible to deuterium exchange (2800–1800 cm^{-1} region: X–D stretching vibration region), and those in hydrophobic environments, which are resistant to deuterium exchange (3800–3000 cm^{-1} region: X–H stretching vibration region). As shown in Fig. 16.7A, the spectral features in the X–H stretching vibration region

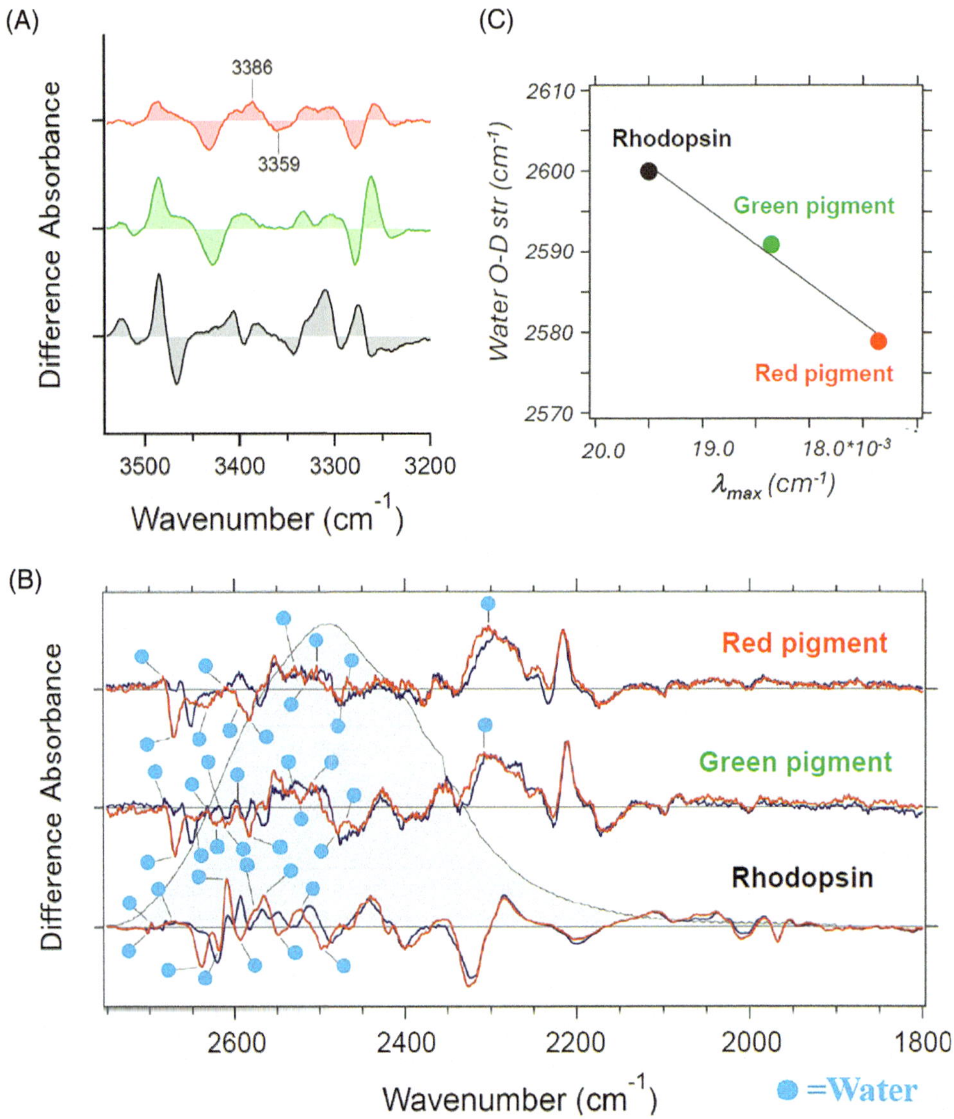

FIGURE 16.7

(A) Light-induced FTIR difference spectra of *red* (top panel), *green* (middle panel) cone pigments, and rhodopsin (bottom panel) in the 3540–3200 cm^{-1} region. Spectra were measured at 77 K in D_2O. Positive and negative bands correspond to the batho intermediate and the initial (dark) state, respectively. One division on the *y*-axis represents 0.00045 absorbance units. (B) Light-induced FTIR difference spectra of *red* (top panel), *green* (middle panel) cone pigments, and rhodopsin (bottom panel) in the 2750–1800 cm^{-1} region. *Red* and *blue* traces indicate spectra measured in D_2O and $D_2{}^{18}O$, respectively. *Blue circles* mark bands assigned to O–D stretching vibrations of water. The *gray* spectrum in the 2700–2000 cm^{-1} region shows the O–D stretching vibrations of bulk D_2O at room temperature. One division on the *y*-axis corresponds to 0.00014 differed significantly among rhodopsin and red and green cone pigments. This observation indicates that the hydrogen-bonding network formed by amino acids in the retinal-binding pocket differs between visual pigments responsible for scotopic vision (rhodopsin) and those involved in color vision (cone pigments).

Interestingly a negative band at 3359 cm^{-1} was observed exclusively in the red cone pigment (Katayama et al., 2010). As this band appeared in the X–H stretching vibration region, it is likely attributed to the O–H or N–H stretching vibrations of polar amino acids located in a hydrophobic environment. Although its precise assignment remains unclear at this stage, this band may serve as a key to understanding the unique retinal-protein interactions in the red cone pigment. Specifically, it is possible that this band originates from one of the three hydroxyl-containing amino acid residues (positions 164, 261, and 269) that are unique to the red cone pigment. Future mutational analyses will be essential for determining the exact assignment of this band.

Protein-bound water molecules actively contribute to spectral tuning by acting as dipoles. Through measurements using D_2O and its isotopically labeled $D_2{}^{18}O$, we successfully identified the O–D stretching vibration of water molecules associated with red and green cone pigments (Fig. 16.7B) (Katayama et al., 2012). In the initial state of these pigments, we detected six and seven distinct O–D stretching vibration bands, respectively, indicating the presence of at least two to three water molecules near the retinal.

Interestingly, when the average vibrational frequencies of water molecules observed in the initial state of red and green cone pigments, as well as rhodopsin, were calculated and plotted against their respective λ_{max} values, a distinct positive correlation was observed (Fig. 16.7C). This finding strongly suggests that these water molecules do more than merely fill voids within the protein—they play an active role in fine-tuning the λ_{max} of red and green cone pigments.

Building on these insights, our experiments conclusively demonstrated that the spectral tuning mechanism of red and green cone pigments is governed primarily by interactions with amino acids and water molecules possessing dipole moments.

16.3.3.5 Fourier-transform infrared spectroscopy of primate blue cone pigment

Blue cone pigments are known to be expressed at approximately one-tenth the level of green and red cone pigments. However, by carefully selecting primate species (Zhang & Fu, 2013) and optimizing protein purification conditions, we successfully measured the FTIR spectra of blue cone pigments in 2017 (Fig. 16.8A) (Katayama et al., 2017). The obtained spectra revealed significant differences in the vibrational modes of the retinal and protein compared to those of green and red cone pigments. Notably the most striking differences were observed in the signals originating from water molecules. FTIR difference spectra between the initial state and the photointermediate formed after photoisomerization were recorded in D_2O at 77 K, revealing a broad and intense O–D stretching vibration band of water molecules around 2500 cm^{-1} with a full width at half maximum exceeding 40 cm^{-1} (Fig. 16.8B). These findings suggest that multiple water molecules form a cluster near the hydrophobic retinal molecule.

Furthermore a decrease in water signals was observed in mutants of amino acids that play a critical role in spectral tuning for blue cone pigment (Fig. 16.8C) (Lin et al., 1998), suggesting that the water

absorbance units. (C) Correlation between the average frequency of water O–D stretching vibration for each pigment and their respective λ_{max} values. *FTIR*, Fourier-transform infrared.

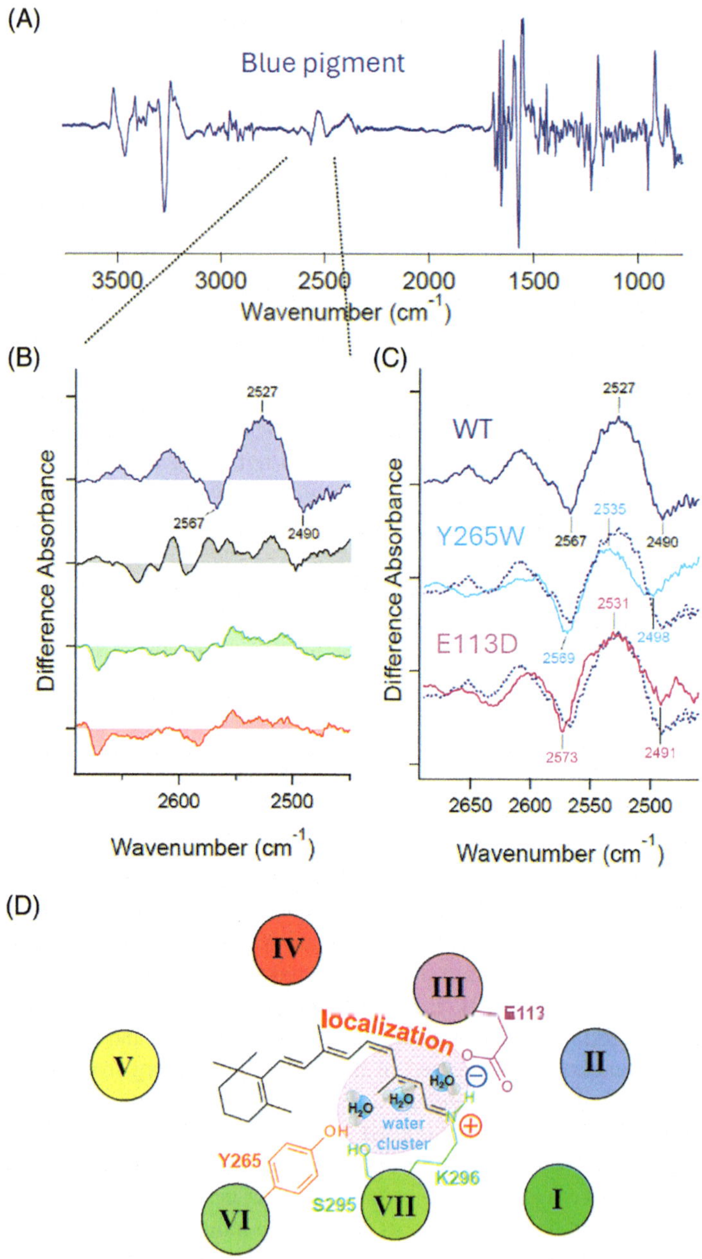

FIGURE 16.8

(A) Light-induced FTIR difference spectra of *blue* cone pigments in the 3540–900 cm^{-1} region. The spectra are measured at 77 K in D_2O. Positive and negative bands originate from the batho intermediate and initial states, respectively. (B) Light-induced FTIR difference spectra of *blue* cone pigment, rhodopsin, *green* cone cluster induces localization of π-electrons along the polyene chain of the retinal, thereby enabling blue light absorption. Although dispersed water molecules in the hydrophobic region contribute to distinguishing between green and red, the water cluster appears to function specifically in sensing blue light (Fig. 16.8D). The flexibility of the protein structure facilitates a delicate balance between water and the hydrophobic retinal, ultimately shaping the "visible light sensitivity" in humans.

16.4 Fourier-transform infrared spectroscopy of nonlight-responsive G protein-coupled receptors

This measurement technique is a form of *stimulus-induced FTIR difference spectroscopy*, which detects structural changes in proteins in response to external stimuli by capturing FTIR difference spectra. Although this approach has been successfully applied to the structural analysis of cone pigments by leveraging "light" as an exceptional stimulus, its application to GPCRs that respond to nonlight stimuli had not been explored. In this context, we successfully measured ion-binding-induced FTIR difference spectra of cone pigments using attenuated total reflection (ATR) FTIR (ATR-FTIR) spectroscopy in Katayama et al. (2018) and Katayama et al. (2019b), marking the first demonstration of structural analysis of a GPCR triggered by nonlight stimuli.

This method employs an evanescent wave, enabling measurements in solution. By introducing ions or ligands that interact with the protein into one of the solutions, it is possible to detect spectral changes induced by chemical stimuli (Doki et al., 2013; Furutani et al., 2011; Kitade et al., 2009; Jiang et al., 2008). In addition, this approach allows membrane proteins reconstituted in lipids to be adsorbed onto the ATR prism (substrate) in a manner that closely resembles their native environment, without requiring specific adsorption onto the substrate (Fig. 16.9).

Building on the findings from ion-binding measurements in cone pigments, we propose that FTIR spectroscopy could serve as a powerful tool for capturing the dynamic structural changes that govern the diverse signal transduction mechanisms of GPCRs.

16.4.1 Fourier-transform infrared spectroscopy of M_2R

We explored the M_2 receptor (M_2R), a subtype of muscarinic acetylcholine receptors (Haga, 2013), as a model GPCR to investigate structural changes induced by ligand binding. By optimizing sample preparation and measurement conditions, we successfully obtained ligand-binding-induced FTIR difference spectra for both agonists and antagonists, revealing distinct structural changes associated with ligands of varying efficacies (Fig. 16.10A) (Katayama et al., 2019a). Notably the amide-I band exhibited a downshift (1666 → 1656 cm^{-1}) upon agonist binding, whereas an upshift (1643 → 1656 cm^{-1}) was observed upon antagonist binding. This finding marks the first demonstration of detecting molecular vibrational changes associated with the functional process of ligand recognition in GPCRs.

pigment, and *red* cone pigment (from top to bottom) in the 2690–2450 cm^{-1} region, measured at 77 K in D_2O. (C) Spectral comparison of the mutant *blue* cone pigments (Y265W and E113D) in the 2690–2450 cm^{-1} region, measured at 77 K in D_2O. (D) Schematic representation of the spectral tuning mechanism in the *blue* cone pigment. *FTIR*, Fourier-transform infrared.

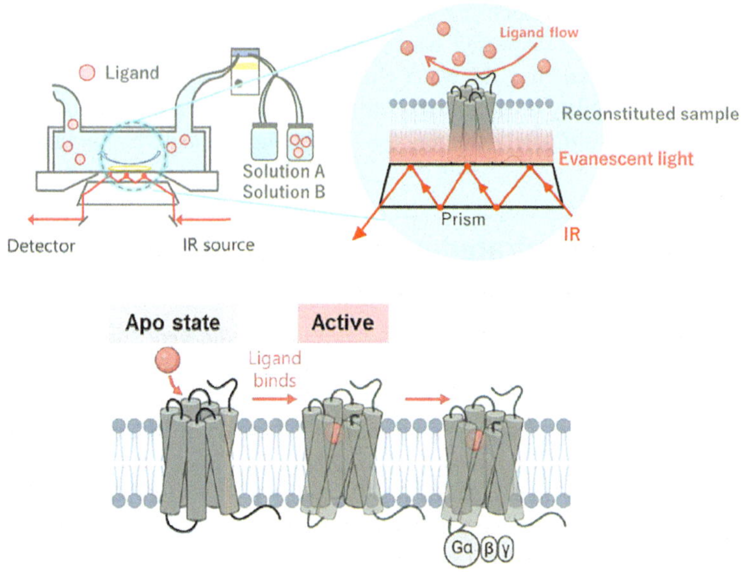

FIGURE 16.9

Schematic diagram of the experimental setup for perfusion-induced ATR-FTIR spectroscopy. *ATR*, Attenuated total reflection; *FTIR*, Fourier-transform infrared.

Moreover, this method enables real-time tracking of ligand binding and dissociation with concentration dependence, allowing the simultaneous acquisition of physicochemical parameters such as binding constants and rate constants (Fig. 16.10B). On the other hand, to accurately extract information on structural changes in both the receptor and ligand from the obtained spectra and to elucidate the mechanisms underlying ligand binding, signaling, and activity regulation, precise assignment of the infrared absorption bands is essential. This can be achieved through experiments utilizing site-directed mutants and isotopic labeling.

16.4.2 Activation mechanism of M_2R based on Fourier-transform infrared spectroscopy

To elucidate the sequence of dynamic structural changes in M_2R at atomic resolution, from ligand binding to signal transduction, we performed measurements using site-directed mutants and isotopically labeled or derivative ligands. These experimental approaches were further complemented by computational analyses to assign infrared signals corresponding to these structural changes.

In particular, we utilized the N404Q mutant, where the asparagine residue at position 404 (Asn404), a key component of the acetylcholine binding site (Haga et al., 2012; Kruse et al., 2013; Xu et al., 2023), was replaced with glutamine (Gln). This substitution enabled us to successfully assign the infrared signals associated with changes in the hydrogen bonding environment of Asn404 upon acetylcholine binding (Fig. 16.11A). In addition, by using acetylcholine labeled with a ^{13}C isotope, we identified the infrared signals of acetylcholine bound to both M_2R wild-type (Suzuki et al., 2021) and the N404Q mutant (Fig. 16.11B).

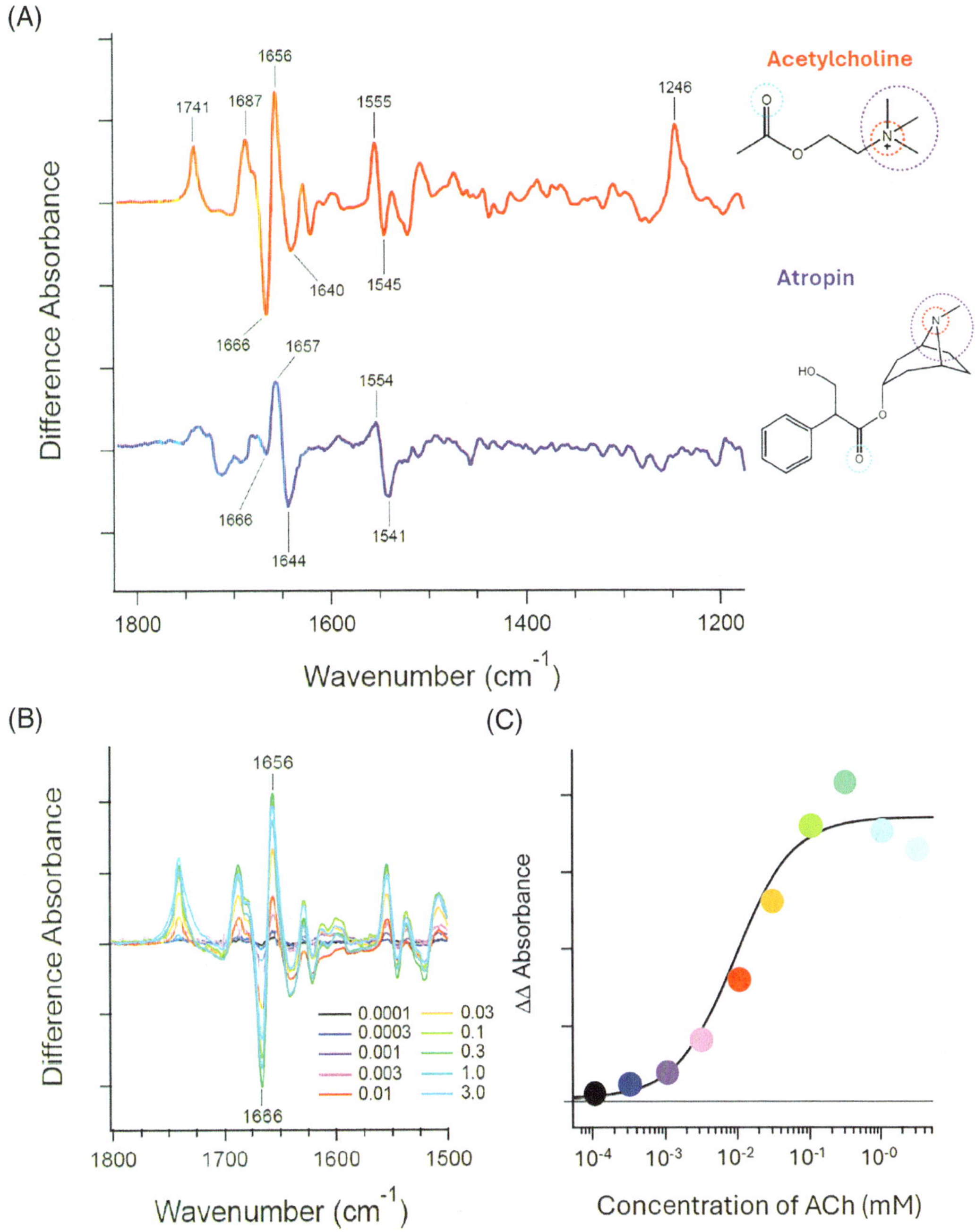

FIGURE 16.10

(A) Ligand binding-induced ATR-FTIR difference spectroscopy of M_2R in the 1820–1200 cm^{-1} region. Positive and negative bands originate from the acetylcholine-bound active state or the atropine-bound inactive state,

Interestingly the N404Q mutant exhibited reduced G protein activation compared to the wild type, despite displaying a 2.2-fold increase in binding affinity for acetylcholine (Fig. 16.12). Typically, mutations in amino acid residues that constitute the ligand-binding site result in a concomitant decrease in both receptor activation and ligand-binding affinity. However, infrared signal assignment of acetylcholine in this study revealed that, in the N404Q mutant, acetylcholine binds in a state that retains a degree of conformational flexibility similar to that in solution. This suggests that structural changes at the ligand-binding site were not effectively propagated to the surrounding α-helices, leading to diminished activation efficiency.

Supporting this interpretation, molecular dynamics (MD) simulations of acetylcholine binding to both the wild type and the N404Q mutant revealed the presence of multiple binding states for acetylcholine in the N404Q mutant, consistent with the results obtained from infrared spectroscopy measurements.

Furthermore, in addition to measurements conducted in D_2O, continued MD simulations of acetylcholine binding to M_2R revealed that a water molecule mediates the interaction between acetylcholine and Asn404, forming a dynamic interaction with an optimal balance of precision and flexibility. This finely tuned hydrogen bond was shown to be essential for M_2R activation (Fig. 16.13).

This study demonstrated that a flexible yet precise hydrogen bond network formed by water molecules plays a crucial role in driving the structural changes and activation of the receptor. Building on this discovery, we hypothesized that fine-tuning the microscopic interactions involving Asn404 could enable precise control of M_2R activity. To test this hypothesis, we synthesized several acetylcholine derivatives with extended side chains and evaluated their activity and FTIR spectra. As expected, acetylcholine derivatives designed to disrupt the water-mediated hydrogen bond network exhibited reduced activation efficiency.

The findings of this study not only provide deeper insights into the activation mechanisms of GPCRs but also establish a solid foundation for the rational design of next-generation ligands with improved efficacy and specificity (Sugiura et al., 2025).

16.4.3 Application of Fourier-transform infrared spectroscopy in drug discovery research

The ligand-binding-induced FTIR difference spectroscopy method, which combines ATR-FTIR spectroscopy with a buffer exchange system, enables repeated measurements by allowing the dissociation and reassociation of ligands with the receptor. This approach is particularly advantageous for analyzing the structural dynamics of low-affinity ligands, which are often difficult to capture using conventional structural biology techniques. In this study, we applied this system to a variety of ligands with different efficacies (agonists, partial agonists, antagonists, and inverse agonists) and monitored the

and the apo state, respectively. (B) Dependence of acetylcholine concentration on ligand-induced infrared spectral changes in M_2R. Spectra were recorded at various acetylcholine concentrations (0.0001–3.0 mM) in the 1800–1500 cm^{-1} region. (C) Plot of peak-to-peak intensity at 1666–1656 cm^{-1} from panel (B) as a function of acetylcholine concentration. The data were fitted using the Hill equation, yielding an apparent K_d value of 10.5 ± 2.7 μM and a Hill coefficient of 1.0. *ATR*, Attenuated total reflection; *FTIR*, Fourier-transform infrared.

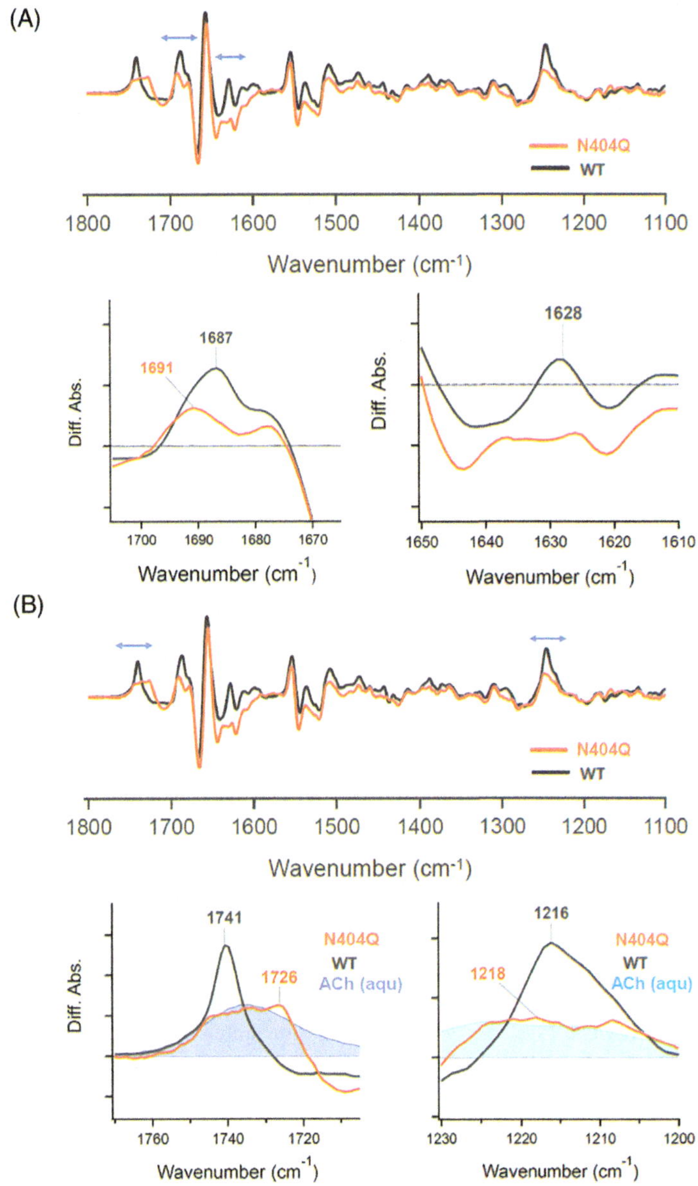

FIGURE 16.11

(A) Comparison of acetylcholine binding-induced FTIR difference spectra in the 1800–1100 cm^{-1} region between WT and N404Q mutant. Enlarged carbonyl C=O stretching vibration (1705–1665 cm^{-1}) (left) region and N–H bending (1650–1610 cm^{-1}) (right) region of the spectra in the panel (A). (B) Left: spectral comparison of the C=O stretching vibration of acetylcholine in WT (*gray*), N404Q mutant (*orange*), and aqueous solution (*light blue*). Right: spectral comparison of the C–O stretching vibration of ^{13}C-labeled acetylcholine in WT (*gray*), N404Q mutant (*orange*), and aqueous solution (*light blue*). *FTIR*, Fourier-transform infrared.

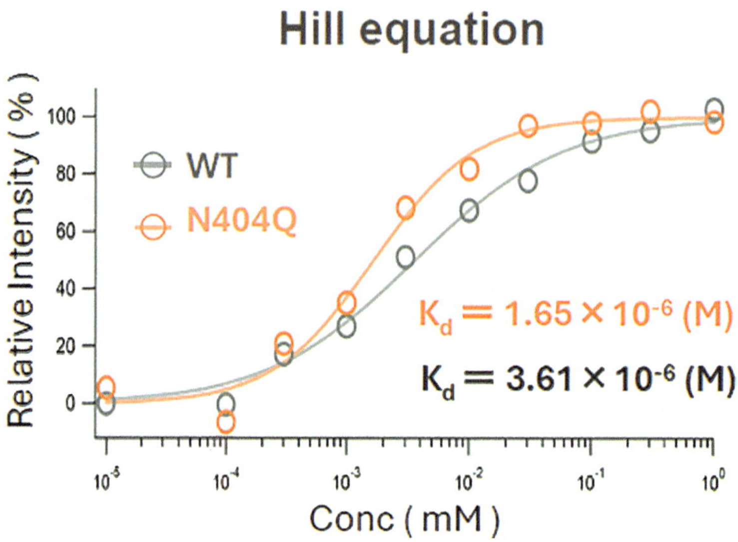

FIGURE 16.12

Dependency of ligand concentration on the acetylcholine-induced infrared bands of WT and N404Q mutant. The peak-to-peak intensity at 1667–1657 cm^{-1} is plotted against the acetylcholine concentration. The data are fitted by Hill equations. The apparent K_d values are estimated as 3.61 ± 0.50 and 1.65 ± 0.24 μM, respectively, and the Hill coefficient is 0.72 and 1.03, respectively.

associated structural changes by detecting alterations in the infrared absorption spectra. Consequently, we successfully obtained FTIR difference spectra that reflected differences in ligand efficacy. Notably, we identified changes in the amide-I band, which is considered an infrared probe for GPCR activation, as a potential indicator for distinguishing between different classes of ligands (Katayama et al., 2021).

Fig. 16.14 presents a quantitative comparative analysis of G protein activation levels inferred from changes in the amide-I band. Specifically, three amide-I bands observed in the agonist-induced difference spectra were designated as ①, ②, and ③, and the sums of their peak intensities (①+② and ②+③) were defined as indicators of the active and inactive states of M_2R, respectively (Fig. 16.14A). The degree of G protein activation was then quantified based on these values. As expected, the ratio (①+②)/(②+③) exceeded 1 only for the agonist, validating the reliability of this approach (Fig. 16.14B).

Furthermore, G protein activity for each ligand was evaluated using a cell-based assay (NanoBiT G protein dissociation assay) (Inoue et al., 2019), with the magnitude of G protein activation plotted on the *y*-axis and the degree of activation inferred from the infrared analysis plotted on the *x*-axis (Fig. 16.14C). As shown in the figure, a clear correlation emerged: as ligand-induced G protein activation increased, the proportion of the activated state also increased. These results suggest that changes in the amide-I band, detected through ligand-binding-induced infrared spectroscopy, can serve as a robust probe for determining the ligand efficacy of M_2R (Katayama et al., 2021).

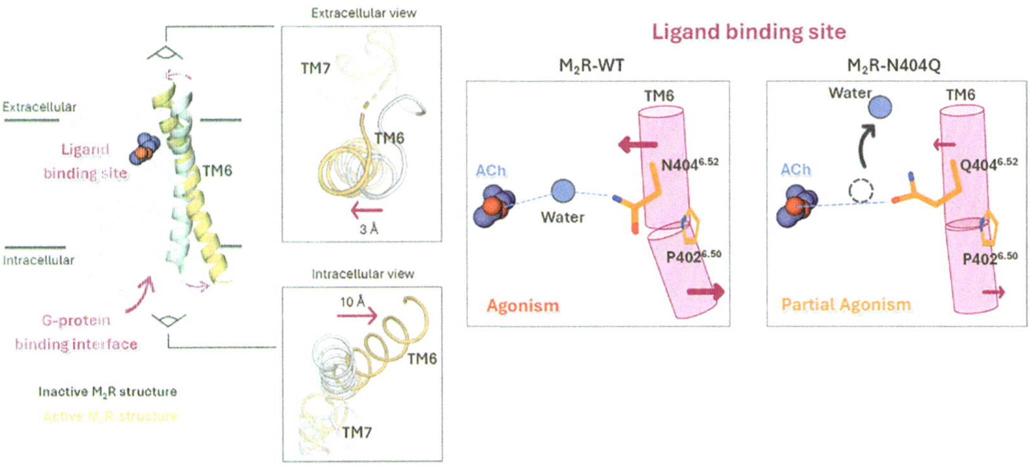

FIGURE 16.13

Differences in the acetylcholine-induced activation mechanism of the WT M_2R and the N404Q mutant. In the WT M_2R the stable binding of acetylcholine induces a sufficient opening motion of TM6. This process is facilitated by water molecules effectively occupying the space between acetylcholine and Asn404, ensuring stable ligand binding. In contrast, in the N404Q mutant, the altered side chain of Gln404 creates a space that is insufficient for water molecules to adequately occupy. This deficiency prevents stable acetylcholine binding and subsequently suppresses the proper opening motion of TM6.

16.5 Concluding remarks

As demonstrated in this study, the application of FTIR difference spectroscopy to elucidate the spectral tuning mechanisms of GPCRs involved in color discrimination has extended beyond its conventional use in light-stimulated systems. This technique has now been successfully applied to chemical stimuli, such as ions and ligands, thereby significantly broadening its potential as a structural analysis method for investigating ligand recognition and activation mechanisms in nonlight-responsive GPCRs. In particular, with recent advances in obtaining atomic-level static structures of GPCRs through X-ray crystallography and single-particle analysis using cryo-electron microscopy (Cryo-EM), the integration of these structural data with the detailed intermolecular interaction information provided by FTIR spectroscopy enables a more comprehensive understanding of ligand recognition and activation mechanisms. This combined approach, leveraging both static and dynamic structural information, offers deeper insights into GPCR function.

Furthermore, this method is expected to establish a novel paradigm in drug design, diverging from conventional structure-based approaches that primarily focus on ligand-binding pockets. By extracting dynamic structural information through interaction analysis, this approach has the potential to enhance control over signal selectivity and drug efficacy. In addition, the applicability of this technique is not limited to GPCRs; it can also be extended to other membrane proteins of pharmacological interest,

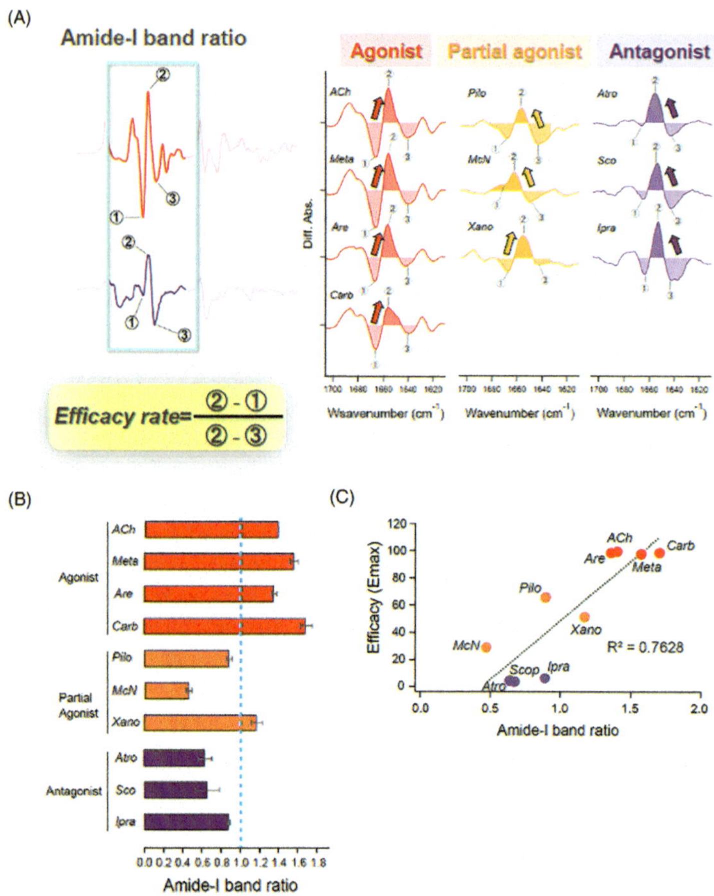

FIGURE 16.14

(A) Left: ligand binding-induced difference ATR-FTIR spectra in the 1800–1200 cm^{-1} region, especially focusing on amide-I band region, which are taken from a. *Red* and *purple lines* correspond to acetylcholine- and atropine-bound spectra, respectively. To quantitatively examine the change in the amide-I band specific to ligand efficacy the ratio between the band strength of (2)1656 (+)/(1)1666 (−) cm^{-1} of acetylcholine-bound spectra at high frequency and (2)1656 (+)/(3)1640 (−) cm^{-1} of acetylcholine-bound spectra at low frequency is calculated and reported as efficacy rate. Right: ligand-dependent spectral changes of amide-I band originating from C=O stretch of α-helix in 1700–1620cm^{-1} region. *Red, orange*, and *purple lines* correspond to agonist-, partial agonist-, and antagonist-bound spectra, respectively. One division of the *y*-axis corresponds to 0.002 absorbance units. (B) Relative populations of active and inactive states upon ligand binding in M_2R derived by the equation of efficacy rate from (A). The error values were calculated from three replicate experiments. *Cyan dotted line* indicates the value = 1. (C) Correlation between ligand efficacy and the relative intensities of the amide-I bands. The relative intensities of the amide-I bands are calculated by the equation of efficacy rate, which is derived from (A). Ligand efficacy is determined by the NanoBiT G protein dissociation assay. Agonists (ACh, Meta, Are, and Carb) are shown as *red circles*, partial agonists (Pilo, McN, and Xano) as *orange circles*, and antagonists (Atro, Scop, and Ipra) as *purple circles. ATR*, Attenuated total reflection; *FTIR*, Fourier-transform infrared.

including ion channels and transporters, thereby offering broader opportunities for advancing drug discovery and development.

Acknowledgment

This study, involving FTIR spectroscopic analysis to elucidate the spectral tuning mechanism in cone pigments, was made possible through the generous support of many collaborators and students. I am especially grateful to Distinguished Professor Hideki Kandori (Nagoya Institute of Technology) for his guidance on the overall research direction. I also thank Professor Hiroo Imai (Kyoto University) and Associate Professor Yuji Furutani (Nagoya Institute of Technology) for their advice on protein expression and FTIR measurements. Appreciation is extended to Professor So Iwata (Kyoto University) and Associate Professor Ryoji Suno (Kansai Medical University) for providing samples and insightful discussions on light-insensitive GPCRs. I am also grateful to Professor Asuka Inoue (Tohoku University) for contributions to GPCR activity assays.

References

Argade, P. V., Rothschild, K. J., Kawamoto, A. H., Herzfeld, J., & Herlihy, W. C. (1981). Resonance Raman spectroscopy of specifically [epsilon-15N]lysine-labeled bacteriorhodopsin. *Proceedings of the National Academy of Sciences, 78*(3), 1643–1646. doi:10.1073/pnas.78.3.1643.

Asenjo, A. B., Rim, J., & Oprian, D. D. (1994). Molecular determinants of human red/green color discrimination. *Neuron, 12*(5), 1131–1138. doi:10.1016/0896-6273(94)90320-4.

Baasov, T., Friedman, N., & Sheves, M. (1987). Factors affecting the C=N stretching in protonated retinal Schiff base: A model study for bacteriorhodopsin and visual pigments. *Biochemistry, 26*(11), 3210–3217. doi:10.1021/bi00385a041.

Barth, A. (2007). Infrared spectroscopy of proteins. *Biochimica et Biophysica Acta (BBA)—Bioenergetics, 1767*(9), 1073–1101. doi:10.1016/j.bbabio.2007.06.004.

Bravaya, K., Bochenkova, A., Granovsky, A., & Nemukhin, A. (2007). An opsin shift in rhodopsin: Retinal S0-S1 excitation in protein, in solution, and in the gas phase. *Journal of the American Chemical Society, 129*(43), 13035–13042. doi:10.1021/ja0732126.

Coto, P. B., Strambi, A., Ferré, N., & Olivucci, M. (2006). The color of rhodopsins at the ab initio multiconfigurational perturbation theory resolution. *Proceedings of the National Academy of Sciences of the United States of America, 103*(46), 17154–17159. doi:10.1073/pnas.0604048103.

Doig, S. J., Reid, P. J., & Mathies, R. A. (1991). Picosecond time-resolved resonance Raman spectroscopy of bacteriorhodopsin's J, K, and KL intermediates. *Journal of Physical Chemistry, 95*(16), 6372–6379. doi:10.1021/j100169a054.

Doki, S., Kato, H. E., Solcan, N., Iwaki, M., Koyama, M., Hattori, M., Iwase, N., Tsukazaki, T., Sugita, Y., Kandori, H., Newstead, S., Ishitani, R., & Nureki, O. (2013). Structural basis for dynamic mechanism of proton-coupled symport by the peptide transporter POT. *Proceedings of the National Academy of Sciences of the United States of America, 110*(28), 11343–11348. http://www.pnas.org/content/110/28/11343.full.pdf+html. doi:10.1073/pnas.1301079110. Japan.

Ernst, O. P., Lodowski, D. T., Elstner, M., Hegemann, P., Brown, L. S., & Kandori, H. (2014). Microbial and animal rhodopsins: Structures, functions, and molecular mechanisms. *Chemical Reviews, 114*(1), 126–163. doi:10.1021/cr4003769.

Eyring, G., & Mathies, R. (1979). Resonance Raman studies of bathorhodopsin: Evidence for a protonated Schiff base linkage. *Proceedings of the National Academy of Sciences of the United States of America, 76*(1), 33–37. doi:10.1073/pnas.76.1.33.

Furutani, Y., Murata, T., & Kandori, H. (2011). Sodium or lithium ion-binding-induced structural changes in the K-ring of V-ATPase from *Enterococcus hirae* revealed by ATR-FTIR spectroscopy. *Journal of the American Chemical Society, 133*(9), 2860–2863. doi:10.1021/ja1116414.

Haga, K., Kruse, A. C., Asada, H., Yurugi-Kobayashi, T., Shiroishi, M., Zhang, C., Weis, W. I., Okada, T., Kobilka, B. K., Haga, T., & Kobayashi, T. (2012). Structure of the human M2 muscarinic acetylcholine receptor bound to an antagonist. *Nature, 482*(7386), 547–551. doi:10.1038/nature10753.

Haga, T. (2013). Molecular properties of muscarinic acetylcholine receptors. *Proceedings of the Japan Academy, Series B, 89*(6), 226–256. doi:10.2183/pjab.89.226.

Hirano, T., Fujioka, N., Imai, H., Kandori, H., Wada, A., Ito, M., & Shichida, Y. (2006). Assignment of the vibrational modes of the chromophores of iodopsin and bathoiodopsin: Low-temperature Fourier transform infrared spectroscopy of 13C- and2H-labeled iodopsins. *Biochemistry, 45*(4), 1285–1294. doi:10.1021/bi0517077.

Hirano, T., Imai, H., Kandori, H., & Shichida, Y. (2001). Chloride effect on iodopsin studied by low-temperature visible and infrared spectroscopies. *Biochemistry, 40*(5), 1385–1392. doi:10.1021/bi001332j.

Hofmann, K. P., & Lamb, T. D. (2023). Rhodopsin, light-sensor of vision. *Progress in Retinal and Eye Research, 93*. www.elsevier.com/locate/preteyeres. doi:10.1016/j.preteyeres.2022.101116.

Imai, H., Terakita, A., & Shichida, Y. (2000). Analysis of amino acid residues in rhodopsin and cone visual pigments that determine their molecular properties. *Methods in Enzymology, 315*, 293–312. doi:10.1016/S0076-6879(00)15850-1.

Imamoto, Y., & Shichida, Y. (2014). Cone visual pigments. *Biochimica et Biophysica Acta—Bioenergetics, 1837*(5), 664–673. www.elsevier.com/locate/bbabio. doi:10.1016/j.bbabio.2013.08.009.

Inoue, A., Raimondi, F., Kadji, F. M. N., Singh, G., Kishi, T., Uwamizu, A., Ono, Y., Shinjo, Y., Ishida, S., Arang, N., Kawakami, K., Gutkind, J. S., Aoki, J., & Russell, R. B. (2019). Illuminating G-protein-coupling selectivity of GPCRs. *Cell, 177*(7), 1933. https://www.sciencedirect.com/journal/cell. doi:10.1016/j.cell.2019.04.044.

Jiang, X., Zaitseva, E., Schmidt, M., Siebert, F., Engelhard, M., Schlesinger, R., Ataka, K., Vogel, R., & Heberle, J. (2008). Resolving voltage-dependent structural changes of a membrane photoreceptor by surface-enhanced IR difference spectroscopy. *Proceedings of the National Academy of Sciences, 105*(34), 12113–12117. doi:10.1073/pnas.0802289105.

Kandori, H., & Katayama, K. (2020). Light-induced difference Fourier-transform infrared spectroscopy of photoreceptive proteins. In *Vibrational spectroscopy in protein research: From purified proteins to aggregates and assemblies* (pp. 23–57). Elsevier. https://www.sciencedirect.com/book/9780128186107. doi:10.1016/B978-0-12-818610-7.00002-5.

Katayama, K., Furutani, Y., Imai, H., & Kandori, H. (2010). An FTIR study of monkey green- and red-sensitive visual pigments. *Angewandte Chemie—International Edition, 49*(5), 891–894. http://www3.interscience.wiley.com/cgi-bin/fulltext/123232975/PDFSTART. doi:10.1002/anie.200903837Japan.

Katayama, K., Okitsu, T., Imai, H., Wada, A., & Kandori, H. (2015). Identical hydrogen-bonding strength of the retinal Schiff base between primate green- and red-sensitive pigments: New insight into color tuning mechanism. *Journal of Physical Chemistry Letters, 6*(7), 1130–1133. http://pubs.acs.org/journal/jpclcd. doi:10.1021/acs.jpclett.5b00291.

Katayama, K., Furutani, Y., Iwaki, M., Fukuda, T., Imai, H., & Kandori, H. (2018). In situ observation of the role of chloride ion binding to monkey green sensitive visual pigment by ATR-FTIR spectroscopy. *Physical Chemistry Chemical Physics, 20*(5), 3381–3387. http://pubs.rsc.org/en/journals/journal/cp. doi:10.1039/c7cp07277e.

Katayama, K., Suzuki, K., Suno, R., Tsujimoto, H., Iwata, S., Kobayashi, T., & Kandori, H. (2019a). Ligand binding-induced structural changes in the M2 muscarinic acetylcholine receptor revealed by vibrational spectroscopy. *Journal of Physical Chemistry Letters, 10*(22), 7270–7276. http://pubs.acs.org/journal/jpclcd. doi:10.1021/acs.jpclett.9b02942.

Katayama, K., Furutani, Y., Imai, H., & Kandori, H. (2012). Protein-bound water molecules in primate red- and green-sensitive visual pigments. *Biochemistry, 51*(6), 1126–1133. doi:10.1021/bi201676y.

Katayama, K., Nakamura, S., Sasaki, T., Imai, H., & Kandori, H. (2019b). Role of Gln114 in spectral tuning of a long-wavelength sensitive visual pigment. *Biochemistry, 58*(26), 2944–2952. http://pubs.acs.org/journal/bichaw. doi:10.1021/acs.biochem.9b00340.

Katayama, K., Nonaka, Y., Tsutsui, K., Imai, H., & Kandori, H. (2017). Spectral tuning mechanism of primate blue-sensitive visual pigment elucidated by FTIR spectroscopy. *Scientific Reports, 7*(1). www.nature.com/srep/index.html. doi:10.1038/s41598-017-05177-4.

Katayama, K., Suzuki, K., Suno, R., Kise, R., Tsujimoto, H., Iwata, S., Inoue, A., Kobayashi, T., & Kandori, H. (2021). Vibrational spectroscopy analysis of ligand efficacy in human M2 muscarinic acetylcholine receptor (M2R). *Communications Biology, 4*(1). https://www.springer.com/journal/11852. doi:10.1038/s42003-021-02836-1.

Kitade, Y., Furutani, Y., Kamo, N., & Kandori, H. (2009). Proton release group of pharaonis phoborhodopsin revealed by ATR-FTIR spectroscopy. *Biochemistry, 48*(7), 1595–1603. http://pubs.acs.org/doi/pdfplus/10.1021/bi801984u. doi:10.1021/bi801984u. Japan.

Kochendoerfer, G. G., Lin, S. W., Sakmar, T. P., & Mathies, R. A. (1999). How color visual pigments are tuned. *Trends in Biochemical Sciences, 24*(8), 300–305. www.elsevier.com/locate/tibs. doi:10.1016/S0968-0004(99)01432-2.

Kochendoerfer, G. G., Wang, Z., Oprian, D. D., & Mathies, R. A. (1997). Resonance Raman examination of the wavelength regulation mechanism in human visual pigments. *Biochemistry, 36*(22), 6577–6587. doi:10.1021/bi970322o.

Kruse, A. C., Ring, A. M., Manglik, A., Hu, J., Hu, K., Eitel, K., Hübner, H., Pardon, E., Valant, C., Sexton, P. M., Christopoulos, A., Felder, C. C., Gmeiner, P., Steyaert, J., Weis, W. I., Garcia, K. C., Wess, J., & Kobilka, B. K. (2013). Activation and allosteric modulation of a muscarinic acetylcholine receptor. *Nature, 504*(7478), 101–106. doi:10.1038/nature12735.

Lin, S. W., Kochendoerfer, G. G., Carroll, K. S., Wang, D., Mathies, R. A., & Sakmar, T. P. (1998). Mechanisms of spectral tuning in blue cone visual pigments. *Journal of Biological Chemistry, 273*(38), 24583–24591. doi:10.1074/jbc.273.38.24583.

Neitz, M., Neitz, J., & Jacobs, G. H. (1991). Spectral tuning of pigments underlying red-green color vision. *Science, 252*(5008), 971–974. doi:10.1126/science.1903559.

Okano, T., Kojima, D., Fukada, Y., Shichida, Y., & Yoshizawa, T. (1992). Primary structures of chicken cone visual pigments: Vertebrate rhodopsins have evolved out of cone visual pigments. *Proceedings of the National Academy of Sciences of the United States of America, 89*(13), 5932–5936. www.pnas.org. doi:10.1073/pnas.89.13.5932.

Palczewski, K. (2012). Chemistry and biology of vision. *Journal of Biological Chemistry, 287*(3), 1612–1619. http://www.jbc.org/content/287/3/1612.full.pdf+html. doi:10.1074/jbc.R111.301150. United States.

Palczewski, K., Kumasaka, T., Hori, T., Behnke, C. A., Motoshima, H., Fox, B. A., Le Trong, I., Teller, D. C., Okada, T., Stenkamp, R. E., Yamamoto, M., & Miyano, M. (2000). Crystal structure of rhodopsin: A G protein-coupled receptor. *Science, 289*(5480), 739–745. doi:10.1126/science.289.5480.739.

Palings, I., van den Berg, E. M. M., Lugtenburg, J., & Mathies, R. A. (1989). Complete assignment of the hydrogen out-of-plane wagging vibrations of bathorhodopsin: Chromophore structure and energy storage in the primary photoproduct of vision. *Biochemistry, 28*(4), 1498–1507. doi:10.1021/bi00430a012.

Pedraza-González, L., Barneschi, L., Padula, D., De Vico, L., & Olivucci, M. (2022). Evolution of the automatic rhodopsin modeling (ARM) protocol. *Topics in Current Chemistry, 380*(3). https://www.springer.com/journal/41061. doi:10.1007/s41061-022-00374-w.

Rajamani, R., Lin, Y. L., & Gao, J. (2011). The opsin shift and mechanism of spectral tuning in rhodopsin. *Journal of Computational Chemistry, 32*(5), 854–865. doi:10.1002/jcc.21663.

Shichida, Y., & Imai, H. (1998). Visual pigment: G-protein-coupled receptor for light signals. *Cellular and Molecular Life Sciences (CMLS), 54*(12), 1299–1315. doi:10.1007/s000180050256.

Stenkamp, R. E., Filipek, S., Driessen, C. A. G. G., Teller, D. C., & Palczewski, K. (2002). Crystal structure of rhodopsin: A template for cone visual pigments and other G protein-coupled receptors. *Biochimica et Biophysica Acta (BBA)—Biomembranes, 1565*(2), 168–182. doi:10.1016/s0005-2736(02)00567-9.

Sugiura, Y., Ikuta, T., Sumii, Y., Tsujimoto, H., Suzuki, K., Suno, R., Ariff, P. N. A. B. M., Iwata, S., Shibata, N., Inoue, A., Kobayashi, T., Kandori, H., & Katayama, K. (2025). Discovering key activation hotspots in the M2 muscarinic receptor. *Journal of the American Chemical Society, 147*(14), 11754–11765. http://pubs.acs.org/journal/jacsat. doi:10.1021/jacs.4c14385.

Suzuki, K., Katayama, K., Sumii, Y., Nakagita, T., Suno, R., Tsujimoto, H., Iwata, S., Kobayashi, T., Shibata, N., & Kandori, H. (2021). Vibrational analysis of acetylcholine binding to the M 2 receptor. *RSC Advances, 11*(21), 12559–12567. doi:10.1039/d1ra01030a.

Wanko, M., Hoffmann, M., Frauenheim, T., & Elstner, M. (2008). Effect of polarization on the opsin shift in rhodopsins. 1. A combined QM/QM/MM model for bacteriorhodopsin and pharaonis sensory rhodopsin II. *Journal of Physical Chemistry B, 112*(37), 11462–11467. http://pubs.acs.org/journal/jpcbfk. doi:10.1021/jp802408g.

Xu, J., Wang, Q., Hübner, H., Hu, Y., Niu, X., Wang, H., Maeda, S., Inoue, A., Tao, Y., Gmeiner, P., Du, Y., Jin, C., & Kobilka, B. K. (2023). Structural and dynamic insights into supra-physiological activation and allosteric modulation of a muscarinic acetylcholine receptor. *Nature Communications, 14*(1). doi:10.1038/s41467-022-35726-z.

Yokoyama, R., & Yokoyama, S. (1990). Convergent evolution of the red- and green-like visual pigment genes in fish, *Astyanax fasciatus*, and human. *Proceedings of the National Academy of Sciences of the United States of America, 87*(23), 9315–9318. www.pnas.org.

Yoshizawa, T., & Kitô, Y. (1958). Chemistry of the rhodopsin cycle. *Nature, 182*(4649), 1604–1605. doi:10.1038/1821604a0.

Zhang, T., & Fu, Y. (2013). A phe-rich region in short-wavelength sensitive opsins is responsible for their aggregation in the absence of 11-cis-retinal. *FEBS Letters, 587*(15), 2430–2434. doi:10.1016/j.febslet.2013.06.012.

CHAPTER

Dynamic interactions in GPCR–G protein coupling: Distinguishing specificity and pluripotency

17

Aditya Manoj[a,b,*] and Durba Sengupta[c]

[a]*CSIR-National Chemical Laboratory, Pune, Maharashtra, India,* [b]*Academy of Scientific and Innovative Research, Ghaziabad, Uttar Pradesh, India,* [c]*Physical Chemistry Division, CSIR-National Chemical Laboratory, Pune, Maharashtra, India*

**Corresponding author: m.aditya@ncl.res.in*

17.1 Introduction

G protein-coupled receptors (GPCRs) are a large family of transmembrane proteins comprised of more than 800 members (Bjarnadóttir et al., 2006). They are critical in a wide range of physiological processes, including neuronal activity (Leung & Wong, 2017), hormonal regulation (Salazar et al., 2007), immunity (Rivera et al., 2008), development, and homeostasis (Wu et al., 2019). The significant roles of GPCRs in controlling numerous cellular signaling pathways have established them as major targets in drug discovery, with approximately one-third of current therapeutic drugs targeting these receptors (Flock et al., 2017). Upon recognition of diverse ligands, including small molecules, peptides, lipids, and proteins, GPCRs initiate the transduction of extracellular physicochemical stimuli into intracellular signaling pathways via the binding of transducers such as the heterotrimeric G proteins. GPCRs are subdivided into different classes (class A to class F), based on their sequence homology and functional similarity (Kolakowski, 1994). Class A comprises the largest number and the best characterized members, such as the adrenergic receptors that function in response to adrenaline and noradrenaline, the serotonin receptors (that bind serotonin), and the dopamine receptors (that bind to dopamine). Although GPCRs are diverse and categorized into different classes, they share a conserved structural architecture (Hilger et al., 2018; Pándy-Szekeres et al., 2018) and a common activation mechanism (Latorraca et al., 2017; Zhou et al., 2019).

Characteristically, these receptors are composed of seven transmembrane helices, which traverse the cell membrane seven times and are consequently also known as seven-(pass)-transmembrane domain receptors (Hollenstein et al., 2014; Venkatakrishnan et al., 2013). A schematic representation of a representative GPCR highlighting its structural characteristics is shown in Fig. 17.1. The seven transmembrane helices are numbered I–VII, and the interconnecting loops are named as intracellular loops (ICL1–3) and extracellular loops (ECL1–3). The lengths of the N- and C-terminal domains exhibit significant variability, often depending on the GPCR subclass, and the C-terminus typically features a surface-aligned helix, named helix-8 (H8). Ligand binding to GPCRs, occurring within the lumen of the transmembrane domain or at the extracellular domain, induces critical conformational transitions that propagate throughout the receptor. These structural changes lead to an "active" state that facilitates the binding of heterotrimeric G protein partners at the intracellular face of the receptor

Protein Dynamics: Experimental and Computational Approaches. DOI: https://doi.org/10.1016/B978-0-443-30042-4.00021-3

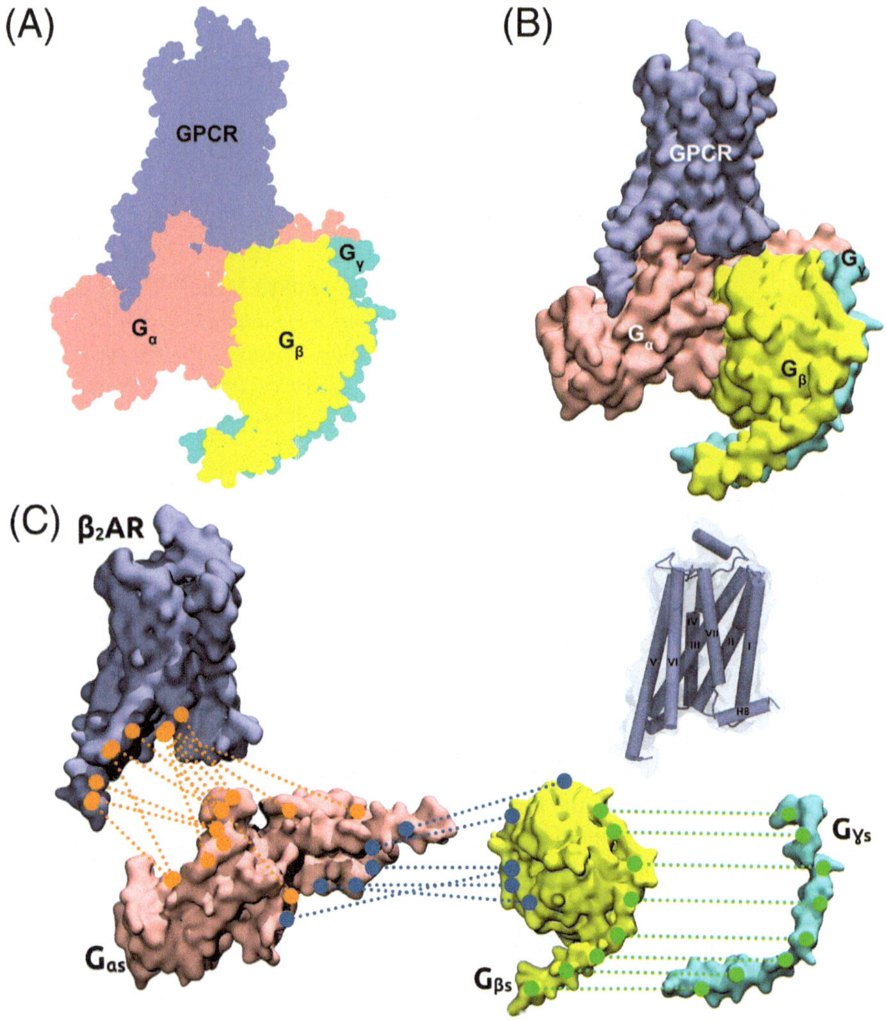

FIGURE 17.1

Structural characteristics of GPCRs bound to G proteins. (A) A schematic representation of the GPCR–G protein complex. The GPCR β_2-adrenergic receptor (β_2AR) is depicted in *purple color.* The G proteins G_α are shown in *pink color*, G_β is shown in *yellow color*, and G_γ is shown in *light blue color.* (B) A surface representation of the complex in which the GPCR β_2-adrenergic receptor (β_2AR) is depicted in *purple*, G_α is shown in *pink*, G_β is shown in *yellow*, and G_γ is shown in *light blue color.* (C) The molecular contacts between the GPCR (β_2AR) complexed with G_s protein. $G_{\alpha s}$ is shown in *pink color*, $G_{\beta s}$ is shown in *yellow color*, and $G_{\gamma s}$ is shown in *light blue color.* The molecular contacts are shown in *dotted line* representation. *Orange dotted lines* indicate molecular contacts between GPCR–$G_{\alpha s}$, *dark blue dotted lines* indicate molecular contacts between $G_{\alpha s}$–$G_{\beta s}$, and *green dotted lines* indicate molecular contacts between $G_{\beta s}$–$G_{\gamma s}$. The structure of a GPCR illustrated in *purple cylinder* representation with transparent surface background is shown in the inset. The transmembrane (TM) helices are numbered in Roman numerals. *GPCR*, G protein-coupled receptor.

(Farrens et al., 1996; Wettschureck & Offermanns, 2005). The β_2-adrenergic receptor (β_2AR) was one of the first receptors whose structural dynamics upon binding of a ligand and G protein were determined by a series of experimentally determined structures (Rasmussen et al., 2011). The active state structure of β_2AR, complexed with its cognate G protein, provides a detailed snapshot of the functional dynamics of the receptor, illustrating the critical interactions that mediate signal transduction (see Fig. 17.1). Over the years, several GPCRs such as serotonin receptors and dopamine receptors have been resolved in the active state bound to the G protein partners (Liu et al., 2021; Özçete et al., 2024). These experimentally resolved structures were instrumental in establishing the molecular mechanisms governing GPCR coupling. For instance the main molecular contacts (shown in *dotted line* representation in the figure) between the GPCR and G protein are relatively few in number. In particular, interactions were reported mainly with one of the proteins of the heterotrimeric G protein (the α subunit) with almost no interactions between the GPCR and the other two proteins (the β and γ subunits). Interestingly, conserved features of G protein coupling are evident across all G protein subtypes, characterized by a similar relative position and orientation of binding to the GPCR. These structural studies suggest a common mechanism for GPCR-induced G protein activation, while also revealing subtle variations in the mechanisms used by different receptor and G protein subtypes. The accumulation of such structural data has significantly advanced our understanding of GPCR signaling, enabling the development of targeted therapeutic interventions.

17.2 G proteins and their subtypes

Guanine nucleotide binding proteins (G proteins) are membrane-associated, heterotrimeric effector proteins composed of three subunits: α, β, and γ (shown in Fig. 17.1) (Preininger & Hamm, 2004). The α subunit of heterotrimeric G proteins is composed of two domains, the GTPase domain and the helical domain (Ahn et al., 2021). The GTPase domain, homologous to other GTPases such as the Ras superfamily, facilitates guanine nucleotide binding and hydrolysis. Crucially, this domain undergoes dynamic conformational shifts upon guanosine diphosphate (GDP)–guanosine triphosphate (GTP) exchange, enabling the α subunit to engage downstream effectors. In the closed inactive state the two domains come together to form a guanine nucleotide-binding pocket but, upon activation, undergo a large rotational opening of the α-helical domain. The β subunit adopts a seven-bladed β-propeller fold, providing a structural scaffold for interaction with the γ subunit. The γ subunit, a comparatively small protein with an extended structure, forms a coiled-coil interaction with the β subunit.

One can consider that ligand-bound active GPCRs behave as guanine nucleotide exchange factors (GEFs), promoting the exchange of bound GDP for GTP on the G_α subunit (Oldham & Hamm, 2008). The trimeric G protein subsequently dissociates from the receptor into the α subunit (GTP-bound) and β/γ dimer as depicted in Fig. 17.1C (Flock et al., 2015; Hauser et al., 2022). The α subunit and the β/γ dimer then activate distinct downstream effectors, such as adenylyl cyclase, phosphodiesterases, phospholipase C, and ion channels. These effectors, in turn, regulate the intracellular concentrations of secondary messengers, such as cAMP, diacylglycerol (DAG), sodium, or calcium cations, which ultimately lead to a physiological response, usually via the downstream regulation of gene transcription (Neves et al., 2002; Svoboda et al., 2004).

G_α proteins are broadly classified into four families, as illustrated in Fig. 17.2. Namely, these are $G_{\alpha s}$ (stimulatory), $G_{\alpha i/o}$ (inhibitory), $G_{\alpha q}$ (toxin-insensitive), and $G_{\alpha 12/13}$ (activation of ρ effector). The various classes represent the diversity of the G proteins in eliciting different cellular responses. $G_{\alpha s}$

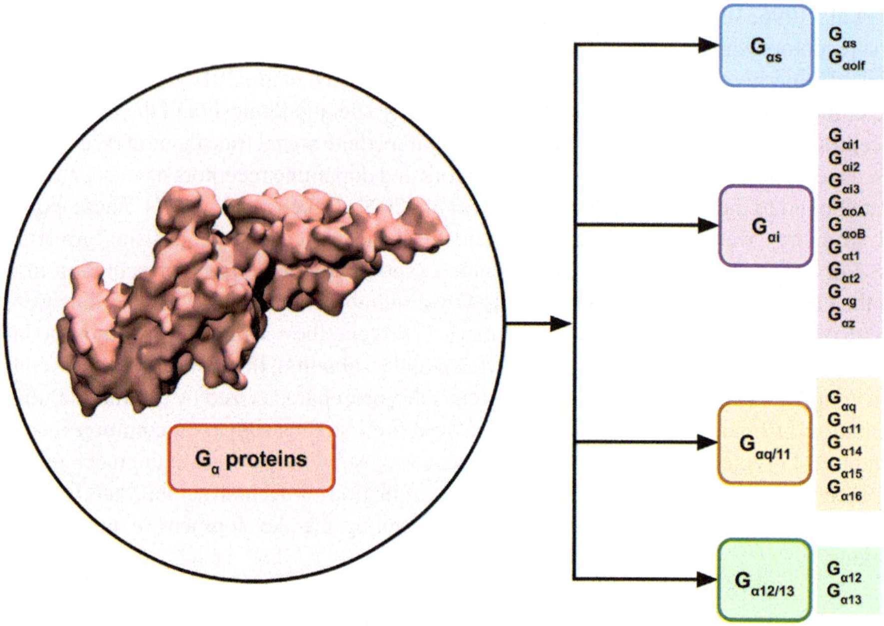

FIGURE 17.2

Classification of G_α proteins: G_α proteins are classified into four families: $G_{\alpha s}$ (stimulatory responses) is shown in *blue color*, $G_{\alpha i/o}$ (inhibitory responses) is shown in *purple color*, $G_{\alpha q}$ (toxin-insensitive responses) is shown in *yellow color*, and $G_{\alpha 12/13}$ (activation of ρ effectors) is shown in *green color*.

stimulates the synthesis of cyclic AMP (cAMP) from ATP, which, in turn, activates the cAMP-dependent pathway. The membrane-associated enzyme adenylate cyclase is directly stimulated. Then, cAMP may function as a second messenger, interacting with and activating protein kinase A (PKA), which has the ability to phosphorylate numerous downstream targets. $G_{\alpha i/o}$ is reported to primarily inhibit adenylyl cyclase and reduce cAMP levels, that is, inhibit the production of cAMP from ATP. $G_{\alpha q}$ stimulates the membrane-bound phospholipase C beta, which then cleaves phosphatidylinositol 4,5-bisphosphate (PIP_2) into two second messengers, inositol trisphosphate (IP_3), and DAG. IP_3 induces calcium release from the endoplasmic reticulum. DAG activates protein kinase C. The inositol phospholipid-dependent pathway is used as a signal transduction pathway for many hormones. $G_{\alpha 12/13}$ are involved in Rho GTPase signaling family involved in control of cell cytoskeleton remodeling and thus in regulating cell migration (Syrovatkina et al., 2016).

These main classes are further subdivided into subtypes or isoforms with specialized functions. For instance, $G_{\alpha i}$ proteins (of the $G_{\alpha i/o}$ class) are widely distributed throughout various tissues and cell types and include $G_{\alpha i1}$, $G_{\alpha i2}$, and $G_{\alpha i3}$ subclasses, whereas $G_{\alpha o}$ subunits are mainly present in the brain and are involved in various neuronal functions. $G_{\alpha z}$ proteins also belong to the $G_{i/o}$ subclass and are involved in the inhibition of adenylyl cyclase but differ from $G_{\alpha i/o}$ proteins in their insensitivity to pertussis toxin. The $G_{\alpha s}$ family includes $G_{\alpha s}$ and $G_{\alpha olf}$ subclasses, both stimulating adenylyl cyclase to increase cAMP levels, activating PKA. The $G_{\alpha q}$ family includes $G_{\alpha q}$, $G_{\alpha 11}$, $G_{\alpha 14}$, $G_{\alpha 15}$, and $G_{\alpha 16}$, all activating PLC-β, leading to IP_3 and DAG production, increasing intracellular calcium levels. The $G_{\alpha 12/13}$ family consists of $G_{\alpha 12}$ and $G_{\alpha 13}$, which activate Rho-GEFs, regulating cytoskeletal dynamics and cell movement (Suzuki et al., 2009).

17.3 G protein coupling: Emerging perspectives

The diversity of G protein partners helps GPCRs to evoke complex intracellular pathways while being able to respond to a broad range of extracellular stimuli (Downes & Gautam, 1999). The conventional model of GPCR signaling considered that each GPCR engages only a particular type of G protein subtype (Jiang et al., 2022; Satagopam et al., 2010). This meant that GPCRs could only couple to their own partner G proteins (cognate G proteins) to elicit the downstream physiological responses. The pluripotency, that is, capacity of GPCRs to couple to multiple G proteins, is likely responsible for the vast number of complex signaling events and cellular responses. Pluripotency has sometimes been suggested to be due to promiscuous coupling, but emerging evidence suggests a specificity that is not consistent with promiscuity (a feature, not a bug!). A prominent example is the β_2-adrenergic receptor (β_2AR), which was shown to couple to $G_{\alpha s}$ (McGraw & Liggett, 2005). However, several studies have highlighted the interactions with $G_{\alpha i}$ (Xiao, 2001), which triggers the opposite response. Similarly, several studies highlight multicoupling of G proteins in the β_1-adrenergic receptor (β_2AR), D1 dopamine receptor (D1R), histamine H2 receptor (H2R), vasopressin receptor 2 (V2R), and adenosine A1 receptor (A1R) (Masuho et al., 2023; Okashah et al., 2019). Till date a large number of GPCRs have been reported to couple to multiple G proteins (Avet et al., 2022; Inoue et al., 2019; Kapolka et al., 2020; Masuho et al., 2023, 2015; Okashah et al., 2020; Olsen et al., 2020; Wright et al., 2024), although with varying strengths and efficacies. A few examples of specific and pluripotent coupling of GPCRs are given in Table 17.1. It was reported that the bradykinin B2 receptor exhibits one of the most pluripotent couplings, whereas purinergic receptor P2Y12 exhibits one of the most specific interactions to Gi subtypes (Sandhu et al., 2022). In general, Gi coupling GPCRs appear to exhibit lower pluripotency. Notably, G protein coupling events may occur at different temporal coupling times, giving GPCRs the ability to periodically stimulate specific responses in waves (Herenbrink et al., 2016; Masuho et al., 2015).

Considering that GPCRs can be pluripotent in G protein coupling, the identification and characterization of the different coupling preferences of GPCRs is critical (Avet et al., 2022; Inoue et al., 2019; Kapolka et al., 2020; Masuho et al., 2015; Okashah et al., 2019; Olsen et al., 2020). Surprisingly the characterization of G protein coupling from different studies has shown low agreement across the studies, suggesting complex protein–protein dynamics that are modulated by several factors (Hauser et al., 2022). An average "promiscuity" index has been proposed by normalizing and attributing weights on the different studies (Sandhu et al., 2022). In general, GPCRs exhibit varying affinities for different G protein subtypes (Suno, 2024). However, this can be further modulated, and the role of agonist binding in the activation of GPCRs and subsequent signaling bias has been well characterized (Che et al., 2018; Wacker et al., 2017). Although the effects of agonist binding events can be an important factor in G protein selectivity (Xu et al., 2021), the molecular determinants are unclear. Further, it has been reported that promiscuous couplings are common across specific G protein families (Alexander et al., 2019).

Table 17.1 GPCR pluripotency: G protein subtype coupling and determining factors.

S. no.	GPCR	G protein subtype	Reported factor	Refs.
1.	β_2-adrenergic receptor (β_2AR)	Gs	Agonist, cardiomyocyte cells	Xiao (2001)
		Gi	High agonist, specific cell types	Xiao (2001)
2.	Angiotensin II type 1 receptor (AT1R)	Gq/11	Angiotensin II, vascular smooth muscle	Hunyady and Catt (2006)
		Gi	Biased agonists, some cell types	Hunyady and Catt (2006)
3.	Dopamine D1 receptor (D1R)	$G_{i/o}$	Dopamine, striatal neurons	Rashid et al. (2007)
		Gq/11	High dopamine, specific brain regions	Rashid et al. (2007)
4.	Chemokine receptor CXCR4	Gi	CXCL12, immune cells	Arang and Gutkind (2020)
		Gq	Specific chemokines, cancer cells	Arang and Gutkind (2020)
5.	Bradykinin B2 receptor	Gq/11	Bradykinin, sensory neurons	Blaukat (2003)
		Gi	Some cell types	Blaukat (2003)
		Gs	–	Blaukat (2003)
		G12	–	Blaukat (2003)
6.	Adhesion G protein-coupled receptor G2 (ADGRG2)	Gq	Selective	Sandhu et al. (2022)
7.	Frizzled-10 (FZD10)	G12/13q	Selective	Sandhu et al. (2022)
8.	Purinergic Receptor P2Y12 (P2RY12)	Gi	Selective	Sandhu et al. (2022)

GPCR, *G protein-coupled receptor.*

The lipid microenvironment has also been reported to contribute to G protein coupling selectivity (Yen et al., 2018). In this light, understanding the molecular mechanism behind the selective and nonspecific coupling of agonist-bound GPCRs with G proteins is crucial.

17.4 Characterizing signatures of G protein coupling in G protein-coupled receptors

Several GPCR–G protein complexes have been structurally characterized and have been investigated to identify features that determine coupling selectivity (Carpenter & Tate, 2016; Draper-Joyce et al., 2018; Garc´ıa-Nafr´ıa et al., 2018; Kang et al., 2018; Koehl et al., 2018; Tsai et al., 2018). A representative structure is shown in Fig. 17.3 in which the main structural domains have been labeled. As can be seen

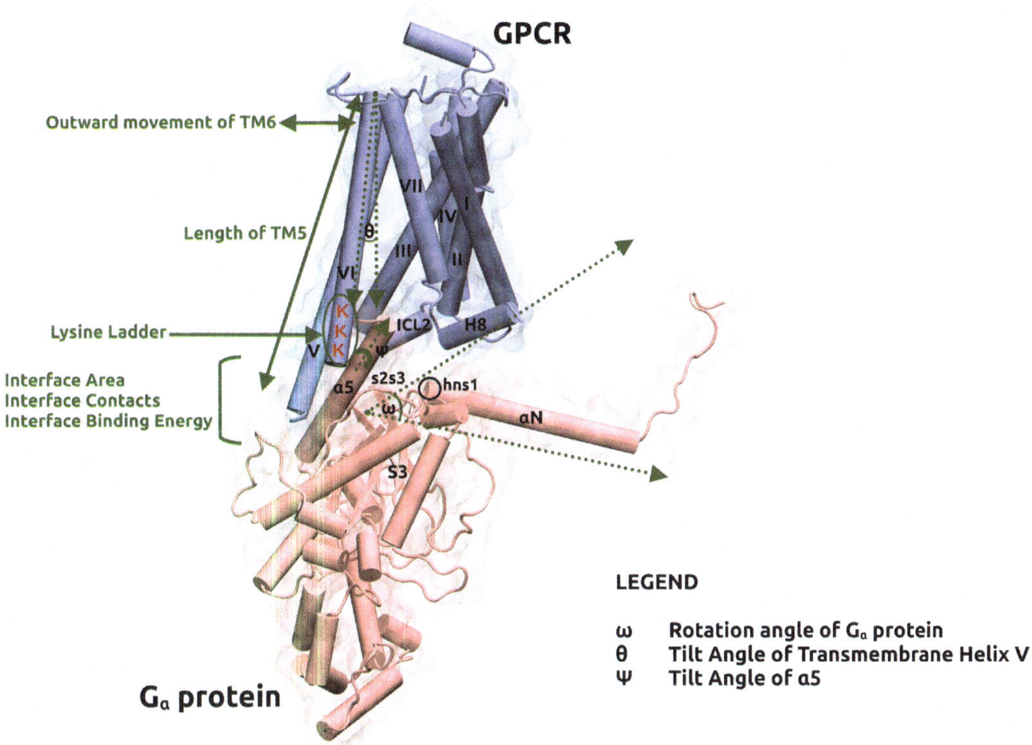

FIGURE 17.3

The structural determinants of G_α protein coupling. The figure shows a GPCR (β_2AR) depicted in *purple cylinder* representation with a *light purple* surface in the background. The G_α protein is also shown in *pink cylinder* representation with a *light pink* surface in the background. *GPCR*, G protein-coupled receptor.

in the figure, the $\alpha5$ helix, also known as the G_α carboxy (C) terminus, is the main site for G protein coupling. The transmembrane (TM) domain, especially TM helices V and VI, envelopes the distal part of the $\alpha5$ helix, promoting the coupling of receptors to the partner G_α proteins (Conklin et al., 1993, 1996; Hamm et al., 1988; Kostenis et al., 1997; Liu et al., 1995; Simonds et al., 1989; Sullivan et al., 1987; West et al., 1985). TM helix VI has been shown to exhibit an outward motion to create a cavity, which facilitates the binding of $\alpha5$ helix. The main interactions were reported with TM helix V, although only a few contacts such as with the lysine ladder have been shown to be important. In all GPCR–G protein complexes that have been experimentally resolved so far, the $\alpha5$ helix is the major structural interaction site, and it has been shown to undergo conformational changes such as helix translocation, rotation, and extension upon binding (Hauser et al., 2022; Semack et al., 2016). The overall orientation of G_α with respect to the GPCR (ω) and the tilt angle of $\alpha5$ helix (Ψ) act as key factors in GPCR–G protein coupling by modulating the interaction between the G protein and the receptor. Further, interactions were reported with other structural domains of GPCRs, including the ICLs and the C-terminal H8. In fact, ICL2 was found to be an important contact hub, interacting broadly with the structural elements, hns1, S1, and S3 of the G_α subunit. Interactions with ICL3 have been suggested from functional studies but are often missing in structural studies because of the high dynamics of this domain. H8 is also a site of posttranslational modifications, such as receptor palmitoylation, and these modifications can modulate the interaction of the G protein with the GPCRs. These extended GPCR and G_α regions are suggested to contribute in the recognition by individual GPCRs (Blahos et al., 2001; Grishina & Berlot, 2000; Kostenis et al., 1997; Lee et al., 1995). However, these receptor-specific and G protein subtype-related structural differences are less distinctive and are discussed below.

17.4.1 Structural determinants of G_α protein subtype coupling

The main features of GPCR–G protein complexes have been analyzed by comparing several experimentally resolved structures with different G protein subtypes (Hauser et al., 2022; Masuho et al., 2023). The structural characteristics observed in different GPCR complexes appear to be more complex and extensive compared to the main $\alpha5$ helix–TMV coupling. We have mapped the main GPCR features reported on a structure of a representative GPCR (Fig. 17.3). As discussed earlier, in the active conformational state of all GPCRs, transmembrane helices V and VI appear to be extended and/or oriented outwards. This allows for the main interaction of transmembrane helices V and VI of GPCRs with $\alpha5$ helix of G_α. Surprisingly, $\alpha5$ helix interactions with TM helices V and VI are important in all subtypes (Conklin et al., 1993; Hamm et al., 1988; Simonds et al., 1989; Sullivan et al., 1987; West et al., 1985). However, it has been reported in the serotonin receptor family that TMV and TMVI may alternately increase or decrease in length as a determining factor for receptor selectivity to Gs and Gi (Huang et al., 2022).

Several other interactions have been suggested to play a role in pluripotent GPCR coupling, particularly in determining the specificity and efficiency of the interaction. Comparison between individual G_α subtypes revealed a wider range of subtype specific interaction elements between GPCRs and G_α subtypes. The importance of ICL2 that was identified from structural studies was validated by functional studies that demonstrated that interactions involving ICL2 are critical for Gi activation by β_1AR (Alegre et al., 2021). The interface area and binding energetics were reported to differ between the subtypes. For instance, in the case of the β_1-adrenergic receptor, a smaller interaction surface was reported between the receptor and Gi compared to Gs (Alegre et al., 2021). These differences are facilitated by differential loop interactions but also the overall orientation of G_α (ω), the tilt of $\alpha5$ helix, differential length of transmembrane helix V, and the outward tilt of transmembrane helix VI. However, robust quantitative measures of these differences have proven challenging. Between the subtypes, $G_{\alpha i/o}$ and $G_{\alpha q}$ were found to follow a similar coupling pattern, only differing in TMVI coupling. The outward movement of the intracellular end of TMVI was suggested to be less in the Gi-bound complexes compared to the Gs complexes (Draper-Joyce et al., 2018), but this was not observed in the β_1-adrenergic receptor (Alegre et al., 2021; Kang et al., 2018). Further, TMVI was found to be a major element in interacting with $G_{\alpha i/o}$ but surprisingly not for $G_{\alpha q}$. The coupling mechanism of $G_{\alpha s}$ was reported to be more complex, and the most distinctive feature of this coupling was reported to be the binding of TMV and ICL3 to the long hgh4 loop. This loop was reported to be absent in other G_α subunits. Several receptors (especially those coupling to $G_{\alpha 12/13}$) also showed prominent TMIV–HN interactions (Hauser et al., 2022; Masuho et al., 2023). These diverse interaction patterns suggest a highly dynamic protein–protein interface in which multiple structural elements govern G protein selectivity through differential engagement mechanisms.

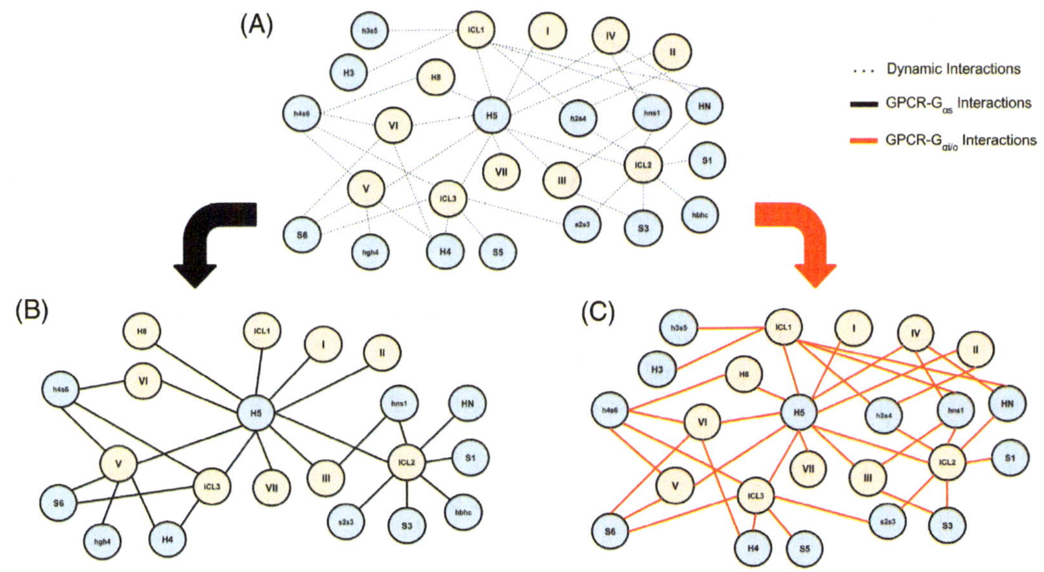

FIGURE 17.4

The dynamic nature of interactions in GPCR–G$_\alpha$ protein coupling. The GPCR elements are shown in *yellow circles*, and G$_\alpha$ protein elements are shown in *blue circles*. (A) The dynamic interactions between GPCRs and G$_{\alpha s}$/G$_{\alpha i/o}$ proteins are shown as *black dotted lines*. The *dotted lines* indicate the transient nature of the interactions. (B) Interactions between GPCRs and G$_{\alpha s}$ proteins (shown in solid *black lines*). (C) Interactions between GPCRs and G$_{\alpha i/o}$ proteins (shown in *solid red lines*). *GPCR*, G protein-coupled receptor.

17.4.2 Differential contacts modulate *G*$_\alpha$ protein subtype coupling

An extensive interaction interface with multiple contact sites has been identified in GPCR–G protein complexes (Avet et al., 2022; Hauser et al., 2022; Inoue et al., 2019; Kapolka et al., 2020; Masuho et al., 2023, 2015; Matic et al., 2023; Okashah et al., 2019; Olsen et al., 2020). To highlight the interaction interfaces in the GPCR–G protein complexes, we have represented the consensus contacts via a protein interaction network, based mainly on the interface contact network analysis reported by Matic and coworkers (Matic et al., 2023) (Fig. 17.4A). In this plot the hubs (nodes) represent the different structural elements of the two proteins, and interconnecting lines (edges) represent a direct contact between them. The interacting sites that were common to the different G protein subtypes are taken as the putative interaction elements required for G$_\alpha$ protein coupling in GPCRs. As several interactions have been reported, not all of which occur simultaneously, we have referred to them as dynamic contacts. A few of these dynamic interactions correspond to step-wise conformational changes with multiple binding conformers during the conformational transition from receptor binding to GDP release (Ahn et al., 2021). These dynamic interactions could also correspond to transient interactions with the complex switching between different sites at a given timescale or as stochastic interactions whose occurrence is determined by external factors such as the membrane environment.

Interestingly, it was observed from the consensus contacts that all transmembrane helices of the GPCRs, except TM IV, interact with the G_α protein, although TM IV contacts have also been reported in some cases (especially with Gq). Similarly, contacts were observed with all ICLs. It can be easily discerned that TMV, TMVI, ICL2, and ICL3 exhibited increased interactions with G_α protein structural elements. Notably, we can see that TMV, TMVI, and ICL3 act as bridging elements between two different G_α protein structural elements. TMV bridges S5 and H6; TMVI bridges H5 and h4s6, and ICL3 bridges H5 and H4. ICL2, on the other hand, acts as a central node in the interaction of G_α with other G_α protein structural elements and forms a bridge to the central H5 helix. ICL2 interactions were observed with G_α protein strutural elements, HN, hns1, S1, s2s3, and S3. From the perspective of the G_α protein, the H5 helix was found to have the highest number of interactions with the GPCR structural elements. The H5 helix interacted with all the structural elements of the GPCR except TMIV and is regarded as the most central and important contacting element. In addition to H5, several other structural elements of the G_α protein interacted with the GPCR, however, to a lesser extent. In particular, G_α protein structural elements h4s6, S6, and H4 were found to have limited interactions with the GPCR (Hauser et al., 2022). Overall, TMV, TMVI, ICL2, and ICL3 can be considered to be the important contacting GPCR structural elements, whereas H5 is the main interaction site of the G proteins.

These interaction networks have been further decomposed for the different G protein subtypes. Fig. 17.4B shows interactions between GPCR–$G_{\alpha s}$ proteins, and Fig. 17.4C shows interactions between GPCR–$G_{\alpha i/o}$ proteins. On comparing the interactions in GPCR–$G_{\alpha s}$ and GPCR–$G_{\alpha i/o}$ complexes, it appears that GPCR–$G_{\alpha s}$ have less interactions than GPCR–$G_{\alpha i/o}$ complexes. Fewer contacts are observed at TMVI coupling to $G_{\alpha s}$ relative to $G_{\alpha i/o}$, and interactions with G protein structural elements such as S6 and H4 are lost. Similarly, contacts with the G protein site, h2s4, are observed only in $G_{\alpha i/o}$. In contrast, hgh4 contacts are observed only in $G_{\alpha s}$. Further, ICL2 makes contacts with hbhc in $G_{\alpha s}$ but not in $G_{\alpha i/o}$. Overall a complex dynamic interaction pattern is observed for G protein subtype specific coupling.

17.4.3 Residues mediating G protein-coupled receptor–G_α interactions

Several residues have been implicated in mediating the contacts between the different GPCR and G protein structural domains discussed above. A large number of studies have highlighted the role of specific residues at the interaction interface in contributing to GPCR–G_α coupling specificity. At the interaction interface, it was reported that ≈100 residues of GPCR interacted with ≈75 residues of the G_α protein (Masuho et al., 2023). However, the actual contacts identified in the experimentally resolved structures (often with truncated loop regions) are far fewer. A consensus network of the residue–residue contacts reported between the GPCRs and G_α proteins (Masuho et al., 2023; Matic et al., 2023) was constructed and is shown in Fig. 17.5. The residues are clubbed based on the structural domains, and a line linking them implies a previously reported direct contact. The residues are numbered as per the Ballesteros–Weinstein numbering that provides a standardized numbering scheme that allows an easy comparison of residue location within the transmembrane helices and identifies similar structural and functional roles.

Analysis of the residue interactions in Fig. 17.5 highlights the residue pairs that form the GPCR–G_α protein complexes. In the experimentally-resolved structures, the maximum interactions are from TMVI, which include five residues, 6.29, 6.32, 6.33, 6.36, and 6.37that interact with the C-terminal H5 residues, H5.24, H5.25, and H5.26. Of these, the charged residues on TMVI (usually lysine "ladder"

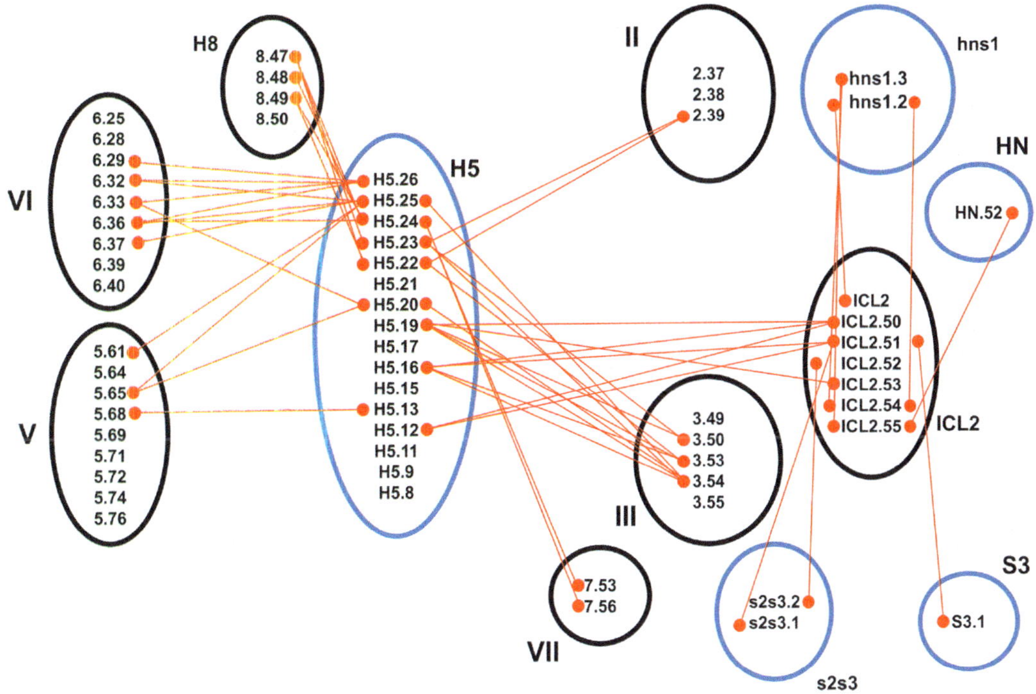

FIGURE 17.5

The consensus contacts of residue–residue interactions between GPCRs and G_α proteins. The GPCR elements are shown in *black outline*, and G_α protein elements are shown in *blue outline*. Within the outline the residues in contact in between the respective elements are shown in *black text*. The residue contacts are shown in *red lines*. The contacts depict a GPCR–G_α protein interaction network at the residue level. The common GPCR numbering system was used to represent the residues of the GPCR, whereas the common G_α protein numbering system was used to represent the residues of the G_α protein. *GPCR*, G protein-coupled receptor.

residues) at positions 6.29 and 6.32 have been shown to be critical. Similarly, TMV contains three residues, 5.61, 5.65, and 5.68, which interact with H5 residues H5.25, H5.20, and H5.13. Surprisingly, these residues are not often conserved and hint at highly dynamic G protein interfaces. In addition to the residues present in the helical regions the ICL2 was reported to contain a significant number of residues, which can interact with G_α protein residues. Of these residues, ICL2.2.50, ICL2.2.51, and ICL2.2.53 were reported to interact with H5 residues H5.12, H5.16, and H5.19. In hns1, residues hns1.2 and hns1.3 interact with ICL2 and ICL2.54 (Avet et al., 2022; Hauser et al., 2022; Inoue et al., 2019; Kapolka et al., 2020; Masuho et al., 2023, 2015; Matic et al., 2023; Okashah et al., 2019; Olsen et al., 2020). Of the remaining transmembrane helices the TMII residue, 2.39 was reported to interact with H5.23 and H5.22, and TMIII residues 3.50, 3.53, and 3.54 were reported to interact with H5 residues, H5.16, H5.19, and H5.20. TMVII, on a similar note, was reported to contain two residues, 7.53 and 7.56, that interact with H5.24 of H5. Interactions of helix H8 were reported via residues 8.47, 8.48, and 8.49, which interact with H5 residues H5.22, H5.23, and H5.24.

17.5 The role of lipid microenvironment in G_α protein coupling

The nanoenvironment of GPCRs is comprised of a diverse set of lipids (Bogdanov & Dowhan, 2021; Levental & Lyman, 2023; Ohki and Miyata, 2018; Tieleman et al., 2021(Ingo´lfsson et al., 2014)). The typical plasma membrane can be considered to be composed of varying percentages of neutral and charged phospholipids, sphingolipids, glycolipids, and sterols (Harayama and Riezman, 2018). The outer membrane leaflet usually consists of sphingolipids, glycolipids, and neutral phospholipids such as phosphatidylcholine and phosphatidylethanolamine. In contrast the inner leaflet is composed of neutral phospholipids as well as charged phospholipids such as phosphatidylserines (POPS) and phosphoinositides (Harayama & Riezman, 2018;). In addition, cholesterol, the plasma membrane sterol in human cells, is present at varying concentrations (Paukner et al., 2022). Membrane lipid composition varies between different cell types and is regulated such that deregulation is often linked to different diseased conditions. Interestingly, this variation of membrane lipid composition has been shown to modulate GPCR function, such as in ligand binding, structural stability, and oligomerization (Baccouch et al., 2022; Jafurulla & Chattopadhyay, 2013; Pucadyil & Chattopadhyay, 2006). Structural (Hanson et al., 2008) and computational studies (Prasanna et al., 2020; Sengupta & Chattopadhyay, 2012; Sengupta et al., 2018) have highlighted specific interaction hotspots that could be important for lipid-mediated allosteric modulation of receptor dynamics.

With regard to G protein coupling, cholesterol and anionic phospholipids have been reported to modulate these interactions (Chao et al., 2024; Kumar et al., 2021; Paila & Chattopadhyay, 2010; Thakur et al., 2023). However, the specificity of anionic lipids is still emerging and could be receptor-dependent. Phosphoinositides, especially phosphatidylinositol 4-phosphate (PIP_1) and phosphatidylinositol-4,5-biphosphate (PIP_2), have been reported in several studies to be important for G protein coupling (Yen et al., 2018). For instance, it was reported that PIP_1 lipids were able to increase G protein activity in the serotonin$_{1A}$ receptor, compared to other charged lipids. The modulation of serotonin$_{1A}$ receptor by PIP_1 lipids was suggested to occur by direct interactions with the receptor because a PIP_1 lipid molecule was resolved in the cryo-electron microscopy (cryo-EM) structure of the serotonin$_{1A}$ receptor–$G_{\alpha i/o}$ complex (PDB ID: 7E2X) (Xu et al., 2021). The experimentally resolved structure with the bound lipids is shown in Fig. 17.6. In this structure a PIP_1 lipid molecule is present at the $5HT_{1A}$–$G_{\alpha i/o}$ interface, together with cholesterol molecules. The head group of the PIP_1 lipid was observed to be buried at the crevice of transmembrane helices VI and VII, whereas the lipid tails were splayed. Adjacent cholesterol molecules appear to help stabilize this complex. In subsequent studies (Warren et al., 2025), it was shown that PIP lipid interactions were implicated in G protein coupling for a few G protein subtypes. The PIP_1 lipid was oriented at the same crevice in a similar conformation. The PIP_1 lipid interacted with R134(3.50), and the mutation of this residue abolished the activation of $G_{\alpha i1}$ and $G_{\alpha z}$ but not $G_{\alpha oA}$. This suggests that PIP_1 lipids could contribute to G protein selectivity.

Similarly, other anionic lipids were reported to stabilize the active states of the β_2-adrenergic receptor (Strohman, 2019)the adenosine A_{2A} receptor ($A_{2A}R$) (Thakur et al., 2023), and the neurotensin receptor 1 (NTSR1)(Inagaki, 2012). Further, high resolution mass spectrometry was used to examine endogenous lipid–receptor interactions (Yen et al., 2018). It was observed that PIP_2 stabilizes the receptors in active states, increases GTPase activity, and enhances G protein coupling selectivity. It was suggested that cytoplasmic surfaces contain hotspots for PIP_2 binding where PIP_2 forms bridging interactions with basic residues on GPCRs. It has been suggested these lipids prime GPCRs to form complexes with

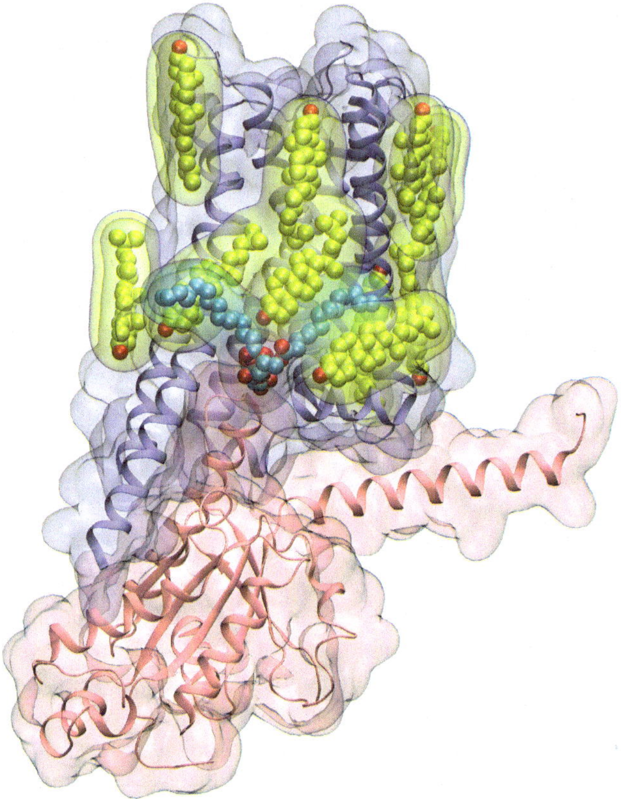

FIGURE 17.6

The lipid microenvironment modulates G_α protein coupling in GPCRs. The figure shows a cryo-electron microscopy (cryo-EM) structure of the apo serotonin 1A $5HT_{1A}$ receptor–$G_{\alpha i/o}$ protein complex (PDB ID: 7E2X). $5HT_{1A}$ is depicted in *purple cartoon* representation with a *light purple* surface in the background. The membrane lipid head groups are shown as *red spheres*. Cholesterol is shown in *yellow sphere* representation with its hydroxyl group in *red color*. A *light yellow* surface background encloses the cholesterol molecules. The $G_{\alpha i/o}$ protein is shown in *pink cartoon* representation with a *light pink* surface in the background. At the $5HT_{1A}$ receptor–$G_{\alpha i/o}$ protein interface a PIP_1 lipid (shown in *cyan colored spheres* with a *cyan-colored* surface background) was crystallized, implicating its modulatory role in G_α protein coupling.

G proteins via a conformational selection process. Further a synergy between anionic lipids and the efficacy of bound drugs was shown in the human A_{2A} adenosine receptor by using lipid nanodiscs to carefully control phospholipid concentration. It was shown that different anionic lipids can have similar effects, and several cationic residues were identified that were shown to be conserved across all receptor classes, indicating a shared trend in the GPCR superfamily (Thakur et al., 2023). Whether anionic lipids compete with each other for G protein coupling or whether it is receptor dependent is still unclear. In general synergistic cobinding and competition between different lipids at the GPCR nanoenvironment have been reported (Mohole et al., 2022) that could modulate the overall lipid effects. Similarly, posttranslational modifications such as receptor palmitoylation have been suggested to modulate the receptor nanoenvironment (Naglekar et al., 2023) that could further modulate G protein coupling. An interesting outcome of this lipid-specific G protein coupling is that it has been suggested that the subcellular location of lipids could promote location-selective coupling (Radoux-Mergault et al., 2023).

17.6 Summary and future perspective

The canonical view of the strict GPCR–G protein specificity has been updated by accumulating evidence demonstrating pluripotent (sometimes suggested to be promiscuous) coupling. This chapter provides a comprehensive overview of recent findings from diverse research groups, highlighting the emerging concepts of GPCR–G protein interactions. In particular the molecular details of GPCR–G protein coupling and the dynamic nature of these receptor–transducer interfaces are discussed. We also highlight factors such as the lipid microenvironment that have been reported to play an important role. A consideration of the structural and dynamic features of GPCR–G protein coupling specificity and pluripotency is important for advancements in healthcare and quality of life through improved GPCR-targeted therapies.

Structural and functional studies have shown that the main features of the GPCR–G protein complex are the interactions of the Gα5 helix of the G protein with TM V and VI. However, an important aspect that is emerging is that a large number of structures from GPCR, including TMI-VII, H8, and all ICLs (ICL1, ICL2, and ICL3), have been implicated. TMIV interactions have often been considered to be absent but shown to be important in Gq coupling. Similarly, several structural domains of the G protein have been suggested to make direct contacts with the GPCRs, such as H5, HN, hns1, S3, and s2s3. Several dynamic interactions are responsible for G subtype specific coupling, such as the binding of ICL3 to the long hgh4 loop in $G_{\alpha s}$. Although there are several reports of the molecular details of the subtype specific interactions, and consensus determinants have been suggested, a robust molecular framework is yet to be identified. Importantly, it has been observed that anionic phospholipids such as PIP_1, PIP_2, and POPS have a vital role in G protein coupling, suggesting that the subcellular location of lipids could promote location-selective coupling. Another important emerging direction is that of receptor priming in GPCRs (Culhane et al., 2022; Gupte et al., 2017) and how they affect G protein coupling events. Further the effects of posttranslational modifications in G proteins that can also further modulate this phenomenon (Chakravorty & Assmann, 2018) need to be better examined. Importantly the phenomenon of biased signaling in GPCRs in the context of tissue specificity is important and is still unclear. Overall a nuanced interplay between GPCR–G protein coupling specificity and pluripotency represents a critical determinant of cellular signaling fidelity and plasticity, with direct implications for human health and affording opportunities for highly targeted pharmacological interventions.

Acknowledgments

The authors express their gratitude to CSIR—National Chemical Laboratory for providing the research infrastructure in preparation of this manuscript. The authors thank Dr. Amit Naglekar, Dr. Siddhanta Nikte, and other members of their research group for their valuable inputs in writing and illustration. Discussions with Prof. Amitabha Chattopadhyay and Dr. Manali Joshi are gratefully acknowledged.

References

Ahn, D., Ham, D., & Chung, K. Y. (2021). The conformational transition during G protein–coupled receptor (GPCR) and G protein interaction. *Current Opinion in Structural Biology, 69*, 117–123.

Alegre, K. O., Paknejad, N., Su, M., Lou, J.-S., Huang, J., Jordan, K. D., Eng, E. T., Meyerson, J. R., Hite, R. K., & Huang, X.-Y. (2021). Structural basis and mechanism of activation of two different families of g proteins by the same GPCR. *Nature Structural & Molecular Biology, 28*(11), 936–944.

Alexander, S. P. H., Christopoulos, A., Davenport, A. P., Kelly, E., Mathie, A., Peters, J. A., Veale, E. L., Armstrong, J. F., Faccenda, E., Harding, S. D., Pawson, A. J., Sharman, J. L., Southan, C., & Davies, J. A.CGTP Collaborators. (2019). The concise guide to pharmacology 2019/20: G protein-coupled receptors. *British Journal of Pharmacology, 176*, S21–S141.

Arang, N., & Gutkind, J. S. (2020). G protein-coupled receptors and heterotrimeric G proteins as cancer drivers. *FEBS Letters, 594*(24), 4201–4232.

Avet, C., Mancini, A., Breton, B., Le Gouill, C., Hauser, A. S., Normand, C., Kobayashi, H., Gross, F., Hogue, M., Lukasheva, V., St-Onge, S., Carrier, M., Héroux, M., Morissette, S., Fauman, E. B., Fortin, J.-P., Schann, S., Leroy, X., Gloriam, D. E., & Bouvier, M. (2022). Effector membrane translocation biosensors reveal G protein and βarrestin coupling profiles of 100 therapeutically relevant GPCRS. *eLife, 11*, e74101.

Baccouch, R., Rascol, E., Stoklosa, K., & Alves, I. D. (2022). The role of the lipid environment in the activity of G protein coupled receptors. *Biophysical Chemistry, 285*, 106794.

Bjarnadóttir, T. K., Gloriam, D. E., Hellstrand, S. H., Kristiansson, H., Fredriksson, R., & Schiöth, H. B. (2006). Comprehensive repertoire and phylogenetic analysis of the G protein-coupled receptors in human and mouse. *Genomics, 88*(3), 263–273.

Blahos, J., Fischer, T., Brabet, I., Stauffer, D., Rovelli, G., Bockaert, J., & Pin, J.-P. (2001). A novel site on the G-protein that recognizes heptahelical receptors. *Journal of Biological Chemistry, 276*(5), 3262–3269.

Blaukat, A. (2003). Structure and signalling pathways of kinin receptors. *Andrologia, 35*(1), 17–23.

Bogdanov, M., & Dowhan, W. (2021). Functional roles of lipids in biological membranes. *Biochemistry of lipids, lipoproteins and membranes* (pp. 1–51). Elsevier.

Carpenter, B., & Tate, C. G. (2016). Engineering a minimal G protein to facilitate crystallisation of G protein-coupled receptors in their active conformation. *Protein Engineering, Design and Selection, 29*(12), 583–594.

Chakravorty, D., & Assmann, S. M. (2018). G protein subunit phosphorylation as a regulatory mechanism in heterotrimeric G protein signaling in mammals, yeast, and plants. *Biochemical Journal, 475*(21), 3331–3357.

Chao, K., Oqua, A., Tomas, A., & Rouse, S. L. (2024). Modulation of class B GPCRs by their lipid environment. *Biophysical Journal, 123*(3), 152A.

Che, T., Majumdar, S., Zaidi, S. A., Ondachi, P., McCorvy, J. D., Wang, S., Mosier, P. D., Uprety, R., Vardy, E., Krumm, B. E., Han, G. W., Lee, M.-Y., Pardon, E., Steyaert, J., Huang, X.-P., Strachan, R. T., Tribo, A. R., Pasternak, G. W., Carroll, F. I., … Roth, B. L. (2018). Structure of the nanobody-stabilized active state of the kappa opioid receptor. *Cell, 172*(1), 55–67.

Conklin, B. R., Farfel, Z., Lustig, K. D., Julius, D., & Bourne, H. R. (1993). Substitution of three amino acids switches receptor specificity of Gqα to that of Giα. *Nature, 363*(6426), 274–276.

Conklin, B. R., Herzmark, P., Ishida, S., Voyno-Yasenetskaya, T. A., Sun, Y., Farfel, Z., & Bourne, H. R. (1996). Carboxyl-terminal mutations of Gq alpha and Gs alpha that alter the fidelity of receptor activation. *Molecular Pharmacology, 50*(4), 885–890.

Culhane, K., Gupte, T., Madhugiri, I., Gadgil, C., & Sivaramakrishnan, S. (2022). Kinetic model of GPCR-G protein interactions reveals allokairic modulation of signaling. *Biophysical Journal, 121*(3), 122a.

Downes, G. B., & Gautam, N. (1999). The G protein subunit gene families. *Genomics, 62*(3), 544–552.

Draper-Joyce, C. J., Khoshouei, M., Thal, D. M., Liang, Y.-L., Nguyen, A. T. N., Furness, S. G. B., Venugopal, H., Baltos, J.-A., Plitzko, J. M., Danev, R., Baumeister, W., May, L. T., Wootten, D., Sexton, P. M., Glukhova, A., & Christopoulos, A. (2018). Structure of the adenosine-bound human adenosine A_1 receptor–G_i complex. *Nature, 558*(7711), 559–563.

Farrens, D. L., Altenbach, C., Yang, K., Hubbell, W. L., & Gobind Khorana, H. (1996). Requirement of rigid-body motion of transmembrane helices for light activation of rhodopsin. *Science, 274*(5288), 768–770.

Flock, T., Hauser, A. S., Lund, N., Gloriam, D. E., Balaji, S., & Madan Babu, M. (2017). Selectivity determinants of GPCR–G-protein binding. *Nature, 545*(7654), 317–322.

Flock, T., Ravarani, C. N. J., Sun, D., Venkatakrishnan, A. J., Kayikci, M., Tate, C. G., Veprintsev, D. B., & Madan Babu, M. (2015). Universal allosteric mechanism for Gα activation by GPCRs. *Nature, 524*(7564), 173–179.

Garc´ıa-Nafr´ıa, J., Nehm´e, R., Edwards, P. C., & Tate, C. G. (2018). Cryo-EM structure of the serotonin 5-HT_{1B} receptor coupled to heterotrimeric G_o. *Nature, 558*(7711), 620–623.

Grishina, G., & Berlot, C. H. (2000). A surface-exposed region of Gsα in which substitutions decrease receptor-mediated activation and increase receptor affinity. *Molecular Pharmacology, 57*(6), 1081–1092.

Gupte, T. M., Malik, R. U., Sommese, R. F., Ritt, M., & Sivara-makrishnan, S. (2017). Priming GPCR signaling through the synergistic effect of two G proteins. *Proceedings of the National Academy of Sciences, 114*(14), 3756–3761.

Hamm, H. E., Deretic, D., Arendt, A., Hargrave, P. A., Koenig, B., & Hofmann, K. P. (1988). Site of G protein binding to rhodopsin mapped with synthetic peptides from the α subunit. *Science, 241*(4867), 832–835.

Hanson, M. A., Cherezov, V., Griffith, M. T., Roth, C. B., Jaakola, V.-P., Chien, E. Y. T., Velasquez, J., Kuhn, P., & Stevens, R. C. (2008). A specific cholesterol binding site is established by the 2.8 A structure of the human beta2-adrenergic receptor. *Structure, 16*, 897–905.

Harayama, T., & Riezman, H. (2018). Understanding the diversity of membrane lipid composition. *Nature Reviews Molecular Cell Biology, 19*(5), 281–296.

Hauser, A. S., Avet, C., Normand, C., Mancini, A., In-oue, A., Bouvier, M., & Gloriam, D. E. (2022). Common coupling map advances GPCR-G protein selectivity. *eLife, 11*, e74107.

Herenbrink, C. K., Sykes, D. A., Donthamsetti, P., Canals, M., Coudrat, T., Shonberg, J., Scammells, P. J., Capuano, B., Sexton, P. M., Charlton, S. J., Javitch, J. A., Christopoulos, A., & Robert Lane, J. (2016). The role of kinetic context in apparent biased agonism at GPCRs. *Nature Communications, 7*(1), 10842.

Hilger, D., Masureel, M., & Kobilka, B. K. (2018). Structure and dynamics of GPCR signaling complexes. *Nature Structural & Molecular Biology, 25*(1), 4–12.

Hollenstein, K., de Graaf, C., Bortolato, A., Wang, M.-W., Marshall, F. H., & Stevens, R. C. (2014). Insights into the structure of class B GPCRs. *Trends in Pharmacological Sciences, 35*(1), 12–22.

Huang, S., Xu, P., Shen, D.-D., Simon, I. A., Mao, C., Tan, Y., Zhang, H., Harpsøe, K., Li, H., Zhang, Y., You, C., Yu, X., Jiang, Y., Zhang, Y., Gloriam, D. E., & Xu, H. E. (2022). GPCRs steer Gi and Gs selectivity via TM5-TM6 switches as revealed by structures of serotonin receptors. *Molecular Cell, 82*(14), 2681–2695.

Hunyady, L., & Catt, K. J. (2006). Pleiotropic AT1 receptor signaling pathways mediating physiological and pathogenic actions of angiotensin II. *Molecular Endocrinology, 20*(5), 953–970.

Inagaki, S., et al. (2012). Modulation of the interaction between neurotensin receptor NTS1 and Gq protein by lipid. *J. Mol. Biol., 417*, 95–111.

Ingo´lfsson, H. I., Melo, M. N., Eerden, F. J. V., Arnarez, C., Lopez, C. A., Wassenaar, T. A., Periole, X., de Vries, A. H., Peter Tieleman, D., & Marrink, S. J. (2014). Lipid organization of the plasma membrane. *Journal of the American Chemical Society, 136*(41), 14554–14559.

Inoue, A., Raimondi, F., Kadji, F. M. N., Singh, G., Kishi, T., Uwamizu, A., Ono, Y., Shinjo, Y., Ishida, S., Arang, N., Kawakami, K., Gutkind, J. S., Aoki, J., & Russell, R. B. (2019). Illuminating G-protein-coupling selectivity of GPCRs. *Cell, 177*(7), 1933–1947.

Jafurulla, M., & Chattopadhyay, A. (2013). Membrane lipids in the function of serotonin and adrenergic receptors. *Current Medicinal Chemistry, 20*(1), 47–55.

Jiang, H., Galtes, D., Wang, J., & Rockman, H. A. (2022). G protein-coupled receptor signaling: Transducers and effectors. *American Journal of Physiology-Cell Physiology, 323*(3), C731–C748.

Kang, Y., Kuybeda, O., de Waal, P. W., Mukherjee, S., Eps, N. V., Dutka, P., Edward Zhou, X., Bartesaghi, A., Er-ramilli, S., Morizumi, T., Gu, X., Yin, Y., Liu, P., Jiang, Y., Meng, X., Zhao, G., Melcher, K., Ernst, O. P., Kossiakoff, A. A., … Xu, H. E. (2018). Cryo-EM structure of human rhodopsin bound to an inhibitory g protein. *Nature, 558*(7711), 553–558.

Kapolka, N. J., Taghon, G. J., Rowe, J. B., Morgan, W. M., Enten, J. F., Lambert, N. A., & Isom, D. G. (2020). DCyFIR: A high-throughput CRISPR platform for multiplexed G protein-coupled receptor profiling and ligand discovery. *Proceedings of the National Academy of Sciences, 117*(23), 13117–13126.

Koehl, A., Hu, H., Maeda, S., Zhang, Y., Qu, Q., Paggi, J. M., Latorraca, N. R., Hilger, D., Dawson, R., Matile, H., Schertler, G. F. X., Granier, S., Weis, W. I., Dror, R. O., Manglik, A., Skiniotis, G., & Kobilka, B. K. (2018). Structure of the μ-opioid receptor–Gi protein complex. *Nature, 558*(7711), 547–552.

Kolakowski, L., Jr. (1994). GCRDb: A G-protein–coupled receptor. *Receptors and Channels, 2*, 1–7.

Kostenis, E., Conklin, B. R., & Wess, J. (1997). Molecular basis of receptor/G protein coupling selectivity studied by coexpression of wild type and mutant m2 muscarinic receptors with mutant Gαq subunits. *Biochemistry, 36*(6), 1487–1495.

Kostenis, E., Degtyarev, M. Y., Conklin, B. R., & Wess, J. (1997). The N-terminal extension of Gαq is critical for constraining the selectivity of receptor coupling. *Journal of Biological Chemistry, 272*(31), 19107–19110.

Kumar, G. A., Sarkar, P., Stepniewski, T. M., Jafurulla, M., Singh, S. P., Selent, J., & Chattopadhyay, A. (2021). A molecular sensor for cholesterol in the human serotonin1A receptor. *Science Advances, 7*(30), Eabh2922.

Latorraca, N. R., Venkatakrishnan, A. J., & Dror, R. O. (2017). GPCR dynamics: Structures in motion. *Chemical Reviews, 117*(1), 139–155.

Lee, C. H., Katz, A., & Simon, M. I. (1995). Multiple regions of G alpha 16 contribute to the specificity of activation by the C5a receptor. *Molecular Pharmacology, 47*(2), 218–223.

Leung, C. C. Y., & Wong, Y. H. (2017). Role of G protein-coupled receptors in the regulation of structural plasticity and cognitive function. *Molecules, 22*(7), 1239.

Levental, I., & Lyman, E. (2023). Regulation of membrane protein structure and function by their lipid nano-environment. *Nature Reviews Molecular Cell Biology, 24*(2), 107–122.

Liu, C., Goel, P., & Kaeser, P. S. (2021). Spatial and temporal scales of dopamine transmission. *Nature Reviews Neuroscience, 22*(6), 345–358.

Liu, J., Conklin, B. R., Blin, N., Yun, J., & Wess, J. (1995). Identification of a receptor/G-protein contact site critical for signaling specificity and G-protein activation. *Proceedings of the National Academy of Sciences, 92*(25), 11642–11646.

Masuho, I., Kise, R., Gainza, P., Von Moo, E., Li, X., Tany, R., Wakasugi-Masuho, H., Correia, B. E., & Martemyanov, K. A. (2023). Rules and mechanisms governing G protein coupling selectivity of GPCRs. *Cell Reports, 42*(10), 113173.

Masuho, I., Ostrovskaya, O., Kramer, G. M., Jones, C. D., Xie, K., & Martemyanov, K. A. (2015). Distinct profiles of functional discrimination among G proteins determine the actions of G protein–coupled receptors. *Science Signaling, 8*(405), ra123.

Matic, M., Miglionico, P., Tatsumi, M., Inoue, A., & Raimondi, F. (2023). GPCRome-wide analysis of G-protein-coupling diversity using a computational biology approach. *Nature Communications, 14*(1), 4361.

McGraw, D. W., & Liggett, S. B. (2005). Molecular mechanisms of β2-adrenergic receptor function and regulation. *Proceedings of the American Thoracic Society, 2*(4), 292–296.

Mohole, M., Sengupta, D., & Chattopadhyay, A. (2022). Synergistic and competitive lipid interactions in the serotonin1A receptor microenvironment. *ACS Chemical Neuroscience, 13*(23), 3403–3415.

Naglekar, A., Chattopadhyay, A., & Sengupta, D. (2023). Palmitoylation of the glucagon-like peptide-1 receptor modulates cholesterol interactions at the receptor–lipid microenvironment. *The Journal of Physical Chemistry B, 127*(51), 11000–11010.

Neves, S. R., Ram, P. T., & Iyengar, R. (2002). G protein pathways. *Science, 296*(5573), 1636–1639.

Ohki, K., & Miyata, H. (2018). Physical properties of biomembranes and cellular functions. *Physical Principles of Biomembranes and Cells* (pp. 81–93). Springer.

Okashah, N., Wan, Q., Ghosh, S., Sandhu, M., In-oue, A., Vaidehi, N., & Lambert, N. A. (2019). Variable g protein determinants of GPCR coupling selectivity. *Proceedings of the National Academy of Sciences, 116*(24), 12054–12059.

Okashah, N., Wright, S. C., Kawakami, K., Mathiasen, S., Zhou, J., Lu, S., Javitch, J. A., Inoue, A., Bouvier, M., & Lam-bert, N. A. (2020). Agonist-induced formation of unproductive receptor-G_{12} complexes. *Proceedings of the National Academy of Sciences, 117*(35), 21723–21730.

Oldham, W. M., & Hamm, H. E. (2008). Heterotrimeric G protein activation by G-protein-coupled receptors. *Nature Reviews Molecular Cell Biology, 9*(1), 60–71.

Olsen, R. H. J., DiBerto, J. F., English, J. G., Glaudin, A. M., Krumm, B. E., Slocum, S. T., Che, T., Gavin, A. C., McCorvy, J. D., Roth, B. L., & Strachan, R. T. (2020). TRUPATH, an open-source biosensor platform for interrogating the GPCR transducerome. *Nature Chemical Biology, 16*(8), 841–849.

Özçete, Ö. D., Banerjee, A., & Kaeser, P. S. (2024). Mechanisms of neuromodulatory volume transmission. *Molecular Psychiatry, 29*, 3680–3693.

Paila, Y. D., & Chattopadhyay, A. (2010). Membrane cholesterol in the function and organization of G-protein coupled receptors. *Sub-Cellular Biochemistry, 51*, 439–466.

Pándy-Szekeres, G., Munk, C., Tsonkov, T. M., Mordalski, S., Harpsøe, K., Hauser, A. S., Bojarski, A. J., & Gloriam, D. E. (2018). GPCRdb in 2018: Adding GPCR structure models and ligands. *Nucleic Acids Research, 46*(D1), D440–D446.

Paukner, K., Lesna´, I. K´´, & Poledne, R. (2022). Cholesterol in the cell membrane—An emerging player in atherogenesis. *International Journal of Molecular Sciences, 23*(1), 533.

Prasanna, X., Mohole, M., Chattopadhyay, A., & Sen-gupta, D. (2020). Role of cholesterol-mediated effects in GPCR heterodimers. *Chemistry and Physics of Lipids, 227*, 104852.

Preininger, A. M., & Hamm, H. E. (2004). G protein signaling: Insights from new structures. *Science's STKE, 2004*(218), re3.

Pucadyil, T. J., & Chattopadhyay, A. (2006). Role of cholesterol in the function and organization of G-protein coupled receptors. *Progress in Lipid Research, 45*, 295–333.

Radoux-Mergault, A., Oberhauser, L., Aureli, S., Luigi Ger-vasio, F., & Stoeber, M. (2023). Subcellular location defines GPCR signal transduction. *Science Advances, 9*(16), eAdf6059.

Rashid, A. J., O-Dowd, B. F., Verma, V., & George, S. R. (2007). Neuronal Gq/11-coupled dopamine receptors: An uncharted role for dopamine. *Trends in Pharmacological Sciences, 28*(11), 551–555.

Rasmussen, S. G. F., DeVree, B. T., Zou, Y., Kruse, A. C., Chung, K. Y., Kobilka, T. S., Thian, F. S., Chae, P. S., Pardon, E., Calinski, D., Mathiesen, J. M., Shah, S. T. A., Lyons, J. A., Caffrey, M., Gellman, S. H., Steyaert, J., Skiniotis, G., Weis, W. I., Sunahara, R. K., & Kobilka, B. K. (2011). Crystal structure of the β2 adrenergic receptor–Gs protein complex. *Nature, 477*(7366), 549–555.

Rivera, J., Proia, R. L., & Olivera, A. (2008). The alliance of sphingosine-1-phosphate and its receptors in immunity. *Nature Reviews Immunology, 8*(10), 753–763.

Salazar, N. C., Chen, J., & Rockman, H. A. (2007). Cardiac GPCRs: GPCR signaling in healthy and failing hearts. *Biochimica et Biophysica Acta (BBA)-Biomembranes, 1768*(4), 1006–1018.

Sandhu, M., Cho, A., Ma, N., Mukhaleva, E., Namkung, Y., Lee, S., Ghosh, S., Lee, J. H., Gloriam, D. E., Laporte, S. A., Babu, M. M., & Vaidehi, N. (2022). Dynamic spatiotemporal determinants modulate GPCR: G protein coupling selectivity and promiscuity. *Nature Communications, 13*(1), 7428.

Satagopam, V. P., Theodoropoulou, M. C., Stampolakis, C. K., Pavlopoulos, G. A., Papandreou, N. C., Bagos, P. G., Schneider, R., & Hamodrakas, S. J. (2010). GPCRs, G-proteins, effectors and their interactions: Human-gpDB, a database employing visualization tools and data integration techniques. *Database, 2010*, bAq019.

Semack, A., Sandhu, M., Malik, R. U., Vaidehi, N., & Sivaramakrishnan, S. (2016). Structural elements in the Gαs and Gαq C termini that mediate selective G protein-coupled receptor (GPCR) signaling. *Journal of Biological Chemistry, 291*(34), 17929–17940.

Sengupta, D., & Chattopadhyay, A. (2012). Identification of cholesterol binding sites in the serotonin1A receptor. *The Journal of Physical Chemistry B, 116*(43), 12991–12996.

Sengupta, D., Prasanna, X., Mohole, M., & Chattopad-hyay, A. (2018). Exploring GPCR–lipid interactions by molecular dynamics simulations: Excitements, challenges, and the way forward. *The Journal of Physical Chemistry B, 122*, 5727–5737.

Simonds, W. F., Goldsmith, P. K., Codina, J., Unson, C. G., & Spiegel, A. M. (1989). Gi2 mediates alpha 2-adrenergic inhibition of adenylyl cyclase in platelet membranes: In situ identification with G alpha C-terminal antibodies. *Proceedings of the National Academy of Sciences, 86*(20), 7809–7813.

Strohman, M. J., et al. (2019). Local membrane charge regulates β2 adrenergic receptor coupling to Gi3. *Nat. Comm., 10*, 2234.

Sullivan, K. A., Tyler Miller, R., Masters, S. B., Beiderman, B., Heideman, W., & Bourne, H. R. (1987). Identification of receptor contact site involved in receptor–G protein coupling. *Nature, 330*(6150), 758–760.

Suno, R. (2024). Exploring diverse signaling mechanisms of G protein-coupled receptors through structural biology. *The Journal of Biochemistry, 175*(4), 357–365.

Suzuki, N., Hajicek, N., & Kozasa, T. (2009). Regulation and physiological functions of g12/13-mediated signaling pathways. *Neurosignals, 17*(1), 55–70.

Svoboda, P., Teisinger, J., Novotny, J., Bourova, L., Drmota, T., Hejnova, L., Moravcova, Z., Lisy, V., Rudajev, V., Stohr, J., Vokurková, A., Svandová, I., & Durchánková, D. (2004). Biochemistry of transmembrane signaling mediated by trimeric G proteins. *Physiological Research, 53*, S141–S152.

Syrovatkina, V., Alegre, K. O., Dey, R., & Huang, X.-Y. (2016). Regulation, signaling, and physiological functions of G-proteins. *Journal of Molecular Biology, 428*(19), 3850–3868.

Thakur, N., Ray, A. P., Sharp, L., Jin, B., Duong, A., Pour, N.-G., Obeng, S., Wijesekara, A. V., Gao, Z.-G., McCurdy, C.-R., Jacobson, K. A., Lyman, E., & Eddy, M. T. (2023). Anionic phospholipids control mechanisms of GPCR-G protein recognition. *Nature Communications, 14*(1), 794.

Tieleman, D. P., Sejdiu, B. I., Cino, E. A., Smith, P., Barreto-Ojeda, E., Khan, H. M., & Corradi, V. (2021). Insights into lipid-protein interactions from computer simulations. *Biophysical Reviews, 13*(6), 1019–1027.

Tsai, C.-J., Pamula, F., Nehm´e, R., Mu¨hle, J., Weinert, T., Flock, T., Nogly, P., Edwards, P. C., Carpenter, B., Gruhl, T., Ma, P., Deupi, X., Standfuss, J., Tate, C. G., & Schertler, G. F. X. (2018). Crystal structure of rhodopsin in complex with a mini-G_0 sheds light on the principles of G protein selectivity. *Science Advances, 4*(9), eaat7052.

Venkatakrishnan, A. J., Deupi, X., Lebon, G., Tate, C. G., Schertler, G.-F., & Madan Babu, M. (2013). Molecular signatures of G-protein-coupled receptors. *Nature, 494*(7436), 185–194.

Wacker, D., Stevens, R. C., & Roth, B. L. (2017). How ligands illuminate GPCR molecular pharmacology. *Cell, 170*(3), 414–427.

Warren, A. L., Zilberg, G., Abbassi, A., Abraham, A., Yang, S., & Wacker, D. (2025). Structural determinants of G protein subtype selectivity at the serotonin receptor 5-HT1A. *Science Advances, 11*(31), eAdu9851.

West, R. E., Moss, J., Vaughan, M., Liu, T., & Liu, T.-Y. (1985). Pertussis toxin-catalyzed ADP-ribosylation of transducin. Cysteine 347 is the ADP-ribose acceptor site. *Journal of Biological Chemistry, 260*(27), 14428–14430.

Wettschureck, N., & Offermanns, S. (2005). Mammalian G proteins and their cell type specific functions. *Physiological Reviews, 85*(4), 1159–1204.

Wright, S. C., Avet, C., Gaitonde, S. A., Muneta-Arrate, I., Gouill, C.-L., Hogue, M., Breton, B., Koutsilieri, S., Diez-Alarcia, R., Héroux, M., Lauschke, V. M., & Bouvier, M. (2024). Conformation-and activation-based BRET sensors differentially report on GPCR–G protein coupling. *Science Signaling, 17*(841), eadi4747.

Wu, V., Yeerna, H., Nohata, N., Chiou, J., Harismendy, O., Raimondi, F., Inoue, A., Russell, R. B., Tamayo, P., & Silvio Gutkind, J. (2019). Illuminating the onco-GPCrome: Novel G protein–coupled receptor-driven oncocrine networks and targets for cancer immunotherapy. *Journal of Biological Chemistry, 294*(29), 11062–11086.

Xiao, R.-P. (2001). β-adrenergic signaling in the heart: Dual coupling of the β2-adrenergic receptor to Gs and Gi proteins. *Science's STKE, 2001*(104), re15.

Xu, P., Huang, S., Zhang, H., Mao, C., Edward Zhou, X., Cheng, X., Simon, I. A., Shen, D.-D., Yen, H.-Y., Robinson, C. V., Harpsøe, K., Svensson, B., Guo, J., Jiang, H., Gloriam, D. E., Melcher, K., Jiang, Y., Zhang, Y., & Xu, H. E. (2021). Structural insights into the lipid and ligand regulation of serotonin receptors. *Nature, 592*(7854), 469–473.

Xu, X., Kaindl, J., Clark, M. J., Hu¨bner, H., Hirata, K., Sunahara, R. K., Gmeiner, P., Kobilka, B. K., & Liu, X. (2021). Binding pathway determines norepinephrine selectivity for the human β1ar over β2ar. *Cell Research, 31*(5), 569–579.

Yen, H.-Y., Hoi, K. K., Liko, I., Hedger, G., Horrell, M. R., Song, W., Wu, D., Heine, P., Warne, T., Lee, Y., Carpenter, B., Plückthun, A., Tate, C. G., Sansom, M. S. P., & Robinson, C. V. (2018). Ptdins(4, 5)p2 stabilizes active states of GPCRs and enhances selectivity of G-protein coupling. *Nature, 559*(7714), 423–427.

Zhou, Q., Yang, D., Wu, M., Guo, Y., Guo, W., Zhong, L., Cai, X., Dai, A., Jang, W., Shakhnovich, E. I., Liu, Z.-J., Stevens, R. C., Lambert, N. A., Babu, M. M., Wang, M.-W., & Zhao, S. (2019). Common activation mechanism of class A GPCRs. *eLife, 8*, e50279.

CHAPTER

Deciphering protein dynamics from experimentally resolved structures: Introductory classroom hands-on exercises

18

Amrutha Keshavamurthy[a,b] and Durba Sengupta[c,*]

[a]CSIR-National Chemical Laboratory, Pune, Maharashtra, India, [b]Academy of Scientific and Innovative Research, Ghaziabad, Uttar Pradesh, India, [c]Physical Chemistry Division, CSIR-National Chemical Laboratory, Pune, Maharashtra, India

**Corresponding author: d.sengupta@ncl.res.in*

18.1 Introduction

Advances in structural biology have helped highlight the molecular basis of life by elucidating the atomic details of key biomolecules, including the diverse repertoire of proteins within cells (Chen et al., 2015; Congreve et al., 2005). However, the representation of protein structures in textbooks or databases can often lead to the perception that they are static, rigid entities. These experimentally-resolved structures provide detailed images of the "folds, twists, and turns," but these snapshots only capture a fleeting moment in the life of a protein. Naturally, this structural blueprint of proteins is crucial, as biological function is usually linked to its structure, but it is becoming increasingly clear that protein conformational flexibility and heterogeneity are equally important, if not more, for their biological activity (Berendsen & Hayward, 2000; Boehr et al., 2009).

Protein dynamics occurs at hierarchical spatio-temporal scales, from localized side-chain fluctuations to large-scale domain motion (Grimaldo et al., 2019), and is fundamental to most biological processes (Bahar et al., 2010). For instance, hemoglobin undergoes allosteric transitions in its quaternary structure, enabling cooperative binding and release of oxygen (Shibayama, 2020). Similarly, lysozyme exhibits pH-dependent conformational changes that optimize its catalytic efficiency in hydrolyzing bacterial cell walls (Bonincontro et al., 1998). G protein-coupled receptors undergo a series of local conformational rearrangements upon ligand binding, and multiple conformational states have been identified (Manglik & Kobilka, 2014). The conformational states of membrane-bound transporters, such as the magnesium transporter CoRA, allow them to selectively transport the metabolite of choice (Johansen et al., 2022). Post-translational modifications such as ubiquitin have been shown to affect the native-state dynamics of the modified protein and could provide a regulatory handle to cells (Gavrilov et al., 2015). Molecular chaperones like HSP70 facilitate proper protein folding and prevent

Protein Dynamics: Experimental and Computational Approaches. DOI: https://doi.org/10.1016/B978-0-443-30042-4.00004-3

aggregation through ATP-driven conformational cycles. Disruptions in these dynamic processes, as seen in amyloid-beta aggregation in Alzheimer's disease, highlight the importance of protein flexibility in maintaining cellular function (Rutledge et al., 2022). Overcoming the challenges of characterizing these dynamic ensembles represents a key frontier in understanding protein function and developing novel therapeutics.

Although proteins are inherently dynamic entities, existing as dynamic conformational ensembles (Wei et al., 2016), it has proven challenging to capture these in atomic-resolution experiments. Experimental techniques, such as X-ray crystallography (Meisburger et al., 2017), nuclear magnetic resonance (NMR) spectroscopy (Ishima & Torchia, 2000) and cryo-electron microscopy (cryo-EM) (Murata & Wolf, 2018), are currently used to gain insights into these dynamic structural transitions, allowing us to observe and characterize atomic-level protein motions in varying conditions. Complementary approaches, such as hydrogen-deuterium exchange mass spectroscopy, electron paramagnetic resonance spectroscopy, and time-resolved spectroscopy, are often used for specific aspects of protein dynamics but lack the atomic-level resolution (Banari et al., 2025; Buhrke & Hildebrandt, 2020; Nesmelov & Thomas, 2010; Zheng et al., 2019). Each of the structural approaches uses distinctive probes and reporters and offers a unique perspective on protein structure. For instance, X-ray crystallography uses X-rays to probe biomolecular crystals, and the resulting diffraction pattern is analyzed to reveal the electron density maps and, subsequently, the atomic arrangement within the crystal (Matthews, 1976; Shi, 2014). NMR spectroscopy utilizes the interaction of atomic nuclei with strong magnetic fields to probe their quantum mechanical energy states (Fernandez-Leiro & Scheres, 2016; Kay, 1998). The resulting nuclear resonance frequencies are measured and analyzed to elucidate molecular structure and dynamics (Ishima & Torchia, 2000). In cryo-EM, biological samples are rapidly cryo-fixed in vitreous ice to preserve their native hydration and structural integrity. Electron microscopy is then used to image these vitrified samples, and the resulting data is processed to reconstruct high-resolution three-dimensional models (Fernandez-Leiro & Scheres, 2016; Kühlbrandt, 2014).

Although X-ray crystallography provides high-resolution, "frozen" snapshots of proteins in crystal form, comparing structures obtained under different conditions reveals hints of the underlying dynamics. The *B*-factor has often been used as a handle to distinguish intraprotein flexibility (Sun et al., 2019). In contrast, NMR spectroscopy offers a more dynamic view of proteins in solution, allowing us to observe an ensemble of conformations that the protein can adopt. Interestingly, in-cell NMR spectroscopy represents an emerging biophysical tool to investigate the conformational and functional characteristics of biomolecules at the atomic level inside live cells (Ito & Selenko, 2010). Cryo-EM captures "near-native" structures of large, complex protein assemblies, providing high-resolution insights into their architecture and conformational states without the need for crystallization. Although traditional cryo-EM analysis is limited to discrete distinct conformations, several approaches have been proposed to analyze conformational heterogeneity and polymorphism (Chang et al., 2021; Toader et al., 2023). More advanced computational analyses, such as deep learning approaches, have been used to extract the dynamics associated with the atomic fluctuations that are hidden in cryo-EM density maps and shown to correlate well with data obtained from molecular dynamics simulations (Matsumoto et al., 2023). Together, these techniques provide a more comprehensive view of the 'life' of a protein, helping us understand how its structure is influenced by its environment and, in turn, how it relates to its function.

The substantial contributions of structural databanks, particularly the Protein Data Bank (PDB), have facilitated the domain of structural bioinformatics and the use of comparative analyses as a means to investigate protein dynamics (Berman et al., 2003). For instance, approaches integrating experimental structures with computational clustering to explore the conformational landscapes directly from the X-ray, cryo-EM, and NMR structures deposited in the PDB have been developed (Ellaway et al., 2024). Similarly an automated calculation of scaled *B*-factor distributions from experimentally resolved protein–DNA complexes demonstrates a correlation between residue flexibility and its neighborhood (Schneider et al., 2014). A bioinformatic description of amino acid dynamic properties, based on residue-specific average *B*-factors from the PDB, has been proposed to map the dynamic space of protein sequences (Rackovsky & Scheraga, 2020). Other computational approaches have also been proposed to extract protein dynamics from experimentally resolved structures (Katebi & Jernigan, 2014). Overall, structural bioinformatics (De Almeida et al., 2022) provides a valuable set of tools and approaches that can be used to understand protein dynamics.

In this chapter, we provide hands-on computational exercises to explore protein dynamics by analyzing experimentally-resolved protein structures. The exercises provided are aimed at beginners in the field of structural bioinformatics and are a first step to analyzing experimentally-resolved structures using visualization software and analysis methods to deconvolute the underlying protein dynamics. Starting with X-ray crystallography structures of lysozyme, we explore the inherent dynamics of the protein. Moving forward, we explore how varied pH conditions modulate the structure of lysozyme. In the next section, we analyze the NMR structure of ubiquitin and observe the various conformations it adopts and examine how the cellular environment influences protein dynamics. And finally, we investigate the cryo-EM structure of the magnesium transporter CorA to identify its structural changes in the presence and absence of magnesium ions. These exercises offer hands-on experience in visualizing and analyzing the inherent dynamics of proteins and interpreting how structure and dynamics together enable proteins to perform their diverse biological roles. We hope these exercises will spark an interest among students to further explore protein dynamics, for instance, by extending these approaches to larger sets of protein families relevant to their studies or research and perhaps including ML/AI tools to identify novel patterns. This journey from static images to dynamic states highlights the fundamental nature of proteins, intrinsically flexible, versatile, and responsive.

Note: Please note that in the following exercises, command lines that need to be executed are provided as inline code lines. All Python scripts required for the exercises are compiled into a single Jupyter notebook, which is available in the GitHub repository. In addition to the notebook, the repository contains selected figures and videos that support the exercises, as well as the Visual Molecular Dynamics (VMD) scripts for molecular visualization. For convenience a direct link to Google Colab is provided on the GitHub page, allowing users to run the analysis interactively without needing to set up a local environment.

Link to GitHub: https://Sengupta-NCL/protein-dynamics-book-chapter

18.2 Exploring protein dynamics from structures resolved with X-ray crystallography

We start with the model protein, lysozyme, a well-studied antimicrobial enzyme that effectively breaks down bacterial cell walls. In this exercise, we consider several structures of lysozyme resolved with X-ray crystallography and examine the intrinsic dynamics hidden in these structures. In the first exercise, we compare different X-ray crystal structures of lysozyme to examine the local motion of the amino acid residues, and in the later exercises, we shall compare the lysozyme structures crystallized and obtained at different pH conditions to probe the pH-sensitive dynamics of the proteins.

18.2.1 Visualizing local conformational changes

Let us begin by visualizing and comparing X-ray crystal structures of lysozyme (PDB IDs: 193L, 194L, 1AKI, 1AZF, and 1BGI) and examine the structural deviations of the amino acid residues, providing insights into protein flexibility.

18.2.1.1 Quantifying structural deviations

Protocol:

- Visit the RCSB PDB database (https://www.rcsb.org/) and search for the specified PDB IDs (193L, 194L, 1AKI, 1AZF, 1BGI). Download the structures in the .pdb format to your local computer.
- Open the PyMOL software to load the protein strcutures. Go to `File > Open` and select the downloaded.pdb files. Open all five files one after the other. Alternatively, use the command given in the following:

  ```
  load/path/194L.pdb
  ```
- Remove all water molecules using the command:

  ```
  remove solvent
  ```
- Remove ions: Open the *SEQ* option in the right-side panel to display the sequences of each PDB ID. Select the ions listed at the end of the sequences. The ions will appear as a "sele" object in the panel; right-click and choose `Action (A) > Remove Atoms`.
- Align all other structures to 193L using the following commands:

  ```
  cmd.align('194L', '193L')
  cmd.align('1AKI', '193L')
  cmd.align('1AZF', '193L')
  cmd.align('1BGI', '193L')
  ```

  Note: These aligned structures will be examined in greater detail in Section 18.2.1.3.

- Plotting **the** root mean square deviation (RMSD) matrix**:** To compare the aligned structures, calculate the RMSD values, which quantifies structural differences (lower values indicate higher similarity). Variations in the RMSD values reflect flexibility and dynamic behavior of the protein. Run the code in Section 2.1.1 of the Colab notebook to compute and visualize the RMSD matrix. The heatmap of the RMSD matrix will highlight conformational variability across the structures (refer to Fig. 18.1).

  Questions:

1. Which pairs of structures show comparatively higher RMSD values?
2. Which pair of protein structures shows the highest structural similarity?
   Hint: Look for low RMSD values excluding the diagonal (0) values in the RMSD matrix.
3. Are any two lysozyme structures the same?

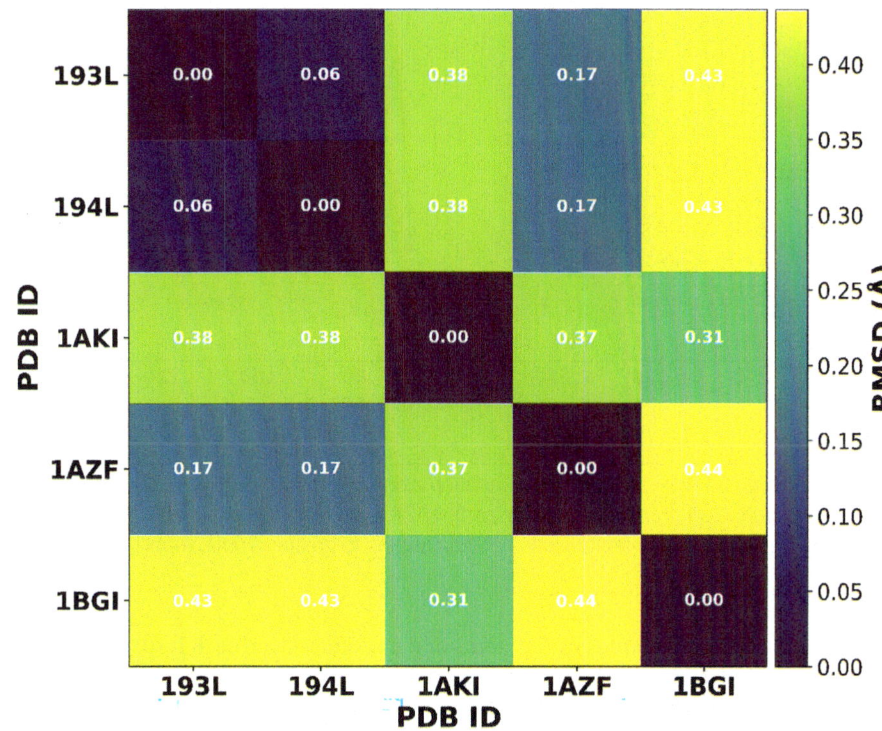

FIGURE 18.1

Structural deviation of lysozyme structures: Heatmap representation of RMSD values (Å) comparing different protein structures (PDB IDs: 193L, 194L, 1AKI, 1AZF, and 1BGI). The diagonal indicates self-comparisons (RMSD = 0), whereas off-diagonal elements highlight structural differences between pairs of proteins. *PDB*, Protein Data Bank; *RMSD*, root mean square deviation.

Notice that no two structures have an RMSD of zero, implying that each lysozyme structure is distinct. But why do these differences exist? These variations arise from the inherent dynamics of amino acids and their subtle movements, which help capture them in different poses in these seemingly static structural snapshots. Although RMSD reveals overall structural deviations, it does not indicate which specific regions of the protein are more flexible or variable. To explore these local differences, we next examine the *B*-factor values associated with each structure. These values, derived from crystallographic models, reflect the extent of atomic motion at each position. Analyzing the *B*-factors helps identify flexible regions within the protein and offers deeper insight into the local dynamics underlying the observed structural differences.

18.2.1.2 *Examining local flexibility through* B*-factor profiles*

Protocol:

- Run the code provided in **Section 2.1.2** of the Colab notebook.

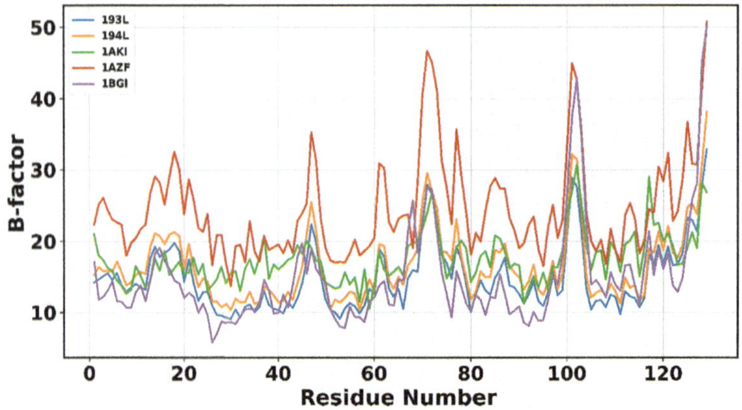

FIGURE 18.2

B-Factor profiles of lysozyme structures: *B*-factor variation along the residue sequence for protein structures with PDB IDs 193L (*blue*), 194L (*orange*), 1AKI (*green*), 1AZF (*red*), and 1BGI (*purple*). The profiles indicate residue-wise flexibility, with higher *B*-factor values corresponding to greater atomic mobility. *PDB*, Protein Data Bank.

- The script extracts ***B*-factor values** of Cα atoms of the residues from each lysozyme structure (PDB IDs: 193L, 194L, 1AKI, 1AZF, 1BGI) and plots them as a function of residue number.
- The resulting plot (Fig. 18.2) overlays the *B*-factor profiles of all the structures, allowing a visual comparison.
- In crystallographic models, **higher *B*-factor values** indicate greater atomic displacement, often reflecting flexibility or disorder in that region.

Questions:

1. Which lysozyme structure consistently exhibits higher *B*-factor values across most residues? ***Hint:*** Try to explore why this structure exhibits increased flexibility. You may refer to the crystallographic or experimental details provided in the PDB entry.
2. Which regions show noticeable *B*-factor peaks, and what does that imply about their flexibility?
3. Do the flexible regions correspond to certain structural elements, such as loops or termini of the protein?

To explore this further, we shall visualize the structures in PyMOL by coloring them according to their *B*-factor values and observe where the flexible regions are located.

- Color each structure based on *B*-factor values:

```
spectrum b, blue_white_red, minimum=20, maximum=50
as cartoon
cartoon putty
```

Note: This representation varies both color and tube thickness based on *B*-factor values, with blue (thin) indicating low *B*-factors (rigid regions) and red (thick) indicating high *B*-factors (flexible regions).

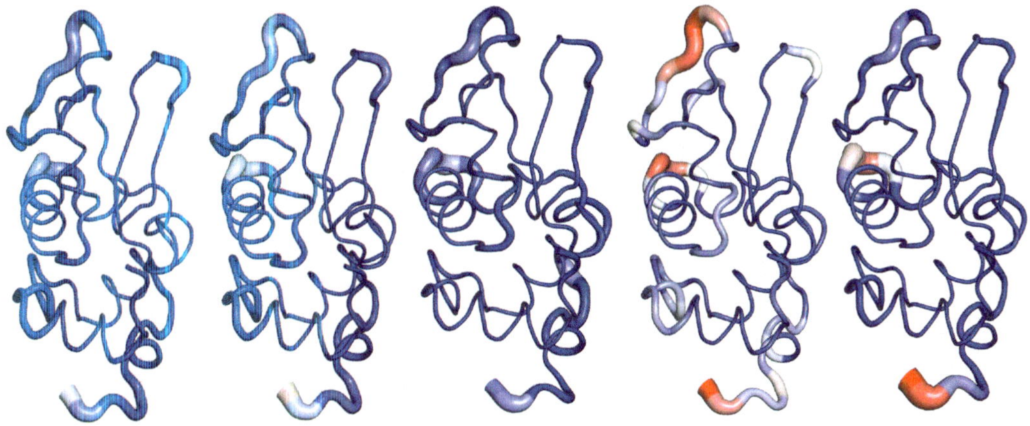

FIGURE 18.3

B-Factor representation of lysozyme structures: Lysozyme structures are shown as cartoons with *B*-factor-dependent coloring and scaling. Rigid regions are represented in *thin blue*, whereas flexible regions are highlighted in *thick red*, corresponding to low and high *B*-factor values, respectively.

- View all structures side by side using grid view:

```
Display > Grid > By Object
```

Now, you should be able to see the structures as shown in Fig. 18.3. This helps you directly observe the flexible and rigid regions in 3D space.

Questions:

1. Which structure(s) show the most prominent red and thickened regions?
2. How do these visually flexible regions correlate with the peaks observed in the *B*-factor profile plot?
3. Can you identify which structural elements (e.g., loops or termini) correspond to the highest flexibility?

18.2.1.3 Visualizing structural differences

In the next step, we examine the aligned protein structures obtained in Section 18.2.1.1 in greater detail to uncover the molecular motions. In PyMOL the default representation is the "cartoon representation," which represents the backbone fold of the protein structure. Carefully examine the aligned protein structures shown in Fig. 18.4 for structural changes.

Question:

1. Did you observe that most of the regions are superimposed across all the structures?
 Hint: Observe how the loop regions, especially around residues 70, 71, and 72, show subtle inward tilts in 1AKI and 1BGI compared to 194L, 193L, and 1AZF.

These movements, though small, raise an intriguing question: could changes in the outer loops alter the overall surface of the protein? Think about how these shifts might influence the external structure or interactions of the protein.

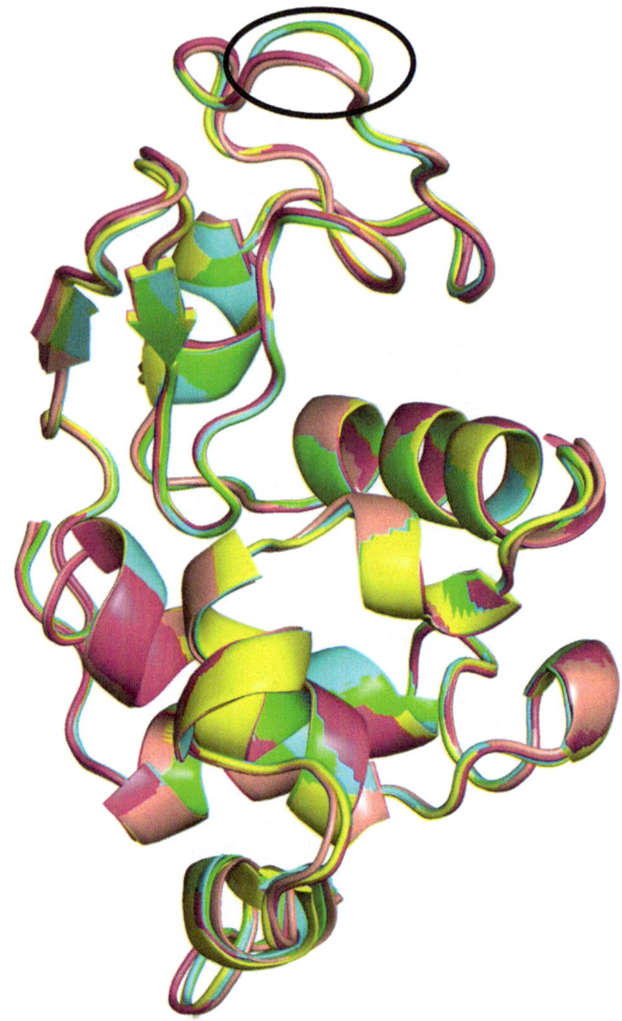

FIGURE 18.4

Comparison of aligned lysozyme structures: Superimposed lysozyme structures of 193L (*green*), 194L (*cyan*), 1AKI (*magenta*), 1AZF (*yellow*), and 1BGI (*light brown*). Loop regions exhibiting major conformational differences are highlighted with a *black circle*.

18.2.1.4 Visualizing surface representation of proteins

Protocol:

Next, let us switch to the surface representation of the proteins. This representation shows the molecular surface, helping us explore how conformational changes, such as those in the loops, may reshape the protein surface.

- Hide cartoon and show surface representation.

```
All > Hide > Cartoon
All > Show > Surface
```

- Color each protein in different colors.

```
All > Color (C) > Auto > By Object
```

- View the structures side by side

```
Display > Grid > By object
```

Question:

1. Did you notice any notable changes in the surface of the proteins?
   Hint: Pay close attention to the tail regions, protruding surfaces , and cavities. (Refer to Figure S1 provided in GitHub).

18.2.1.5 Visualizing local side-chain motion

Protocol:

Now, consider the possible reasons for these surface changes. Such movements often arise from the side-chain dynamics of amino acids. To explore this further, let us switch to the sticks representation, which provides a clearer view of individual side chains and their contributions to these conformational changes.

- Hide surface and show line representation

```
All > Hide > Surface
All > Show > Sticks
```

- Disable the Grid view option if it is enabled.

```
Display > Grid > Disable
```

- Zoom into the sites and try to locate the regions where you observed changes in the surface representation.

Hint: You can select specific residues (R21, R45, P70, R73, and R128) across all the structures and observe the orientation of their side chain. These differences are related to the intrinsic protein dynamics sampled. Refer to Figure S2 provided in GitHub.

18.2.2 Visualizing the proteins crystallized under different pH conditions

In this exercise, we explore lysozyme structures crystallized under varying pH conditions. Lysozyme is highly sensitive to pH, with an optimal activity around pH 5. This exercise aims to investigate how pH alterations influence the orientation of amino acids, the overall shape of the protein, and its associated dynamics.

We consider three X-ray crystallographic structures with the following PDB IDs: **1REX, 1I20**, and **1BB4**, crystallized at pH values of **4.5, 6**, and **8.5**, respectively. Let us examine how these differences manifest at the structural level and aim to identify structural differences between them.

Protocol:

- Load the three structures in PyMOL:

```
fetch 1rex
fetch 1i20
fetch 1bb4
```

- Remove water from these structures:

```
remove solvent
```

- Protein with PDB ID 1BB4 has a homodimer. So, remove the B chain of 1BB.

```
remove 1bb4 and chain B
```

- Set the background to transparent for better visibility.

```
bg_color white
```

- Align 1i20 and 1bb4 structures to 1rex structure:

```
align 1i20,1rex
align 1bb4,1rex
```

- Grid option can be used to view all the objects side by side, zoom into all, and compare the structures.
- Carefully examine the protein structures looking for changes in the overall shape of the protein, secondary structural elements, loop movements, and so on.

```
Go to Display > Grid > By object
zoom all
```

- Rotate the protein in all possible ways, looking for the structural differences.

Questions:

1. Did you observe any changes in the secondary structural elements under different pH conditions? (Refer to Exercise 18.2.2.1)
2. Did you notice any change in the overall surface of the protein? (Refer to Exercise 18.2.2.2)
3. Knowing that the side chains of these residues are flexible, can you think of how to measure these movements? (Refer to Exercise 18.2.2.3 and 18.2.2.4)
4. Given the role of phi and psi angles in determining protein structure, can you think of how to capture the differences in these dihedral angles to understand protein dynamics under different pH conditions? (Refer to Exercise 18.2.2.5)

18.2.2.1 Secondary structural changes

In this exercise, we identify the regions where the secondary structural elements of the proteins are altered because of changes in pH.

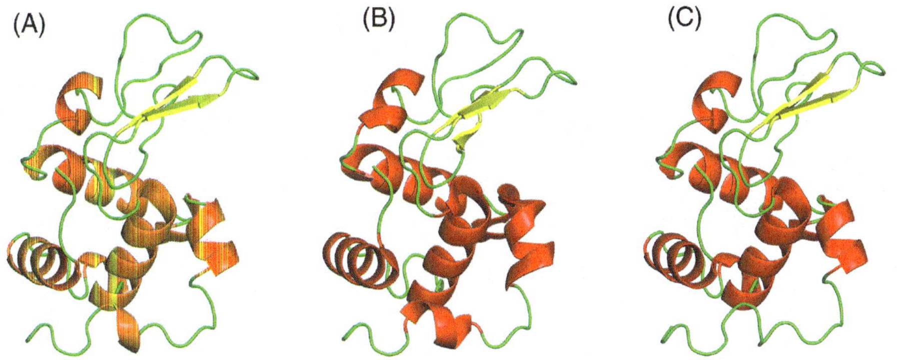

FIGURE 18.5

Secondary structure representations of lysozyme at different pH conditions: α-helices are shown in *red*, β-sheets in *yellow*, and loops in *green*: (A) 1REX at pH 4.5, (B) 1I20 at pH 6.0, and (C) 1BB4 at pH 8.5.

Table 18.1 Overview of secondary structural elements in the crystal structures chosen.

ss/PDB ID	1REX (pH 4.5)	1I20 (pH 6)	1BB4 (pH 8.5)
Helices			
Sheets			

PDB, *Protein Data Bank.*

Protocol:

- Continue with the grid view of these structures set in Exercise 18.2.2.
- Color the structures by secondary structure information:

```
util.cbss("all","red","yellow","green",_self=cmd)
```

This will color the helices in red, sheets in yellow, and loops in green, as shown in Fig. 18.5.

- Highlight the residues of the helices:

```
select helices, ss H
```

- Open the sequence tab by clicking the SEQ button at the bottom right corner of the PyMOL window. Respective residues get highlighted. Now, count the number of helices.
- Highlight the residues of the sheets:

```
select sheets, ss S
```

- Follow the previous step to count the sheets.
- After counting the number of helices and sheets, try to fill out Table 18.1.

Note: You may refer to the answers provided in the GitHub repository (Table S1).

Questions:

1. Did you observe the decrease in the number of helices in 1I20 (pH 6) when compared to structures of 1REX (pH 4.5) and 1BB4 (pH 8.5)? Also, look for the change in the number of sheets.
2. Can you locate the region that changes from a loop at pH 4.5 to a sheet at pH 6, then back to a loop at pH 8.5? ***Hint:*** Observe the secondary structure of residues 59 and 60 in all three structures.
3. Can you identify the helix region present at pH 4.5 and 8.5, but which transitions to a loop at pH 6? ***Hint:*** Observe the secondary structure of residues 20, 21, and 22 in all three structures.

You may refer to Figure S3 provided on GitHub to observe these secondary structure differences highlighted.

18.2.2.2 Highlighting the protein in surface representation

So far, we have visualized the protein using cartoon representation, which highlights its secondary structural elements. Now, let us switch to surface representation to get a clear view of the overall shape of the protein. This will allow us to observe whether there are any noticeable changes in the protein surface and overall structure under different pH conditions.

Protocol:

- Set "surface representation" for all the proteins and transparency to 80%, keeping cartoon representation as such:

```
show surface, all
set transparency, 0.8
```

- Color the three structures in three different colors:

```
color green, 1rex
color yellow, 1i20
color cyan, 1bb4
```

- Orient the molecules to get the view as shown in Fig. 18.6.

In this view, you can see that the surface at the top (highlighted in circles) has been modulated. The narrowed region got altered, with a reduced surface area. Now, let us calculate the solvent-accessible surface area (SASA) for the selected residues (residues 46–50, Sequence-NAGDR) present at that highlighted region.

- Set the dot_solvent option on:

```
set dot_solvent, on
```

- Select the residues from 46 to 50 and calculate SASA by executing the commands individually for each structure and note down the SASA values that show up in the PyMOL command console.

```
For 1REX: select sele-1rex, 1rex and resi 46-50
get_area sele-1rex
For 1I20: select sele-1i20, 1i20 and resi 46-50
get_area sele-1i20
For 1BB4: select sele-1bb4, 1bb4 and resi 46-50
get_area sele-1bb4
```

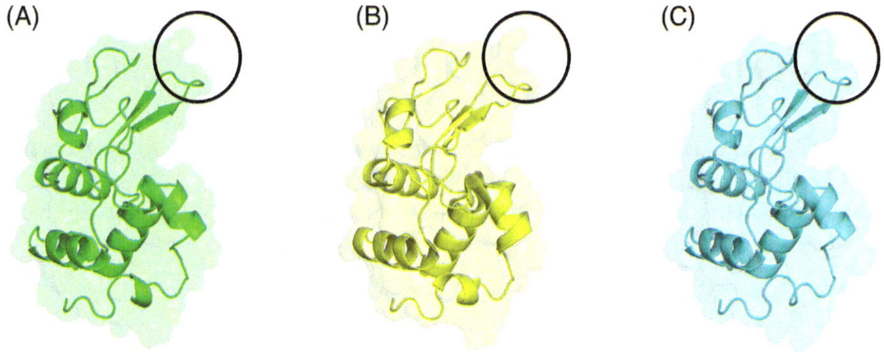

FIGURE 18.6

Lysozyme structures shown in cartoon with transparent surface representation: (A) 1REX at pH 4.5 (*green*), (B) 1I20 at pH 6.0 (*yellow*), and (C) 1BB4 at pH 8.5 (*cyan*). Altered surface regions across pH conditions are highlighted with *black circles*.

Table 18.2 Overview of solvent-accessible solvent area in the crystal structures chosen.

pH	1REX (pH 4.5)	1I20 (pH 6)	1BB4 (pH 8.5)
SASA (Å^2)			

- Tabulate the observed SASA values for residues 46–50 into Table 18.2 for each structure.

Note: The answers have been provided on GitHub (Table S2).
Question:

1. Did you observe a decrease in the SASA values for the selected residues as the pH increases from 4.5 to 8.5?

The reduction in SASA suggests a move toward a more compact conformation, in which the residues are increasingly shielded within the protein core instead of being solvent-exposed. Let us look at the residues present in this region to gain further insights about its dynamics.

18.2.2.3 ***Exploring side-chain motion***

In this exercise, we shall consider arginine (R50), which is present in the region analyzed in the previous example. We observe that the arginine residue (R50), which is a pH-sensitive amino acid, appears to be more buried with increasing pH.
Protocol:

- Visualize the movement of arginine (R50) in different pH conditions.
- Overlap all three crystal structures and view only arginine 50:
- Turn off the grid display:

```
Display > Grid > Disable
```

- Hide everything and view only arginine 50 from the three structures:

```
hide everything
show sticks, (resi 50)
```

- Zoom at residue 50, that is, Arg50 (use Zoom All option), and orient them to have a better view.
- Color the residue by element and label them with the atom element:
- Select the R50 in all three structures so that this selection is stored as sele group. Next, use the commands given in the following:

```
Coloring: util.cnc("sele",_self=cmd)
Labeling: cmd.label("sele","name")
```

Question:

1. Did you notice the side chain movement of arginine 50 across structures in different pH conditions? ***Hint:*** Refer to Figure 18.7A to observe side chain movement differences of arginine 50.

The change in the pH influences the orientation of the amino acid residues, particularly their side chain groups. At the lower pH of 4.5 the side chain of arginine is more extended. As the pH increases to 6 the side chain of arginine shows a slight bend. At pH 8.5 the side chain bends further inward, closer to the core. Now, let us quantify the side chain movement of residue arginine 50 by measuring the angle.

18.2.2.4 ***Quantifying side-chain motion***

For measuring the angle, we shall consider 1REX as the reference. The first and second atoms would be atoms NE and CD of 1REX, respectively. The third atom would be atom NE of 1I20 or 1BB4.

Protocol:

- Set the angle measurement option:

```
wizard measurement
cmd.get_wizard().set_mode("angle")
```

- For measuring 1I20 side chain movement:

```
angle angle_1, /1rex///50/NE, /1rex///50/CD, /1i20///50/NE
```

Note down the angle_1 value.

- For measuring 1BB4 side chain movement:

```
angle angle_2, /1rex///50/NE, /1rex///50/CD, /1bb4///50/NE
```

Note down the angle_2 value.

Arginine 50 exhibits a more extended conformation at pH 4.5, and starts bending to an angle of 56.3° as the pH varies from 4.5 to 6. Further, it bends toward the inner core by 120.0° relative to the reference structure (1REX, pH 4.5) as the pH increases to 8.5 (Fig. 18.7).

The bending of arginine toward the inner core of protein (with an increase in pH) could arise due to the ionization state of its side chain and its interactions with the surrounding environment. At a lower pH of 4.5, well below its p*K*a of 12.5, the side chain is fully protonated, maintaining a more extended conformation because of favorable hydrogen bonding with the surrounding solvent. As pH increases to

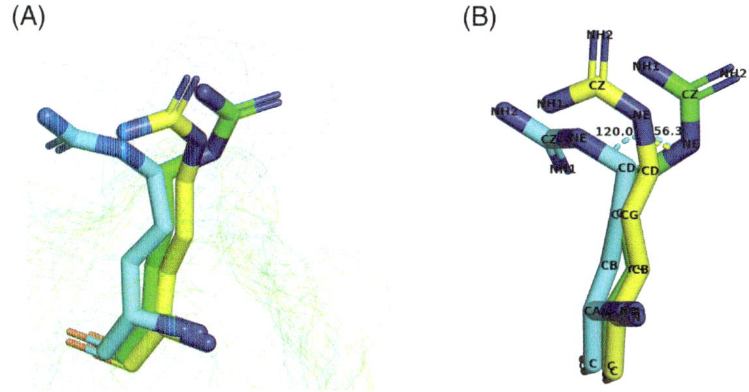

FIGURE 18.7

Conformational changes of arginine 50 of lysozyme under different pH conditions: (A) Mesh and stick representation of arginine 50 of lysozyme structures at pH 4.5 (*green*, 1REX), pH 6.0 (*yellow*, 1I20), and pH 8.5 (*cyan*, 1BB4). (B) Inward bending of the residue is observed with increasing pH.

6 with p*K*a 12.5, then to pH 8.5 with lowered p*K*a 12.43, the side chain gets deprotonated leading to a more compact or inward-bending conformation.

Note: pKa values are calculated using the PropKa server (Li et al., 2005). You can access the PropKa server at https://www.ddl.unimi.it/vegaol/propka.htm

18.2.2.5 Analysis of backbone dihedral angles using the Ramachandran plot

The altered orientation of the side chain results in changes to the backbone dihedral angles (Φ and Ψ). These variations can be visualized using a Ramachandran plot. In this exercise the PPBuilder module from the Bio.PDB package in Biopython will be used to calculate the peptide bond dihedral angles.

Protocol:

- Execute the code snippet provided as Section 2.2.5 in the Google Colab notebook.
- The code fetches the PDB files, extracts the Φ and Ψ angles for arginine 50, and plots them in a color-coded Ramachandran plot as shown in Fig. 18.8.

Fig. 18.8A shows the backbone structure of an amino acid, highlighting the **phi (Φ)** and **psi (Ψ)** torsion angles. These angles represent the rotations around the N–C_α bond (phi) and the C_α–C bond (psi) within the peptide backbone. These angles are critical for understanding the conformational flexibility of proteins, as they dictate the folding and spatial arrangement of the polypeptide chain. The phi and psi angles are constrained by steric hindrance, and their allowed combinations give rise to different secondary structures such as α-helices and β-sheets. Fig. 18.8B is a Ramachandran plot, which maps the phi (x axis) and psi (y axis) angles of each residue in a protein. Each dot on the plot represents the phi–psi combination for arginine 50 in the protein structure. In this case, the dihedral angles of the residue were 66.6° and 22.8° at pH 4.5, but shifted to 71.1° and 14.0° at pH 6, and 58.3° and 24.7° at pH 8. This led to the slightly altered conformation of the protein as a whole.

Explore further!

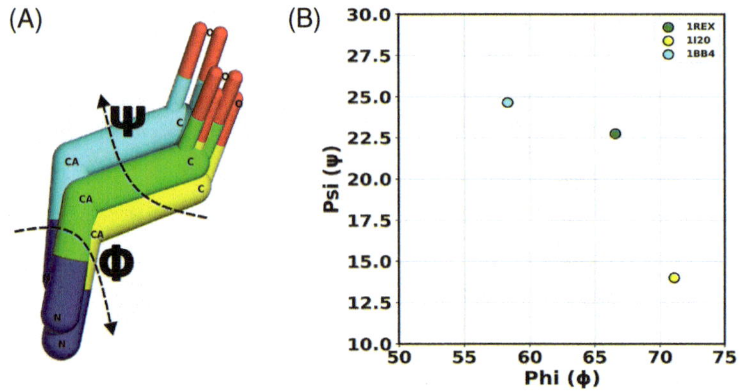

FIGURE 18.8

Dihedral angle plot of arginine 50 of lysozyme under varying pH conditions: (A) Arginine backbone illustrating the phi (Φ) and psi (Ψ) torsion angles within the peptide backbone. (B) Ramachandran plot showing the distribution of these torsion angles for arginine 50, with *dots colored green* (pH 4.5, PDB ID: 1REX), *yellow* (pH 6.0, PDB ID: 1I20), and *cyan* (pH 8.5, PDB ID: 1BB4).

You may extend the same analysis to other residues (Arg44, Tyr45, and Gly72) in the three structures to examine the possible gains or losses of interactions. In addition, explore structures solved under varying temperature conditions by using the advanced search feature in the PDB.

18.3 Exploring the dynamics of proteins from nuclear magnetic resonance structures

Probing a protein structure with NMR allows us to see a range of different conformations that a protein can adopt. Because NMR structures are determined in solution, they closely mimick the natural environment of a protein, making it ideal for studying protein flexibility. In this section, we will work through a series of practical exercises designed to explore protein dynamics. This includes visualization of protein conformations in PyMOL and analysis of the dynamics of the proteins in the NMR ensemble using Python codes.

18.3.1 Visualizing protein conformational flexibility

This exercise guides you through using PyMOL to explore the concept of protein flexibility by examining the NMR structure of a protein with multiple conformations deposited in the PDB. The NMR structure of ubiquitin (PDB ID: 1D3Z) has been selected for the exercise.

Protocol:

- Access the PDB website using the provided link https://www.rcsb.org/, search for the entry with PDB ID **1D3Z**, and check the number of conformers deposited for this entry.

- Open PyMOL and fetch the protein structure using the PDB ID by executing the command given in the following:

```
fetch 1d3z
```

- Visualize and animate through all states to explore the ensemble of conformations deposited for this entry
- Display all states at once: `set all_states, 1`
- Display only the first state: `frame 1; set all_states, 0`
- Set the frame rate to 1 frame per second (fps) to observe the transitions at a controlled speed: `set movie_fps, 1`
- Play the animation and visualize the transitions between states: `mplay`

  Refer to the video provided on GitHub (Video 1).

- Grid option can be used to view all the states side by side and compare each state.

```
Go to Display > Grid > By state
```

- Try using surface representation and compare

```
as surface, all
color green, all
```

Questions:

1. Did you observe any significant transitions between the different conformational states? What stood out to you? ***Hint:*** While viewing the protein conformers in the grid layout, you can focus on residues 71–76 to observe differences across conformers. Select residues 71–76 (`select region, resi 71-76`) and color them red (`color red, region`). This will highlight the specific region in all conformers within the grid, helping you observe any notable changes. Feel free to try out other representations (such as cartoon or stick) for further insights.
2. Do you think there could be a conformation that is revisited or repeated in the ensemble of states? ***Hint:*** Think of how to quantify the pairwise RMSD across all the conformations of the NMR structure. [Refer to Figure S4 provided on GitHub]. You can use the code provided in Section 3.1 of the notebook to calculate the RMSD.

Additional Note: For more practice, try using the PDB IDs **2RSC** and **2LXI**, which have multiple conformations. To find more PDB entries with multiple conformations, use the PDB advanced search options.

18.3.2 Finding a representative structure from the ensemble of protein structures

We have now observed that structures resolved with NMR consist of an ensemble of structures. A natural question that arises is whether we can choose a representative structure. A common practice could be to derive an "average" (mean) structure by averaging the atomic positions of the structure. However, it is crucial to check if this average is realistic or if it introduces artifacts. The code provided in Colab notebook downloads an NMR PDB file, calculates the average atomic coordinates from all the models, and saves this average as a new PDB file. The resulting mean structure can be visualized to identify any potential artifacts introduced by the averaging process.

Protocol:

- Execute the code provided in Section 3.2 of the notebook to compute the average structure for the NMR ensemble of structures. This will generate and save the PDB file with the mean coordinates to your local computer
- Open PyMOL and load the average PDB file:

  `load 1D3Z_mean.pdb` OR `Go to File > Open > Downloads > 1D3Z_mean.pdb`

- Fetch the 1D3Z structure from the RCSB PDB:

```
fetch 1D3Z
```

- Set the representation of the average structure to "lines":

```
hide all
show lines, 1D3Z_mean
show lines,1D3Z
```

- Set different colors for both objects:

```
color red, 1D3Z_mean
color green, 1D3Z
```

- Examine the average structure and the original conformations for any noticeable structural artifacts. Zoom in on areas with significant differences to assess how well the average structure represents the protein and identify any issues from averaging the coordinates.

Questions:

1. How does the average structure look compared to the individual conformations of the protein?
2. When visualizing the average structure in line representation, did you notice any unrealistic features, such as unusually short or stretched bonds, or regions where atoms are packed too closely? ***Hint:*** You may refer to Video S2 provided on GitHub.
3. Considering the observations above, how do you interpret these findings? Is the average structure a realistic protein conformation?

18.3.2.1 ***Is the mode structure the representative structure?***

From the previous exercise, it is clear that the mean structure can introduce structural discrepancies. This exercise aims to identify a more representative structure from the NMR ensemble, termed the "mode structure," as it represents the most frequently occurring conformation in the ensemble.

The code provided downloads an NMR PDB file, extracts Cα atom coordinates, calculates pairwise RMSD values, and clusters the models using hierarchical clustering. It then identifies the most frequent cluster to determine the mode structure, which is saved as 1D3Z_mode.pdb.

Protocol:

- Execute Section 3.2.1 code provided in the Notebook to compute the mode structure for the NMR ensemble of structures. Running the code will download the mode PDB file to your local computer.

- Open PyMOL and load the average PDB file:

```
load 1D3Z_mode.pdb   OR Go to File > Open > Downloads > 1D3Z_mode.pdb
```

- Fetch the 1D3Z structure from the RCSB PDB:

```
fetch 1D3Z
```

- Set the representation to lines, and color the objects in different colors as was done in the previous exercise.
- Consider animating through the states:

```
set movie_fps, 1
mplay
```

- In a similar way, animate through all the states with the mean structure.

Note: Refer to the videos uploaded on GitHub to observe all the states aligned to the mean structure (Video S3) and the mode structure (Video S4).

Questions:

1. Did you observe any structural discrepancy in the mode structure? (Refer to video S5 provided on GitHub)
2. After visualizing both the mode and mean structures with the individual states in the ensemble, which structure do you think better captures the variability of the ensemble?

18.3.3 Is protein dynamics modulated in a cellular environment?

Imagine a protein inside a cell surrounded by a myriad of other biomolecules—does this environment make the protein more or less dynamic compared to a controlled, *in vitro* setting? In this exercise, you will explore this question by analyzing protein dynamics in both environments using visualization and basic metrics such as RMSD and root mean square fluctuation (RMSF), focusing on two selected proteins: heavy metal-binding protein S and immunoglobulin G-binding protein G (GB1). This comparison will help determine how the cellular context affects protein behavior.

18.3.3.1 Visualizing the dynamics of heavy metal-binding proteins in cellular environments

Protocol:

- Consider the PDB structures of the heavy metal-binding protein of *Thermus thermophilus*, one solved in an *in vitro* environment (PDB ID-2ROE) and the other in a cellular environment (PDB ID:2ROG).
- Load and visualize protein states in PyMOL:

```
fetch 2roe
fetch 2rog
```

- Align the two structures:

```
align 2ROE, 2ROG
```

- To compare states side-by-side using a grid layout:

```
Go to Display > Grid > By object
```

- Visualize all states simultaneously:

```
set all_states, 1
```

Question:

1. Upon comparing the ensemble of structures (2ROE and 2ROG), could you identify in which environment the protein samples a wider set of conformations?

Try animating these states to gain a better understanding of the dynamics of the proteins in both environments.

```
frame 1
set all_states, 0
mplay
```

Note: You may refer to the video S6 provided on GitHub to compare the dynamics of heavy metal binding protein *in vitro* (left side) and in the cellular environment (right side).

18.3.3.2 Assessing protein flexibility in cellular environments

We will consider basic analyses such as RMSD and RMSF, which are key metrics used to assess the structural dynamics of proteins. RMSD measures the average distance between atoms of superimposed protein structures, indicating how much a structure deviates from a reference, which helps in assessing structural changes and comparing conformations. On the other hand, RMSF quantifies the average fluctuation of each atom or residue around its mean position over time, revealing the flexibility of protein residues and providing insights into dynamic regions.

Protocol:

- Execute the codes provided in Section 3.3.2 in the notebook.
- The section is divided into Part A (RMSD analysis) and Part B (RMSF analysis); in both cases, the states are aligned to mean and mode structures as reference.
- Analyze the comparative plots of RMSD and RMSF of the protein obtained *in vitro* and in cell conditions, illustrated in Fig. 18.9.

Question:

1. What did you infer from the comparative plots of RMSD and RMSF about the dynamics of the protein in the two environments? ***Hint:*** Check in which environment did you observe higher RMSD and RMSF values. Higher RMSD and RMSF imply higher structural variability of the protein.

18.3.4 Investigating the dynamics of immunoglobulin G-binding protein G in cellular environments

Protocol:

- Consider the two PDB structures of the Immunoglobulin G-binding protein G (GB1), with one structure determined *in vitro* (PDB ID: 2N9K) and the other in living sf9 cells (cellular environment) (PDB ID: 5Z4B)

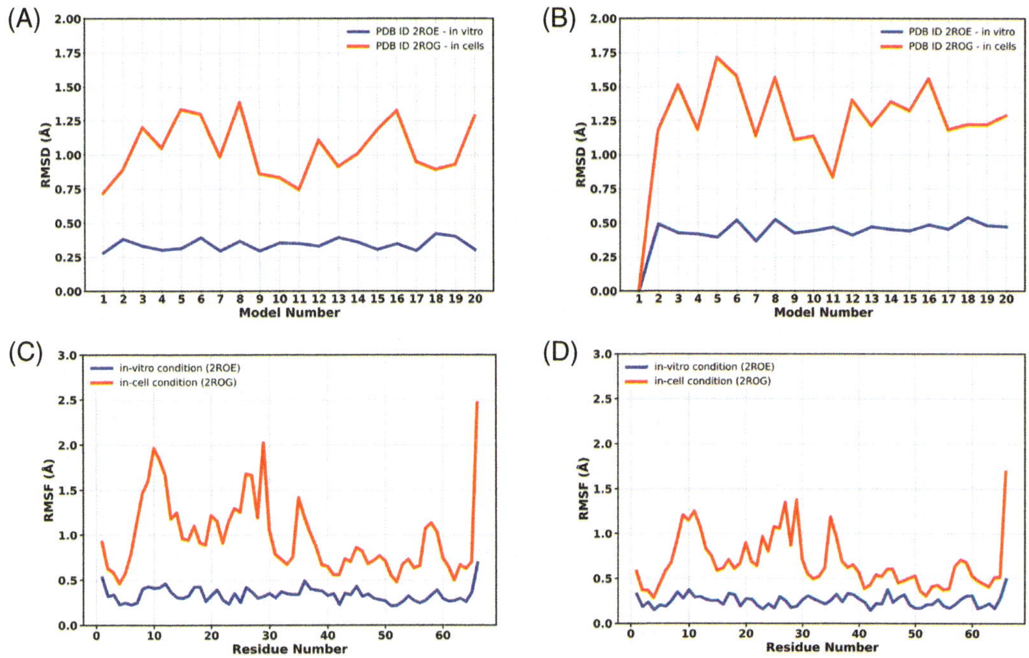

FIGURE 18.9

Comparative RMSD and RMSF profiles of the protein in *in vitro* and in-cell conditions: RMSD (top) and RMSF (bottom) profiles for the protein *in vitro* (PDB ID: 2ROE) and in-cell conditions (PDB ID: 2ROG). Panels (A) and (C) use the mean structure as the reference, whereas panels (B) and (D) use the mode structure as the reference. *PDB*, Protein Data Bank; *RMSD*, root mean square deviation; *RMSF*, root mean square fluctuation.

- Load and visualize protein states in PyMOL:

```
fetch 2n9k
fetch 5z4b
```

- Align the two structures:

```
align 2n9k, 5z4b
```

- Repeat the steps we performed for the heavy metal binding protein and check if the same behavior is observed in this case.
- Focus on the alpha helix of the GB1 protein and compare its position in both structures.

Question:

1. Did you notice the alpha-helix tilting away from the beta-sheet in the cellular environment (5Z4B)? [Refer to video S7 provided on GitHub]

2. If yes, can you think of a way to measure this movement of the alpha helix?

18.3.4.1 Quantifying alpha-helix tilt in the cellular environment

Protocol:

- Focus on the alpha-helix region in both the 2N9K (*in vitro* condition) and 5Z4B (cellular environment) structures in PyMOL.
- Select the following key residues: Asp23 and Asp37 in 2N9K and Asp23 in 5Z4B.
- Color the selected residues in different colors (may be red).
- Go to angle measurement option:
- In PyMOL, go to the **Wizard** menu.
- From the menu, select **Measurement**.
- In the right-side panel, from the dropdown options, select **Angle Measurement**.
- Select the atoms for angle measurement:
- Represent Asp23 and Asp37 of 2N9K and Asp23 of 5Z4B in stick representation and label their atoms.
- First atom: Select the CA (alpha carbon) of Asp23 in 2N9K.
- Second atom: Select the CA (alpha carbon) of Asp37 in 2N9K.
- Third atom: Select the CA (alpha carbon) of Asp23 in 5Z4B.
- Measure the angle.
- PyMOL will now calculate the angle between the two alpha helices, using the selected residues and atoms.
- Move the frames one by one using the movie options provided in the right-side panel. Note down the values of the angle measured for each frame and plot those values using the code provided in the notebook. You should be able to get a plot as shown in Fig. 18.10.
- Analyze the plot to understand the dynamics of the alpha helix modulated by the cellular environment.
- You may find the average helix angle to see how far it deviates from its original position in the cellular environment.

Explore further!

To further enhance your understanding of protein dynamics, we encourage you to apply the exercises presented here to other proteins discussed in the review article by Hu et al. (2021). This will not only deepen your knowledge but also expand your insights into the dynamic behavior of various proteins studied using NMR approaches.

18.4 Exploring the dynamics of proteins resolved with cryo-electron microscopy

In this section, we explore protein dynamics from cryo-EM structures. Cryo-EM is particularly suited for capturing the structural variability of large, flexible proteins and complexes, often under conditions

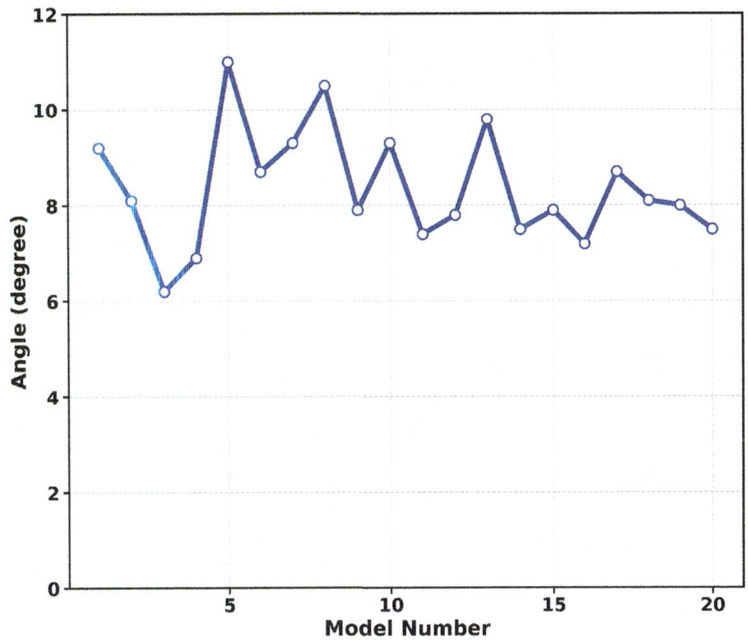

FIGURE 18.10

Movement of the alpha helix of the GB1 protein in the cellular environment: Line graph showing frame-by-frame changes in the α-helix orientation of the GB1 protein under cellular conditions.

closer to their native states. Its ability to resolve multiple conformational states provides valuable insights into protein dynamics that are challenging to achieve with other techniques.

In this exercise, we will analyze the dynamics of CorA, a magnesium transporter protein, using its cryo-EM structures in different functional states. By comparing the protein structures with and without magnesium ions, we aim to understand how its domains behave and identify structural changes that may influence its function. Specifically, we will examine the following:

- The magnesium-bound, closed symmetric state (PDB ID: 3JCF)
- The magnesium-free, asymmetric open state I (PDB ID: 3JCG)
- The magnesium-free, asymmetric open state II (PDB ID: 3JCH)

These structures will help us investigate the relationship between structural transitions and functional mechanisms of the protein.

18.4.1 Visualization of conformational transitions in CorA protein

Protocol:

- Download the structures for the above-stated PDB IDs from the RCSB PDB database in PDB format.
- Load the protein structures in PyMOL and visualize the states.

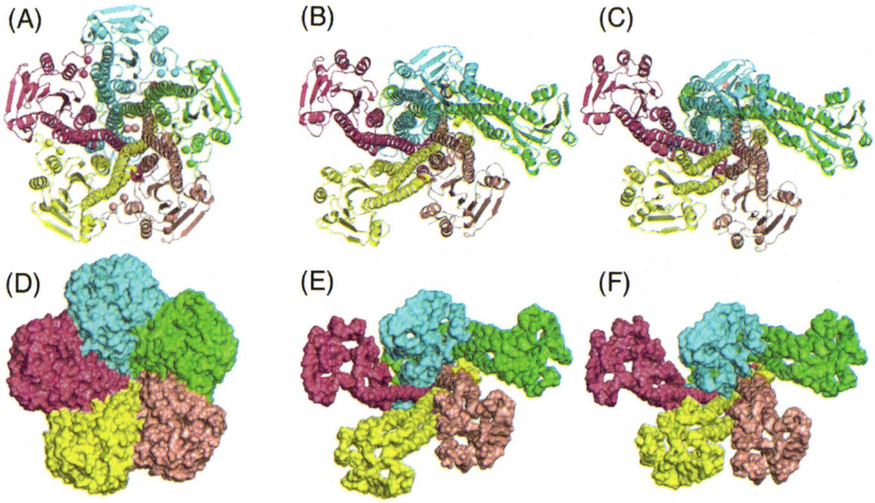

FIGURE 18.11

Structural comparison of closed and open states of CorA protein: (A–C) Top view in cartoon representation and (D–F) surface representation. Proteins are colored by chain: chain A (*green*), chain B (*blue*), chain C (*magenta*), chain D (*yellow*), and chain E (*light brown*).

- Consider 3JCF, which is the closed state of the CorA, as the reference structure.
- Align the structures to 3JCF:

```
align 3jcg, 3jcf
align 3jch,3jcf
```

- Color the protein structures by chain

```
All > Color > by chain
```

Question:

1. Did you notice changes between the closed and open states? Which chains showed the maximal motion?
   - View the structures side-by-side using the grid display by object option in PyMOL. You should observe that chains A and B move outward in the open states.
   - Top view comparison: Rotate the proteins to get a top view. You should clearly see the differences in the channel path, with chain C moving inward in the open states as illustrated in Fig. 18.11A–C.
   - Surface representation: Switch to the surface representation. You should be able to clearly see the loss of symmetry in the open states as shown in Fig. 18.11D–F. The closed state is symmetrical,

whereas the open states show some chains moving outward and others moving inward, resulting in a loss of symmetry.

2. Can you think of a way to quantify the differences in symmetry observed between the closed and open states?

   Hint: The next exercise will guide you through the steps required for the calculation.

18.4.2 Quantifying protomer motion and symmetry loss in CorA open states

The overall challenge in this section is to capture how the symmetry of the CorA pentamer changes between closed and open states. As the protein is a homo-pentamer, each of its five chains is structurally identical. To represent the relative position of these chains, we focus on the residues located at the top region of each chain and compute their center of mass (COM). This gives one representative COM per chain, resulting in five COMs in total. By connecting these COMs, we obtain a pentagon that reflects the geometry of the protein from the top view. Comparing these pentagons across different conformational states allows us to visualize and quantify symmetry loss during channel opening.

For this analysis, we select the following residues from each chain: 32, 36, 37, 38, 39, 57, 58, 80, 81, 103, 104, 133, 134, 160, 163, 164, 165, 239, 241, 243, 244, and 245. This set of residues roughly corresponds to the top region of the chains and was chosen because it is consistently present across all conformational states of CorA. These residues will serve as the basis for calculating the COM of each chain in the subsequent steps.

Protocol:

- Download all the necessary pdb files and vmd scripts in the local directory.
- Now, launch the VMD in the same local directory path:
  `vmd`
- In the VMD interface, go to `File > New Molecule`.
- Click `Browse`, select your *3jcf.pdb* file, and hit `Load`.
- To visualize the protein, follow these steps:
- Navigate to `Display > Orthographic` to adjust the view.
- Go to `Graphics > Representation`.
- Set `Drawing Method > New Cartoon` to visualize the structure in cartoon representation.
- For a clearer view, set the VMD background to white:
- Go to `Graphics > Colors`
- Under `Category`, select `Display`
- Under `Name`, choose `Background`
- Set the `Color` to `White`
- Open the *Tk Console* by navigating to `Extensions > Tk Console`.
- **Calculation of the center of mass:** Run the following script to compute the COM of the selected residues in each chain and represent them as labeled spheres (A–E). Type the following command in the console and press enter:

```
source com-3jcf.tcl
```

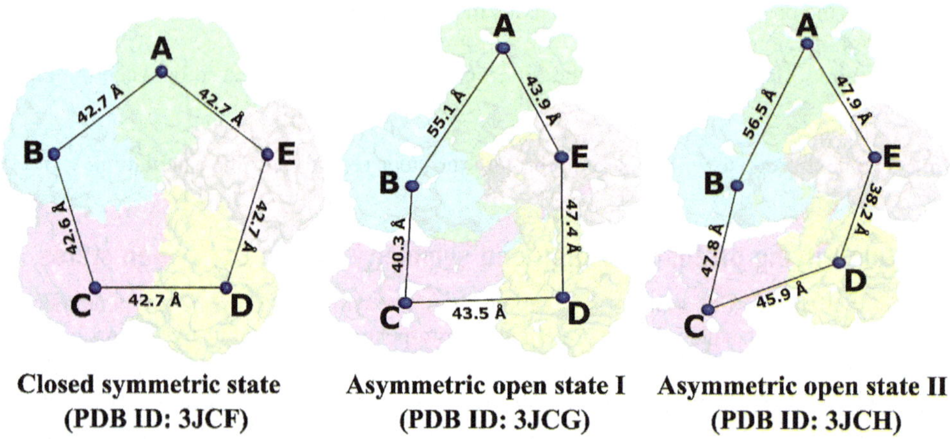

FIGURE 18.12

Conformational states of the protein complex represented by pentagons formed from centers of mass of chains A–E: Centers of mass (COM) of chains A–E are shown as blue spheres and connected to form pentagons for three conformational states of the protein complex. Distances (Å) between COMs are indicated along the edges of each pentagon. The geometry clearly demonstrates a well-defined symmetric pentagon in the closed state, whereas clear distortions and loss of symmetry are evident in the open states.

- **Pentagon construction and distance measurement:** Run the following script to construct a pentagon by joining the COMs, compute the distances between them, and display the values along each edge. Type the following command in the console and press enter:

```
source pentgon-3jcf.tcl
```

- Repeat the steps for 3JCG and 3JCH also. Use the respective scripts provided.
- The resultant pentagons obtained from these steps are illustrated in Fig. 18.12.
- On comparing the pentagons obtained by three PDB ids (3JCF, 3JCG, and 3JCH), the loss of symmetry can be observed.

See the channel in motion!

You may wonder how the transition between the closed and open states occurs. The authors of the study (Matthies et al. (2016)) provide a detailed schematic model of CorA, showing how Mg^{2+} binding and release drives asymmetric movements within the pentamer. For a dynamic view of this process, refer to Supplementary Movie S4 in the same study (Matthies et al., 2016), which vividly illustrates the stepwise conformational changes.

Explore further!

We encourage readers to explore additional studies where protein dynamics have been examined using cryo-EM structures. Such studies often illustrate state transitions, domain motion, and related dynamic behavior. By delving into these examples, readers can deepen their understanding of how cryo-EM enhances our knowledge of protein dynamics.

18.5 Conclusion

In this chapter, we have probed protein dynamics, moving beyond static structures to dynamic ensembles that reveal the flexibility of proteins that are essential for their biological function. Using protein structures resolved by X-ray crystallography, NMR spectroscopy, and cryo-EM, we analyzed different aspects of protein dynamics through structured exercises. At first, we examined lysozyme, a pH-sensitive enzyme, through X-ray structures crystallized at varying pH levels. This exercise provided a foundation for understanding how proteins alter their shape, secondary structure, and solvent accessibility in response to environmental changes. Through hands-on visualization in PyMOL, we observed specific changes in the structural elements of lysozyme, laying the groundwork for analyzing dynamic responses in proteins. In the next section, we explored the use of NMR spectroscopy to capture conformational flexibility ubiquitin in solution. By analyzing ensembles of structures and calculating RMSD and RMSF values, we quantified structural variability and identified flexible regions. The comparison of protein structures *in vitro* and in cellular environments further demonstrated how cellular conditions influence protein dynamics, adding complexity to their behavior. In the last section, we focused on cryo-EM structures, examining the magnesium transporter CorA to investigate conformational transitions. By comparing the closed symmetric state with open asymmetric states the analysis revealed significant domain movements, including outward shifts of specific chains and changes in the channel path. These structural changes provided insights into the functional mechanisms of CorA and highlighted the ability of cryo-EM to capture dynamic transitions in large, flexible protein complexes.

In summary, this chapter provided a comprehensive, hands-on approach to explore protein dynamics from experimentally resolved protein structures. Through these exercises, we have developed a foundational understanding of how proteins function as flexible, adaptable molecules, essential for life's complexity. These exercises are only a first step toward exploring protein dynamics through structural bioinformatics approaches, and we recommend the reader to explore further.

Abbreviations

cryo-EM cryo-electron microscopy
NMR nuclear magnetic resonance

References

Bahar, I., Lezon, T. R., Bakan, A., & Shrivastava, I. H. (2010). Normal mode analysis of biomolecular structures: Functional mechanisms of membrane proteins. *Chemical Reviews, 110*(3), 1463–1497. doi:10.1021/cr900095e.

Banari, A., Samanta, A. K., Munke, A., Laugks, T., Bajt, S., Grünewald, K., Marlovits, T. C., Küpper, J., Maia, F. R. N. C., Chapman, H. N., Oberthür, D., & Seuring, C. (2025). Advancing time-resolved structural biology: Latest strategies in cryo-EM and X-ray crystallography. *Nature Methods, 22*(7), 1420–1435. https://www.nature.com/nmeth/. doi:10.1038/s41592-025-02659-6.

Berendsen, H. J., & Hayward, S. (2000). Collective protein dynamics in relation to function. *Current Opinion in Structural Biology, 10*(2), 165–169. http://www.elsevier.com/locate/sbi. doi:10.1016/S0959-440X(00)00061-0.

Berman, H., Henrick, K., & Nakamura, H. (2003). Announcing the worldwide Protein Data Bank. *Nature Structural & Molecular Biology, 10*(12), 980. doi:10.1038/nsb1203-980.

Boehr, D. D., Nussinov, R., & Wright, P. E. (2009). The role of dynamic conformational ensembles in biomolecular recognition. *Nature Chemical Biology, 5*(11), 789–796. http://www.nature.com/nchembio. doi:10.1038/nchembio.232.

Bonincontro, A., De Francesco, A., & Onori, G. (1998). Influence of pH on lysozyme conformation revealed by dielectric spectroscopy. *Colloids and Surfaces B: Biointerfaces, 12*(1), 1–5. doi:10.1016/s0927-7765(98)00048-4.

Buhrke, D., & Hildebrandt, P. (2020). Probing structure and reaction dynamics of proteins using time-resolved resonance Raman spectroscopy. *Chemical Reviews, 120*(7), 3577–3630. doi:10.1021/acs.chemrev.9b00429.

Chang, W. H., Huang, S. H., Lin, H. H., Chung, S. C., & Tu, I. P. (2021). Cryo-EM analyses permit visualization of structural polymorphism of biological macromolecules. *Frontiers in Bioinformatics*. Www.frontiersin.org/journals/bioinformatics. doi:10.3389/fbinf.2021.788308.

Chen, X., Salerno, M., Kramer, C. M., Mehta, B. B., Yang, Y., Shaw, P., & Epstein, F. H. (2015). Prospectively accelerated first-pass myocardial perfusion imaging in patients using motion-compensated compressed sensing exploiting regional low-rank sparsity. *Journal of Cardiovascular Magnetic Resonance, 1*, O40. doi:10.1186/1532-429x-17-s1-o40.

Congreve, M., Murray, C. W., & Blundell, T. L. (2005). Keynote review: Structural biology and drug discovery. *Drug Discovery Today, 10*(13), 895–907. doi:10.1016/S1359-6446(05)03484-7.

De Almeida, A. J. P. O., De Oliveira, J. C. P. L., Da Silva Pontes, L. V., De Souza Júnior, J. F., Gonçalves, T. A. F., Dantas, S. H., De Almeida Feitosa, M. S., Silva, A. O., & De Medeiros, I. A. (2022). ROS: Basic concepts, sources, cellular signaling, and its implications in aging pathways. *Oxidative Medicine and Cellular Longevity, 202*. http://www.hindawi.com/journals/oximed/. doi:10.1155/2022/1225578.

Ellaway, J. I. J., Anyango, S., Nair, S., Zaki, H. A., Nadzirin, N., Powell, H. R., Gutmanas, A., Varadi, M., & Velankar, S. (2024). Identifying protein conformational states in the Protein Data Bank: Toward unlocking the potential of integrative dynamics studies. *Structural Dynamics, 11*(3). http://scitation.aip.org/content/aca/journal/sdy. doi:10.1063/4.0000251.

Fernandez-Leiro, R., & Scheres, S. H. W. (2016). Unravelling biological macromolecules with cryo-electron microscopy. *Nature, 537*(7620), 339–346. http://www.nature.com/nature/index.html. doi:10.1038/nature19948.

Gavrilov, Y., Hagai, T., & Levy, Y. (2015). Nonspecific yet decisive: Ubiquitination can affect the native-state dynamics of the modified protein. *Protein Science, 24*(10), 1580–1592. doi:10.1002/pro.2688.

Grimaldo, M., Roosen-Runge, F., Zhang, F., Schreiber, F., & Seydel, T. (2019). Dynamics of proteins in solution. *Quarterly Reviews of Biophysics, 5*. doi:10.1017/s0033583519000027.

Hu, Y., Cheng, K., He, L., Zhang, X., Jiang, B., Jiang, L., Li, C., Wang, G., Yang, Y., & Liu, M. (2021). NMR-based methods for protein analysis. *Analytical Chemistry, 93*(4), 1866–1879. http://pubs.acs.org/journal/ancham. doi:10.1021/acs.analchem.0c03830.

Ishima, R., & Torchia, D. A. (2000). Protein dynamics from NMR. *Nature Structural Biology, 7*(9), 740–743. doi:10.1038/78963.

Ito, Y., & Selenko, P. (2010). Cellular structural biology. *Current Opinion in Structural Biology, 20*(5), 640–648. doi:10.1016/j.sbi.2010.07.006.

Johansen, N. T., Bonaccorsi, M., Bengtsen, T., Larsen, A. H., Tidemand, F. Grøæk, Pedersen, M. C., Huda, P., Berndtsson, J., Darwish, T., Yepuri, N. R., Martel, A., Pomorski, T. Gü, Bertarello, A., Sansom, M., Rapp, M., Crehuet, R., Schubeis, T., Lindorff-Larsen, K., Pintacuda, G., & Arleth, L. (2022). Mg^{2+}-dependent conformational equilibria in CorA and an integrated view on transport regulation. *eLife, 1*. doi:10.7554/elife.71887.

Katebi, A. R., & Jernigan, R. L. (2014). The critical role of the loops of triosephosphate isomerase for its oligomerization, dynamics, and functionality. *Protein Science, 23*(2), 213–228. http://onlinelibrary.wiley.com/journal/10.1002/(ISSN)1469-896X. doi:10.1002/pro.2407.

Kay, L. E. (1998). Protein dynamics from NMR. *Biochemistry and Cell Biology, 76*(2–3), 145–152. doi:10.1139/bcb-76-2-3-145.

Kühlbrandt, W. (2014). Cryo-EM enters a new era. *eLife*, e03678. doi:10.7554/eLife.03678.

Li, H., Robertson, A. D., & Jensen, J. H. (2005). Very fast empirical prediction and rationalization of protein p*K*a values. *Proteins: Structure, Function and Genetics, 61*(4), 704–721. doi:10.1002/prot.20660.

Manglik, A., & Kobilka, B. (2014). The role of protein dynamics in GPCR function: Insights from the β2AR and rhodopsin. *Current Opinion in Cell Biology, 27*(1), 136–143. doi:10.1016/j.ceb.2014.01.008.

Matsumoto, S., Ishida, S., Terayama, K., & Okuno, Y. (2023). Quantitative analysis of protein dynamics using a deep learning technique combined with experimental cryo-EM density data and MD simulations. *Biophysics and Physicobiology, 20*(2). n/a. doi:10.2142/biophysico.bppb-v20.0022.

Matthews, B. W. (1976). X-ray crystallographic studies of proteins. *Annual Review of Physical Chemistry, 27*(1), 493. doi:10.1146/annurev.pc.27.100176.002425.

Matthies, D., Dalmas, O., Borgnia, M. J., Dominik, P. K., Merk, A., Rao, P., Reddy, B. G., Islam, S., Bartesaghi, A., Perozo, E., & Subramaniam, S. (2016). Cryo-EM structures of the magnesium channel CorA reveal symmetry break upon gating. *Cell, 164*(4), 747–756. https://www.sciencedirect.com/journal/cell. doi:10.1016/j.cell.2015.12.055.

Meisburger, S. P., Thomas, W. C., Watkins, M. B., & Ando, N. (2017). X-ray scattering studies of protein structural dynamics. *Chemical Reviews, 117*(12), 7615–7672. http://pubs.acs.org/journal/chreay. doi:10.1021/acs.chemrev.6b00790.

Murata, K., & Wolf, M. (2018). Cryo-electron microscopy for structural analysis of dynamic biological macromolecules. *Biochimica et Biophysica Acta (BBA) - General Subjects, 1862*(2), 324–334. doi:10.1016/j.bbagen.2017.07.020.

Nesmelov, Y. E., & Thomas, D. D. (2010). Protein structural dynamics revealed by site-directed spin labeling and multifrequency EPR. *Biophysical Reviews, 2*(2), 91–99. http://www.springer.com/life+sciences/biochemistry+%26+biophysics/journal/12551. doi:10.1007/s12551-010-0032-5.

Rackovsky, S., & Scheraga, H. A. (2020). The structure of protein dynamic space. *Proceedings of the National Academy of Sciences of the United States of America, 117*(33), 19938–19942. http://www.pnas.org/lookup/doi/10.1073/pnas.2008873117. doi:10.1073/pnas.2008873117.

Rutledge, B. S., Choy, W. Y., & Duennwald, M. L. (2022). Folding or holding?—Hsp70 and Hsp90 chaperoning of misfolded proteins in neurodegenerative disease. *Journal of Biological Chemistry, 298*(5). https://www.jbc.org/article/S0021-9258(22)00345-3/fulltext. doi:10.1016/j.jbc.2022.101905.

Schneider, B., Gelly, J. C., De Brevern, A. G., & Černý, J. (2014). Local dynamics of proteins and DNA evaluated from crystallographic B factors. *Acta Crystallographica Section D: Biological Crystallography, 70*(9), 2413–2419. http://journals.iucr.org/d/journalhomepage.html. doi:10.1107/S1399004714014631.

Shi, Y. (2014). A glimpse of structural biology through X-ray crystallography. *Cell, 159*(5), 995–1014. doi:10.1016/j.cell.2014.10.051.

Shibayama, N. (2020). Allosteric transitions in hemoglobin revisited. *Biochimica et Biophysica Acta (BBA)—General Subjects, 1864*(2), 129335. doi:10.1016/j.bbagen.2019.03.021.

Sun, Z., Liu, Q., Qu, G., Feng, Y., & Reetz, M. T. (2019). Utility of B-factors in protein science: Interpreting rigidity, flexibility, and internal motion and engineering thermostability. *Chemical Reviews, 119*(3), 1626–1665. http://pubs.acs.org/journal/chreay. doi:10.1021/acs.chemrev.8b00290.

Toader, B., Sigworth, F. J., & Lederman, R. R. (2023). Methods for Cryo-EM single particle reconstruction of macromolecules having continuous heterogeneity. *Journal of Molecular Biology, 435*(9). https://www.journals.elsevier.com/journal-of-molecular-biology. doi:10.1016/j.jmb.2023.168020.

Wei, G., Xi, W., Nussinov, R., & Ma, B. (2016). Protein Ensembles: How does nature harness thermodynamic fluctuations for life? The diverse functional roles of conformational ensembles in the cell. *Chemical Reviews, 116*(11), 6516–6551. doi:10.1021/acs.chemrev.5b00562.

Zheng, J., Strutzenberg, T., Pascal, B. D., & Griffin, P. R. (2019). Protein dynamics and conformational changes explored by hydrogen/deuterium exchange mass spectrometry. *Current Opinion in Structural Biology, 5*, 305–313. http://www.elsevier.com/locate/sbi. doi:10.1016/j.sbi.2019.06.007.

CHAPTER

G protein–coupled receptors: An introductory laboratory class

19

Ioana-Daniela Dumitru[a], Alina-Maria Halaicu[a] and Ana-Nicoleta Bondar[a,b,*]
[a]*Faculty of Physics, University of Bucharest, Bucharest, Romania,* [b]*Faculty of Interdisciplinary Sciences, University of Bucharest, Bucharest, Romania*
**Corresponding author: a.bondar@fz-juelich.de*

19.1 Introduction

G protein–coupled receptors (GPCRs) are seven-helical transmembrane (TM) proteins that mediate signal transduction across cellular membranes. A GPCR binds an activating ligand, called an agonist, at the extracellular (ECL) side of the membrane, changes conformation and becomes activated, and binds and activates the G protein partner at the other, cytoplasmic side of the membrane. The human genome encodes about 800 distinct GPCRs (Sakmar, 2017; Yang et al., 2021). These proteins have essential physiological roles, have been implicated in numerous human diseases, and are the target of about one third of the drugs approved by the FDA (Yang et al., 2021). There are six different classes of GPCRs, labeled A to F; class A is also known as rhodopsin-like (Deupi, 2014; Yang et al., 2021).

In this chapter we use three class A GPCRs as model systems for an introductory laboratory class aimed at undergraduate and graduate students. We introduce briefly the GPCRs used and present possible tasks for a practical computer class. The tasks can be assigned such that students work on each of the three GPCRs or, alternatively, three teams of students work in parallel on one of the GPCRs and then discuss the results. To facilitate the assignment of tasks that involve working with protein sequences and structures, our brief introduction of the three GPCRs is largely focused on the architecture and functional motifs of the receptor. For reviews on GPCRs we refer the reader to Cheng et al. (2023), Deupi (2014), Deupi et al. (2012), Hilger (2021), Latorraca et al. (2017), Lorente et al. (2025), Sakmar (2017), Schertler (2005), Tan et al. (2004), Trzaskowski et al. (2012), Weis and Kobilka (2018), Yang et al. (2021), Zhang et al. (2024), and Zhou et al. (2021). The GPCRs used as model systems are as follows:

1. The *visual pigment rhodopsin* found in the rod cells of the outer retina (Athanasiou et al., 2018) is among the best studied GPCRs. The projection structure of rhodopsin, solved in 1993 at a resolution of 9 Å using cryo-electron microscopy (cryo-EM), revealed the membrane orientation of the seven TM helices (Schertler et al., 1993). The first high-resolution (2.8 Å) three-dimensional (3D) structure of bovine rhodopsin, solved with X-ray crystallography in 2000, showed protein interactions that could help explain the absorption maximum of the retinal chromophore (Palczewski et al., 2000). The 2.45 Å X-ray structure solved in 2004 identified ordered water molecules that hydrogen bond at key functional sites of the receptor (Li et al., 2004). The retinal chromophore is covalently

Protein Dynamics: Experimental and Computational Approaches. DOI: https://doi.org/10.1016/B978-0-443-30042-4.00007-9

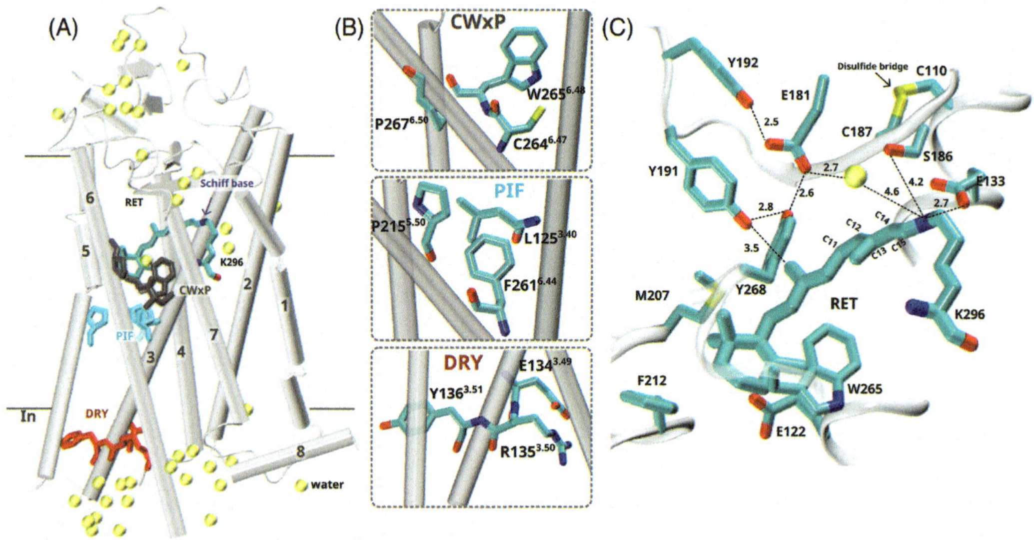

FIGURE 19.1

Architecture and functional motifs of bovine rhodopsin. The image was prepared using visual molecular dynamics, VMD (Humphrey et al., 1986) version 1.9.3, and the protein structure PDB ID: 7ZBC solved at a resolution of 1.80 Å (Gruhl et al., 2023). (A) The receptor's seven transmembrane helices are typically labeled 1–7. The retinal (RET) binding site is at the extracellular side of the receptor. We show with small *yellow spheres* the oxygen atoms of the water molecules that are within 12 Å of the interhelical region of the receptor. The retinal and the covalently bound K296 are shown in atom colors, with carbon atoms, *cyan*; nitrogen, *blue*; and oxygen, *red*. Residues of the CWxP motif are colored *black*, PIF motif, *light cyan*; NPxxY motif, *yellow*; and DRY, *dark red*. (B and C) Close view of the functional motifs of class A GPCRs illustrated for bovine rhodopsin (panel B) and of the retinal binding pocket (panel C).

bound via a protonated Schiff base to a conserved Lys residue on the seventh helix of the receptor (Fig. 19.1A). The retinal, initially in an isomeric state that keeps the receptor inactive, changes geometry upon absorption of a photon. This triggers a change in the receptor conformation, which becomes activated, and it binds and activates its cytosolic heterotrimeric G protein partner, initiating a signaling cascade in the cell. A number of rhodopsin mutations are associated with autosomal dominant retinitis pigmentosa, a degenerative condition that leads to blindness (Athanasiou et al., 2018; Deupi, 2014).

2. The chemokine receptors are GPCRs that bind chemokines, small proteins whose name is derived from "chemotactic cytokines," and are essential for the movement of immune cells (Griffith et al., 2014), with implications in human diseases such as inflammation, asthma, and multiple sclerosis (Proudfoot, 2002). There are about 20 signaling chemokine receptors and 50 chemokines (Griffith et al., 2014; Proudfoot, 2002). CXCR1 and CXCR2 bind CXC chemokines with an ELR (Glu–Leu–Arg) motif next to their CXC sequence. Human CXCL8 [interleukin-8 (IL-8)] and CXCL6 can bind to CXCR1, and CXCL1 to CXCL7 bind to CXCR2 (Pelus & Fukuda, 2006; Tsai et al., 2002). CCR5 is a receptor for CC chemokines and a coreceptor for HIV-1 entry into host cells (Arimont et al., 2017). A timeline of experimental structural biology studies of chemokine receptors

up to 2023 and the corresponding Protein Data Bank (PDB; Burley et al., 2024) entries of the structures are presented in Kayastha et al. (2024). Here, we will use the structure of CXCR1 bound to the G-protein (Fig. 19.2).

3. The *proton-sensing GPCRs* (GPR4, GPR65, GPR68, and GPR132) are class A receptors that sense the ECL acidic pH. The pH-sensing mechanisms of these receptors have been proposed to involve histidine and/or carboxylic sidechains of the receptor (Kapur et al., 2023; Ludwig et al., 2003; Rowe et al., 2021). pH-sensing GPCRs have been implicated in, for example, inflammation (Okajima, 2013; Wang et al., 2018), ischemia (Russell et al., 2012), and cancer (Nayak & Penn, 2020). GPR68, also known as the ovarian cancer G protein–coupled receptor 1, OGR1 (Show et al., 2025), has been implicated in prostate cancer and medulloblastoma (Wiley et al., 2019). GPR68 is inactive at pH = 7.8 and activated at acidic pH = 6.8 (Ludwig et al., 2003). 3D structures of GPR4, GPR65, GPR68, and GPR132 bound to mini-Gα constructs were solved very recently with cryo-EM at resolutions of 2.8–3.0 Å (Howard et al., 2024). In this chapter we will discuss the architecture and key residues of GPR68 (Fig. 19.3).

In what follows we will illustrate how tools for molecular visualization, molecular modeling, and bioinformatics analyses could be used for an introductory computer class on GPCRs to teach students how to

1. perform a sequence alignment and evaluate the conservation of key residues of interest;
2. use molecular visualization software to inspect the 3D structure of a membrane receptor;
3. perform a homology modeling computation for a sequence of interest and evaluate basic structural features of the model;
4. work with molecular graphics software to generate images of a biomolecule and understand structure–function relationships.

The laboratory class assumes that students have prior basic knowledge of protein structure and folding and of the physical–chemical properties of the amino acid residues from which proteins are constituted. All software and databases used for the class are available free of charge to academic users. The tools and protocols we present are generally applicable to other membrane proteins.

This chapter is organized as follows: First, we discuss briefly the structure and key functional residues of each of the three model GPCRs. To prepare the student for the analysis of the sequence alignments and of the homology models, we include in the discussion of receptor sequences several examples of tasks that could be performed by the students. Second, we present examples of computations that can be performed in a computer class—each computation presented could be a computer laboratory class. Each section concludes with suggestions of tasks that could be assigned to students taking the laboratory class. The chapter ends with a conclusion section where we discuss potential usage of these types of computations for other membrane proteins of interest for a computer class.

19.2 Architecture and essential interactions of class A G protein–coupled receptors

To introduce the student to key features of a GPCR, we discuss briefly the 3D structure of three class A GPCRs, with a focus on the conserved functional motifs of these receptors—conserved residues at

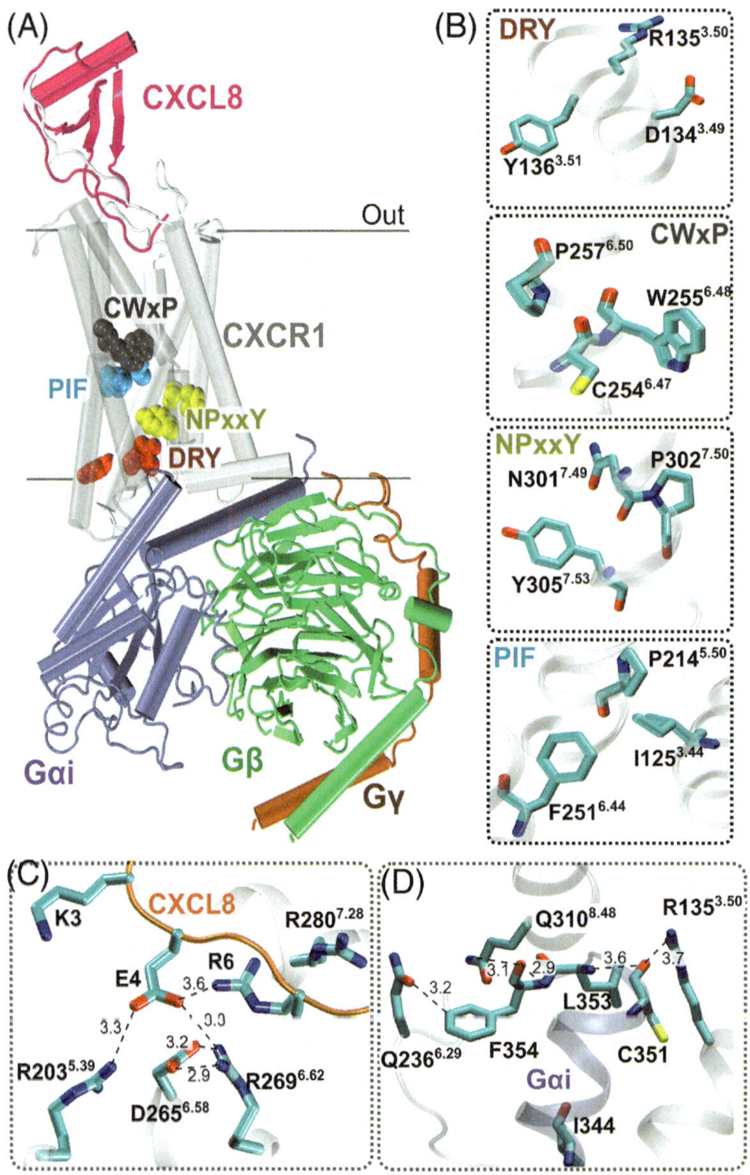

FIGURE 19.2

Architecture of CXCR1 bound to CXCL8 and G protein. (A) Structure of the complex from the cryo-EM structure PDB ID: 8IC0 (Ishimoto et al., 2023). (B) Functional motifs of CXCR1: CWxP, C254$^{6.47}$, W255$^{6.48}$, and P257$^{6.50}$; DRY, D134$^{3.49}$ and R135$^{3.50}$, Y136$^{3.51}$; PIF, P214$^{5.50}$, I125$^{3.40}$, and F251$^{6.44}$; NPxxY, N301$^{7.49}$, P302$^{7.50}$, and Y305$^{7.53}$. (C and D) Illustration of hydrogen-bonding and hydrophobic interactions between CXCR1 and CXCL8 (panel C) and between CXCR1 and Gαi (panel D). In panel D, Gαi's I344, C351, L353, and F354 are within 3.5 Å distance of CXCR1 residues R135 of DRY motif, Q236, and Q310, respectively.

sites where important interactions and structural rearrangements occur during the reaction cycle of the receptor. Thus the tasks to inspect the sequences, 3D structures, and key sites of the receptors could be assigned to three groups of students working in parallel, with a common discussion at the end of the class.

It is common to refer to conserved residues of class A GPCRs using the Ballesteros Weinstein (BW) scheme, in which residues that belong to TM α-helical segments are assigned two numbers separated by a dot: the helix number (from 1 to 7), followed by a number that indicates the position of that residue relative to the most conserved residue of the helix, which is assigned the number 50 (Ballesteros & Weinstein, 1995). For example, $R135^{3.50}$ is the Arg residue with number 135 in the amino acid residue sequence, and the upper script indicates its BW number = 3.50; that is, it is an Arg residue on helix 3, and it is the most conserved residue of TM helix 3. In what follows we will refer to protein residues using residue sequence numbers and, for specific functionally important residues, we will also use the BW numbers. Although using BW numbers could be somewhat advanced for an introductory class, the BW numbers have the advantage of simpler comparisons between residue interactions in the three GPCRs used. Tools for generic residue numbers are available (Pándy-Szekeres et al., 2018; Munk et al., 2016).

19.2.1 Bovine rhodopsin

Bovine rhodopsin has 348 residues. The retinal chromophore is covalently bound to a conserved Lys residue on the seventh TM helix, K296, via a protonated Schiff base (Fig. 19.1). Carbon atoms of the retinal are commonly labeled as C1, C2, and so on, with C15 the carbon atom covalently bounded to the Schiff base nitrogen atom, which is typically labeled N16. In the recent crystal structure of the dark state of the receptor from Gruhl et al. (2023) the retinal is 11-*cis*, and the Schiff base is within 2.8 Å from the negatively charged E113 (Fig. 19.1). Several other charged and polar residues near the retinal and water molecules participate in the protein–water hydrogen-bond network at the ligand binding site (Fig. 19.1).

Fig. 19.1 shows the residues of the functional motifs of bovine rhodopsin based on the dark-state structure PDB ID: 7ZBC (Gruhl et al., 2023). The conserved *E/DRY motif* consists of $E134^{3.49}$, $R135^{3.50}$, and $Y136^{3.51}$ (Fig. 19.1B). In the inactive receptor a salt bridge between E3.49 and R3.50, known as the "ionic lock," helps stabilize the conformation of the receptor (Trzaskowski et al., 2012); the ionic lock breakes when the ligand binds and the receptor becomes activated (Hauser et al., 2021; Trzaskowski et al., 2012). The CWxP motif consists of $C264^{6.47}$, $W265^{6.48}$, and $P267^{6.50}$; W265 changes its rotameric state during receptor activation (Standfuss et al., 2011; Trzaskowski et al., 2012). The PIF motif ($P215^{5.50}$, $L125^{3.40}$, and $F261^{6.44}$) participates in conformational dynamics of TM helix 6 (Hilger, 2021; Pope et al., 2020). The ligand-binding pocket contains several charged and polar residues, some of which have hydrogen bonds with backbone or sidechain groups (Fig. 19.1). For example, the carboxyl group of the negatively charged E181 is within hydrogen-bond distance of the hydroxyl oxygen atoms of Y192 and Y268, while Y191 is within hydrogen-bond distance of Y268. Adjacent in the sequence to S186, C187 is disulfide-bridged to C110.

19.2.2 The CXC chemokine receptor 1

The CXC chemokine receptor 1 (CXCR1) is expressed by neutrophils, macrophages, and endothelial cells (Mishra et al., 2021), and its function has been implicated in, for example, the progression of

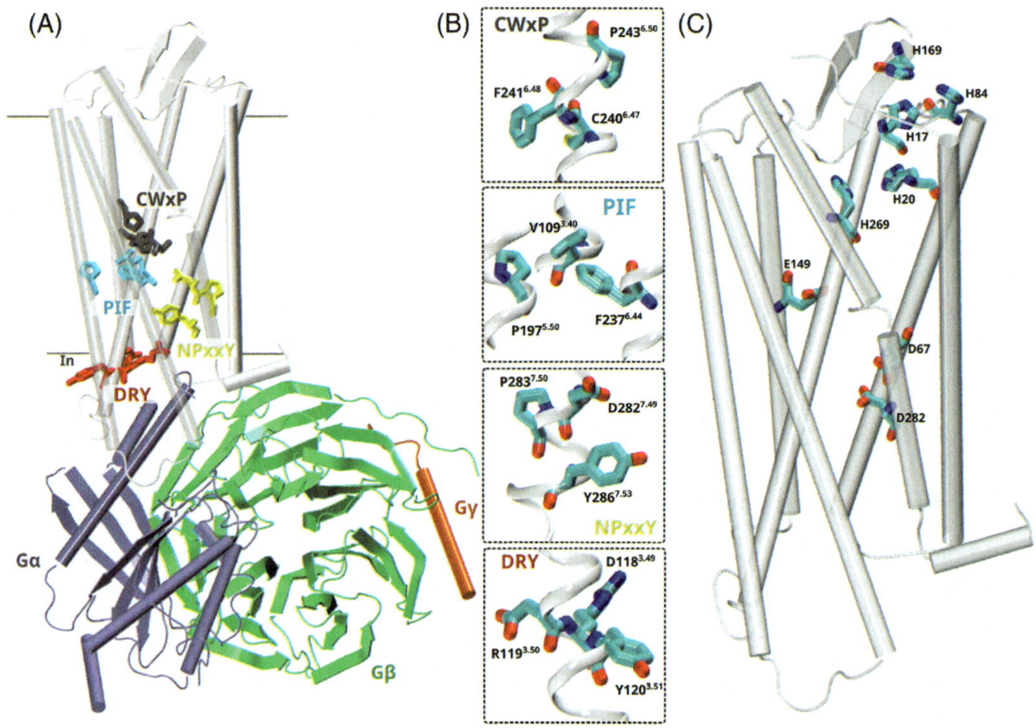

FIGURE 19.3

Structure of GPR68 bound to miniGs/q protein complex solved at pH = 7.5. The figure is based on the cryo-EM structure PDB ID: 9BI6 (Howard et al., 2024). (A) The GPR68-G protein complex and the functional motifs of GPR68. (B) Functional motifs of GPR68: CWxP (C240$^{6.47}$, F241$^{6.48}$, and P243$^{6.50}$), PIF (P197$^{5.50}$, V109$^{3.40}$, and F237$^{6.44}$), NPxxY (D282$^{7.49}$, P283$^{7.50}$, and Y286$^{7.53}$) and DRY (D118$^{3.49}$, R119$^{3.50}$, and Y120$^{3.51}$). (C) The histidine and carboxylic side chains are proposed to play a role in pH-sensing. *PDB*, Protein Data Bank.

breast cancer (Mishra et al., 2021). The cryo-EM structure of the CXCR1 bound to the G protein and IL-8 ligand (also known as C–X–C motif ligand 8, CXCL8; Mishra et al., 2021), solved at a resolution of 3.4 Å, shows that the C-terminal helix of Gαi inserts into the TM region of the receptor (Ishimoto et al., 2023).

Interactions between the receptor and the G protein, and between the receptor and the ligand, are illustrated in Fig. 19.2. E4 and R6 of the ligand are within 3.6 Å of each other, which is close to the 3.5 Å distance typically used as a cutoff for hydrogen bond distances. E4 is also within 3.5 Å of CXCR1-D269$^{6.62}$ and R203$^{5.39}$, and D265 is within 3.5 Å of R269, such that residues of the ligand and of the receptor are part of a local hydrogen-bond network (Fig. 19.2A). The Cα atom of D265 is within 11 Å of the CWxP motif and within 33 Å of DRY. The interface between the receptor and the G protein is stabilized by hydrogen bonding and hydrophobic contacts (Fig. 19.2D).

19.2.3 The proton-sensing GPR68

The proton-sensing GPR68 and the other proton-sensing GPCRs (GPR4, GPR65, and GPR132) detect changes in the ECL pH. GPR68 is inactive at pH = 7.8, starts to activate at pH = 7.4, and is fully activated at pH = 6.8 (Ludwig et al., 2003). The mechanism by which GPR68 binds protons from the ECL side and becomes activated is largely unclear. The two main scenarios that had been proposed consider distinct structural elements as the pH sensing unit, which is either (1) a cluster of five histidine residues located at the ECL side of the receptor (H17, H20, H84, H169, and H269 in Fig. 19.3C) or (2) a triad of buried acidic residues (Rowe et al., 2021) (see D67, E149, and D282 in Fig. 19.3C). More recently, atomic-level simulations and graph-based analyses have demonstrated that the ECL histidine residues and the internal carboxylic triad are in fact part of a common protein–water hydrogen-bond network with exquisite sensitivity to protonation change (Kapur et al., 2023).

While we were working on this chapter the experimental cryo-EM structures of pH-sensing GPCRs, including GPR68, became available (Howard et al., 2024). The structure of GPR68 shows that the ECL histidine cluster is accessible from the bulk (Fig. 19.3C). At the center of the receptor the carboxyl oxygen atoms of D67 and D282 of the carboxylic triad are within 3.3 Å distance from each other, and D67 is also within 3.5 Å distance from the hydroxyl group of S108 (Fig. 19.3C). Interhelical hydrogen bonds between Asp/Glu carboxyl groups and Ser/Thr hydroxyl groups are found at proton-binding sites of membrane transporters (del Val et al., 2014). The hydrogen-bond distances between D67 and D282, and between D67 and S108, suggest that D67 or D282 could be protonated in the structure. The third carboxylic group of the internal triad, E149, is within hydrogen-bond distance (2.7 Å) of the backbone amide group of N104, which is one helical turn before S108 and adjacent in the sequence to E103—whose carboxyl group hydrogen bonds to R189. That is, each of the four internal carboxylic side chains is hydrogen-bonded in the structure, and the short distance between two of the carboxylic groups is compatible with one of the residues being neutral (protonated) in the experimental structure.

19.2.4 Tasks for a laboratory class on three-dimensional structures of G protein–coupled receptors

Interactions between residues with different locations in the sequence and distances between functionally important residues. The student may note from the discussion above that the key residues of the functional motifs are from different parts of the protein polypeptide chain. Fig. 19.1 shows that the functional motifs are in the interhelical region of bovine rhodopsin. The distance between the Cα atoms of W265 (CWxP motif) and R135 (DRY motif) is 22.6 Å, and that between the retinal Schiff-based nitrogen atom and R135-Cα is 31.1 Å. In Fig. 19.2A the distance between the ligand and the G protein is about 7.1 Å. In Fig. 19.3C the ECL histidine cluster is within ~34.8 Å from the GPR68-G protein interface. For comparison the length of a C–C bond is 1.5 Å, the average hydrogen-bond distance between two oxygen atoms in crystal structures of biomolecules is ~2.8 Å (Mitra & Ramakrishnan, 1977), and the average thickness of an eukaryotic plasma membrane is 40 Å (Mitra et al., 2004). Fig. 19.3C illustrates internal hydrogen bonding that could be interpreted to suggest protonation of an internal carboxylic sidechain (D67 or D282).

Hydrogen bonds and clusters of hydrogen bonds. A possible task during the laboratory class is for students to use a molecular visualization software such as visual molecular dynamics (VMD) to inspect the 3D structure of the protein, locate key functional sites, and measure distances between these sites.

These distances can be compared with the length of a C–C covalent bond (see the previous task). In the case of the bovine rhodopsin binding pocket, students additionally measure the distance between the sulfur atoms of the disulfide bridge.

Water molecules interacting with protein residues that are known to be important for function. The VMD software can be used to display water molecules within hydrogen-bond distance of the functionally important groups or all water molecules inside the receptor; identify the types of atoms that are within hydrogen-bond distance of these water molecules; and visualize the network of protein–water hydrogen bonds in the interhelical region of the receptor (see example in Fig. 19.1).

19.3 Structural modeling of G protein–coupled receptors

In section two above we discussed the architecture of three distinct GPCRs based on 3D protein structures that were solved experimentally. As of June 2023, more than 1100 structures for 180 unique GPCRs had been solved experimentally (Chen et al., 2024). For GPCRs whose 3D structures have not yet been solved, computational approaches can be used to derive structural models. AlphaFold-generated models of proteins can be downloaded from the Universal Protein Resource (UniProt) database. For 422 human nonolfactory GPCRs, 844 state-specific structural models of the receptors in active or inactive conformations have been derived using AlphaFold2 Multistate (Pándy-Szekeres et al., 2023). To illustrate the usefulness of AlphaFold GPCR structural modeling for a laboratory class, here we use CCR4 and GPR68 as test systems. For GPR68, whose structure was published while this chapter was being prepared, we also compare the AlphaFold structural model with the experimental structure.

This section is organized as follows. We first briefly discuss the public repositories for protein sequences and protein structures, how schematic diagrams of a GPCR sequence can be drawn and inspected, and sequence alignments of GPCRs. In the second half of this section, we discuss homology modeling using multiple templates and AlphaFold structural modeling of CCR4 and GPR68. Each of these subsections could constitute a task in the practical computer class.

19.3.1 UniProt searches for basic information about the protein of interest

UniProt is a web-based database that contains a wealth of information about proteins, including protein function, expression, homologues, whether 3D structures are available for the protein (Boutet et al., 2007), and, more recently, links to AlphaFold structural predictions that can be downloaded and used in the class. UniProt searches of the three GPCRs we use as model systems can be used in the laboratory class to find out, for example, where each of the receptors is normally expressed, proteins that share high residue similarity with the receptor of interest, and mutations that could be linked to disease conditions.

Bovine rhodopsin (UniProt ID: P02699) is expressed in rod-shaped photoreceptor cells in the retina. Rhodopsins are conserved throughout distinct species: bovine, human, mouse, and dog rhodopsins share >90% sequence identity, for example. Mutations, such as M207R in bovine and human rhodopsins, are linked with retinitis pigmentosa, a degenerative disease that causes loss of vision over time. M207 is thought to influence the hydrogen bond network in the vicinity of the retinal chromophore (Hernández-Rodríguez et al., 2012).

For CXCR1 (UniProt ID: P25024) there are in UniProt no fewer than 17 entries of proteins that share identity with CXCR1; for example, apes such as chimpanzees (UniProt ID: P55920) and gorillas CXCR1 (UniProt ID: P55919) share at least 90% sequence identity with the human receptor, with more distant species such as rodents showing lower identity (~75%–85%). Several naturally occurring CXCR1 variants, such as M31R, S276T, and R335C, have been linked to altered receptor expression or signaling and are associated with inflammatory and infectious diseases such as asthma or HIV-1 progression.

GPR68 (UniProt ID: Q15743) is expressed in the nervous system and other cell types. According to UniProt, there are >60 entries for proteins that share at least 90% identity with human GPR68, for example, mouse GPR68, monkey GPR68, and horse GPR68. GPR68 mutations are linked to several diseases, such as cancer and teeth problems. Mutations at the carboxylic triad, D282N or E149Q, disrupt the pH sensing of the receptor, which could result in cancer progression (Rowe et al., 2021).

19.3.2 Finding protein structures of interest in the Protein Data Bank

Protein structures solved with structural biology are deposited in the public repository PDB, available online at www.rcsb.org. Here, one can search for 3D structures of proteins of interest by keywords or PDB identifier, or ID, which is an alphanumeric code of four characters. Alternatively, one can type in the search field of the main PDB window the name of the protein of interest.

The example illustrated in Fig. 19.4 shows screen snapshots of the main PDB window for PDB ID: 5UIW (Zheng et al., 2017), which is the structure of the CC chemokine receptor 5, CCR5, in complex with an HIV entry inhibitor. On November 11, 2024 a query with the keyword "CCR5" returned 131 structures, some of which are of CCR5, some, of related proteins; options for the order in which these results are to be displayed include the resolution of the structure, the year when it was published, and entry IDs. The PDB information (Fig. 19.4B) includes (1) the classification of the protein, here, a signaling protein; (2) the organism from which the protein was taken; (3) the experimental method used to solve the structure; (4) the resolution at which the structure was solved; (5) general information about the protein, including its mass and the number of atoms. The "3D report" tab opens a separate, interactive window that displays the sequence of the protein(s), a view of the structure, and a number of tabs one can use to display the structure and derive information from it (Fig. 19.4C). One type of analysis that could be included as a task in the laboratory class is to select, in the interactive interface, residues of interest, for example, residues of functional motifs introduced in Section 19.2.

19.3.3 Using GPCRdb to draw and inspect the topology of a G protein–coupled receptor

GPCRdb is an open-access database that provides a wealth of information, including the availability of experimental structures, mutations, ligand-binding sites, specific ligand–receptor interactions (Munk et al., 2016; Pándy-Szekeres et al., 2018). We used GPCRdb to draw snake plots and helix-box diagrams of human CXCR1 and of GPR65 and to identify the location in the sequence of residues whose mutation impairs receptor function.

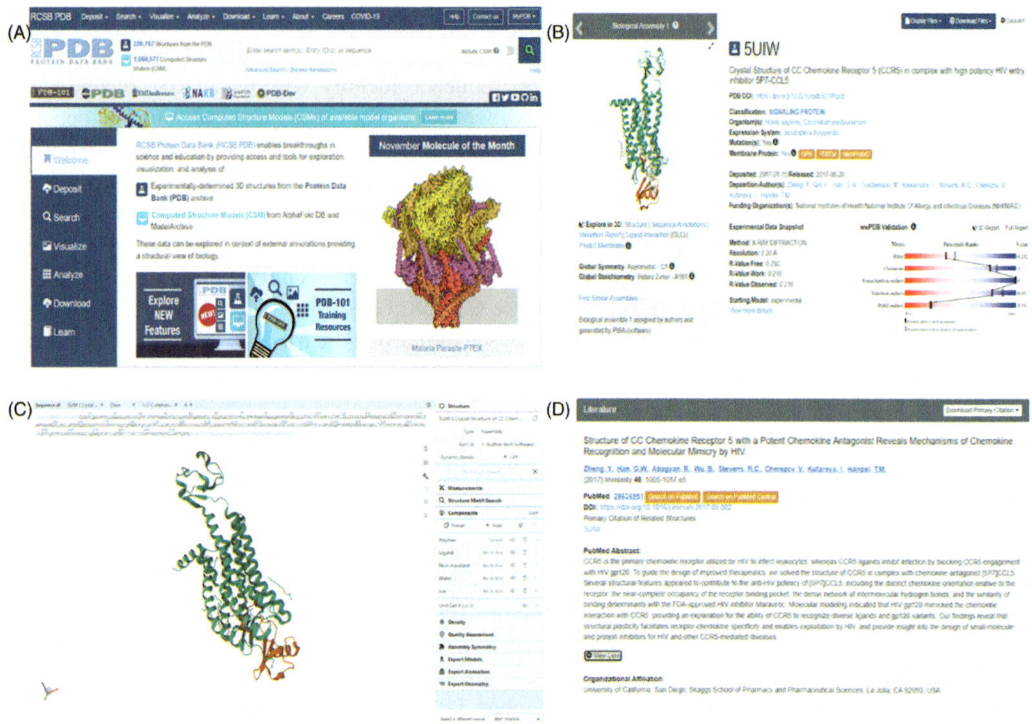

FIGURE 19.4

Using the Protein Data Bank (PDB), rcsb.org, to search for protein structures of interest. (A) Screenshot of the main window of rcsb.org. In the search tab, one can enter the name of the protein, the identification code (PDB ID) of the protein structure, or other queries. (B) The protein of interest is found in rcsb.org. For structures solved with X-ray crystallography or cryo-EM the resolution at which the structure was solved is given in Å. (C) The "Biological Assembly" tab gives a molecular representation of the protein of interest. (D) The "Literature" tab provides information about the publication that reported the structure of interest.

The *snake plot* shown in Fig. 19.5A represents (1) the seven TM helices labeled TM1–TM7, the ECL and intracellular loops, and the posttranslational modifications (methylation) at R297 and R280; and (2) residues associated with natural variations. The *helical wheel diagrams* represent cross-sections of the TM helices with residues color-coded based on information about mutations. In Fig. 19.5B, variations are color coded green (tolerated mutations) or red and violet (deleterious mutations). Fig. 19.5C shows in vitro mutations, with red indicating loss of function and yellow, neutral or unknown effects. For example, the S276T mutation (on the seventh helix) is a natural but tolerated variation, whereas Y46C and N57D (on the first helix) are predicted as deleterious. These residues are shown in Fig. 19.5G.

In Fig. 19.5D and E we show the shake plots and ECL view of the helical section of GPR65. I231L (Fig. 19.5H) is a mutation associated with inflammatory bowel disease (Lassen et al., 2016) and is predicted as deleterious together with the E295K mutation (Fig. 19.5H).

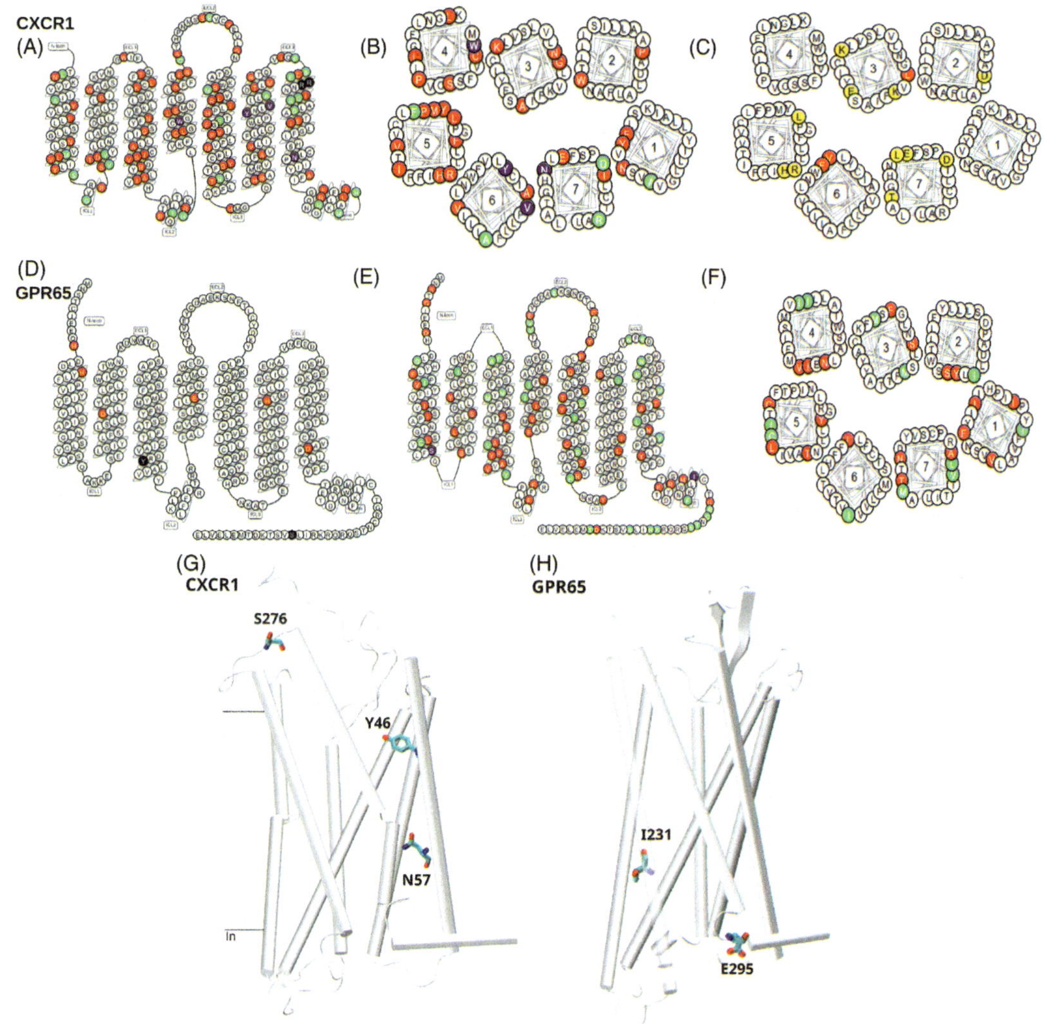

FIGURE 19.5

Using GPCRdb to draw schematic representations of the topology and key sites of a GPCR. (A) Snake plot of CXCR1 with marked residues showing natural variations. (B and C) Schematic representation of the transmembrane helices viewed from the extracellular side, with residues colored according to the toleration predicted effect: *green*, tolerated mutation; *red* and *violet*, deleterious mutation (panel B); *red*, loss of function, *yellow*, neutral or unknown effect (panel C). (D) Snake plots of GPR65 with key functional residues and posttranslational modification sites marked *red* and *black*, respectively. (E) Snake plot of GPR65 with residues where natural variation can occur colored *green*, tolerated mutations, *red*, and predicted deleterious mutation effects, *purple*. (F) Schematic representation showing sites where natural variation can occur. (G and H) Molecular graphics of CXCR1 and GPR65 showing selected mutation sites.

19.3.4 Protein sequence alignments

The EMBL–EBI online tutorial defines sequence alignment as "a way of arranging the primary sequences of a protein to identify regions of similarity that may be a consequence of functional, structural, or evolutionary relationships between the sequences" (https://www.ebi.ac.uk/training/). A sequence similarity analysis reports statistically relevant similarity between protein sequences (Pearson, 2013); the more stringent sequence identity reports the percentage of residues that are the same in two aligned sequences. These scores are of direct interest to homology modeling, as proteins with high sequence identity tend to have similar structures (for an in-depth discussion, see Koehl & Levitt, 2002).

We used ClustalOmega (Sievers et al., 2011) to align and inspect the sequences for human chemokine receptors and, separately, for GPR68 and the angiotensin II type 1 receptor, AT1R. We include AT1R in the sequence alignment, as this receptor is part of the default templates selected by AlphaFold to model GPR68. The color-coded sequence alignments presented in Figs. 19.6 and 19.7 illustrate the overall conservation of the residues, including of the residues that are part of the conserved functional motifs.

The sequence alignments served to verify the conservation of some of the residues discussed above for their involvement in potentially important interactions and mutations. We verified (and suggest as tasks for the class) that the conserved functional motifs are conserved (with Tyr–Phe mutations in the DRY motifs of the chemokine receptors and pH-sensing GPCRs and Cys to Phe in the CWxP motif of chemokine receptors). The hydrogen bond residues discussed in Section 19.2.3—residues N104, S108, and R189—are conserved.

To quantify the sequence similarity and sequence identity, in Table 19.1 we collect these scores for pairs of GPCR sequences from the datasets we used. This summary illustrates the high sequence identity (>30%) between the CXCRs of our data set, as high as 76% between CXCR1 and CXCR2. The sequence identity between GPR68 and bovine rhodopsin, CXCR1, and AT1 is comparatively lower, although still reasonably high—and with accompanying sequence similarity values above 40% Structures of CXCR1, AT1R, and bovine rhodopsins could be used in the dataset of templates for structural modeling of GPR68 during a laboratory class for students.

19.3.5 G protein–coupled receptor homology modeling with Swiss-Model and Phyre2

Swiss-Model (Benkert et al., 2011) and Phyre2 (Protein Homology/Analogy Recognition Engine; Kelley et al. (2015) are online tools for homology modeling that can be easily incorporated into an introductory laboratory class. Swiss-Model ranks homology models according to the Global Model Quality Estimation (GMQE) score that can take values between 0 and 1 (Waterhouse et al., 2018), with "0" indicating the lowest score and "1," the highest score. Phyre2 uses the sequence identity score to show the accuracy between the protein and the template—the higher the identity score, the better the model.

Here we illustrate the usefulness of Swiss-Model and Phyre2 to generate homology models of the CCR4 receptor (Fig. 19.8A–C) and of GPR68 (Fig. 19.8D–I). The GMQE scores of the Swiss-Model CCR4 models are about 0.6 when the full-length sequence is used (Fig. 19.8A). When we recomputed the homology models without the N- and C-termini, we obtained better scores of around 0.8 (Fig. 19.8B). The model predicted by Phyre2 shows 52% sequence identity between the target CCR4 sequence and CCR1 chemokine receptor (PDB ID 7VL9), with 100% confidence score for the TM

```
sp|P25024|CXCR1_HUMAN      ----------MSNITDPQMWDFDDLN----FTGMPPADEDYSPCMLET-ETLNKYVVIIA 45
sp|P25025|CXCR2_HUMAN      -----MEDFNMESDSFEDFWKGEDLSNYSYSSTLPPFLLDAAPCEPES-LEINKYFVVII 54
sp|P51681|CCR5_HUMAN       ------MDYQVSSPIYD---------I---------NYYTSEPCQKINVKQIAARLLPPL 36
sp|P51679|CCR4_HUMAN       MNPTDIADTTLDESIYS---------NY------YLYESIPKPCTKEGIKAFGELFLPPL 45
                                     :..                             **       :   .:

sp|P25024|CXCR1_HUMAN      YALVFLLSLLGNSLVMLVILYSRVGRSVTDVYLLNLALADLLFALTLPIWAASKVNGWIF 105
sp|P25025|CXCR2_HUMAN      YALVFLLSLLGNSLVMLVILYSRVGRSVTDVYLLNLALADLLFALTLPIWAASKVNGWIF 114
sp|P51681|CCR5_HUMAN       YSLVFIFGFVGNMLVILILINCKRLKSMTDIYLLNLAISDLFFLLTVPFWAHYAAAQWDF 96
sp|P51679|CCR4_HUMAN       YSLVFVFGLLGNSVVVLVLFKYKRLRSMTDVYLLNLAISDLLFVFSLPFWGYYAADQWVF 105
                           *:***::.::** :*:*:::  :  :*:**:******::**:* :::*:*.   .  * *

sp|P25024|CXCR1_HUMAN      GTFLCKVVSLLKEVNFYSGILLLACISVDRYLAIVHATRTLTQKRHLV-KFVCLGCWGLS 164
sp|P25025|CXCR2_HUMAN      GTFLCKVVSLLKEVNFYSGILLLACISVDRYLAIVHATRTLTQKRYLV-KFICLSIWGLS 173
sp|P51681|CCR5_HUMAN       GNTMCQLLTGLYFIGFFSGIFFIILLTIDRYLAVVHAVFALKARTVTFGVVTSVITWVVA 156
sp|P51679|CCR4_HUMAN       GLGLCKMISWMYLVGFYSGIFFVMLMSIDRYLAIVHAVFSLRARTLTYGVITSLATWSVA 165
                           *   :*:::: :   :.*:***:::   :::*****:***. :*  :       . .:  * ::

sp|P25024|CXCR1_HUMAN      MNLSLPFFLFRQAYHPNNSSPVCYEVLGNDTAKWRMVLRILPHTFGFIVPLFVMLFCYGF 224
sp|P25025|CXCR2_HUMAN      LLLALPVLLFRRTVYSSNVSPACYEDMGNNTANWRMLLRILPQSFGFIVPLLIMLFCYGF 233
sp|P51681|CCR5_HUMAN       VFASLPGIIFTRSQKEGLHYTCSSHFPYSQYQFWKNFQTLKIVILGLVLPLLVMVICYSG 216
sp|P51679|CCR4_HUMAN       VFASLPGFLFSTCYTERNHTYCKTKYSLN-STTWKVLSSLEINILGLVIPLGIMLFCYSM 224
                           :  :** ::*              .    .    *: .  :     :*:::** :*::**.

sp|P25024|CXCR1_HUMAN      TLRTLFKAHMGQK-HRAMRVIFAVVLIFLLCWLPYNLVLLADTLMRTQVIQESCERRNNI 283
sp|P25025|CXCR2_HUMAN      TLRTLFKAHMGQK-HRAMRVIFAVVLIFLLCWLPYNLVLLADTLMRTQVIQETCERRNHI 292
sp|P51681|CCR5_HUMAN       ILKTLLRCRNEKKRHRAVRLIFTIMIVYFLFWAPYNIVLLLNTFQEFFGLN-NCSSSNRL 275
sp|P51679|CCR4_HUMAN       IIRTLQHCKNEKK-NKAVKMIFAVVVLFLGFWTPYNIVLFLETLVELEVLQ-DCTFERYL 282
                            ::** :.:  :* ::*:::**:::::::  * ***:**: :*: .   ::  *   . :

sp|P25024|CXCR1_HUMAN      GRALDATEILGFLHSCLNPIIYAFIGQNFRHGFLKILAMHGLVS---------KEFLARH 334
sp|P25025|CXCR2_HUMAN      DRALDATEILGILHSCLNPLIYAFIGQKFRHGLLKILAIHGLIS---------KDSLPKD 343
sp|P51681|CCR5_HUMAN       DQAMQVTETLGMTHCCINPIIYAFVGEKFRNYLLVFFQKHIA-KRFCKCCSIFQQEAPER 334
sp|P51679|CCR4_HUMAN       DYAIQATETLAFVHCCLNPIIYFFLGEKFRKYILQLFKTCRGLFVLCQYCGLLQIYSADT 342
                           . *::.** *.: *.*:**:** *:*::**: :* ::                    :

sp|P25024|CXCR1_HUMAN      RVTSYTSSSVNVSSNL--   350
sp|P25025|CXCR2_HUMAN      SRPSFVGSSSGHTSTTL-   360
sp|P51681|CCR5_HUMAN       ASSVYTRSTGEQEISVGL   352
sp|P51679|CCR4_HUMAN       PSSSYTQSTMDHDLHDAL   360
                               :. *:
```

FIGURE 19.6

Sequence alignment of the human chemokine receptors CXCR1, CXCR2, CCR5, and CCR4. The sequence alignment was generated using ClustalOmega and sequences with UniProt IDs P25024 (CXCR1), P25025 (CXCR2), P51681 (CCR5), and P51679 (CCR4). The following color codes are used: hydrophobic residues, *red;* hydrophilic residues, *green;* acidic residues (Asp and Glu), *blue;* and basic residues (Lys, Arg, and His), *magenta*. The "*" symbol indicates residues present in all four sequences; ":", conservative substitutions; ".", semiconserved substitutions; and "-", insertions and deletions. The functional residues were marked as follows: CWxP motif, colored *dark gray;* PIF motif, *light cyan;* NPxxY motif, *yellow;* DRY, *dark red;* and key residues, *black*.

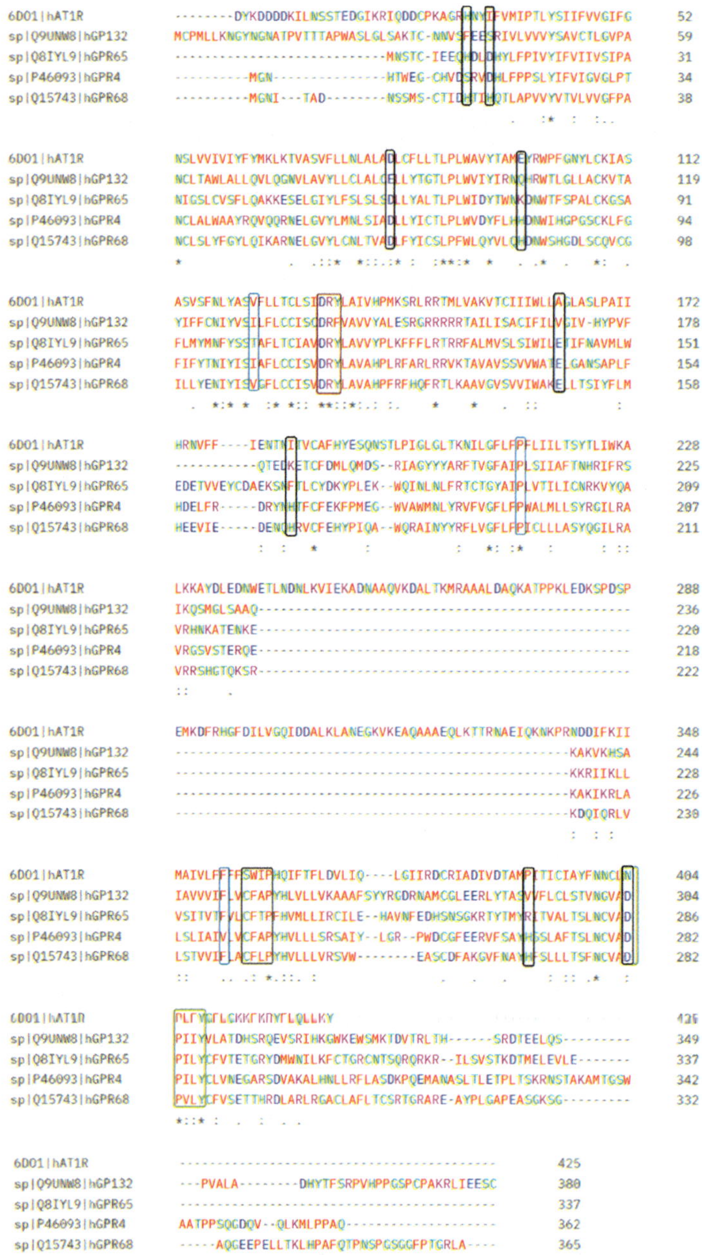

FIGURE 19.7

Sequence alignment for four proton-sensing GPCRs and angiotensin II type 1 receptor, AT1R. We aligned the sequences of human GP132 (UniProt ID: Q9UNW8), GPR65 (Q8IYL9), GPR4 (P46093), GPR68 (Q15743), and AT1R (PDB ID: 6DO1). The color coding and symbols are the same as in Fig. 19.6. *GPCRs*, G protein–coupled receptors; *PDB*, Protein Data Bank.

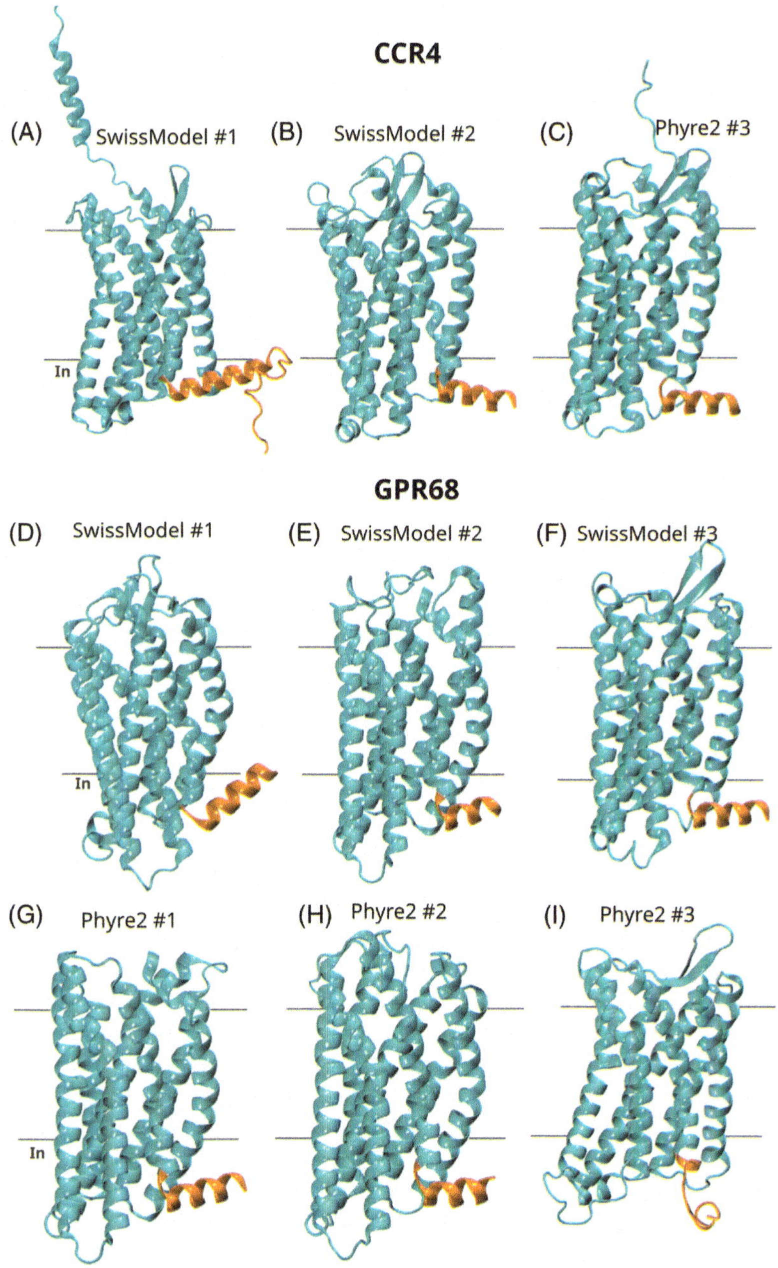

FIGURE 19.8

Homology models of CCR4 and GPR68.(A–C) Homology models of CCR4 obtained with Swiss-Model (panels A and B) versus Phyre2 (panel C). (D–I) Homology models of GPR68 obtained with Swiss-Model (panels D–F) versus Phyre2 (panels G–I). The amphipathic is colored *orange*.

Table 19.1 Sequence similarity and sequence identity values for GPCRs of our dataset, as reported by Clustal Omega.

GPCR pair		Sequence similarity (%)	Sequence identity (%)
Chemokine receptors			
CXCR1	CXCR2	87	76
	CCR4	60	40
	CCR5	55	34
CXCR2	CCR4	61	41
	CCR5	58	35
CCR4	CCR5	68	49
pH-sensing GPCRs			
GPR68	GPR4	60	49
	GPR65	56	36
	GPR132	49	29
GPR65	GPR4	62	41
	GPR132	51	30
GPR4	GPR132	59	38
Additional sequence alignments for GPR68 and CXCR1			
GPR68	ATR	48	31
	CCR5	45	24
	Bovine rhodopsin	41	25
CXCR1	Bovine rhodopsin	44	26

GPCRs, *G protein–coupled receptors.*

domain. The three homology models shown in Fig. 19.8A–C have overall similar structures of the TM region of the receptor and noticeable differences in the relative orientation of the short amphipathic helix (see the helical segment shown as an orange ribbon). We obtained Swiss-Model homology model scores between 0.6 and 0.9; model 1 with GMQE score of 0.9 is shown in Fig. 19.8D.

The Swiss-Model and Phyre2 homology modeling of CCR4 and GPR68 took about 0.5 and 1 hour, respectively; thus homology modeling exercises using either of these online servers can be integrated into a practical computer class. We also suggest that, for a practical computer class aiming to illustrate homology modeling of GPCRs, a useful task would be verifying the templates used to generate the models. This can be important to verify, for example, the resolution at which template structures had been solved experimentally and the conformational state of the receptor. In Table 19.2 we present the templates selected during our tests with Swiss-Model and Phyre2.

For CCR4 the default template selection with Swiss-Model included experimental CXCR1 and CCR5 structures solved at resolutions between 2.60 and 3.15 Å, but also a computational AlphaFold model; the Phyre2 default template selection included an antagonist-bound structure and structures bound to G protein constructs. Likewise, for GPR68, the Swiss-Model default template selection included an AlphaFold model of the protein. The different default structural templates selected by the two online servers contribute to the differences in structural details of the resulting models.

Table 19.2 Summary of the templates selected by Swiss-Model and Phyre2 to model CCR4 and GPR68. For each experimental structure we indicate the PDB ID, the resolution in Å, and pairwise sequence identity scores. PDB IDs with italic fonts are templates selected by both servers.

Web server	Template				Sequence identity
	Protein/protein complex	Organism	PDB ID	Resolution (Å)	
CCR4, UniProt ID: P51679					
Swiss-Model	CCR5-rubredoxin chimera	human	*4mbs*	2.71	49
	CCR5-HIV inhibitor		5uiw	2.20	49
	CCR1-Gi-CCL15 complex		7vla	2.70	50
	CCR5-G1-CCL5 complex		7o7f	3.15	50
	CCR1-Gi-CCL15 complex		*7vl9*	2.60	51
	CCR4 AF P51679.1.A[a]		–	–	100

(continued on next page)

Table 19.2 Summary of the templates selected by Swiss-Model and Phyre2 to model CCR4 and GPR68. For each experimental structure we indicate the PDB ID, the resolution in Å, and pairwise sequence identity scores. PDB IDs with italic fonts are templates selected by both servers—cont'd

Web server	Template				Sequence identity
Phyre2	CCR1-Gi-CCL15 complex	human	*7vl9*	2.60	52
	CCR2-antagonists chimera		5t1a	2.81	55
	CCR5-MIP-1a complex		7f1t	2.60	49
	CCR5-rubredoxin chimera		*4mbs*	2.71	52
	Apo CCR3-Gi complex		7 × 9y	3.10	51
GPR68, UniProt ID: Q15743					
Swiss-Model	GPR68 AF D3ZSW1.1.A	rat	–	–	95.2
	GPR4-Gs complex	human	*8zfj*	3.1	50.2
	CCL15-Gi-CCR1 complex		7vla	2.7	27.2
Phyre2	GPR4-Gs complex	human	*8zfj*	3.1	48
	GPR132-Gi complex		8hqe	2.9	31
	BLT1-Agonist Complex		7k15	2.9	30

[a]*AlphaFold models.*PDB, *Protein Data Bank.*

Table 19.3 Summary of the templates selected automatically by AlphaFold for structural modeling of GPR68. For each template we indicate the PDB ID, the protein, and the resolution (in Å). The templates were used for GPR68 without the N- and C-termini present (Fig. 19.9).

PDB ID	Structure	Resolution (Å)
6do1	Angiotensin II type 1 receptor bound to peptide	2.90
7f6g	Human angiotensin receptor in complex with G proteins and agonist	2.90
7l1u	Orexin receptor 2 in complex with G protein and peptide-agonist	3.20
7l1v	Yeast THO-Sub2 complex	3.70
7sqo	Orexin-2 receptor bound to G protein, orexin 2 receptor agonist	3.17
7sr8	Orexin receptor bound to TAK-925	3.30
7t11	Somatostatin receptor 2 in complex with octreotide and Gi protein	2.70
7wic	Somatostatin 14-bound human SSTR2-Gi1 complex	2.80
7wig	Somatostatin 14 bound to nonpeptide agonist	2.70
7wq3	Galanin-bound galanin receptor 1 in complex with G protein	2.70
7wq4	Galanin-bound galanin receptor 2 in complex with G protein	2.60
7xat	Somatostatin receptor 2 bound with SST14	2.85
7xau	Somatostatin receptor 2 bound with octreotide	2.97
7xav	Somatostatin receptor 2 bound with lanreotide	2.87
7xjl	Spexin-bound GALR2-miniGq complex	3.50
7y24	Octreotide-bound SSTR2-miniGo-scFv16 complex	3.25
7y26	Octreotide-bound SSTR2-miniGq-scFv16 complex	3.30
7y27	SST-14-bound SSTR2-miniGq-scFv16 complex	3.48
7yae	Octreotide-bound SSTR2-Gi complex	3.37
7yit	Kappa opioid receptor	3.30

19.3.6 Protein structural modeling with AlphaFold

AlphaFold2 (Mirdita et al., 2022) structural modeling tasks can easily be incorporated into a class focused on GPCRs. Here we used ColabFold (Jumper et al., 2021) to generate structural models of GPR68, inspected the default template selections obtained, and compared the results with the Swiss-Model and Phyre2 structural models.

AlphaFold2 uses multiple sequence alignments and pairwise distance distributions to model structures whose reliability is evaluated using the predicted local distance difference test (pLDDT), which gives the confidence score of the generated structure (Tunyasuvunakool et al., 2021). The higher the pLDDT score, the more reliable the structural model.

We checked the PDB IDs of all 20 default templates selected by the ColabFold computation (Table 19.3). Structures of distinct GPCRs, in different complexes, were part of this selection. All five

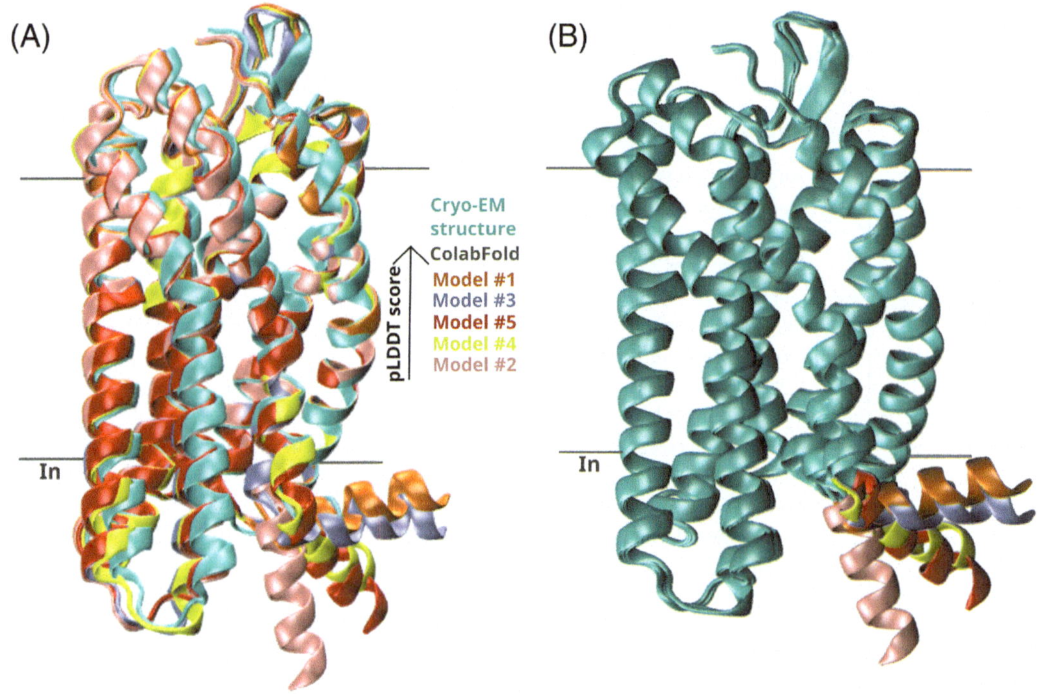

FIGURE 19.9

AlphaFold structural modeling of GPR68. (A) Overlap of the cryo-EM GPR68 structure (Howard et al., 2024); (PDB ID: 9BHM) and the five ColabFold-generated structural models. Note that the cryo-EM GPR68 structure lacks the short amphipathic helix. (B) Overlap of five ColabFold GPR68 structural models obtained without the N- and C-termini. The five structures have almost identical transmembrane regions but differ in the relative orientation of the short amphipathic helix. This amphipathic helix is color-coded according to the structure it belongs to and to its pLDDT rank. Model 1 (*orange*) is the highest ranked with pLDDT = 88.2%, followed by structural models 3 (*blue*, 88.1%), 5 (*red*, 87.1%), 4 (*yellow*, 86.5%), and 2 (*pink*, 86.2%). The figure was generated with VMD. *PDB*, Protein Data Bank; *VMD*, visual molecular dynamics.

structural models obtained have good pLDDT scores >80, indicating structures are reasonable. We have also overlapped each of the ColabFold models with the experimental cryo-EM structure (Fig. 19.9A) and found that the root-mean-square distances for the TM domain are between 1.73 and 2.03 Å; that is, the models are reasonable. By contrast the short amphipathic helical segment has largely distinct orientations in each structure (Fig. 19.9B).

The ColabFold calculation of the five GPR68 models took 0.5 hours using GPU, which is reasonable for a typical 2-hour lab class. ColabFold computations could also be assigned for individual study and tasks.

Table 19.4 PROPKA estimation of pK_a values measured in GPR68 that are notably different from the nominal values of each residue. The nominal pK_a values were taken from Li et al. (2005) for Asp, Glu, His, and Tyr.

Residue	PROPKA pK_a	Nominal pK_a value	ΔpK_a (PROPKA—nominal)	Location of the residue
D67	4.7	3.8	0.9	Internal, H-bonded to S108 and D282 (Fig. 19.3C)
D118	5.1		1.3	Internal
D282	11.0		6.5	Internal, H-bonded to D67 (Fig. 19.3C)
E103	3.4	4.5	−1.1	Internal, H-bonded to R189 (Fig. 19.3C)
E174	6.1		1.6	Internal
H17	8.5	6.5	2.0	External (Fig. 19.3C)
H20	4.7		−1.8	External (Fig. 19.3C)
H175	4.1		−2.4	External
H245	1.9		−4.6	External
Y102	16.2	10	6.2	Internal
Y176	16.3		6.3	External
Y268	19.6		9.6	External

19.3.7 pK_a estimations using PROPKA

The pK_a value (acid dissociation constant) is a fundamental parameter in chemistry that indicates the pH at which a molecule's ionizable group is equally protonated and deprotonated (Krieger et al., 2012). PROPKA3 (Olsson et al., 2011; Søndergaard et al., 2011) is a widely used software tool designed to predict the pK_a values of ionizable groups in proteins and protein–ligand complexes based on their 3D structures. The pK_a values of the titratable residues are estimated according to an equation that includes desolvation effects, hydrogen bonding interactions, and electrostatic interactions between ionizable groups (Li et al., 2005). We used PROPKA3 on the online server https://www.ddl.unimi.it/vegaol/propka.htm for the GPR68 structure PDB ID: 9BHM. Compatible with our discussion above based on visual inspection of the structure (Fig. 19.9A), PROPKA3 predicts for D282 a high pK_a value indicative of this residue potentially being protonated. By contrast, for E103 the predicted pK_a (=3.4) is close to the nominal value (Table 19.4, Fig. 19.3C).

19.4 Conclusion

In this chapter, we presented an introductory overview of three GPCRs that could be used as model systems for an introductory lab class on membrane receptors. We chose bovine rhodopsin as a representative class A GPCR. Bovine rhodopsin is also an example of a light-activated receptor—its retinal ligand absorbs a photon and changes geometry from 11-*cis* to all-*trans*. Chemokine receptor CXCR1, which we chose as a GPCR whose ligand is a small protein (chemokine) and for which a structure of the G protein–bound receptor existed at the time when we started to prepare this introductory chapter, could be of broad interest to students given the relevance of chemokine receptors to human health. GPR68 we chose as an example of a receptor activated by the binding of protons. Thus, via the three receptors we used as model systems, the students can learn about different ligands and activation mechanisms of class A GPCRs. For each of the GPCRs we discussed basic analyses of structure–function relationships, including distances between functionally important sites—by taking the length of a hydrogen bond and the average membrane thickness as a reference. The sequence analyses and structural modeling can be integrated into introductory bioinformatics computer classes on protein structural modeling. The introductory discussion we presented could also be used as starting point for more advanced tasks, such as molecular modeling and simulations of membrane proteins.

Acknowledgments

I-DD and A-MH are supported by a doctoral scholarship from the Doctoral School of the Faculty of Physics of the University of Bucharest.

References

Arimont, M., Sun, S. L., Leurs, R., Smit, M., De Esch, I. J. P., & De Graaf, C. (2017). Structural analysis of chemokine receptor-ligand interactions. *Journal of Medicinal Chemistry, 60*(12), 4735–4779. http://pubs.acs.org/jmc. doi:10.1021/acs.jmedchem.6b01309.

Athanasiou, D., Aguila, M., Bellingham, J., Li, W., McCulley, C., Reeves, P. J., & Cheetham, M. E. (2018). The molecular and cellular basis of rhodopsin retinitis pigmentosa reveals potential strategies for therapy. *Progress in Retinal and Eye Research, 62*, 1–23. Www.elsevier.com/locate/preteyeres. doi:10.1016/j.preteyeres.2017.10.002.

Ballesteros, J. A., & Weinstein, H. (1995). Integrated methods for the construction of three-dimensional models and computational probing of structure-function relations in G protein-coupled receptors. *Methods in Neurosciences, 25*(C), 366–428. doi:10.1016/S1043-9471(05)80049-7.

Benkert, P., Biasini, M., & Schwede, T. (2011). Toward the estimation of the absolute quality of individual protein structure models. *Bioinformatics, 27*(3), 343–350. doi:10.1093/bioinformatics/btq662.

Boutet, E., Lieberherr, D., Tognolli, M., Schneider, M., & Bairoch, A. (2007). In *UniProtKB/Swiss-Prot: 406* (pp. 89–112). Springer Science and Business Media LLC. doi:10.1007/978-1-59745-535-0_4.

Burley, S. K., Piehl, D. W., Vallat, B., & Zardecki, C. (2024). RCSB Protein Data Bank: Supporting research and education worldwide through explorations of experimentally determined and computationally predicted atomic level 3D biostructures. *IUCrJ, 11*(3), 279–286. http://www.iucrj.org/m/. doi:10.1107/S2052252524002604.

Chen, Z., Ren, X., Zhou, Y. u, & Huang, N. (2024). Exploring structure-based drug discovery of GPCRs beyond the orthosteric binding site. *hLife, 2*(5), 211–226. doi:10.1016/j.hlife.2024.01.002.

Cheng, L., Xia, F., Li, Z., Shen, C., Yang, Z., Hou, H., Sun, S., Feng, Y., Yong, X., Tian, X., Qin, H., Yan, W., & Shao, Z. (2023). Structure, function and drug discovery of GPCR signaling. *Molecular Biomedicine, 4*(1). doi:10.1186/s43556-023-00156-w.

Deupi, X. (2014). Relevance of rhodopsin studies for GPCR activation. *Biochimica et Biophysica Acta (BBA)—Bioenergetics, 1837*(5), 674–682. doi:10.1016/j.bbabio.2013.09.002.

Deupi, X., Standfuss, J. ö, & Schertler, G. (2012). Conserved activation pathways in G-protein-coupled receptors. *Biochemical Society Transactions, 40*(2), 383–388. doi:10.1042/bst20120001.

Griffith, J. W., Sokol, C. L., & Luster, A. D. (2014). Chemokines and chemokine receptors: Positioning cells for host defense and immunity. *Annual Review of Immunology, 32*, 659–702. http://arjournals.annualreviews.org/loi/immunol. doi:10.1146/annurev-immunol-032713-120145.

Gruhl, T., Weinert, T., Rodrigues, M. J., Milne, C. J., Ortolani, G., Nass, K., Nango, E., Sen, S., Johnson, P. J. M., Cirelli, C., Furrer, A., Mous, S., Skopintsev, P., James, D., Dworkowski, F., Båth, P., Kekilli, D., Ozerov, D., Tanaka, R., … Panneels, V. (2023). Ultrafast structural changes direct the first molecular events of vision. *Nature, 615*(7954), 939–944. https://www.nature.com/nature/. doi:10.1038/s41586-023-05863-6.

Hauser, A. S., Kooistra, A. J., Munk, C., Heydenreich, F. M., Veprintsev, D. B., Bouvier, M., Babu, M. M., & Gloriam, D. E. (2021). GPCR activation mechanisms across classes and macro/microscales. *Nature Structural and Molecular Biology, 28*(11), 879–888. http://www.nature.com/nsmb/index.html. doi:10.1038/s41594-021-00674-7.

Hernández-Rodríguez, E. W., Sánchez-García, E., Crespo-Otero, R., Montero-Alejo, A. L., Montero, L. A., & Thiel, W. (2012). Understanding rhodopsin mutations linked to the Retinitis pigmentosa disease: A QM/MM and DFT/MRCI study. *Journal of Physical Chemistry B, 116*(3), 1060–1076. http://pubs.acs.org/journal/jpcbfk. doi:10.1021/jp2037334.

Hilger, D. (2021). The role of structural dynamics in GPCR-mediated signaling. *The FEBS Journal, 288*(8), 2461–2489. doi:10.1111/febs.15841.

Howard, M. K., Hoppe, N., Huang, X. P., Macdonald, C. B., Mehrota, E., Grimes, P. R., Zahm, A., Trinidad, D. D., English, J., Coyote-Maestas, W., & Manglik, A. (2024). Molecular basis of proton-sensing by G protein-coupled receptors. *bioRxiv*. https://www.biorxiv.org. doi:10.1101/2024.04.17.590000.

Humphrey, W., Dalke, A., & Schulten, K. (1986). VMD: visual molecular dynamics. *Journal of Molecular Graphics, 14*, 33–38. doi:10.1016/0263-7855(96)00018-5.

Ishimoto, N., Park, J. H., Kawakami, K., Tajiri, M., Mizutani, K., Akashi, S., Tame, J. R. H., Inoue, A., & Park, S. Y. (2023). Structural basis of CXC chemokine receptor 1 ligand binding and activation. *Nature Communications, 14*(1). https://www.nature.com/ncomms/. doi:10.1038/s41467-023-39799-2.

Jumper, J., Evans, R., Pritzel, A., Green, T., Figurnov, M., Ronneberger, O., Tunyasuvunakool, K., Bates, R., Žídek, A., Potapenko, A., Bridgland, A., Meyer, C., Kohl, S. A. A., Ballard, A. J., Cowie, A., Romera-Paredes, B., Nikolov, S., Jain, R., Adler, J., … Hassabis, D. (2021). Highly accurate protein structure prediction with AlphaFold. *Nature, 596*(7873), 583–589. http://www.nature.com/nature/index.html. doi:10.1038/s41586-021-03819-2.

Kapur, B., Baldessari, F., Lazaratos, M., Nar, H., Schnapp, G., Giorgetti, A., & Bondar, A.-N. (2023). Protons taken hostage: Dynamic H-bond networks of the pH-sensing GPR68. *Computational and Structural Biotechnology Journal, 21*, 4370–4384. doi:10.1016/j.csbj.2023.08.034.

Kayastha, K., Zhou, Y., & Brünle, S. (2024). Structural perspectives on chemokine receptors. *Biochemical Society Transactions, 52*(3), 1011–1024. doi:10.1042/bst20230358.

Kelley, L. A., Mezulis, S., Yates, C. M., Wass, M. N., & Sternberg, M. J. E. (2015). The Phyre2 web portal for protein modeling, prediction and analysis. *Nature Protocols, 10*(6), 845–858. http://www.natureprotocols.com/. doi:10.1038/nprot.2015.053.

Koehl, P., & Levitt, M. (2002). Sequence variations within protein families are linearly related to structural variations. *Journal of Molecular Biology, 323*(3), 551–562. doi:10.1016/s0022-2836(02)00971-3.

Krieger, E., Dunbrack, R. L., Hooft, R. W. W., & Krieger, B. (2012). Assignment of protonation states in proteins and ligands: Combining p*K*a prediction with hydrogen bonding network optimization. *Methods in Molecular Biology, 819*, 405–421. https://www.springer.com/series/7651. doi:10.1007/978-1-61779-465-0_25.

Lassen, K. G., McKenzie, C. I., Mari, M., Murano, T., Begun, J., Baxt, L. A., Goel, G., Villablanca, E. J., Kuo, S. Y., Huang, H., Macia, L., Bhan, A. K., Batten, M., Daly, M. J., Reggiori, F., Mackay, C. R., & Xavier, R. J. (2016). Genetic coding variant in GPR65 alters lysosomal pH and links lysosomal dysfunction with colitis risk. *Immunity, 44*(6), 1392–1405. Www.immunity.com. doi:10.1016/j.immuni.2016.05.007.

Latorraca, N. R., Venkatakrishnan, A. J., & Dror, R. O. (2017). GPCR dynamics: Structures in motion. *Chemical Reviews, 117*(1), 139–155. http://pubs.acs.org/journal/chreay. doi:10.1021/acs.chemrev.6b00177.

Li, H., Robertson, A. D., & Jensen, J. H. (2005). Very fast empirical prediction and rationalization of protein p*K*a values. *Proteins: Structure, Function, and Bioinformatics, 61*(4), 704–721. doi:10.1002/prot.20660.

Li, J., Edwards, P. C., Burghammer, M., Villa, C., & Schertler, G. F. X. (2004). Structure of bovine rhodopsin in a trigonal crystal form. *Journal of Molecular Biology, 343*(5), 1409–1438. https://www.journals.elsevier.com/journal-of-molecular-biology. doi:10.1016/j.jmb.2004.08.090.

Lorente, J. S., Sokolov, A. V., Ferguson, G., Schiöth, H. B., Hauser, A. S., & Gloriam, D. E. (2025). GPCR drug discovery: New agents, targets and indications. *Nature Reviews Drug Discovery, 24*(6), 458–479. https://www.nature.com/nrd/. doi:10.1038/s41573-025-01139-y.

Ludwig, M. G., Vanek, M., Guerini, D., Gasser, J. A., Jones, C. E., Junker, U., Hofstetter, H., Wolf, R. M., & Seuwen, K. (2003). Proton-sensing G-protein-coupled receptors. *Nature, 425*(6953), 93–98. doi:10.1038/nature01905.

Mirdita, M., Schütze, K., Moriwaki, Y., Heo, L., Ovchinnikov, S., & Steinegger, M. (2022). ColabFold: Making protein folding accessible to all. *Nature Methods, 19*(6), 679–682. doi:10.1038/s41592-022-01488-1.

Mishra, A., Suman, K. H., Nair, N., Majeed, J., & Tripathi, V. (2021). An updated review on the role of the CXCL8-CXCR1/2 axis in the progression and metastasis of breast cancer. *Molecular Biology Reports, 48*(9), 6551–6561. Www.springer.com/journal/11033. doi:10.1007/s11033-021-06648-8.

Mitra, J., & Ramakrishnan, C. (1977). Analysis of O-H…O hydrogen bonds. *International Journal of Peptide and Protein Research, 9*(1), 27–48. doi:10.1111/j.1399-3011.1977.tb01834.x.

Mitra, K., Ubarretxena-Belandia, I., Taguchi, T., Warren, G., & Engelman, D. M. (2004). Modulation of the bilayer thickness of exocytic pathway membranes by membrane proteins rather than cholesterol. *Proceedings of the National Academy of Sciences, 101*(12), 4083–4088. doi:10.1073/pnas.0307332101.

Munk, C., Isberg, V., Mordalski, S., Harpsøe, K., Rataj, K., Hauser, A. S., Kolb, P., Bojarski, A. J., Vriend, G., & Gloriam, D. E. (2016). GPCRdb: The G protein-coupled receptor database – An introduction. *British Journal of Pharmacology, 173*(14), 2195–2207. doi:10.1111/bph.13509.

Nayak, A. P., & Penn, R. B. (2020). The proton-sensing receptor ovarian cancer G-protein coupled receptor 1 (OGR1) in airway physiology and disease. *Current Opinion in Pharmacology, 51*, 1–10. https://www.journals.elsevier.com/current-opinion-in-pharmacology. doi:10.1016/j.coph.2020.03.004.

Okajima, F. (2013). Regulation of inflammation by extracellular acidification and proton-sensing GPCRs. *Cellular Signalling, 25*(11), 2263–2271. doi:10.1016/j.cellsig.2013.07.022.

Olsson, M. H. M., SØndergaard, C. R., Rostkowski, M., & Jensen, J. H. (2011). PROPKA3: Consistent treatment of internal and surface residues in empirical p*K*a predictions. *Journal of Chemical Theory and Computation, 7*(2), 525–537. doi:10.1021/ct100578z.

Palczewski, K., Kumasaka, T., Hori, T., Behnke, C. A., Motoshima, H., Fox, B. A., Le Trong, I., Teller, D. C., Okada, T., Stenkamp, R. E., Yamamoto, M., & Miyano, M. (2000). Crystal structure of rhodopsin: A G protein-coupled receptor. *Science, 289*(5480), 739–745. doi:10.1126/science.289.5480.739.

Pándy-Szekeres, G., Munk, C., Tsonkov, T. M., Mordalski, S., Harpsøe, K., Hauser, A. S., . . . Gloriam, D. E. (2018). GPCRdb in 2018: adding GPCR structures and ligands. *Nucleic Acids Research, 46*, D440–D446. doi:10.1093/nar/gkx1109.

Pándy-Szekeres, G., Munk, C., Tsonkov, T. M., Mordalski, S., Harpsøe, K., Hauser, A. S., Bojarski, A. J., & Gloriam, D. E. (2018). GPCRdb in 2018: Adding GPCR structure models and ligands. *Nucleic Acids Research, 46*(1), D440. http://nar.oxfordjournals.org/. doi:10.1093/nar/gkx1109.

Pándy-Szekeres, G., Caroli, J., Mamyrbekov, A., Kermani, A. A., Keseru, G. M., Kooistra, A. J., & Gloriam, D. E. (2023). GPCRdb in 2023: State-specific structure models using AlphaFold2 and new ligand resources. *Nucleic Acids Research, 51*(1), D395. https://academic.oup.com/nar/issue. doi:10.1093/nar/gkac1013.

Pearson, W. R. (2013). An introduction to sequence similarity ("homology") searching. *Current Protocols in Bioinformatics*, (42). http://onlinelibrary.wiley.com/store/10.1002/0471250953.bi0301s42/asset/bi0301.pdf?v=1&t=hl30qnfd&s=b7cbeec4e490370a549d91d7ddb9a8506b9fcb78. doi:10.1002/0471250953.bi0301s42. United States.

Pelus, L. M., & Fukuda, S. (2006). Peripheral blood stem cell mobilization: The CXCR2 ligand GROβ rapidly mobilizes hematopoietic stem cells with enhanced engraftment properties. *Experimental Hematology, 34*(8), 1010–1020. doi:10.1016/j.exphem.2006.04.004.

Pope, A. L., Sanchez-Reyes, O. B., South, K., Zaitseva, E., Ziliox, M., Vogel, R., Reeves, P. J., & Smith, S. O. (2020). A conserved proline hinge mediates helix dynamics and activation of rhodopsin. *Structure, 28*(9), 1004. http://www.journals.elsevier.com/structure/. doi:10.1016/j.str.2020.05.004.

Proudfoot, A. E. I. (2002). Chemokine receptors: Multifaceted therapeutic targets. *Nature Reviews Immunology, 2*(2), 106–115. http://www.nature.com/nri/index.html. doi:10.1038/nri722.

Rowe, J. B., Kapolka, N. J., Taghon, G. J., Morgan, W. M., & Isom, D. G. (2021). The evolution and mechanism of GPCR proton sensing. *Journal of Biological Chemistry, 296*, 100167. doi:10.1074/jbc.ra120.016352.

Russell, J. L., Goetsch, S. C., Aguilar, H. R., Coe, H., Luo, X., Liu, N., Van Rooij, E., Frantz, D. E., & Schneider, J. W. (2012). Regulated expression of pH sensing G protein-coupled receptor-68 identified through chemical biology defines a new drug target for ischemic heart disease. *ACS Chemical Biology, 7*(6), 1077–1083. doi:10.1021/cb300001m.

Sakmar, T. P. (2017). Introduction: G-protein coupled receptors. *Chemical Reviews, 117*(1), 1–3. http://pubs.acs.org/journal/chreay. doi:10.1021/acs.chemrev.6b00686.

Schertler, G. F. X., Villa, C., & Henderson, R. (1993). Projection structure of rhodopsin. *Nature, 362*(6422), 770–772. doi:10.1038/362770a0.

Schertler, G. (2005). Structure of rhodopsin and the metarhodopsin I photointermediate. *Current Opinion in Structural Biology, 15*(4), 408–415. doi:10.1016/j.sbi.2005.07.010.

Show, S., Mukherjee, A., Nongthomba, U., & Mahadesh Prasad, A. J. (2025). Guardian of tissue, accomplice of cancer: The two faces of a proton sensor-G protein-coupled receptor 68. *Discover Medicine, 2*(1). doi:10.1007/s44337-025-00196-x.

Sievers, F., Wilm, A., Dineen, D., Gibson, T. J., Karplus, K., Li, W., Lopez, R., McWilliam, H., Remmert, M., Söding, J., Thompson, J. D., & Higgins, D. G. (2011). Fast, scalable generation of high-quality protein multiple sequence alignments using Clustal Omega. *Molecular Systems Biology, 7*. doi:10.1038/msb.2011.75.

Søndergaard, C. R., Olsson, M. H. M., Rostkowski, M., & Jensen, J. H. (2011). Improved treatment of ligands and coupling effects in empirical calculation and rationalization of p*K*a values. *Journal of Chemical Theory and Computation, 7*(7), 2284–2295. doi:10.1021/ct200133y.

Standfuss, J., Edwards, P. C., D'Antona, A., Fransen, M., Xie, G., Oprian, D. D., & Schertler, G. F. X. (2011). The structural basis of agonist-induced activation in constitutively active rhodopsin. *Nature, 471*(7340), 656–660. doi:10.1038/nature09795.

Tan, C. M., Brady, A. E., Nickols, H. H., Wang, Q., & Limbird, L. E. (2004). Membrane trafficking of G protein-coupled receptors. *Annual Review of Pharmacology and Toxicology, 44*, 559–609. doi:10.1146/annurev.pharmtox.44.101802.121558.

Trzaskowski, B., Latek, D., Yuan, S., Ghoshdastider, U., Debinski, A., & Filipek, S. (2012). Action of molecular switches in GPCRs—Theoretical and experimental studies. *Current Medicinal Chemistry, 19*(8), 1090–1109. doi:10.2174/092986712799320556.

Tsai, H. H., Frost, E., To, V., Robinson, S., Ffrench-Constant, C., Geertman, R., Ransohoff, R. M., & Miller, R. H. (2002). The chemokine receptor CXCR2 controls positioning of oligodendrocyte precursors in developing spinal cord by arresting their migration. *Cell, 110*(3), 373–383. doi:10.1016/S0092-8674(02)00838-3.

Tunyasuvunakool, K., Adler, J., Wu, Z., Green, T., Zielinski, M., Žídek, A., Bridgland, A., Cowie, A., Meyer, C., Laydon, A., Velankar, S., Kleywegt, G. J., Bateman, A., Evans, R., Pritzel, A., Figurnov, M., Ronneberger, O., Bates, R., Kohl, S. A. A., ... Hassabis, D. (2021). Highly accurate protein structure prediction for the human proteome. *Nature, 596*(7873), 590–596. http://www.nature.com/nature/index.html. doi:10.1038/s41586-021-03828-1.

del Val, C., Bondar, L., & Bondar, A.-N. (2014). Coupling between inter-helical hydrogen bonding and water dynamics in a proton transporter. *Journal of Structural Biology, 186*(1), 95–111. doi:10.1016/j.jsb.2014.02.010.

Wang, Y., de Vallière, C., Imenez Silva, P. H., Leonardi, I., Gruber, S., Gerstgrasser, A., Melhem, H., Weber, A., Leucht, K., Wolfram, L., Hausmann, M., Krieg, C., Thomasson, K., Boyman, O., Frey-Wagner, I., Rogler, G., & Wagner, C. A. (2018). The proton-activated receptor GPR4 modulates intestinal inflammation. *Journal of Crohn's and Colitis, 12*(3), 355–369. https://academic.oup.com/ecco-jcc/issue. doi:10.1093/ecco-jcc/jjx147.

Waterhouse, A., Bertoni, M., Bienert, S., Studer, G., Tauriello, G., Gumienny, R., Heer, F. T., De Beer, T. A. P., Rempfer, C., Bordoli, L., Lepore, R., & Schwede, T. (2018). SWISS-MODEL: Homology modelling of protein structures and complexes. *Nucleic Acids Research, 46*(1), W296. https://academic.oup.com/nar/issue. doi:10.1093/nar/gky427.

Weis, W. I., & Kobilka, B. K. (2018). The molecular basis of G protein-coupled receptor activation. *Annual Review of Biochemistry, 87*, 897–919. http://www.annualreviews.org/journal/biochem. doi:10.1146/annurev-biochem-060614-033910.

Wiley, S. Z., Sriram, K., Salmerón, C., & Insel, P. A. (2019). GPR68: An emerging drug target in cancer. *International Journal of Molecular Sciences, 20*(3). https://www.mdpi.com/1422-0067/20/3/559/pdf. doi:10.3390/ijms20030559.

Yang, D., Zhou, Q., Labroska, V., Qin, S., Darbalaei, S., Wu, Y., Yuliantie, E., Xie, L., Tao, H., Cheng, J., Liu, Q., Zhao, S., Shui, W., Jiang, Y. i, & Wang, M.-W. (2021). G protein-coupled receptors: Structure- and function-based drug discovery. *Signal Transduction and Targeted Therapy, 6*(1). doi:10.1038/s41392-020-00435-w.

Zhang, M., Chen, T., Lu, X., Lan, X., Chen, Z., & Lu, S. (2024). G protein-coupled receptors (GPCRs): Advances in structures, mechanisms and drug discovery. *Signal Transduction and Targeted Therapy, 9*(1). doi:10.1038/s41392-024-01803-6.

Zheng, Y., Han, G. W., Abagyan, R., Wu, B., Stevens, R. C., Cherezov, V., Kufareva, I., & Handel, T. M. (2017). Structure of CC chemokine receptor 5 with a potent chemokine antagonist reveals mechanisms of chemokine recognition and molecular mimicry by HIV. *Immunity, 46*(6), 1005. Www.immunity.com. doi:10.1016/j.immuni.2017.05.002.

Zhou, Y., Meng, J., Xu, C., & Liu, J. (2021). Multiple GPCR functional assays based on resonance energy transfer sensors. *Frontiers in Cell and Developmental Biology, 9*. https://doi.org/10.3389/fcell.2021.611443.

Author Index

C

D

E

F

G

H

I

J

K

M

N

O

P

Q

R

S

W

X

Y

Z

Index

Page numbers followed by "*f*" and "*t*" indicate, figures and tables respectively.

D

E

F

G

H

I

L

M

N

O

P

T

U

V

W

X

Printed and bound by CPI Group (UK) Ltd, Croydon, CR0 4YY
14/09/2026
02216593-0001